ECOLOGY AND MANAGEMENT OF NEOTROPICAL MIGRATORY BIRDS

ECOLOGY AND MANAGEMENT OF NEOTROPICAL MIGRATORY BIRDS

A Synthesis and Review of Critical Issues

Edited by

THOMAS E. MARTIN

and

DEBORAH M. FINCH

New York Oxford

OXFORD UNIVERSITY PRESS

1995

Oxford University Press

Oxford New York
Athens Auckland Bangkok Bombay
Calcutta Cape Town Dar es Salaam
Delhi Florence Hong Kong Istanbul
Karachi Kuala Lumpur Madras Madrid
Melbourne Mexico City Nairobi Paris
Singapore Taipei Tokyo Toronto

and associated companies in
Berlin Ibadan

Published by Oxford University Press Inc.,
198 Madison Avenue, New York, New York 10016

Library of Congress Cataloging-in-Publication Data
Ecology and management of neotropical migratory birds:
a synthesis and review of critical issues/
edited by Thomas E. Martin and Deborah M. Finch.
p. cm. Includes bibliographical references and index.
ISBN 0-19-508452-7 (pbk.).— ISBN 0-19-508440-3 (cloth)
1. Birds, Protection of—America. 2. Birds—Migration—America.
3. Birds—Ecology—America. 4. Birds—Habitat—America.
I. Finch, Deborah M. II. Martin, Thomas E.
QL680.E28 1995
598.252'5'097—dc20 95-35551

9 8 7 6 5 4 3 2 1

Printed in the United States of America
on acid-free paper

CONTENTS

Introduction: Importance of Knowledge and its Application in Neotropical Migratory Birds xiii
Thomas E. Martin and Deborah M. Finch

PART I POPULATION TRENDS

1 Population Trends from the North American Breeding Bird Survey 3
Bruce G. Peterjohn, John R. Sauer, and Chandler S. Robbins

2 The Strength of Inferences about Causes o Trends in Populations 40
Frances C. James and Charles E. McCulloch

PART II TEMPORAL PERSPECTIVES ON POPULATION LIMITATION AND HABITAT USE

3 When and How are Populations Limited? The Roles of Insect Outbreaks, Fire, and Other Natural Perturbations 55
John T. Rotenberry, Robert J. Cooper, Joseph M. Wunderle, and Kimberly G. Smith

4 Summer versus Winter Limitation of Populations: What are the Issues and What is the Evidence? 85
Thomas W. Sherry and Richard T. Holmes

5 Habitat Requirements During Migration: Important Link in Conservation 121
Frank R. Moore, Sidney A. Gauthreaux, Jr, Paul Kerlinger, and Theodore R. Simons

6 Habitat Use and Conservation in the Neotropics 145
Daniel R. Petit, James F. Lynch, Richard L. Hutto, John G. Blake, and Robert B. Waide

PART III FOREST MANAGEMENT

7 Impacts of Silviculture: Overview and Management Recommendations 201
Frank R. Thompson, III, John R. Probst, and Martin G. Raphael

8 Effects of Silvicultural Treatments in the Rocky Mountains 220
Sallie J. Hejl, Richard L. Hutto, Charles R. Preston, and Deborah M. Finch

9 Silviculture in Central and Southeastern Oak–Pine Forests 245
James G. Dickson, Frank R. Thompson, III, Richard N. Conner, and Kathleen E. Franzreb

PART IV GENERAL HUMAN EFFECTS

10 Effects of Agricultural Practices and Farmland Structures 269
Nicholas L. Rodenhouse, Louis B. Best, Raymond J. O'Connor, and Eric K. Bollinger

11 An Assessment of Potential Hazards of Pesticides and Environmental Contaminants 294
Nicholas W. Gard and Michael J. Hooper

12 Livestock Grazing Effects in Western North America 311
Victoria A. Saab, Carl E. Bock, Terrell D. Rich, and David S. Dobkin

PART V SCALE PERSPECTIVES

13 Habitat Fragmentation in the Temperate Zone 357
John Faaborg, Margaret Brittingham, Therese Donovan, and John Blake

14 A Landscape Ecology Perspective for Research, Conservation, and Management 381
Kathryn E. Freemark, John B. Dunning, Sallie J. Hejl, and John R. Probst

15 Ecology and Behavior of Cowbirds and their Impact on Host Populations 428
Scott K. Robinson, Stephen I. Rothstein, Margaret C. Brittingham, Lisa J. Petit, and Joseph A. Grzybowski

16 Single-Species versus Multiple-Species Approaches for Management 461
William M. Block, Deborah M. Finch, and Leonard A. Brennan

17 Summary: Model Organisms for Advancing Understanding of Ecology and Land Management 477
Thomas E. Martin

Index 485

CONTRIBUTORS

Louis B. Best, Department of Animal Ecology, 124 Science II, Iowa State University, Ames, Iowa 50011

John G. Blake, Department of Biology, University of Missouri–St Louis, 8001 Natural Bridge Road, St. Louis, Missouri 63121

William M. Block, USDA Forest Service, Rocky Mountain Forest and Range Experiment Station, 2500 S. Pine Knoll Drive, Flagstaff, Arizona 86001

Carl E. Bock, Department of EPO Biology, University of Colorado, Boulder, Colorado 80309-0334

Eric R. Bollinger, Department of Zoology, Eastern Illinois University, Charleston, Illinois 61920

Leonard A. Brennan, Tall Timbers Research Station, Route 1, Box 678, Tallahassee, Florida 32312

Margaret C. Brittingham, School of Forest Resources, 320 Forest Resources Laboratory, Pennsylvania State University, University Park, Pennsylvania 16802

Richard N. Conner, Southern Forest Experiment Station, Box 7600, SFA Station, Nacogdoches, Texas 75962

Robert J. Cooper, Department of Biology, Memphis State University, Memphis, Tennessee 38152

James G. Dickson, US Forest Service, Box 7600, SFA Station, Nacogdoches, Texas 75962

David S. Dobkin, The High Desert Ecological Research Institute, 15 SW Colorado Avenue Suite 300, Bend, Oregon 97702

Therese Donovan, Division of Biological Sciences, University of Missouri–Columbia, Columbia, Missouri 65211

John B. Dunning, University of Georgia, Institute of Ecology, 724 Biological Sciences Building, Athens, Georgia 30602

John Faaborg, Division of Biological Sciences, University of Missouri–Columbia, Columbia, Missouri 65211

Deborah M. Finch, USDA Forest Service, Rocky Mountain Forest and Range Experiment Station, 2205 Columbia SE, Albuquerque, New Mexico 87106

Kathleen E. Franzreb, Department of Forest Resources, Clemson University, Clemson, South Carolina 29634-1003

Kathryn E. Freemark, National Wildlife Research Centre, Environment Canada, Ottawa, Canada K1A 0H3; c/o US Environmental Protection Agency, 200 SW 35th Street, Corvallis, Oregon 97333

Nicholas W. Gard, The Institute of Wildlife and Environmental Toxicology, and Department of Environmental Toxicology, Clemson University, PO Box 709, Pendleton, South Carolina 29670

Sidney A. Gauthreaux, Jr, Department of Biological Sciences, Clemson University, Clemson, South Carolina 29634

Joseph A. Grzybowski, Department of Biology, University of Central Oklahoma, Edmond, Oklahoma 73034

Sallie J. Hejl, USDA Forest Service, Intermountain Research Station, PO Box 8089, Missoula, Montana 59807

Richard T. Holmes, Department of Biological Sciences, Dartmouth College, Hanover, New Hampshire 03755

Michael J. Hooper, The Institute of Wildlife and Environmental Toxicology, and Department of Environmental Toxicology, Clemson University, PO Box 709, Pendleton, South Carolina 29670

Richard L. Hutto, Division of Biological Sciences, University of Montana, Missoula, Montana 59812

Frances C. James, Department of Biological Science, Florida State University, Tallahassee, Florida 32306-2043

Paul Kerlinger, Cape May Bird Observatory, Cape May Point, New Jersey 08212

James F. Lynch, Smithsonian Environmental Research Center, PO Box 28, Edgewater, Maryland 21037

Thomas E. Martin, US National Biological Service, Montana Cooperative Wildlife Research Unit, University of Montana, Missoula, MT 59812

Charles E. McCulloch, Biometrics Unit, 337 Warren Hall, Cornell University, Ithaca, New York 14853

Frank R. Moore, Department of Biological Sciences, University of Southern Mississippi, Hattiesburg, Mississippi 39406

Raymond J. O'Connor, Department of Wildlife, 240 Nutting Hall, University of Maine, Orono, Maine 04469

Bruce G. Peterjohn, US National Biological Service, Office of Migratory Bird Management, Patuxent Wildlife Research Center, Laurel, Maryland 20708

Daniel R. Petit, Smithsonian Environmental Research Center, PO Box 28, Edgewater, Maryland 21037. Current address: US Fish and Wildlife Service, Office of Migratory Bird Management, 4401 North Fairfax Drive, Arlington, Virginia 22203

Lisa J. Petit, Smithsonian Migratory Bird Center, National Zoological Park, Washington, DC 20008

Charles L. Preston, Department of Zoology, Denver Museum of Natural History, Denver, Colorado 80205

John R. Probst, USDA Forest Service, North Central Forest Experiment Station, Forestry Sciences Laboratory, PO Box 898, Rhinelander, Wisconsin 54501-0898

Martin G. Raphael, Pacific Northwest Forest and Range Experiment Station, 3625 93rd Avenue SW, Olympia, Washington 98502

Terrell D. Rich, USDI Bureau of Land Management, 3380 Americana Terrace, Boise, Idaho 83706

Chandler S. Robbins, US National Biological Service, Patuxent Wildlife Research Center, Laurel, Maryland 20708

Scott R. Robinson, Illinois Natural History Survey, 607 East Peabody Drive, Champaign, Illinois 61820

Nicholas L. Rodenhouse, Department of Biological Sciences, Wellesly College, Wellesly, Massachusetts 02181

John T. Rotenberry, Natural Reserve System and Department of Biology, University of California, Riverside, California 92521

Stephen I. Rothstein, Department of Biological Sciences, University of California, Santa Barbara, California 93106

Victoria A. Saab, USDA Forest Service, Intermountain Research Station, 316 E. Myrtle Street, Boise, Idaho 83702 and Department of EPO Biology, University of Colorado, Boulder, Colorado 80309-0334

John R. Sauer, US National Biological Service, Patuxent Wildlife Research Center, Laurel, Maryland 20708

Thomas W. Sherry, Department of Ecology, Evolution, and Organismal Biology, 310 Dinwiddie Hall, Tulane University, New Orleans, Louisiana 70118

Theodore R. Simons, CPSU/NBS, North Carolina State University, Raleigh, North Carolina 27695

Kimberly G. Smith, Department of Biological Sciences, University of Arkansas, Fayetteville, Arkansas 72701

Frank R. Thompson, III, North Central Forest Experiment Station, 1/26 Agriculture Building, University of Missouri, Columbia, Missouri 65211

Robert B. Waide, Center for Energy and Environment Research, University of Puerto Rico, GPO Box 3682, San Juan, Puerto Rico 009 36

Joseph M. Wunderle, International Institute of Tropical Forestry, USDA Forest Service, PO Box B, Palmer, Puerto Rico 00721

INTRODUCTION: IMPORTANCE OF KNOWLEDGE AND ITS APPLICATION IN NEOTROPICAL MIGRATORY BIRDS

THOMAS E. MARTIN AND DEBORAH M. FINCH

The pursuit of knowledge has always been a noble occupation, but when knowledge is applied to improve conditions for living organisms, whether they are plants, animals, or humans, the value of knowledge expands to a new dimension involving ethics and responsibility. Knowledge upon which application is based originates from research; advances in any applied field, whether it is medicine, horticulture, or biology, depend on innovative and vigorous research programs. If we relied solely on existing knowledge to manage ourselves and our environment, rather than generating new information to address new problems, the world as we know it, including its parts, processes, and functions, could quickly spiral out of control. The continuous need for new knowledge is exemplified in cases when diseases emerge in new forms, rendering previous cures ineffective, when agricultural plant varieties suddenly become susceptible to newly evolved diseases, and when deadly diseases remain incurable. Yet, research results may have no immediate or obvious impact because their use in solving problems is dependent on experimental applications several steps removed from the research itself. Research results need to be conveyed to practitioners (e.g., physicians, resource managers) to indicate the meaning and value of research knowledge for recommending applications in usable forms.

The transfer of research knowledge to practitioners such as resource managers and conservationists is particularly critical for sustaining ecological systems. Ecological systems are currently undergoing stress as natural resources such as water, timber, and minerals are depleted by growing human populations, resulting in the development of new problems. To find answers to new problems, the need for vigorous ongoing research is just as important as the application of existing knowledge. Ecological systems are difficult to understand and manage because of the complexity created by the intricate web of interactions among myriad species and their environments. Moreover, problems facing ecological systems are constantly evolving and expanding, owing to the extent and manner of environmental changes caused by humans interacting with nature. These ecological webs and evolving problems must be studied to understand how they work; how they respond to human activities; and how humans can adaptively manage them given that they are constantly changing. We cannot simply wait for further understanding because ecological systems can become irreversibly damaged while we wait as species decline to extinction or near-extinction and cause cascading effects. By acting now, we may preserve options for the future, and at the same time, we can develop a vigorous research program to improve our understanding and future approaches. Our intent with this volume is to distill research information on Neotropical migratory birds for use and application by practitioners (e.g., natural resource managers) and researchers alike.

Existing knowledge is made available most readily by critical reviews and syntheses. Their importance is clearly recognized by the existence of publications like *Annual Reviews*, *Biological Reviews*, *Current Ornithology*, and *Trends in Ecology and Evolution*. Such reviews are particularly valuable when a field is unusually active due to its importance, or when it has many disciplines that make it difficult to stay current with all relevant work. The study of Neotropical migratory birds (birds that breed in North America and migrate to wintering grounds south of the United States) represents both—a field that is unusually active due to its importance and one with many subdisciplines. Thus, increased access to existing information on issues that are critical to the conservation and study of Neotropical migratory birds is needed, which this volume accomplishes.

This volume differs from other edited volumes in three ways. First, we asked authors to write chapters that were not simply descriptions of individual research studies, but rather reviews and syntheses of critical conceptual and topic areas. Second, we contacted multiple authors to work together to write each chapter to enhance the probability that multiple sides of issues were covered and possible biases of individuals were minimized. Finally, we asked authors to extract management and conservation implications from the reviews they prepared to make this information more readily available to resource managers.

Neotropical migratory birds are particularly appropriate for a volume addressing conservation of ecological systems. Recognition of the need to develop methods for managing entire assemblages of species rather than single resources is growing rapidly. Birds have already had a major influence on the understanding of major areas of biology (Konishi et al. 1989). Knowledge of birds can provide additional critical insights into understanding and sustaining the integrity and functioning of ecological systems because birds are speciose, because species differ in environmental requirements and tolerances and can serve as sensitive indicators of environmental health, and because demographic parameters (i.e., breeding productivity, survival) that provide the strongest assay of population health and habitat suitability are more easily monitored for birds than for any other group of vertebrate taxa (Martin 1992, Martin and Geupel 1993). In addition, birds are of wide public interest; a survey in 1985 revealed that 46% of the United States population over 16 years of age (82.5 million people) purchase food for wild birds and even more watch them (US Department of the Interior 1988). Finally, some evidence suggests that a variety of Neotropical migratory birds may be showing long-term population declines (e.g., Robbins et al. 1989) and, thus, in particular need of conservation attention.

The potential existence of population problems and the ensuing public and research interest in Neotropical migratory birds provided the catalyst for development of an international interagency program, *Partners in Flight—Aves de las Americas*. This conservation program is based on cooperative partnerships among a full array of federal and state agencies, nongovernmental organizations, and academic institutions. The vigor of this program and the interest in Neotropical migratory bird conservation was reflected by attendance (more than 700 people) at the National Training Workshop, Status and Management of Neotropical Migratory Birds, held 21–25 September 1992, at Estes Park, Colorado. Authors of this book also worked closely together to produce an earlier volume that focused specifically on priorities for managing populations of Neotropical migratory birds on public and private lands in North America and Mexico (Finch and Stangel 1993). As members of the Research Working Group for *Partners in Flight*, authors felt that scientists could make a critical contribution to the *Partners in Flight* program by providing ecologically based management recommendations to natural resource managers. However, authors also concurred that the scientific basis underlying those guidelines should be presented in a synthesized form to allow informed decision-making. This resulting book is a companion to the volume by Finch and Stangel (1993).

Part I (Population Trends) begins with a chapter that examines potential population problems detected using Breeding Bird Survey (BBS) data. BBS results showed spatial and temporal variation in population trends across species, habitats, and ecological conditions. When population data show long-term population declines for some species, an important step for reversing a negative trend is to determine and interpret the cause of the decline. The second chapter in Part I evaluates and develops experimental designs that affect the strength of inference when examining causes of population trends.

Population trends can be complicated by temporal variation in environmental conditions and resources causing strong variation in populations. Such variation must be recognized and incorporated into considerations of population trends. In Part II (Temporal Perspectives on Population Limitation and Habitat Use), the first chapter begins by examining the role of natural perturbations (e.g., insect outbreaks, fire, hurricanes) on bird populations and how variations in natural disturbances across years and space cause population fluctuations and limitations. In addition to year-to-year fluctuations, populations may also respond differentially across seasons (i.e., breeding, migration, or winter). In the next chapter, the authors conclude that populations can be limited in all seasons and, thus, determination of the habitat conditions needed in each season is necessary for conservation. Consequently, the final two chapters in Part II examine conceptual bases of the importance of habitats during migration and winter seasons and summarize existing information on habitat use during these periods.

Initial concerns about population trends of Neotropical migrants arose from concerns of birds of eastern deciduous forests and much attention focused on forest systems. A major land use factor altering the quality and quantity of forest habitats at local, landscape, and regional scales is logging and its geographic variability. The kinds of logging practices, the size of cuts, and the distribution of cuts all can influence habitat quality and distribution for Neotropical migrants. Part III (Forest Management) begins with an overview and description of kinds of silvicultural practices, their potential consequences for Neotropical migrants, and the importance of within-habitat and landscape-level considerations. The overview is followed by two more detailed case studies in two different forest systems: Rocky Mountains versus Central and Southeastern oak–pine forests.

Clearly, human activities can have direct influences on habitat quality and demography within habitats by affecting the landscape, and/or habitats within the landscapes. In Part IV (General Human Effects), effects of grazing, agriculture and pesticides on quality of habitats and consequences for Neotropical migratory birds are examined. Agriculture can affect habitats of grassland birds directly, but can also affect habitats of Neotropical migrants indirectly through landscape effects or pesticides. Pesticides and contaminants can be a larger general problem and can negatively affect Neotropical migrants by causing decreased breeding productivity. Grazing affects habitats by modifying vegetation structure within habitats, affecting distribution of Brown-headed Cowbirds, and needs to be considered in terms of spatial distribution of grazed and ungrazed lands.

The chapters throughout the volume emphasize the need to understand and examine the influence and importance of differing spatial scales on populations. This perspective is developed further in Part V (Scale Perspectives), which begins by focusing on fragmentation issues. Fragmentation of habitats can cause loss of some species from small fragments, by affecting presence and abundance of required microhabitats, and distribution and abundance of nest predators and brood parasites (i.e., Brown-headed Cowbirds). Local populations using fragments are not necessarily isolated and may interact with other nearby populations through dispersal and migration. Hence, analyses of demography must account for mobile populations and metapopulations interacting with complex environments at larger spatial scales, such as landscapes and habitat mosaics. The next chapter addresses the topic of landscape ecology not only because of the growing call to manage ecological systems at spatial scales larger than those traditionally used, but also because we now know that habitat use and demography of Neotropical migratory birds is influenced by large-scale processes and factors. Composition and pattern of land use and habitat within landscapes can influence the abundance and distribution of predators and brood parasites, thereby influencing processes such as nest predation and brood parasitism that act on the demography of populations. Concerns about impacts of cowbird parasitism, in particular, have been increasing. As a result, spatial and environmental influences on Brown-headed Cowbird abundance and distribution, and resulting rates of parasitism, are considered in detail in the next chapter. This chapter highlights the need to consider spatial scales at the level of fragments and landscapes, but also at even larger scales pertinent to species geographic ranges because the role and importance of some processes such as Brown-headed Cowbird parasitism can vary geographically. Spatial scale is only one scale perspective, however. Land managers are faced with a bewildering diversity of bird species to manage across habitats, and the appropriate biological scale (single species vs assemblage vs community vs ecosystem) for management decisions is equally important. In the final chapter in Part V, the value, ecological basis, and decision-making approach for managing habitats and lands for multiple species are discussed.

The book ends with a short summary chapter that summarizes many of the book's key points and which simply points out that: (1) we have learned a great deal, but even so, habitats for many neotropical migrants continue to disappear or decline in quality; (2) some populations are at risk of endangerment; and (3) that we need to act immediately to apply the knowledge we have. Nonetheless, many conservation problems have yet to be solved, and active pursuit of knowledge and solutions is needed to sustain the parts and processes of ecological systems, including its biological components such as Neotropical migratory birds. To maximize our effectiveness at conserving migratory birds in all their beauty, mystery, and function, it is imperative that land managers and researchers work closely together to develop scientifically based management strategies for ensuring the sustainability of migrant populations and their habitats.

LITERATURE CITED

Finch, D. M., and P. W. Stangel (eds). 1993. Status and management of Neotropical migratory birds. USDA Forest Serv. Rocky Mt. Forest Range Exp. Sta. Fort Collins, CO, Gen. Tech. Rep. RM-229. 422 pp.

Konishi, M., S. T. Emlen, R. E. Ricklefs, and J. C. Wingfield. 1989. Contributions of bird studies to biology. Science 246:465–472.

Martin, T. E. 1992. Breeding productivity considerations: what are the appropriate habitat features for management? Pp. 455–473 *in* Ecology and conservation of Neotropical migrant landbirds (J. M. Hagan, III and D. W. Johnston, eds). Smithsonian Institution Press, Washington.

Martin, T. E., and G. R. Geupel. 1993. Nest-monitoring plots: Methods for locating nests and monitoring success. J. Field Ornithol. 64:507–519.

US Department of the Interior. 1988. 1985 National survey of fishing, hunting, and wildlife associated recreation. USDI Fish Wildl. Serv. Rep.

Robbins, C. S., J. R. Sauer, R. S. Greenberg, and S. Droege. 1989. Population declines in North American birds that migrate to the Neotropics. Proc. Natl Acad. Sci. USA 86:7658–7662.

PART I

POPULATION TRENDS

1

POPULATION TRENDS FROM THE NORTH AMERICAN BREEDING BIRD SURVEY

BRUCE G. PETERJOHN, JOHN R. SAUER, AND CHANDLER S. ROBBINS

INTRODUCTION

Most Neotropical migrant birds are difficult to count accurately and are moderately common over large breeding distributions. Consequently, little historical information exists on their large-scale population changes, and most of this information is anecdotal. Surveys begun in this century such as Breeding Bird Censuses and Christmas Bird Counts have the potential to provide this information, but only the North American Breeding Bird Survey (BBS) achieves the extensive continental coverage necessary to document population changes for most Neotropical migrant birds. Conservationists and ecologists have begun to use BBS data to estimate population trends, but there is still widespread confusion over exactly what these data show regarding population changes.

In this chapter, we review the current state of knowledge regarding population changes in Neotropical migrant birds and the methods used to analyze these changes. The primary emphasis is on the BBS (Robbins et al. 1986) because this survey provides the best available data for estimating trends of Neotropical migrants on a continental scale. To address questions about methods of analyzing survey data, we review and compare some alternative methods of analyzing BBS data. We also discuss the effectiveness of the BBS in sampling Neotropical migrant species, and review possibilities for use of alternative data sets to verify trends from the BBS.

HISTORICAL SUMMARY

Since colonial times, writers have commented on the decline in bird numbers. Birds used for food were the first species of concern, as exemplified by the disappearance of Wild Turkeys from most of their original range (Wright 1915a, b). Coues and Prentiss (1862, 1883), were among the first to mention declines of some Neotropical migrants (Table 1-1). They documented the disappearance of some species as the human population of the District of Columbia increased from 60,000 to 180,000 between 1862 and 1882 while woodland and field habitats decreased in extent.

In the New York City area, Chapman (1894) reported reduced numbers of a few Neotropical migrants during the late 19th century. An expanded list of declining species was reported by Griscom (1923) (Table 1-1). Elsewhere in the eastern United States, William Brewster (1903) listed many species of Neotropical migrants in the Cambridge, Massachusetts area whose populations had declined since 1870. Griscom (1949) compared his observations with unpublished 19th century counts by William Brewster, and described the effects of habitat changes on trends in bird populations near Concord, Massachusetts. West of the Appalachians, Trautman (1940) reported that the greatest changes in habitats and bird populations in central Ohio occurred between 1821 and 1890.

An inspection of Table 1-1 demonstrates little regional consistency in the Neotropical

Table 1-1. List of Neotropical migrants experiencing regional population declines during the 19th and early 20th centuries. See American Ornithologists' Union (1983) for scientific names of these species.

Species	Sources					
	1	2	3	4	5	6
Broad-winged Hawk					×	
Common Nighthawk					×	
Whip-poor-will	×		×			×
Chimney Swift			×			
Ruby-throated Hummingbird					×	
Olive-sided Flycatcher			×			
Eastern Wood-Pewee						×
Yellow-bellied Flycatcher					×	
Acadian Flycatcher						×
Willow Flycatcher						×
Purple Martin					×	
Bank Swallow					×	
Cliff Swallow		×	×	×		
Barn Swallow	×		×	×		
Blue-gray Gnatcatcher	×					
Gray-cheeked Thrush					×	
Swainson's Thrush					×	
Wood Thrush						×
White-eyed Vireo			×	×		
Solitary Vireo					×	
Yellow-throated Vireo				×	×	
Warbling Vireo					×	
Nashville Warbler					×	
Northern Parula					×	
Yellow Warbler						×
Magnolia Warbler					×	
Black-throated Blue Warbler					×	
Black-throated Green Warbler					×	
Blackburnian Warbler					×	
Blackpoll Warbler					×	
Cerulean Warbler						×
Black-and-white Warbler					×	
American Redstart						×
Ovenbird					×	
Northern Waterthrush					×	
Kentucky Warbler						×
Canada Warbler					×	
Summer Tanager				×		
Scarlet Tanager					×	
Indigo Bunting					×	
Dickcissel	×	×	×	×		
Grasshopper Sparrow				×		
Bobolink		×	×			

References: 1 = Coues and Prentiss (1862, 1883); 2 = Chapman (1894); 3 = Brewster (1903); 4 = Griscom (1923); 5 = Griscom (1949); 6 = Trautman (1940).

migrants experiencing declines during the 19th and early 20th centuries. Similar inconsistent patterns of decline have been documented later in the 20th century. For example, Wilcove (1988) found no evidence of a widespread decline in Neotropical migrant birds in the Smoky Mountains between 1947 and 1983, while Hall (1984) noted some increasing and some decreasing species in West Virginia during the same decades. In the west, Marshall (1988) reported the disappearance of Olive-sided Flycatcher and Swainson's Thrush from study sites in the Sierra Nevada Mountains of California while most other Neotropical migrants remained fairly stable in habitats that did not change substantially over a 50-year period. Numerous additional studies have described trends in Neotropical migrants in recent years (such as Graber and Graber 1963, Ambuel and Temple 1982, Johnson and Cicero 1985, Sharp 1985), and some studies associated changes in bird populations with habitat changes (Robbins 1980, Askins and Philbrick 1987).

In general, Neotropical migrants have shown few consistent regional and temporal patterns in their trends during the past century. These inconsistent patterns are not too surprising, since they probably reflect regional differences in habitat availability, habitat changes, weather, and many other factors. However, widespread concern did not begin to focus on massive habitat alterations, pesticide use, and other factors as threats to Neotropical migrants until the 1970s and 1980s (Keast and Morton 1980, Hagan and Johnston 1992). This concern became more focused during the late 1980s after Robbins et al. (1989a) demonstrated consistent declines in most Neotropical species in the northeastern United States after 1980, perhaps indicating that a factor or several factors are operating to adversely effect the populations of these species.

BREEDING BIRD SURVEY

During the 1960s, the widespread use of dichloro-diphenyl-trichloro ethane (DDT) and other chlorinated hydrocarbon pesticides was associated with anecdotal accounts of mortality of songbirds, raising serious questions as to whether continental bird populations were detrimentally affected by this pesticide use. Responding to the need for a continental monitoring program, the US

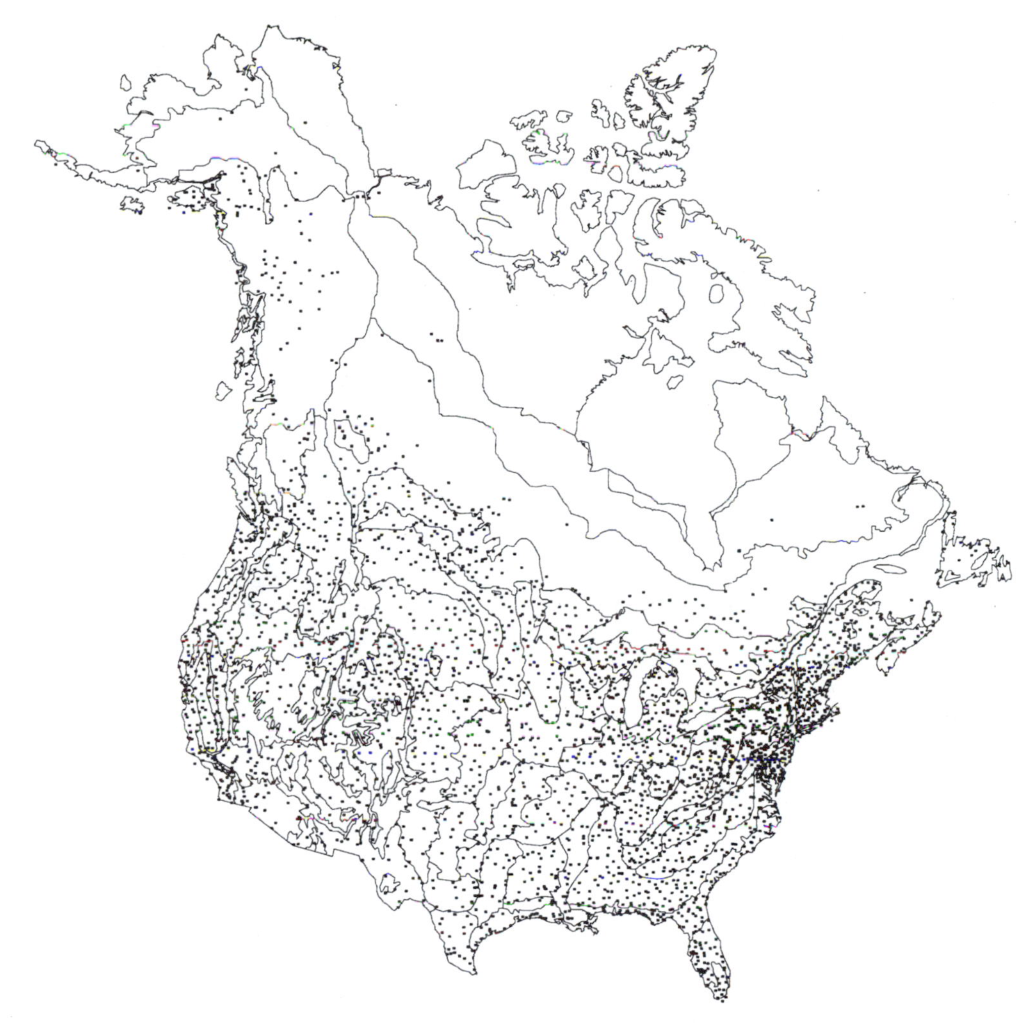

Figure 1-1. Map of BBS route locations in North America. Physiographic strata boundaries are indicated as solid lines.

Fish and Wildlife Service developed and tested the Breeding Bird Survey. Following trials on routes in Maryland (50 routes) and Delaware (ten routes) in the summer of 1965, the BBS was launched in 1966 throughout the eastern United States and Canada in cooperation with the Canadian Wildlife Service. Over the next 2 years it was expanded to include the continental United States and southern Canada. Routes in Alaska, Yukon, and Northwest Territories were added in the 1980s.

The BBS currently comprises approximately 3400 randomly located permanent survey routes established along secondary roads (Fig. 1-1). Each route is 24.5 miles (39.4 km) long, consisting of 50 stops spaced at 0.5 mile (0.8 km) intervals. These routes are surveyed once annually during the peak of the breeding season by volunteers knowledgeable in the identification of birds by sight and sound. All birds detected within 0.25 mile (0.4 km) of each stop during a 3 minute observation period are recorded. More detailed information on the BBS methodology is provided by Robbins et al. (1986).

ESTIMATION OF TRENDS

We estimated population trends and annual indices using the route-regression methods described by Geissler and Sauer (1990) and Sauer and Geissler (1990). In this analysis, trend, a consistent change in counts of birds on a route is estimated. Annual indices of abundance are used to assess other patterns in the data in the context of the estimated trend.

Route trends are estimated for individual routes from linear regression of the natural logarithm of counts (plus 0.5 to accommodate zero counts) on year of survey and observer covariables. The observer covariables allow the location of the estimated trend line to shift higher or lower depending on the observer. Covariables are necessary because observers show significant differences in their ability to count birds on BBS routes, and the quality of new observers on routes tends to be better than the observers they replace. Failure to include covariables causes a positive bias in trend estimates (Sauer et al. 1994). The slope estimate for year, when transformed to the multiplicative scale, provides the trend estimate for the route.

Regional trends are computed as weighted averages of route trends. Besides inherent variability in the counts on the route, missing counts (from years when the route was not surveyed) and observer changes (that modify the quality of the data) both tend to make route data less reliable. Consequently, it is necessary to weight the route trends by a measure of the consistency of counting on the route. The measure we use is the variance of the slope divided by the mean square error from the route-specific regression. This measure is proportional to the number of years run and number of observer changes, but as it does not contain the mean square error of the count data, it provides no information on variation in counts. We additionally weight route trend estimates by mean route count (Geissler and Sauer 1990). Owing to geographic variation in the sample intensity, we also weight by areas of the physiographic strata within states. This weighting makes the stratum-within-state regions the smallest regions for estimating trends, and trends from larger regions are composed of area-weighted averages of trend estimates from the component stratum-within-state regions. Bootstrapping is used to estimate variances of trends (Efron 1982, Geissler and Link 1988, Geissler and Sauer 1990) and because of the area weighting the degrees of freedom for a trend estimate is the number of routes minus the number of strata-within-state regions. We used z-tests to evaluate the null hypothesis that trends were equal to 0, and discuss as significant those where $P < 0.1$ in statistical tests.

Annual Indices of Abundance

Using the regional trend estimate and a regional average count, a curve can be drawn that depicts the predicted trend in counts over time. We define annual indices as deviations from this regional predicted trend. These annual indices can be thought of as residual variation remaining after the trend is modeled.

For each route in the region, we estimate observer effects using the regression described above but, instead of estimating a route trend, we incorporate the regional estimate of trend into the regression. Residuals are calculated for each year the route was run. Residuals from all routes are averaged by year and the yearly averages are then added to the yearly counts from the regional predicted trend. Resulting estimates of composite annual indices are then exponentiated to produce the yearly annual indices (Sauer and Geissler 1990). This approach allows direct observation of yearly indices to see abrupt changes in indices, and the indices can be smoothed using a procedure such as LOESS (Cleveland and Devlin 1988) to allow observation of patterns in the data.

Regions and Time Periods of Analysis

We estimated population trends for all species of Neotropical migrant birds for three time periods: 1966–1991 (the entire survey interval); 1966–1979 (an "early" period); and 1980–1991 (a "recent" period). The 1966–1979 period corresponds to the first published summary of the BBS (Robbins et al. 1986) and was used to be directly comparable with the results provided in that summary.

Regional trends were estimated for states and provinces, physiographic strata (Fig. 1-1), Eastern, Central, and Western BBS Regions (Robbins et al. 1986), and the entire survey area. Annual indices of abundance were estimated for selected regions for the entire survey interval.

Summary of Regional Patterns of Population Changes

One important aspect of BBS analyses is that we can evaluate the consistency of population changes among regions. If the annual indices of abundance for a species exhibit a consistent pattern across a number of physiographic strata, this pattern provides evidence that some common factor is affecting regional population trends. Consistency of pattern among species within a region suggests that some common factor may be influencing these species. Also, variation in patterns of population change within a species may reflect an association of population changes with an environmental feature such as regional changes in land use. BBS data are appropriate to evaluate bird–environmental associations (Robbins et al. 1986, Sauer and Droege 1990, 1992), within the constraints of the sample design (Barker and Sauer 1992). We present two different analyses of these patterns, one emphasizing regional trend patterns within species and the other emphasizing regional trend patterns in groups of species.

Species-specific Regional Analyses

We present detailed regional analyses for a few selected species of Neotropical migrants to illustrate that regional patterns in population changes exist in the BBS data. These species exemplify some of the regional patterns apparent within the entire Neotropical migrant guild.

Regional Analyses of Guilds of Neotropical Migrants

We summarized the trends for Neotropical migrant species within each of the regions by determining the percentages of species that had positive estimates of trend. Binomial tests were used to examine the null hypothesis that the percentages did not differ from 50%. If these percentages significantly differ from 50%, they suggest that the physiographic stratum is experiencing a disproportionate number of increasing or decreasing species. We conducted *z*-tests to determine whether the proportion of increasing species in each region for each guild differed between the 1966–1979 and 1980–1991 time periods.

We estimated the percentage of increasing species by physiographic strata, the BBS regions, and continentally for all Neotropical migrant birds and for several species groups. For our analysis, Neotropical migrants were defined as the Category A species on the list of Neotropical migrant birds developed by Gauthreaux (1991). This list excludes all waterbirds and shorebirds that regularly migrate to the tropics, but otherwise includes all widely distributed North American species whose winter ranges predominantly extend into the Neotropics.

We divided these species into guilds associated with breeding habitats (urban, grassland, wetland, scrub, or woodland), nest type (cavity nesting or open cup), and nest location (ground/low or mid-story/canopy), based on characteristics described by Ehrlich et al. (1988) and Harrison (1978). Species were placed into breeding habitat guilds based on the habitat they most frequently use for nesting. The nest location guild reflects the variability exhibited by some species; for example, some tree-nesting species exhibit a wide range of variability in the heights of their nests while some ground-nesting species will also regularly place their nests in low shrubs. Assigning guild membership can be critical to the results of any analysis, and we recognize that some of our definitions are arbitrary and may be questioned by others.

We estimated the percentages of increasing species for each of these guilds. In all analyses, only trends based on >14 degrees of freedom and weighted average counts of at least 1.0 bird/route were used to estimate the percentages of increasing species. These relatively conservative criteria were chosen to minimize possible positive bias associated with small samples and low relative abundances (Geissler and Link 1988, Barker and Sauer 1992).

Use of broadly defined guilds for comparative studies has been questioned when heterogeneity exists among members of the guild (Jaksic 1981, Mannan et al. 1984). Life history traits may be interdependent, so their importance may not be distinguished by separate analysis of each characteristic. Given these limitations, we urge caution interpreting the results of our guild analysis.

Evaluation of Efficiency of the BBS Sample

How good is the BBS in detecting population changes of Neotropical migrant birds? This question must be answered before we can decide which species need additional monitoring by other surveys or by extension of the BBS into new regions. To answer the question properly we must consider two aspects of sampling: (1) which species must be considered a priori to be poorly monitored by the BBS because the BBS does not adequately cover their breeding ranges; and (2) which species are not sampled appropriately with the existing survey?

Examination of BBS route locations (Fig. 1-1) indicates that the survey has relatively few routes in some northern habitats, especially portions of the boreal forest, northern Rocky Mountains, and most of Alaska. Neotropical migrants that are widely distributed in these habitats may not be accurately sampled by the survey.

Sauer and Droege (unpublished) conducted an extensive analysis of the validity of the BBS sample. Regional variation in sample efficiency was documented, and certain states and provinces (such as Maine and Idaho) had sparse samples that make long-term trend estimates of questionable validity. They also documented the precision of trend estimates by state and larger geographic regions. We present a revised version of this analysis for Neotropical migrant species.

We suggest that for a species to be well sampled in the BBS, trend estimates should be of sufficient precision to detect a population change of 50% over a 25-year period with a probability of 0.9. We use the existing trend estimates to determine which species currently meet this criterion. However, trend estimates and their variances do not by themselves provide sufficient information to allow us to judge the effectiveness of the BBS sample. Species that are seen on few BBS routes, or are recorded in relatively low numbers, also tend to be poorly sampled (Geissler and Link 1988). Therefore, we also note relative abundances and degrees of freedom for trend analyses. As noted above, trends from species seen at relative abundances of <1.0 bird/route or with fewer than 14 degrees of freedom must be viewed with caution, as the variances may be underestimated and the trend estimates may be positively biased (Barker and Sauer 1992).

RESULTS AND DISCUSSION

Guild Analysis

For all Neotropical migrants during the entire survey, increasing species significantly outnumbered decreasing species in the Western ($P < 0.1$) Region. Decreasing species outnumbered increasing species only in the Central Region ($P < 0.05$) (Table 1-2). However, the "early" and "recent" periods exhibited markedly different patterns in their proportions of increasing species. During 1966–1979, the Eastern ($P < 0.01$), Western ($P < 0.1$), and Continental ($P < 0.01$) Regions all had significantly more species increasing than decreasing. In contrast, only the Western Region ($P < 0.1$) had significantly more increasing species during 1980–1991, while the Eastern ($P < 0.01$) and Central ($P < 0.05$) had significantly more decreasing species. A comparison of these two intervals indicated a significant reduction in the number of increasing species in the Eastern and Continental Regions after 1980 (Table 1-2).

Similar patterns were evident in the guilds based on breeding habitats (Table 1-2). In the Woodland guild more species increased than decreased in the Eastern ($P < 0.01$) and Continental ($P < 0.01$) Regions during 1966–1979. No region had significantly more increasing species during 1980–1991, when decreasing species outnumbered increasing species in the Eastern Region ($P < 0.05$). A significant reduction in the number of

Table 1-2. Proportion of increasing species by guilds for Neotropical migrant birds. Proportions (*Prop.*), the statistical significance of the test that the proportion does not differ from 0.5 (A = $P < 0.10$; B = $P < 0.05$; C = $P < 0.01$), and the number of species in the guild are presented for BBS regions and the entire survey range for three time intervals: 1966–1979, 1980–1991, and 1966–1991. To appear in this list, the weighted relative abundance of the species had to be $\geqslant 1.0$ for the region, and the degrees of freedom had to be $\geqslant 14$. We also present a test for the significance of the differences between the 1966–1979 and 1980–1991 intervals (P_{diff}).

Region	Interval						P_{diff}
	1966–1979		1980–1991		1966–1991		
	N	*Prop.*	*N*	*Prop.*	*N*	*Prop.*	
Successional/scrub							
Eastern	14	0.86^B	14	0.36	13	0.54	C
Central	16	0.31	14	0.29	15	0.27	
Western	20	0.75^B	21	0.71^A	21	0.76^B	
Continent	27	0.74^B	29	0.52	28	0.54	
Woodland							
Eastern	29	0.86^C	34	0.29^B	33	0.67^A	C
Central	14	0.50	14	0.29	13	0.15^B	
Western	19	0.58	21	0.52	20	0.50	
Continent	42	0.76^C	44	0.41	44	0.61	C
Open cup							
Eastern	45	0.80^C	50	0.32^B	48	0.58	C
Central	35	0.34^A	33	0.33^A	33	0.21^C	
Western	40	0.65^A	43	0.63	42	0.60	
Continent	71	0.72^C	75	0.44	74	0.51	C
Ground/low-nesting							
Eastern	23	0.78^B	26	0.27^B	24	0.54	C
Central	22	0.27^A	20	0.15^C	21	0.14^C	
Western	21	0.76^B	22	0.64	22	0.64	
Continent	38	0.71^B	41	0.34^A	39	0.46	C
Mid-story/canopy							
Eastern	30	0.83^C	33	0.39	33	0.64	C
Central	24	0.58	24	0.46	23	0.39	
Western	28	0.57	30	0.67^A	29	0.62	
Continent	46	0.74^C	47	0.57	48	0.65^A	
All Neotropical migrants							
Eastern	55	0.78^C	61	0.33^C	59	0.58	C
Central	54	0.48	52	0.33^B	51	0.35^B	
Western	56	0.63^A	60	0.63^A	58	0.62^A	
Continent	94	0.70^C	100	0.46	97	0.55	C

increasing species occurred in the Eastern and Continental Regions after 1980 ($P < 0.01$; Table 1-2). For the entire survey period, increasing species significantly outnumbered decreasing species in the Eastern Region ($P < 0.1$) while the opposite was true in the Central Region ($P < 0.05$).

Successional/scrub nesting birds included more increasing than decreasing species in the Eastern ($P < 0.05$), Western ($P < 0.05$), and Continental ($P < 0.05$) Regions during 1966–1979 and in the Western Region ($P < 0.1$) during 1980–1991 (Table 1-2). Decreasing species did not significantly outnumber increasing species in any region, although fewer species were increasing after 1980 in the Eastern Region.

Open-cup nesters had significantly more decreasing than increasing species in the Central Region ($P < 0.01$) for the entire survey period while no region had significantly more increasing species (Table 1-2). During 1966–1979, the Eastern ($P < 0.01$), Western ($P < 0.1$), and Continental ($P < 0.01$) regions had significantly more increasing species while the Central Region ($P < 0.1$) had more decreasing species. After 1980, both the Eastern ($P < 0.05$) and Central ($P < 0.1$) regions had more decreasing than increasing species.

Similar patterns were apparent for the ground/low-nesting guild. The only significant result during the entire survey period was for decreasing species in the Central Region ($P < 0.01$) (Table 1-2). During 1966–1979, increasing species outnumbered decreasing species in the Eastern ($P < 0.05$), Western ($P < 0.05$), and Continental ($P < 0.05$) regions while more decreasing than increasing species were evident in the Central Region ($P < 0.1$). During 1980–1991, the Eastern ($P < 0.05$), Central ($P < 0.01$), and Continental ($P < 0.1$) regions all had significantly more decreasing than increasing species.

In the mid-story/canopy nesting guild, no region had significantly more decreasing species during any period (Table 1-2). Significantly more increasing species were noted in the Eastern ($P < 0.01$) and Continental ($P < 0.01$) Regions during 1966–1979, Western Region ($P < 0.1$) during 1980–1991, and the Continental Region ($P < 0.1$) for the entire survey. However, significantly fewer species increased in the Eastern Region after 1980.

Several general patterns are evident in this guild analysis. During 1966–1979, significantly more increasing species generally occurred within guilds in the Eastern,

Western, and Continental Regions. These results contrasted with the 1980–1991 results when the Eastern and Central Regions tended to have significantly more decreasing species in most guilds. The Western Region still had significantly more increasing species in a few guilds, while the Continental guilds had nearly equal proportions of increasing and decreasing species. As a result of these conflicting patterns, relatively few results were significant for the entire survey period, except for the Central Region which had significantly more decreasing species in most guilds.

In the Eastern Region, the tendency for guilds to have more increasing species during 1966–1979 and more decreasing species after 1980 differs from the Central Region, where guilds tend to have more decreasing species throughout the survey, and the Western Region where they generally have more increasing species in both periods. These regional differences may indicate that different factors are responsible for the observed patterns of increasing species in guilds. Habitat degradation in their temperate breeding range, destruction of tropical wintering habitats, and increased predation and nest parasitism during the breeding season in North America have been implicated in the recent declines of Neotropical migrants (Askins et al. 1990), but the relative importance of each factor in the various regions of the continent remains to be determined. In any event, the regional guild analysis indicates that a single continent-wide approach toward the conservation of Neotropical migrants may not be realistic, and that conservation programs should address the specific factors influencing population trends in each region.

Sample Efficiency

Only 16 species of Neotropical migrants are not sampled with sufficient intensity in the BBS to meet our minimum precision criteria (Table 1-3). Three species are not included in this table owing to their extremely small populations (Black-capped Vireo, Golden-cheeked Warbler, and Kirtland's Warbler). However, a number of species have small sample sizes and low relative abundances at the continental level, indicating their results should be viewed with caution.

Species Analysis

During 1966–1979, 34 species of Neotropical migrants exhibited significant increases while 19 experienced significant decreases (Table 1-3). This pattern was reversed in 1980–1991 when 35 species had significant decreases and only 15 underwent significant increases. For the entire survey period, the number of significantly increasing and decreasing species was 18 and 23, respectively.

Relatively few Neotropical migrants exhibited significant trends that consistently increased or decreased during the entire survey period. The only species with significant increases during 1966–1979 and 1980–1991 at the continental level were Upland Sandpiper, House Wren, Solitary Vireo, Warbling Vireo, Red-eyed Vireo, and Blue Grosbeak. Continental populations of Chimney Swift, Eastern Wood-Pewee, Lark Sparrow, and Grasshopper Sparrow significantly decreased during both periods.

The Eastern Region had nearly equal numbers of significantly increasing and decreasing Neotropical migrant species with 13 and 15, respectively (Table 1-4). The regional analysis demonstrated a low proportion of increasing species in the Central Region, results consistent with the guild analysis. Only six species significantly increased as compared with 17 that significantly declined. This pattern was reversed in the Western Region where 22 species significantly increased and only eight significantly decreased, reflecting the relatively high proportion of increasing species apparent in the guild analysis.

Very few Neotropical migrants exhibited consistent significant trends across their breeding ranges. For example, of the ten species experiencing significant increases or decreases throughout the entire survey period (Table 1-3), only Eastern Wood-Pewee (decreases), Solitary Vireo (increases), and Grasshopper Sparrow (decreases) exhibit similar significant trends in every region where they breed. Other species showing significant decreases across two or more regions include Yellow-billed Cuckoo, Veery,

Table 1-3. Continental population trends for Neotropical migrant species over the intervals 1966–1979 and 1980–1991 and the entire BBS period, 1966–1991. Trends are presented for all species that were seen on 14 or more BBS routes during 1966–1991. For the three intervals, we present the percentage change per year (%/year), sample size (in df = number of routes − number of state-stratum regions), statistical significance (A = $P < 0.10$; B = $P < 0.05$; C = $P < 0.01$) and the mean count on BBS routes over the 1966–1991 interval (relative abundance, RA). Guilds are denoted by letters: s = successional/scrub nesting species; w = woodland species; c = cavity nesting species; o = open-cup nesting species; g = ground–low nesting species; m = mid-story/canopy nesting species. See American Ornithologists' Union (1983) for scientific names of bird species.

Species	Guild	1966–1979		1980–1991		1966–1991		
		%/year	DF	%/year	DF	%/year	DF	RA
American Swallow-tailed Kite	w	−0.3	9	1.8	34	1.8	40	0.11
Mississippi Kite		1.0	56	−1.2	100	0.3	117	0.77
Broad-winged Hawk	w	2.0^{B}	516	-1.8^{A}	570	0.8	818	0.15
Swainson's Hawk		0.2	280	2.9^{B}	400	1.2	507	0.78
Merlin	w	0.2	37	0.8	67	1.3^{A}	109	0.04
Mountain Plover	g	6.3^{A}	6	−3.9	23	-3.7^{C}	25	0.39
Upland Sandpiper	g	3.5^{C}	369	2.7^{B}	427	3.6^{C}	557	1.90
Long-billed Curlew	g	3.0	76	−0.4	136	−0.5	170	1.48
Band-tailed Pigeon	w	−0.1	125	-10.4^{C}	121	-3.1^{C}	156	1.74
Black-billed Cuckoo	wog	11.6^{B}	833	1.4	898	0.6	1141	0.64
Yellow-billed Cuckoo	wog	3.2^{C}	1130	-4.5^{C}	1224	-1.3^{C}	1432	3.79
Burrowing Owl		1.8	163	1.6	199	−0.2	285	0.61
Lesser Nighthawk	s	−5.4	62	2.7	57	5.2^{B}	90	1.18
Common Nighthawk		1.0	850	-2.8^{B}	1000	−0.1	1349	2.12
Chuck-will's-widow	w	−0.1	327	−0.1	386	−0.9	445	1.45
Whip-poor-will	w	−0.4	371	-2.7^{C}	314	−1.0	506	0.27
Black Swift[a]		25.5	38	1.8	41	3.2	60	1.72
Chimney Swift		-1.1^{A}	1308	-2.0^{C}	1413	−0.8	1591	5.85
Vaux's Swift	wc	1.1	60	-5.0^{A}	97	−0.7	120	0.52
White-throated Swift[a]		−0.1	69	−5.8	91	−3.2	133	0.91
Ruby-throated Hummingbird	w	1.0	903	2.3^{B}	1016	1.4	1271	0.32
Black-chinned Hummingbird	w	−0.9	89	−0.5	98	0.9	147	0.21
Costa's Hummingbird[a]	s	5.3	34	0.9	27	2.0	48	0.35
Calliope Hummingbird	w	−0.9	42	−1.4	61	1.2	82	0.19
Broad-tailed Hummingbird	w	−1.2	25	1.2	61	0.3	80	1.84
Rufous Hummingbird	w	0.6	101	−2.0	115	-3.1^{B}	152	1.14
Allen's Hummingbird		2.8	32	-3.7^{A}	25	−0.5	36	0.47
Olive-sided Flycatcher	wom	−1.6	421	-2.2^{B}	471	-2.5^{C}	622	1.32
Western Wood-Pewee	wom	−1.5	317	0.4	439	−0.4	526	2.76
Eastern Wood-Pewee	wom	-1.9^{C}	1256	-1.4^{C}	1346	-1.6^{C}	1529	2.64
Yellow-bellied Flycatcher	wog	10.2^{C}	125	5.5	134	5.1^{C}	203	0.94
Acadian Flycatcher	wom	1.0	503	0.0	581	0.5	720	1.29
Least Flycatcher	wom	−0.9	743	−0.3	796	−0.7	1001	3.61
Hammond's Flycatcher	wom	0.3	83	3.5^{B}	150	1.3	175	2.58
Dusky Flycatcher	wog	10.2^{C}	88	−3.2	184	0.8	214	1.85
Gray Flycatcher[a]	sog	−4.6	21	9.1^{C}	54	2.9	64	0.81
Vermilion Flycatcher	wom	−3.1	36	0.6	26	-2.8^{A}	43	0.33
Ash-throated Flycatcher	scg	5.2^{C}	224	0.7	240	2.4^{C}	304	3.81
Great Crested Flycatcher	wcm	0.9	1285	−0.4	1448	0.0	1612	3.09
Cassin's Kingbird	om	0.4	57	−4.9	73	−2.4	104	2.01
Western Kingbird	om	1.8^{A}	518	0.8	660	1.5^{B}	799	5.37
Eastern Kingbird	om	−0.7	1530	0.0	1785	−0.1	1999	4.19
Scissor-tailed Flycatcher	om	-4.8^{C}	177	1.0	177	−0.4	207	9.49
Purple Martin	cm	4.0^{C}	1092	-1.3^{A}	1123	0.9	1404	3.99
Violet-green Swallow	cm	−0.7	250	2.1	342	0.8	420	4.67
Northern Rough-winged Swallow	m	3.7^{C}	1263	−0.3	1485	1.1	1832	1.58
Bank Swallow	m	0.5	761	4.9	787	0.2	1104	3.39
Cliff Swallow	m	4.9^{C}	964	0.8	1238	1.1	1485	16.57
Barn Swallow	om	4.5^{C}	1780	-2.1^{C}	2153	0.6	2405	13.72
House Wren	scm	2.4^{C}	1205	1.6^{C}	1439	1.4^{C}	1662	3.86

(continued)

Table 1-3 (*cont.*)

Species	Guild	1966–1979		1980–1991		1966–1991		
		%/year	DF	%/year	DF	%/year	DF	RA
Blue-gray Gnatcatcher	wom	0.9	692	0.6	876	0.8	1033	1.49
Veery	wog	1.2	606	−2.1[C]	687	−1.0[B]	821	3.96
Gray-cheeked Thrush[a]	wog	−9.3[B]	8	−0.5	6	0.7	14	0.24
Swainson's Thrush	wog	3.0[B]	380	−1.0	440	0.2	581	12.93
Wood Thrush	wom	0.8	1072	−2.5[C]	1169	−2.0[C]	1339	4.58
Gray Catbird	sog	0.6	1339	−1.1[B]	1484	−0.5	1698	2.25
Phainopepla	som	4.8	62	0.9	61	3.2[B]	85	1.36
White-eyed Vireo	sog	0.3	586	−0.6	699	−0.2	815	3.44
Bell's Vireo	sog	−5.4[B]	183	−1.4	160	−2.3[B]	244	0.84
Gray Vireo[a]	sog	6.6	4	6.7	7	5.4	16	0.34
Solitary Vireo	wom	3.2[B]	490	4.5[C]	625	3.3[C]	791	1.17
Yellow-throated Vireo	wom	0.0	709	1.0	844	0.7	1039	0.60
Warbling Vireo	wom	2.0[B]	1061	1.4[A]	1263	1.4[C]	1504	2.63
Philadelphia Vireo	wom	1.9	68	4.3[B]	101	1.5	151	1.05
Red-eyed Vireo	wom	2.7[C]	1378	1.4[C]	1533	1.4[C]	1780	9.07
Blue-winged Warbler	sog	1.4	255	−1.8[A]	292	−0.1	399	0.31
Golden-winged Warbler	sog	−2.1[A]	195	0.2	185	−2.4[C]	271	0.25
Tennessee Warbler[a]	wog	11.9[A]	172	−9.1[C]	189	4.5	278	5.59
Orange-crowned Warbler	sog	0.4	176	1.9[B]	238	−0.6	284	2.12
Nashville Warbler	sog	0.7	407	0.2	480	1.7	589	4.94
Virginia's Warbler	sog	0.3	11	4.3[C]	16	3.0	22	1.00
Northern Parula	wom	2.1[B]	553	−1.0	632	1.0	812	1.10
Yellow Warbler	som	0.4	1337	0.6	1571	0.9[A]	1859	3.84
Chestnut-sided Warbler	sog	1.8[A]	531	−0.9	569	−0.6	698	5.24
Magnolia Warbler	wom	3.5[C]	309	0.6	346	2.9[C]	449	4.97
Cape May Warbler	wom	16.5	113	−7.8[A]	136	3.2[A]	203	1.18
Black-throated Blue Warbler	wom	−2.6[B]	249	0.2	289	0.4	391	0.52
Black-throated Gray Warbler	som	1.7	95	0.6	124	2.1[B]	152	1.19
Townsend's Warbler[a]	wom	7.0[C]	60	3.0	97	2.9	121	3.23
Hermit Warbler	wom	3.8	49	1.0	55	1.2	66	3.02
Black-throated Green Warbler	wom	0.4	376	−1.1	430	−0.1	535	2.41
Blackburnian Warbler	wom	2.1	305	2.5[B]	345	1.0	448	1.16
Yellow-throated Warbler	wom	0.8	234	0.4	309	0.4	409	0.42
Grace's Warbler[a]	wom	−8.3[C]	8	1.3	10	−1.2	15	1.51
Prairie Warbler	sog	−4.5[C]	482	−0.6	522	−2.4[C]	650	1.39
Palm Warbler	sog	1.3	29	4.8[A]	34	1.8	58	0.08
Bay-breasted Warbler[a]	wom	11.1	126	−19.3[C]	124	1.1	193	2.40
Blackpoll Warbler[a]	wom	18.9[C]	84	−10.0[C]	39	3.0	120	3.15
Cerulean Warbler	wom	−3.1[C]	151	−0.6	144	−2.7[C]	237	0.21
Black-and-white Warbler	wog	0.9	684	0.9	735	1.1	952	1.31
American Redstart	wom	1.2	828	−1.5[B]	818	−0.6	1095	2.54
Prothonotary Warbler	wcm	1.9[A]	284	−1.0	291	−0.5	380	0.90
Worm-eating Warbler	wog	0.6	168	0.6	217	0.6	291	0.19
Swainson's Warbler	wog	6.6[C]	34	−1.1	55	1.4	76	0.11
Ovenbird	wg	1.5[B]	819	−0.6	901	0.7	1088	5.32
Northern Waterthrush	wog	2.2	338	−0.9	361	0.8	511	1.56
Louisiana Waterthrush	wog	−0.2	309	−1.4[A]	374	0.1	493	0.21
Kentucky Warbler	wog	0.2	400	−1.4	453	−0.7	575	1.09
Connecticut Warbler	sog	−3.7	43	−2.1	72	1.8	90	0.46
Mourning Warbler	sog	1.9	299	−2.0[C]	357	0.5	450	4.09
Macgillivray's Warbler	sog	0.2	145	−2.0	207	−0.8	253	3.05
Common Yellowthroat	sog	0.8	1562	−2.1[C]	1829	−0.5	2068	7.18
Hooded Warbler	wog	1.1	315	1.1	391	1.3	505	1.06
Wilson's Warbler	sog	1.7	270	−2.3[A]	311	0.9	419	1.34
Canada Warbler	wog	−1.3	313	−3.0[C]	328	0.4	438	1.12
Yellow-breasted Chat	sog	−3.3[C]	780	0.5	854	−0.6	1070	2.85
Hepatic Tanager[a]	wom	2.8	7	0.2	13	5.2[C]	15	0.51
Summer Tanager	wom	0.6	503	−1.7[C]	557	−0.2	647	2.15
Scarlet Tanager	wom	2.3[C]	820	−1.1[A]	928	0.2	1100	1.17
Western Tanager	wom	−2.9[B]	251	0.8	323	−0.5	390	3.09

Table 1-3 (*cont.*)

Species	Guild	1966–1979		1980–1991		1966–1991		
		%/year	DF	%/year	DF	%/year	DF	RA
Rose-breasted Grosbeak	wom	5.1[C]	763	−3.8[C]	874	0.0	1005	2.21
Black-headed Grosbeak	wom	−2.2	253	1.1	358	−0.1	423	1.85
Blue Grosbeak	sog	3.5[C]	612	1.9[C]	742	2.0[C]	860	2.37
Lazuli Bunting	sog	−1.9	180	2.4	279	0.1	333	1.03
Indigo Bunting	sog	0.3	1223	−1.1[B]	1332	−0.5	1514	9.64
Painted Bunting	sog	−2.4[B]	183	−0.4	179	−3.4[C]	224	4.77
Dickcissel	og	−6.0[C]	548	0.9	576	−1.7[C]	667	12.39
Green-tailed Towhee	sog	1.7	68	0.6	146	0.3	166	2.42
Chipping Sparrow	om	−1.2[C]	1510	0.3	1732	−0.1	2018	6.64
Clay-colored Sparrow	sog	1.4	218	−1.1	316	−1.4[B]	379	6.09
Brewer's Sparrow	sog	−2.6	124	−4.9[C]	247	−4.1[C]	294	6.89
Lark Sparrow	sog	−4.6[C]	518	−2.1[B]	607	−3.4[C]	771	4.06
Lark Bunting	og	−6.2[C]	156	−0.4	229	−3.1[B]	284	30.24
Baird's Sparrow	og	−3.3	55	−0.5	84	−1.9	110	1.49
Grasshopper Sparrow	og	−4.3[C]	925	−2.7[C]	1003	−4.6[C]	1248	3.30
Lincoln's Sparrow	sog	5.5[A]	154	−2.1	249	4.1[B]	320	2.28
Bobolink	og	1.3[A]	754	−4.6[C]	853	−1.1[B]	1002	5.61
Yellow-headed Blackbird	mog	4.8[B]	283	1.7	436	3.0[B]	531	7.80
Orchard Oriole	m	−2.2[C]	817	0.4	957	−1.3[B]	1125	2.78
Hooded Oriole[a]	m	0.3	35	0.3	28	0.8	49	0.24
Bullock's Oriole	m	−1.1	320	0.5	422	−0.8	511	1.67
Baltimore Oriole	m	2.6[C]	1126	−1.3[B]	1204	0.4	1406	2.65
Scott's Oriole	sm	1.3	54	0.3	55	2.7	79	1.49

[a] Species not monitored with sufficient intensity to detect a 50% decline in population over a 25-year period with probability 0.9.

Wood Thrush, Prairie Warbler, Painted Bunting, and Brewer's Sparrow. House Wren is the only other species to exhibit significant increases in at least two regions.

Geographic Patterns in Trends of Individual Species

Neotropical migrants experiencing regionally and temporally significant decreases in population trends may be in need of conservation measures, while those undergoing significant increases may have relatively secure populations. However, most Neotropical migrants exhibit a geographical mosaic of increases and decreases as a result of regional and local factors influencing their populations. A detailed discussion of the geographic patterns in the trends of each Neotropical migrant species is beyond the scope of this paper. However, we analyze these patterns for four species to exemplify the geographic variation in the trends of Neotropical migrants, recognizing that the BBS is useful for developing an understanding of the long-term fluctuations in bird populations but does not identify the factors responsible for these fluctuations.

Acadian Flycatcher

The Acadian Flycatcher is a Neotropical migrant that does not show a significant trend for any time period or geographic region in this analysis (Tables 1-3 and 1-4). Within its breeding range, the physiographic strata exhibit a mosaic of increases and decreases without any apparent geographic pattern (Fig. 1-2). Significant decreases occur in the Ozark–Ouachita Plateau, Blue Ridge Mountains, and Osage Plain–Cross Timbers physiographic strata. Acadian Flycatchers experience significant increases in the Highland Rim and Allegheny Plateau strata.

Examination of temporal patterns in trends does not reveal any consistent results (Fig. 1-3A–E). For strata exhibiting significant trends (Fig. 1-3A and B), these trends are fairly consistent over the entire survey. All strata show some year-to-year variation in the annual indices, reflecting regional and annual population fluctuations as well as sampling inconsistencies.

Table 1-4. Population trends (%/year), sample size (in DF = number of routes − number of state-strata), relative abundances (RA) (birds/route) and statistical significances (A = $P < 0.01$, B = $P < 0.05$, C = $P < 0.10$) for Eastern, Central, and Western BBS Regions, the continental United States, Canada, and the survey area for Neotropical migrant birds for the interval 1966–1991.

Species	Eastern			Central			Western		
	%/yr	DF	RA	%/yr	DF	RA	%/yr	DF	RA
American Swallow-tailed Kite	1.8	40	0.13						
Mississippi Kite	1.9	50	0.07	0.3	67	1.23			
Broad-winged Hawk	0.9	744	0.18	0.1	69	0.10			
Swainson's Hawk				0.3	283	0.97	2.6^{B}	224	0.68
Merlin	1.9^{C}	35	0.03				1.4	63	0.06
Mountain Plover							−3.8	16	0.27
Upland Sandpiper	2.0	238	0.32	3.8^{A}	275	3.73	2.5	44	0.72
Long-billed Curlew				-4.8^{C}	60	1.58	4.2^{A}	110	1.43
Band-tailed Pigeon							-3.2^{A}	156	1.79
Black-billed Cuckoo	0.9	841	0.69	0.5	249	0.67	−1.1	51	0.37
Yellow-billed Cuckoo	-1.7^{A}	1041	3.32	-1.0^{B}	376	5.08	−3.2	15	0.10
Burrowing Owl				−1.8	141	0.83	3.9^{B}	134	0.44
Lesser Nighthawk				3.1	24	1.07	7.9^{C}	66	1.24
Common Nighthawk	-2.4^{A}	518	0.61	0.6	411	4.02	−1.0	420	2.14
Chuck-will's-widow	-1.7^{B}	311	1.68	0.8	134	1.22			
Whip-poor-will	-1.0^{C}	452	0.26	−1.3	54	0.32			
Black Swift							3.0	60	1.72
Chimney Swift	-1.1^{B}	1248	7.27	0.1	343	4.21			
Vaux's Swift							−0.9	120	0.52
White-throated Swift							-5.4^{C}	127	0.90
Ruby-throated Hummingbird	0.6	1078	0.32	4.0	177	0.37	−0.2	16	0.04
Black-chinned Hummingbird				1.3	38	0.35	0.4	109	0.17
Costa's Hummingbird							2.1	48	0.35
Calliope Hummingbird							1.4	82	0.22
Broad-tailed Hummingbird							0.3	80	1.95
Rufous Hummingbird							-3.1^{A}	152	1.14
Allen's Hummingbird							−0.4	36	0.47
Olive-sided Flycatcher	−0.6	317	1.08				-3.5^{A}	297	1.67
Western Wood-Pewee				1.2	60	0.39	−0.5	466	3.73
Eastern Wood-Pewee	-1.3^{A}	1248	3.24	-2.6^{A}	279	1.72			
Yellow-bellied Flycatcher	5.2^{A}	200	1.08						
Acadian Flycatcher	0.7	608	1.65	−1.6	112	0.62			
Least Flycatcher	-1.7^{A}	749	4.93	−1.5	126	1.30	2.6^{A}	126	3.13
Hammond's Flycatcher							1.2	175	2.58
Dusky Flycatcher							1.0	210	2.03
Gray Flycatcher							2.7	64	0.81
Vermilion Flycatcher				-5.0^{A}	34	0.33			
Ash-throated Flycatcher				1.9	49	1.27	2.6^{A}	255	4.67
Great Crested Flycatcher	0.0	1249	4.16	−0.1	342	2.20	4.5	21	0.36
Cassin's Kingbird							−2.9	93	3.01
Western Kingbird				1.7^{A}	360	8.09	0.8	434	3.99
Eastern Kingbird	−0.6	1297	3.57	0.3	483	7.33	0.7	219	1.54
Scissor-tailed Flycatcher				−0.3	200	11.16			
Purple Martin	−0.1	971	5.63	2.4^{A}	350	5.02	2.2	83	0.44
Violet-green Swallow							0.7	410	5.40
Northern Rough-winged Swallow	2.0^{A}	980	1.06	1.0	416	1.32	0.3	436	2.23
Bank Swallow	0.7	681	5.36	−0.3	229	1.78	−1.1	194	2.47
Cliff Swallow	1.0	598	3.48	0.0	360	23.77	1.9^{C}	527	23.01
Barn Swallow	0.2	1305	14.56	1.7^{A}	527	18.70	−0.5	573	9.06
House Wren	0.5	932	3.06	1.9^{A}	318	4.82	2.3^{A}	412	3.94
Blue-gray Gnatcatcher	1.4^{C}	732	2.16	−1.0	185	1.76	4.3	116	0.40
Veery	-1.1^{C}	678	7.01	-3.4^{A}	35	0.29	−0.5	108	1.42
Gray-cheeked Thrush	0.6	14	0.24						
Swainson's Thrush	0.8	299	20.37				−0.4	272	10.21
Wood Thrush	-2.0^{A}	1199	5.86	-1.8^{B}	140	1.15			
Gray Catbird	−0.5	1229	4.19	−1.6	325	1.03	1.6	144	0.48
Phainopepla							3.3^{B}	85	1.36

Table 1-4 *(cont.)*

Species	Eastern			Central			Western		
	%/yr	DF	RA	%/yr	DF	RA	%/yr	DF	RA
White-eyed Vireo	0.2	646	3.48	−1.9[A]	169	3.45			
Bell's Vireo	−1.0	41	0.11	−3.5[A]	178	0.77	1.2	25	1.56
Gray Vireo							5.5	16	0.38
Solitary Vireo	4.0[A]	489	1.15				2.5[A]	297	1.35
Yellow-throated Vireo	1.0[C]	914	0.75	−1.5	125	0.31			
Warbling Vireo	0.1	799	1.20	0.4	293	1.23	2.3[A]	412	5.31
Philadelphia Vireo	1.5	135	1.27				−1.0	16	0.08
Red-eyed Vireo	1.6[A]	1306	15.99	−0.3	305	1.98	1.3	169	4.38
Blue-winged Warbler	−0.2	377	0.36	2.8	22	0.15			
Golden-winged Warbler	−2.3[A]	271	0.28						
Tennessee Warbler	4.2	217	9.12				1.0	52	0.48
Orange-crowned Warbler							−0.6	280	2.73
Nashville Warbler	1.6	451	7.74				2.0[C]	129	1.17
Virginia's Warbler							2.9	22	1.00
Northern Parula	1.4[B]	708	1.25	−1.7	104	0.78			
Yellow Warbler	1.4[A]	1077	3.85	−0.4	285	1.82	0.4	497	5.08
Chestnut-sided Warbler	−0.6	679	6.50						
Magnolia Warbler	3.0[A]	426	6.75				−6.2	20	0.50
Cape May Warbler	3.1[C]	202	1.32						
Black-throated Blue Warbler	0.3	391	0.52						
Black-throated Gray Warbler							2.2[B]	152	1.19
Townsend's Warbler							2.8	121	3.23
Hermit Warbler							1.1	66	3.02
Black-throated Green Warbler	−0.1	533	2.64						
Blackburnian Warbler	1.0	442	1.25						
Yellow-throated Warbler	0.3	348	0.54	0.3	61	0.19			
Grace's Warbler							−1.0	15	1.51
Prairie Warbler	−2.2[A]	578	1.54	−4.0[A]	72	0.89			
Palm Warbler	1.9	56	0.09						
Bay-breasted Warbler	1.0	193	2.58						
Blackpoll Warbler	3.1	111	3.78						
Cerulean Warbler	−2.7[A]	219	0.25	−4.4	18	0.05			
Black-and-white Warbler	1.3	847	1.69	−0.9	93	0.46			
American Redstart	−0.7	928	4.14	1.8	86	0.24	−0.1	81	1.24
Prothonotary Warbler	0.0	293	0.83	−2.2	87	1.05			
Worm-eating Warbler	0.9	264	0.24	−0.2	27	0.10			
Swainson's Warbler	1.5	62	0.11	1.1	14	0.10			
Ovenbird	0.8	983	8.67	−4.3[A]	69	0.58	−1.0	36	0.36
Northern Waterthrush	0.0	435	1.53				2.7[B]	69	1.78
Louisiana Waterthrush	0.0	436	0.24	0.5	57	0.12			
Kentucky Warbler	−0.1	475	0.99	−2.1	100	1.34			
Connecticut Warbler	1.1	61	0.38				4.8[B]	21	0.66
Mourning Warbler	0.5	408	5.33	0.3	15	1.34	3.3	27	0.52
Macgillivray's Warbler							−0.7	251	3.13
Common Yellowthroat	−0.6	1332	13.00	−1.0	412	6.33	2.7[B]	324	1.43
Hooded Warbler	2.3[A]	450	1.01	−4.6	55	1.22			
Wilson's Warbler	2.8	135	0.91				0.4	283	1.75
Canada Warbler	0.5	436	1.22						
Yellow-breasted Chat	−0.9	690	5.97	−0.1	218	2.72	0.5	162	0.62
Hepatic Tanager							5.1[A]	15	0.62
Summer Tanager	−0.3	454	3.18	−0.2	179	1.97	2.7[C]	14	0.20
Scarlet Tanager	0.3	1024	1.47	−1.4	76	0.22			
Western Tanager							−0.5	381	3.26
Rose-breasted Grosbeak	−0.1	807	3.14	0.1	159	0.90	2.7	39	0.78
Black-headed Grosbeak				2.5[B]	64	0.15	−0.1	359	2.52
Blue Grosbeak	2.8[A]	454	4.51	−0.0	306	1.70	2.9[A]	100	1.23
Lazuli Bunting				−2.0	34	0.26	0.1	299	1.29
Indigo Bunting	−0.5	1185	14.86	−0.6	318	5.63			
Painted Bunting	−2.8[A]	32	0.59	−3.6[A]	183	6.57			

(continued)

Table 1-4 (*cont.*)

Species	Eastern			Central			Western		
	%/yr	DF	RA	%/yr	DF	RA	%/yr	DF	RA
Dickcissel	−3.8[C]	267	3.08	−1.3	398	19.36			
Green-tailed Towhee							0.1	157	2.71
Chipping Sparrow	0.3	1239	10.76	1.3	315	1.78	−2.7[A]	464	6.13
Clay-colored Sparrow	−3.3	125	0.80	−1.7	132	5.22	−1.2[B]	122	11.86
Brewer's Sparrow				−7.3[A]	60	2.48	−3.4[A]	234	8.08
Lark Sparrow	−8.1	47	0.07	−3.8[A]	389	7.08	−1.2	335	2.57
Lark Bunting				−3.5[B]	204	51.16	9.3	80	5.01
Baird's Sparrow				−2.8	61	1.66	−0.1	49	1.46
Grasshopper Sparrow	−5.3[A]	733	1.29	−4.3[A]	402	8.02	−2.7[B]	113	0.36
Lincoln's Sparrow	1.8	177	3.36				12.4[A]	140	1.56
Bobolink	−0.8	703	8.14	−3.7[A]	213	5.86	4.2	86	0.62
Yellow-headed Blackbird	2.5	39	0.55	0.3	213	13.77	4.7[B]	279	5.98
Orchard Oriole	−0.1	703	3.04	−3.5[A]	414	2.79			
Hooded Oriole							0.9	40	0.29
Bullock's Oriole				−1.0	143	1.70	−0.8	368	1.66
Baltimore Oriole	−0.2	970	2.44	−0.0	353	2.79	4.6[A]	83	3.00
Scott's Oriole							2.7	75	1.69

Wood Thrush

The BBS shows a fairly extensive geographic decline for the Wood Thrush since 1966 (Fig. 1-4), with significant declines in 16 strata while significant increases occur only in the Dissected Till Plains stratum. Many other strata show nonsignificant declines.

Wood Thrushes exhibit an interesting temporal pattern in their trends within physiographic strata (Fig. 1-5A–F). A few strata show fairly consistent declines throughout the survey (Fig. 1-5A and B). In most strata, however, Wood Thrush populations were relatively stable through the late 1970s and have undergone noticeable declines since 1980 (Fig. 1-5C–F).

A number of factors have been implicated in the recent declines in Wood Thrush populations. On the breeding grounds, Wood Thrushes occupy mature deciduous woodlands and disappear from localities where wooded habitats become extensively fragmented (Robbins et al. 1989b). Some populations experience high levels of nest parasitism by Brown-headed Cowbirds (Robinson 1992). In their winter range, Wood Thrushes are restricted to mature tropical forests, and avoid agricultural and scrub habitats (Rappole et al. 1989). Hence, deforestation in Central America may contribute to their population declines. The increased rate of decline on BBS routes after 1980 may indicate these factors have had a greater influence on Wood Thrush populations in recent years.

Black-and-white Warbler

Black-and-white Warblers exhibit nonsignificant trends at the continental and regional levels (Tables 1-3 and 1-4), but have experienced significant declines along the Appalachian Mountains in the Ridge and Valley, Blue Ridge Mountains, Ohio Hills, and Allegheny Plateau strata (Fig. 1-6). Significant increases are widely scattered in the Coastal Flatwoods, Mississippi Alluvial Plain, Lexington Plain, and St. Lawrence River Plain strata, although the greatest proportion of significant and nonsignificant increases occurs at the northern edge of its range.

The physiographic strata show a variety of temporal patterns in the trends of Black-and-white Warblers (Fig. 1-7A–F). Along the Appalachian Mountains declines are fairly consistent after 1966 (Fig. 1-7A–C). Other strata show nearly stable or slightly increasing trends (Fig. 1-7D and E), while the Cumberland Plateau stratum shows an increase through 1974 followed by a decrease (Fig. 1-7F). The factors responsible for the significant decrease in Black-and-white

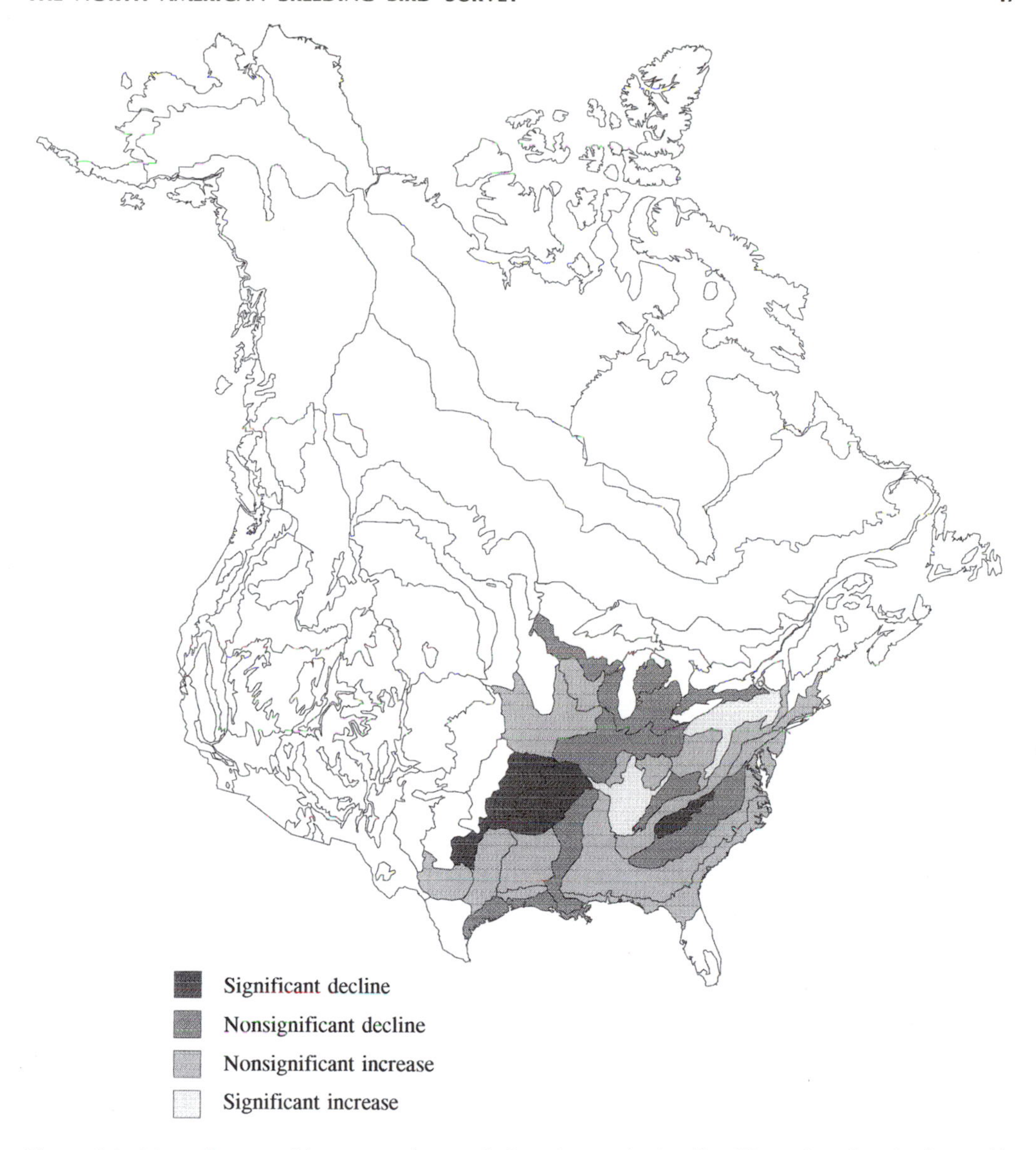

Figure 1-2. Map of geographic patterns in population changes in Acadian Flycatchers by physiographic stratum, 1966–1991.

Warblers along the Appalachian Mountains have not been identified, but may be different from those influencing population trends elsewhere in its range.

Scarlet Tanager

At the continental level, Scarlet Tanagers increased ($P < 0.01$) during 1966–1979 and decreased ($P < 0.1$) after 1980 (Table 1-3). For the entire period, neither the continental nor regional population trends are significant. Their trends are also nonsignificant within most physiographic strata (Fig. 1-8). Significant trends are limited to increases in the Highland Rim, Lexington Plain, Great Lakes Transition, and Cumberland Plateau strata and decreases in the Upper Coastal Plain stratum.

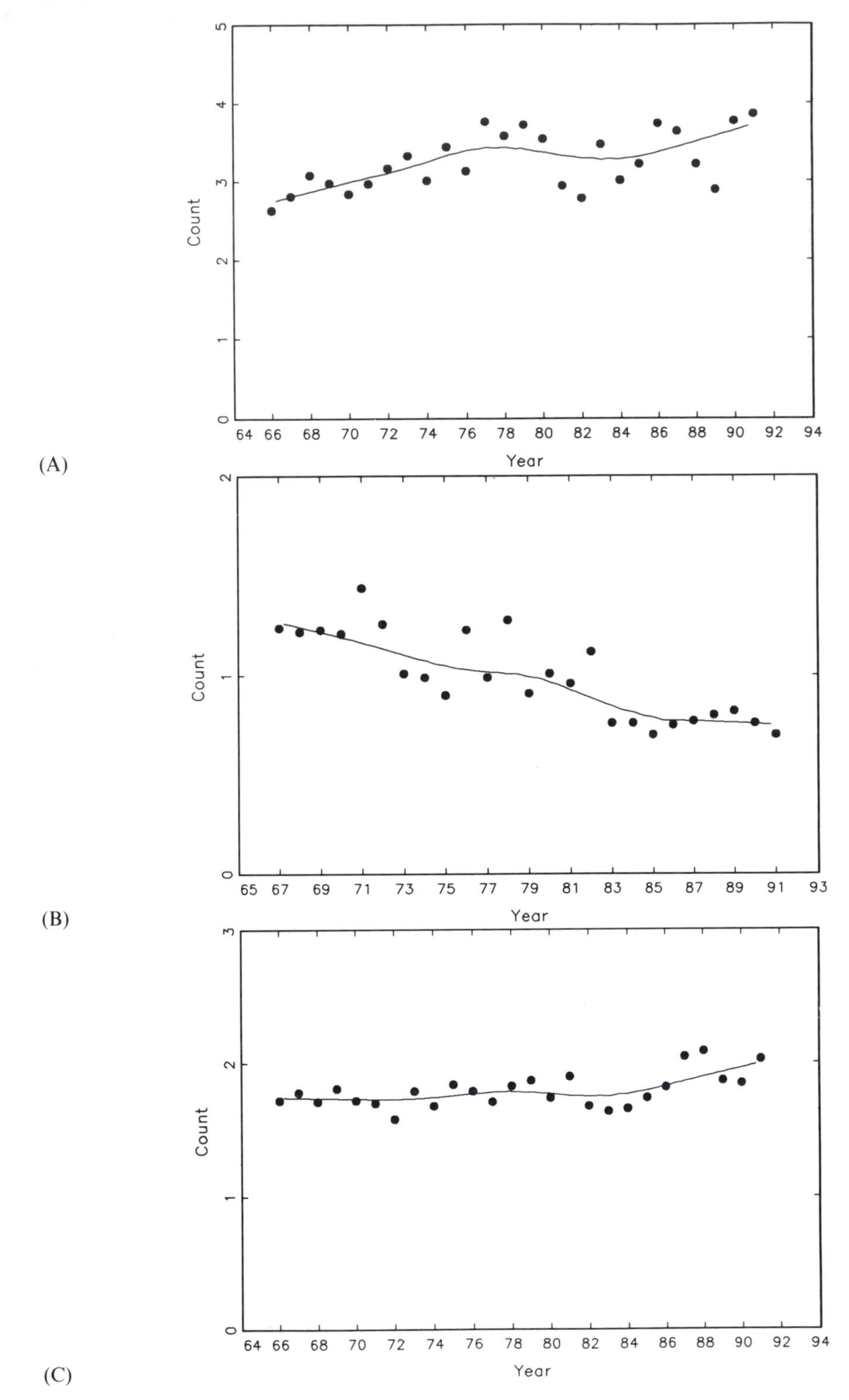
Count
Year
(A)
(B)
(C)

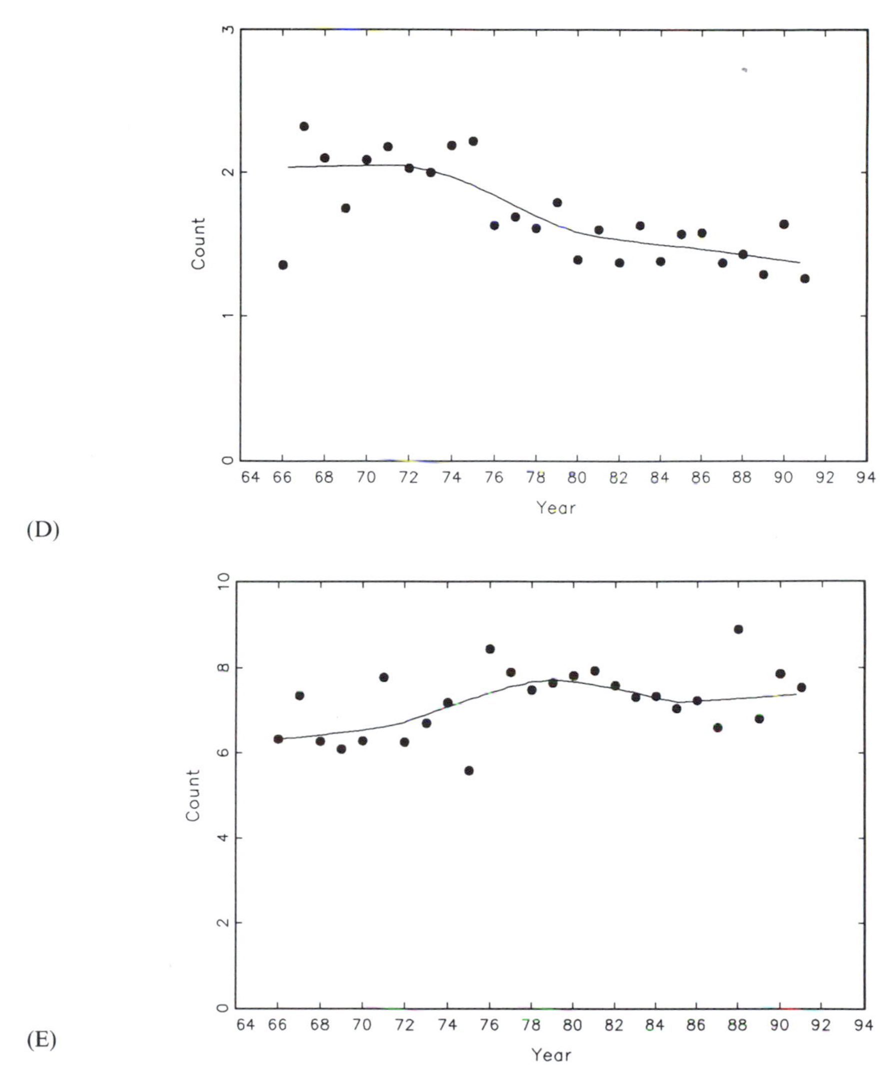

Figure 1-3. Annual indices of abundance of Acadian Flycatchers for (A) Highland Rim (14), (B) Ozark-Ouachita Plateau (19), (C) Upper Coastal Plain (4), (D) Southern Piedmont (11), and (E) Ohio Hills (22) strata.

The temporal pattern of Scarlet Tanager trends varies considerably among strata (Fig. 1-9A–E). Their trends tend to increase in most strata during the first years of the BBS. In some strata, these increases are fairly consistent through 1991 (Fig. 1-9A and B). Other strata exhibit declines during the 1980s that negate the increases of earlier years and result in stable long-term trends (Fig. 1-9C–E).

Potential Biases of the BBS

Because BBS data are based on roadside survey counts, a number of potential biases are associated with them and the trend estimates are always subject to question. These potential biases fall into two general categories, observer biases and habitat biases.

Most attention has been directed towards

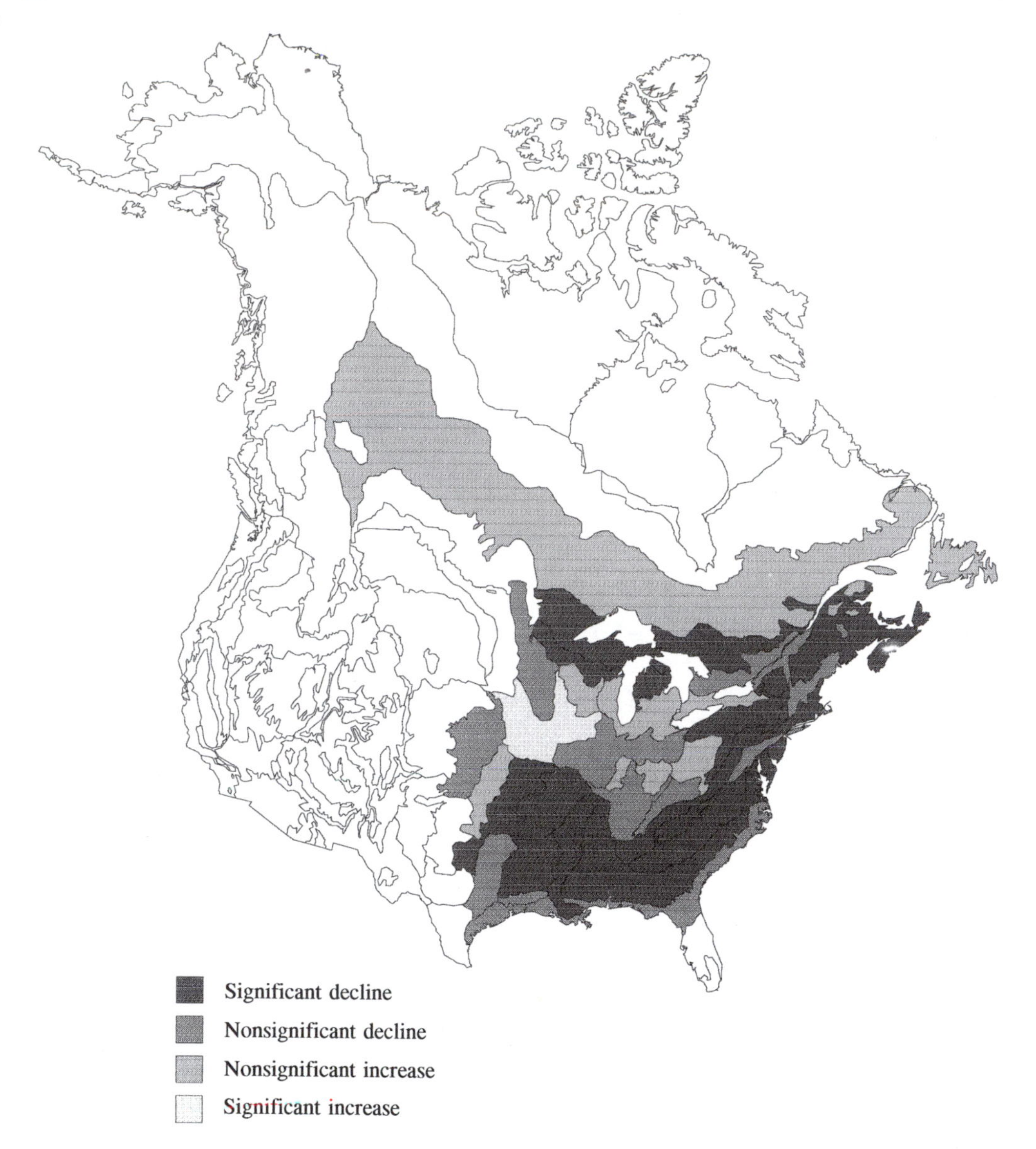

Figure 1-4. Map of geographic patterns in population changes in Wood Thrush by physiographic stratum, 1966–1991.

potential observer biases within the BBS and observer quality is a major factor affecting counts on routes. Sauer et al. (1994) analyzed observer differences in counts of bird species along BBS routes, and documented significant interobserver effects for over 50% of the species seen on the BBS. More importantly, observers have improved in quality, and a new observer on a route may count a higher proportion of the birds that are present than earlier observers. Consequently, population changes should not be calculated on BBS routes for years when observers change, and covariables should be included in the analysis to accommodate the differences in observer quality.

Other observer biases have been identified that cannot be accommodated through

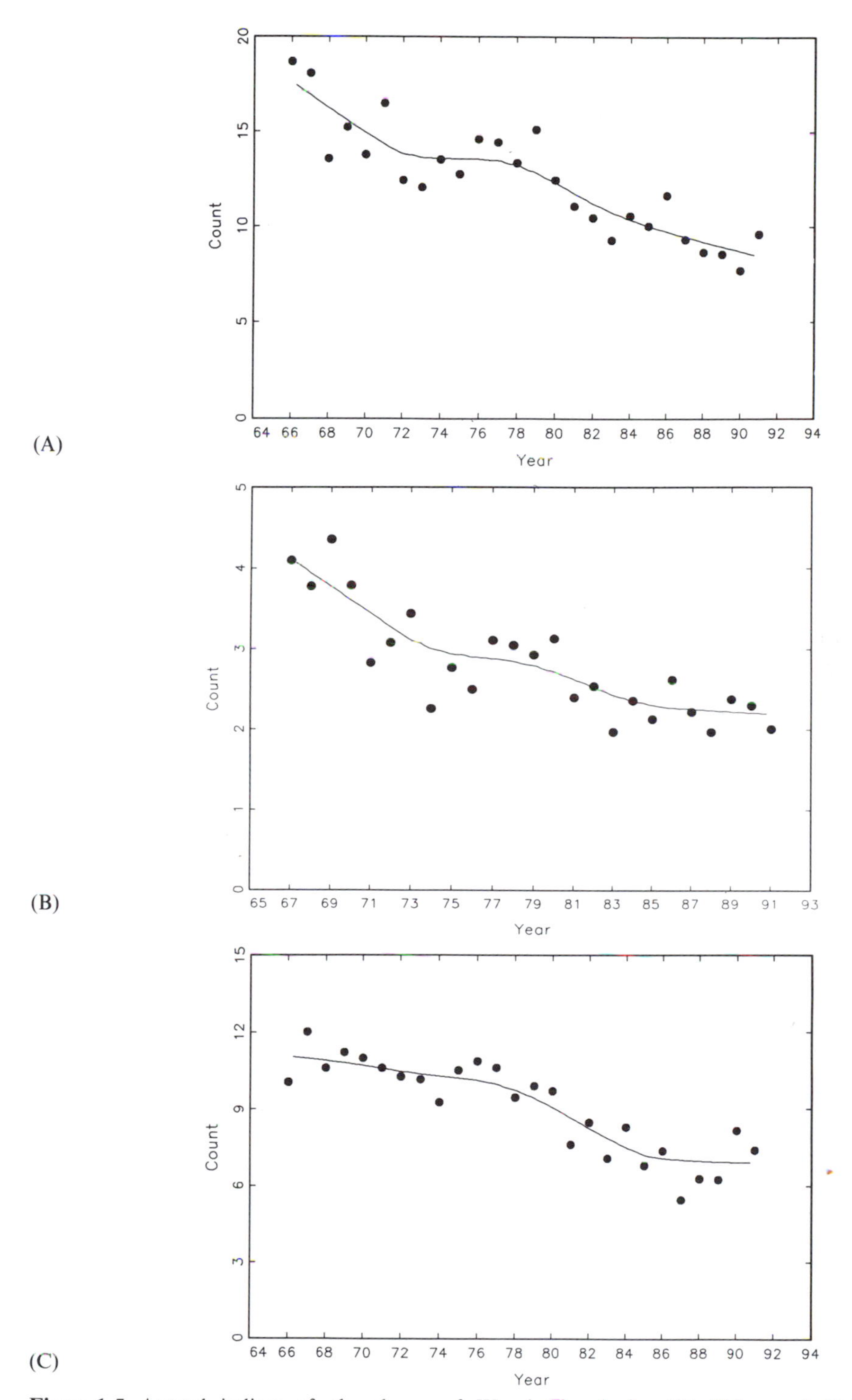

Figure 1-5. Annual indices of abundance of Wood Thrush for (A) Ridge and Valley (13), (B) Ozark-Ouachita Plateau (19), (C) Upper Coastral Plain (4) strata.

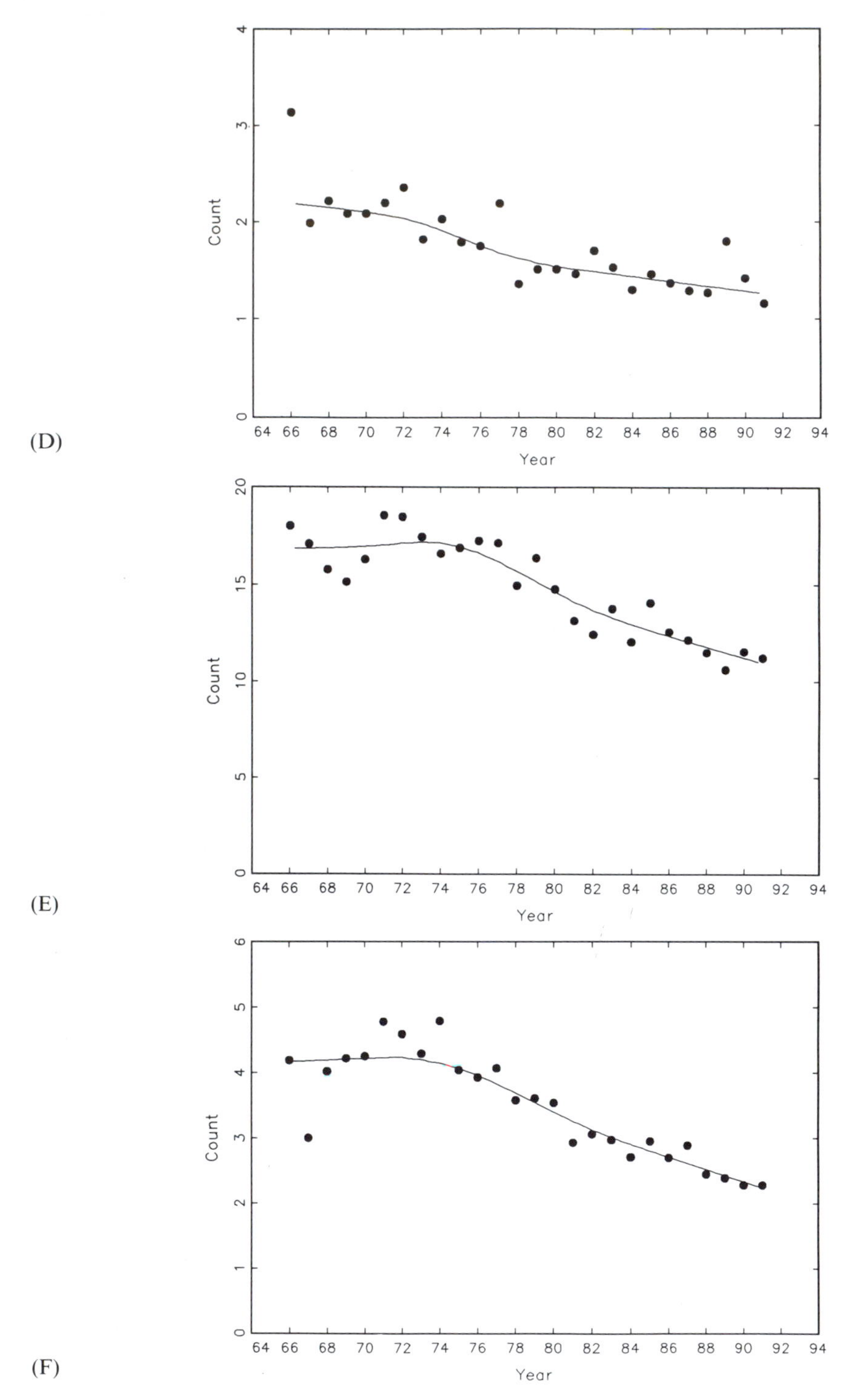

Figure 1-5 (*cont.*). (D) Great Lakes Transition (20), (E) Allegheny Plateau (24), and (F) Northern Spruce-Hardwoods (28) strata.

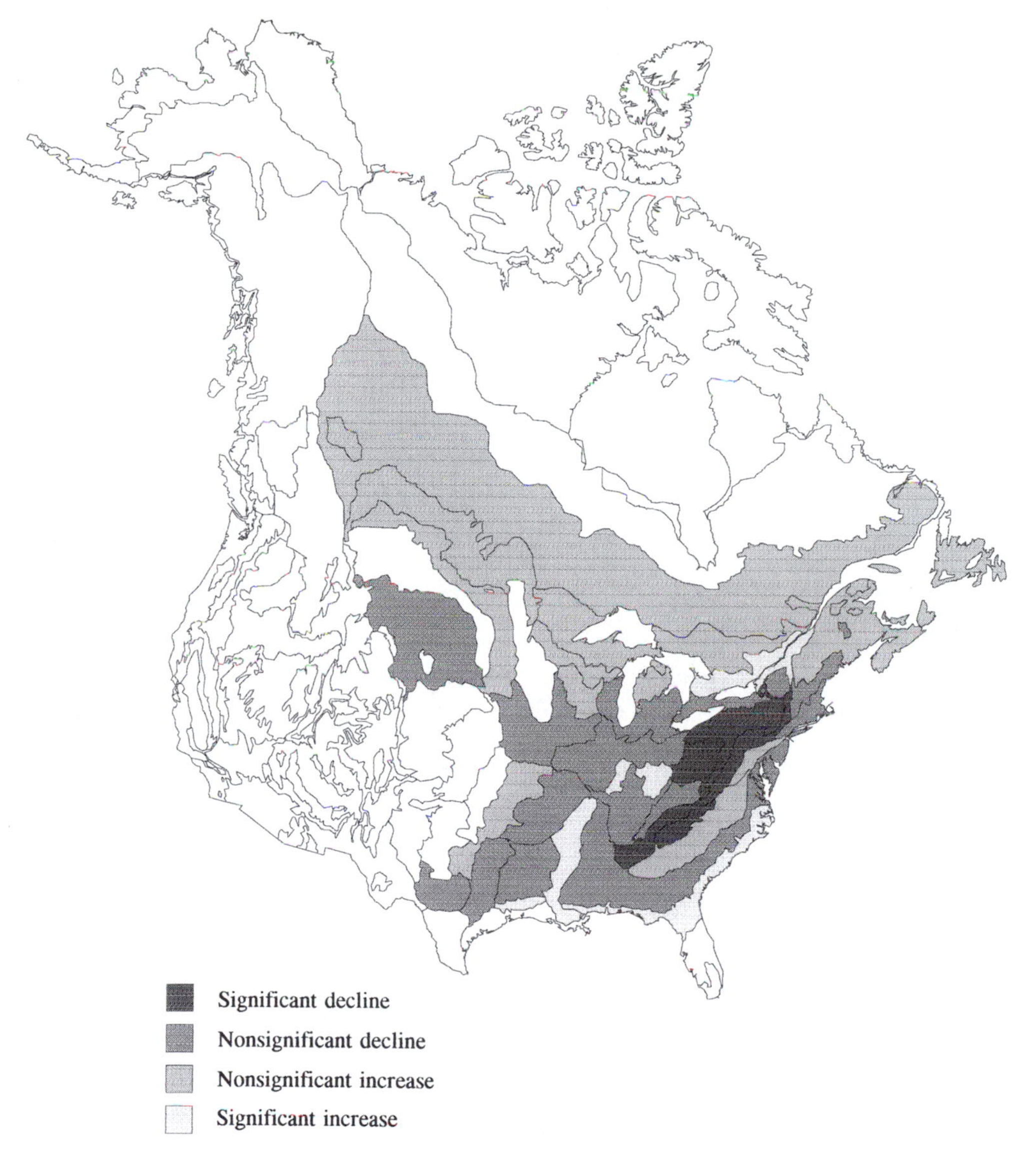

Figure 1-6. Map of geographic patterns in population changes in Black-and-white Warblers by physiographic stratum, 1966–1991.

employment of appropriate statistical methodologies. Bart and Schoultz (1984) reported that BBS observers tend to underestimate the abundance of all commonly occurring species along the routes, which could result in underestimating changes in population abundance by as much as 25–33% for these species. Erskine (1978) noted the tendency for observer skills to improve between the first and second years they conduct surveys, resulting in increased counts of individuals and species during the second year. However, this "learning effect" does not markedly influence BBS trend estimates as they are currently estimated (Kendall and Peterjohn, unpublished data).

Actions of the observers also introduce a degree of uncertainty into the analysis of the

BBS data. Each observer employs slightly different approaches to counting birds, and different observers may use slightly different stop locations on the routes. Consequently, each observer tends to perceive a slightly different population of birds and tends to count at different levels of efficiency along the same route. Year-to-year differences in survey dates, weather conditions, and presence of a second person to record data can influence counts along BBS routes. While these latter factors may not introduce unidirectional biases into the data base, they may partially obscure trends that are occurring within populations of some species.

With regard to habitat biases, whether or not changes in habitats along the roadsides surveyed by the BBS are representative of regional habitat changes has been questioned. Habitat changes along BBS routes have been discussed by Witham and Hunter (1992) for a portion of coastal Maine and are the subject of a number of current research projects. The importance of potential habitat biases probably varies regionally, as suburban development following road corridors is a more important factor in some portions of the continent than in others.

A related problem is whether or not a roadside survey can adequately sample populations of species that avoid the edge habitats created along roads. Even if counts of individuals were similar in roadside and off-road habitats, whether or not the trends of populations in these habitats are similar has never been examined.

Comparison with Other Surveys

All of these biases complicate analyses of the BBS data and raise questions concerning the accuracy of the BBS trend estimates. One means of examining the validity of the BBS trends is to compare them with results obtained from independent surveys of bird populations. Because all surveys provide estimates with unknown biases, we can never make strong conclusions about the validity of a survey from comparison with other surveys as they may all have similar biases.

Of the potential data sets for use in documenting population trends of Neotropical migrant birds in North America, the Breeding Bird Census (BBC) may be the best for direct comparison with the BBS. The BBC is composed of a series of sites censused using standardized territory-mapping procedures requiring eight or more visits to the site each year (Johnston 1990). The indices derived from territory mapping tend to be consistent within but not between observers, hence, the BBC has sampling deficiencies similar to those of the BBS. The BBC sites are much less consistently surveyed than BBS routes, site selection is not random in the BBC, and the sample of BBC sites is relatively small. Consequently, analyses of population changes from BBC data are usually conducted on either single sites or for several sites within a localized region (Terborgh 1989, Taub 1990). The BBC may provide some information on long-term trends of Neotropical migrants, but must be viewed strictly within the constraints of the geographic and temporal extent of the sample. The relevance of the resulting population trend estimates to regional bird populations is difficult to assess.

Data from BBC sites in the eastern United States were summarized for Wood Thrush and Red-eyed Vireo by C. S. Robbins (unpublished data), and the route-regression procedure was used to estimate population trends. Because BBC data were not categorized by physiographic strata, we weighted the results by areas within states and provinces rather than by physiographic stratum areas. Trends and annual indices were estimated for the interval 1966–1989.

For Wood Thrush, trends were estimated from 60 BBC sites in 16 states and provinces. The estimated trend was -3.8% per year ($P < 0.05$), which is not significantly different from the BBS trend of -2.1% per year ($P < 0.05$) for the species in the Eastern region. Annual population changes were quite similar between the two surveys (Fig. 1-10), although the BBC exhibited a slightly greater decline during the early 1980s. However, the regional trends from the BBC are based on results from relatively few states and provinces, with 38 of the sites in Connecticut, Delaware, and West Virginia. Thus, similarity in regional trends should be viewed cautiously.

For Red-eyed Vireo, the BBC trend

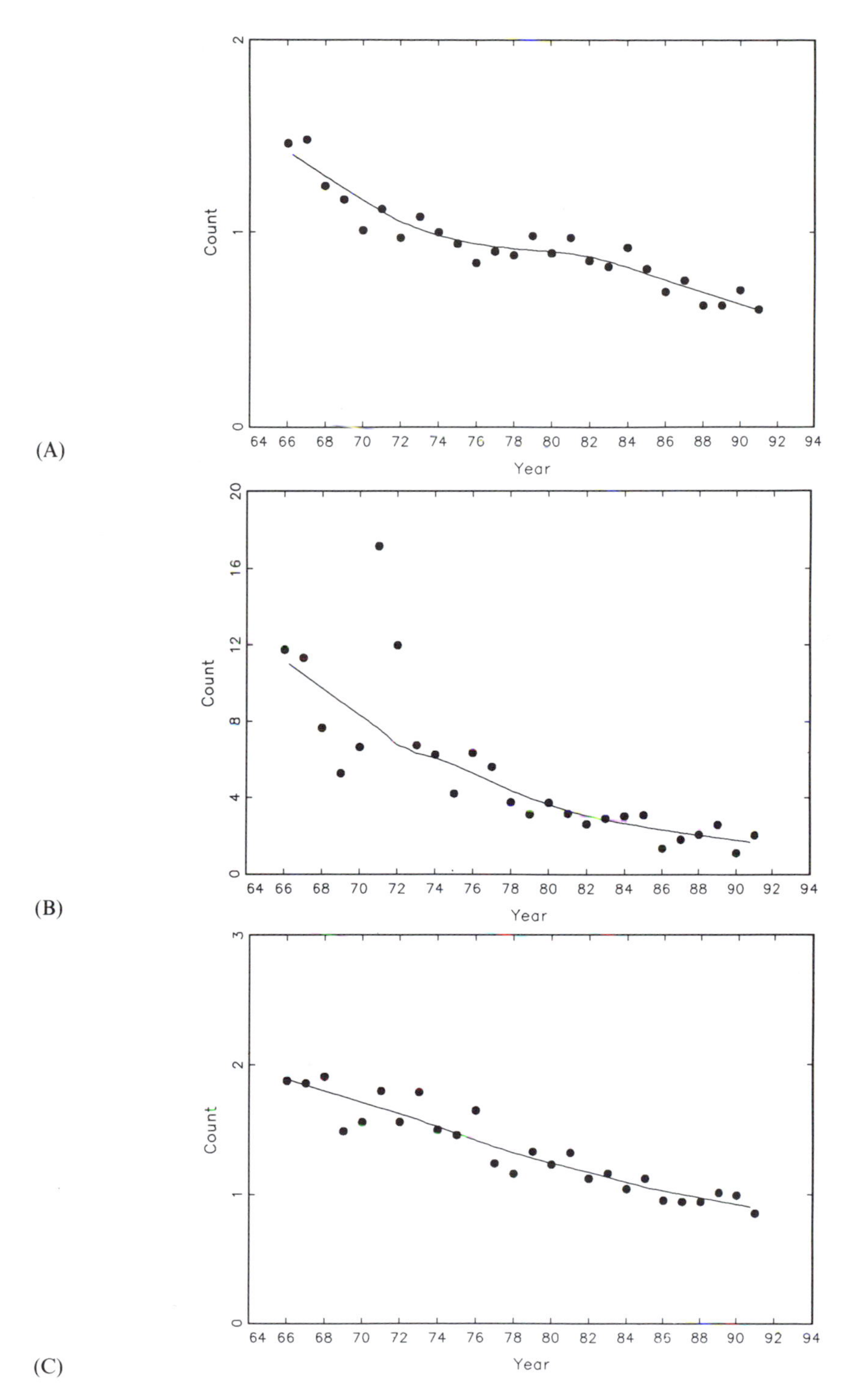

Figure 1-7. Annual indices of abundance of Black-and-white Warblers for (A) Ridge and Valley (13), (B) Blue Ridge Moutains (23), (C) Allegheny Plateau (24) strata.

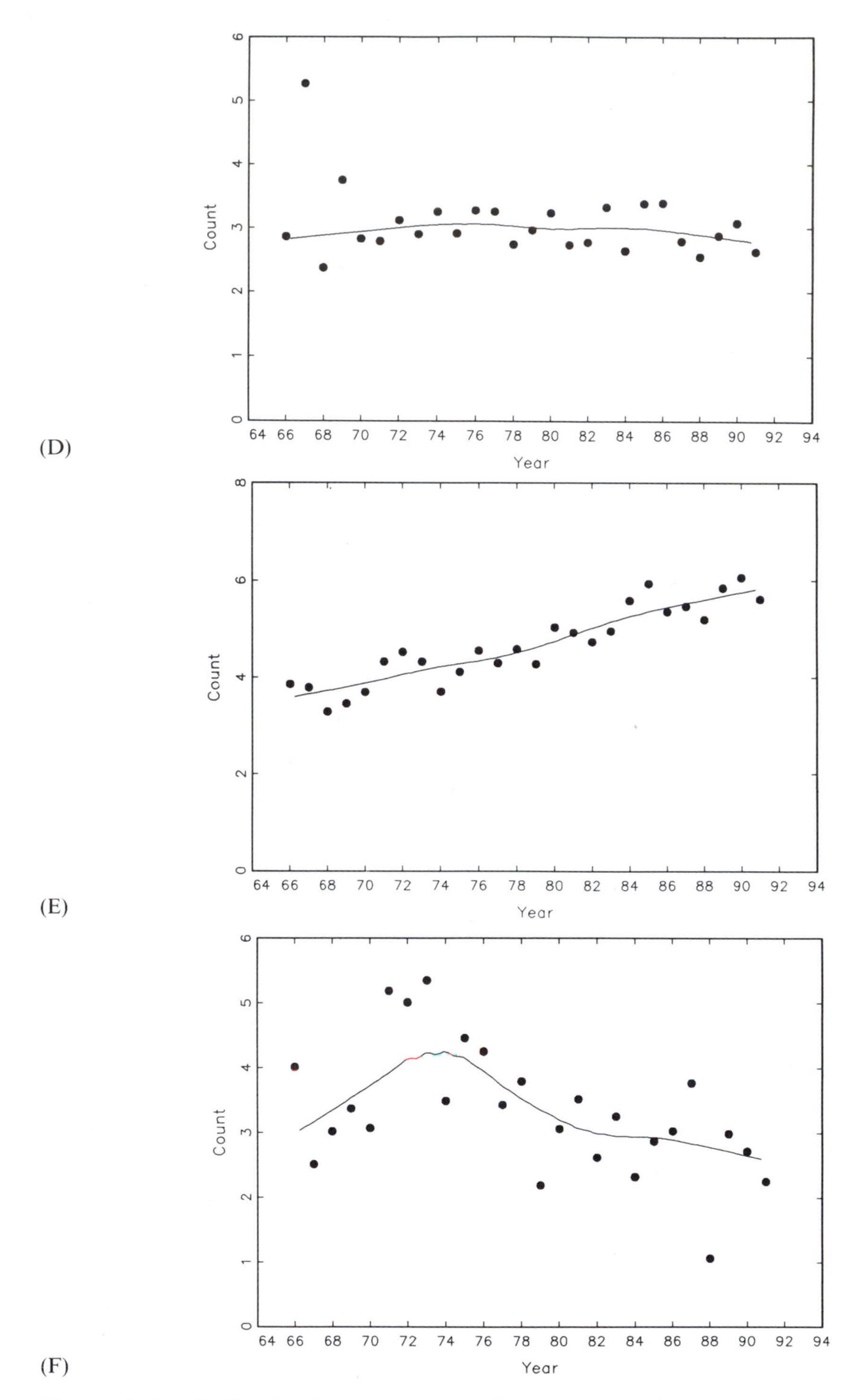

Figure 1-7 (*cont.*). (D) Southern New England (12), (E) Northern Spruce-Hardwoods (28), and (F) Cumberland Plateau (21) strata.

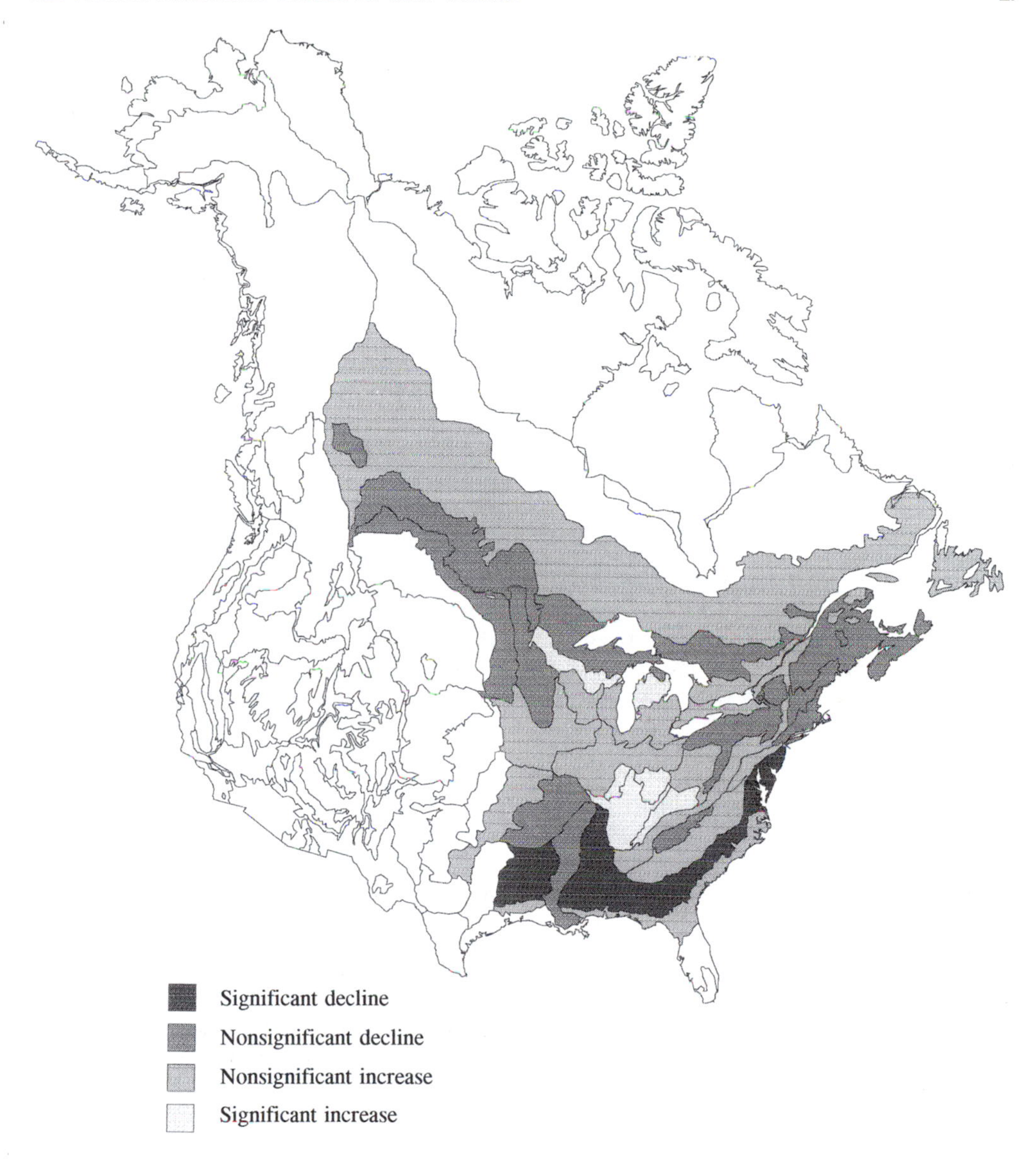

Figure 1-8. Map of geographic patterns in population changes in Scarlet Tanagers by physiographic stratum, 1966–1991.

estimate is −0.9% per year based on data from 70 sites in 22 states or provinces. This trend is not significantly different from 0.0. The BBS estimate is 1.5% per year ($P < 0.05$), and the difference between these results is of borderline significance ($P < 0.08$). The pattern in annual population changes is fairly different between these two surveys (Fig. 1-11). The BBS exhibits a gradual increase throughout the period, contrasting with the variability in the BBC data including a decline between 1978 and 1983 followed by a noticeable increase in numbers during 1984.

Results from other surveys have also been compared with BBS trend estimates and have frequently produced similar estimates of population trends. Butcher and Fuller (1986) and Butcher (1990) analyzed population trends for selected species using BBS and Christmas Bird Count data and found general agreement in trends, especially for species exhibiting significant increases and decreases. Checklist data from Wisconsin were compared with BBS trend estimates by Temple and Cary (1990) for 143 bird species. They found a strong positive correlation between the two data sets for both rare and common species including many Neotropical migrants. Robbins et al. (1989c) reviewed the use of repeated atlases in assessing population trends of breeding birds including Neotropical migrants. They demonstrated a general consistency between BBS and atlas-based trends, and suggested that repeated atlas data would be useful for species that are relatively rare or locally distributed and not well sampled by the BBS.

Migration banding projects and counts provide a potential source of information on population trends, but there are many methodological difficulties in estimating a yearly index of abundance. Also, the population sampled during migration is poorly defined, and migration pathways may change from year-to-year (Dawson 1990). Several attempts have been made to correlate trends in migration counts with BBS population trend estimates. Hagan et al. (1992) analyzed migration count data from coastal Massachusetts and western Pennsylvania, finding similarities between these counts and BBS indices from strata in which the species breed. Hussell et al. (1992) performed a similar analysis using data from Long Point, Ontario, and detected trends in population change for 45 species including a number of Neotropical migrants that were similar to the BBS trend estimates for their breeding populations.

Some authors have evaluated long-term changes at a specific site or for a single species and compared these changes with BBS trends. For example, Holmes and Sherry (1988) compared population trends from an undisturbed 10 ha forest in New Hampshire with the statewide BBS trends, and observed that 12 of 19 species exhibited similar trends. Clark et al. (1983) found a significant positive correlation between counts of male Red-winged Blackbirds along a BBS route with the known density derived from an independent survey. While these studies provide some additional validation of the accuracy of BBS trend estimates, the results should be viewed cautiously since site-specific results are difficult to generalize into regional patterns without some regional replication of random samples.

Alternative Methods of BBS Analysis

Correlation of population trends from the BBS with results of independent surveys increases our confidence that our BBS trend estimates are realistic. However, other researchers have employed various statistical procedures to analyze the BBS data set and have obtained results that contradict the route-regression trend estimates. While a detailed statistical analysis of the various methodologies used to analyze BBS data is beyond the scope of this paper, we summarize the major issues involved in the analysis of BBS data, compare some results from methods that have been used recently to analyze these data, and discuss the strengths and weaknesses of each methodology.

A fundamental constraint exists on all analyses of BBS data: we do not count every bird on each route. Instead, a portion of the birds are not counted, and the count on a route represents the product of the actual number of birds present and the proportion of detected birds, which we denote as p_{ij}. Within p_{ij}, i represents the year and j represents the route. Any analysis of BBS data makes assumptions about the consistency of p_{ij} over time and space. Unfortunately, it is very likely that p is less than 0.5 for most species, and p is known to vary both among years within a route (i) and among routes (j). In general, we feel that variation in these detection probabilities can severely limit the analyses of BBS data and, assuming that the count data are accurate and precise estimates of the actual population may result in incorrect conclusions about changes in populations over time (Barker and Sauer 1992).

Time series have several components that are of interest to ecologists. Consistent

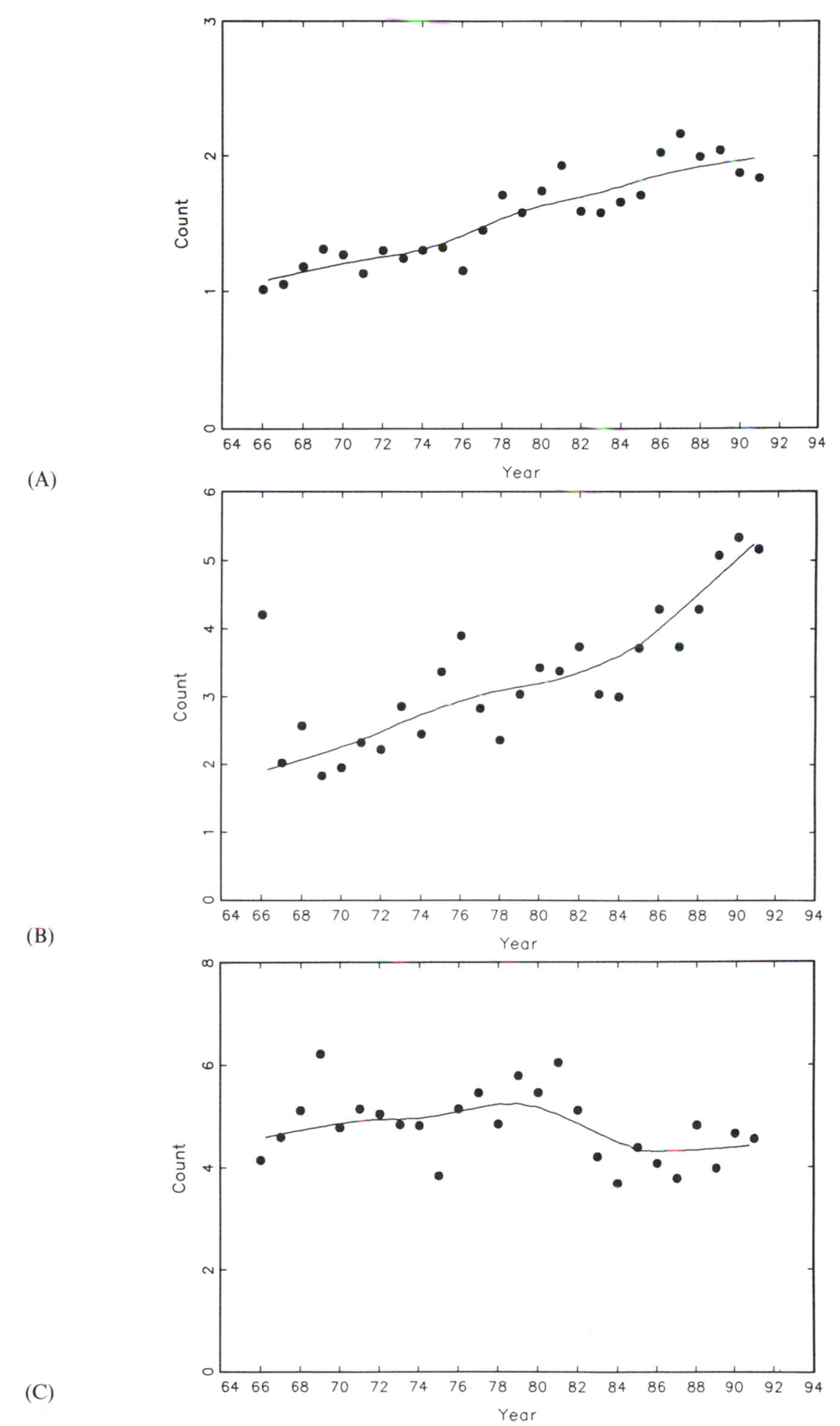

Figure 1-9. Annual indices of abundance of Scarlet Tanagers for (A) Highland Rim (14), (B) Great Lakes Transition (20), (C) Southern New England (12) strata.

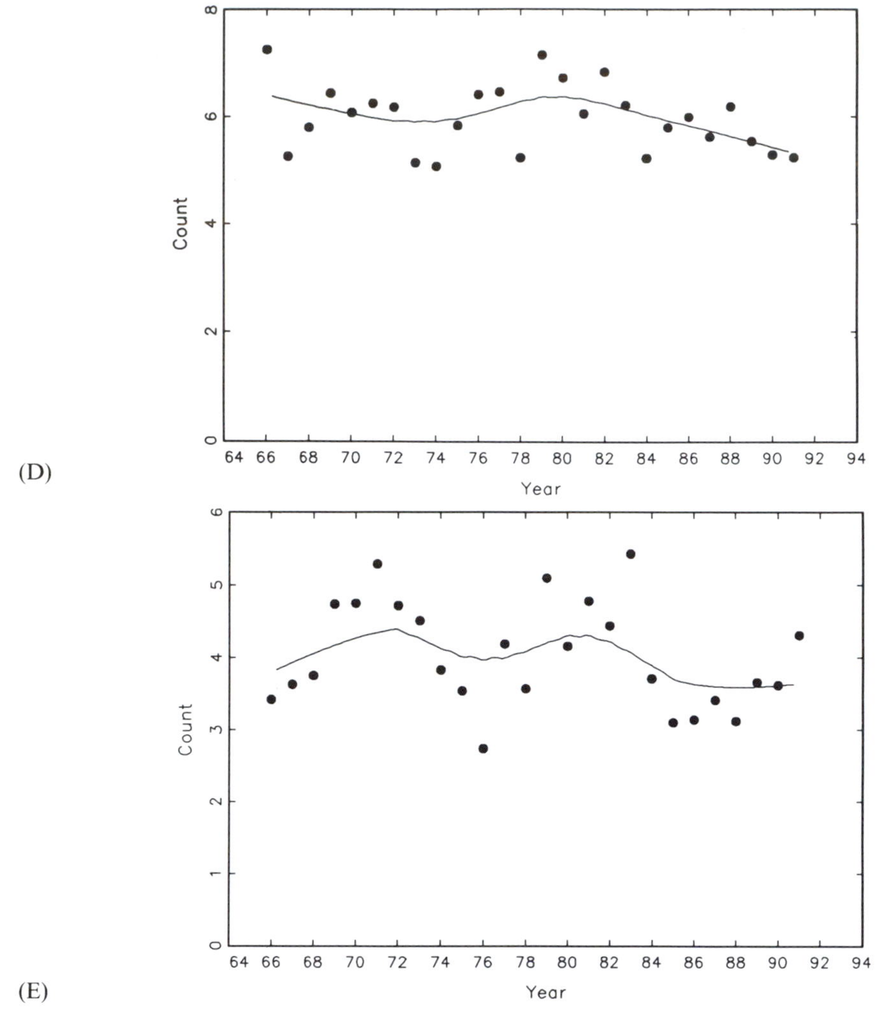

Figure 1-9 (*cont.*). (D) Allegheny Plateau (24), and (E) Northern New England (27) strata.

long-term changes in the populations, or trends, are perhaps the most fundamental component of a time series. Given the constraints in analyzing BBS data, the trend may be the only time-series component that can easily be modelled from the data. Other sources of pattern such as population cycles or other autocorrelations are also of interest to ecologists, and can confuse the interpretation of trends. For example, a negative trend from a short time series may only be a part of a long-term cycle that cannot be detected from the time series. Additionally, some forms of autocorrelation could be interpreted as density dependence (Bulmer 1975). Finally, abrupt changes in the level of time series may be associated with environmental changes. Any analysis of BBS data should accommodate pattern related to trend, population autocorrelation, and irregular changes due to the environment.

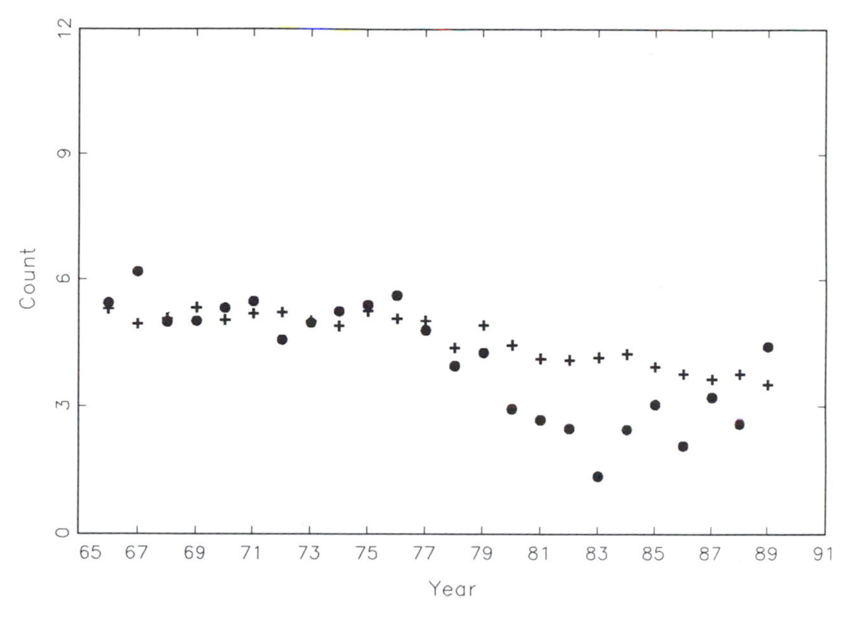

Figure 1-10. Annual indices of abundance for Wood Thrush from BBS (circles) and Breeding Bird Census (crosses) data.

The BBS as a Sample of Bird Populations

The BBS is an extremely ambitious program for monitoring birds at a continental scale. The sampling procedure and route locations are compromises between statistical accuracy (and precision) and the practical constraints of sampling North America.

Regional Differences in Sample Intensity

At a regional level, the BBS sample is random along roadsides. However, at a continental scale the sample locations are not random, with many more routes located in the eastern United States than Canada and the western United States (Fig. 1-1). The variation in sample effort also occurs within physiographic strata, as portions of strata located near urban centers tend to have more routes than portions located in rural areas. For example, Stratum 4 (Upper Coastal Plain) extends from New Jersey to Texas, and the density of routes among states within Stratum 4 varies from 0.056 routes/100 square miles in Texas to 0.54 routes/100 square miles in Maryland, indicating that routes are 9.65 times more dense in Maryland. Simple averages of route data over the entire stratum would be weighted toward the parts of the stratum with the larger sample sizes, which could result in biased estimates of trends and regional relative abundances. To avoid this bias, stratification within states (or provinces) is necessary, and regional estimates of trends and relative abundances can be calculated as area-weighted averages of the state-stratum average trends.

Temporal Differences in Sample Intensity

Most BBS routes have not been surveyed each year of the 1966–1991 period. The survey was not established in the western United States and southern Canada until 1968, and gaps occur in coverage on most routes. Missing data on routes have been identified as a major constraint on BBS analyses, and much of the development of trend estimation methods has been designed to adjust counts to a subset of "comparable" routes (e.g., Geissler and Noon 1981, Geissler and Sauer 1990, James et al. 1990).

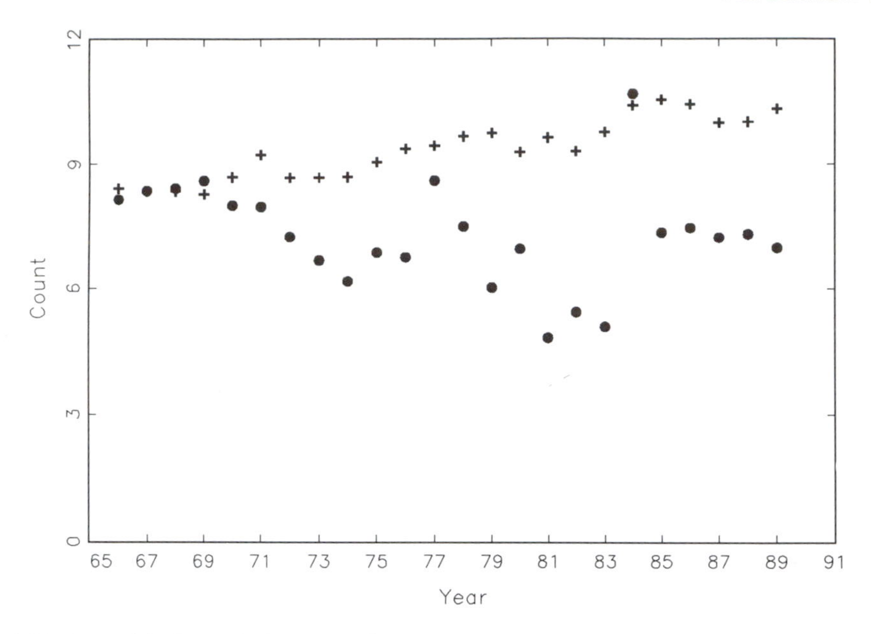

Figure 1-11. Annual indices of abundance for Red-eyed Vireo from BBS (circles) and Breeding Bird Census (crosses) data.

Methods of Analysis for BBS Data

BBS analyses generally attempt to estimate trend and yearly population indices for a geographic region. They must combine data from the component routes, taking into account the incomplete counts on the routes, the missing data and observer effects, and the geographic inequities in the sample. No analysis is fully successful in dealing with these problems, but certain analyses are better than others in accommodating the vagaries of the BBS sample.

Regional Mean Counts

Simply averaging counts by regions and year for a species is the "naive" way of analyzing BBS data. Most investigators would agree that average counts are not a valid yearly index to bird abundance because of missing data (e.g., Geissler and Noon 1981, James et al. 1990). Some patterns of change are adequately modeled by average counts (Robbins et al. 1986). However, many spurious patterns are introduced into the yearly average counts by changes in coverage of routes and other artifacts of sampling, and real population changes cannot always be separated from these artifacts of sampling.

Linear Model Fits

One way of modeling regional yearly indices is to fit a linear model to the count data. These analysis of variance type models estimate year effects in the context of other factors, and yearly expected values can be used as the regional annual indices. Several variations on this general approach have been used on BBS data (Dolbeer and Stehn 1979, Sauer and Geissler 1990, Moses and Rabinowitz 1990). These models tend to be very complex when factors are included to accommodate year, route, observer, and area effects, and the models are only infrequently used (Sauer and Geissler 1990). Some investigators fit simplified models to BBS data that do not contain observer effects and omit routes with incomplete data (e.g., Bohning-Gaese et al. 1992). Unfortunately, indices estimated without observer effects tend to contain additional variability and produce positive bias in trend estimates, making them imprecise and inaccurate.

Base-year Methods

The base-year method was used to analyze population changes in the BBS during the early years of the survey, and is still used for the Common Birds Census in the United Kingdom (Marchant et al. 1990). In this method, counts on "comparable" routes (e.g., routes surveyed by the same observer in both years) are summed for each 2-year period, and the ratios of these sums are used to predict future and past counts from some "base-year" or starting point. Sometimes the base-year count is set at 100, but other applications use the mean of all the counts as the base-year count. This method has many flaws and we do not recommend its use. Geissler and Noon (1981) noted that error accumulates in base-year indices, making them very imprecise. It is also important to note that ratios tend to be positively biased, and the incomplete counts from survey data such as the BBS accentuate this bias. Consequently, base-year indices are positively biased, especially when the ratios are based upon few routes or species with low counts (Barker and Sauer 1992). We also note that this positive bias exists in any analysis of averages of counts, and may account for the large population changes observed by O'Connor (1992) in his analysis of yearly population changes of Neotropical migrant birds.

Mountford's Method

Mountford (1985) suggested a method of estimating year effects that uses comparable route counts from a series of years. The logarithms of ratios of counts for comparable years from all possible sets of years in the interval are calculated from the data, and year effects (deviations from a mean value) are estimated using a least-square fitting procedure. He also provided a test for homogeneity of changes. Baillie and Peach (1992) have prepared a FORTRAN program for fitting Mountford's model to survey data.

Our use of Mountford's method indicates that the approach will not generally be useful in estimating large-scale trends from BBS data. It appears that the model generally does not fit population time series of more than 11 years for most survey datasets (Baillie and Peach 1992). Of course, poor fit of a model does not necessarily invalidate its use, and we fit route-regression trends to nonlinear time series. However, when the model does not fit, the precision of the yearly indices of abundance, and statistical hypothesis tests of changes over time, are not valid. Our experience suggests that the assumption of homogeneity will only infrequently be met in the BBS. We have repeatedly demonstrated temporal and geographic variation in trends of Neotropical migrant birds in the eastern United States (e.g., Robbins et al. 1989b, Sauer and Droege 1992), which would clearly invalidate use of Mountford's model at this scale.

To avoid the geographic heterogeneity, Mountford's model could still be fit to data from small geographic areas, such as strata within states. Unfortunately, there is evidence that technical and pr·ctical aspects of model fitting also invalidate its use at smaller scales. For example, data for our four example species was not homogeneous within Maryland, suggesting that Mountford's model would not be applicable. Even more importantly, the model cannot be applied in regions with missing comparable year data.

LOESS Smooths

LOESS, or locally weighted least squares, is a statistical procedure in which a time series is "smoothed" by a weighted average of the counts, in which the counts near the point to be smoothed are weighted more heavily than counts farther from the point. It has been suggested that problems associated with missing data on BBS routes could be minimized by using LOESS on each BBS route, and using the LOESS-smoothed data for each year to calculate regional mean counts (James et al. 1990). Trend and patterns in the population could then be modelled from these smoothed regional mean counts. Because of the recent use of these methods to model population changes, we have calculated LOESS-smoothed annual indices of abundance at the continental level for selected species.

Data analyzed in this way generally provide population time series that appear much less variable than annual indices from the residual method owing to the smoothing of variability at the route level (Fig. 1-12). The smoothing removes information about interventions, or abrupt changes in the level of the time series. For example, the pattern introduced by the smoothing can obscure year effects of interest, such as the decline in 1973 in Purple Martins associated with severe weather (Sauer et al. 1987). Also, LOESS smooths do not extend beyond the end points of count data from any route, so sample sizes decrease near the end of the time series. In our analysis, the decrease in sample sizes at the ends of the time series caused clearly spurious patterns for all four species during the first 6 years of the time series. Also, a LOESS smooth can provide a misleading view of population changes when observer differences are not taken into account. We suggest that the LOESS smoothed time series be used as visual aids in identifying patterns in the data, but that the patterns be verified by rigorous statistical estimation procedures.

Route Regression and Residual Indices

In this paper, we use the route regression procedure to estimate trends, and the residual indices to estimate annual indices of abundances. We feel that this procedure mitigates most factors that can bias trend estimation, and the bias and precision of the procedure has been evaluated (Geissler and Link 1988). Even this method has its deficiencies. For example, route-regression trend analyses suggest that Tennessee Warblers had extreme increases 1966–1979, but declines 1980–1991 (Table 1-3). The residual indices or abundance fail to catch the extreme population changes of the Tennessee Warbler (Fig. 1-12). However, the change in trend for this species is well modeled by the interval trend analysis, suggesting that a combination of interval trend estimates and annual indices should be used in evaluating the significance of observed population changes from any indices.

Within the general structure of route-regression analysis, certain aspects of the analysis can be improved. For example, the linear regression model used to estimate trends has been criticized because a constant must be added to the counts to prevent the occurrence of 0 data (Geissler and Sauer 1990). Recently, Link and Sauer (1994) developed a procedure to directly estimate trends that does not require the regression model but is instead based on estimating equations. We are investigating the application of this procedure in BBS analyses (Peterjohn et al., 1994). Preliminary analyses suggest that estimating equation estimates provide generally similar results to the linear regression estimates for abundant species, but low abundance species are estimated with less precision. See Peterjohn et al. (1994) for commentary regarding the future of route regression (and other) trend analyses.

Analysis of Population Changes with the BBS

It has been suggested that several methods should be used to evaluate population changes. Unfortunately, we have shown that some methods when applied to BBS data can produce biased results. We suggest that, rather than use inappropriate methods to evaluate trends, it may be better to evaluate trends at several levels of scale, to see if route data are summarized appropriately in regional analyses.

A number of additional potential biases have been identified within the BBS. Because it is a roadside survey, the BBS most thoroughly samples species occupying habitats found along roads. Species that are rare, locally distributed, or occupy habitats that are seldom traversed by roads, such as bogs and alpine areas, tend to be poorly represented in the BBS. If roadside habitats are changing at a different rate than habitats away from roads, or if the roadside habitats are not representative of regional habitat availability, then the BBS population trend estimates may not accurately reflect regional trends of some species.

Traffic noise has become an increasing problem along BBS routes since the mid-1960s, especially in the more populated portions of the continent. As traffic noise becomes worse, it tends to drown out bird

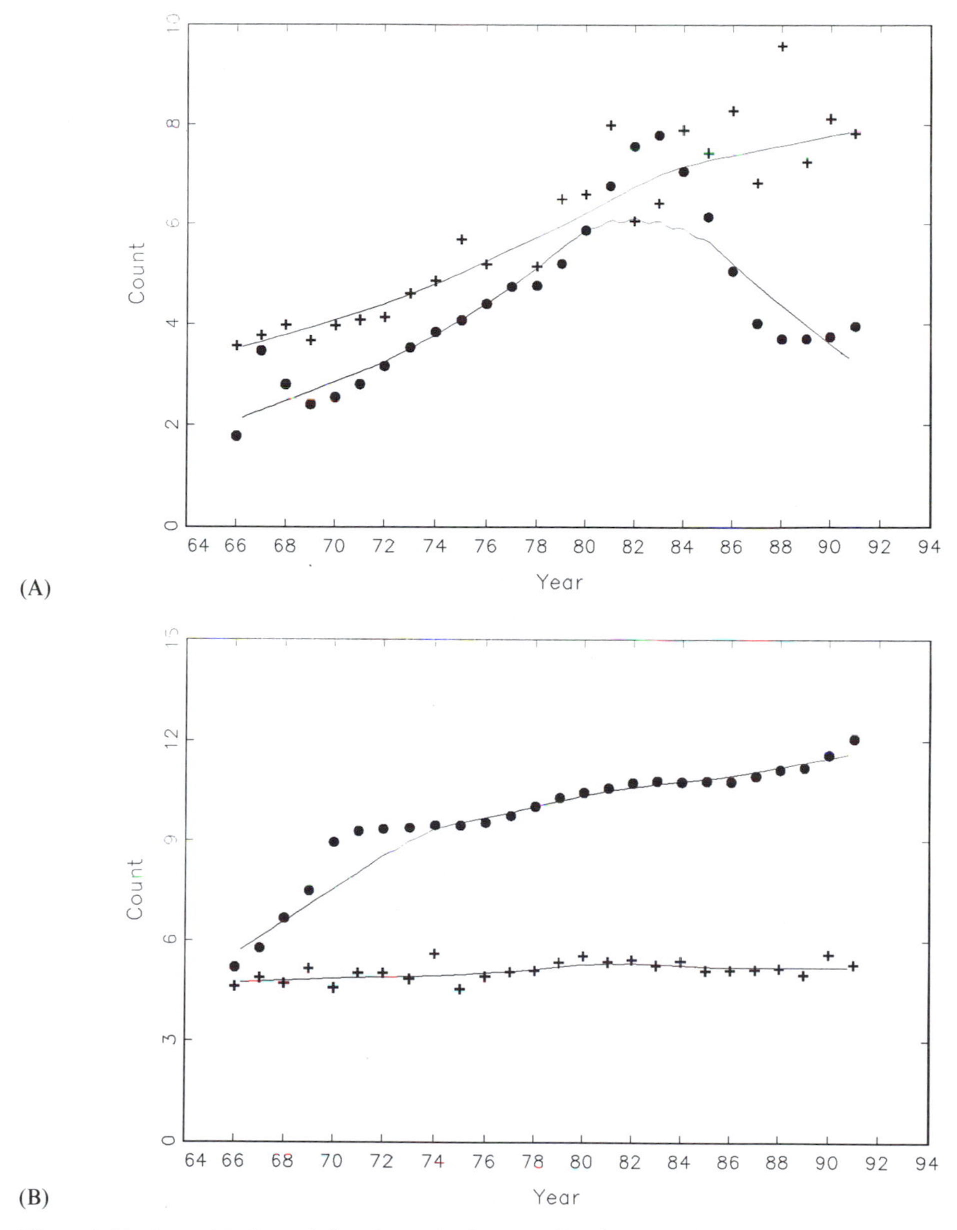

Figure 1-12. Annual indices of abundance for four species of Neotropical migrants from residual indices (crosses) and LOESS-smoothed counts (circles): (A) Tennessee Warbler, (B) Ovenbird.

song, and consequently reduce the numbers of individuals and species recorded along a route. Because traffic volume has not been routinely recorded along BBS routes, the extent of this problem remains to be determined.

No two observers have exactly the same experiences and abilities in bird identification. In order to reduce potential biases resulting from differing observer capabilities, observer covariables are routinely used during the analysis of the BBS data set.

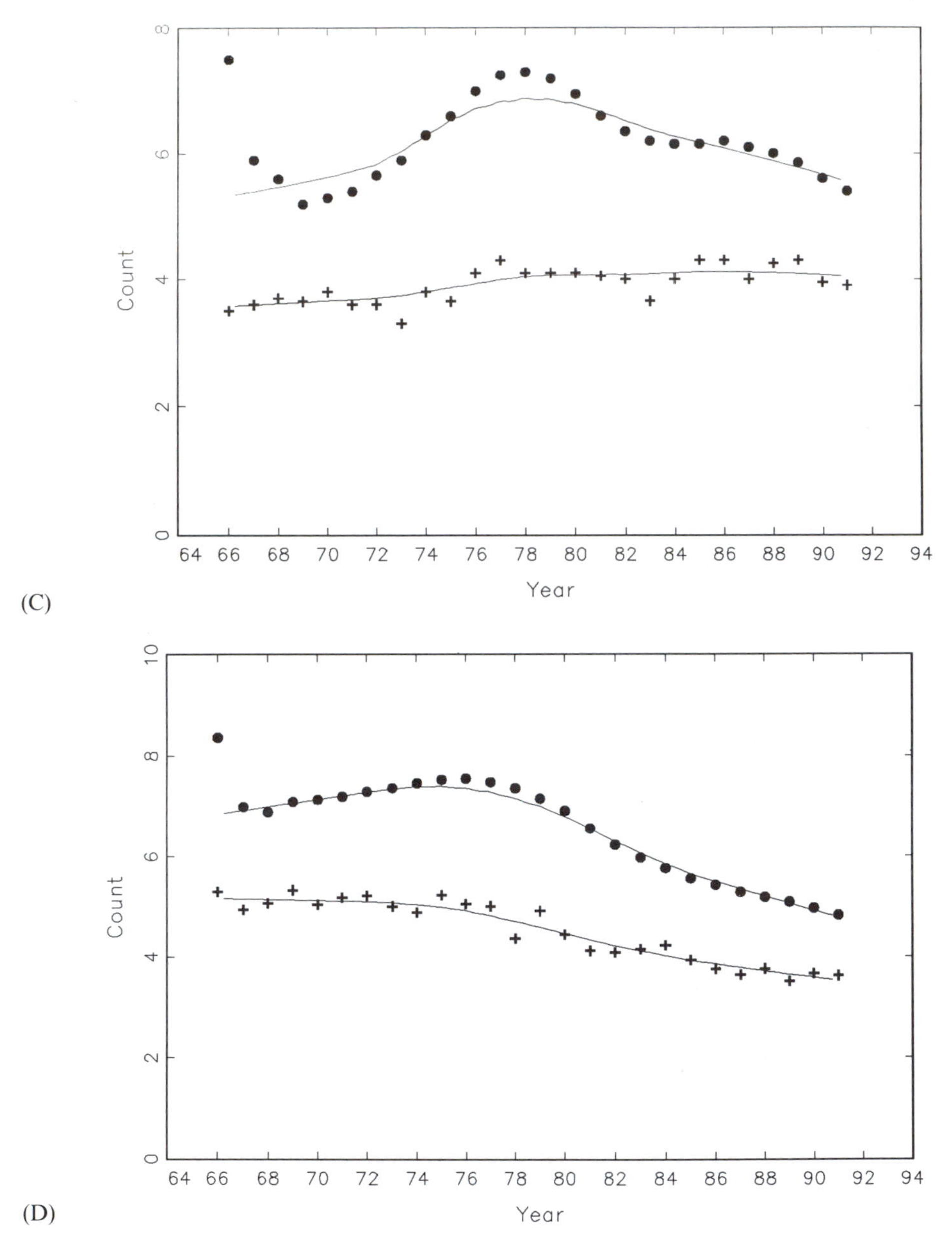

Figure 1-12 (*cont.*). (C) Purple Martin, (D) Wood Thrush.

However, use of observer covariables does not eliminate all sources of observer biases. For example, as observers become more familiar with the BBS methodology, they may record increased numbers of species and individuals during the first few years they conduct a survey. Improved identification skills over time can produce similar results. Conversely, hearing loss as an observer grows older can result in fewer species and individuals recorded along a BBS route. These intraobserver biases exist within the

BBS data set, and their influence on the population trend estimates is currently under evaluation.

ACKNOWLEDGMENTS

We are very appreciative of the efforts of the thousands of volunteers who have participated in the BBS over the years. We thank S. Orsillo for assistance in preparing the figures. D. Dawson, C. Moore, and J. Nichols provided very useful comments on earlier drafts of the manuscript.

LITERATURE CITED

Ambuel, B., and S. A. Temple. 1982. Songbird populations in southern Wisconsin forests. J. Field Ornithol. 53:149–158.

American Ornithologists' Union. 1983. Check-list of North American birds, 6th ed. Allen Press, Lawrence, KS.

Askins, R. A., and M. J. Philbrick. 1987. Effect of changes in regional forest abundance on the decline and recovery of a forest bird community. Wilson Bull. 99:7–21.

Askins, R. A., J. F. Lynch, and R. S. Greenberg. 1990. Population declines in migratory birds in eastern North America. Curr. Ornithol. 7:1–57.

Baillie, S., and W. Peach. 1992. Program MOUNTF. Unpublished user documentation.

Barker, R. J., and J. R. Sauer. 1992. Modelling population change from time series data. Pp. 182–194 *in* Wildlife 2001: populations (D. R. McCullough and R. Barrett, eds). Elsevier, New York.

Bart, J., and J. D. Schoultz. 1984. Reliability of singing bird surveys: changes in observer efficiency with avian density. Auk 101:307–318.

Böhning-Gaese, K., M. L. Taper, and J. H. Brown. 1992. Are declines in North American insectivorous songbirds due to causes on the breeding range? Conserv. Biol. 7:76–86.

Brewster, W. 1903. The birds of the Cambridge region of Massachusetts. Nuttall Ornithol. Club Memoirs 4, Cambridge, MA.

Bulmer, M. G. 1975. The statistical analysis of density dependence. Biometrics 31:901–911.

Butcher, G. S. 1990. Audubon Christmas bird counts. Pp. 5–13 *in* Survey designs and statistical methods for the estimation of avian population trends (J. R. Sauer and S. Droege, eds). US Fish Wildl. Serv. Biol. Rep. 90(1).

Butcher, G. S., and M. R. Fuller. 1986. Bird populations in winter and summer: an evaluation of the Christmas Bird Count and a comparison with the Breeding Bird Survey. Report to US Fish Wildl. Serv., Washington, DC.

Chapman, F. M. 1894. Visitors' guide to the local collection of birds in the American Museum of Natural History, New York City, with an annotated list of the birds known to occur within 50 miles of New York City. American Museum of Natural History, New York.

Clark, R. G., P. J. Weatherhead, and R. D. Titman. 1983. On the relationship between breeding bird survey counts and estimates of male density in the Red-winged Blackbird. Wilson Bull. 95:453–459.

Cleveland, W. S., and S. J. Devlin. 1988. Locally weighted regression, an approach to regression analysis by local fitting. J. Amer. Stat. Assoc. 83:596–610.

Coues, E., and D. W. Prentiss. 1862. List of birds ascertained to inhabit the District of Columbia, with the times of arrival and departure of such as are non-residents, and brief notes of habits, etc. Pp. 399–421 *in* Ann. Rept. Board of Regents of the Smithsonian Inst... for 1861. 37th Congress 2d Session, House of Rep., Misc. Doc. 77.

Coues, E., and D. W. Prentiss. 1883. Avifauna Columbiana, being a list of birds ascertained to inhabit the District of Columbia,... etc. 2nd ed. US Natl. Mus. Bull. 26.

Dawson, D. K. 1990. Migration banding data: a source of information on bird population trends? Pp. 37–40 *in* Survey designs and statistical methods for the estimation of avian population trends (J.R. Sauer and S. Droege, eds). US Fish Wildl. Serv. Biol. Rep. 90(1).

Dolbeer, R. A., and R. A. Stehn. 1979. Population trends of blackbirds and starlings in North America, 1966–76. US Fish Wildl. Serv. Spec. Sci. Rep. Wildl. 214.

Ehrlich, P. R., D. S. Dobkin, and D. Wheye. 1988. The birder's handbook: A field guide to the natural history of North American birds. Simon & Schuster, New York.

Efron, B. 1982. The jackknife, the bootstrap and other resampling plans. Soc. Indust. Appl. Math., Philadelphia, PA.

Erskine, A. J. 1978. The first ten years of the cooperative breeding bird survey in Canada. Can. Wildl. Serv. Rep. Ser. 42:1–61.

Gauthreaux, S. A., Jr. 1991. Preliminary lists of migrants for Neotropical migrant bird conservation program. Partners in Flight Program, Washington, DC.

Geissler, P. H., and W. A. Link. 1988. Bias of animal population trend estimates. Pp. 755–759 *in* Proc. 20th Symp. Interface Comput. Sci. Stat. American Statistical Association, Arlington, VA.

Geissler, P. H., and B. R. Noon. 1981. Estimates of avian population trends from the North American Breeding Bird Survey. Pp. 42–51 *in* Estimating numbers of terrestrial birds. (C. J. Ralph and J. M. Scott, eds). Stud. Avian Biol. 6.

Geissler, P. H., and J. R. Sauer. 1990. Topics in route regression analysis. Pp. 54–57 *in* Survey designs and statistical methods for the estimation of avian population trends (J. R. Sauer and S. Droege, eds). US Fish Wildl. Serv. Biol. Rep. 90(1).

Graber, R. R., and J. W. Graber. 1963. A comparative study of bird populations in Illinois, 1906–1909 and 1956–1958. Illinois Natural Hist. Surv. Bull. 28(3):383–528.

Griscom, L. 1923. Birds of the New York City region. Amer. Mus. Natural Hist. Handbook Ser. no. 9.

Griscom, L. 1949. The birds of Concord—a study in population trends. N. Engl. Bird Studies 2, Harvard University, Cambridge, MA.

Hagan, J. M., III, and D. W. Johnston. 1992. Ecology and conservation of Neotropical migrant landbirds. Smithsonian Institution Press, Washington, DC.

Hagan, J. M., III, T. L. Lloyd-Evans, and J. L. Atwood. 1992. Long-term changes in migratory landbirds in the northeastern United States: Evidence from migration capture data. Pp. 115–130 *in* Ecology and conservation of Neotropical migrant landbirds (J. M. Hagan, III and D. W. Johnston, eds). Smithsonian Institution Press, Washington, DC.

Hall, G. A. 1984. Population decline of Neotropical migrants in an Appalachian forest. Amer. Birds 38:14–18.

Harrison, C. 1978. A field guide to the nests, eggs, and nestlings of North American birds. William Collins Sons & Co., Glasgow, UK.

Holmes, R. T., and T. W. Sherry. 1988. Assessing population trends of New Hampshire forest birds: local vs. regional patterns. Auk 105:756–768.

Hussell, D. J. T., M. H. Mather, and P. H. Sinclair. 1992. Trends in numbers of tropical- and temperate-wintering migrant landbirds in migration at Long Point, Ontario. Pp. 101–114 *in* Ecology and conservation of Neotropical migrant landbirds (J. M. Hagan, III and D. W. Johnston, eds). Smithsonian Institution Press, Washington, DC.

Jaksic, F. M. 1981. Abuse and misuse of the term "guild" in ecological studies. Oikos 37:397–400.

James, F. C., C. E. McCulloch, and L. E. Wolfe. 1990. Methodological issues in the estimation of trends in bird populations with an example: the Pine Warbler. Pp. 84–97 *in* Survey designs and statistical methods for the estimation of avian population trends (J. R. Sauer and S. Droege, eds). US Fish Wildl. Serv. Biol. Rep. 90(1).

Johnson, N. K., and C. Cicero. 1985. The breeding avifauna of San Benito Mountain, California: evidence for change over one-half century. West. Birds 16:1–23.

Johnston, D. W. 1990. Descriptions of Surveys: Breeding Bird Censuses. Pp. 33–36 *in* Survey designs and statistical methods for the estimation of avian population trends (J. R. Sauer and S. Droege, eds). US Fish Wildl. Serv. Biol. Rep. 90(1).

Keast, A., and E. S. Morton. 1980. Migrant birds in the Neotropics: ecology, behavior, distribution, and conservation. Smithsonian Institution Press, Washington DC.

Link, W. A., and J. R. Sauer. 1994. Estimating equation estimates of trends. Bird Populations 2:23–32.

Mannan, R. W., M. L. Morrison, and E. C. Meslow. 1984. The use of guilds in forest bird management. Wildl. Soc. Bull. 12: 426–430.

Marchant, J. H., R. Hudson, S. P. Carter, and P. Whittington. 1990. Population trends in British breeding birds. British Trust for Ornithology, Tring, Hertfordshire, UK.

Marshall, J. T. 1988. Birds lost from a giant Sequoia forest during fifty years. Condor 90:359–372.

Moses, L. E., and D. Rabinowitz. 1990. Estimating (relative) species abundance from route counts of the Breeding Bird Survey. Pp. 71–79 *in* Survey designs and statistical methods for the estimation of avian population trends (J. R. Sauer and S. Droege, eds). US Fish Wildl. Serv. Biol. Rep. 90(1).

Mountford, M. D. 1985. An index of population change with application to the Common Birds Census. Appl. Stat. 31:135–143.

O'Connor, R. J. 1992. Population variation in relation to migrancy status in some North American birds. Pp. 64–74 *in* Ecology and conservation of Neotropical migrant landbirds (J. M. Hagan and D. W. Johnston, eds). Smithsonian Institution Press, Washington, DC.

Peterjohn, B. G., J. R. Sauer, and W. A. Link.

1994. The 1992–1993 Summary of the North American Breeding Bird Survey. Bird Populations 2:46–61.

Rappole, J. H., M. A. Ramos, and K. Winker. 1989. Wintering Wood Thrush movements and mortality in southern Veracruz. Auk 106:402–410.

Robbins, C. S. 1980. Effect of forest fragmentation on breeding bird populations in the Piedmont of the mid-Atlantlc region. Atlantic Natur. 33:31–36.

Robbins, C. S., D. Bystrak, and P. H. Geissler. 1986. The Breeding Bird Survey: its first fifteen years, 1965–1979. US Fish Wildl. Serv. Resource Publ. 157.

Robbins, C. S., J. R. Sauer, R. S. Greenberg, and S. Droege. 1989a. Population declines in North American birds that migrate to the neotropics. Proc. Natl Acad. Sci. USA 86:7658–7662.

Robbins, C. S., D. K. Dawson, and B. A. Dowell. 1989b. Habitat area requirements of breeding forest birds of the Middle Atlantic States. Wildl. Monogr. 103.

Robbins, C. S., S. Droege, and J. R. Sauer. 1989c. Monitoring bird populations with Breeding Bird Survey and atlas data. Ann. Zool. Fennici 26:297–304.

Robinson, S. K. 1992. Population dynamics of breeding Neotropical migrants in a fragmented Illinois landscape. Pp. 408–418 *in* Ecology and conservation of Neotropical migrant landbirds (J. M. Hagan, III and D. W. Johnston eds). Smithsonian Institution Press, Washington, DC.

Sauer, J.R., and S. Droege. 1990. Recent population trends of the Eastern Bluebird. Wilson Bull. 102:239–252.

Sauer, J. R., and S. Droege. 1992. Geographic patterns in population trends of Neotropical migrants in North America. Pp. 26–42 *in* Ecology and conservation of neotropical migrant landbirds (J. M. Hagan, III and D. W. Johnston, eds). Smithsonian Institution Press, Washington, DC.

Sauer, J. R., and P. H. Geissler. 1990. Estimation of annual indices from roadside surveys. Pp. 58–62 *in* Survey designs and statistical methods for the estimation of avian population trends (J. R. Sauer and S. Droege, eds). US Fish Wildl. Serv. Biol. Rep. 90(1).

Sauer, J. R., M. K. Klimkiewicz, and S. Droege. 1987. Population trends of the purple martin in North America, 1966–1986. Nature Soc. News 22:1–2.

Sauer, J. R., B. G. Peterjohn, and W. A. Link. 1994. Observer differences in the North American Breeding Bird Survey. Auk 111: 50–62.

Sharp, B. 1985. Avifaunal changes in central Oregon since 1899. Western Birds 16:63–70.

Taub, S. R. 1990. Smoothed scatterplot analysis of long term Breeding Bird Census data. Pp. 80–83 *in* Survey designs and statistical methods for the estimation of avian population trends. (J. R. Sauer and S. Droege, eds). US Fish Wildl. Serv. Biol. Rep. 90(1).

Temple, S. A., and J. R. Cary. 1990. Using checklist records to reveal trends in bird populations. Pp. 98–104 *in* Survey designs and statistical methods for the estimation of avian population trends (J. R. Sauer and S. Droege, eds). US Fish Wildl. Serv. Biol. Rep. 90(1).

Terborgh, J. 1989. Where have all the birds gone? Princeton University Press, Princeton, NJ.

Trautman, M. B. 1940. The birds of Buckeye Lake, Ohio. Univ. Michigan Mus. Zool. Misc. Publ. 44.

Wilcove, D. S. 1988. Changes in the avifauna of the Great Smoky Mountains: 1947–1983. Wilson Bull. 100:256–271.

Witham, J. W., and M. L. Hunter, Jr. 1992. Population trends of Neotropical migrant landbirds in northern coastal New England. Pp. 85–95 *in* Ecology and conservation of Neotropical migrant landbirds. (J. M. Hagan, III and D. W. Johnston, eds.) Smithsonian Institution Press, Washington, DC.

Wright, A. M. 1915a. Early records of the Wild Turkey, part 3. Auk 32:61–81.

Wright, A. M. 1915b. Early records of the Wild Turkey, part 4. Auk 32:207–224.

2

THE STRENGTH OF INFERENCES ABOUT CAUSES OF TRENDS IN POPULATIONS

FRANCES C. JAMES AND CHARLES E. McCULLOCH

INTRODUCTION

The idea that land birds that breed in the United States and Canada and spend the winter in the New World tropics have been suffering severe population declines is the basis for the United States government-sponsored program *Partners in Flight* and is the rationale for the publication of this volume. We believe that the evidence of an overall decline in populations of Neotropical migrants is weak (James et al., in press). Analyses of the best data for population trends in North American land birds for the past 25 years, the Breeding Bird Survey (BBS), indicate general declines in some species of Neotropical migrants and increases in others, but no overall pattern that pertains to the group as a whole (Peterjohn et al., Chapter 1, this volume). However, documentation of population trends is not the issue in this paper. Instead we will address methods for the confirmation of hypothesized causes of population trends, once population trends have been established. The confirmation of causes is more a matter of logic than of data analysis. It involves the design of the study, how comparisons are made among observations, rather than the significance of statistical tests. Of course, both steps, documentation of trends and determination of their causes, should precede decisions about management.

Controlled experiments are usually considered to be the only way to determine cause-and-effect relationships, but true field experiments (including random assignment of experimental units to treatments) are often not possible, and one must be satisfied with either quasi-experiments or with observational studies. In this sense, most field experiments are, in fact, quasi-experiments. (For definitions of terms, see Table 2-1, and for notation that makes the logic of causal reasoning explicit, see Table 2-2.) In addition, the external validity of field experiments, that is, the extrapolation of the results to domains larger than that of the experiment, is a matter of judgment rather than rigorous statistical inference (Eberhardt and Thomas 1991). What we recommend is a further step, that principles of quasi-experimental design be applied to comparative observational studies. Although such efforts will provide even weaker causal inferences than would quasi-experiments, they have the important advantage that they can incorporate the entire domain of interest. Just as with the new nonmanipulative comparative methods in evolutionary biology (Harvey and Pagel 1991), studies of population regulation could benefit from advances in the development of comparative methods. Carefully designed comparative observational studies (the analytical sampling of Eberhardt and Thomas 1991) should not be viewed as second-class experiments. They have their own justification (Kish 1987).

Referring to the field of epidemiology, another area that seeks to find the causes of broad-scale phenomena, Rothman (1988) wrote that, without such effort toward falsification, we risk forming a consensus based on shared irrational belief buoyed by an accumulation of supporting observations. Maclure (1988) added that, without attention to alternative explanations, we are left with untested beliefs and uncritical convictions about what should be done. Our argument here is that any real confirmation of a cause

Table 2-1. Definitions, based largely on Campbell and Stanley (1966), Cook and Campbell (1979), and Cochran (1983).

Comparative observational study A study involving sampling of naturally occurring events but employing elements of quasi-experimental design such as control groups to gain inferences about causal relationships; the same as the "*ex post facto* experiments" of Campbell and Stanley (1966) and "analytical sampling" by Eberhardt and Thomas (1991); also includes the "natural experiments" of Diamond (1986) in those cases in which rival hypotheses can be rendered implausible.

Control group An experimental unit that did not receive the treatment; in a comparative observational study, a group to be compared with another group that was subjected to a natural intervention.

Controlled experiment A study in which the researcher applies a treatment, as opposed to an observational study or a survey.

Descriptive survey A sampling study not planned for the analysis of causes; nevertheless the results could be analyzed using quasi-experimental designs, especially if the effects of distinct perturbations are evident; the BBS is an example.

Experimental design An arrangement of observational units and treatments planned to detect the effect of a treatment or perturbation.

Intervention An event that occurred within a series of observations of the same group; may be a treatment applied by a researcher or a naturally occurring event.

Multiple-time-series design A comparative design involving more than one group, each of which has several observations before and after a treatment.

Observational study A study consisting of sampling; the study is uncontrolled in the sense that the researcher did not apply a treatment (Cochran 1983); it can be simply a descriptive survey without any intent toward causal inference or a comparative observational study planned to analyze causes; it can encompass the entire domain of interest more easily than can most experiments.

Quasi-experiment An experiment in which there is no randomization of the assignment of experimental units to treatments (Cook and Campbell 1979); same as the "pseudoexperiments" of Cochran (1983).

Randomization In experimental design, the random assignment of experimental units to treatments, often not possible in field experiments.

Replication Repeated cases of the application of treatments to experimental units, whether or not they are under the control of the observer.

Time-series design with intervention A design consisting of several observations before and after an intervening treatment.

True experiment An experiment whose design includes control groups, randomization and replication; for an example of a true field experiment with birds, see Walters et al. (1990).

of a population trend would have to be made on the basis of a combination of the results of controlled field experiments, which would provide inference about a local situation, and the results of comparative observational studies, which would provide information about processes in the large intact ecosystem.

RESEARCH DESIGNS FOR CAUSAL ANALYSIS

Our definitions (Table 2-1) and notation (Table 2-2) are based primarily on those of Campbell and Stanley (1966), Cook and Campbell (1979), and Cochran (1983), and we cite alternative terms used by other authors. Campbell and Stanley (1966) used the following notation:

X = exposure of a group to a treatment or an event
O = an observation
R = random assignment of experimental units to treatments.

Rows of Xs and Os are for single groups, and columns are for simultaneous treatments and observations. Replication (multiple experimental units per group) would ordinarily be included and is not discussed here. It affects the likely statistical significance of effects but not the strength of causal inference that can be made once effects have been detected. The first three categories in Table 2-2 are experimental in that the researcher applies the treatment. The fourth category, comparative observational studies, involves making planned comparisons of sets of observations of naturally occurring events. The random assignment of experimental units to treatments is only possible with category 2, an example of a true experimental design. We estimate the strength of inference possible with each design on a subjectively determined scale from 0 to 5.

Table 2-2. Examples of four categories of research designs used for causal analysis (from Campbell and Stanley 1966).

Design	Notation[a]	Treatment Controlled by Observer	Randomization	Control Group	Strength of Causal Inference (0–5)[b]	Domain (1–3)[c]
1. Inadequate designs						
a. One-shot case study	X O	+	—	—	0	1–2
b. One-group pretest–posttest	O X O	+	—	—	1	1–2
c. Static-group comparison	X O O	+	—	+	1	1–2
2. A true experimental design	R O X O R O O	+	+	+	5	1
3. Quasi-experimental designs (nonrandomized control)						
a. Nonequivalent-control-group design (removes effect of time)	O X O O O	+	—	+	3	2
b. Single time series with intervention	O O O O X O O O O	+	—	—	3	2
c. Multiple time series with intervention	O O O O X O O O O O O O O O O O O	+	—	+	4	2
4. Comparative observational studies[d]						
a. Nonequivalent-control-group design	O X O O O	—	—	+	1	2–3
b. Single time series with natural intervention	O O O O X O O O O	—	—	—	2	2–3
c. Multiple time series with natural intervention	O O O O X O O O O O O O O O O O O	—	—	+	3	2–3

[a] X = treatment; O = observation; R = units randomly assigned to treatments or controls; rows are for single groups; columns are for simultaneous events.
[b] Categories, assigned by the authors, range from 0 for none to 5 for maximal causal inference.
[c] Categories range from 1 for the usual restricted geographical (or temporal) domain of a field experiment to 3 for the entire domain of interest.
[d] Same designs as in category 3, quasi-experimental designs, but Xs are naturally occurring events.

Inadequate Designs

We define controlled experiments as those cases in which the researcher is in control of the application of treatment. The first three experimental designs in Table 2-2, the one-shot case study,

X O,

the one group pretest–posttest,

O X O,

and the static group comparison,

X O
O,

are inadequate for causal inference, even if the work is based on controlled experiments, because they offer insufficient control over extraneous processes. If the researcher wished to conclude that the observation after the treatment was affected by the treatment, that conclusion would have to be confirmed from information beyond the experiment, because the logic of the design was inadequate for causal inference. In the first case, one group was studied once, so there was no control and no comparison. In the second case, one group was studied before and after a treatment. This design controls for initial inequalities by using the group as its own control group, but does not control for alternate causes that occurred at the same time as X. In neither case is there any assurance that only the treatment caused the effect. In the third case, there is a control group that did not receive the treatment, but because there was no random assignment, there is no assurance that the initial groups did not differ before the treatment.

A True Experimental Design

We give only one example of a true experimental design (Table 2-2, category 2):

R O X O
R O O,

the pretest–posttest control-group design. Here the Rs signify that, before the experiment, experimental units were randomly assigned to treatments. A comparison of the differences between observations before and after the treatment and differences without the treatment is a test for the effect of the treatment.

Quasi-experimental Designs

The term quasi-experiment refers to an experiment in which full experimental control is lacking because the random assignment of experimental units to treatments was not performed (Table 2-1). Inferences from such experiments are necessarily weaker than would be those from a true experiment. According to Cook and Campbell (1979, p. 6) the basic issue with quasi-experiments is that

> comparisons depend on non-equivalent groups that differ from each other in many ways other than the presence of a treatment whose effects are being tested. The task confronting persons who try to interpret the results from quasi-experiments is basically one of separating the effects of a treatment from those due to initial noncomparability between average units in each treatment group; only the effects of the treatment are of research interest. To achieve this separation of effects, the researcher has to explicate the specific threats to valid causal inference that a random assignment rules out and then in some way deal with these threats. In a sense, quasi-experiments require making explicit the irrelevant causal forces hidden within the *ceteris paribus* of random assignment.

With the nonequivalent-control-group design, in which both groups are observed pretest and posttest,

O X O
O O,

the only difference from the example of a true experimental design is that there is no pre-experimental randomization to assure sampling equivalence. Because of the simultaneous observations made of the two groups, the design removes the effect of time. However, because of the lack of randomization, even if a statistical comparison similar to the one for the true experiment is significant, it does not necessarily test the effect of the treatment (see Hurlbert 1984).

A single time series of observations into which the observer has introduced an

intervention,

is another type of quasi-experimental design. In this case there is no control group, but the multiple observations pretest and posttest help the researcher judge the effect of the treatment. The weakness of this design is that it does not remove the effects of competing causes occurring at the same time as X.

With the inclusion of a control group that is not subjected to the intervention, the single-time-series design becomes the multiple-time-series design,

O O O O X O O O O

O O O O   O O O O,

and substantial strength of inference is gained. If a researcher is unable to perform a true experiment, the multiple-time-series design is a particularly strong alternative.

It is possible to extend all of these designs to cover tests of alternative causes. For example, with the multiple-time-series design and treatments X and Y, the notation would be

O O O O X O O O O

O O O O Y O O O O

O O O O   O O O O

O O O O XY O O O O.

The notation XY means that both treatments X and Y occurred together.

Comparative Observational Studies

Observational studies are not experimental because the researcher is not in control of treatments. Even so, naturally occurring events can be viewed as treatments. In some cases principles of quasi-experimental design, because they help the researcher gain some control over irrelevant causal forces, can be used to strengthen causal inferences (Cook and Campbell 1979; Table 2-2, category 4). Of course inadequate experimental designs (Table 2-2, category 1) will be inadequate in these cases as well. Many causal hypotheses about naturally occurring events are first addressed through inadequate designs, so the question becomes how to move them forward through a progression of designs that provide stronger causal inference. The category of comparative observational studies addresses this issue, how some rival hypotheses about causes can be rendered implausible by the design of the comparisons (Cook and Campbell 1979). This category includes the "natural experiments" of Diamond (1986).

When there have been no abrupt natural events, the analysis of causes (treatments) in comparative observational studies is the most difficult of all, yet this is the situation in many cases of the analysis of trends in bird populations. Even if multiple observations are available, the analysis may require quantification of static levels of hypothesized causes. For two hypothesized causes, the design is likely to be

X O O O O

Y O O O O

XY O O O O

   O O O O

which would have to be interpreted as a static group comparison with multiple observations, a weak design.

Suppose we wish to separate the effects of changing levels of cowbird parasitism (X) and habitat fragmentation (Y) on population trends of the Wood Thrush. We would need measurements of the degree of parasitism and fragmentation for each of a number of sites. Our actual design would be complicated because parasitism and fragmentation would probably be continuous measures rather than just absent (control) or present (X or Y). So with population size measurements (O) in consecutive years (e.g., as in BBS data) and continuous (measured) levels of parasitism ($X_1, X_2, \ldots, X_n$) and fragmentation ($Y_1, Y_2, \ldots, Y_n$) the design would be:

O X_1Y_1 O

O X_2Y_2 O

O X_3Y_3 O

⋮   ⋮   ⋮

O X_nY_n 0

With treatments that were merely present or absent, and observations before and after treatments, it would have been easy to calculate the effect of parasitism (by comparing the change in the XY group to the change in the Y group and/or X to control) or fragmentation (by comparing XY to X and/or Y to control), nonequivalent-control-group designs. However, because the levels of each factor vary among groups, a multiple-regression model must be used to try to infer the effect of each factor independently of the other. Unfortunately, this is a common situation, and the validity of the inferences "will depend not only on the study design but also on the skills and ability of the investigator to develop sound subjective interpretations of the test results" (Skalski and Robson 1992). Incorrect assumptions incorporated in the regression model (which is necessarily a simplification of the true situation) may result in biased estimates of causes.

Correlation and Causation

The arguments in this paper are relevant to the commonly stated idea that one cannot infer causation from correlations among variables or, more properly, that correlational analysis alone is an insufficient basis for inferring causation. With observational studies, the problem is that it may be difficult to control possibly confounding factors. The way to improve such control is less a matter of the method of analysis (e.g., correlation, regression) than of the design of the study (e.g., single time series, multiple time series). Correlational analyses based on a strong design can provide causal inference. For discussion of appropriate data analysis to accompany various experimental designs, see Scheiner and Gurevitch (1993).

EXAMPLES OF QUASI-EXPERIMENTS AND COMPARATIVE OBSERVATIONAL STUDIES WITH COMMENTS ON THE STRENGTH OF THE CAUSAL INFERENCES THAT THEY ALLOW

In this section we present four examples of studies that attempt to identify causes of population trends in birds, and we show how their designs fit into the categories in Table 2-2. The first two examples are local quasi-experiments, the third and fourth examples are broad-scale comparative observational studies. The example of the Kirtland's Warbler uses a single-time-series design with two interventions due to two different types of treatments. The example of the Great Tit demonstrates the added strength of inference that might have been gained with a multiple-time-series design.

With comparative observational studies, much larger geographic areas can be considered. The third example, a comparative observational study of population trends in the Ovenbird, uses BBS data. It begins with an inadequate design but is somewhat strengthened by an independent analysis. The fourth example, the Carolina Wren, uses BBS data in a comparative observational study with a multiple-time-series design. Results for different physiographic strata are viewed as groups, and the effects of a naturally occurring perturbation, severe winter weather, had different levels of impact in different strata.

1. A Quasi-experimental Design, Single Time Series: Kirtland's Warbler

The endangered Kirtland's Warbler has only one known breeding population, the one in early successional stands of jack pines (*Pinus banksiana*) mainly in two counties in central Michigan. Field surveys showed that this population declined by 50% in the 1960s and that brood parasitism by the Brown-headed Cowbird was severe. With the establishment of a sustained intensive cowbird control program in the 1970s, the number of singing male warblers stabilized at about 200 individuals, but no major increases in the breeding population occurred (Probst 1986). In 1980 a prescribed fire got out of hand and burned several thousand acres, setting into motion an extensive natural regeneration of jack pines. By the late 1980s the young forest in the Mack Lake Burn had become suitable habitat for the Kirtland's Warbler, and since that time the breeding population has increased dramatically (Fig. 2-1). The continuing cowbird control program is probably necessary, and the conservation of the species will probably be well served by a program of habitat manipulation that involves

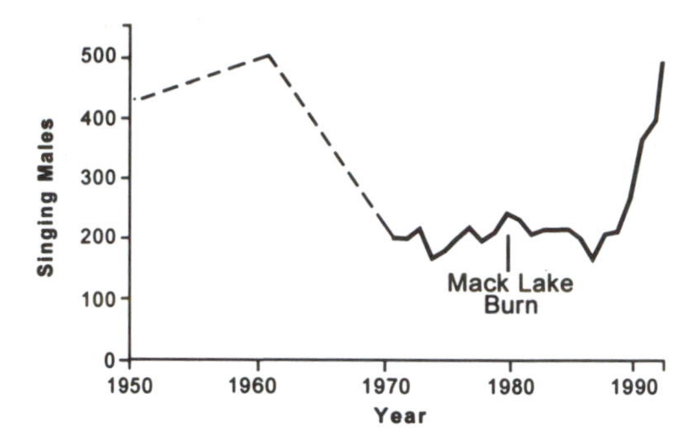

Figure 2-1. The number of singing male Kirtland's Warblers stabilized after the Brown-headed Cowbird control program began in 1970 and rose sharply after 1990, when ecological succession following the Mack Lake Burn provided new suitable habitat.

exceptionally hot fire and natural regeneration of jack-pine forest. This single time series, a quasi-experimental design with no control and two interventions,

O O O X O O Y O O O,

allows moderately strong inferences even though it does not remove the effects of a competing cause occurring at the same time as X or Y. We rank its strength of inference a 3 (Table 2-2, category 3b). Had there been a comparison group that had not been subjected to either cowbird control or the hot fire, stronger inference would have been possible.

2. A Quasi-experimental Design, Single Time Series: Great Tit

Tinbergen et al. (1985) provide an example of the danger of misinterpreting the results of field experiments that offer only moderate strength of inference and that do not consider alternative causes. The data are from the classic long-term studies of fluctuations in a population of Great Tits using nest boxes on the island of Vlieland in The Netherlands. Kluyver and coworkers had experimentally reduced the numbers of fledglings in the breeding seasons of 1960–1963 and 1967–68 and found that adult survival was increased in the years following manipulations. This result was interpreted by Kluyver (1971, as cited by Tinbergen et al. 1985) as evidence of density-dependent population regulation. Tinbergen et al. (1985) reanalyzed the data, adding information up to 1978 that included further manipulations from 1970 to 1974 (Fig. 2-2). They also used data for annual variation in the seed crop available to birds for food in winter. The reanalysis showed that years when there were manipulations also happened to be years with abundant winter food. Adult survival was, in fact, positively related to the seed stock in the previous winter and the number of fledglings produced per pair in the previous breeding season. Thus, winter food supply was as valid an explanation of variation in adult survival as was density-dependent population regulation. Had there been a control population that was not manipulated, and thus a multiple-time-series rather than a single-time-series design,

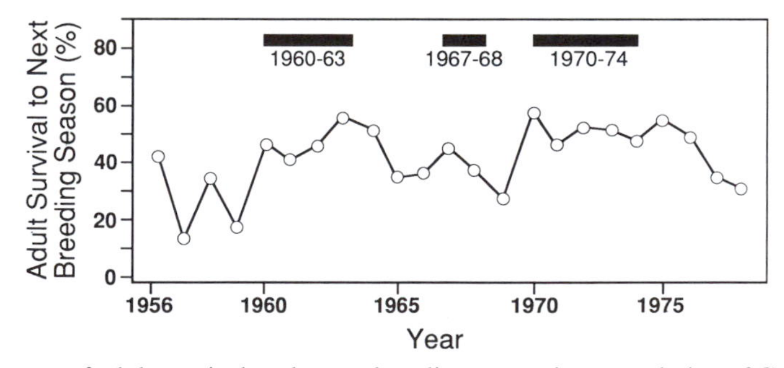

Figure 2-2. The rate of adult survival to the next breeding season in a population of Great Tits using nest boxes on the island of Vlieland, The Netherlands. The black bars show years when juvenile recruitment was reduced by removal of eggs or birds. Based on Tinbergen et al. (1985).

the presence or lack of causal association between the experimental conditions and adult survival would have been clear from the beginning. This example shows two things: (1) the danger, even with experiments, of not considering possibly confounding factors; and (2) the difference in strength of inference between two quasi-experimental designs (Table 2-2, categories 3b and c, inference levels 3 and 4).

3. A Comparative Observational Study: Static Group Comparisons with Multiple Observations Plus Independent Evidence: Ovenbird

Cases in which there are no clear environmental perturbations and no data sets for underlying environmental variation among strata call for the examination of population trends among populations to identify patterns and then the development of hypotheses about causes. For example, in BBS data for 20 physiographic strata for the Ovenbird, the overall 25-year trends are estimated to be substantial (an average change of more than one bird per route) in nine strata (Table 2-3). Of these nine strata, the Blue Ridge Mountains and the Adirondack Mountains are both highland areas. The design of comparisons between highland and lowland strata is basically a static group comparison with multiple observations where X is highland areas,

X O O O O

O O O O,

by itself an inadequate design (Table 2-2, category 1c). However, in a recent multispecies analysis using BBS data for the Ovenbird and 19 other species of wood warblers, we showed that highland strata in the eastern and central United States tend to have more declining species generally than do other physiographic strata (James et al. in press). This result is too weak for causal inference, but it suggests that research is warranted to determine what correlates of altitude might be affecting populations of warblers. The hypothesized treatment (X) is something that is present in highland areas that is not present in lowland areas. The next

Table 2-3. Population trends in the Ovenbird from 1970–1972 to 1986–1988 by 20 physiographic strata, as analyzed by semiparametric route regression limited to 300 iterations.[a] All strata are reported that had differences of an average of more that one bird per route. An asterisk indicates statistical significance at $\rho < 0.05$.

Physiographic stratum	df	Average Change in Birds per Route
Blue Ridge Mountains	2	−5.6*
Adirondack Mountains	4	−2.8
Southern Upper Coastal Plain	28	1.1
Spruce Hardwoods and Boreal Forest	67	1.2
Greak Lakes Transition	19	1.5
St Lawrence River Plain	19	2.1*
Southern New England and Glaciated Coastal Plain	23	2.3
Allegheny Plateau	36	2.4*
Northern New England	20	3.6

[a] With semiparametric route regression, yearly data of counts for a species on a route are smoothed with LOESS after additive observer effects are removed, then the yearly values are aggregated by strata and smoothed again to give an estimate of the trend. Tests of the statistical significance of a trend for a stratum are based on paired t-tests between averages of the smoothed counts for the first and last five years of the period under consideration. (See James et al., in press, for further details.)
df = degrees of freedom.

step should be analyses for each species that explore different combinations of levels of hypothesized causes. For example, if fragmentation, cowbird parasitism, or atmospheric pollution are the most likely suspected causes, one should construct data sets for different levels of these factors among strata, and possibly changes in levels within the BBS period, and then strengthen the design from a static group comparison by treating it as a multiple-time-series design in which different groups have different combinations and levels of these potential causes.

4. A Comparative Observational Study, Multiple Time Series with Different Levels of a Natural Intervention: Carolina Wren

Declines of populations of the nonmigratory Carolina Wren after severe winter weather is well illustrated in BBS data. In many places changes in the size of the breeding population

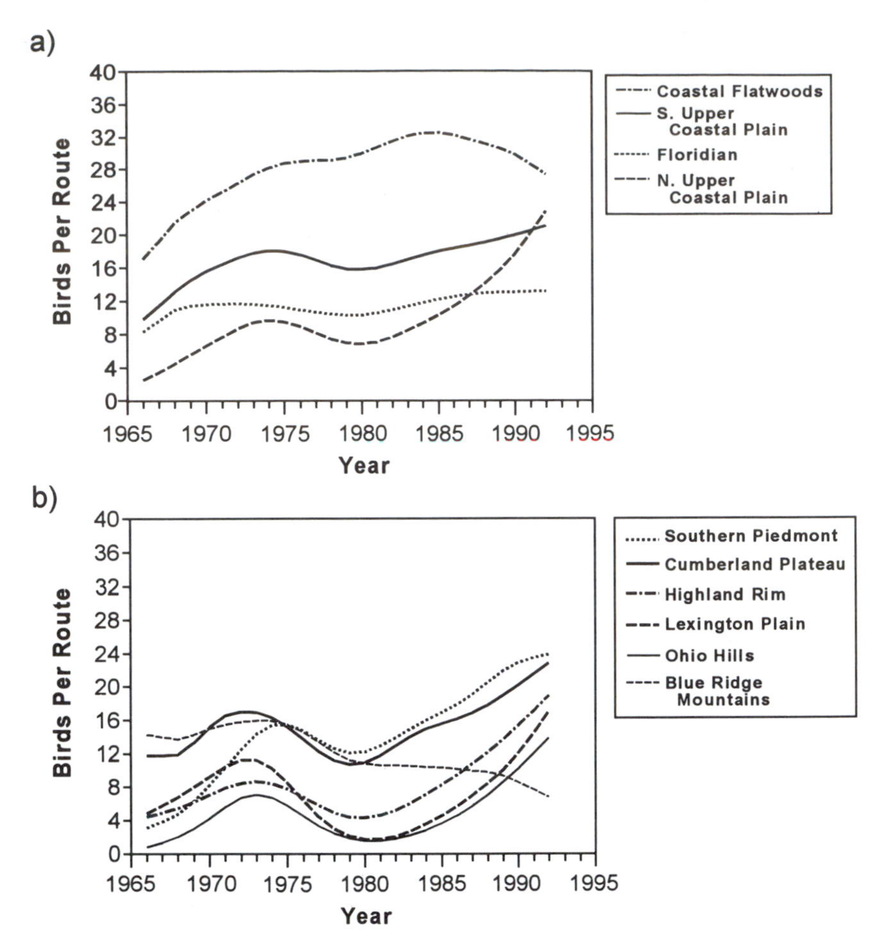

Figure 2-3. Population trends in the Carolina Wren in ten physiographic strata as estimated by nonlinear semiparametric route regression (James et al., in press) applied to data from the BBS, but without smoothing.

seem to be primarily a function of the severity of the previous winter (Robbins et al. 1986, Fig. 2-36). The particularly severe snow and ice storms in the winters of 1976–1977 and 1977–1978 were associated with sharp declines as far south as the Coastal Plain (Fig. 2-3a) and were most dramatic in three physiographic strata, the Ohio Hills, Lexington Plain, and Highland Rim (Fig. 2-3b). If BBS data are interpreted as a comparative observational study, the natural intervention, with its different levels of severity in different physiographic strata, llows the researcher to compare nonlinear population trends among strata and to make inferences about the relative effects of the two harsh winters (combined in this analysis) in strata that were affected to differing degrees. Apparently, populations in the strata most affected did not fully recover until the 1990s. This example has multiple naturally selected treatment and control groups, all with a set of pretest and posttest observations. As organized here, each group (row) is the average number of Carolina Wrens per BBS route in a stratum across years. The design has groups that received different levels of the same treatment (X), the two severe winters, which varied in intensity in different strata

$$\begin{array}{ccccccccc} O & O & O & O & X_0 & O & O & O & O \\ O & O & O & O & X_1 & O & O & O & O \\ O & O & O & O & X_2 & O & O & O & O \\ O & O & O & O & X_3 & O & O & O & O \end{array}$$

CAUSAL INFERENCES ABOUT POPULATION TRENDS IN NEOTROPICAL MIGRANTS

What are the implications of the previous discussion for the conservation of Neotropical migrant birds? Concern for Neotropical migrants increased in the 1980s with several reports of declines in suburban forest study plots in the eastern United States (Terborgh 1989). Vague correlations with tropical deforestation were interpreted as causal by some authors, correlations with forest fragmentation on the breeding grounds by others (see review by Askins et al. 1990). Brittingham and Temple (1983) inferred a causal basis for a purported negative correlation between the abundance of the brood-parasitic Brown-headed Cowbird and the abundances of Neotropical migrants. Askins and Philbrick (1987) incorporated alternative causal factors into a multiple-regression analysis of census data for forest-interior, long-distance migrants in a preserve in Connecticut and concluded that the degree of habitat isolation of the study area was the main factor involved in regulating their populations. Their design was a one-group pretest–posttest. On the basis of field experiments, Wilcove (1985) found that predation on artificial nests was higher in edge than in forest-interior habitats and inferred that general population declines were due to predation associated with forest fragmentation.

Most researchers acknowledge that the best source of data for the documentation of regional population trends in North American land birds is the Breeding Bird Survey, the roadside census program started in 1966 and sponsored by the US Fish and Wildlife Service (FWS) and the Canadian Wildlife Service (Robbins et al. 1986). The most recent FWS analyses (Robbins et al. 1989; Peterjohn et al. Chapter 1, this volume) of BBS data suggest that a major feature of the data is that populations of Neotropical migrants increased in the 1970s and decreased in the 1980s. This phenomenon has received no attention in terms of causal analysis. However, we have not been able to find this peak in our own analyses of BBS data for wood warblers (James et al., in press).

Robbins et al. (1989) and Bohning-Gaese et al. (1992) demonstrated associations between population trends and causes using regression-type analyses with BBS data. We have reservations about their analyses, however, because they assumed that data on trends in different species were independent. This is unlikely to be true because BBS data for all species come from the same routes run at the same times.

In all of the above literature, purported multispecies declines were attributed to a preferred alternative cause, and the design of the research did not adequately rule out alternative explanations. In addition to their weak designs, the validity of extrapolations from population trends detected in local studies to scales larger than their immediate surroundings has rarely been checked. For example, there is no evidence in BBS data that cowbirds have been increasing in abundance in the eastern United States in the last 25 years (John Sauer, David Wiedenfeld, personal communications) or even that Neotropical migrant forest birds as a group are declining in New England (Smith et al. 1993). Another relevant issue is whether observed variation in survival (for example, due to predation) actually affects the size of the population in the following breeding season (Newton 1980). Unfortunately, we do not have complete information about the causes of population limitation for even one species.

At least three major problems must be addressed before valid inferences can be developed about what factors affect population trends of Neotropical migrants. The first, of course, is accurate documentation of their fluctuations. This is fairly easily accomplished at a local scale but determination of regional population trends requires sampling at that scale. The second problem, which is the one discussed in this paper, is how to determine the major causes of observed population trends. If experiments are feasible, they may show what environmental factors, if modified, would cause a change in the number of individuals in the next breeding season in a local population. Even so, their results can be misleading. The validity of causal inferences derived from experiments depends upon their design. In addition, extrapolation of the results of local field experiments to large geographic areas may not be justified. We think that carefully designed comparative observational studies are called for. The third problem is that confirmed sources of mortality are not necessarily causes of population regulation. There is a clear need for well-designed experiments that allow for the identification of processes that, when modified, result in changes in the sizes of the breeding populations.

RECOMMENDATIONS

1. Study individual species that have different population trends in different areas. The geographic variation in population trends within species can be used to advantage in causal analyses. Construct data sets for different levels of potential causes in the different areas. Seek examples in which levels of treatments change during a time series. Analyze the data for trends and causes as comparative observational studies that use the strongest possible levels of inference, preferably with the multiple-time-series design. The BBS is an appropriate source of data for the population trends. In statistical analyses of BBS data, do not treat species as independent units.
2. If feasible, conduct field experiments to test whether purported causes actually affect the size of the breeding population.
3. Be explicit about the level and scope of causal inference allowed by the design of a study, including the extent to which alternative explanations have been accounted for.

CONCLUSION

At present, ecological research intended to discover the causes of broad-scale ecological phenomena, such as long-term population trends in birds, is poorly developed, and there is a large gap in knowledge at the research–management interface. Until sound inferences can be developed about what environmental factors, when modified, will make a difference, conservation and management will be inefficient. An important part of the development of such causal inferences should be the incorporation of principles of quasi-experimental design into broad-scale observational studies. This type of research deserves more attention.

ACKNOWLEDGMENTS

We thank James Cox, Duncan Evered, Charles Hess, Don Levitan, Arie van Noordwijk, Noel Wamer, David Wiedenfeld, and two anonymous reviewers for their helpful criticism of earlier drafts of the manuscript.

LITERATURE CITED

Askins, R. A., and M. J. Philbrick. 1987. Effects of changes in regional forest abundance on the decline and recovery of a forest bird community. Wilson Bull. 99:7–21.

Askins, R. A., J. F. Lynch, and R. Greenberg. 1990. Population declines in migratory birds in eastern North America. Pp. 1–57 *in* Current ornithology, Vol. 7 (D. M. Power, ed.). Plenum Press, New York.

Bohning-Gaese, K., M. L. Taper, and J. H. Brown. 1992. Are declines in North American insectivorous songbirds due to causes on the breeding range? Conserv. Biol. 7:6–86.

Brittingham, M. C., and S. A. Temple. 1983. Have cowbirds caused forest songbirds to decline? BioScience 33:31–35.

Campbell, D. T., and J. C. Stanley. 1966. Experimental and quasi-experimental designs for research. Rand McNally, Chicago, IL.

Cochran, W. G. 1983. Planning and analysis of observational studies. Wiley, New York.

Cook, T. D., and D. T. Campbell. 1979. Quasi-Experimentation: design and analysis issues for field settings. Rand McNally, Chicago, IL.

Diamond, J. 1986. Overview: laboratory experiments, field experiments, and natural experiments. Pp. 3–22 *in* Community ecology (J. Diamond and T. J. Case, eds). Harper and Row, New York.

Eberhardt, L. L., and J. M. Thomas. 1991. Designing environmental field studies. Ecol. Monogr. 61:53–73.

Harvey, P. H., and M. D. Pagel. 1991. The comparative method in evolutionary biology. Oxford University Press, New York.

Hurlbert, S. H. 1984. Pseudoreplication and the design of ecological field experiments. Ecol. Monogr. 54:187–211.

James, F. C., C. E. McCulloch, and D. A. Wiedenfeld. In press. New approaches to the analysis of population trends in land birds. Ecology.

Kish, L. 1987. Statistical design for research. Wiley, New York.

Maclure, M. 1988. Refutation in epidemiology: why else not? Pp. 130–138 *in* Causal inference (K. J. Rothman, ed.). Epidemiology Resources, Chestnut Hill, MA.

Newton, I. 1980. The role of food in limiting bird numbers. Ardea 68:11–30.

Probst, J. R. 1986. A review of factors limiting the Kirtland's Warbler on its breeding grounds. Amer. Midl. Natur. 116:87–100.

Robbins, C. S., D. Bystrak, and P. H. Geissler. 1986. The Breeding Bird Survey: its first fifteen years, 1965–1977. US Fish Wildl. Service, Resource Publ. 157, Washington, DC.

Robbins, C. S., J. R. Sauer, R. S. Greenberg, and S. Droege. 1989. Population declines in North American birds that migrate to the Neotropics. Proc. Natl. Acad. Sci. USA 86:7654–7662.

Rothman, K. J. 1988. Causal inference. Epidemiology Resources, Chestnut Hill, MA.

Scheiner, S. M., and J. Gurevitch, eds. 1993. Design and analysis of ecological experiments. Chapman and Hall, New York.

Skalski, J. R., and D. S. Robson. 1992. Techniques for wildlife investigations: design and analysis of capture data. Academic Press, San Diego, CA.

Smith, C. R., D. M. Pence, and R. J. O'Connor. 1993. Status of Neotropical migratory birds in the Northeast: a preliminary assessment. Pp. 172–188 *in* Status and management of Neotropical migratory birds (D. M. Finch and P. W. Stangel, eds). USDA Rocky Mountain Forest and Range Experiment Station, Ft Collins, CO.

Terborgh, J. 1989. Where have all the birds gone? Essays in the biology and conservation of birds that migrate to the American tropics. Princeton University Press, Princeton, NJ.

Tinbergen, J. M., J. H. Van Balen, and H. M. Van Eck. 1985. Density dependent survival in an isolated Great Tit population: Kluyver's data reanalyzed. Ardea 73:3–48.

Walters, J. R., C. K. Copeyon, and J. H. Carter, III. 1990. Test of the ecological basis of cooperative breeding in Red-cockaded Woodpeckers. Auk 109:90–97.

Wilcove, D. S. 1985. Nest predation in forest tracts and the decline of migratory songbirds. Ecology 66:1211–1214.

PART II

TEMPORAL PERSPECTIVES ON POPULATION LIMITATION AND HABITAT USE

3

WHEN AND HOW ARE POPULATIONS LIMITED? THE ROLES OF INSECT OUTBREAKS, FIRE, AND OTHER NATURAL PERTURBATIONS

JOHN T. ROTENBERRY, ROBERT J. COOPER, JOSEPH M. WUNDERLE, AND KIMBERLY G. SMITH

INTRODUCTION

Populations of migrant birds are affected by natural changes in weather, climate, and habitat, including those brought about by unpredictable events like drought, floods, fire, and insect infestations. This chapter will review effects of potentially "catastrophic" natural events on bird populations and discuss methods for understanding, interpreting, and managing (or at least mitigating) effects of such disturbances. In most cases we will focus on the proximate, rather than ultimate, consequences of these events, although we note where appropriate if a particular species appears adapted to a recurring disturbance. Likewise, we emphasize extrinsic rather than intrinsic (e.g., the relationship between food abundance and reproductive success) factors relating to population limitations, the latter being amplified by Sherry and Holmes (Chapter 4, this volume).

Sousa (1984) provides an excellent review of potentially catastrophic natural disturbances that may occur in ecological systems. He points out that the old view of disturbance as an uncommon, irregular event that causes an abrupt, structural change in a natural community and moves it away from a static, near-equilibrium condition, probably has little utility. Evidence from long-term censuses suggests that few natural populations or communities persist at or near an equilibrium on a *local scale* (Connell and Sousa 1983), and that the "damage" caused by any force can vary from negligible to extreme, depending on the intensity of the force and the vulnerability of target organisms. That events such as wildfire, drought, and outbreaks of insects are often viewed as "natural disasters" thus remains a matter of perspective. We prefer to view these events as natural phenomena that may, depending on one's particular management goal, have catastrophic consequences. As natural phenomena, however, they are integral features of every ecosystem (more frequent in some than in others), and the question of coping with their occurrence in any managed ecosystem is not a question of "if" they occur, but of "when."

Perhaps more important from the perspective of a land manager charged with the responsibility of preserving a particular species or habitat type is not whether a population's abundance is "regulated" by some particular disturbance, but whether such a disturbance may possibly cause the extinction of that population. As we document below, such can clearly be the case. Indeed, we may have just witnessed the loss of potentially 5–6 species of Hawaiian Islands endemics, whose last known populations (most less than 50 individuals; Scott et al. 1986) occurred in the Alakai Swamp on Kauai, an island devastated by Hurricane Iniki on 11 September 1992. However, apart from the observation that smaller populations are more at risk than are larger ones, theoretical generalizations are few. Most models that examine relationships between population characteristics and persistence in the face of localized environmental change tend to be demographic simulations specific to a particular species and place (e.g., Gilpin 1987, Pulliam et al. 1992). The few generalized analytical models that exist (e.g., Ewens et al. 1987, Shaffer 1987) suggest that median time to extinction for single populations (given birth, death, and catastrophe occurrence rates) rises only as the log of initial population size, not linearly or exponentially, as it does for environmental and demographic stochasticity, respectively. This implies that catastrophe can be a more serious threat to population persistence than either demographic or environmental stochasticity (e.g., Gilpin and Soulé 1986, Shaffer 1987).

In the next several sections we discuss more specific examples of the effects of natural disturbances due to insect outbreaks, climate, and wildfires on populations of birds, with emphasis on Neotropical migrant species, especially those breeding in eastern North America and wintering in the Caribbean.

CLIMATE

Variation in climate (long-term meteorological expectation) or weather (short-term actual experience), as well as their effects on ecological systems, can be expressed over a variety of spatial and temporal scales (Michaels and Hayden 1987). Drought, for example, is only manifest over a period of months or even years (and may affect tens of thousands of km^2), whereas a tropical storm may intensify to hurricane strength, make landfall, then dissipate within a week. A large thunderstorm with torrential rain or hail may form and disperse in the course of a few hours, spawning tornados or downbursts whose existence may be measured in minutes or even seconds (and may cause destruction localized to a few hectares). All of these (and other weather-related) phenomena can affect bird populations both directly, by killing individuals or destroying their nests, and indirectly by their effect on food and habitat. The diversity of effects possible precludes a complete review in the space available. Instead, we choose to focus on two areas with which we are familiar: drought and hurricanes.

Drought

Drought represents a climatic condition, or long-term expression of weather, expressed over one or more seasons or annual cycles. Its effects on birds are especially pronounced (or at least well-studied) in arid and semi-arid regions. In such ecosystems there is a strong association between various population parameters of birds (such as survival and reproductive success) and annual levels of precipitation (Marr and Raitt 1983, Petersen et al. 1986, Rotenberry and Wiens 1991). This relationship appears to be indirect, mediated by the direct effects of the availability of moisture on primary and secondary production (Noy-Meir 1973, Seely and Louw 1980, Cody 1981, Fuentes and Campusano 1985). For example, in arid shrubsteppe habitats avian reproductive success, which is substantially influenced by food availability (e.g., Martin 1987), was reduced during drier years, although there appeared to be little effect of drought on overall avian density or that of any particular species (Rotenberry and Wiens 1980, 1989, 1991). George et al. (1992) also noted drought effects on reproductive success of birds in grassland habitats. Unlike Rotenberry and Wiens, however, they observed transient changes in population abundances (decline during a drought year followed by recovery the next year) of six of eight common species.

Along a more mesic montane gradient, Smith (1982) observed that drought effects on avian population density and community structure were most evident in deciduous forest compared to coniferous forests or meadows. His data suggested that, although individuals of some species may have suffered from water stress, insectivores and nectarivores were likely more adversely affected by the indirect effect of drought on food resources. This is consistent with Martin's (1987) more general observation that insectivores (particularly aerial foragers such as swallows) appeared more sensitive than other species because of the dependence of their food source on the vagaries of weather.

Blake et al. (1992) also noted that, during a period of moderate to extreme drought (1986–1988) in the Upper Midwest, birds breeding in upland deciduous forest were more strongly adversely affected than those in other wooded habitats. Eight of 11 long-distance migrants examined in Wisconsin and eight of 13 in Michigan significantly declined in abundance, compared to a much lower proportion of short-distance migrants and permanent resident species. They speculated that long-distance migrants might have been more affected by drought than other species because most long-distance migrants nested in June, when the effects of drought were most severe, whereas other species bred earlier in the year before local environmental conditions had deteriorated.

Because drought operates over relatively large temporal and spatial scales (e.g., DeSante et al. 1993), it has the additional characteristic that it may constrain the manifestation of effects due to weather phenomena that operate over much shorter or smaller scales (Allen and Starr 1982). For example, day-to-day variation in weather during a breeding season may impinge directly on daily time and energy budgets of parents and their young (e.g., Johnson and Best 1982). In many systems, survivorship and reproductive success are higher during periods of mild weather than under extremes of temperature or precipitation (e.g., Bryant 1975, 1978, Johnson and Best 1982, Petersen et al. 1986, Hussell and Quinney 1987, Rotenberry and Wiens 1991). Again, insectivorous species are likely more sensitive than others, as the short-term availability of food may be directly related to weather; arthropods are simply less active under cooler, rainier conditions, reducing their vulnerability to avian predators (e.g., Bryant 1973, Craig 1978, Avery and Krebs 1984). However, the expression of these effects is context-sensitive; although nestlings might be reared during a period of otherwise benign, favorable weather, their mass gain and survival rates can still be relatively low if overall food resources are low due to poor precipitation during the preceding rainy season (Marr and Raitt 1983, Blancher and Robertson 1987, Jarvinen 1989, Rotenberry and Wiens 1991).

Lowered reproductive success as a result of climate or weather during the breeding season can have direct repercussions on population abundance of migrant species. For example, Sherry and Holmes (1992) noted that over a 9-year period the breeding densities of American Redstarts was significantly correlated with yearling male recruitment, which in turn was related to fledging success the previous summer. Fledging success depended primarily on nest depredation, but also on within-season weather-related starvation. Furthermore, starvation may not reflect the full impact of food limitation on populations, because such limitation may also be expressed through reduced clutch sizes or fewer breeding attempts, rather than simply mortality (e.g., Martin 1987, Rotenberry and Wiens 1991, Rodenhouse and Holmes 1992).

Average annual rainfall is also an important factor influencing overall migrant abundance for birds wintering in the Caribbean. Xeric woodlands consistently had the lowest total counts of migrants of any habitats in surveys of the Greater Antilles (Wunderle and Waide 1993), as previously observed throughout the Caribbean (Lack and Lack 1972, Terborgh and Faaborg 1980, Askins et al. 1992), but in contrast to parts of Mexico (Hutto 1980, Waide 1980, Waide et al. 1980). This is attributed to the fact that the migrants wintering in the Caribbean breed in the mesic eastern forests of North America, whereas many species that winter in Mexico also breed in xeric sites in the western United States (Terborgh 1989). The low abundance of migrants wintering in xeric Caribbean habitats is interesting, given that seasonally dry habitats predominated in the Caribbean in the Pleistocene (Pregill and Olson 1981), presumably providing sufficient time for adaptation to xeric conditions. Possibly as a result, xeric habitats contain the highest species richness of resident species on an island (Kepler and Kepler 1970). Yet no migrant species winters exclusively in dry habitats, nor do any migrants have their maximum abundance in xeric Caribbean habitats despite the historic abundance of this habitat type.

Xeric habitats in the Caribbean are highly seasonal with a distinct wet (usually May–November) and dry season (usually December–April), which causes fluctuations in food resources, as found elsewhere for insects (Wolda 1978, Hespenheide 1980) and fruit (Morton 1980). The seasonal decline in food supplies in xeric habitats may make it difficult for some species to obtain adequate fat reserves for spring migration (Orejuela et al. 1980, Bosque and Lentino 1987). These habitats are also characterized by severe droughts during which both resident and migrant populations may decline, although the degree of response varies among species (Orejuela et al. 1980, Faaborg et al. 1984). Kirtland's Warblers may be drought sensitive on their Bahamian wintering grounds, as illustrated by a major decline on their breeding grounds following a severe drought in the Bahamas during the winter of 1970–1971 (Radabaugh 1974), and by a positive correlation between annual breeding counts and rainfall in the Bahamas (Mayfield 1993). Drought on wintering grounds has also been demonstrated to cause population declines on breeding grounds, at least in one European species (Winstanley et al. 1974, Batten and Marchant 1977).

The likelihood that winter food supplies will deteriorate faster in xeric than mesic habitats may make them less suitable for species establishing winter territories (Bosque and Lentino 1987). Individuals with territories in xeric habitats may be subordinates who have been relegated to this habitat via territorial exclusion by dominant individuals occupying more mesic sites. This occurs in American Redstarts in which females (subordinates) are most abundant in dry habitats while older males exclude them from territories in nearby mangroves on Jamaica (Parrish and Sherry 1991, Marra et al. 1991). This, too, may occur in Black-and-white Warblers, in which females predominate in samples from the Guanica dry forest where the species was not territorial, in contrast to more mesic habitats (Faaborg and Arendt 1984).

Finally, opportunistic species, such as Cape May and Prothonotary warblers appear to take advantage of drought-induced declines in resident insectivores and, in their absence, move into dry forest (Faaborg et al. 1984). Thus, tropical xeric habitats may commonly be inhabited by subordinate individuals displaced by territorial exclusion from more mesic sites or by opportunistic species capable of using temporarily available or drought-resistant food resources.

Mesic or moist habitats may provide the most reliable food supplies for wintering migrants, but they are also the most likely to be converted into agriculture in many tropical regions. Therefore, wintering migrants may be forced into those habitats in which effects of periodic drought are most severe, or at least denied mesic refugia under drought conditions. This is the case in the Bahamas where agricultural development has occurred primarily on sites with the deepest soils, where effects of drought would normally be less severe (Sealey 1985). Some mesic sites in the Bahamas can be extremely rich in migrants. Mangroves, which might otherwise be inhabited by migrants in dry areas elsewhere, are often of diminished stature (< 1.5 m) in the Bahamas, making them unsuitable for overwintering or as drought refugia for most migrants. Thus, as mesic habitats are lost, wintering migrants are more exposed to the effects of drought.

The most obvious management strategy to lessen the impact of drought on wintering migrants in the Caribbean is to preserve moist and mesic forests, particularly in drought-prone regions. In xeric coastal areas, mangrove forests may serve as refugia during drought. Mangrove forests are often rich in migrants (as well as other species) and should be a focus of migrant conservation efforts in the tropics.

Unfortunately, an important feature of drought and within-season weather variation is that these disturbances occur over a variety of spatial and temporal scales (but often large and long), thus precluding any simple generalization of their effects or of methods for mitigating those effects (Rotenberry et al. 1993). An additional complication, not addressed here, is that even larger scale climate change (e.g., "global warming;" Schneider 1989) may have profound effects on both the wintering and breeding areas of Neotropical migrant landbirds, with different effects on different species in different areas (Rodenhouse 1992). We briefly address "management" of such potential problems below and elsewhere (Rotenberry et al. 1993).

Hurricanes

Hurricanes have been implicated as a factor causing declines in migrants during passage through or while wintering in the southeastern United States and Caribbean (e.g., Walkinshaw 1983, Hamel 1986). Some aspects of this issue were studied after Hurricane Gilbert passed across 10 sites that had been previously sampled in Jamaica. Four months after the hurricane those 10 sites were resampled to document changes in both vegetation structure and bird populations (Wunderle et al. 1992). Overall, it was found that montane populations were more likely to decline than lowland populations. Montane population declines were apparently related to diet as population declines were most common in nectarivores and fruit/seed eaters, and less common in insectivores (resident or migrant), as also observed by others (Askins and Ewert 1991, Lynch 1991, Waide 1991, Wauer and Wunderle 1992). The fact that avian population declines are related to diet indicates that a hurricane's greatest stress occurs after its passage rather than during its impact. Population increases in some lowland habitats were attributed to increased wandering, migration, or the presence of canopy dwellers at ground level.

Since most hurricanes occur early in the migratory period, migrant birds are not likely to suffer directly from hurricanes, but are more likely to suffer from secondary effects. Migrants as a group showed no consistent pattern in response to recent hurricanes in Mexico (Lynch 1991) and the islands of St John (Askins and Ewert 1991) and Jamaica (Wunderle et al. 1992). Most migrants in these studies were insectivores and, therefore, the inconsistent response of migrants was not related to diet. More likely, the response of migrants to hurricanes can be attributed primarily to changes in foraging substrate or vegetation structure (Wunderle et al. 1992). For example, loss of high canopy foraging substrates likely explains the absence of Black-and-white Warblers from montane pine plantations and their decline in shade coffee plantations on Jamaica. Overstory trees in both these habitats suffered substantial losses of branches and twigs from which this species normally gleans insects. Blackthroated Blue Warbler populations declined in sun coffee plantations in which an estimated 60–80% of the coffee trees were blown over, but remained unchanged in nearby montane broadleaf forest and pine plantations. Extensive damage to mimosaceous trees in the overstory of shade coffee, pasture, and in dry limestone scrub undoubtedly contributed to the decline of overwintering migrant Prairie Warblers, which regularly glean insects from their leaves (Lack and Lack 1972). Thus, structural damage to vegetation produced by hurricanes could eliminate foraging substrates.

Hurricane damage to vegetation could also eliminate or change the vegetation characteristics used by some species (or sexes) as primary cues for habitat selection or segregation. This has been demonstrated in Hooded Warblers in which males and females show habitat segregation based on vegetation structure (Lynch et al. 1985). Wintering Hooded Warbler males defended exclusive territories mostly in closed canopy forest of moderate to tall stature, whereas females defended exclusive territories mostly in lower, more open vegetation. Laboratory experiments with hand-reared birds indicated that females preferentially responded to two-dimensional arrays of oblique black lines, in contrast to males, which preferentially responded to vertical strips (Morton 1990). After a hurricane struck a previously studied area on the Yucatan peninsula, females moved into storm-damaged sites previously occupied by males (Morton et al. 1993). Post-hurricane analysis of vegetation angles, sizes, and density of stems showed that females were found on territories where the verticality of the vegetation was reduced (i.e., more oblique lines) compared to the verticality of vegetation still defended by males. Thus by reducing the verticality of stems, the hurricane increased the habitat cues used by females and decreased those used by males, thereby changing the sex ratio in the area.

As with resident species, wintering migrants showed population increases in some habitats following the passage of hurricanes. In some instances, population increases may be attributed to movement from badly damaged sites into new habitats, as observed in Prairie Warblers on Jamaica (Wunderle et al. 1992). Here Prairie Warblers showed significant declines in three lowland habitats, but increased markedly in nearby mangroves. In other cases, migrant increases may be attributed simply to the presence of displaced canopy dwellers foraging at ground level where they are easier to detect following a hurricane. This was the case in the lower montane broadleaf forest in Puerto Rico after the passage of Hurricane Hugo, which destroyed most of the canopy. Possibly as a result, populations of canopy or subcanopy dwelling migrants such as Black-throated Blue Warblers, Black-and-white Warblers, American Redstarts, and Northern Parulas showed major increases after the hurricane, but returned to normal levels after refoliation in the following year (Wunderle 1990, Waide 1991). Thus, hurricanes may cause temporary shifts in the use of habitats or foraging sites.

Hurricane-induced habitat shifts are more unlikely as habitats become more fragmented, thereby exposing populations to local extinction. This was undoubtedly the case in the Puerto Rican Bullfinch on St Kitts after two hurricanes swept through the only remaining montane forest fragment in the late 1800s (Raffaele 1977). Compounding this species' risk was a reliance on seeds and fruit, a resource highly susceptible to hurricane damage. Similarly, the fruit/seed diet, low population size, and restriction to remnant forest fragments of the Bridled Quail-dove may explain the significant population decline of this species and its disappearance from traditional sites on St Croix (Wauer and Wunderle 1992). Such a scenario may have been a contributing factor to the demise of Bachman's Warbler, a migrant species whose wintering grounds included Cuba and nearby Isla de Juventude (Hamel 1986). Agricultural development of these wintering grounds since the 1500s (Rappole et al. 1983) undoubtedly produced habitat fragmentation, thereby increasing the species' vulnerability to hurricanes. Moreover, this vulnerability may have been enhanced if wintering Bachman's Warblers relied heavily on a food resource highly susceptible to hurricane damage, such as nectar as some observers suggest (see citations in Hamel 1986). In contrast, this fate has not occurred in the highly nectarivorous Cape May Warbler, possibly because of its widespread distribution among a variety of islands.

The effects of hurricanes on the temperate breeding grounds are less well documented than those in the tropics. However, it is well known that species requiring large, old trees for nesting or roosting are particularly susceptible to hurricane effects. This was documented in the aftermath of Hurricane Hugo, in which 44% (24/54) of South Carolina's Bald Eagle nests were destroyed (Cely 1991). In the breeding season following the hurricane, nest building got underway too late in the season to produce any young on some territories, thereby temporarily setting back eagle recovery in the state.

Another species found to be highly vulnerable to hurricane destruction of nest/roost sites is the Red-cockaded Woodpecker, a species endemic to old growth pine forests in the southeastern United States. This small, colonial woodpecker is restricted to old growth pine forest in which it constructs its nest and roost cavity in living older pines (60–90 years old), often infected with red heart, a fungal disease (Baker 1971). These old cavity trees were especially vulnerable to Hurricane Hugo's high winds, which destroyed 87% of 1765 active cavities used by woodpeckers prior to Hugo in the Francis Marion National Forest, South Carolina (Hooper et al. 1990). Although only 10 dead woodpeckers were found afterwards, Hooper et al. (1990) estimated that 63% of the woodpeckers in the forest were killed or missing, based on the mean number of surviving birds per colony (1.5 after vs 4.0 before Hugo). Loss of cavity trees undoubtedly contributed to this mortality, because woodpeckers forced to roost outside of cavities were more exposed to predators and inclement weather (Engstrom and Evans 1990). Moreover, the long-term effects are likely to be substantial as potential cavity trees will continue to show high mortality rates as a result of increased likelihood of disease, insect outbreaks, and catastrophic wildfires.

Hurricanes may also cause the loss of materials needed for nest construction, as observed for Swallow-tailed Kites, which abandoned nest sites in parts of south Florida after hurricanes destroyed an important component (Spanish moss, *Tillandsia usneoides*) of kite nests (Bent 1937).

The finding that the greatest stress of a hurricane occurs after its passage rather than during its impact indicates that several vegetation management strategies might lessen the impact of hurricanes on migrants wintering in the tropics (and on resident species as well). From these studies it is obvious that effective montane forest reserves require habitat corridors to lowland forest reserves to which montane forest inhabitants can migrate after hurricanes. Lowland vegetation, because of its rapid growth rate, can recover faster than montane vegetation, and thus can serve as a refugia for montane species in a storm's aftermath.

Because of its reduced likelihood of structural damage, and rapid recovery after storms, young second growth vegetation (sapling stage) is essential for post-hurricane survival of many species. While natural treefall gaps and landslides normally provide second-growth habitat patches, their size and frequency may at times be insufficient to provide adequate-sized patches. This is most likely to occur in reserves of primary forests (i.e., national parks) and where the recurrence rate of hurricanes is low. Low levels of cutting for charcoal production may actually be beneficial in providing second growth patches that can contribute to post-hurricane survival.

Hooper et al. (1990) have suggested that it might be possible to "hurricane-proof" pine stands by having more densely stocked stands of pine than is currently practiced, based on a direct relationship between pine stand density and hurricane survival in South Carolina. Although this may be appropriate for pine stands in South Carolina, for other forests it is unclear if comparatively high stem density makes forests more or less vulnerable to damage (Brokaw and Walker 1991, Reilly 1991). Recently thinned stands were more severely affected by a hurricane than unthinned stands on Puerto Rico, although age/size structure also influenced stand vulnerability (Wadsworth and Englerth 1959). Obviously, no single management prescription will exist to "hurricane-proof" forests by manipulating stand density; management will instead depend upon a variety of factors, including site, species, and age composition.

INSECT OUTBREAKS

An insect outbreak can be defined as an explosive increase in the abundance of a species, or a complex of species, that occurs over a relatively short period of time, usually including several generations of the insect. Thus, a normal emergence of an insect, such as a mayfly hatch or cicada emergence, although perhaps explosive, would not constitute an outbreak. When an outbreak has a deleterious effect on human society, such as monetary or esthetic losses through tree mortality, those insect species are referred to as pests. Pest species tend to exhibit strongly bimodal trends; in any one year or place, they may be so rare as to be nearly undetectable, or so abundant as to defoliate their hosts completely. Although we will use examples of other kinds of insects, this section will focus upon interactions among forest pest species and Neotropical migratory birds. Key pest species in North America include the eastern and western spruce budworms (*Choristoneura fumiferana* and *C. occidentalis*; Tortricidae: Lepidoptera), the gypsy moth (*Lymantria dispar*; Lymantriidae: Lepidoptera), the Douglas fir tussock moth (*Orgyia pseudotsugata*; Lymantriidae: Lepidoptera), and the southern pine beetle (*Dendroctonus frontalis*; Scolytidae: Coleoptera). We justify this focus because: (1) forests are richer in Neotropical migratory bird species than most other habitats, (2) most forest insect outbreak management will be directed towards several economically important species; and (3) most of the literature on insect outbreaks is devoted to those species. We intend this to be a brief overview on that subject, concentrating instead on how best to manage insect outbreaks with birds in mind. Several large, important publications exist on insect outbreaks, especially of economically important species (Brookes et al. 1978, 1987, Doane and McManus 1981, Thatcher et al. 1982, USDA 1985).

The Nature of Insect Outbreaks

Despite the economic importance of insect outbreaks, there are still many basic unanswered questions regarding outbreaks (Barbosa and Schultz 1987). For example, when attempting to identify a cause of an outbreak, researchers have had difficulty even defining spatial and temporal limits to the outbreak, much less its cause. Somehow, environmental conditions favoring sudden, rapid population growth occur, and populations that are normally limited by natural controls, such as predation, irrupt (Berryman 1987, Crawford and Jennings 1989, Elkinton and Liebhold 1990). Climate seems to be a major factor in irruptions of many species of forest pests (Martinat 1987).

Many native species exhibit cyclic outbreaks at more or less regular intervals (Berryman 1987, 1988). Outbreaks commonly last 3–4 years, but may be longer. Sometimes outbreaks are less severe; a reported population increase may only mean that a few otherwise very rare larvae are observed in foliage samples. Because they are exotic, gypsy moths tend to cause greater forest damage upon initial contact with a forest stand (Campbell and Sloan 1977). After initial contact, population irruption, and subsequent crash, gypsy moth populations tend to behave more like endemic caterpillar species, with cyclic outbreaks of less severity.

Birds generally do not have much effect on an insect population during an outbreak, but may help to control insect populations between outbreaks (Otvos 1979, Takekawa et al. 1982, Crawford and Jennings 1989). Exclosure studies have demonstrated suppression of insect populations by birds in nonoutbreak years in both forest (Holmes et al. 1979, Greene 1989) and grassland ecosystems (Joern 1986, Fowler et al. 1991).

Frequency of insect outbreaks of any kind over time has been poorly studied. Three caterpillar outbreaks of any consequence occurred during the 16 year study by Holmes et al. (1986), and only the saddled prominent (*Heterocampa guttivitta*; Notodontidae: Lepidoptera) had a major effect on birds in that study. Other than invasion by the gypsy moth, over a 12 year period in West Virginia three major insect events occurred. One was an outbreak of a geometrid caterpillar complex; the others were emergences of two different broods of periodical cicadas, which emerged in different years for one year only. Other than those two data sets, there have been few long-term studies of defoliating Lepidoptera or other forest insect populations and their influence on life-history parameters of birds.

Effects of Outbreaks on Neotropical Migratory Birds

Beneficial Aspects

Many caterpillars are spring defoliating species, emerging when new foliage appears on trees. While perhaps not influencing reproduction, they undoubtedly affect adult survival during migration and first arrival on breeding grounds. Caterpillars, including early-instar gypsy moth larvae, were the most abundant prey in stomachs of 25 different Neotropical migratory species collected while migrating through West Virginia (R.J. Cooper, unpublished data).

Most Neotropical migratory birds are insectivorous during the breeding season. Although diets of forest birds consist of arthropod types most commonly encountered during characteristic searching and attach behaviors, birds are also opportunistic, taking advantage of abundant or profitable prey types (Robinson and Holmes 1982, Cooper 1988). Chief among these in forests are caterpillars, which are low in chitin content but high in fat and water content. Nearly every study of the dietary ecology of insectivorous birds in eastern deciduous forests has demonstrated the importance of caterpillars to birds (e.g., Robinson and Holmes 1982, Holmes et al. 1986, Cooper 1988, Rodenhouse and Holmes 1992). It is therefore not surprising that many species of Neotropical migrants have been shown to respond functionally (change in diet) and/or numerically (density, productivity) to caterpillar outbreaks (e.g., Kendeigh 1947, Morris et al. 1958, Zach and Falls 1975, Morse 1978, Sealy 1979, Holmes et al. 1986, Crawford and Jennings 1989). Similar responses have been found to emergences of periodical cicadas (*Magicicada* spp.; Homoptera: Cicadidae) (reviewed in Stephen et al. 1990), although only certain species can use them since birds must be large enough to capture and eat those relatively large prey (Karban 1982).

When an outbreak or major emergence of a palatable, profitable insect, such as a large caterpillar, occurs, many bird species switch to feed on that insect. Birds may immigrate into an area experiencing an outbreak (Morris et al. 1958, Sealy 1979). If the outbreak coincides with the nestling or postfledging phase of the birds' breeding cycle, then reproductive success, fecundity, and perhaps first year survivorship increase (Holmes et al. 1986).

Low reproductive success in several species led Holmes et al. (1986) to suggest that several forest species were normally subjected to a limited availability of food except during caterpillar outbreaks, when release from food limitation occurred. Holmes et al. (1986) found that fluctuations in abundance of caterpillars accounted for some of the major fluctuations in bird populations observed on a New Hampshire study site. Similarly, low densities of caterpillars caused poor reproductive success of Black-throated Blue Warblers (Rodenhouse and Holmes 1992).

Detrimental Effects

It would seem from the limited evidence discussed above that insect outbreaks are not only beneficial but perhaps necessary to maintain viable forest bird populations. Based on historical observations, insect outbreaks apparently are a natural component of most ecosystems (Nothnagle and Schultz 1987). Few systems, however, remain "natural" (i.e., unaltered by humans). As with agricultural systems, humans have created vast forest monocultures that have greater predisposition to outbreaks of insects such as the southern pine beetle, which is not generally eaten by Neotropical migrants due to its woodboring habit. Introduced insect pests, such as the gypsy moth, also cause greater damage to forests in North America than in their native Eurasia (Wallner and McManus 1988). Few North American bird species other than cuckoos eat gypsy moths to any extent (Smith 1985).

The most significant negative effects of insect outbreaks are likely manifested through changes in habitat suitability due to the removal of vegetation and tree mortality. For example, nest predation may increase in areas with less foliage (Martin 1992); indeed, artificial nests placed in gypsy moth-infested areas had greater predation than stands with little defoliation (D. Thurber and R. Whitmore, unpublished data). It also seems likely that breeding birds in forests opened up by defoliation or tree mortality are likely to experience increased nest parasitism, although this has not yet been demonstrated. Of greatest long-term concern, however, is that relatively few large tracts of unfragmented forest remain in North America. Excessive tree mortality caused by an insect outbreak further contributes to that fragmentation, and may also result in increasing edge habitat in temperate forests.

To our knowledge, the only long-term study of forest bird response to habitat alteration brought about by an insect outbreak is being conducted in West Virginia (Thurber 1992). In that study, initiated in 1984, gypsy moths reached defoliating levels in 1987. The outbreak ended in 1990; since that time gypsy moth populations have been at low densities. Birds were censused from 1984 to 1992. As expected, numerous edge species, including some species of Neotropical migrants such as the Indigo Bunting, increased in abundance over this period. Also, several deep forest species declined in abundance during that period, although only three species, Acadian Flycatcher, Eastern Wood-pewee, and Blue-gray Gnatcatcher, showed significant declines. Another species, the Black-throated Green Warbler, which was never very abundant on the study area, disappeared entirely by 1990. Lack of a suitable control site, a problem in this type of study, precludes strong inference regarding exact cause for the declines. However, statewide Breeding Bird Survey (BBS) data indicate that none of the four species were declining significantly statewide from 1980 to 1989 (Droege and Sauer 1990), lending indirect support to the hypothesis that habitat changes caused the declines.

In summary, most bird species tend to respond favorably to an outbreak of a small, palatable insect such as spruce budworm

Table 3-1. Numerical response of birds to insect outbreaks.

Insect	Nature of Impact	Reference	Bird	Nature of Response
Larch sawfly (*Pristiphora erichsonii*)	Increased food	Buckner and Turnock (1965)	Yellow-bellied Sapsucker	D+
			Eastern Phoebe	D+
			Least Flycatcher	D+
			Gray Jay	D+
			Golden-crowned Kinglet	D+
			American Robin	D+
			Cedar Waxwing	D+
			Red-eyed Vireo	D+
			Tennessee Warbler	D+
			Nashville Warbler	D+
			Magnolia Warbler	D+/−
			Yellow-rumped Warbler	D+
			Common Yellowthroat	D+
			Rose-breasted Grosbeak	D+
			Song Sparrow	D+
			Chipping Sparrow	D+
			White-throated Sparrow	D+
			Purple Finch	D+
			Evening Grosbeak	D+
Spruce budworm	Increased food	Morris et al. (1958)	Tennessee Warbler	B+D+
			Blackburnian Warbler	B+D+
			Bay-breasted Warbler	B+D+
		Jennings and Crawford (1985)	Yellow-rumped Warbler	D+
			Black-throated Green Warbler	D+
		Crawford and Jennings (1989)	Red-breasted Nuthatch	D+
			Canada Warbler	D+
Periodical cicadas	Increased food	Nolan and Thompson (1975)	Yellow-billed Cuckoo	D+
		Anderson (1977)	House Sparrow	B+
			Tree Sparrow	B+
Forest tent caterpillar (*Malacosoma disstria*)	Increased food	Sealy (1979)	Bay-breasted Warbler	B+D+
Gypsy moth	Habitat alteration	DeGraaf (1987)	Blue Jay	D−
			Black-capped Chickadee	D−
			House Wren	D+
			Wood Thrush	D−
			Northern Oriole	D−
		Thurber (1992)	Mourning Dove	D+
			Red-bellied Woodpecker	D+
			Downy Woodpecker	D+
			Hairy Woodpecker	D+
			Northern Flicker	D+
			Eastern Wood-pewee	D−
			Acadian Flycatcher	D−
			Blue Jay	D−
			Black-capped Chickadee	D−
			Tufted Titmouse	D+
			Brown Creeper	D+
			Carolina Wren	D+
			Blue-gray Gnatcatcher	D−
			Eastern Bluebird	D+
			Black-throated Green Warbler	D−
			Black-and-white Warbler	D+
			Ovenbird	D+
			Hooded Warbler	D+
			Rose-breasted Grosbeak	D+
			Indigo Bunting	D+
			Rufous-sided Towhee	D+
			Chipping Sparrow	D+
			Brown-headed Cowbird	D−

B = breeding response; D = density response; + = increase; − = decrease.

(Table 3-1). Fewer species tend to respond to cicadas simply because of their size. Unpalatable species such as the gypsy moth are not eaten to any great extent, so the effect is mostly one of habitat alteration through defoliation and tree mortality. Most, but not all, species responding favorably to this type of alteration are resident, edge-favoring species.

Insect Outbreak Management and Neotropical Migrants

Integrated Pest Management

Modern pest management involves an integration of all available technologies and methods to reduce damage caused by an insect pest. Emphasis is given to the augmentation of natural controls and the manipulation of forest conditions to make them less favorable for outbreaks and to ameliorate losses on a long-term, continuing basis (Waters and Stark 1980). Quantitative methodology related to the measurement, analysis, and modeling of forest growth and development, and the dynamics of insect populations, is also important. In theory, while pesticides have a place in integrated pest management, their use is played down.

Pesticides

In reality, pesticide application remains the most common method of dealing with insect outbreaks. This is due in part to the fact that pesticides are the only known way to protect a forest from severe timber losses; alternative methods have an element of uncertainty involved. Other methods, discussed below, will reduce losses to an acceptable level. However, there often is public or governmental pressure to suppress a pest population actively, and pesticides are the only way to do that for most insect species.

Modern pesticides used for control of forest insect outbreaks are generally more benign than their earlier counterparts such as dichloro-diphenyl-trichloroethane (DDT). While a variety of pesticides are used, the trend is towards biological pesticides such as the bacterium, *Bacillus thuringiensis* (Bt). The insect growth regulator diflubenzuron is also gaining favor for use on some pests such as gypsy moth. Both pesticides are specific to larval insects, and neither has much demonstrated direct effect on vertebrates. Also, they are generally effective enough that they need not be applied every year for effective control of gypsy moth damage (Cooper et al. 1990).

However, there are still problems with widespread use of those and other pesticides. First, once committed to pesticide use, one must continue application, even if it is only once every few years, indefinitely. Second, although direct effects on vertebrates are minimal, indirect effects through elimination of important prey, such as nontarget caterpillars, may be severe. Rodenhouse and Holmes (1992) experimentally reduced nesting attempts of Black-throated Blue Warblers by eliminating caterpillar prey with Bt. Both diflubenzuron (Martinat et al. 1988) and Bt (Rodenhouse and Holmes 1992) reduced numbers of caterpillars compared with control areas. Five bird species had significantly fewer caterpillars in their diets in areas treated with diflubenzuron than birds in untreated areas (Sample et al. 1993). Also, body condition (fat content) of six species of Neotropical migrants was less in areas treated with diflubenzuron (Whitmore et al. 1993). However, reproductive success of Acadian Flycatchers was unaffected, probably because flycatchers normally do not eat large numbers of caterpillars, and can switch to flying prey without much energetic cost (Shearer 1990).

Each of the above studies was conducted during the year of pesticide application only. To our knowledge, effects in subsequent years, or long-term effects of pesticide use during every other year or so on bird populations has never been studied. Given the above results, it would seem likely that frequent reduction of important prey, such as caterpillars, would affect survivorship, especially of young birds, and fecundity.

Alternative Management Techniques

A wide variety of alternative techniques exist for reducing the damage caused by insect outbreaks, although many are still in experimental phases. These include introduction of predators, parasites, and diseases, predator habitat management, sterile male introductions, and various silvicultural techniques.

Perhaps the most promising alternative techniques, both from the standpoint of minimizing forest damage and maintaining viable populations of Neotropical migrants, involve silvicultural prescriptions. Before 1975, most silvicultural prescriptions ignored insect pests. Because the effectiveness of silvicultural prescriptions can only be assessed over a relatively long period of time, little data exist on the success of those techniques, which are to some extent still experimental. Prescriptions are based on sound forest ecology, however, and data that do exist indicate promising results.

While silvicultural techniques vary, the general goals are the same: eliminate trees that are likely to die in an outbreak, and increase vigor of remaining trees. In this way, stand vigor and resistance to severe insect losses are increased. For example, overstocked pine stands are more susceptible to infestation by southern pine beetle than stands of lower stem density (Hicks 1982). Overstocked stands, therefore, should be thinned to reduce stand susceptibility. Also, loblolly pine (*Pinus taeda*) has shown higher infestation rates than other pines; other species should be planted instead. Forest stand susceptibility to western spruce budworm infestation is also related to tree species composition, seral stage, and stand density (Wulf and Cates 1987). Among other prescriptions, thinnings, increasing density of nonhost species, and harvesting to achieve an intermediate seral stage are recommended. For gypsy moth, a variety of prescriptions also exist (Gansner et al. 1987). Most involve either selectively harvesting preferred host species, such as oaks, or removing trees of low vigor. Early results indicate that defoliation in untreated stands was higher than in silviculturally treated stands in several West Virginia locations (US Forest Service, unpublished data).

The advantages of silvicultural options are obvious. First, they are more environmentally acceptable than pesticides. Second, they cost less. Third, while the insect outbreak is not halted, and some trees die, pest populations do not achieve the explosive growth patterns observed in some outbreaks. A pulse of food is still provided for birds, which are part of the natural control of most insect pest populations anyway. Finally, the outbreak is allowed to run its course, usually in an abbreviated time frame, without stepping onto the endless pesticide treadmill. A disadvantage of silvicultural methods is that prior planning is involved. Treatments should occur at least 5 years, and frequently longer, before the outbreak occurs, to give the forest stand time to respond to the treatment. Silvicultural methods generally will not work if they are applied at the time of infestation.

Implications for Bird Population Dynamics

Population dynamics models have been developed for passerine birds (Emlen and Pikitch 1989, Lebreton and Clobert 1991, Noon and Sauer 1992, Meyer and Boyce, in press). Each of those studies supported the notion that population growth rates of short-lived bird species are most sensitive to fecundity and prereproductive survival. Those results have some interesting implications for managing insect outbreaks, because management decisions regarding insect pests will directly affect Neotropical migrant populations.

On the one hand, if an outbreak of a palatable, available insect species is allowed to progress, many forest bird populations will be able to exploit that superabundant prey source, and those populations should increase during these outbreaks. Also, egg size (Murphy 1986), fledging success (Strehl and White 1986), and juvenile survival (Holmes et al. 1986) can increase as a result of increased prey availability. On the other hand, widespread application of pesticides to control insect outbreaks has the opposite effect. Even if direct negative effects on birds are minimal, loss of important prey, such as nontarget caterpillars, can decrease nesting attempts or reproductive success (Rodenhouse and Holmes 1992). Adult and juvenile survivorship may decrease as well due to the energetic effects of prey scarcity (Cooper et al. 1990). Again, the obvious implication is that forest bird population growth rates are at least temporarily reduced (Young 1968, Meyer and Boyce, in press). However, if pesticides are applied every 2–3 years over a long period of time, as they are for gypsy moth control, effects are likely to be more than temporary. With depressed populations, the probability of local extinctions is increased through stochastic variation in population parameters (Meyer and Boyce, in press).

Management Implications and Research Needs

Neotropical migrant bird populations may be negatively affected both if an insect outbreak is allowed to proceed with no intervention (through habitat alteration) and if pesticides are used to suppress the pest population. We believe there is greater overall risk to neotropical migrants in widespread application of pesticides, which should have only a limited role in modern forest pest management. For example, spraying a buffer area around a silviculturally treated forest will help prevent emigration of spruce budworm to the forest from high populations in surrounding areas. Insects that irrupt only occasionally and are not economically important, such as *Heterocampa guttivitta* and various Geometridae, should not be sprayed at all, since their impact on forests is minimal but their importance to birds is great. If pesticides are to be used, biological treatments such as Bt are preferred over diflubenzuron or conventional synthetics.

Of the various options available to forest managers for managing insect pests, we believe silvicultural treatments provide the most sound combination of timber protection and conservation of Neotropical migratory birds. Research must continue in this area.

A particularly disturbing finding is that no long-term studies assessing effects of alternative strategies for managing insect pests on forest birds exist. To manage a resource like Neotropical migratory birds intelligently, we must understand what the long-term consequences are of applying pesticides, silviculturally altering stands, or taking no action. We further note that most studies cited in our review are unreplicated, and that most lack any sort of experimental controls. Many do not use inferential statistics. Thus, it is difficult to claim unambiguously a positive or negative response, and the effects described above and in Table 3-1 must be considered as tentative.

These research needs are especially critical for managing the gypsy moth, since its range is still expanding and will eventually cover most of the United States. Monitoring efforts for Neotropical migrants, now in the initial stages of development, should take pest management into consideration, since nearly every study area involved in this program will experience an insect outbreak of some consequence at some point in time.

FIRE

The role of fire in maintaining diversity in ecosystems of North America would appear to have come full circle during this century. Prior to the presence of many humans, fires caused by lightning were natural disturbances that increased the diversity of habitats in most ecosystems (with the possible exception of deserts) (Wright and Bailey 1982). Human use of fire, both purposefully and carelessly (e.g., Smith and Petit 1988), greatly intensified fire's effects and frequency, particularly in grasslands and surrounding habitats (Wright and Bailey 1982), and in coniferous forests (Van Lear and Waldrop 1989, Baker 1992). The practice of fire suppression in North American forests began early this century, and is attributed to European-trained foresters who thought fire was bad because it killed trees (Wright and Bailey 1982). Its application spread to every plant community in North America with the exception of the tallgrass prairie of eastern Kansas, where burning continued in non-drought years (Zimmerman 1993).

Biologists began taking a constructive view of fire in North America starting in the early 1960s with the release of the Leopold Report (Leopold et al. 1963). Reintroduction of fire as a management tool began in the Southeast and Northwest, and fire is now used on at least a limited scale in all areas of North America (Wright and Bailey 1982). Prescribed burning, whereby fire is applied in a knowledgeable manner to wildland fuels on a specific land area under selected weather conditions to accomplish predetermined, well-defined management objectives, has become an important forest management tool (Biswell 1989, Wade and Lundsford 1990). Although suppression of natural fires ("wildfires") continues today, since 1972 some natural fires in specific situations have been allowed to continue burning under the Forest Service's prescribed natural fire program (Daniels 1991).

The renewed interest in the role of fires in the environment has led to many publications appearing over the last 35 years. The annual Tall Timbers Fire Ecology Conference, started in 1962, has been a forum for forest and wildlife managers concerned with the role of fires in ecosystems throughout North America. However, probably more so than any other topic in avian ecology, much information concerning fire is not readily available in the primary literature, but is found in sources such as the proceedings of numerous symposia on fire that have been sponsored by the US Forest Service (e.g., Mooney et al. 1981, Nodvin and Waldrop 1991). In the past, the main focus of this literature has been primarily on effects of fire on soils and vegetation, and only recently has research addressed the impact of fire on animals other than game species (e.g., Bradley et al. 1992).

Effects of Fire and Fire Suppression

Fire can have many impacts upon an ecosystem, including changes in local climatic and microclimatic conditions (i.e., changes in light, temperature, humidity, and wind), changes in the structure of vegetation, and changes in animal abundances and distribution (Bendell 1974). However, a major problem in making general statements about fire is the wide variation that occurs in fires and their effects. Fires may vary in intensity, duration, frequency, location, shape, and extent, and their effects may differ with season, nature of fuel, and properties of the site and soil. Fires rarely kill birds directly, but rather affect population levels indirectly by altering habitat structure, abundance of competing species, and food levels (Dickson 1981). Some birds, particularly birds of prey, are attracted to fire and smoke, and many species are attracted to blackened burns and "greening" burns (Komarek 1969, Lyon and Marzluff 1984, Tomback 1986). Fires in forests may initially benefit cavity-nesting birds by causing large numbers of snags, but, over time, burned areas will become poor nesting areas because snags will be lost, and recruitment of new snags, which occurs in unburned forests when trees die, will not occur (Taylor and Barmore 1980, Raphael and White 1984).

In mixed-conifer forests in western Montana, there was little evidence that fire affected overall numbers of birds, numbers of species, or species diversity (Lyon and Marzluff 1984). However, individual species varied in the extent to which they were affected; ground-feeding species increased and tree-trunk foragers decreased, with little change in canopy feeding species. Along a postfire sequence in western hemlock (*Tsuga heterophylla*)/Douglas-fir (*Pseudotsuga menziesii*) in the Pacific Northwest, nearly the same number and density of species were present the first 2 years after burning as in a nearby 515-year-old forest (Huff et al. 1984). Similar results (little change in density) were found in Wyoming in postfire successional habitats (Taylor and Barmore 1980), and in the Sierra Nevada (Raphael et al. 1987). Raphael et al. (1987) further noted that different foraging guilds responded differently; ground- and brush-foraging birds were more numerous on a burned plot (where shrub cover was twice as great) whereas foliage searching birds were more abundant on an unburned plot. Overall, changing vegetation structure resulted in predictable trends in abundance related to the foraging and nesting habits of the birds they studied.

Without fires, many forested communities become monocultures with inadequate reproduction of tree species and high fuel accumulation (Wright and Bailey 1982). Grasslands suffer lower productivity and are invaded by shrubs and trees (Zimmerman 1993), and shrublands may become impenetrable thickets (Biswell 1989). Since censuses of nongamebird populations were not carried out prior to the turn of the century, it is impossible to assess quantitatively the impact that fire suppression has had on most Neotropical migratory birds, with the exception of those that have become extinct or endangered due to fire suppression. [The role of fire in maintaining higher populations of gamebirds, such as quail (Stoddard 1931), turkey, and qrouse (Bendell 1974, Wright and Bailey 1982), seemed easier to establish.] The nearly extinct Bachman's Warbler may have bred in large expanses of cane (*Arundinaria gigantea*) that no longer exist in the southeastern United States due, in part, to fire suppression (Remsen 1986). The endangered Kirtland's Warbler is dependent on the fire-maintained jack pine (*Pinus banksiana*) forests for breeding (Mayfield 1960), and the population is currently increasing, apparently in response to a large fire (a prescribed burn that escaped control) that occurred about 15 years ago (James and McCulloch, chapter 2, this volume).

Three non-migratory birds apparently also have been adversely affected. Fire suppression contributed to the decline and extinction of the Heath Hen (Thompson and Smith 1970) and is one factor contributing to the disappearance of the Red-cockaded Woodpecker (Ligon et al. 1986). Fire suppression in oak scrub renders habitats unusable for nesting by Florida Scrub Jays (Woolfenden and Fitzpatrick 1984).

Fire suppression and the concomitant increase in understory vegetation may also have caused habitat shifts in species associated with recently burned areas. For example, Bachman's Sparrow was once found on burned areas north of the Arkansas River in Arkansas, but now occurs commonly on young (1–5-year-old) pine plantations in the southern part of the state (Haggerty 1986).

In some instances, effects of naturally occurring fires may be mimicked by prescribed burning, whereas in others they may be quite different (Biswell 1989, Baker 1992), particularly when wildfires occur infrequently (perhaps as a consequence of suppression) and as a result may become high-intensity canopy fires. When wildfires do occur in such circumstances, as a result of extra fuel buildup, they may escape to burn a larger area and at a greater intensity than they would otherwise (e.g., Minnich 1983), transforming a normal ecological event into a catastrophe. For example, high-intensity canopy fires probably occurred less frequently before settlement of the western United States (Baker 1992).

Even under a normal fire regime, habitat conditions are altered and may become temporarily unsuitable for some species. Additionally, fire may enhance the invasion of exotic plants, such as cheatgrass (*Bromus tectorum*) in the western shrubsteppe, which may, in turn, prevent the normal post-fire successional pathway or recovery process (e.g., West 1979).

Prescribed Burning and Birds

Controlled or prescribed burning has numerous management applications: improvement of wildlife habitat, reduction of hazardous fuels, disposal of logging debris, site preparation for seeding or planting, regeneration of both pines and hardwoods, management of competing vegetation, control of insects and disease, improvement of forage for grazing, enhancement of appearance, improved stand access, perpetuation of fire-dependent species, and improved working conditions (Dickson 1981, Van Lear and Waldrop, 1989, Wade and Lundsford 1990).

The basic management questions relevant to a decision to burn are: where, when, under what conditions, how, and what frequency (Stoddard 1962)? Most of these questions could be asked in relation to maximizing the benefit to wildlife, particularly birds, but a review of the literature (presented below) suggests that few studies have actually addressed those issues.

Where?

Burning is commonly used in three kinds of habitats within North America: conifers, grasslands and prairies, and shrublands. Although burning can be used as a management tool in some hardwood stands (e.g., Van Lear and Waldrop 1989), usually the bark of those trees is too thin to withstand the heat of fire (Black 1979).

Conifers. Burning is commonly used in coniferous forests of the Southeast; most species evolved in the presence of fire and are adapted to periodic burning (Van Lear and Waldrop 1989). In longleaf (*Pinus palustris*) and slash (*P. elliottii*) pines, impacts are: (1) control of the understory reduces bird species diversity, (2) reduction of the litter exposes seeds that would not otherwise be available for foraging, and (3) destruction of dead trees eliminates substrate for dead-tree feeders and cavity nesters (Wood and Niles 1978).

Periodic burning was also important in maintaining the "pinebarrens" of the Northeast and the jack-pine forests around the Great Lakes. In the New Jersey Pine Barrens, numbers of Rufous-sided Towhees and Brown Thrashers decline after burning, but numbers of Eastern Bluebirds and Common Nighthawks increase (Leck 1979).

Prescribed burning is commonly recommended for ponderosa pine (*Pinus ponderosa*) to reduce the chances of canopy fires (e.g., Blake 1982). Such burning reduces the number of snags >15 cm dbh by about 45%, but increases the number of snags <15 cm dbh by 20-fold (Horton and Mannan 1988). No species disappeared in the first breeding season after burning, but Northern Flicker and Violet-green Swallows decreased in number, while numbers of Mountain Chickadees increased.

However, not all conifers are resistant to fire, and fires in these forests can affect bird populations negatively. The lowland conifers of southern Canada are one example: Cape May and Magnolia Warblers, Ovenbirds, and Golden-crowned Kinglets were missing from a burned black spruce (*Picea mariana*) stand (Dawson 1979).

Grasslands and prairies. Historically, fires were common on the grasslands and prairies of North America and most of the plants and animals found in these regions are adapted to frequent fires (e.g., Risser et al. 1982). Fire suppression in these areas leads to invasion by woody vegetation, which may totally destroy the nature of the grasslands. For example, fire limits the extension of tree islands and mangroves in the Everglades of south Florida (Robertson 1962), and fire suppression there allows the development of a palmetto understory, which leads to the disappearance of songbirds requiring prairie-like conditions (Komarek 1963).

In the most comprehensive study conducted to date on the effects of fire on prairie bird communities, Zimmerman (1992, 1993) demonstrated that fires caused a decrease in species richness, primarily due to the decrease in Henslow's Sparrows and Common Yellowthroats, both of which rely on litter and standing dead material for nesting. Numbers of other species that are dependent on woody vegetation also decreased due to fire, but effects varied as a function of available moisture, being more severe in years of low moisture. This study clearly demonstrated that impacts of burning may be species-specific and that interactions between fire and other environmental factors must be considered to truly understand the role of fire in a given ecosystem.

In native and exotic grasslands in Arizona, numbers of primarily seed-eating birds using burned areas increased dramatically in fall, but the only species attracted to the burned areas during summer were Mourning Doves, Horned Larks, and Lark Sparrows (Bock and Bock 1992). Fire may have made these grasslands more suitable to these species by reducing heavy accumulations of litter. Grasshopper Sparrow, Botteri's Sparrow, Cassin's Sparrow, and Eastern Meadowlark, on the other hand, generally avoided burned areas. In Saskatchewan, only Vesper Sparrows occurred on a fescue prairie the first year after burn, and densities of Savannah and Clay-colored sparrows were still low after 3 years compared to densities in unburned areas (Pylypec 1991).

Marsh burning is a common practice to improve conditions for wintering waterfowl (Givens 1962, Schlichtemeier 1967, Ward 1968, Kirby et al. 1988), and burning of breeding areas may increase densities of nesting waterfowl (Kirsch and Kruse 1973). Fire can also be used to maintain salt marshes (Wade 1991), but no research has been conducted on the effects of those practices on migratory passerine populations.

Shrublands. Fire is commonly used as a sagebrush (*Artemisia* spp.) control technique due to its effectiveness and cheap cost.

McGee (1976) found that total bird density was reduced in sagebrush after burning, but returned to preburn levels within 3 years. Total bird species diversity was higher during the first postburn year, but returned to control levels after 2 years. Other studies (Castrale 1982, Petersen and Best 1987) found that Sage Sparrow and Sage Thrasher populations were unaffected by fire, Brewer's Sparrow numbers decreased markedly after fire, whereas Horned Larks and Vesper Sparrows appeared after fire. Those studies suggest that the nongamebird community found in sagebrush seemed resilient to fire. However, Rotenberry and Wiens (1978) observed a highly significant decrease in Sage Sparrow density on a burned plot compared to an unburned control during the first 2 years postfire. Likewise, after a fire that completely eliminated sagebrush, Bock and Bock (1987) found that only Western Meadowlarks were equally common in burned and unburned; Lark Buntings, Lark Sparrows, and Brewer's Sparrows completely avoided burned areas, and Grasshopper Sparrows were more abundant on unburned areas. It therefore seems likely that the response of birds to fire in sagebrush shrublands will depend upon the degree of initial shrub removal and subsequent recovery, as well as the differing relationships of individual species to the presence of shrub cover (Rotenberry and Wiens 1978, Bock and Bock 1987). These results also could be due to intraspecific geographical variation in response to fire and the changes in habitat that it produces, a phenomenon which surely exists in some species, but which has received very little attention to date (cf., Collins 1983, Knopf et al. 1990).

In chaparral, no species were eliminated after fire, but the brush-dwelling species declined, while predatory and seed-eating species increased (Lawrence 1966). The greatest changes occurred during the first year postburn (Wirtz 1979, 1982). Birds apparently repopulated burned areas very rapidly, and were limited primarily by availability of nesting sites (Biswell 1989). Periodic prescribed burning in chaparral can cause nearly cyclic changes in bird community structure: after burning, oak-savanna or even grassland species increase and chaparral species decrease, but, as the vegetation recovers, chaparral species increase and the others decline (Biswell 1989).

Other habitats. In some cases, fires may maintain pre-existing openings within forests that may be important to birds. Over 80 species of nongamebirds are associated with forest openings or edges of forest openings in northern Michigan (Taylor and Taylor 1979). If bogs and swamps burn in Michigan, they may attract Savannah Sparrows, which are not ordinarily found in bogs with heavy shrub cover (Taylor and Taylor 1979). Many wading birds in the Okefenokee Swamp are dependent on shallow marshes that are maintained by fire (Cypert 1973). Some agricultural crops, such as sugar cane and rice, are commonly burned after harvest, but impacts of that kind of burning on migratory birds remain to be determined.

How and When?

Firing methods in prescribed burning generally fall into the following categories: backing fire, strip-heading fire, flanking (parallel to wind) fire, point-source fires, center and circular fires, and pile and windrow burning (Biswell 1989, Wade and Lundsford 1990). Each of these may have different effects on the area being burned. For example, fire intensity can determine the number of snags present in southern coniferous forests (Wood and Niles 1978).

Fires can be prescribed in winter, spring, summer or fall, depending on the management objectives. Managing for forest birds usually is associated with winter burns, but differing management options may dictate other burning schedules (Biswell 1989). A relatively cool fire during the dormant season could greatly increase food sources and leave adequate nest sites for ground and brush-foraging birds; an intermediate fire might have a similar effect for ground and brushforaging birds, but would also create more openness for timber drilling and flycatching birds and raptors; a severe fire would seriously reduce the number and diversity of tree-foliage-searching and timber-gleaning birds (Wright and Bailey 1982). For example, the best time to kill understory vegetation with fire in southern conifers may be spring and summer, but burning at that time would destroy bird nests (Wood and Niles 1978) and may disrupt or destroy important bird foods (Givens 1962). The difference between summer and winter burning may only be meaningful for repeated fires over many years; little difference occurred in the effectiveness of one-time summer versus winter burning on woody vegetation in Florida grasslands (Fitzgerald and Tanner 1992).

What Frequency?

The frequency at which burns occur is important in determining effects of fire in a habitat. In southern pines, 3–5 years is recommended; more frequent burning eliminates the understory and its associated avifauna. Burning loblolly (*Pinus taeda*) and shortleaf (*P. echinata*) pine at 3–4 year interval is good for control of hardwood understory. Bachman's Sparrow may benefit from more frequent burning in that habitat, but other bird species do not. Burning every year eliminates many bird species in pines (Myers and Johnson 1978), tallgrass prairies (Zimmerman 1993), and Florida coastal scrub (Breininger and Smith 1992). Petersen and Best (1987) suggested that burning to reduce sagebrush by 40–50% probably would have little effect on birds in sagebrush. Clearly, more research is needed on the effects of different length burning rotations and spot-burning on bird populations.

Alternatives to Burning

Some mechanical management practices, such as chopping, disking, mowing, plowing, chaining, and spraying of herbicides, produce effects on vegetation that are superficially similar to those produced by fire (e.g., Schroeder and Sturges 1975, Castrale 1982, Wiens et al. 1986). On prairie grasslands in Florida, roller chopping for shrub suppression produced faster prairie restoration than did burning (Fitzgerald and Tanner 1992). However, birds were seen on the summer burn plots the day after burning, but did not return to summer chopped plots for 5 months post-treatment! Bird species richness also was lower in winter on chopped plots compared to burned and control plots. Birds in that habitat have evolved and adapted to changes caused by fire, but chopping flattened vegetation to such a degree that it was unusable. Likewise, grasslands are commonly burned to increase food for gamebirds, such as quail, but the effect may not last more than 2 years and disking may be a better treatment (Blakely et al. 1990). Like fire, mowing in grasslands suppresses woody vegetation, but mowing leads to an accumulation of litter, rather than a destruction of litter found with fire, which can have important influences on breeding birds in those habitats (Bollinger and Gavin 1992).

In forested systems, clearcutting timber initiates a successional process that superficially resembles the physiognomy of vegetation recovering from fire. However, some bird species that are adapted to early postfire habitats do not occur in areas that have been clearcut, implying that clearcutting is not completely suitable as a management technique to mimic habitat that would otherwise be created by fire (Hejl et al., chapter 8, this volume).

Summary

Several works have presented long lists of bird species with the predicted effect of fire on populations of each. Because of the length and diversity of those lists, we chose not to repeat them here, but instead summarize briefly their contents. Wright and Bailey (1982) listed birds, many of which are Neotropical migrants, that are favored by open plant communities caused by fires and those that are more common in unburned areas in forests of the Southwest, in sagebrush-grass habitats, in grasslands, and in western coniferous forests. More recently, regional lists and reviews have appeared, e.g., Southeastern pine forests (Landers 1987) and Pacific Northwest forests (McMahon and Calesta 1990). Based on computer modeling, Kerpez and Stauffer (1989) predicted effects of burning on all bird species found in Southeastern pine-hardwood forests, commenting that they could not locate any published field studies of effects of controlled burning on birds in that ecosystem.

Bendell (1974) attempted to synthesize the effects of fire on birds but concluded that it may be too early to generalize about the effects. Rather pessimistically, he stated that few studies at that time were quantitative, had adequate controls, and had been carried on long enough to assess the real effect of a particular fire on birds. A few years later, Dickson (1981) summarized all major field studies on birds and fire available at that time and attempted to predict which passerines would most benefit and which would suffer from burning in southern pine forests. He based his analysis on known effects of fire on vegetation and habitat associations of birds, stating that the amount of real data on effects of prescribed burning on songbirds in southern forests was "meager."

CONCLUDING REMARKS

It is apparent that there are strong interrelationships among such phenomena as climate variation, insect outbreaks, and fire. For example, in forested ecosystems, periods of extended drought may be associated with an increased likelihood of fires; vegetation is more likely to die and less likely to decompose, thus increasing fuel load, and dry vegetation and litter are more likely to ignite and carry a fire than that which carries a higher moisture content. Likewise, drought-stressed vegetation may have reduced resistance to insect infestation, and epidemic outbreaks have been associated with prolonged drought (Mattson and Haack 1987). Conversely, insect outbreaks in forests can contribute to development of a fuel complex that makes fires more probable in future drought years (Knight 1987). In turn, trees scarred by fire serve as epicenters for outbreaks of insects such as mountain pine beetles (Knight and Wallace 1989 and references therein). Finally, changes in vegetation cover, particularly loss of vegetation (due to fire or drought, as well as agricultural clearing or overgrazing), may affect regional patterns of climate, especially precipitation (e.g., Shukla et al. 1990). In this way, disturbances of one kind have an influence on the spread of disturbances of other kinds—a primary theme in landscape ecology (Turner 1987). Obviously, the dynamics of these interactions, including as they do the presence of multiple feedbacks, are quite complex.

Clearly, each of these phenomena has the ability to alter the number of individuals present in a local or regional population of birds, either through their direct effects on survivorship and fecundity, or indirectly by modifying the abundance of prey and/or suitable habitat. While these phenomena are often viewed from a human perspective as catastrophes, their effects on birds may be either positive or negative. It is difficult to determine, however, the degree to which this represents a natural "regulation" of population size (see Sherry and Holmes 1993, this volume). In the case of large-scale climatic patterns (as opposed to localized "disasters"), the answer is almost trivially "yes"—species do not occur where the climate is routinely unfavorable for their particular life history and physiology, or creates a habitat type to which they are not adapted (e.g., Root 1988). This feature is exploited by the model BIOCLIM, which apparently quite accurately predicts the geographical distribution of Australian fauna from temperature, precipitation, and topographic data (Busby 1991). To the extent that a species is dependent upon a fire-maintained habitat type (e.g., Kirtland's Warbler; Mayfield 1960) or any "subclimax," or early successional habitat (e.g., Bachman's Sparrow; Pulliam et al. 1992), then its distribution, if not its actual population size, is "regulated" by fire or disturbance. Prior to extensive conversion of native habitats to agricultural systems beginning in the early 19th century, the boundary between central prairie grasslands and the eastern deciduous forest in the American Mid-west (and presumably the bird fauna associated with each) was largely controlled by fire frequency (e.g., Risser et al. 1982).

For several of the types of disturbances we have discussed we have proffered specific management recommendations to mitigate their effects. We would, however, also like to provide a more general overview in managing for disturbance. Unquestionably, the abundance of any particular species, and hence the composition of bird assemblages, changes with the successional stage of the vegetation (e.g., Shugart and James 1973, Smith and MacMahon 1981). Furthermore, the patterns that arise reflect species-specific responses to changes in vegetation structure and composition, food availability, and interspecific interactions that accompany successional shifts in the plant community. This presents clear difficulties in the development of generalizations, and precludes the provision of formulaic solutions to any particular management problem. To further complicate the issue, even though the "average" migratory species might differ from the "average" resident species in, say, the range of habitats it uses during the nonbreeding season, or in its response to habitat disturbance, those averages conceal considerable variation among species within both groups (Hutto 1992). Because of this variation, Neotropical migrants as a whole may not comprise a useful management group for the purposes of planning conservation action.

We recognize that certain natural phenomena that adversely affect bird populations will inevitably occur even in the most intensively managed ecosystem. Indeed, all of these "catastrophes" occur despite our efforts to prevent them, and while everybody talks about the weather, no one has yet been able to do anything about it in any meaningful way. So, how do we manage the adverse effects on bird populations of these intrinsically unmanageable events?

We would like to discuss a general approach to managing for disturbance based on emerging theory from landscape ecology (e.g., Forman and Godron 1986, Turner 1987, Urban et al. 1987), part of the "new paradigm" in ecology described by Maurer (1993).

We begin by returning to the issue of scale, noting that all of the phenomena we have discussed tend to occur over characteristic spatial and temporal scales (e.g., Michaels and Hayden 1987). Understanding the scale over which any of these events characteristically occurs is important for understanding how to cope with each when it is viewed as a potential catastrophe or disaster. As pointed out above, drought, for example, can affect vegetation over a region of tens of thousands of square kilometers, whereas a microburst downdraft during a thunderstorm may cause intense destruction (e.g., a forest blowdown) over a very localized area, perhaps as little as a hectare.

Likewise, the scope and frequency of fires and insect outbreaks may vary over several orders of magnitude, from less than one hectare to over many tens of thousands, despite our best efforts to limit their size. And although basic and applied biologists tend to focus on the consequences of large scale events, even short-lived, small scale phenomena can have profound effects on community structure (e.g., Turner 1935).

As a result of these disturbances, ecosystems will (or should) consist of a mosaic of patch types (of an average size reflecting the characteristic spatial scale of disturbances) that vary depending on the time since the last disturbance—after some period of time, depending on the type of disturbance, a patch will "recover," or be returned to its predisturbance state (assuming it is possible to do so). Thus the management goal is to have as a managed ecosystem a sufficiently large area (not necessarily contiguous) to contain some minimum number of patches (i.e., some minimum area, which will depend on the life history of the species of management interest) that will be in a recovered state. In other words, the disturbance is "incorporated" as a natural part of the managed ecosystem (Urban et al. 1987). Not only should this provide protection for species characteristic of undisturbed or recovered areas, but allows for the persistence of other species that may depend upon disturbed areas or ecotones.

A bounded or managed landscape that is large enough to incorporate factors that disturb its component patches will have a relatively constant frequency distribution of patches of all types at all times; a smaller landscape unit that is unable to incorporate a disturbance has a transient frequency distribution of patch types, which changes in response to each disturbance event (Pickett and White 1985). The greatest likelihood of a "steady state" of a spatially shifting mosaic of patch types occurs in systems in which disturbance is small in scale relative to an otherwise homogeneous area of habitat, and in which feedback mechanisms influence disturbance frequency (Pickett and White 1985). One example of such a feedback is the increasing flammability of forest stands with time since the last fire. Landscape steady state is perturbed not only by disturbances that are too large, but also by any that are too frequent (Michaels and Hayden 1987).

It is important to note that the "steady state" referred to does not represent a homeostatic landscape, but instead one in which patches are in more or less constant flux. This quasi-stable steady state will shift in response to changing conditions, representing the "flux of nature" rather than the "balance of nature" (Maurer 1987)

The management implications are clear—a management unit that is large enough to incorporate a disturbance will have a constant and predictable (in time, not necessarily space) proportion of disturbed, recovering, and "climax" vegetation or habitat types, presumably containing most of their appropriate bird species as well. In the simplest case, incorporation can be passive—a disturbance is incorporated simply by increasing the scale of reference (Urban et al. 1987). The management question becomes: can incorporation be realized within a particular bounded system (e.g., a park, a reserve, a forest district, or a management area)—that is, given a geographically defined region and the perturbation affecting it, is the region of sufficient scale to incorporate the disturbance? One example relevant to some habitats containing Neotropical migrants has been developed from forest simulation modelling (Shugart and West 1981, Urban et al. 1987). Their models, based on gap-phase patch dynamics driven by tree falls and subsequent regrowth, suggest that a relatively stable dynamic steady state occurs when the area of the bounded landscape is about 50 times the average size of the disturbance of interest. Obviously, any sort of ratio would be site, species, and disturbance specific.

Perhaps the greatest effect of humans on landscapes has been to rescale patterns of disturbance in time and space (Urban et al. 1987). Foremost of these has been to reduce the size of "bounded landscapes" via habitat fragmentation, rendering them less able to incorporate natural disturbances of a given size and/or frequency. Less apparent has been the effect of control of fire. Fire suppression retards the natural frequency of burns in systems that have incorporated fire (e.g., Minnich 1983). When wildfires do occur as a result of fuel buildup, they may escape to burn over a larger area and at a greater intensity than they would otherwise. By altering the size of fires, humans may transform a landscape from a steady state that supports a full mosaic of patch types (and their associated avifauna) to one that contains a substantially smaller subset.

Another example is the establishment of forest plantation monocultures, such as pines, which are not only more susceptible to pine beetle infestations (a species of relatively low palatability to birds), but also spread those impacts over a larger area. Pesticide use may eliminate the immediate disturbance only to replace it with another, perhaps more damaging one, in the long run. By considering insect outbreaks in silvicultural prescriptions, however, individual stands will become more resistant to infestation. Because some stands still will be more or less resistant than others, some tree loss will occur, usually in patches. The landscape, however, has a greater probability of remaining near the original steady state, and insect pests, even exotics, are incorporated.

In conclusion, resource management should be scaled to mimic natural patch dynamics, recognizing that in some cases these dynamics may be influenced by disturbance events (e.g., drought) that occur over continental scales. Thus, in many cases increasing the size of a management unit will not be feasible. Nonetheless options do exist to reduce the size of some natural disturbances relative to a particular management unit. Foresters could use prescriptive scaling, for example, in determining the size, frequency, and distribution of clearcuts, in a way that mimics the size, frequency, and distribution of windthrow or blowdowns that occur naturally. As another example, a land manager could prescribe numerous small burns in a bounded area that has been reduced in size and has become thereby more susceptible to larger natural fires (Urban et al. 1987). In this case, block or patch burning may be better than one large burn (Givens 1962), although in some cases it may not be possible (or desirable) to mimic the patch dynamics of natural fires (e.g., Baker 1992). Likewise, rather than attempting to eradicate a pest infestation throughout its entire extent, insect suppression activities could be concentrated at the periphery of the outbreak, limiting its spread. An integrated approach involving several pest management techniques should produce a landscape with a relatively small area of highly disturbed habitat.

Clearly, the most pressing research need is to determine the scale of disturbances that are important and relevant to the species and habitats that we have an interest in preserving.

ACKNOWLEDGMENTS

We thank B. Maurer, J. Zimmerman, and the editors for thoughtful review and comments on an earlier draft of this chapter. In addition, B. Maurer helped clarify several points relating to landscape ecology. Smith's research on effects of fire on bird community structure has been supported by funds from the US Forest Service through the Bayou Ranger District of the Ozark National Forest and the Arkansas Audubon Society Trust.

LITERATURE CITED

Allen, T. F. H., and T. B. Starr. 1982. Hierarchy: perspectives for ecological complexity. University of Chicago Press, Chicago, IL.

Anderson, T. R. 1977. Reproductive responses of sparrows to a superabundant food supply. Condor 79:205–208.

Askins, R. A., and D. N. Ewert. 1991. Impact of Hurricane Hugo on bird populations on St John, US Virgin Islands. Biotropica 23(4a):481–487.

Askins, R. A., D. N. Ewert, and R. L. Norton. 1992. Abundance of wintering migrants in fragmented and continuous forests in the US Virgin Islands. Pp. 197–206 *in* Ecology and conservation of Neotropical migrant landbirds (J. M. Hagan, III and D. W. Johnston, eds). Smithsonian Institution Press, Washington, DC.

Avery, M. I., and J. R. Krebs. 1984. Temperature and foraging success of Great Tits *Parus major* hunting for spiders. Ibis 1126:33–38.

Baker, W. L. 1992. Effects of settlement and fire suppression on landscape structure. Ecology 73:1879–1887.

Baker, W. W. 1971. Progress report on life history studies of the Red-cockaded Woodpecker at Tall Timbers Research Station. Pp. 44–59 *in* Symposium on the Red-cockaded Woodpecker (R. L. Thompson, ed.). Florida Bur. Sport Fish. Wildl. US Dept. Int. Tall Timbers Res. Sta., Tallahassee, FL.

Barbosa, P., and J. C. Schultz (eds). 1987. Insect outbreaks. Academic Press, San Diego, CA.

Batten, L. A., and J. H. Marchant. 1977. Bird population changes for the years 1974–75. Bird Study 24:55–61.

Bendell, J. F. 1974. Effects of fire on birds and mammals. Pp. 73–138 *in* Fire and ecosystems (T. T. Kozlowski and C. E. Ahlgeren, eds). Academic Press, New York.

Bent, A. C. 1937. Life histories of North American birds of prey, Part 1. Dover Publications, New York, 409 pp.

Berryman, A. A. 1987. The theory and classification of outbreaks. Pp. 3–30 *in* Insect outbreaks (P. Barbosa and J. C. Schultz, Academic Press, San Diego, CA.

Berryman, A. A. (ed.). 1988. Dynamics of forest insect populations: patterns, causes, implications. Plenum, New York.

Biswell, H. H. 1989. Prescribed burning in California wildlands vegetation management. University of California Press, Berkeley, CA.

Black, G. N. 1979. Avian communities and management guidelines of the aspen-birch forest. Pp. 67–79 *in* Proceedings of the workshop management of northcentral and northeastern forests for nongame birds (R. M. DeGraaf, tech. coord.). USDA Forest Serv. Gen. Tech. Rep. NC-51.

Blake, J. G. 1982. Influence of fire and logging on nonbreeding bird communities of ponderosa pine forests. J. Wildl. Manag. 46:404–415.

Blake, J. G., G. J. Niemi, and J. M. Hanowski. 1992. Drought and annual variation in bird populations. Pp. 419–430 *in* Ecology and conservation of Neotropical migrant landbirds (J. M. Hagan, III and D. W. Johnston, eds). Smithsonian Institution Press, Washington, DC.

Blakely, K. L., J. A. Crawford, R. S. Lutz, and K. M. Kilbride. 1990. Response of key foods of California Quail to habitat manipulations. Wildl. Soc. Bull. 18:240–245.

Blancher, P. J., and R. J. Robertson. 1987. Effect of food supply on the breeding biology of Western Kingbirds. Ecology 68: 723–732.

Bock, C. E., and J. H. Bock. 1987. Avian habitat occupancy following fire in a Montana shrubsteppe. Prairie Nat. 19:153–158.

Bock, C. E., and J. H. Bock. 1992. Response of birds to wildfire in native versus exotic Arizona grassland. Southwest. Natur. 37:73–81.

Bollinger, E. K., and T. A. Gavin. 1992. Eastern Bobolink populations: ecology and conservation in an agricultural landscape. Pp. 497–506 *in* Ecology and conservation of Neotropical migrant landbirds J. M. Hagan, III and D. W. Johnson, eds). Smithsonian Institution Press, Washington, DC.

Bosque, C., and M. Lentino. 1987. The passage of North American migratory land birds through xerophytic habitats on the west coast of Venezuela. Biotropica 19:267–273.

Bradley, A. F., N. V. Noste, and W. C. Fischer. 1992. Fire ecology of forests and woodlands in Utah. USDA Forest Serv. Gen. Tech. Rep. INT-287.

Breininger, D. R., and R. B. Smith. 1992. Relationships between fire and bird density in coastal scrub and slash pine flatwoods in Florida. Amer. Midl. Natur. 127:233–240.

Brookes, M. H., R. W. Stark, and R. W. Campbell. 1978. The Douglas-fir tussock moth: a synthesis. USDA Forest Serv. Tech. Bull. 1958.

Brookes, M. H., R. W. Campbell, J. J. Colbert, R. G. Mitchell, and R. W. Stark (eds). 1987. Western spruce budworm. USDA Forest Serv. Tech. Bull. 1694.

Brokaw, N. V. L., and L. R. Walker. 1991. Summary of the effects of Caribbean hurricanes on vegetation. Biotropica 23: 442–447.

Bryant, D. M. 1973. Factors affecting the selection of food by the House Martin (*Delichon urbica* (L.)). J. Anim. Ecol. 42:539–564.

Bryant, D. M. 1975. Breeding biology of House Martins *Delichon urbica* in relation to aerial insect abundances. Ibis 117:180–216.

Bryant, D. M. 1978. Environmental influences on growth and survival of nestling House Martins *Delichon urbica*. Ibis 120:271–283.

Buckner, G. H., and W. J. Turnock. 1965. Avian predation on the larch sawfly, *Pristiphora erichsonii* (HTG), Hymenoptera: Tenthredinidae. Ecology 46:223–236.

Busby, J. R. 1991. BIOCLIM—a bioclimatic analysis and prediction system. Pp. 64–68 *in* Nature conservation: cost effective biological surveys and data analysis (C. R. Margules and M. P. Austin, eds). CSIRO, Melbourne.

Campbell, R. W., and R. J. Sloan. 1977. Forest stand responses to defoliation by the gypsy moth. For. Sci. Monogr. 19.

Castrale, J. S. 1982. Effects of two sagebrush control methods on nongame birds. J. Wildl. Manag. 46:945–952.

Cely, J. E. 1991. Wildlife effects of Hurricane Hugo. J. Coastal Res. 8:319–326.

Cody, M. L. 1981. Habitat selection in birds: the roles of vegetation structure, competition, and productivity. Bioscience 31:107–113.

Collins, S. L. 1983. Geographical variation in habitat structure of the Black-throated Green Warbler (*Dendroica virens*). Auk 100:382–389.

Connell, J. H., and W. P. Sousa. 1983. On the evidence needed to judge ecological stability or persistence. Amer. Nature. 121:789–824.

Cooper, R. J. 1988. Dietary relationships among insectivorous birds of an eastern deciduous forest. PhD dissertation, West Virginia University, Morgantown, WV.

Cooper, R. J., K. M. Dodge, P. J. Martinat, S. B. Donahoe, and R. C. Whitmore. 1990. Effect of diflubenzuron application on eastern deciduous forest birds. J. Wildl. Manag. 54:486–493.

Craig, R. B. 1978. An analysis of the predatory behavior of the Loggerhead shrike. Auk 95:221–234.

Crawford, H. S., and D. T. Jennings. 1989. Predation by birds on spruce budworm *Choristoneura fumiferana*: functional, numerical, and total responses. Ecology 70:152–163.

Cypert, E. 1973. Plant succession on burned areas in Okefenokee Swamp following fires of 1954 and 1955. Tall Timbers Fire Ecol. Conf. 12:199–217.

Daniels, O. L. 1991. A Forest Supervisor's perspective on the prescribed natural fire. Tall Timbers Fire Ecol. Conf. 17:361–366.

Dawson, D. K. 1979. Bird communities associated with succession and management of lowland conifer forests. Pp. 120–131 *in* Proceedings of the workshop on management of northcentral and northeastern forests for nongame birds (R. M. DeGraaf, tech. coord.). USDA Forest Ser. Gen. Tech. Rep. NC-51.

DeGraaf, R. M. 1987. Breeding birds and gypsy moth defoliation: short-term responses of species and guilds. Wildl. Soc. Bull. 15:217–221.

DeSante, D. F., O. E. Williams, and K. M. Burton. 1993. The Monitoring Avian Productivity and Survival (MAPS) program: overview and progress. Pp. 208–222 *in* Status and management of Neotropical migratory birds (D. M. Finch and P. W. Stangel, eds). USDA Forest Serv. Gen. Tech. Rep. RM-229, Fort Collins, CO.

Dickson, J. G. 1981. Effects of forest burning on songbirds. Pp. 67–72 *in* Prescribed fire and wildlife in southern forests (G. W. Wood, ed.). Belle W. Baruch For. Sci. Inst., Clemson University, Georgetown, SC.

Doane, C. C., and M. L. McManus (eds). 1981. The gypsy moth: research toward integrated pest management. USDA Forest Serv. Tech. Bull. 1584.

Droege, S., and J. R. Sauer. 1990. North American Breeding Bird Survey annual summary, 1989. USDI Fish Wildl. Serv. Biol. Rept. 90(8).

Elkinton, J. S., and A. M. Liebhold. 1990. Population dynamics of gypsy moth in North America. Ann. Rev. Entomol. 35:571–596.

Emlen, J. M., and E. K. Pikitch. 1989. Animal population dynamics: identification of critical components. Ecol. Model. 44:253–273.

Engstrom, R. T., and G. W. Evans. 1990. Hurricane damage to Red-cockaded Woodpecker (*Picoides borealis*) cavity trees. Auk 107: 608–610.

Ewens, W. J., P. J. Brockwell, J. M. Gani, and S. I. Resnick. 1987. Minimum viable population size in the presence of catastrophes. Pp. 59–68 *in* Viable populations for conservation (M. E. Soulé, ed.). Cambridge University Press, Cambridge, England.

Faaborg, J. R., and W. J. Arendt. 1984. Population sizes and philopatry of winter resident warblers in Puerto Rico. J. Field Ornithol. 55:376–378.

Faaborg, J. R., W. J. Arendt, and M. S. Kaiser. 1984. Rainfall correlates of bird population fluctuations in a Puerto Rican dry forest: a nine year study. Wilson Bull. 96:575–693.

Fitzgerald, S. M. and G. W. Tanner. 1992. Avian community response to fire and mechanical shrub control in south Florida. J. Range Manag. 45:396–400.

Forman, R. T. T., and M. Godron. 1986. Landscape ecology. John Wiley and Sons, New York.

Fowler, A. C., R. L. Knight, T. L. George, and L. C. McEwen. 1991. Effects of avian predation on grasshopper populations in North Dakota grasslands. Ecology 72:1775–1781.

Fuentes, E. R., and C. Campusano. 1985. Pest outbreaks and rainfall in the semi-arid region of Chile. J. Arid Environ. 8:67–72.

Gansner, D. A., O. W. Herrick, G. N. Mason, and K. W. Gottschalk. 1987. Coping with the gypsy moth on new frontiers of infestation. South. J. Appl. For. 11:201–209.

George, T. L., A. C. Fowler, R. L. Knight, and L. C. McEwen. 1992. Impacts of a severe drought on grassland birds in western North Dakota. Ecol. Appl. 2:275–284.

Gilpin, M. E. 1987. Spatial structure and population vulnerability. Pp. 125–139 *in* Viable populations for conservation (M. E. Soulé, ed.). Cambridge University Press, Cambridge, England.

Gilpin, M. E., and M. E. Soulé. 1986. Minimum viable populations: the process of species extinctions. Pp. 13–34 *in* Conservation biology: the science of scarcity and diversity (M. E. Soulé, ed.). Sinauer Associates, Sunderland, MA.

Givens, L. S. 1962. Use of fire on southeastern wildlife refuges. Tall Timbers Fire Ecol. Conf. 1:121–126.

Greene, E. 1989. Food resources, interspecific agression, and community organization in a guild of insectivorous birds. PhD dissertation, Princeton University, Princeton, NJ.

Haggerty, T. M. 1986. Reproductive ecology of Bachman's Sparrow (*Aimophila aestivalis*) in central Arkansas. PhD dissertation, University of Arkansas, Fayetteville, AR.

Hamel, P. B. 1986. Bachman's Warbler: a species in peril. Smithsonian Institution Press, Washington, DC.

Hespenheide, H. A. 1980. Bird community structure in two Panama forests: residents, migrants, and seasonality during the non-breeding season. Pp. 227–237 *in* Migrant birds in the Neotropics: ecology, behavior, distribution, and conservation (A. Keast and E. S. Morton, eds). Smithsonian Institution Press, Washington, DC.

Hicks, R. R. 1982. Climatic, site, and stand factors. Pp. 55–70 *in* The southern pine beetle (R. C. Thatcher, J. L. Searcy, J. E. Coster, and G. D. Hertel, eds). USDA Forest Ser. Rech. Bull. 1631.

Holmes, R. T., J. C. Schultz, and P. Nothnagle. 1979. Bird predation on forest insects: an exclosure experiment. Science 206:462–463.

Holmes, R. T., T. W. Sherry, and F. W. Sturges. 1986. Bird community dynamics in a temperate deciduous forest: long-term trends at Hubbard Brook. Ecol. Monogr. 56:201–220.

Hooper, R. G., J. C. Watson, and R. E. F. Escano. 1990. Hurricane Hugo's initial effects on Red-cockaded Woodpeckers in the Francis Marion National Forest. Trans. 55th North Amer. Wildl. Nat. Resources Conf. 55:220–224.

Horton, S. P., and R. W. Mannan. 1988. Effects of prescribed fire on snags and cavity-nesting birds in southeastern Arizona pine forests. Wildl. Soc. Bull. 16:37–44.

Huff, H. H., J. K. Agee, and D. A. Manuwal. 1984. Postfire succession of avifauna in the Olympic Mountains, Washington. Pp. 8–15 *in* Fire's effects on wildlife habitat—symposium proceedings (J. E. Lotan and J. K. Brown, compilers). USDA Forest Serv. Gen. Tech. Rep. WO-16.

Hussell, D. J. T., and T. E. Quinney. 1987. Food abundance and clutch size of Tree Swallows *Tachycineta bicolor*. Ibis 129:243–258.

Hutto, R. L. 1980. Winter habitat distribution of migratory land birds in western Mexico, with special reference to small foliage-gleaning insectivores. Pp. 181–203 *in* Migrant birds in the Neotropics: ecology, behavior, distribution, and conservation (A. Keast and E. S. Morton, eds). Smithsonian Institution Press, Washington, DC.

Hutto, R. L. 1992. Habitat distributions of migratory landbird species in western Mexico. Pp. 221–239 *in* Ecology and conservation of Neotropical migrant landbirds (J. M. Hagan, III and D. W. Johnson, eds). Smithsonian Institution Press, Washington, DC.

Järvinen, A. 1989. Geographical variation in temperature variability and predictability and their implications for the breeding strategy of the Pied Flycatcher *Ficedula hypoleuca*. Oikos 54:331–336.

Jennings, D. T., and H. S. Crawford. 1985. Predators of the spruce budworm. USDA Forest Serv. Coop. State Res. Serv. Agric. Handbook no. 644.

Joern, A. 1986. Experimental study of avian predation on coexisting grasshopper populations (Orthoptera: Acrididae) in a sandhills grassland. Oikos 46:243–249.

Johnson, E. J., and L. B. Best. 1982. Factors affecting feeding and brooding of Gray Catbird nestlings. Auk 99:148–156.

Karban, R. 1982. Increased reproductive success at high densities and predator satiation for periodical cicadas. Ecology 63:321–328.

Kendeigh, S. C. 1947. Bird population studies in the coniferous forest biome during a spruce budworm outbreak. Ontario Dept. Lands For., Biol. Bull. 1:1–100.

Kepler, C. B., and A. K. Kepler. 1970. Preliminary comparison of bird species diversity and density in Luquillo and Guanica forests. Pp. 183–191 *in* A tropical rain forest (H. T. Odum and R. F. Pigeon, eds). National Technical Information Service, Springfield, VA.

Kerpez, T. A. and D. F. Stauffer. 1989. Avian communities of pine-hardwood forests in the southeast: characteristics, management, and modeling. Pp. 156–168 *in* Proceedings of pine-hardwood mixtures: a symposium on management and ecology of the type (T. A. Waldrop, ed.). USDA Forest Serv. Gen. Tech. Rep. SE-58.

Kirby, R. E., S. J. Lewis, and T. N. Sexson. 1988. Fire in North American wetland ecosystems and fire wildlife relations: an annotated bibliography. US Fish Wildl. Serv. Biol. Rep. 88(1), 146 pp.

Kirsch, L. M. and A. D. Kruse. 1973. Prairie fires and wildlife. Tall Timbers Fire Ecol. Conf. 12:289–303.

Knight, D. H. 1987. Parasites, lightning, and the vegetation mosaic in wilderness landscapes. Pp. 59–83 *in* Landscape heterogeneity and disturbance (M. G. Turner, ed.). Springer-Verlag, Berlin.

Knight, D. H., and L. L. Wallace. 1989. The Yellowstone fires: issues in landscape ecology. Bioscience 39:700–706.

Knopf, F. L., J. A. Sedgwick, and D. B. Inkley. 1990. Regional correspondence among shrub-steppe bird habitats. Condor 92:45–53.

Komarek, E. V. 1969. Fire and animal behavior. Tall Timbers Fire Ecol. Conf. 9:161–207.

Komarek, R. 1963. Fire and the changing wildlife habitat. Tall Timbers Fire Ecol. Conf. 2:35–43.

Lack, D., and P. Lack. 1972. Wintering warblers in Jamaica. Living Bird 11:129–153.

Landers, J. L. 1987. Prescribed burning for managing wildlife in southeastern pine forests. Pp. 19–27 *in* Managing southern forests for wildlife and fish—a proceedings (J. G. Dickson and O. E. Maugh, eds). USDA Forest Serv. Gen. Tech. Rep. SO-65.

Lawrence, G. E. 1966. Ecology of vertebrate animals in relation to chaparral fire in the Sierra Nevada foothills. Ecology 47:278–291.

Lebreton, J.-D., and J. Clobert. 1991. Bird population dynamics, management, and conservation: the role of mathematical modelling. Pp. 105–125 *in* Bird population studies: their relevance to conservation and management (C. M. Perrins, J.-D. Lebreton, and G. J. M. Hirons, eds). Oxford University Press, Oxford, England.

Leck, C. F. 1979. Birds of the pine barrens. Pp. 457–466 *in* Pine barrens: ecosystem and landscape (R. T. T. Forman, ed.). Academic Press, New York.

Leopold, A. S., S. A. Cain, C. M. Cottam, I. N. Gabrielson, and T. L. Kimball. 1963. Wildlife management in the national parks. Amer. For. 69(4):32–35, 61–63.

Ligon, J. D., P. B. Stacey, R. N. Conner, C. E. Bock, and C. S. Adkisson. 1986. Report of the American Ornithologists' Union Committee for the conservation of the Red-cockaded Woodpecker. Auk 103:848–855.

Lynch, J. F. 1991. Effects of Hurricane Gilbert on birds in a dry tropical forest in the Yucatan Peninsula. Biotropica 23(4a):488–496.

Lynch, J. F., E. S. Morton, and M. E. Van der Voort. 1985. Habitat segregation between the sexes of wintering Hooded Warblers. Auk 102:714–721.

Lyon, L. J. and J. M. Marzluff. 1984. Fire's effects on a small bird population. Pp. 16–22 *in* Fire's effects on wildlife habitat—symposium proceedings (J. E. Lotan and J. K. Brown, comps). USDA Forest Serv. Gen. Tech. Rep. WO-16.

Marr, T. G., and R. J. Raitt. 1983. Annual variations in patterns of reproduction of the Cactus Wren (*Campylorhynchus brunneicapillus*). Southwest. Natur. 28:149–156.

Marra, P. P., T. W. Sherry, and R. T. Holmes. 1991. Territorial exclusion by older males in a Neotropical migrant in winter: removal experiments in American Redstarts. Abstract, American Ornithologists Union Meeting, Montreal.

Martin, T. E. 1987. Food as a limit on breeding birds: a life history perspective. Ann. Rev. Ecol. Syst. 18:453–487.

Martin, T. E. 1992. Breeding productivity considerations: what are the appropriate habitat features for management. Pp. 455–473 *in* Ecology and conservation of Neotropical migrant landbirds (J. M. Hagan, III and D. W. Johnson, eds). Smithsonian Institution Press, Washington, DC.

Martinat, P. J. 1987. The role of climatic variation and weather in forest insect outbreaks. Pp. 241–268 *in* Insect outbreaks (P. Barbosa and J. C. Schultz, eds). Academic Press, San Diego, CA.

Martinat, P. J., C. C. Coffman, K. M. Dodge, R. J. Cooper, and R. C. Whitmore. 1988. Effect of diflubenzuron on the canopy arthropod community in a central Appalachian forest. J. Econ. Entomol. 81:261–267.

Mattson, W. J., and R. A. Haack. 1987. The role of drought in outbreaks of plant-eating insects. Bioscience 37:110–118.

Maurer, B. 1987. Scaling of biological community structure: a systems approach to community complexity. J. Theoret. Biol. 127:97–110.

Maurer, B. A. 1993. Biological diversity, ecological integrity, and neotropical migrants: new perspectives for wildlife management. Pp. 24–31 *in* Status and management of Neotropical migratory birds (D. M. Finch and P. W. Stangel, eds). USDA Forest Serv. Gen. Tech. Rep. RM-229, Fort Collins, CO.

Mayfield, H. F. 1960. The Kirtland's Warbler. Bull. Cranbrook Inst. Sci. 40:1–242.

Mayfield, H. F. 1993. Kirtland's Warblers benefit from large forest tracts. Wilson Bull. 105:351–353.

McGee, J. M. 1976. Some effects of fire suppression and prescribed burning on birds and small mammals. PhD dissertation, University of Wyoming, Laramie, WY.

McMahon, T. E. and D. S. de Calesta. 1990. Effects of fire on fish and wildlife. Pp. 233–250 *in* Natural and prescribed fire in Pacific Northwest forests (J. D. Walstad, S. R. Radosevich, and D. V. Sandberg, eds). Oregon State University Press, Corvallis, OR.

Meyer, J. S., and M. S. Boyce. In press. Life historical consequences of pesticides and other insults to vital rates. *In* The population ecology and wildlife toxicology of agricultural pesticide use: a modelling initiative for avian species (T. E. Lacher, ed.). Soc. Environm. Toxicol. Chem. Spec. Publ., Lewis Publishers, Boca Raton, FL.

Michaels, P. J., and B. P. Hayden. 1987. Modeling the climate dynamics of tree death. Bioscience 37:603–610.

Minnich, R. A. 1983. Fire mosaics in southern California and northern Baja California. Science 219:1287–1294.

Monney, H. A., T. M. Bonnicksen, N. L. Christensen, J. E. Lotan, and W. A. Reines (tech. coords.). 1981. Proceedings of the conference on fire regimes and ecosystem properties. USDA Forest Ser. Gen. Tech. Rep. WO-26.

Morris, R. F., W. F. Cheshire, C. A. Miller, and D. G. Mott. 1958. The numerical response of avian and mammalian predators during a gradation of the spruce budworm. Ecology 39:487–494.

Morse, D. H. 1978. Populations of bay-breasted and Cape May warblers during an outbreak of the spruce budworm. Wilson Bull. 90:404–413.

Morton, E. S. 1980. Adaptations to seasonal changes by migrant land birds in Panama Canal Zone. Pp. 437–453 *in* Migrant birds in the Neotropics: ecology, behavior, distribution, and conservation (A. Keast and E. S. Morton, eds). Smithsonian Institution Press, Washington, DC.

Morton, E. S. 1990. Habitat segregation by sex in the Hooded Warbler: experiments on proximate causation and discussion of its evolution. Amer. Natur. 135:319–333.

Morton, E. S., M. Van der Voort, and R. Greenberg. 1993. How a warbler chooses its habitat: field support for laboratory experiments. Anim. Behav. 46:47–53.

Murphy, M. T. 1986. Temporal components of reproductive variablity in Eastern Kingbirds (*Tyrannus tyrannus*). Ecology 67:1483–1492.

Myers, J. M. and A. S. Johnson. 1978. Bird communities associated with succession and management of loblolly-shortleaf pine forests. Pp. 50–65 *in* Proceedings of the workshop management of southern forests for nongame birds (R. M. DeGraaf, tech. coord.). USDA Forest Serv. Gen. Tech. Rep. SE-14.

Nodvin, S. C. and T. A. Waldrop (eds). 1991. Fire and the environment: ecological and cultural perspectives. USDA Forest Serv. Gen. Tech. Rep. SE-69.

Nolan, V., and C. F. Thompson. 1975. The occurrence and significance of anomalous reproductive activities in two North American non-parasitic cuckoos, *Coccyzus* spp. Ibis 117:496–503.

Noon, B. R., and J. R. Sauer. 1992. Population models for passerine birds: structure, parametrization, and analysis. Pp. 441–464 *in* Wildlife 2001:populations (D. R. McCullough and R. Barrett, eds). Elsevier Applied Science, New York.

Nothnagle, P. J., and J. C. Schultz. 1987. What is a forest pest? Pp. 59–80 *in* Insect outbreaks (P. Barbosa and J. C. Schultz, eds). Academic Press. San Diego, CA.

Noy-Meir, I. 1973. Desert ecosystems: environment and producers. Ann. Rev. Ecol. Syst. 4:25–51.

Orejuela, J. E., R. J. Raitt, and H. Alvarez. 1980. Differential use by North American migrants of three types of Colombian forests. Pp. 253–264 *in* Migrant birds in the Neotropics: ecology, behavior, distribution, and conservation (A. Keast and E. S. Morton, eds). Smithsonian Institution Press, Washington, DC.

Otvos, I. S. 1979. The effects of insectivorous bird activities in forest ecosystems: an evaluation. Pp. 341–374 *in* The role of insectivorous birds in forest ecosystems (J. G. Dickson, R. R. Fleet, J. A. Jackson, and J. C. Kroll, eds). Academic Press, New York.

Parrish, J. D., and T. W. Sherry. 1991. Sexual habitat segregation in American Redstarts wintering in Jamaica. Abstract, Wilson and Cooper Ornithological Societies Joint Meeting, Norman, OK.

Petersen, K. L. and L. B. Best. 1987. Effects of prescribed burning on nongame birds in a sagebrush community. Wildl. Soc. Bull. 15:317–329.

Petersen, K. L., L. B. Best, and B. M. Winter. 1986. Growth of nestling Sage Sparrows and Brewer's Sparrows. Wilson Bull. 98:535–546.

Pickett, S. T. A., and P. S. White. 1985. Patch dynamics: a synthesis. Pp. 371–384 *in* The ecology of natural disturbance and patch dynamics (S. T. A. Pickett and P. S. White, eds). Academic Press, New York.

Pregill, G. K., and S. L. Olson. 1981. Zoogeography of West Indian vertebrates in relation to Pleistocene climatic cycles. Annu. Rev. Ecol. Syst. 12:75–98.

Pulliam, H. R., J. B. Dunning, and J. Liu. 1992. Population dynamics in complex landscapes: a case study. Ecol. Appl. 2:165–177.

Pylypec, B. 1991. Impacts of fire on bird populations in a fescue prairie. Can. Field Natur. 105:346–349.

Radabaugh, B. E. 1974. Kirtland's Warbler and its Bahamas wintering grounds. Wilson Bull. 86:374–383.

Raffaele, H. A. 1977. Comments on the extinction of *Loxigilla portoricensis grandis* in St Kitts, Lesser Antilles. Condor 79:389–390.

Raphael, M. G., and M. White. 1984. Use of snags by cavity-nesting birds in the Sierra Nevada. Wildl. Monogr. 86:1–66.

Raphael, M. G., M. L. Morrison, and M. P. Yoder-Williams. 1987. Breeding bird populations during twenty-five years of postfire succession in the Sierra Nevada. Condor 89:614–626.

Rappole, J. H., E. S. Morton, T. E. Lovejoy, and J. L. Ruos. 1983. Nearctic avian migrants in the neotropics. USDI Fish Wildl. Serv., Washington, DC.

Reilly, A. E. 1991. The effects of Hurricane Hugo in three tropical forests in the US Virgin Islands. Biotropica 23:414–419.

Remsen, J. V. 1986. Was Bachman's Warbler a bamboo specialist? Auk 103:216–219.

Risser, P. G., E. C. Birney, H. D. Blocker, S. W. May, W. J. Paton, and J. A. Wiens. 1982. The true prairie ecosystem. US/IBP Synthesis Ser. 16. Hutchinson Ross, Stroudsburg, PA.

Robertson, W. B. 1962. Fire and vegetation in the Everglades. Tall Timbers Fire Ecol. Conf. 1:67–80.

Robinson, S. K., and R. T. Holmes. 1982. Foraging behavior of forest birds: the relationships among search tactics, diet, and habitat structure. Ecology 63:1918–1931.

Rodenhouse, N. L. 1992. Potential effects of climatic change on a Neotropical migrant landbird. Conserv. Biol. 6:263–272.

Rodenhouse, N. L., and R. T. Holmes. 1992. Results of experimental and natural food reductions for breeding Black-throated Blue Warblers. Ecology 73:357–372.

Root, T. 1988. Environmental factors associated with avian distributional boundaries. J. Biogeogr. 15:489–505.

Rotenberry, J. T., and J. A. Wiens. 1978. Nongame bird communities in northwestern rangelands. Pp. 32–46 *in* Proceedings of a workshop on nongame bird habitat management in the coniferous forests of the western United States (R. M. DeGraaf, tech. coord.). USDA Forest Serv. Gen. Tech. Rep. PNW-64.

Rotenberry, J. T., and J. A. Wiens. 1980. Temporal variation in habitat structure and shrubsteppe bird dynamics. Oecologia 47:1–9.

Rotenberry, J. T., and J. A. Wiens. 1989. Reproductive biology of shrubsteppe passerine birds: geographical and temporal variation in clutch size, brood size, and fledging success. Condor 91:1–14.

Rotenbery, J. T., and J. A. Wiens. 1991. Weather and reproductive variation in shrubsteppe sparrows: a hierarchical analysis. Ecology 72:1325–1335.

Rotenberry, J. T., R. J. Cooper. J. M. Wunderle, and K. G. Smith. 1993. Incorporating effects of natural disturbances in managed ecosystems. Pp. 103–108 *in* Status and management of Neotropical migratory birds (D. M. Finch and P. W. Stangel, eds). USDA Forest Serv. Gen. Tech. Rep. RM-229, Fort Collins, CO.

Sample, B. E., R. J. Cooper, and R. C. Whitmore. 1993. Dietary shifts among songbirds from a diflubenzuron-treated forest. Condor 95:616–624.

Schlichtemeier, G. 1967. Marsh burning for waterfowl. Tall Timbers Fire Ecol. Conf. 6:41–46.

Schneider, S. H. 1989. The greenhouse effect: science and policy. Science 243:771–781.

Schroeder, M. H., and D. L. Sturges. 1975. The effect on the Brewer's sparrow of spraying big sagebrush. J. Wildl. Manag. 28:294–297.

Scott, J. M., S. Mountainspring, F. L. Ramsey, and C. B. Kepler. 1986. Forest bird communities of the Hawaiian Islands: their dynamics, ecology, and conservation. Stud. Avian Biol. 6:1–431.

Sealey, N. E. 1985. Bahamian landscapes: an introduction to the geography of the Bahamas. Collins Caribbean, London.

Sealy, S. G. 1979. Extralimital nesting of bay-breasted warblers: response to forest tent caterpillars? Auk 96:600–603.

Seely, M. K., and G. N. Louw. 1980. First approximation of the effects of rainfall on the ecology and energetics of a Namib Desert dune ecosystem. J. Arid Environ. 3:25–54.

Shaffer, M. 1987. Minimum viable populations: coping with uncertainty. Pp. 69–86 *in* Viable populations for conservation (M. E. Soulé, ed.). Cambridge University Press, Cambridge, England.

Shearer, J. A. 1990. Studies of the Acadian flycatcher (*Empidonax virescens*): responses to Dimilin treatments and nest site selection. MS thesis, West Virginia University, Morgantown, WV.

Sherry, T. W., and R. T. Holmes. 1992. Population fluctuations in a long-distance Neotropical migrant: demographic evidence for the importance of breeding season events in the American Redstart. Pp. 431–442 *in* Ecology and conservation of Neotropical migrant landbirds (J. M. Hagan, III and D. W. Johnston, eds). Smithsonian Institution Press, Washington, DC.

Sherry, T. W., and R. T. Holmes. 1993. Are populations of Neotropical migrant birds limited in summer or winter? Implications for management. Pp. 47–57 *in* Status and management of neotropical migratory birds (D. M. Finch and P. W. Stangel, eds). USDA Forest Serv. Gen. Tech. Rep. RM-229, Fort Collins, CO.

Shugart, H. H., and D. James. 1973. Ecological succession of breeding bird populations in northwest Arkansas. Auk 90:62–77.

Shugart, H. H., and D. C. West. 1981. Long-term dynamics of forest ecosystems. Amer. Sci. 69:647–652.

Shukla, J., C. Nobre, and P. Sellers. 1990. Amazon deforestation and climate change. Science 247:1322–1325.

Smith, H. R. 1985. Wildlife and the gypsy moth. Wildl. Soc. Bull. 13:166–174.

Smith, K. G. 1982. Drought-induced changes in avian community structure along a montane sere. Ecology 63:952–961.

Smith, K. G., and J. A. MacMahon. 1981. Bird communities along a montane sere: community structure and energetics. Auk 98:8–28.

Smith, K. G., and D. R. Petit. 1988. Breeding birds and forestry practices in the Ozarks: past, present, and future relationships. Bird Conserv. 3:23–49.

Sousa, W. P. 1984. The role of disturbance in natural communities. Annu. Rev. Ecol. Syst. 353–391.

Stephen, F. M., G. W. Wallis, and K. G. Smith. 1990. Bird predation on periodical cicadas in Ozark forests: ecological release for other canopy arthropods? Stud. Avian Biol. 13:369–374.

Stoddard, H. L. 1931. The bobwhite quail, its habits, preservation, and increase. Scribner's New York.

Stoddard, H. L. 1962. Use of fire in pine forests

and game lands of the Deep Southeast. Tall Timbers Fire Ecol. Conf. 1:31–42.

Strehl, C. E., and J. White. 1986. Effects of superabundant food on breeding success and behavior of the Red-winged Blackbird. Oecologia 70:178–186.

Takekawa, J. Y., E. O. Garton, and L. A. Langelier. 1982. Biological control of forest insect outbreaks: the use of avian predators. Trans. North Amer. Wildl. Nat. Resources Conf. 47:393–402.

Taylor, C. M. and W. E. Taylor. 1979. Birds of upland openings. Pp. 189–197 *in* Proceedings of the workshop on management of northcentral and northeastern forests for nongame birds (R. M. DeGraaf, tech. coord.). USDA Forest Serv. Gen. Tech. Rep. NC-51.

Taylor, D. L., and W. J. Barmore, Jr. 1980. Post-fire succession of avifauna in coniferous forests of Yellowstone and Grand Teton National Parks, Wyoming. Pp. 130–145 *in* Proceedings of the workshop on management of western forests and grasslands for nongame birds (R. M. DeGraaf, (tech. coord.). USDA Forest Serv. Gen. Tech. Rep.

Terborgh, J. W. 1989. Where have all the birds gone? Princeton University Press, Princeton, NJ.

Terborgh, J. W., and J. R. Faaborg. 1980. Factors affecting the distribution and abundance of North American migrants in the eastern Caribbean region. Pp. 145–155 *in* Migrant birds in the Neotropics: ecology, behavior, distribution, and conservation. (A. Keast and E. S. Morton, eds). Smithsonian Institution Press, Washington, DC.

Thatcher, R. C., J. L. Searcy, J. E. Coster, and G. D. Hertel (eds). 1982. The southern pine beetle. USDA Forest Serv. Tech. Bull. 1631.

Thompson, D. Q., and R. H. Smith. 1970. The forest primeval in the Northeast—a great myth? Tall Timbers Fire Ecol. Conf. 10:255–265.

Thurber, D. K. 1992. Impacts of a gypsy moth outbreak on bird habitats and populations. PhD dissertation, West Virginia University, Morgantown, WV.

Tomback, D. F. 1986. Post-fire regeneration of krummholz whitebark pine: a consequence of nutcracker seed caching. Madrono 33:100–110

Turner, L. M. 1935. Catastrophes and pure stands of southern shortleaf pine. Ecology 16:213–215.

Turner, M. G. (ed.). 1987. Landscape heterogeneity and disturbance. Springer-Verlag, Berlin.

Urban, D. L., R. V. O'Neill, and H. H. Shugart. 1987. Landscape ecology. Bioscience 37:119–127.

USDA 1985. Spruce-fir management and spruce budworm. USDA Gen. Tech. Rep. NE-99.

Van Lear, D. H., and T. A. Waldrop. 1989. History, uses and effects of fire in the Appalachians. US Forest Serv. Gen. Tech. Rep. SE-54.

Wade, D. D. 1991. High intensity prescribed fire to maintain Spartina marsh at the urban-wildland interface. Tall Timbers Fire Ecol. Conf. 17:211–216.

Wade, D. D., and J. Lundsford. 1990. Fire as a forest management tool: prescribed burning in the southern United States. Unasylva 41:28–38.

Wadsworth, F. H., and G. W. Englerth. 1959. Effects of the 1956 hurricane on forests in Puerto Rico. Carib. For. 20:38–51.

Waide, R. B. 1980. Resource partitioning between migrant and resident birds: the use of irregular resources. Pp. 337–352 *in* Migrant birds in the Neotropics: ecology, behavior, distribution, and conservation (A. Keast and E. S. Morton, eds). Smithsonian Institution Press, Washington, DC.

Waide, R. B. 1991. The effect of Hurricane Hugo on bird populations in the Luquillo Experimental Forest, Puerto Rico. Biotropica 23(4a):475–480.

Waide, R. B., J. T. Emlen, and E. J. Tramer. 1980. Distribution of migrant birds in the Yucatan Peninsula: a survey. Pp. 165–171 *in* Migrant birds in the Neotropics: ecology, behavior, distribution, and conservation (A. Keast and E. S. Morton, eds). Smithsonian Institution Press, Washington, DC.

Walkinshaw, L. H. 1983. Kirtland's Warbler. Bull. Cranbrook Inst. Sci. 58:1–207.

Wallner, W. E., and K. A. McManus (eds). 1988. Lymantriidae: a comparison of features of New and Old World tussock moths. USDA Forest Serv. Gen. Tech. Rept. NE-123.

Ward, P. 1968. Fire in relation to waterfowl habitat of the Delta marshes. Tall Timbers Fire Ecol. Conf. 8:255–267.

Waters, W. E., and R. W. Stark. 1980. Forest pest management: concept and reality. Annu. Rev. Entomol. 25:479–509.

Wauer, R. H., and J. M. Wunderle. 1992. The effect of Hurricane Hugo on bird populations on St Croix, US Virgin Islands. Wilson Bull. 104:656–669.

West, N. E. 1979. Basic synecological relationships of sagebrush-dominated lands in the Great Basin and the Colorado Plateau. Pp. 33–41 *in* The sagebrush ecosystem: a symposium. Utah State University, College of Natural Resources, Logan, UT.

Whitmore, R. C., R. J. Cooper, and B. E. Sample.

1993. Bird fat reductions in forests treated with Dimilin. Environ. Toxicol. Chem. 12:2059–2064.

Wiens, J. A., J. T. Rotenberry, and B. Van Horne. 1986. A lesson in the limitations of field experiments: shurbsteppe birds and habitat alteration. Ecology 67:365–376.

Winstanley, D., R. Spencer, and K. Williamson. 1974. Where have all the Whitethroats gone? Bird Study 21:1–14.

Wirtz, W. O. 1979. Effects of fire on birds in chaparral. Calif. Nevada Wildl. Trans. 1979:114–124.

Wirtz, W. O. 1982. Postfire community structure of birds and rodents in southern California chaparral. Pp. 241–246 *in* Proceedings of the symposium on dynamics and management of Mediterranean-type ecosystems (C. E. Conrad and W. C. Oechel, tech. coord.). USDA Forest Serv. Gen. Tech. Rep. PSW-58.

Wolda, H. 1978. Fluctuations in abundance of tropical insects. Amer. Natur. 112:1017–1045.

Wood, G. W., and L. J. Niles. 1978. Effects of management practices on nongame bird habitat in longleaf-slash pine forests. Pp.40–49 *in* Proceedings of the workshop on management of southern forests for nongame birds (R. M. DeGraaf, tech. coord.). USDA Forest Serv. Gen. Tech. Rep. SE-14.

Woolfenden, G. E., and J. W. Fitzpatrick. 1984. The Florida Scrub Jay. Monogr. Pop. Biol. 20. Princeton University Press, Princeton, NJ.

Wright, H. A., and A. W. Bailey. 1982. Fire ecology. John Wiley & Sons, New York.

Wulf, N. W., and R. G. Cates. 1987. Site and stand characteristics. Pp. 90–114 *in* Western spruce budworm (M. H. Brookes, R. W. Campbell, R. G. Mitchell, and R. W. Stark, eds). USDA Forest Serv. Tech. Bull. 1694.

Wunderle, J. M. 1990. The effects of Hurricane Hugo on bird populations in a Puerto Rican rainforest. Abstract, Wilson Ornithological Society Meeting, Norton, MA.

Wunderle, J. M., and R. B. Waide. 1993. Distribution of overwintering nearctic migrants in the Bahamas and Greater Antilles. Condor 95:904–933.

Wunderle, J. M., D. J. Lodge, and R. B. Waide. 1992. Short-term effects of Hurricane Gilbert on terrestrial bird populations on Jamaica. Auk 109:148–166.

Young, H. 1968. A consideration of insecticide effects on hypothetical avian populations. Ecology 49:991–993.

Zach, R., and J. B. Falls. 1975. Response of the ovenbird (Aves: Parulidae) to an outbreak of the spruce budworm. Can. J. Zool. 53:1669–1672.

Zimmerman, J. L. 1992. Density-independent factors affecting the avian diversity of the tallgrass prairie. Wilson Bull. 104: 85–94.

Zimmerman, J. L. 1993. The birds of Konza. University Press of Kansas, Lawrence, KS.

4

SUMMER VERSUS WINTER LIMITATION OF POPULATIONS: WHAT ARE THE ISSUES AND WHAT IS THE EVIDENCE?

THOMAS W. SHERRY AND RICHARD T. HOLMES

INTRODUCTION

Migratory passerine birds that breed in the temperate zone and winter at tropical latitudes comprise one of the grand spectacles of nature, in terms of their prodigious flying and navigating abilities, abundance of individuals, diversity of species traveling up to tens of thousands of kilometers annually, and the emotional impact of their seasonal song choruses. Beyond their esthetic value, these migratory birds consume vast quantities of insects and other arthropods, and perform other ecosystem functions, although the potential ecosystem and economic consequences of the migratory bird phenomenon remain virtually unstudied. The possibility that many of these migrant species may be declining due to global environmental problems is thus alarming (reviewed by Terborgh 1989, 1992, Askins et al. 1990, Finch 1991). Ultimately the most important foundation for the development of sound conservation strategies for these species is a sound scientific understanding of their ecology and evolution. Improved understanding of migrant birds, and of their recent population declines, requires intensive studies of diverse populations and the factors that limit them, including reproductive success, survivorship, habitat suitability, and the effects of food resources and predator abundance. Moreover, because long-distance migrants typically spend only about 3–4 months in their temperate breeding areas (Keast 1980), studies of their population structure, demography, and habitat use must also take place in winter and along migratory routes.

When during the annual cycle are populations of long-distance migratory passerine birds limited? The answers to this question have been slow in coming, due largely to: (1) the global spatial scales over which these species travel annually; (2) their long-distance dispersal capability; (3) the diverse ecological factors potentially influencing their populations; and (4) their potentially flexible responses to seasonally and annually variable resources or habitats. One frequently held view is that these migrant species are limited principally by events affecting overwintering survival, i.e., in the winter grounds (Lack 1968, Fretwell 1972, 1986, Morse 1980b, Alerstam and Hogstedt 1982, Baillie and Peach 1992, Morton 1992, Rappole et al. 1992). This view is derived largely from: theoretical arguments (e.g., Fretwell and Lucas 1970, Fretwell 1972); extrapolations from studies of non-migratory species, which tend to be limited by mortality in winter (e.g., Dhondt 1971, Perrins 1980, van Balen 1980, Ekman 1984a,b, Arcese et al. 1992); and observations that events where migratory birds winter sometimes influence numbers of breeders in subsequent summers (e.g., Winstanley et al. 1974, Morse 1980b, Fretwell 1986, van der Have 1991, Baillie and Peach 1992). Robbins et al. (1989) noted that several migrant species showing the most significant declines were those that depend on forest habitat in winter but utilize nonforested habitats for breeding, implying that forest loss in winter might be the critical factor in population declines (see also Askins et al. 1990, Morton 1992, Rappole et al. 1992). Winter limitation has also been inferred from an apparent absence of resource limitation during summer breeding periods (e.g., Wiens 1977, Pulliam and Dunning 1987), suggesting that it must then occur in winter.

An alternative view is that events on the breeding grounds may be at least as important, if not more so, than those in winter (e.g., Holmes et al. 1986, Probst 1986, Martin 1987, Holmes and Sherry 1988, Hutto 1988, Robinson 1992, Sherry and Holmes 1992). Fragmentation of forest habitats in eastern North America, for instance, has been strongly implicated as one cause of reduced breeding success, and consequently lowered breeding densities of some songbird populations (Robbins 1979, Whitcomb et al. 1981, Ambuel and Temple 1983, Wilcove and Whitcomb 1983, Lynch and Whigham 1984, Wilcove and Robinson 1990). In this case, lowered songbird densities are thought to be due to reductions in the amount of undisturbed forest interior breeding habitat and/or to an increase in nest predation and nest parasitism along forest edges (Mayfield 1978, Brittingham and Temple 1983, Wilcove 1985, Wilcove et al. 1986, Andrén and Angelstam 1988, Wilcove and Robinson 1990, Robinson 1992, Robinson et al. 1993). In addition, variable food abundance during the breeding season, even within unfragmented landscapes, influences songbird reproductive success and survival (Enemar et al. 1984, Holmes et al. 1986, 1991, Tomialojac and Wesolowski 1990, Rodenhouse and Holmes 1992) as does summer weather (Holmes et al. 1986, Hejl et al. 1988, Blake et al. 1992).

Based on the foregoing review, populations of migratory birds could be limited over global spatial scales, and could be influenced by myriad ecological factors and circumstances. Despite numerous recent studies and reviews (e.g., Keast and Morton 1980, Rappole et al. 1983, Terborgh 1989, Finch 1991, Hagan and Johnston 1992), understanding of migratory birds' population dynamics remains at a primitive stage relative to the scope of the problem. For example, we do not know for any single Neotropical migrant bird species how mortality varies throughout the annual cycle. Nor do we know typical dispersal distances for individuals of most populations, nor where individuals within a particular breeding population spend the winter and vice versa. We cannot assess rigorously with the available information the relative importance of wintering or breeding areas for any single migratory species. Students of migratory bird populations are thus much like Rudyard Kipling's blind men, each of whom tries to understand the same elephant by examining a piece of the beast. Understanding the whole migratory phenomenon is essential for effective conservation and management.

Despite the lack of knowledge about many aspects of the ecology of most migrant species, particularly in winter, reasonable inferences can now be made concerning some of the most important ecological factors influencing and limiting their populations. In this paper, we review the evidence for limitation during the annual cycle, and conclude that migrant bird populations appear to be limited contemporaneously by their need for quality habitats in which to maintain high fecundity in summer and in which to maintain high survival in winter, i.e., we develop the idea of summer-and-winter limitation for migrant species, an idea described previously by Morse (1980b) and Cox (1985). We focus on patterns of habitat use because bird–habitat relationships are becoming relatively well understood conceptually, because information is becoming increasingly available on habitat-specific demography, and because habitats are often manageable units of landscapes. We provide only a brief section here of potential limiting factors during migration because relatively few data are available to evaluate its impact on population dynamics, but also because this topic is reviewed in depth by Moore et al. elsewhere in this volume (Chapter 5).

Our review and synthesis of how habitat use by migrant birds influences population dynamics year-round is organized into four sections: (1) general conceptual issues relevant to population dynamics and population regulation; (2) examples of how habitat quality and quantity limits populations in both summer and winter; (3) life-history characteristics indicating how migrants are adapted to use habitats and resources in both summer and winter; and (4) a set of recommendations for migrant habitat conservation and population management (see also Sherry and Holmes 1993). We try to understand population dynamics of migrants as a function of their age- and sex-specific reproductive and survival characteristics, dispersal, site faithfulness from one season to the next, and all of these in the typical arrays of habitats occupied throughout the annual cycle. The results of this review imply that the continued existence of the Neotropical–Nearctic migratory bird phenomenon depends critically on effective preservation of substantial quantities of high-quality habitats throughout the annual cycle.

We acknowledge at the outset that habitat is necessary, but may not be sufficient, to safeguard already threatened or endangered populations. Genetic phenomena such as inbreeding depression (Gilpin and Soulé 1986) and demographic problems such as inability to disperse to new habitat or find mates (Lande 1988, Simberloff 1988) may cause precipitous declines of small or isolated populations to critically low levels. Some of these problems probably contributed to the now probable extinction of Bachman's Warbler (Terborgh 1989), and to recent declines of Kirtland's Warbler (Probst 1986). Geographically restricted populations are also particularly susceptible to chance ecological (or human) disturbances that can eliminate all remaining individuals of a species even when suitable habitat remains. We will not say any more about these problems peculiar to small populations, other than to acknowledge their obvious importance for managing particular species.

CONCEPTUAL ISSUES CONCERNING POPULATION LIMITATION

Definition of Terms

To clarify what we mean by population limitation, we begin by defining several terms, following the usage of Begon and Mortimer (1986) and Sinclair (1989; see also Arcese et al. 1992). Population regulation is the dynamic process by which a population remains at an environmental carrying capacity, because of a tendency for the population to decrease at densities above this level and to increase at densities below it. Thus regulation involves "density-dependent" (= regulatory) factors, such as decreasing per capita fecundity with increased population density due to increased competition for food or safe nesting sites. The magnitude of density-independent factors, by contrast, does not depend on population density, as might happen through the effects of a severe drought or a hurricane. Population limitation (= determination of abundance) refers to the sum of factors, density-dependent and density-independent, causing a population to decline or remain below some level.

We envision the life cycle of Neotropical–Nearctic migrant birds as comprised, in simplest terms, of two major seasons of selecting a habitat and either breeding (summer) or surviving (summer and winter—Fig. 4-1). Between these two seasons is a migration period that also can involve selection of habitat and certainly mortality, which we discuss briefly below (see also Moore et al., Chapter 5, this volume). We concentrate here on the summer and winter periods, which jointly occupy the majority of these birds' annual cycles. For young birds, the postfledging period in mid- to late summer can be particularly hazardous (e.g., Sullivan 1989), as can migratory periods. Disappearance of birds is a function of both mortality and dispersal, but the two are difficult to distinguish and dispersal has rarely been measured in these highly vagile animals.

Population Regulation, Habitat Use, and Migrant Birds

Much evidence has accumulated in support of the idea that fitness of many animals, especially vertebrates, declines with increased density (see reviews by Sinclair 1989, Blondel et al. 1990, Perrins et al. 1991, Murdoch 1994). For instance, populations of Pied and Collared Flycatchers (respectively, both Palearctic–Paleotropical migrants) are cited by Sinclair (1989) as examples of where regulation has been demonstrated.

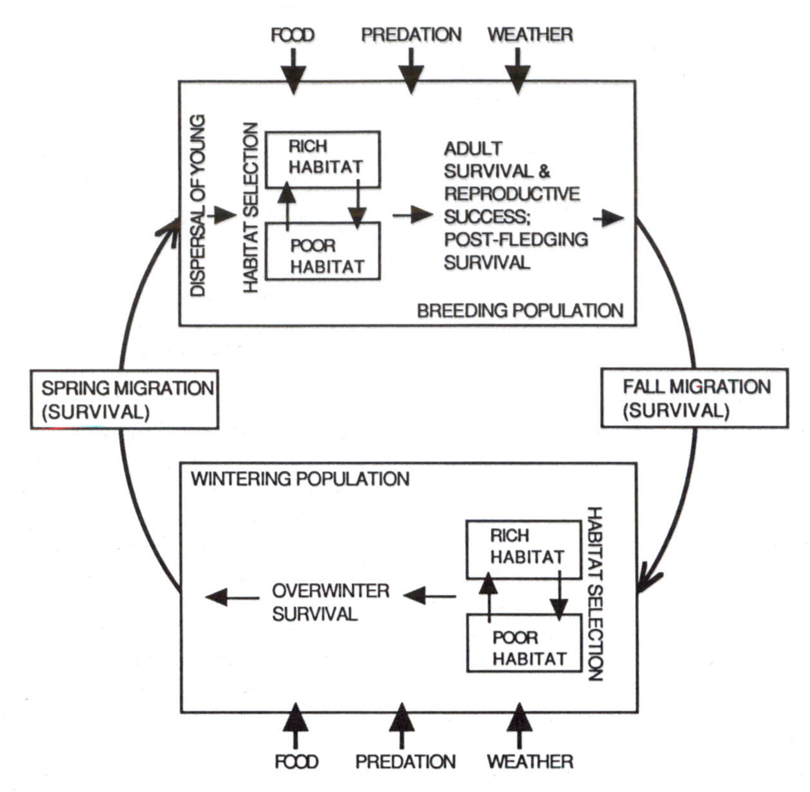

Figure 4-1. Simplified scheme illustrating key population processes facing migratory birds. Following each migration, birds confront the problem of defending or selecting habitat anew, often from a mosaic of habitat patches (after Sherry and Holmes 1991).

Several methods are used to demonstrate regulation (Sinclair 1989). First, experiments in which a population returns after disturbance to predisturbance levels provide evidence for regulation in some animals. Such experiments are not simple, and have never been performed, to our knowledge, with highly mobile animals such as migratory birds. Second, several kinds of statistical analyses are used to detect density-dependence, and thus infer regulation. One such approach is key-factor analysis (e.g., Baillie and Peach 1992), in which mortality ("killing-power") is partitioned into different stages of the life cycle. The key factor is defined as that one event or period of the annual cycle best correlated with total annual mortality. Baillie and Peach (1992) used this approach with seven species of Palearctic-African migrant bird species, based on long-term data sets of the British Trust for Ornithology, and concluded that: (1) migrant species populations tended to fluctuate most in response to events in nonbreeding seasons when density dependence was also greatest; (2) winter mortality was significantly associated with rainfall in the Sahel region of sub-Saharan Africa; and (3) some species populations tended to be limited by competition for winter food resources and other species by competition for nesting sites. Presently we lack the kinds of long-term, wide-scale demographic data necessary for a key-factor analysis using any Neotropical–Nearctic migrant population.

Other methods to identify population regulation are generally less satisfactory. Removal experiments during the breeding season (e.g., Hensley and Cope 1951, Stewart and Aldrich 1951, Sherry and Holmes 1989) suggest that populations contain floater individuals, which are kept from reproducing by territorial behavior of conspecifics. To our knowledge, however, females in migratory species have never been shown to be prevented from breeding by territorial behavior, and thus published removal experiments have little bearing on population limitation. Removal experiments in winter are also scarce and small in scale (Marra et al. 1993, Stutchbury 1994), although they do suggest that some individuals are kept from occupying preferred habitat, which could reduce overall survival and help regulate a population. Fitness should be reduced when densities increase, if populations are regulated, but we know of no augmentation experiments involving migrant bird species.

Neotropical–Nearctic migrant birds often occupy multiple habitats and disperse widely (e.g., Payne 1990) compared to most other organisms, suggesting that population dynamics must be examined at multiple-habitat or greater spatial scales, a relatively neglected topic in many past studies of population regulation. Models of how animals use habitats assume that, at times of low population density individuals will occupy those habitats in which they can achieve greatest fitness, referred to here as the primary habitats. Fretwell and Lucas (1970; Fretwell 1972) and Brown (1969b) devised explicit graphical models for the decline in fitness ("suitability") as the density of animals increases within the primary habitat. Furthermore, they noted that continued increase in density will depress suitability of primary habitat to the level achieved by individuals settling in a secondary habitat, at which point they should settle in both habitats. Increasing density should cause further suitability declines in both habitats as animals continue to settle both areas. At some point, further increases in density cause suitability to decline to such a level that individuals cannot reproduce or even maintain themselves, at which point they become "floaters," searching for unoccupied patches or newly created habitat vacancies (Fig. 4-2). Fretwell and Lucas also distinguished two habitat-settlement mechanisms with correspondingly different patterns of density-dependent distributions of abundances and suitabilities among the habitats. In the "ideal-free" case, individuals settle independently (free) of each other in the optimum habitat, such that, at any given density, fitnesses and suitabilities are equalized among all occupied habitats. In the "ideal-despotic" case, individuals settling first in preferred habitats constrain the settlement of subsequent individuals, e.g., via territorial behavior, such that fitness of individuals in secondary habitats is less than that of individuals in primary habitats. These general theories have motivated much empirical research as well as refinements to the theory (e.g.. Rosenzweig 1985, 1991; Morris 1987, 1989; Pulliam 1987, Bernstein et al. 1991).

Theoretically, density-dependent habitat use in vertebrates leads to potentially powerful regulatory mechanisms, which could operate widely in Neotropical–Nearctic migrant birds. Specifically, despotic habitat-selection could stabilize population dynamics in multiple-habitat landscapes, both via increased dispersal at greater population densities and via declining per capita fitness as a population is constrained at greater densities to occupy increasingly less suitable habitats (Lomnicki 1978, 1980; O'Connor 1985; Newton 1986; Andrén 1990; Dhondt et al. 1992). Dhondt et al. (1992), for example, showed that clutch size in European tits declines with density not because average clutch size declines within good habitat, but because small clutches are produced in secondary habitat. This idea of reduced average per capita fitness at higher densities (Fig. 4-2) leads to an asymptotic relationship between total number of young produced (or survival, depending on season) and population density. In other words, above some density at which suitable habitats become saturated, either no more young will be produced and/or adult survival will decline—a crucial relationship for understanding the joint influence of summer and winter habitats on population size, as we discuss below. The other point to note from the model in Fig. 4-2 is that per capita fitness declines with increased density, which is the essence of population regulation.

This mechanism of density-dependent fitness variation may be widespread in Neotropical–Nearctic migrants if despotic habitat-use patterns cause fitness to decline in secondary habitats. Some evidence suggests that fitness does depend on habitat, often in

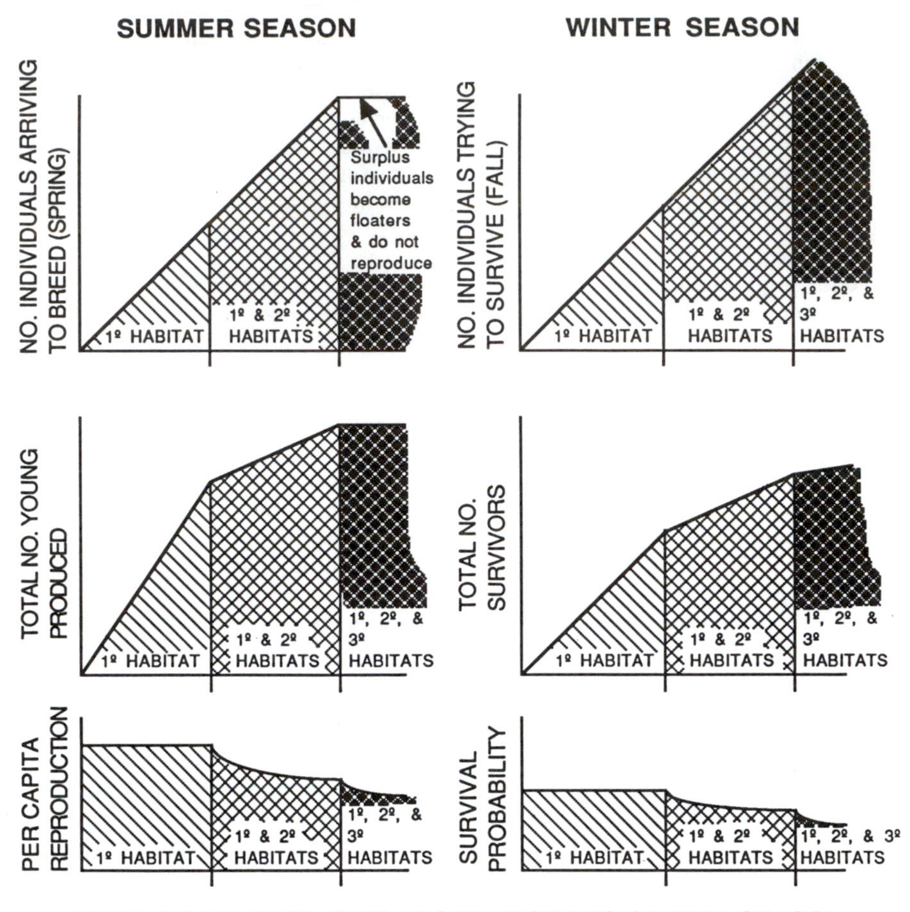

Figure 4-2. Density-dependent total and per capita productivity (summer season) and mortality (winter season) for organisms occupying multiple-habitat landscapes. As population increases, it expands from occupying just primary to occupying primary plus secondary habitats, and so on, with the consequence that total productivity and survival tends to reach an asymptote as suitable habitat(s) become saturated, and per capita fitness declines, consistent with the idea of regulation of the population to the total habitat available (see text for detailed explanation).

relation to age of bird in migrant species (e.g., Carey and Nolan 1975, Gauthreaux 1978, Lanyon and Thompson 1986, Sherry and Holmes 1995 and unpublished data, Holmes et al. 1995), and in a variety of other types of birds (e.g., Best 1977, Reese and Kadlec 1985, Newton 1986, Matthyson 1990). Such a mechanism could operate both in the breeding season and during winter, particularly if birds compete for limited amounts of the highest quality habitats. In a later section we review evidence suggesting that Neotropical–Nearctic migrant birds compete for habitat in both summer and winter, and emphasize the possibility that such density-dependent habitat selection may have a major impact on population dynamics.

Winter vs Summer Population Limitation: Theoretical Considerations

To draw some inferences about the breadth of factors contributing to year-round population limitation and regulation, we can view a hypothetical migrant population simplistically as oscillating between summer and winter grounds (i.e., ignoring the migratory period for the moment; Fig. 4-1). Most explicit theoretical discussions of seasonal populations model: (1) population size in autumn as a function of that in spring (based on summer habitat-selection processes, related primarily to nest site choice and reproduction); (2) spring population size as a function of that in autumn (taking into consideration winter habitat-selection processes, related to winter survival); and (3) combine these two sets of events mathematically or graphically (e.g., Fretwell 1972, Pulliam 1987, LeBreton and Clobert 1991). We follow Pulliam's (1987) treatment here (Fig. 4-3) to deduce the possible shapes of the curve of autumn population size as a function of that in spring. His ideal-dominance curve (Fig. 4-3A) seems most appropriate for Neotropical–Nearctic migrants, based on their widespread territoriality and intraspecific competition for preferred breeding and wintering sites (see below). A similar ideal-dominance curve is probably most appropriate for spring populations as a function of those in the fall (Fig. 4-3B). The "ideal" part of the model assumes both that birds can disperse widely among an array of habitats and can assess their quality, both of which are probably reasonable (if simplistic) assumptions for birds under most circumstances. These ideal-dominance curves are combined into one graph (Fig. 4-3C), from which several deductions about Neotropical–Nearctic migrant populations are possible.

These two curves approach zero slope asymptotically with increased density (Fig. 4-3A and B), a qualitative pattern that is crucial for our deductions that follow, and so we wish to know to what extent this pattern is reasonable. We noted that ideal despotic habitat selection models lead to this pattern (Fig. 4-2), and that there is some empirical support for the idea that Neotropical–Nearctic migrant populations occupy habitat in such a manner, e.g., winter territoriality in diverse species suggests that despotic habitat use is widespread (although certainly not universal, see below). Declining slope in the curve of fall population size as a function of that in spring (Fig. 4-3A) means that per capita productivity of young birds declines with density, and declining slope in the spring vs fall numbers (Fig. 4-3B) refers to declining per capita survival over winter. Even in the ideal-free case, in the absence of aggression, individuals will be increasingly likely to use up limited resources (food, safe nesting sites), causing a decline in fitness with density. This latter scenario might apply to the many nonterritorial Palearctic–African and Neotropical–Nearctic species in which individuals flock or wander nomadically in search of unpredictable or seasonally localized winter resources (Karr 1976, Leisler 1990, Levey and Stiles 1992, Sherry and Holmes 1995). Even in such opportunistic species, however, dominance hierarchies may form within flocks or at local resource concentrations, which could have survival or population consequences.

We now make several deductions from the graphical model (Fig. 4-3C–F). First, such populations are regulated (Pulliam 1987). This is because their populations tend to increase at population sizes below the equilibrium point (denoted by asterisks, which represent the projections of the equilibrium point on each axis), and decrease if they exceed the equilibrium point (Fig. 4-3C). It makes no sense to speak of populations approaching this equilibrium limited in either summer or winter—they are regulated by circumstances in both summer and winter because circumstances in both summer and winter contribute to the location of the equilibrium point. To look at it another way, any change in the total habitat available (caused by a change in either quantity or quality of the habitat) will change the height of at least one of the curves in Fig. 4-3A and B, which will translate to a different equilibrium point (Fig. 4-3C) in both seasons. Several recent long-term population studies of birds and other animals document the assertion that multiple factors often limit a population (Sinclair 1989, Arcese et al. 1992). Provisioning of nest boxes in local populations of many cavity-nesting species, for example, increases populations (Martin 1992) to a new level at which food or other resources then become limiting. Severe loss of habitat in either winter or summer—corresponding with a flattened curve of numbers in one season as a function of

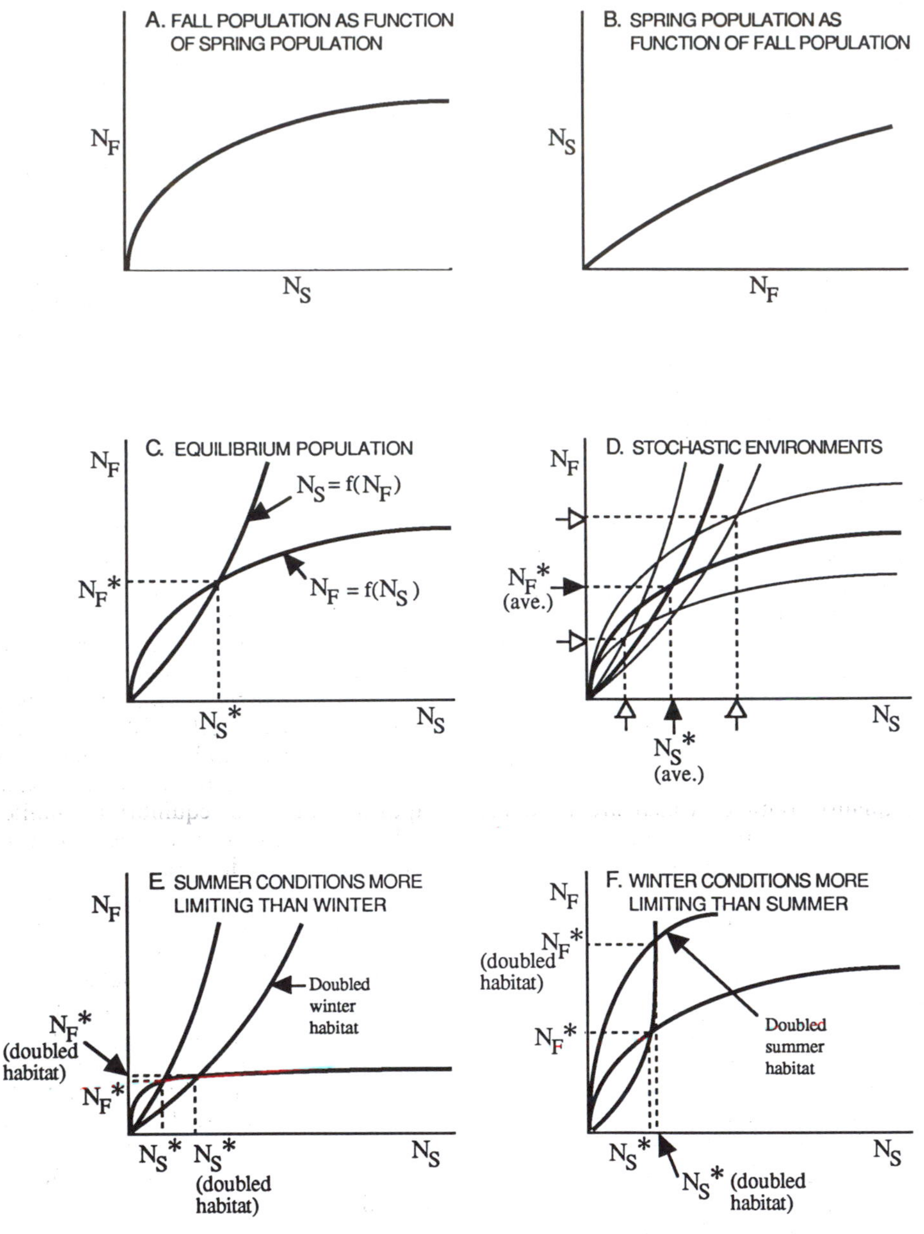

Figure 4-3. Regulation of migratory population as a function of events in both summer and winter season. Numbers of individuals in fall increase asymptotically with the total population starting the reproductive season in spring (A), as do numbers in spring as a function of those starting the fall survival season (B). Combining these two curves on one graph with the same axes (C) is possible by rotating the curve in (B) about the origin (i.e., transposing the axes), which leads to an equilibrium population, designated by N_F^* and N_S^*. Stochastically varying environments lead to a cloud of curves (D) with a mean equilibrium population, but extensive variation about this mean for both N_F^* and N_S^*. Relatively restrictive conditions for habitat selection in either summer (E) or winter (F) can place the effective limit on total population size in either season, although even in these cases the final populations are at least in part a function of events in both seasons (see text for further explanation).

numbers in the other—would have a devastating effect on a population, leading in extreme cases to extinction. This extreme condition clarifies the idea that both habitats are necessary to maintain a migratory species population (Cox 1985).

The second point is that most, if not all, populations occupy variable environments with some degree of unpredictability in resources. We illustrate this environmental stochasticity with families of curves (Fig. 4-3D), indicating that the population may fluctuate within a wide range of values due to changing conditions (carrying capacity of the environment) in either summer or winter habitats. The effects of environmental variability are represented by density-independent effects, which combine additively with density-dependent effects to influence the population (Sinclair 1989). As either the frequency and/or severity of environment changes increase, the relative importance of density-independent factors relative to density-dependent ones increases. A point is reached, potentially in some migrant species, in which the effects of droughts, fires, abundance of nest predators, or breeding season food supply fluctuate frequently and with sufficient amplitude that the "carrying capacity" changes before a population can approach it through density-regulating factors. Many migrant species, especially Paleotropical–Palearctic species wintering in African habitats, depend on periodically superabundant, ephemeral food supplies (Leisler 1990), illustrating effects of stochastic environments. Similarly, the migrants we have studied intensively for over two decades in the Hubbard Brook Experimental Forest in New Hampshire have undergone dramatic fluctuations due to changing food abundance (caterpillar irruptions), weather, nest predator populations, and possibly changing conditions on the wintering grounds (Holmes et al. 1986, Holmes 1990, Sherry and Holmes 1992). The point we wish to emphasize here is that such populations in stochastic environments might appear to be unregulated, and may show strong correlations with weather or other seemingly density-independent factors, even though regulation was operating sufficiently to maintain the population bounded within a range of abundances (e.g., Begon and Mortimer 1986). This stabilization renders a population less susceptible to extinction than one undergoing a random walk (Murdoch 1994), or than one that declines to the level of genetic and demographic extinction vortices (see Introduction). Stochastic variability, although exaggerated here (Fig. 4-3D), is likely the norm, and is omitted from other figures only for practical purposes.

The third point concerning winter vs summer limitation is that a population may in theory be more severely limited by habitat availability or other conditions in one season than in another (Fig. 4-3E and F). Summer limitation is exemplified by restriction to one or a few specialized summer (but not winter) habitats, in which fecundity is high, but outside of which it drops abruptly (Fig. 4-3E). Difficulty in foraging, attracting mates, or building (or otherwise acquiring) safe nest sites could explain why a particular species might have higher fitness in its primary breeding habitat. Prothonotary Warblers, for example, prefer a swamp nesting habitat because of the relatively high food availability compared to surrounding habitats, and reproductive success is also constrained there by availability of nesting cavities (Petit 1991a,b). Availability of cavities and other safe nesting sites may often limit bird populations that have such requirements (O'Connor 1985, Martin 1992). Ultimately, restriction to a specialized habitat or nesting site could be enforced through interspecific competition with one or more other migratory or temperate resident species. In theory birds may be more sensitive to changes in breeding than wintering habitat, since breeding requirements (finding mates and nest sites; rearing, protecting, feeding additional dependent individuals; and often molting immediately after the breeding season) may be more restrictive and constraining than wintering (purely self-maintenance) activities (O'Connor 1985, Martin 1992).

Alternatively, Terborgh (1980, 1989) and others have argued that migrants are more sensitive to winter habitat change because they are purported to occur at greater density in winter. Few data exist, however, to assess this argument. A population may be relatively more limited in winter than summer habitat (Fig. 4-3F). Cerulean Warblers provide a possible example, since they forage with mixed-species flocks within a relatively narrow elevational range of tall, relatively undisturbed South American forests, primarily in Eastern Andean foothills (Robbins et al. 1992b). Restriction to a narrow elevational range in this species may be influenced by competition with Blackburnian Warblers (another Neotropical migrant), and possibly with resident insectivores (Robbins et al. 1992b). The utility of our graphical model (Fig. 4-3E and F) from the perspective of wildlife management is that any effort to increase the amount of habitat in the less limiting season (e.g., doubled winter habitat when summer habitat is most limiting—Fig. 4-3E) will have relatively little effect on population size in the most limiting season.

We can now rephrase the underlying question behind this review: Instead of asking whether Neotropical–Nearctic migrants are limited in summer or winter, we hypothesize that such populations are probably limited by the interaction of ecological events in both summer and winter (Fig. 4-3). We should thus be asking questions about the relative quantitative effect on Neotropical–Nearctic migrant populations of any change in the availability of habitat in summer or winter, a presently unanswerable question for any species. Answers to this question could come, in theory, from data on areal extent of different habitats (perhaps using geographic information systems and satellite-scale earth imagery—Powell et al. 1992) coupled with estimates of fitness performance for migrants as a function of habitat type; see Sherry and Holmes (1995) for the latter. Landscape configuration (e.g., degree of fragmentation of preferred habitats, their connectedness, and barriers to dispersal between suitable habitat fragments) is another factor that could influence success in either season, as we discuss briefly below.

Morse (1980b, 1989) recognized the possibility that ecological conditions in winter or summer, or both seasons could limit populations, and Cox (1985; see also Svennson 1985) proposed an explicit hypothesis in which migrant birds' populations are limited simultaneously in summer and winter by a dynamic equilibrium between fecundity and mortality in a changing array of habitats. Cox argued that a population in which overwinter survival temporarily increased would expand into a greater array of breeding habitats (because the greater winter survival would compensate for decreased fecundity in the newly added, but inherently less preferred breeding habitats). Conversely, improved breeding season fecundity would allow increased range of habitats in which individuals could survive in winter. We illustrate this hypothesis schematically in Fig. 4-4, using a particular form of relationship between fecundity and survival (namely the product of the two = constant), in which changes in one parameter are compensated by changes in the other so as to maintain a constant population—implicit in our discussion of regulated populations as discussed above. An inverse relationship between fecundity and survival among species has been frequently documented (e.g., O'Connor 1985, Martin 1995).

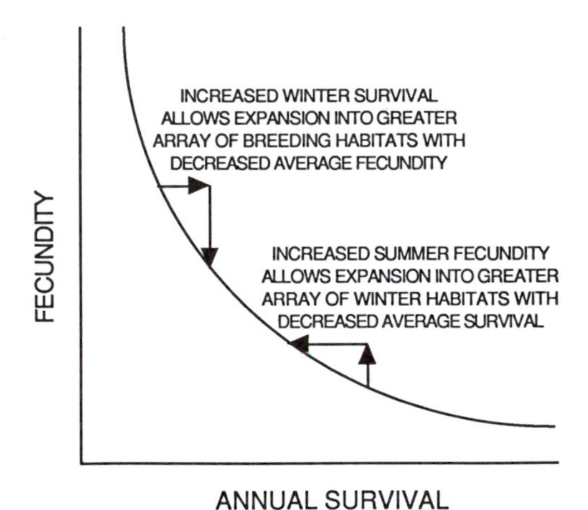

Figure 4-4. Summer-and-winter limitation hypothesis for Neotropical–Nearctic migrant birds illustrated in terms of tradeoff between fecundity and annual survival in hypothetical constant population. An increase in fitness of animals in either season allows expansion of the population into worse habitats in the other season, given the maintenance of the same overall population (determined by the product of annual survival and per capita reproduction).

Habitat Configuration

We have treated population processes of migrants thus far as though the population in a particular season were strictly proportional to the total quantity and quality of habitat available. We think that this assumption is reasonable as a first approximation for highly vagile organisms such as many, if not all, migrants. However, the spatial arrangement of these habitats and their quality are potentially important to population dynamics, particularly in landscapes in which habitats are fragmented, e.g. by human activities. High-quality habitats could be available, for example but remain unused in a fragmented landscape if birds cannot find them, or populations occupying a habitat with insufficient reproductive success to maintain it locally (a "sink" habitat), could be maintained by continual immigration from distant habitats producing an excess of individuals ("source" habitats—Temple and Carey 1988, Pulliam 1988, Pulliam and Danielson 1991, Bernstein et al. 1991, Robinson 1992). Moreover, a population fragmented into somewhat isolated subpopulations across a landscape (collectively referred to as a "metapopulation"), may repeatedly colonize, then become extinct from habitat patches depending on their size, isolation, and other variables. The importance of the configuration of habitats within a landscape and the dynamics of metapopulations are clearly important topics in understanding bird populations, and are discussed in detail elsewhere (e.g. Bernstein et al. 1991, Hunter 1991, Temple 1991, Freemark and Collins 1992, Villard et al. 1992, Faaborg et al., Chapter 13, this volume, Freemark et al., Chapter 14, this volume). Habitat configuration may have a proportionally larger effect than habitat quantity in species with specialized habitat needs and poor dispersal ability (Pulliam and Danielson 1991). Also, effects of habitat fragmentation seem to be less important to Neotropical–Nearctic migrants in winter than in summer (Robbins et al. 1987, 1992a, Freemark et al., Chapter 14, this volume, Petit et al., Chapter 6, this volume), although the reasons for this are unclear.

The Migration Period

Although less is known about habitat requirements and limitation of birds during migratory than breeding or wintering seasons, events during migration must also influence population abundances and population dynamics. Some shorebirds concentrate major proportions of their population in a few "staging areas" during migration, so that these areas take on considerable conservation importance (Myers 1987). Also, when landbird migrants confront large physical barriers, such as the Gulf of Mexico, certain habitats become particularly important to their welfare because of the need of restoring energy reserves for subsequent travel (Moore et al., Chapter 5, this volume). The potential for limitation and food competition in passage habitats is indicated by territoriality in some species at stopover sites (Rappole and Warner 1976), by the results of exclosure experiments showing depletion of insect abundance on foliage (Moore and Yong 1991), and by the dispersion of birds among available stopover habitats (Martin 1980), which may at least in part be due to food abundance (Moore et al., Chapter 5, this volume).

Migrant birds may also be effectively limited by food during migration because of time constraints to find suitable habitats (Moore and Simons 1992, Moore et al., Chapter 5, this volume). Such limitation could cause increased mortality of individuals in proportion to the size of migrants' populations, i.e., to density-dependent mortality during migration, although this phenomenon has yet to be demonstrated at this phase of the annual cycle. Loss of staging and refueling habitats along a migratory route and long-term weather patterns could both contribute to long-term population declines of migrants (Moore and Simons 1992, Moore et al., Chapter 5, this volume).

A final consideration in understanding how the migration period influences population dynamics is the phenomenon of extended, or multiple-stage migrations. Eastern Kingbirds, for example, undergo large-scale movements throughout the winter as availability of fruits and other food resources change in South America (Morton

1971). Eastern Wood-pewees spend up to several months wintering in Central American habitats in the fall before proceeding further south later in the winter (F. G. Stiles, personal communication). Various paruline warblers appear to shift habitats and regions during the winter in northern South America (Lefebvre et al. 1992). Such gradual movements throughout the winter are common in African winter residents (see references in Leisler 1990, Rappole 1991). In all these cases the migration period becomes indistinguishable from the wintering period, resulting in longer time periods for ecological conditions in the tropics to influence nonbreeding season survival. Clearly more research attention needs to be focused on the frequency and extent to which habitat might be limiting for migrants during passage (or during winter in the case of multiple-stage migrants), and on the environmental resources or circumstances that determine habitat quality during these sometimes extended periods of the birds' annual cycle.

EVIDENCE FOR LIMITATION OF HABITAT QUANTITY AND QUALITY IN SUMMER AND WINTER

Competition for Habitat

If habitat is limiting to migrant birds, competition for preferred areas should be evident. Such competition could be based on variability in habitat quality, if individuals occupying certain habitats reproduce more successfully or survive better than individuals in other habitats. Territorial behavior resulting in an exclusively occupied area is usually considered as strong evidence for competition (Brown 1969a), and Morse (1980b) mentioned evidence for aggressive territorial behavior, and thus habitat limitation, in both summer and winter. Additional evidence for habitat competition comes from different age- or sex-specific habitat segregation, and from experimental removals of territorial individuals. We review here such evidence for habitat limitation and competition in both breeding and wintering grounds of Neotropical migrant landbirds.

Intraspecific Competition for Breeding Habitat

Territoriality among songbirds breeding in the temperate zone is widespread, suggesting strong intraspecific competition for habitat and the resources that accrue to habitat defense (e.g., mates, food, safe nesting sites). In most species, a male establishes a territory immediately upon arrival in the breeding area, and advertises himself and the territory using song and visual displays. Although the function or ultimate adaptive value of such behavior in birds is still debated (e.g., Morse 1980a, Gauthier 1987, Möller 1987), the proximate result in many cases is to overdisperse individuals throughout available habitat (Sherry and Holmes 1985), and to prevent some individuals from settling in an otherwise preferred habitat. If there is a limit to which territories can be compressed, i.e., if the rubber disc analogy of Kluijver and Tinbergen (1953) holds, then territoriality limits the number of individuals that can settle in a given patch of habitat. Results of removal experiments in a variety of long-distance migrant species during the breeding season (e.g., Hensley and Cope 1951, Stewart and Aldrich 1951, Thompson 1977, Arvidsson and Klaesson 1984, Sherry and Holmes 1989; see review by Newton 1992) suggest that territorial behavior does limit access by some individuals, usually males, to preferred habitat. However, we know of no removal studies in migrant species showing that both females and males are prevented from breeding, a condition that must be shown to conclude that territoriality limits populations (Morse 1980a). Thus, more research is needed to verify the frequency of population limitation by territorial behavior.

Another line of evidence for habitat limitation comes from age-specific distribution patterns, and fitness consequences thereof. If older males were dominant socially, or arrived earlier to claim territories in the best habitats, then one would expect to find younger individuals in the least preferred habitats experiencing relatively lower reproductive success (e.g., lower mating success, smaller clutch sizes, lower fledging succcess). This pattern has been documented in several Neotropical migrant birds during the breeding season (Carey and

Nolan 1975, Lanyon and Thompson 1986, Holmes et al. 1995), as well as in a variety of European species (reviewed by Andrén 1990).

Intraspecific Competition for Winter Habitat

Much less is known about competition for habitat by migrants in their Neotropical winter grounds. Although some migrants occur in mixed- or single-species flocks or even as pairs (see reviews by Greenberg 1985, 1986), they are frequently found as solitary individuals, either defending localized resources, such as flowering or fruiting trees, or larger patches of habitat for more extended periods of time (Rappole and Warner 1980, Greenberg 1985). Evidence for long-term defense of winter territory, which suggests some form of habitat limitation, derives from: (1) patterns of overdispersion of individuals through winter habitat (Elgood et al. 1966, Sliwa and Sherry 1992); (2) local site fidelity of wintering individuals (Price 1981, Holmes et al. 1989, Holmes and Sherry 1992); (3) the occurrence of intraspecific aggressive interactions (Willis 1966, Gorski 1969, 1971, Emlen 1973, Morton 1980, Stacier 1992); or (4) some combination of the above (Schwartz 1964, 1980, Nisbet and Medway 1972, Rappole and Warner 1980, Greenberg 1984, Holmes et al. 1989, Winker et al. 1990, Bates 1992, Stutchbury 1994). Because no breeding occurs during the winter period, such territoriality cannot serve the functions of defense of paternity (e.g., Gauthier 1987, Möller 1987) or of spacing to reduce nest predation (Patterson 1985). Rather, it seems to function primarily for self-maintenance, probably for securing a patch of habitat with dependable food (Price 1981, Hutto 1985, Greenberg 1986, Kelsey 1989, Marra et al. 1993) or possibly for providing refuge from predators and inclement weather (Winker et al. 1990).

Although social intolerance among wintering migrants appears to disperse individuals among available habitats, does it limit the numbers of birds that can occupy particular habitats? Preliminary data indicate that it can have such effect. Rappole and Warner (1980) reported that color-marked Hooded Warblers that disappeared from Mexican habitats were replaced by new, unmarked individuals, as was also the case for Black-throated Blue Warblers and American Redstarts in Jamaica (Holmes et al. 1989). Experimental removals of wintering migrants have been conducted only four times, three with Hooded Warblers (Rappole and Warner 1980, Morton et al. 1987, Stutchbury 1994) and once with American Redstarts (Marra et al. 1993). In all these studies, individuals removed were replaced, mostly within 2 days, except for some Hooded Warblers in "female" habitat (Morton et al. 1987). From these results, it appears that winter habitats may be saturated for some species, forcing some individuals to become floaters, which fill vacant territories as they occur.

Winter social organization of these long-distance migrant species differs fundamentally from that in summer. Whereas only adult males are territorial in summer, both males and females as well as juveniles and yearlings defend territories in winter (Schwartz 1964, Rappole and Warner 1980, Greenberg, 1985, 1986, Holmes et al. 1989, Sliwa and Sherry 1992). In some habitats, the sexes and age classes defend territories against each other (see Holmes et al. 1989, Marra et al. 1993). In other cases, the sexes are segregated by habitat (Lynch et al. 1985, Lopez and Greenberg 1990, Wunderle 1992, Parrish and Sherry 1994). It is not clear whether sex-specific segregation among habitats is due to inherent differences in habitat preferences (Morton et al. 1987, Morton 1990) or to the effects of behavioral dominance (Marra et al. 1993), or both. Whatever the mechanism, the important question from the perspective of population limitation is whether these habitat differences affect survival or other fitness-related parameters. Preliminary evidence supports the idea that fitness of birds does differ among winter habitats, and that ages and sexes differ in winter habitat-specific fitness (Greenberg 1992, Sherry and Holmes 1995). Greater mortality of females in winter could help explain the male-biased sex ratio during the breeding season in many migrant species (Breitwisch 1989, Gibbs and Faaborg 1990).

Finally, the occurrence of territoriality and behavioral dominance in both breeding and wintering grounds suggests that the population dynamics in many of these long-distance migrant species are structured around density-dependent mechanisms acting in both seasons. A dominance-based system in winter, which differentially affects survival by sex or age may play a role in structuring events on the breeding grounds. The reverse is almost certainly true, in that reproductive success and survival in the breeding period will influence winter densities, habitat settlement patterns and perhaps mortality rates (Sherry and Holmes 1991, 1992). These empirical findings of competitive, socially constrained habitat selection in winter support earlier theoretical models in which dominance-mediated habitat selection in one season influences population size in the subsequent season (Fretwell 1972, Pulliam 1987). The evidence reviewed above concerning competitive intraspecific relationships in these birds therefore supports the idea that habitats could be limiting populations both in summer and in winter.

Changes in Habitat Quality and Quantity Affect Neotropical Migrant Bird Populations

Here we review available evidence indicating that habitat quality and quantity available to Neotropical migrant populations are correlated with population changes in these birds.

Habitat Quality in Summer

Evidence from four studies indicates that yearling recruitment in breeding populations of long-distance migrants is determined by the previous summer's fledging success (Nolan 1978, Virolainen 1984, Sherry and Holmes 1992, Holmes et al. 1992). This relationship appears to hold despite variation in mortality occurring on migration or in winter, demonstrating the paramount importance of breeding productivity. Thus, events in summer breeding periods are clearly important to population dynamics.

Various aspects of habitat quality strongly affect reproductive success. Those factors most important are adequate food supply, cover, and nest sites. Food has been shown to be important for many temperate breeding passerines (reviewed by Martin 1987). For instance, in New Hampshire forests, Black-throated Blue Warblers fledged more young per season when caterpillars were more abundant (Rodenhouse and Holmes 1992). Furthermore, between 1986 and 1990, the abundances of all bird species, of all foliage gleaning species, and of Black-throated Blue Warblers, in particular, were positively correlated with caterpillar abundances, not in the same season, but in the previous season (Holmes et al. 1991). This suggests that food in one year affects bird abundances in the following year via the mechanism of variation in nestling productivity and possibly also fledgling survival. Several species of spruce-woods warblers also respond numerically to population irruptions (rapid population increase followed by crash) of spruce budworm caterpillars (*Choristoneura fumiferana*), indicating that food abundance influences the carrying capacity, or quality of their boreal habitats in North America. The Bay-breasted Warbler, Cape May Warbler, and Tennessee Warbler are budworm specialists insofar as their populations increase dramatically in abundance in budworm outbreak areas, in part because of increased reproductive success due to larger clutch sizes (MacArthur 1958; reviewed by Morse 1989). Interspecific competition in these species, e.g. the decline in Blackburnian Warblers in response to increased local populations of budworms (Morse 1989), is yet another line of evidence for habitat saturation, although the carrying capacity of the habitat may change by orders of magnitude in response to changing food abundance.

Weather during the breeding season also affects habitat quality directly by influencing bird survivorship (Zumeta and Holmes 1978) and fledging success (Rodenhouse and Holmes 1992). Weather also affects habitat quality indirectly by influencing food availability (Hejl et al. 1988, Blake et al. 1992) or nest site selection (Martin 1993b).

Nest predation is the main factor affecting nesting success in most passerine birds (Ricklefs 1969, Holmes et al. 1992, Martin 1992, 1993a, Sherry and Holmes 1992), and this factor may have caused recent population declines in many migrant species

(Böhning-Gaese et al. 1993). Nest predation is particularly likely to be increased by fragmentation of forested habitat, since many predators benefit from the increased ratio of edge habitat to forest interior, particularly whenever fragmentation is accompanied by human residential developments and the attendant food subsidies for nest predator populations such as Blue Jays, skunks, and raccoons (Wilcove 1985, Temple and Carey 1988, Yahner 1988, Small and Hunter 1988, Terborgh 1989, Finch 1991). Forest fragmentation decreases habitat quality of Neotropical–Nearctic migrants by increasing the predator populations that decrease nesting success for open-cup nesting species, and that may decrease renesting attempts of shrub and ground nesters (Martin 1993a).

Similarly, Brown-headed Cowbirds increase with forest fragmention and with regional extent of agriculture, and these brood parasites significantly decrease annual reproductive success of many open-nesting Neotropical migrant birds (e.g., Brittingham and Temple 1983, Terborgh 1989, Robinson 1992, Robinson et al. 1993). Cowbird parasitism is implicated as at least one cause of the endangered status of Kirtland's Warbler, Black-capped Vireo, and Least Bell's Vireo (Robinson et al. 1993). Golden-cheeked Warblers are parasitized heavily by cowbirds, but renest when parasitized, thus reducing the impact compared with other species (Morse 1989). The Shiny Cowbird, which is expanding its geographic range from Florida into neighboring southeastern states, is another brood parasite threatening-habitat quality for open-cup nesting migrants.

Finally, nest addition experiments have increased breeding populations of cavity nesters (e.g., Brawn and Balda 1988, review in Martin 1992), suggesting the possibility that nesting opportunities were limiting local population sizes via habitat quality.

Habitat Quantity in Summer

The quantity of North American breeding habitat has probably changed greatly and continuously over many centuries, as humans have modified the landscape to increase food production (Morse 1989). Moreau (1972) hypothesized that many Old-World migrant bird populations may have been decimated in Europe over the last 1000 years. Likewise, in North America, clearing of the primeval forests by European colonists in the 17th and 18th centuries must have resulted in major changes in migratory bird populations. J. J. Audubon purportedly saw very few Chestnut-sided Warblers during his lifetime, whereas now the species is widespread in forest/field-edge habitats, which abound throughout eastern North America (Morse 1989).

Several specific examples of how habitat loss probably caused population declines or changes directly may be cited. A prime example is the Kirtland's warbler, which requires large (> 32 ha), 6–22 year-old stands of regenerating jack pine, especially in association with fire for nesting (Morse 1989). This species' population increased dramatically within its specialized breeding habitat when new stands were created, mostly in Michigan, about 10 years after each of a series of large fires between 1871 and 1980—reviewed by Mayfield 1993) that initiated secondary succession of the pines. Conditions for nesting subsequently became unsuitable as jack pine stands aged, and the population declined—to a low total population of 167 males in 1974 (Probst 1986, Mayfield 1993). Similarly, the reduction of canebreaks in the southeastern United States may have led to the demise of Bachman's Warbler (Remsen 1986), although this may also have been caused by loss of lowland native forests in Cuba where this species wintered (Terborgh 1989). The loss of suitable breeding habitat for Cerulean Warblers has been considered to be a possible cause of their continuing decline (Robbins et al. 1991b). Loss of habitat has contributed to the decline of the Black-capped Vireo to endangered status, although cowbird parasitism, habitat fragmentation (loss of quality), and habitat modification via grazing have also contributed in this species (Grzybowski et al. 1986).

Habitat Quality in Winter

Likewise, changes in winter habitat quality may result in population changes (reviewed by Sherry and Holmes 1995). Food shortage induced by drought in the Sahel correlated with major population declines of White-Throats and several other species of Palearctic–African migrants (Baillie and Peach 1992). The importance of winter habitat suitability for Kirtland's Warbler is indicated by positive correlations between winter rainfall in the Bahamas, where this species winters, and minor but significant fluctuations in summer adult censuses in Michigan (Ryel 1981; unpublished analyses by Lawrence Ryel and Duncan Evered, cited by Mayfield 1993). Thus, wet winter conditions have apparently contributed significantly to increased overwinter survival of Kirtland's Warbler, presumably by augmenting winter food supplies or total area of suitable wintering habitat (for discussions of the relation between rainfall and food supply in northern tropical latitudes, see Baillie and Peach 1992, Greenberg 1992, Parrish and Sherry 1994, Sherry and Holmes 1995).

Habitat Quantity in Winter

Finally, habitat loss in the winter may also be involved, although its effects are very difficult to quantify. Bachman's Warbler is often cited as an example of how winter habitat loss can cause population decline to extinction (e.g., Terborgh 1989), although summer habitat loss is an alternative explanation (see above). Marshall (1988) attributed reduced population sizes of breeding Olive-sided Flycatchers in the Sierra Nevada, California, to changes in its winter grounds. Similarly, winter habitat loss is blamed for declines of Cerulean Warblers (Robbins et al. 1992b) and Wood Thrushes (Morton 1992, Rappole et al. 1992). Shorebirds wintering on the Argentinian pampas may also be declining due to habitat loss in winter (Terborgh 1989). The evidence that winter habitat loss caused these particular population declines is circumstantial, and thus weak, indicating the need for serious demographic work on wintering populations (Sherry and Holmes 1995).

Evidence reviewed above suggests that populations of Neotropical–Nearctic migrants may be limited in breeding areas for some species, wintering areas for others, or both areas in a few cases. A fundamental issue here is that very few species have been studied intensively enough to detect the effects of habitat change on population size in any season, let alone in both, but the absence of such empirical evidence does not mean that populations may not be limited in both seasons. Unfortunately, we know of no empirical data set for any Neotropical–Nearctic migrant species to assess quantitatively whether limitation is occurring in summer, winter, or both—such as was done in a preliminary way for several Old-World migrants using key-factor analysis (Baillie and Peach 1992).

LIFE HISTORIES OF MIGRANTS VS RESIDENTS

Life-history traits of an organism are the result of natural selection and other evolutionary forces shaping the organism to particular environmental characteristics. In the case of Neotropical–Nearctic migratory birds, these environments are diverse because these birds comprise many species and occupy diverse latitudes and habitats, ranging from boreal conifer forests to tropical rainforests and scrublands. The migratory phenomenon raises many evolutionary questions. For example, how did long-distance migration evolve in the first place? Are migrants' demographic characteristics, habitat-use patterns, or feeding adaptations unique, and, if so, what characteristics of the environment have shaped these traits? Are migrants simply opportunists exploiting seasonal or disturbed habitats, or are migrants uniquely specialized to native habitats currently threatened by humans? Are migrants relatively specialized for summer or winter ecological conditions, or for both? Answers to these questions have important management implications, for example, informing us about the potential of migrants to survive in human-disturbed landscapes. Thus we review the demography, habitat-use patterns, and feeding adaptations of migrants in comparison with both temperate and tropical resident birds. We mention several controversial and polarized views of migrant life histories, which indicate the need for much further work in this area, and we stress simultaneously the generalization that migrants as a group may be particularly well adapted to exploit highly seasonal resources and the seemingly contradictory point that migrants are evolutionarily so diverse as to make any generalizations challenging to discover.

The very act of migration, taking advantage of seasonally changing environments across an array of latitudes, is a form of ecological opportunism. The traditional view of migrants (e.g., Morel and Bourlierre 1952, Karr 1976, Morse 1980b) describes ecological opportunists, avoiding the most mature forested habitats and thus avoiding resident competitors both in tropical winter grounds and higher latitude breeding areas, and as vagrants in search of seasonal or ephemeral resources that resident species cannot dominate. Some more recent scholarship also supports this view of migrants as opportunists (Lack 1986, Leisler 1990, 1992, O'Connor 1990). For example, the highly seasonal flushes of vegetation and foliage insects in the broad-leaved temperate forests are dominated by up to 70–80% migratory birds (MacArthur 1959, Rabenold 1993). Migrants such as the abundant small-bodied warblers, flycatchers, and vireos are well adapted to exploit these small foliage arthropods opportunistically because of shallower and shorter beaks, shorter wings and legs, and shorter toes than found in either temperate or tropical resident bird species (Greenberg 1981, Ricklefs 1992). O'Connor (1990) found a significant relationship between low densities of British residents following a severe winter and numbers of migrants occupying habitats the following summer, suggesting that temperate residents outcompete migrants, probably for safe habitats in which to begin breeding early in the season. O'Connor (1985, 1990) concluded that migrants were relatively "r-selected" insofar as migrants tended to be smaller and tended to vary more in abundance than temperate residents. North American migrant species tend to have smaller total (absolute) populations that are less tightly regulated around a carrying capacity than resident species, although there is much overlap between the two groups (O'Connor 1992). Mostly anecdotal observations from the tropical wintering season also suggest that migrants are occasionally socially subordinate to, and outcompeted by, resident tropical species of birds (Willis 1966, Leck 1972, Leisler 1990), and some evidence supports the idea that migrants compete with tropical residents on large spatial scales (Slud 1976, Ricklefs 1992).

Alternatively, migrants are viewed not simply as opportunists relegated to the ecological fringes of tropical communities, but rather as competitive habitat and even resource specialists, derived from tropical ancestors, and thus well adapted to tropical communities where they still spend more than half the year (Schwartz 1980, Stiles 1980, Rappole et al. 1983, Ramos 1988, Morton and Greenberg 1989, Rappole 1991, Levey and Stiles 1992, Powell et al. 1992, Rappole et al. 1992). As ecological specialists in native habitats such as rainforests, migrants face ecological calamity when Neotropical forests are converted to agriculture and other land uses (Terborgh 1989, Rappole 1991, Morton 1992).

Evolutionary opportunism is often associated with "r-selected" life-history characteristics, so-called because of high population growth potential (high "r" in population growth models) and good dispersal ability to facilitate occupation of ephemeral and often disturbed environments. Typical r-selected species are "weeds," effective invaders of disturbed or otherwise temporary habitats occupied by few competitors. Stable habitats tend to contain species with "K-selected" life-history characteristics, including competitive ability under conditions in which populations live constantly at the environmental carrying capacity, designated by "K" in population models. The traditional view of migrants as r-selected opportunists implies that they should need less management than if they were K-selected specialists (Morton and Greenberg 1989, Rappole 1991, Morton, 1992, Powell et al. 1992, Rappole et al. 1992), and yet, as we have indicated above, some migrants can be considered as specialized occupants of tropical rainforests and other species-rich, predictable habitats. In general we find the dichotomy of r- vs K-selected life histories too simplistic for helping to understand adaptations of migratory birds, in that they exhibit mixtures of both and vary greatly from species to species and among regions of the globe. We review some of these life-history features below.

Demographic Comparisons of Migrants and Residents

Annual survival of long-distance migrants is often greater than that of temperate zone residents, usually exceeding 50% (Greenberg 1980, Morse 1980b, 1989, O'Connor 1981, 1985, 1990, Ricklefs 1992; but see Karr et al. 1990, Martin 1993b, 1995). Mortality of temperate residents such as woodpeckers, nuthatches, titmice, and chickadees tends to occur most often during harsh winter months (reviewed by Arcese et al. 1992), which may also be the case for some migrants wintering in drought-prone tropical regions (Baillie and Peach 1992). The greatest survival difference between migrants and residents may be in yearling birds. Young of migrant species appear to survive their first nonbreeding period about half as well as their respective adults, whereas yearling residents survive only about one-quarter as well as their adults (Greenberg 1980; see also Karr et al. 1990). Data on survivorship, however, are sparse.

Fecundity, which depends on both the number of broods and the number of offspring produced per brood, is often higher in temperate residents than in Neotropical migrants (Greenberg 1980, O'Connor 1981, 1985, 1990, Morse 1989, Ricklefs 1992, Martin 1995), although there is considerable overlap (e.g., O'Connor 1985, 1990). Some temperate resident species achieve relatively high fecundity by producing large broods in safe nesting sites (such as tree cavities). Other residents begin reproduction relatively early in the season, which allows time to rear two or more broods (Greenberg 1980, O'Connor 1981, 1985, 1990), although this may depend more on habitat than migration status (O'Connor 1985, Martin 1995). Production of multiple broods annually is not unusual in migratory species (O'Connor 1985, Keast 1990, Holmes et al. 1992, Rodenhouse and Holmes 1992), and, according to Martin (1995), number of broods does not differ significantly between migrants and residents, suggesting that brood size does differ so as to make overall productivity greater in residents than in migrants.

Relatively few studies have been conducted on tropical residents' demography, but the few available suggest little if any difference in survival compared with migrants (Karr et al. 1990, Ricklefs 1992) and, if annual fecundity is any different in the two groups, then it is slightly higher in migrants than tropical residents (Ricklefs 1992). The greater demographic similarity of Neotropical migrants to Neotropical residents than to Nearctic residents is consistent with the idea that many Neotropical migrants evolved from close relatives that are resident year-round in tropical or subtropical latitudes (Stiles 1980, Rappole et al. 1983, Cox 1985, Rappole 1991, Levey and Stiles 1992).

These demographic studies, particularly by showing the lack of strong differences between migrants and tropical residents, suggest that Neotropical–Nearctic migrants are highly adapted to tropical habitats, and can hardly be considered simply as weedy interlopers in tropical systems, as traditionally believed. Relatively low seasonal productivity of migrants compared to temperate residents, if confirmed to be generally true, would be consistent with relatively high survival in the former, insofar as survival and fecundity are often related inversely (e.g., O'Connor 1985, Martin 1995; see also Fig. 4-4).

Habitat-Use Patterns

What are the ecological conditions to which migrants are adapted? We now consider habitats occupied to assess whether or not migrants are uniquely adapted compared to residents. In summer, many migratory landbirds are typically found in specialized breeding habitats. For example, some migrant paruline warblers are specialized for breeding in either conifer or hardwood forests (Greenberg 1979), and others are associated with particular seral stages or floristic associations (references in Morse 1989): e.g., Kirtland's Warbler in 6–21-year-old jack pine stands in Michigan's Upper Peninsula, Chestnut-sided Warbler at edges of mature hardwoods forests; Louisiana Waterthrush along streams, and Northern Waterthrush in marshes and swamps; Black-throated Blue Warblers in relatively old-growth Eastern deciduous forests with a well-developed shrub understory; Pine Warbler in pines; Yellow-throated Warbler in cypress swamps, and other relatively old-growth conifer forests of the Southeast; Prothonotary Warbler in swamp forest; and Mourning Warbler in 5–10-year-old seral stages of Northern Hardwoods. Breeding habitat is specialized enough in the bewilderingly similar species of *Empidonax* flycatchers that it is an important field clue to species identification. Black-capped Vireos depend on scrub-oak shrublands (Grzybowski et al. 1986).

Migrants breeding in Britain tend to be more specialized in habitat use than resident species, especially those that travel furthest between breeding and wintering sites (O'Connor 1985, 1990). The longest-distance migrants tend to return late to breed, and may be particularly dependent on finding relatively safe and productive nesting sites, which may restrict the range of available nesting habitats compared with resident species. O'Connor (1985) argues, moreover, that habitat breadths are narrower in summer than winter because conditions for reproduction (safe nesting sites, song posts, and food supply to provision broods) are more restrictive than conditions for surviving winter (feeding and self-maintenance). Thus migrants tend to be highly dependent on specialized breeding sites in which to maintain high fecundity, often in habitats of native vegetation (as opposed to exotic, human-modified habitats). Where migrants are behaviorally plastic enough to breed in agricultural landscapes, they largely depend on uncultivated areas that probably resemble native habitats in terms of foods and safe nesting sites (Rodenhouse et al. 1993). Thus the importance of specific breeding habitat for migrants is clear, whether this is native forest, grassland, shrubland, fire-maintained seral stages or certain floristic associations. Even this generalization, however, does not apply to all migrants, some of which (e.g. Red-eyed Vireo and American Redstart) breed in a wide variety of floristic associations and successional stages of vegetation.

In winter, migrant species are rarely habitat specialists. Migrants are distributed among at least as many habitats as are residents (Stiles 1980, Lynch 1992, Hutto 1992), and essentially all tropical habitats contain migrants. When migrants are specialized in the use of forested tropical habitats (Terborgh 1980, Rappole and Morton 1985, Morton and Greenberg 1989, Rappole 1991), this specialization may vary regionally (Hutto 1992). Rappole and others (Morton 1992, Powell et al. 1992, Rappole et al. 1992) observed Wood Thrushes and Kentucky Warblers to be largely restricted to rainforest interior habitats in winter, whereas Lynch (1992) found these species in the Yucatan Peninsula in a wide variety of habitats, although their abundances were greatest in wetter forests. Terborgh (1989) listed 54 species of North American migrants wintering in mature tropical habitats, which implies that the remainder, and by far the majority of species, do not winter in mature tropical forest. Recent reviews of Neotropical–Nearctic migrant habitat use (Hutto 1992, Lynch 1992, Petit et al. 1993, Chapter 6, in this volume) indicate a clear preference for early successional or disturbed habitats. Human residential and agricultural habitats such as citrus and coffee are used frequently as well by many migrant species (Robbins et al. 1992a).

Similarly, European–African migrants overlap resident species, but tend to be habitat generalists in winter, especially abundant in relatively dry or edge habitats, where rainfall resources are highly seasonal or where residents are relatively less abundant (Lack 1986, Leisler 1990, 1992). In this respect, Neotropical–Nearctic migrants may differ from Old-World migrants (Rappole 1991, Mönkkönen et al. 1992) in that the former include many species that occur in forest habitats in the winter while the latter tend to be in more open habitats, such as savanna and scrub. Seasonally dry habitats in the northern Neotropics tend to have the greatest concentration of migrant individuals and species (Terborgh 1989, Petit et al. 1993), and this seasonality may explain why Neotropical migrants wintering in Mexico tend to be less specialized in use of habitats than resident species (Lynch 1989, Hutto 1992).

Inclement weather systems ("nortes") may force many migrants (and even some residents) out of mountainous areas in the northern Neotropics, which may have been a factor in their evolution of broad habitat tolerances (Ramos 1988, Rappole et al. 1989). Periodic hurricanes may have had a similar effect on both residents and migrant species in the Caribbean region (Wunderle et al. 1992).

To summarize habitat use, migrants in summer are often specialized on particular seral stages or floristic associations, where safe nesting sites and usually dependably abundant food may be found. In winter, some migrant species are as restricted to particular habitats such as wet tropical forests as are some residents, but the majority occupy the most seasonal or disturbed habitats including edge and secondary growth. The ability of many Neotropical–Nearctic migrants to use a variety of native winter habitats, including often highly seasonal ones, explains why migrants tend to be widespread and abundant in human-modified tropical habitats (Petit et al. 1993). Considering the number of Neotropical–Nearctic migrant species that depend on particular native habitats, especially in summer, we conclude that habitat maintenance will continue to be an important management strategy (see recommendations below).

Resource-Use Patterns

An essential aspect of migratory behavior is the seasonal movement between habitats or regions, suggesting that an important aspect of the resource base of migrants might be resources that vary to such an extent seasonally or ephemerally that residents cannot dominate them completely. A number of studies tentatively support this idea. For example, the greatest breeding densities of migrants occur in those habitats and latitudes in which insect food resources undergo the greatest seasonal increases (MacArthur 1959, Willson 1976, Herrera 1978, O'Connor 1981), and where winter foods are seasonally reduced enough to decrease the carrying capacity of temperate residents (Herrera 1978, O'Connor 1981, 1990). The seasonal flush of spring foliage and attendant herbivorous arthropods in temperate deciduous forests is so dramatic that it supports a reversal of the typical latitudinal species-diversity gradient, i.e., the diversity of foliage-gleaning bird species in eastern North American forests increases to the north in the most seasonal forests (Rabenold 1993). In addition, several species of spruce woods warblers in North America have apparently evolved specializations to take advantage of periodic, spatially unpredictable, summer population irruptions of spruce budworm and other caterpillars (MacArthur 1958, Morse 1989). Migratory bird populations respond dramatically to caterpillar outbreaks (Zach and Falls 1975, Sealy 1979, Enemar et al. 1984, Holmes et al. 1986, 1991, Holmes 1988, 1990, Crawford and Jennings 1989, Rodenhouse and Holmes 1992).

In winter, many migrants forage at temporarily abundant, and often spatially unpredictable food resources such as large fruiting or flowering trees, reproductive flights of social insects such as termites, and recently burned grasslands where ants and termites are conspicuous (Willis 1966, Leck 1972, Karr 1976, Rappole et al. 1983, Leisler 1990, 1992). Quantitative dietary studies suggest that migratory flycatchers are more opportunistic than tropical resident species in Costa Rica where they coexist in rainforest (Sherry 1984, 1990). Greenberg et al. (1993) describe winter territories of Yellow-rumped Warblers established opportunistically around honeydew food sources (carbohydrate exudates from homopteran scale insects) in subtropical Mexican forests. These Yellow-rumped Warbler territories protect food from birds of several species, thereby establishing a relatively clear connection between food productivity and presence of migrant birds in winter. Many migrants that are insectivorous in summer switch to omnivorous diets in winter, including fruit, nectar, and many kinds of small insects opportunistically, especially in rainforest canopy, in dry forests, and otherwise disturbed habitats (Schwartz 1980, Stiles 1980, Rappole et al. 1983, 1992, Martin 1985, Morton and Greenberg 1989, Morse 1989, Levey and Stiles 1992).

On the other hand, some migrants are both behaviorally and even morphologically specialized in winter to exploit distinctive tropical resources (Morton 1971, 1992, Morton and Greenberg 1989). Examples include arthropods in dead leaf clusters (Worm-eating and Blue-winged Warblers), nectar at particular species of flowering trees and vines (Cape May and Tennessee Warblers), and fruit (Eastern Kingbirds, vireos, tanagers, many warblers). This apparent controversy over whether migrants are either opportunistic or specialized dietarily in winter, is resolved if we recognize that seasonal foods can be predictable, and that migrant species differ in their degree of specialization. Wintering migrants tend to feed on temporally abundant (i.e., seasonal) foods including nectar, fruit, and insects, some of which are seasonally predictable enough for evolution of morphological specializations (Sherry 1990), e.g., the fringed tongue of Cape May Warblers for collecting nectar, which occurs only in winter. A corollary of such morphological and even psychological specializations (Morton and Greenberg 1989, Greenberg 1990) is that particular migrants are constrained at least to some degree in their responses to changing environments, which makes them susceptible to habitat loss in either winter or summer. Once again, we must add that some species are more specialized dietarily in winter than others, and without a thorough review it may be difficult to make broad generalizations.

Increasingly, ecologists recognize that even the "stable tropics" vary dramatically, both seasonally and annually, and migrants often respond to these seasonally changing resources in similar ways to many residents (Schwartz 1980, Stiles 1980, Rappole et al. 1983). Levey and Stiles (1992) propose that a diet of fruit or nectar, which tends to be unpredictable in both space and time compared with many insect and other resources, may have preadapted diverse tropical resident species in diverse families to evolve opportunistic dietary and life-history strategies. Such strategies include movements between habitats and elevationally within a region, at one extreme, to the dramatic latitudinal movements of long-distance migrants with nonoverlapping breeding and wintering ranges at the other extreme (see also Ramos 1988). By this view, long-distance Neotropical–Nearctic migrants are simply the endpoint of a spectrum of movement strategies evolved to take advantage of seasonally predictable, but temporally and spatially patchy resources.

Thus even though migrants are ecological opportunists at a diverse range of spatial scales, they are nonetheless well adapted to tropical conditions, often competing effectively with closely related residents for seasonal foods. Migratory life-histories should be favored by seasonally changing environments, both within the tropics and temperate zone, which preclude resident populations in either region from expanding totally each season (in part because of their low fecundity) into available environments. The late dry season, which occurs typically from January to March in the northern Neotropics, could be particularly important in influencing habitat preferences of wintering migrants because some habitats maintain resources and suitable foraging conditions better than other habitats (Poulin et al. 1993, Parrish and Sherry 1994, Lefebvre et al. 1994, Sherry and Holmes 1995). Hurricanes and periodic droughts in the Caribbean region (see above) may reduce residents to populations too low to exclude the hordes of migrants arriving seasonally at a time when resources are often relatively abundant in the late rainy season (September–November).

To summarize life-history characteristics of migrant birds, we need to revise our conception of what these birds are based on their evolved traits such as their life histories. Many migrant species can be characterized as demographically similar to, and evolutionarily derived from, resident tropical species. They thus spend part of their annual cycle (winter) in tropical habitats where they can maintain sufficiently high survival. They then migrate to breeding habitats where they can maintain a higher fecundity than they could as tropical residents, but probably not as high as that of many temperate residents. The latter have relatively more opportunities for laying large clutches or double-brooding by defending the best early-season nesting and feeding sites. Migrants appear to reach greatest abundance, both in summer and winter, where resources (food year-round, safe nesting sites in summer) vary seasonally to such an extent as to preclude exploitation by resident bird communities. These resources are often, but not always, unpredictable in space, necessitating broad movements on the part of the migrants even within a season, and in some cases necessitating habitat specialization and morphological specializations to exploit these seasonal resources efficiently. Migrants' niches (use of both resources and habitats) are probably shaped in general by competition from both temperate residents and tropical residents (Bennett 1980, Ricklefs 1992). Migrants are thus uniquely adapted to depend on the most seasonal and perhaps unpredictable resources on a seasonal planet. Even though our review reveals a few tentative distinctions between migrants and residents in terms of life histories, it also emphasizes the considerable overlap of migrants with both tropical and higher-latitude resident birds in terms of morphological characteristics, demography, habitat-use patterns, and resource specialization (Levey and Stiles 1992, Ricklefs 1992). This difficulty of distinguishing long-distance migrants as a group from other birds results from the diversity of migrants' life histories, making "migrants" a poor ecological category for simple management strategies (Hutto 1992). Both these images of migrants—as a unique life-history solution to seasonal environments and as a diverse assemblage of unique species adapted to diverse habitats and resources—have important management implications.

MANAGEMENT RECOMMENDATIONS

The information reviewed above about populations of long-distance migratory bird species supports the hypothesis that habitats for Neotropical migrants may be limiting in both summer and winter, as well as during migration. These findings have important implications for conservation and management efforts, which we develop below in the form of seven recommendations (see also Sherry and Holmes 1993).

1. Most importantly, management policy must include habitat necessary to maintain populations of migrant birds in summer breeding areas, in wintering quarters, and along migratory routes. Migrants compete for high-quality habitat at essentially all times of year, and thus significant loss or deterioration of habitat at any major part of the annual cycle could lead to population declines.

2. Management should emphasize habitats as landscapes where migrants can sustain their own populations based on fecundity and survival. These species cannot be treated as independent entities distinguishable from the habitat in which they evolved. Viewing a species outside the context of its habitat is particularly dangerous when it leads to arbitrary management targets (such as numbers of nest boxes or snags). Such targets too often address the symptoms of a population decline (e.g., loss of nest sites) rather than underlying ecological causes such as increased nest parasitism, increased nest predators due to habitat fragmentation, or insufficient time for regeneration of new habitat. Instead, management should focus on sustainable habitat quality and quantity necessary to support particular species. If a population has already reached threatened or endangered status, any action to increase the population should of course be considered, but actions such as putting out nest boxes or cowbird control programs should be viewed as temporary measures until a sufficient amount of new habitat can be made available.

Our review of migrants' life histories leads to a view of their habitats as entire landscapes characterized by inherent variability and continuous change. Change is a part of normal ecosystem processes. Habitats for migrant birds should not be viewed simply as assemblages of snags or plants with distinctive floristic and physiognomic characteristics, but as ecosystems capable of sustaining complex processes of disturbance, regeneration, and seral development in various ways. Many migrant species populations have evolved the ability to respond to variability in their habitats, including dramatic seasonal and year-to-year changes in food abundance (e.g., emergences of cicadas or aquatic insects; and irruptions of defoliating insects such as spruce budworm, saddled prominent caterpillar, douglas-fir tussock moth, and various loopers), nest predator populations, fires, and weather anomalies such as droughts, floods, and hurricanes. Whereas some resident species populations even in the tropics fluctuate dramatically (e.g., seabird populations responding to effects of the El Niño-Southern Oscillation), we expect such fluctuations to be the norm in migrant populations. Birds in general, but migrants in particular, have extraordinary dispersal capabilities to find newly created habitats, or to move from deteriorating ones. Migratory birds can move over global scales to exploit completely different breeding versus winter survival niches (Alerstam and Hogstedt 1982). This opportunistic use of seasonal environments may in some migrant species require the availability of continually changing habitats as a refuge from resident populations.

We have deliberately emphasized habitat management for Neotropical–Nearctic migrants, rather than population management. The former is certainly not a new concept, but it does emphasize what we hope is a continued shift in conservation policy away from managing species, such as the Spotted Owl or Red-Cockaded Woodpecker, as if they were somehow disconnected from their habitats. Migrant birds are intimately tied, both evolutionarily and ecologically, to their habitats, and habitat management will help preserve not only populations of these particular species, but also other inhabitants of these habitats including some species with greater home range requirements than migratory birds.

Research has hardly begun to explore another aspect of migrants, namely their possible roles in ecosystem function, e.g., their potential role in cropping insect populations to levels that facilitate plant growth (e.g., Marquis and Whelan 1994), or the role of migrant birds in seed dispersal and plant pollination (Morton and Greenberg 1989). Moreover, habitats are units of ecosystems, which function as refuges for biological control agents, and modify atmospheric and hydrological chemistry within limits suitable for life—to name just a couple of functions. Understanding such ecosystem functions of migratory birds will be necessary to manage habitats within the broader ecosystem context.

3. Management should identify a normal range of migrant population sizes, rather than target any one level of abundance. Temporarily low populations might be acceptable as long as new habitats become available quickly enough to rebuild populations, and as long as genetic factors such as inbreeding depression do not inhibit successful reproduction. Thus management strategies must be sufficiently flexible to accommodate the continuous changes inherent in the habitats exploited by migrants, and the resulting, but normal, population fluctuations. Important in this regard is the need for management flexibility, especially in view of global climate change projected during coming decades. Global change scenarios suggest that habitats we recognize at present will not only shift in location (potentially crossing present political boundaries and regional mandates), but will in some cases become completely unrecognizeable due to independently shifting ranges of plant species comprising those habitats (e.g., Botkin 1990). Successful management will require the ability to anticipate such changes, to reorganize management guidelines and priorities, to transfer responsibilities for management among political entities, and possibly even to help mobilize the political support necessary to reorganize some landscapes (e.g., adding to current park boundaries to increase the range of elevations and habitats, or to establish wildlife corridors). For example, if global warming eliminates present jack pine stands in Michigan, they will not simply be shifted northward, because of lack of suitable sandy soils (Botkin 1990). In this case, managers will need to consider relocating birds to radically new areas where large burns may be established to regenerate the young jack pine stands that Kirtland's Warblers need to reproduce. More than anything else, managers must come to accept uncertainty and change as natural and acceptable aspects of the ecosystems in which most animals thrive (Botkin 1990). Variability in habitat characteristics, such as the frequency and extent of environmental disturbances, may be the best argument yet for managing habitat-species complexes rather than managing any particular species per se. In a few cases it may be possible to use an indicator approach, i.e., manage

particular habitats for several species (e.g., bottomlands hardwoods in the southeastern United States, containing Swainson's Warbler, Yellow-throated Warbler, Northern Parula Warbler, Prothonotary Warbler, and Cerulean Warbler). Managing habitats rather than species has the benefit of conserving many other kinds of organisms as well as ecosystem processes.

4. Sufficient migrant habitat must be maintained to buffer against temporary, local habitat loss or disturbance. This point is derived from a knowledge of the demographic and life-history characteristics of migrants. Many migrant populations may be less resilient than populations of temperate resident species to temporary declines, i.e. slower to recover from declines (O'Connor 1992). Recent research on how migrants use habitats, particularly those in winter, has shown that not all migrants can be viewed as behaviorally plastic ecological generalists, i.e. temporary invaders of disturbed tropical habitats. In some cases, migrants are habitat specialists, using specific microhabitats (such as dead leaf clusters containing concealed arthropods) or specialized food resources such as nectar (Morton and Greenberg 1989). Environmental disasters such as habitat loss, drought, or hurricanes may thus devastate some migrant populations as much as, if not more than, resident bird populations. Managers must therefore manage habitats and landscapes such that migrant populations remain at a great enough total size and can spread across multiple landscape units, so that "normal" and natural, yet potentially catostrophic, local habitat disturbances do not annihilate entire species. This recommendation is particularly important considering the rapidity and scope of human-influenced habitat changes.

5. Management must often be tailored to the needs of individual species. Although seemingly contrary to the points above, there are times where some species are so rare that we must take every effort to boost their populations, even if this means targeting those particular habitats or even breeding birds in captivity. Migrants include a diverse set of species, each with its own particular habitat requirements, demography, and life history. Thus, some species may behave very differently from others in response to habitat changes and may require fundamentally different management considerations. Some migrant species are extraordinarily flexible in terms of habitats and diet (e.g., Yellow-rumped Warblers and American Redstarts) and these probably require little concern from managers at present. Other species are presently endangered in part because of stereotypical dependence on particular safe nesting sites, nesting materials, foods, or other resources (Morse 1989). A good example may be Bachman's Warbler, as we mentioned above, even though we cannot presently resolve whether summer or winter habitats were more important (see Remsen 1986, Terborgh 1989).

6. Managers must distinguish between quantity and quality of habitat available for Neotropical–Nearctic migrants. Assessment of habitat availability or quantity is of primary importance. On a first approximation, the total abundance of a migrant species is roughly proportional to the total area of suitable habitat available. Whereas quantities of some North American habitats have stabilized (e.g., eastern deciduous forest; Terborgh 1992), other habitats important to migrants are either declining at an alarming rate at present, or are already nearly annihilated (e.g., Texas plateau woodlands occupied by Golden-Cheeked Warblers); large, recently burned jack pine stands occupied by Kirtland's Warbler; old-growth riparian hardwoods occupied by Cerulean Warblers—Morse 1989). Habitat quantity will probably be monitored most efficiently in the future using remote sensing coupled with geographic information system technology, which are then associated with accurate ground-based data such as censuses (e.g. Powell et al. 1992).

Quality of habitats must be considered in addition to quantity. In the breeding season, some habitats are more suitable than others, as indicated by bird densities, age structures, mating success, and particularly breeding productivities, as discussed previously. Likewise, variation in quality of winter habitat for migrants is suggested by differences in density, sex- and age-distribution patterns, variation in body mass, and competition for "good" sites (Sherry and Holmes 1995). Information about habitat quality is crucial for management purposes because it helps establish priorities among habitats, and such information is necessary for models of how habitat conversion within a landscape will alter total population size of a species. Differences in population abundances among habitats tell us how populations change, but demographic information will tell why population densities change (food, predator or brood parasite abundance, etc.). Thus, management policies must also be based on demographic information (Table 4-1). We emphasize that a habitat is more than just a particular floristic, physiognomic, or seral stand of plants. It is a place where birds feed, locate mates, nest, and survive. Knowledge of diets, such as the use of fruit or nectar in winter, leads to the idea of manipulating densities of food plants such as fruit-bearing plants in the family Melastomataceae in montane forests of Central America or the Caribbean region. Thus managers must not forget the biology of the organisms involved, and in many species breeding and feeding ecology are poorly known (e.g., Cerulean Warbler, see Robbins et al. 1992b).

Habitat quality raises the issue of "source" and "sink" populations (see Pulliam 1988, Pulliam and Danielson 1991). A source population produces sufficient individuals to maintain local abundance and produce potential colonists of newly available habitat, whereas a sink population is maintained only by continual immigration of individuals from other, more productive habitats. Some populations of Neotropical migrants occupying fragmented woodland habitats, at least in parts of the midwestern United States, are presently sink populations (Gibbs and Faaborg 1990, Robinson 1992). Migrant species breeding in the White Mountains of New Hampshire, on the other hand, appear to be source populations, benefitting from extensive, nonfragmented stands of northern hardwoods forest (Holmes et al. 1992, Sherry and Holmes 1992). Using Black-throated Blue Warblers in the Hubbard Brook Experimental Forest as an example, and assuming annual survival is between 50% and 70% and that fledglings survive to the start of the next breeding season about half as well as adults (see life-history section for references), then 1.7–4 fledglings are needed to replace adults lost (Fig. 4-5). Holmes et al. (1992) found that Black-throated Blue Warblers fledged an average of 4.3 young per female per season (range = 3.5–4.9) over a period of 4 years, indicating more than enough fledglings were produced to maintain the local population. However, even in the White Mountains, habitats are patchy, and in some, species such as the Black-throated Blue Warbler produce insufficient young to maintain themselves locally (Holmes et al. 1995). Ideally, managers need to be aware of which habitats are sources and which are sinks, so as to increase the proportion of source habitats in a landscape in which a population is declining. However, even sink habitats can help stabilize population dynamics by maintaining a pool of individuals that will rapidly colonize nearby preferred habitats when individuals there disappear for whatever reason (Bernstein et al. 1991).

Another reason to distinguish quality from quantity of habitats is their potentially independent effects on bird populations. A cowbird-elimination program begun in Kirtland's Warbler habitat in 1971, for example, dramatically increased nestling productivity (see Morse 1989, Fig. 11-2), but

Table 4-1. Important demographic parameters of bird populations useful in assessing habitat quality.

Fecundity

Clutch size
Number of annual broods (many migrants single-brooded but some double-brooded)
Nesting (fledging) success
Mass at fledging, or other index to postfledging survival
Age of first breeding (by sex)
Mating success

Survival

Annual survival (summer-to-summer, or winter-to-winter; see Holmes and Sherry 1992)
Oversummer survival
Overwinter survival

Other related parameters

Age structure (e.g., proportion of yearlings: adults)
Sex ratio
Dispersal distances (by age or sex)

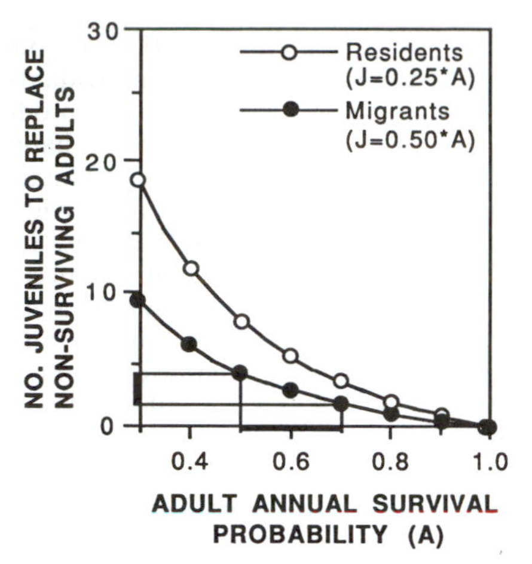

Figure 4-5. Number of juveniles needed to replace adults lost annually in a steady-state population of typical temperate resident and Neotropical migrant species. Annual mortality for juveniles is assumed to be one-quarter that of adults for residents and one-half that of adults for migrants (after Greenberg 1980). The heavy line on the ordinate is the projected range of fledgling productivity (1.7–4 fledglings per pair per season) of a "source" population, in a migrant such as Black-throated Blue Warbler with annual adult survival of 50–70% as shown by the heavy line projected on to the abscissa. Ordinate given by expression $2 * (1 - A)/J$, where A = adult annual survival probability, J = juvenile annual survival probability (see Rodenhouse and Holmes 1992, and text).

had little effect on the population. Several factors probably contributed to this result, including the difficulty of many birds finding mates (Probst 1986), but the biggest factor was almost certainly the limited quantity of habitat, as shown by the effect of newly burned jack pine stands that are currently boosting the population (Mayfield 1993).

Finally, it is becoming increasingly clear that many migrants compete for quality (source) habitats in both summer and winter (e.g., Sherry and Holmes 1989, Marra et al. 1993), suggesting that they are limited in supply relative to demand year-round. In such habitats, individuals behaviorally limit the density of birds sustainable per unit of habitat (Newton 1992). In such cases, management plans that focus on expanding the availability of habitat will be far more effective than trying to push densities above these habitat-specific carrying capacities.

Habitat quality is best assessed by monitoring a population's demography (Table 4-1), but this requires a substantial commitment of time, effort, and trained personnel. Ralph et al. (1992) provide an up-to-date manual on standardized methods and information necessary to study bird populations in the field, such as methods to quantify seasonal productivity of offspring using nest-monitoring studies. It may be possible for managers to involve scientists and amateur bird enthusiasts in the process of monitoring migrants demographically, and not just numerically. It may also be possible to develop effective short cuts in assessing habitat quality. Density of birds may be on average an accurate index of habitat quality, even if some exceptions to this generalization are known (Sherry and Holmes 1995).

7. Safeguarding the annual range of habitats necessary to maintain viable migrant populations will require extraordinary communication and coordination among managers, scientists, and the public across international borders. Migrants illustrate particularly well the adage that a chain is only as strong as its weakest link. We thus cannot overemphasize the importance of habitat preservation and management year-round, i.e., throughout the entire breeding, wintering, and migratory passage ranges, since most migrant populations will be sensitive to loss of habitat at any part of the annual cycle. No matter how much money goes into preserving habitat in the breeding range, a population could still go extinct due to deterioration of its wintering habitat, and vice versa. What little we know presently about the biology of these birds clearly emphasizes the importance of simultaneous breeding and wintering season population limitation for these species.

We thus urge that those vested with managing these populations increase communication and collaboration with scientists and land managers in all countries where these species spend parts of the year. Partnerships organized to span the geographical range of particularly threatened or endangered migrant species, involving North American and Latin American/Caribbean governments, resource-management agencies, private conservation organizations, or scientists in tropical countries must be encouraged. Such formal partnerships, including "sister forests," would be particularly productive if they coupled groups working on the same threatened or endangered species at different times of the year, such as the Kirtland Warbler Recovery Team with agencies in the Bahamas.

We have stressed repeatedly the importance of conserving large tracts of quality habitat throughout the year to safeguard healthy populations of Neotropical–Nearctic migrants. This is, of course, a far more complicated task than it sounds for a variety of reasons besides just the global extent of the habitats under consideration. This task will require maintaining not only large quantities of habitat, but also quality habitat in terms of the potential fecundity and survival probabilities of the birds. Thus, monitoring of abundances and demographic characteristics of populations must be extensive and accurate. Radically creative solutions may be required to act on the information so gathered because much habitat is already occupied or under pressure to be used by humans in ways that are not necessarily compatible with the birds' requirements. Conservation of habitats in the wintering ranges of migrant birds will be particularly important, because loss of habitat in the tropics continues unabated. Thus it is difficult to escape the conclusion that tropical wintering habitat will become limiting to at least some species in the near future, if it is not occurring already (Terborgh 1980, 1989, Morton 1992, Rappole et al. 1992, Robbins et al. 1992b, Sherry and Holmes 1995). Declines in habitat quality in North America have also severely affected Neotropical migrant populations and will continue to do so as land use changes. Preliminary efforts are underway to assess the potential vulnerability of species and habitats most threatened in winter (Morton 1992) and year-round (Terborgh 1989, Reed 1992), and such efforts need to be refined and expanded.

ACKNOWLEDGMENTS

Our research on Neotropical migrant birds has been supported by the National Science Foundation, Tulane University, and Dartmouth College; and was facilitated by cooperation of the Hubbard Brook Experimental Forest of the Northeast Forest Experiment Station, US Forest Service, and various individuals and government agencies in Jamaica, West Indies. We gratefully acknowledge all of their help and support. The manuscript was written in part while TWS was on sabbatical leave at the Wildlife Department, University of Maine, where the support and discussions with Raymond O'Connor and William Glanz were particularly helpful. We also thank Deborah Finch, Russ Greenberg, Dick Hutto, Pete Marra, Tom Martin, and an anonymous reviewer for helpful comments on the manuscript.

LITERATURE CITED

Alerstam, T., and G. Hogstedt. 1982. Bird migration and reproduction in relation to habitats for survival and breeding. Ornis. Scand. 13:25–37.

Ambuel, B., and S. A. Temple. 1983. Area-dependent changes in the bird communities and vegetation of southern Wisconsin forests. Ecology 64:1057–1068.

Andrén, H. 1990. Despotic distribution, unequal reproductive success, and population regulation in the jay *Garrulus glandarius* L. Ecology 71:1796–1803.

Andrén, H., and P. Angelstam. 1988. Elevated predation rates as an edge effect in habitat islands: experimental evidence. Ecology 69: 544–547.

Arcese, P., J. N. M. Smith, W. M. Hochachka, C. M. Rogers, and D. Ludwig. 1992. Stability, regulation, and the determination of abundance in an insular Song Sparrow population. Ecology 73:805–822.

Arvidsson, B. L., and P. Klaesson. 1984. Removal experiments in a population of willow warblers *Phylloscopus trochilus* in mountain birch forest, in Ammarnäs, Swedish Lapland. Ornis Scand. 15:63–66.

Askins, R. A., J. F. Lynch, and R. Greenberg. 1990. Population declines in migratory birds in eastern North America. Curr. Ornithol. 7:1–57.

Baillie, S. R., and W. J. Peach. 1992. Population limitation in Palearctic–African migrant passerines. Ibis 134 (Suppl. 1): 120–132.

Bates, J. M. 1992. Winter territorial behavior of gray vireos. Wilson Bull. 104:425–433.

Begon, M., and M. Mortimer. 1986. Population ecology: a unified study of animals and plants, 2nd ed. Blackwell Scientific Publications, Oxford, United Kingdom.

Bennett, S. E. 1980. Interspecific competition and the niche of the American Redstart (*Setophaga ruticilla*) in wintering and breeding communities. Pp. 319–335 *in* Migrants in the Neotropics: ecology, behavior, distribution and conservation (A. Keast and E. S. Morton, eds). Smithsonian Institution Press, Washington DC.

Bernstein, C., J. R. Krebs, and A. Kacelnik. 1991. Distribution of birds amongst habitats: theory and relevance to conservation. Pp. 317–345 *in* Bird population studies: relevance to conservation and management (C. M. Perrins, J.-D. Lebreton, and G. J. M. Hirons, eds). Oxford University Press, New York.

Best, L. B. 1977. Territory quality and mating success in the field sparrow (*Spizella pusilla*). Condor 79:192–204.

Blake, J. G., G. J. Niemi, and J. M. Hanowski. 1992. Drought and annual variation in bird populations. Pp. 419–430 *in* Ecology and conservation of Neotropical migrant landbirds (J. M. Hagan, III and D. W. Johnston, eds). Smithsonian Institution Press, Washington, DC.

Blondel, J., G. Gosler, J.-D. Lebreton, and R. McCleery (eds). 1990. Population biology of passerine birds. Springer-Verlag, Berlin.

Böhning-Gaese, K., M. L. Taper, and J. H. Brown. 1993. Are declines in North American insectivorous songbirds due to causes on the breeding range? Conserv. Biol. 7:76–86.

Botkin, D. 1990. Discordant harmonies: a new ecology for the twenty-first century. Oxford University Press, New York.

Brawn, J. D., and R. P. Balda. 1988. Population biology of cavity nesters in northern Arizona: do nest sites limit breeding densities? Condor 90:61–71.

Breitwisch, R. 1989. Mortality patterns, sex ratios, and parental investment in monogamous birds. Curr. Ornithol. 6:1–50.

Brittingham, M. C., and S. A. Temple. 1983. Have cowbirds caused forest songbirds to decline? Bioscience 33:31–35.

Brown, J. L. 1969a. Territorial behavior and population regulation in birds. Wilson Bull. 81:293–329.

Brown, J. L. 1969b. The buffer effect and productivity in tit populations. Amer. Natur. 103:347–354.

Carey, M., and V. Nolan. 1975. Polygyny in indigo buntings: a hypothesis tested. Science 190: 1296–1297.

Cox, G. W. 1985. The evolution of avian migration systems between temperate and tropical regions of the New World. Amer. Natur. 126:451–474.

Crawford, H. S., and D. T. Jennings. 1989. Predation by birds on spruce budworm *Choristoneura fumiferana*: functional, numerical, and total responses. Ecology 70:152–163.

Dhondt, A. A. 1971. The regulation of numbers in Belgian populations of Great Tits. Proc. Adv. Study Inst. (Osterbeek 1970): 532–547.

Dhondt, A. A., B. Kempenaers, and F. Adriaensen. 1992. Density-dependent clutch size caused by habitat heterogeneity. J. Anim. Ecol. 61:643–648.

Ekman, J. 1984a. Density-dependent seasonal mortality and population fluctuations of the temperate-zone willow tit (*Parus montanus*). J. Anim. Ecol. 53:119–134.

Ekman, J. 1984b. Stability and persistence of an age-structured avian population in a seasonal environment. J. Anim. Ecol. 53:135–146.

Elgood, J. H., R. E. Sharland, and P. E. Ward. 1966. Palearctic migrants in Nigeria. Ibis 108:84–116.

Emlen, J. 1973. Territorial aggression in wintering warblers at Bahama Agave blossoms. Wilson Bull. 85:71–74.

Enemar, A., L. Nilsson, and B. Sjostrand. 1984. The composition and dynamics of the passerine bird community in a subalpine birch forest, Swedish Lapland: a 20-year study. Ann. Zool. Fennica 21:321–338.

Finch, D. 1991. Population ecology, habitat requirements, and conservation of Neotropical migratory birds. Gen. Tech. Rep. RM-205. USDA, Forest Serv. Rocky Mt. Forest Range Exp. Sta., Fort Collins, CO, 26 p.

Freemark, K., and B. Collins. 1992. Landscape ecology of birds breeding in temperate forest fragments. Pp. 443–454 *in* Ecology and conservation of Neotropical migrant landbirds (J. M. Hagan, III and D. W. Johnston, eds). Smithsonian Institution Press, Washington, DC.

Fretwell, S. 1986. Distribution and abundance of the Dickcissel. Curr. Ornithol. 4:211–242.

Fretwell, S., and H. L. Lucas, Jr. 1970. On territorial behavior and other factors influencing habitat distribution in birds. I. Theoretical development. Acta Biotheor. 19:16–36.

Fretwell, S. D. 1972. Populations in a seasonal environment. Princeton University Press, Princeton, NJ.

Gauthier, G. 1987. The adaptive significance of territorial behavior in breeding Buffleheads: a test of three hypotheses. Anim. Behav. 35:348–360.

Gauthreaux, S. A., Jr. 1978. The ecological significance of behavioural dominance. Perspec. Ethol. 3:17–54.

Gibbs, J. P., and J. Faaborg. 1990. Estimating the viability of Ovenbird and Kentucky Warbler populations in forest fragments. Conserv. Biol. 4:193–196.

Gilpin, M. E., and M. E. Soulé. 1986. Minimum viable populations: processes of species extinctions. Pp. 19–34 *in* Conservation biology: the science of scarcity and diversity (Soulé, M. E., ed). Sinauer Associates, Sunderland, MA.

Gorski, L. J. 1969. Traill's Flycatchers of the 'fitz-bew' songform wintering in Panama. Auk 86:745–747.

Gorski, L. J. 1971. Traill's Flycatchers of the 'fee-bee-o' songform wintering in Peru. Auk 88:429–431.

Greenberg, R. 1979. Body size, breeding habitat, and winter exploitation systems in *Dendroica.* Auk 96:756–766.

Greenberg, R. 1980. Demographic aspects of long-distance migration. Pp. 493–504 *in* Migrants in the Neotropics: ecology, behavior, distribution and conservation (A. Keast and E. S. Morton, eds). Smithsonian Institution Press, Washington DC.

Greenberg, R. 1981. Dissimilar bill shapes in New World tropical versus temperate forest foliage-gleaning birds. Oecologia (Berlin) 49:143–147.

Greenberg, R. 1984. The winter exploitation systems of Bay-breasted and Chestnut-sided warblers in Panama. Univ. Calif. Publ. Zool. 116:1–106.

Greenberg, R. 1985. The social behavior and foraging ecology of Neotropical migrants in the non-breeding season. Acta XVIII Congr. Internat. Ornithol: 648–653.

Greenberg, R. 1986. Competition in migrant birds in the nonbreeding season. Curr. Ornithol. 3:281–307.

Greenberg, R. 1990. Ecological plasticity, neophobia, and resource use in birds. Stud. Avian Biol. 13:431–437.

Greenberg, R. 1992. Forest migrants in non-orest habitats on the Yucatan Peninsula. Pp. 273–286 *in* Ecology and conservation of Neotropical migrant landbirds (J. M. Hagan, III and D. W. Johnston, eds). Smithsonian Institution Press, Washington, DC.

Greenberg, R., C. M. Caballero, and P. Bichier. 1993. Defense of homopteran honeydew by birds in the Mexican highlands and other warm temperate forests. Oikos 68:519–524.

Grzybowski, J. A., R. B. Clapp, and J. T. Marshall, Jr. 1986. History and population status of the Black-capped Vireo in Oklahoma. Amer. Birds 40:1151–1161.

Hagan, J. M., and D. W. Johnston (eds). 1992. Ecology and conservation of Neotropical migrant landbirds. Smithsonian Institution Press, Washington, DC.

Hejl, S. J., J. Verner, and R. P. Balda. 1988. Weather and bird populations in true fir forests of the Sierra Nevada, California. Condor 90:561–574.

Hensley, M. M., and J. B. Cope. 1951. Further data on removal and repopulation of the breeding birds in a spruce-fir forest community. Auk 68:483–493.

Herrera, C. 1978. Ecological correlates of residence and non-residence in a mediterranean passerine bird community. J. Anim. Ecol. 47: 871–890.

Holmes, R. T. 1988. Community structure, population fluctuations, and resource dynamics of birds in temperate deciduous forests. Acta XIX Congr. Internat. Ornithol.: 1318–1327.

Holmes, R. T. 1990. The structure of a temperate deciduous forest bird community: variability in time and space. Pp. 121–140 *in* Biogeography and ecology of forest bird communities (A. Keast, ed.). SPB Academic Publications, The Hague.

Holmes, R. T., and T. W. Sherry. 1988. Assessing population trends of New Hampshire forest birds: local vs. regional patterns. Auk 105:756–768.

Holmes, R. T., and T. W. Sherry. 1992. Site fidelity of migratory warblers in temperate breeding and Neotropical wintering area: implications for population dynamics, habitat selection, and conservation. Pp. 563–578 *in* Ecology and conservation of Neotropical migrant landbirds (J. M. Hagan, III and D. W. Johnston, eds). Smithsonian Institution Press, Washington, DC.

Holmes, R. T., T. W. Sherry, and F. W. Sturges. 1986. Bird community dynamics in a temperate deciduous forest: long-term trends at Hubbard Brook. Ecol. Monogr. 56:201–220.

Holmes, R. T., T. W. Sherry, and L. Reitsma. 1989. Population structure, territoriality, and overwinter survival of two migrant warbler species in Jamaica. Condor 91:545–561.

Holmes, R. T., T. W. Sherry, and F. W. Sturges. 1991. Numerical and demographic responses of temperate forest birds to annual fluctuations in their food resources. Acta XX Congr. Internat. Ornithol.: 1559–1567.

Holmes, R. T., T. W. Sherry, P. P. Marra, and K. E. Petit. 1992. Multiple-brooding and productivity of a Neotropical migrant, the Black-throated Blue Warbler (*Dendroica caerulescens*) in an unfragmented temperate forest. Auk 109:321–333.

Holmes, R. T., P. P. Marra, and T. W. Sherry. 1995. Habitat-specific demography of breeding Black-throated Blue Warblers (*Dendroica caerulescens*): implications for population dynamics. J. Anim. Ecol. (accepted for publication).

Hunter, M. L., Jr 1991. Introductory remarks: bird conservation at a landscape scale: seeing the world from a birds'-eye view. Acta XX Congr. Internat. Ornithol.: 2283–2285.

Hutto, R. T. 1985. Habitat selection by non-breeding migratory landbirds. Pp. 455–476 *in* Habitat selection in birds (M. L. Cody, ed.). Academic Press, New York.

Hutto, R. L. 1988. Is tropical deforestation responsible for the reported declines in Neotropical migrant populations? Amer. Birds 42:375–379.

Hutto, R. L. 1992. Habitat distributions of migratory landbird species in western Mexico. Pp. 221–239 *in* Ecology and conservation of Neotropical migrant landbirds (J. M. Hagan, III and D. W. Johnston, eds). Smithsonian Institution Press, Washington, DC.

Karr, J. R. 1976. On the relative abundance of migrants from the north temperate zone in tropical habitats. Wilson Bull. 88:433–458.

Karr, J. R., J. D. Nichols, M. K. Klimkiewicz, and J. D. Brawn. 1990. Survival rates of birds of tropical and temperate forests: will the dogma survive? Amer. Natur. 136:277–291.

Keast, A. 1980. Migratory Parulidae: what can species co-occurrence in the north reveal about ecological plasticity and wintering patterns? Pp. 457–476 *in* Migrants in the Neotropics: ecology, behavior, distribution and conservation (A. Keast and E. S. Morton, eds). Smithsonian Institution Press, Washington, DC.

Keast, A. 1990. Biogeography and ecology of forest bird communities. SPB Academic Publications, The Hague.

Keast, A., and E. S. Morton (eds). 1980. Migrants in the Neotropics: ecology, behavior, distribution and conservation. Smithsonian Institution, Washington, DC.

Kelsey, M. G. 1989. A comparison of the song and territorial behaviour of a long-distance migrant, the Marsh Warbler *Acrocephalus palustris*, in summer and winter. Ibis 131: 403–414.

Kluijver, H. N., and L. Tinbergen. 1953. Territory and regulation of density in titmice. Arch. Neerl. Zool. 10:265–289.

Lack, D. 1968. Bird migration and natural selection Oikos 19:1–9.

Lack, P. 1986. Ecological correlates of migrants and residents in a tropical African savanna. Ardea 74:111–119.

Lande, R. 1988. Genetics and demography in biological conservation. Science 241:1455–1460.

Lanyon, S. M., and C. F. Thompson. 1986. Site fidelity and habitat quality as determinants of settlement pattern in male painted buntings. Condor 88:206–210.

LeBreton, J.-D., and J. Clobert. 1991. Bird population dynamics, management, and conservation: the role of mathematical modelling. Pp. 105–125 *in* Bird population studies: relevance to conservation and management (C. M. Perrins, J.-D. Lebreton, and G. J. M. Hirons, eds). Oxford University Press, New York.

Leck, C. 1972. The impact of some North American migrants at fruiting trees in Panama. Auk 89:842–850.

Lefebvre, G., B. Poulin, and R. McNeil. 1992. Abundance, feeding behavior, and body condition of Nearctic warblers wintering in Venezuelan mangroves. Wilson Bull. 104: 400–412.

Lefebvre, G., B. Poulin, and R. McNeil. 1994. Temporal dynamics of mangrove bird communities in Venezuela with special reference to migrant warblers. Auk 111:405–415.

Leisler, B. 1990. Selection and use of habitat of wintering migrants. Pp. 156–174 *in* Bird migration (E. Gwinner, ed.). Springer-Verlag, Berlin.

Leisler, B. 1992. Habitat selection and coexistence of migrants and afrotropical residents. Ibis 134 (Suppl. 1): 77–82.

Levey, D., and F. G. Stiles. 1992. Evolutionary precursors of long-distance migration: resource availability and movement patterns in Neotropical landbirds. Amer. Natur. 140: 447–476.

Lomnicki, A. 1978. Individual differences between animals and the natural regulation of their numbers. J. Anim. Ecol. 47:461–475.

Lomnicki, A. 1980. Regulation of population density due to individual differences and patchy environment. Oikos 35:185–193.

Lopez Ornat, A. L., and R. Greenberg. 1990. Sexual segregation by habitat in migratory warblers in Quintana Roo, Mexico. Auk 107:539–543.

Lynch, J. F. 1989. Distribution of overwintering Nearctic migrants in the Yucatan Penninsula. 1. general patterns of occurrence. Condor 91:515–544.

Lynch, J. F. 1992. Distribution of overwintering Neartic migrants in the Yucatan Peninsula, II: use of native and human-modified vegetation. Pp. 178–196 *in* Ecology and conservation of Neotropical migrant landbirds (J. M. Hagan, III and D. W. Johnston, eds). Smithsonian Institution Press, Washington, DC.

Lynch, J. F., and D. F. Whigham. 1984. Effects of forest fragmentation on breeding bird communities in Maryland, USA. Biol. Conserv. 28:287–324.

Lynch, J. F., E. S. Morton, and M. E. Van der Voort. 1985. Habitat segregation between the sexes of wintering Hooded Warblers (*Wilsonia citrina*). Auk 102:714–721.

MacArthur, R. H. 1958. Population ecology of some warblers of northeastern coniferous forests. Ecology 39:599–619.

MacArthur, R. H. 1959. On the breeding distribution patterns of North American migrant birds. Auk 76:318–325.

Marquis, R. J., and C. J. Whelan. 1994. Insectivorous birds increase growth of white oak through consumption of leaf-chewing insects. Ecology 75:2007–2014.

Marra, P. P., T. W. Sherry, and R. T. Holmes. 1993. Territorial exclusion by a long-distance migrant warbler in Jamaica: a removal experiment with American redstarts (*Setophaga ruticilla*). Auk 110:565–572.

Marshall, J. T. 1988. Birds lost from a giant sequoia forest during fifty years. Condor 90:359–372.

Martin, T. E. 1980. Diversity and abundance of spring migratory birds using habitat islands on the Great Plains. Condor 82:430–439.

Martin, T. E. 1985. Selection of second-growth woodlands by frugivorous migrating birds in Panama: an effect of fruit size and plant density. J. Tropical Ecol. 1:157–170.

Martin, T. E. 1987. Food as a limit on breeding birds: a life history perspective. Ann. Rev. Ecol. Syst. 19:453–487.

Martin, T. E. 1988. Processes organizing open-nesting bird assemblages: competition or nest predation? Evol. Ecol. 2:37–50.

Martin, T. E. 1992. Breeding productivity considerations: what are the appropriate habitat features for management? Pp. 455–473 *in* Ecology and conservation of Neotropical migrant landbirds (J. M. Hagan, III and D. W. Johnston, eds). Smithsonian Institution Press, Washington, DC.

Martin, T. E. 1993a. Nest predation among vegetation layers and habitat types: revising the dogmas. Amer. Natur. 141:897–913.

Martin, T. E. 1993b. Nest predation and nest sites: new perspectives on old patterns. BioScience 43:1–18.

Martin, T. E. 1995. Variation and covariation of life history traits in birds in relation to nest sites, nest predation, and food. Ecology (in press).

Matthysen, E. 1990. Behavioral and ecological correlates of territory quality in the Eurasian nuthatch (*Sitta europea*). Auk 107:86–95.

Mayfield, H. F. 1978. Brood parasitism: reducing interactions between Kirtland's Warblers and Brown-headed Cowbirds. Pp. 85–91 *in* Endangered birds: management techniques for preserving threatened species (S. A. Temple, ed.). University of Wisconsin Press, Madison, WI.

Mayfield, H. F. 1993. Kirtland's warblers benefit from large forest tracts. Wilson Bull. 105: 351–353.

Mönkkönen, M., P. Helle, and D. Welsh. 1992. Perspectives on Palaearctic and Nearctic bird migration; comparisons and overview of life-history and ecology of migrant passerines. Ibis 134 (Suppl. 1): 7–13.

Möller, A. P. 1987. Intruders and defenders in avian breeding territories: the effect of sperm competition. Oikos 48:47–54.

Moore, F. R., and T. R. Simons. 1992. Habitat suitability and stopover ecology of Neotropical landbird migrants. Pp. 345–355 *in* Ecology and conservation of Neotropical migrant landbirds (J. M. Hagan, III and D. W. Johnston, eds). Smithsonian Institution Press, Washington, DC.

Moreau, R. E. 1972. The Palaearctic–African bird migration systems. Academic Press, London, England.

Morel, G., and F. Bourlière. 1952. Relations écologiques des avifaunes sédentaires et migratrices dans une savane sahélienne du bas Sénégal. Terre Vie 190:371–393.

Morris, D. W. 1987. Tests of density-dependent habitat selection in a patchy environment. Ecol. Monogr. 57:269–281.

Morris. D. W. 1989. Density-dependent habitat selection: testing the theory with fitness data. Evol. Ecol. 3:80–94.

Morse, D. H. 1980a. Behavioral mechanisms in ecology. Harvard University Press, Cambridge, MA.

Morse, D. H. 1980b. Population limitation: breeding or wintering grounds? Pp. 505–516 *in* Migrants in the Neotropics: ecology, behavior, distribution and conservation (A. Keast and E. S. Morton, eds). Smithsonian Institution Press, Washington, DC.

Morse, D. H. 1989. American Warblers. Harvard University Press, Cambridge, MA.

Morton, E. S. 1971. Food and migration habits of the Eastern Kingbird in Panama. Auk 88: 925–926.

Morton, E. S. 1980. Adaptations to seasonal changes by migrant landbirds in the Panama Canal Zone. Pp. 437–453 *in* Migrants in the Neotropics: ecology, behavior, distribution and conservation (A. Keast and E. S. Morton, eds). Smithsonian Institution Press, Washington, DC.

Morton, E. S. 1990. Habitat segregation by sex in the hooded warbler: experiments on proximate causation and discussion of its evolution. Amer. Natur. 135:319–333.

Morton, E. S. 1992. What do we know about the future of migrant landbirds? Pp. 579–589 *in* Ecology and conservation of Neotropical migrant landbirds (J. M. Hagan, III and D. W. Johnston, eds). Smithsonian Institution Press, Washington, DC.

Morton, E. S., and R. Greenberg. 1989. The outlook for migratory birds: "future shock" for birders. Amer. Birds 43:178–183.

Morton, E. S., J. F. Lynch, K. Young, and P. Mehlhop. 1987. Do male hooded warblers exclude females from nonbreeding territories in tropical forest? Auk 104: 133–135.

Murdoch, W. W. 1994. Population regulation in theory and practice. Ecology 75:271–287.

Myers, J. P. 1987. Conservation strategy for migratory species. Amer. Sci. 75:18–26.

Newton, I. 1986. The Sparrowhawk. T&AD Poyser, Calton, England.

Newton, I. 1992. Experiments on the limitation of bird numbers by territorial behaviour. Biol. Rev. 67:129–174.

Nisbet, E. C. T., and L. Medway. 1972. Dispersion, population ecology and migration of Eastern Great Reed Warblers *Acrocephalus orientalis* wintering in Malaysia. Ibis 114: 452–494.

Nolan, V. Jr. 1978. The ecology and behavior of the Prairie Warbler *Dendroica discolor*. Ornithol. Monogr. 26:1–595.

O'Connor, R. J. 1981. Comparisons between migrant and non-migrant birds in Britain. Pp. 167–195 *in* Animal migration (Aidley, D. J., ed.). Cambridge University Press, Cambridge England.

O'Connor, R. J. 1985. Behavioural regulation of bird populations: a review of habitat use in relation to migration and residency. Pp. 105–142 *in* Behavioural ecology: ecological consequences of adaptive behaviour (R. M. Sibly, and R. H. Smith, eds). Blackwell Scientific Press, Oxford, England.

O'Connor, R. J. 1990. Some ecological aspects of migrants and residents. Pp. 175–182 *in* Bird migration (E. Gwinner, ed.). Springer-Verlag, Berlin, Heidelberg.

O'Connor, R. J. 1992. Population variation in relation to migrancy status in some North American birds. Pp. 64–74 *in* Ecology and conservation of Neotropical migrant landbirds (J. M. Hagan, III and D. W. Johnston, eds). Smithsonian Institution Press, Washington, DC.

Parrish, J. D., and T. W. Sherry. 1994. Sexual habitat segregation by American Redstarts wintering in Jamaica: importance of resource seasonality. Auk 111:38–49.

Patterson, I. J. 1985. Territorial behaviour and the limitation of bird populations. Acta XVIII Congr. Internat. Ornithol. 770–773.

Payne, R. B. 1990. Natal dispersal and population structure in a migratory songbird, the Indigo Bunting. Evolution 45:49–62.

Perrins, C. M. 1980. Survival of young great tits. *Parus major*. Proc. XVII Internat. Ornithol. Congr.: 159–174.

Perrins, C. M., J.-D. Lebreton, and G. J. M. Hirons (eds). 1991. Bird population studies: relevance to conservation and management. Oxford University Press, New York.

Petit, D. R., J. F. Lynch, R. L. Hutto, J. G. Blake, and R. B. Waide. 1993. Management and conservation of migratory landbirds overwintering in the Neotropics. Pp. 70–92 *in* Status and management of Neotropical migratory birds (D. M. Finch and P. W. Stangel, eds). General Technical Report RM-229. US Department of Agriculture, Forest Service, Rocky Mountain Forest and Range Experiment Station, Fort Collins, CO.

Petit, L. J. 1991a. Adaptive tolerance of cowbird parasitism by Prothonotary Warblers: a consequence of nest-site limitation? Anim. Behav. 41:425–432.

Petit, L. J. 1991b. Experimentally induced polygyny in a monogamous bird species: prothonotary warblers and the polygyny threshold. Behav. Ecol. Sociobiol. 29:177–187.

Poulin, B., G. Lefebvre, and R. McNeil. 1993. Variations in bird abundance in tropical arid and semi-arid habitats. Ibis 135:432–441.

Powell, G. V. N., J. H. Rappole, and S. A. Sader. 1992. Neotropical migrant landbird use of lowland Atlantic habitats in Costa Rica: a test of remote sensing for identification of habitat. Pp. 287–298 *in* Ecology and conservation of Neotropical migrant landbirds (J. M. Hagan, III and D. W. Johnston, eds). Smithsonian Institution Press, Washington, DC.

Price, T. 1981. The ecology of the greenish warbler

Phylloscopus trochiloides in its winter quarters. Ibis 123:131–144.

Probst, J. R. 1986. A review of factors limiting the Kirtland's Warbler on its breeding grounds. Amer. Midl. Natur. 116:87–100.

Pulliam, H. R. 1987. On the evolution of density-regulating behavior. Perspect. Ethol. 7:99–124.

Pulliam, H. R. 1988. Sources, sinks, and population regulation. Amer. Natur. 132: 652–661.

Pulliam, H. R., and B. J. Danielson. 1991. Sources, sinks, and habitat selection: a landscape perspective on population dynamics. Amer. Natur. 137 (Suppl. 1): S50–S66.

Pulliam, H. R., and J. B. Dunning. 1987. The influence of food supply on local density and diversity of sparrows. Ecology 68:1009–1014.

Rabenold, K. N. 1993. Latitudinal gradients in avian species diversity and the role of long-distance migration. Curr. Ornithol. 10:247–274.

Ralph, C. J., G. R. Geupel, P. Pyle, T. E. Martin, and D. F. DeSante. 1992. Field methods for monitoring landbirds. USDA Forest Service, Arcata, CA.

Ramos, M. A. 1988. Eco-evolutionary aspects of bird movements in the northern Neotropical region. Proc. Internat. Ornithol. Congr. 19:251–293.

Rappole, J. H. 1991. Migrant birds in Neotropical forest: a review from a conservation perspective. Pp. 259–277 *in* Conserving migratory birds. (T. Salathé, ed.). ICBP technical publication no. 12. Cambridge, England.

Rappole, J. H., and E. S. Morton. 1985. Effects of habitat alteration on a tropical avian forest community. Pp. 1013–1021 *in* Neotropical ornithology (P. A. Buckley, M. S. Foster, E. S. Morton, R. S. Ridgely, and F. G. Buckley, eds). Ornithological Monographs 36, American Ornithologists' Union, Washington, DC.

Rappole, J. H., and D. W. Warner. 1976. Relationships between behavior, physiology, and weather in avian transients at a migration stopover site. Oecologia (Berlin) 26: 193–212.

Rappole, J. H., and D. W. Warner. 1980. Ecological aspects of migrant bird behavior in Veracruz, Mexico. Pp. 353–393 *in* Migrants in the Neotropics: ecology, behavior, distribution and conservation (A. Keast and E. S. Morton, eds). Smithsonian Institution Press, Washington, DC.

Rappole, J. H., E. S. Morton, T. E. Lovejoy, and J. L. Ruos. 1983. Nearctic avian migrants in the Neotropics. US Department of the Interior, Fish and Wildlife Service, Washington, DC.

Rappole, J. H., M. A. Ramos, and K. Winker, 1989. Wintering wood thrush movements and mortality in southern Veracruz. Auk 106: 402–410.

Rappole, J. H., E. S. Morton, and M. A. Ramos. 1992. Density, philopatry, and population estimates for songbird migrants wintering in Veracruz. Pp. 337–344 *in* Ecology and conservation of Neotropical migrant landbirds (J. M. Hagan, III and D. W. Johnston, eds). Smithsonian Institution Press, Washington, DC.

Reed, J. M. 1992. A system for ranking conservation priorities for Neotropical migrant birds based on relative susceptibility to extinction. Pp. 524–536 *in* Ecology and conservation of Neotropical migrant landbirds (J. M. Hagan, III and D. W. Johnston, eds). Smithsonian Institution Press, Washington, DC.

Reese, K. P., and J. A. Kadlec. 1985. Influence of high density and parental age on the habitat selection and reproduction of black-billed magpies. Condor 87:96–105.

Remsen, J. V., Jr. 1986. Was Bachman's Warbler a bamboo specialist? Auk 103:216–219.

Ricklefs, R. E. 1969. An analysis of nesting mortality in birds. Smithsonian Contrib. Zool. 9:1–48.

Ricklefs, R. E. 1992. The megapopulation: a model of demographic coupling between migrant and resident landbird populations. Pp. 537–548 *in* Ecology and conservation of Neotropical migrant landbirds (J. M. Hagan, III and D. W. Johnston, eds). Smithsonian Institution Press, Washington, DC.

Robbins, C. S. 1979. Effect of forest fragmentation on bird populations. Pp. 33–48 *in* Management of northcentral and northeastern forests for non-game birds (R. M. DeGraaf and K. E. Evans, eds). Gen. Tech. Rep. NC-51, North Central Forest Exp. Stn., US Forest Serv, St. Paul, MN.

Robbins, C. S., B. A. Dowell, D. K. Dawson, J. Colon, F. Espinoza, J. Rodriguez, R. Sutton, and T. Vargas. 1987. Comparison of Neotropical winter bird populations in isolated patches versus extensive forest. Acta Oecologica 8:285–292.

Robbins, C. S., J. R. Sauer, R. Greenberg, and S. Droege. 1989. Population declines in North American birds that migrate to the Neotropics. Proc. Natl. Acad. Sci. USA 86:7658–7662.

Robbins, C. S., B. A. Dowell, D. K. Dawson, J. A. Colón, R. Estrada, A. Sutton, R. Sutton, and D. Weyer. 1992a. Comparison of Neotropical migrant landbird populations wintering in tropical forest, isolated forest fragments, and agricultural lands. Pp. 207–220 *in* Ecology and conservation of Neotropical migrant landbirds (J. M. Hagan, III and D. W. Johnston, eds). Smithsonian Institution Press, Washington, DC.

Robbins, C. S., J. W. Fitzpatrick, and P. B. Hamel. 1992b. A warbler in trouble: *Dendroica cerulea*. Pp. 549–562 *in* Ecology and conservation of Neotropical migrant landbirds (J. M. Hagan, III and D. W. Johnston, eds). Smithsonian Institution Press, Washington, DC.

Robinson, S. K. 1992. Population dynamics of breeding birds in a fragmented Illinois landscape. Pp. 408–418 *in* Ecology and conservation of Neotropical migrant landbirds (J. M. Hagan, III and D. W. Johnson, eds.) Smithsonian Institution Press, Washington, DC.

Robinson, S. K., J. A. Grzybowski, S. I. Rothstein, M. C. Brittingham, L. J. Petit, and F. R. Thompson. 1993. Management implications of cowbird parasitism on Neotropical migrant songbirds. Pp. 93–102 *in* Status and management of Neotropical migratory birds (D. M. Finch and P. W. Stangel, eds), Gen. Tech. Rep. RM-229. USDA Forest Serv. Rocky Mt. Forest Range Exp. Sta., Fort Collins, CO.

Rodenhouse, N. L., and R. T. Holmes. 1992. Results of experimental and natural food reductions for breeding Black-throated Blue Warblers. Ecology 73:357–372.

Rodenhouse, N. L., L. B. Best, R. J. O'Connor, and E. K. Bollinger. 1993. Effects of temperate agriculture on Neotropical migrant landbirds. Pp. 280–295 *in* Status and management of Neotropical migratory birds (D. M. Finch and P. W. Stangel, eds), General Technical Report RM-229. USDA Forest Serv. Rocky Mt. Forest Range Exp. Sta., Fort Collins, CO.

Rosenzweig, M. L. 1985. Some theoretical aspects of habitat selection. Pp. 517–540 *in* Habitat selection in birds (M. Cody, ed.). Academic Press, New York.

Rosenzweig, M. L. 1991. Habitat selection and population interactions: the search for mechanism. Amer. Natur. 137 (Suppl.): S5–S28.

Ryel, L. A. 1981. Population change in the Kirtland's Warbler. Jack-Pine Warbler 59: 77–90.

Schwartz, P. 1964. The Northern Waterthrush in Venezuela. Living Bird 2: 169–184.

Schwartz, P. 1980. Some considerations on migratory birds. Pp. 31–36 *in* Migrants in the Neotropics: ecology, behavior, distribution and conservation (A. Keast and E. S. Morton, eds). Smithsonian Institution Press, Washington, DC.

Sealy, S. G. 1979. Extralimital nesting of Bay-breasted Warblers: response to Forest Tent Caterpillars? Auk 96:600–603.

Sherry, T. W. 1984. Comparative dietary ecology of sympatric insectivorous Neotropical flycatchers (Tyrannidae). Ecol. Monogr. 54: 313–338.

Sherry, T. W. 1990. When are birds dietarily specialized? Distinguishing ecological from evolutionary approaches. Stud. Avian Biol. 13:337–352.

Sherry, T. W., and R. T. Holmes. 1985. Dispersion patterns and habitat responses of birds in northern hardwoods forests. Pp. 283–309 *in* Habitat selection in birds (M. L. Cody, ed). Academic Press, New York.

Sherry, T. W., and R. T. Holmes. 1989. Age-specific social dominance affects habitat use by breeding American redstarts (*Setophaga ruticilla*): a removal experiment. Behav. Ecol. Sociobiol. 25:327–333.

Sherry, T. W., and R. T. Holmes. 1991. Population age structure of long-distance migratory passerine birds: variation in time and space. Acta XX Congr. Internat. Ornithol.: 1542–1556.

Sherry, T. W., and R. T. Holmes. 1992. Population fluctuations in a long-distance Neotropical migrant: demographic evidence for the importance of breeding season events in the American Redstart. Pp. 431–442 *in* Ecology and conservation of Neotropical migrant landbirds (J. M. Hagan, III and D. W. Johnston, eds). Smithsonian Institution Press, Washington, DC.

Sherry, T. W., and R. T. Holmes. 1993. Are populations of Neotropical migrant birds limited in summer or winter? Implications for management. Pp. 47–57 *in* Status and management of Neotropical migratory birds (D. M. Finch and P. W. Stangel, eds). Gen. Tech. Rep. RM-229. USDA Forest Serv. Rocky Mt. Forest Range Exp. Sta., Fort Collins, CO.

Sherry, T. W., and R. T. Holmes. 1995. Winter habitat limitation in Neotropical–Nearctic migrant birds: implications for population dynamics and conservation. Ecology 76 (in press).

Simberloff, D. 1988. The contribution of population and community biology to conservation science. Annu. Rev. Ecol. System. 19:473–511.

Sinclair, A. R. E. 1989. Population regulation in animals. Pp. 197–241 *in* Ecological concepts, the contribution of ecology to an understanding of the natural world. (J. M. Cherrett, ed.). 29th Symposium of the British Ecological

Society. Blackwell Scientific Publications, Oxford, England.

Sliwa, A., and T. W. Sherry. 1992. Surveying wintering warbler populations in Jamaica: point counts with and without broadcast vocalizations. Condor 94:924–936.

Sliwa, A., and T. W. Sherry, in preparation.

Slud, P. 1976. Geographic and climatic relationships of avifaunas with special reference to comparative distribution in the Neotropics. Smithsonian Contrib. Zool. 212:1–149.

Small, M. F., and M. L. Hunter. 1988. Forest fragmentation and avian nest predation in forested landscapes. Oecologica (Berlin) 76:62–64.

Stacier, C. A. 1992. Social behavior of the Northern Parula, Cape May Warbler, and Prairie Warbler wintering in second-growth forest in southwestern Puerto Rico. Pp. 308–320 *in* Ecology and conservation of Neotropical migrant landbirds (J. M. Hagan, III and D. W. Johnston, eds). Smithsonian Institution Press, Washington, DC.

Stewart, R. E., and J. W. Aldrich. 1951. Removal and repopulation of breeding birds in a spruce-fir forest community. Auk 68:471–482.

Stiles, F. G. 1980. Evolutionary implications of habitat relations between permanent and winter resident landbirds in Costa Rica. Pp. 421–436 *in* Migrants in the Neotropics: ecology, behavior, distribution and conservation (A. Keast and E. S. Morton, eds). Smithsonian Institution Press, Washington, DC.

Stutchbury, B. J. 1994. Competition for winter territories in a Neotropical migrant: the role of age, sex, and color. Auk 111: 63–69.

Sullivan, K. A. 1989. Predation and starvation: age-specific mortality in juvenile juncos. J. Anim. Ecol. 58:275–286.

Svensson, S. E. 1985. Effects of changes in tropical environments on the North European avifauna. Ornis Fenn. 62:56–63.

Temple, S. A. 1991. The role of dispersal in the maintenance of bird populations in a fragmented landscape. Acta XX Congr. Internat. Ornithol.: 2298–2305.

Temple, S. A., and J. R. Cary. 1988. Modeling dynamics of habitat-interior bird populations in fragmented landscapes. Conserv. Biol. 2:340–347.

Terborgh, J. W. 1980. The conservation status of Neotropical migrants: present and future. Pp. 21–30 *in* Migrants in the Neotropics: ecology, behavior, distribution and conservation (A. Keast and E. S. Morton, eds). Smithsonian Institution Press, Washington, DC.

Terborgh, J. W. 1989. Where have all the birds gone? Princeton University Press, Princeton, NJ.

Terborgh, J. W. 1992. Perspectives on the conservation of Neotropical migrant landbirds. Pp. 7–12 *in* Ecology and conservation of Neotropical migrant landbirds (J. M. Hagan, III and D. W. Johnston, eds). Smithsonian Institution Press, Washington, DC.

Thompson, C. F. 1977. Experimental removal and replacement of territorial male Yellow-Breasted Chats. Auk 94:107–113.

Tomialojc, L., and T. Wesolowski. 1990. Bird communities of the primeval temperate forest of Bialowieza, Poland. Pp. 141–165 *in* Biogeography and ecology of forest bird communities (A. Keast, ed.). SPB Academic Publishers, The Hague.

van Balen, J. H. 1980. Population fluctuations of the Great Tit and feeding conditions in winter. Ardea 68:143–164.

van der Have, T. M. 1991. Conservation of Palearctic–African migrants: are both ends burning? Trends Ecol. Evol. 6:308–310.

Villard, M.-A., K. Freemark, and G. Merriam. 1992. Metapopulation theory and Neotropical migrant birds in temperate forests: an empirical investigation. Pp. 474–482 *in* Ecology and conservation of Neotropical migrant landbirds (J. M. Hagan, III and D. W. Johnston, eds). Smithsonian Institution Press, Washington, DC.

Virolainen, M. 1984. Breeding biology of the Pied Flycatcher *Ficedula hypoleuca* in relation to population density. Ann. Zool. Fennica 21:187–197.

Whitcomb, R. F., C. S. Robbins, J. F. Lynch, F. L. Whitcomb, M. K. Klimkiewicz, and D. Bystrack. 1981. Effects of forest fragmentation on the avifauna of the eastern deciduous forest. Pp. 125–205 *in* Forest island dynamics in man-dominated landscapes (R. L. Burgess and D. M. Sharpe, eds). Springer-Verlag, New York.

Wiens, J. A. 1977. On competition and variable environments. Amer. Sci. 65:590–597.

Wilcove, D. S. 1985. Nest predation in forest tracts and the decline of migratory songbirds. Ecology 66: 1211–1214.

Wilcove, D. S., and S. K. Robinson. 1990. The impact of forest fragmentation on bird communities in Eastern North America. Pp. 319–331 *in* Biogeography and ecology of forest bird communities (A. Keast, ed.). SPB Academic Publishing, The Hague.

Wilcove, D. S., and R. F. Whitcomb. 1983. Gone with the trees. Natural History 9/83:82–91.

Wilcove, D. S., C. H. McLellan, and A. P. Dobson.

1986. Habitat fragmentation in the temperate zone. Pp. 237–256 *in* Conservation biology: the science of scarcity and diversity (M. E. Soulé, ed.). Sinauer Associates, Sunderland, MA.

Willis, E. O. 1966. The role of migrant birds at swarms of army ants. Living Bird 5:187–231.

Willson, M. F. 1976. The breeding distribution of North American migrant birds: a critique of MacArthur (1959). Wilson Bull. 88:582–587.

Winker, K., J. H. Rappole, and M. A. Ramos. 1990. Population dynamics of the Wood Thrush in southern Veracruz, Mexico. Condor 92:444–460.

Winstanley, D., R. Spencer, and K. Williamson. 1974. Where have all the Whitethroats gone? Bird Study 2:1–14.

Wunderle, J. M., Jr. 1992. Sexual habitat segregation in wintering Black-throated Blue Warblers in Puerto Rico. Pp. 299–307 *in* Ecology and conservation of Neotropical migrant landbirds (J. M. Hagan, III and D. W. Johnston, eds). Smithsonian Institution Press, Washington, DC.

Wunderle, J. M., Jr., D. J. Lodge, and R. B. Waide. 1992. Short-term effects of Hurricane Gilbert on terrestrial bird populations on Jamaica. Auk 109:148–146.

Yahner, R. H. 1988. Changes in wildlife communities near edges. Conserv. Biol. 2:333–339.

Zach, R., and J. B. Falls. 1975. Response of the ovenbird (Aves: Parulidae) to an outbreak of the spruce budworm. Can. J. Zool. 53:1669–1672.

Zumeta, D., and R. T. Holmes. 1978. Habitat shift and roadside mortality of Scarlet Tanagers during a cold, wet New England spring. Wilson Bull. 90:575–283.

5

HABITAT REQUIREMENTS DURING MIGRATION: IMPORTANT LINK IN CONSERVATION

FRANK R. MOORE, SIDNEY A. GAUTHREAUX, JR, PAUL KERLINGER, AND THEODORE R. SIMONS

STATEMENT OF PROBLEM

Conservation of Neotropical landbird migrants is complicated by the very life-history characteristic that permits these birds to exploit seasonal environments, namely migration. Choice of habitat must be made in Neotropical wintering quarters, in temperate breeding areas, and repeatedly during migration. Each of the habitats encountered during the migrant's annual cycle faces different threats of degradation and destruction (Gradwohl and Greenberg 1989, Askins et al. 1990).

Stopover habitat is defined here as "... an area with the combination of resources (like food, cover, water) and environmental conditions (temperature, precipitation, presence and absence of predators and competitors) that promotes occupancy by individuals of a given species (or population) and allows those individuals to survive ..." (Morrison et al. 1992) during passage. The importance of habitat during migration has been largely overlooked in our developing conservation strategy (Moore and Simons 1992, Berthold and Terrill 1991). Migration is a period of exceptional energy demand, and small landbird migrants generally do not deposit enough fat to fly without stopping between breeding and wintering areas (Berthold 1975, Blem 1980). Hence, availability of suitable habitats where depleted fat stores can be safely and rapidly replenished becomes critical to a successful migration. Yet, we know little about what types of habitats are most important during migration, where they occur, and how their distribution and abundance are changing as a result of development and land conversion. Nor do we know much about migrant–habitat relations (*sensu* Morrison et al. 1992).

Unless habitat requirements during migration are met, conservation measures that focus on temperate breeding grounds or Neotropical wintering areas will be compromised. Our objectives are threefold: (1) to document the importance of habitat as a critical link in the conservation of Neotropical landbird migrants, (2) to examine the importance of scale in understanding the ecology of landbird migrants during passage, and (3) to identify management guidelines specific to the migratory phase of the annual cycle.

CONSEQUENCES OF HABITAT USE DURING MIGRATION

Migration allows individuals to take advantage of different habitats as life-history requirements alter or as environments change seasonally (e.g., Greenberg 1980, Cox 1985, Levey and Stiles 1992), and "... will be favored if survival and reproduction are greater in a new habitat in spite of the risks of migrating" (Dingle 1980). The mortality associated with intercontinental migration, though difficult to estimate, may be substantial (Lack 1946, Moreau 1972, Ketterson and Nolan 1982), and yearling migrants suffer greater mortality than adult (after hatching year) migrants (Nisbet and Medway 1972, Alerstam 1978, Greenberg 1980, Ketterson and Nolan 1983, 1985, DeSante 1983, Ramos 1988). Potential costs of migration include: (1) high energy demand of transport (Berthold 1975, Blem 1980); (2) adjustment to unfamiliar habitats (Morse 1971); (3) conflicting demands between predator avoidance and food acquisition (Lindström 1989, Moore 1994); (4) competition with other migrants and resident birds for limited resources (Moore and Yong 1991, Carpenter 1993a,b); (5) unfavorable weather (Richardson 1978), and (6) orientation errors (Ralph 1978, Moore 1984, 1990, Alerstam 1990).

Although the effect of habitat use on fitness (i.e., survival and reproductive success) is difficult to estimate during migration, more immediate consequences of habitat use can be measured in relation to the energetic requirements of migration; how successfully migrants satisfy energy demand and meet contingencies en route depends in large measure on the habitat quality. Birds mobilize stored lipid (and sometimes protein reserves) for energy during migratory flight, so many individuals arrive in stopover habitat in a fat-depleted condition (e.g., Rappole and Warner 1976, Moore and Kerlinger 1987, Kuenzi et al. 1991, Winker et al. 1992a). Individuals that quickly restore fat loads to levels appropriate for the next stage of their passage will minimize time spent on migration.

Consider the consequences of loss of stopover habitat on landbird migrant populations (cf. Evans et al. 1991): densities will increase in remaining areas, which may intensify competition. Increased competition may reduce food availability and increase interference, lowering food intake rates and the rate of fat deposition, thereby slowing migration. Increased competition may also redistribute birds among habitats, with younger, less experienced migrants forced into poorer sites where mortality rates can be expected to be higher. A migrant that departs a stopover site with lower than usual fat stores has a smaller "margin of safety" to buffer the effect of adverse weather on the availability of food supplies at the next stopover (see Moore and Kerlinger 1991). If a migrant stays longer than usual at a stopover site and does not make up lost time, arrival on the wintering or breeding grounds is necessarily delayed (cf. Francis and Cooke 1986; Lavee and Safriel 1989). Migrants that arrive late on the breeding grounds may jeopardize opportunities to secure a territory or a mate. If a bird expects to "catch-up" with the overall time schedule of migration, it must refuel faster than average during its next stopover. Yet, a penalty may be attached to late arrival at the next stopover site if resource levels have been depressed by earlier migrants (cf. Schneider and Harrington 1981, Moore and Yong 1991)—a delay in passage may cause a "domino effect" (Piersma 1990).

EFFECT OF SCALE ON HABITAT USE DURING MIGRATION

Habitat selection during migration is a hierarchical decision-making process (*sensu* Johnson 1980, Hutto 1985a; see Moore and Simons 1992). To understand how migrants respond to "problems" during passage we must realize that birds make decisions at different scales and that different factors, some extrinsic to habitat per se, operate at these different scales. Intrinsic constraints on habitat use are those factors thought to determine habitat quality and upon which migrants made decisions about habitat use. As the spatial scale broadens, factors intrinsic to habitat give way to factors largely unrelated to habitat (extrinsic constraints), such as synoptic weather patterns (see Hutto 1985a).

We begin with an emphasis on the large scale (intercontinental and continental) movements of birds between their breeding grounds and wintering grounds. Most individuals are "programmed" to follow a migratory pathway between their breeding and nonbreeding areas. The intrinsic factors (e.g., amount of resource, protection from predation) that determine the suitability of a stopover habitat may have influenced the evolution of the currently used migratory routes and wintering areas (Hutto 1985a). Pathways that are not frequented may have high extrinsic costs associated with travelling them rather than high intrinsic costs associated with the use of habitat along the way. Conversely, the use of some pathways may result from extrinsic benefits (short distance or favorable wind patterns) rather than the intrinsic quality of associated habitats (see Gauthreaux 1980b).

Extrinsic Constraints on Habitat Selection

Geographic Pattern of Migration

A wealth of information on the geographical patterns of bird migration in North America can be obtained from state bird books and checklists; state, regional, and national bird periodicals (e.g., *American Birds*); and even the range maps in some popular identification field guides (e.g., Robbins et al. 1983, National Geographic Society 1983). A very broad picture of migration in North America can be obtained from analyses of the total number of landbird species breeding versus wintering in quadrats of 500 km per side in different parts of North and Central America (Fig. 5-1). The greatest change in species numbers from summer to winter occurs in Canada, while little change can be found in the southern United States. In the latter region departing migrants are replaced by arriving migrants from further north. Numbers of migrants are higher for quadrats south of the border during the winter than during the summer because Neotropical migrants vacate their breeding grounds in late summer and fall, and move into Central and northern South America to overwinter. However, not all of the changes in species numbers over North America can be attributed to Neotropical migrants, because short-distance migrants do not contribute to changes. Information on the breeding distribution of Neotropical migrants is also needed to understand the geographical pattern of Neotropical migration.

In general considerably more Neotropical landbird migration occurs in the eastern two-thirds of the United States than in the West (see also Lowery 1951, Lowery and Newman 1955, 1966). One explanation for the greater amount of migration in the East comes from the fact that the breeding ranges of several "eastern" species of Neotropical migrant extend considerably further west and north of the eastern forests of the United States, but the birds migrate through the East. The breeding and migration range of the Philadelphia Vireo (Fig. 5-2) illustrates this point. Approximately 33 species of Neotropical migrants conform to this pattern. Another basis for the pattern of more migration in the East is that more Neotropical migrants (species and individuals) breed in the East (MacArthur 1959). For example, among North American wood warblers that migrate to the Neotropics, 40 species occur east of the Rocky Mountains and 15 species are found west of the mountains. The western species winter almost entirely within a narrow strip of west Mexico from Sonora south to Guatemala, while the eastern warblers generally winter in geographically separate areas of the Bahamas, West Indies, eastern Mexico, Central America, and northern South America (Hutto 1985a).

When drawing inferences about continent-wide patterns, it should be kept in mind that information on the spatial and temporal pattern of migration, not to mention migration volume ("traffic rate"), is not readily available for the southwestern United States or the West in general. Radar and direct visual (ceilometer and moon watching) studies must be conducted to fill that gap in our knowledge. Yet, it is clear that riparian or riverine habitats in the southwestern United States are vital to landbird migrants, notably woodland species (Sprunt 1975). Similarly, shelter-belts on the Great Plains represent "islands" of suitable habitat for woodland migrants (Martin 1980).

For the most part the longitudinal separation or species and populations of migratory landbirds characterizing the breeding season persists during the migration phase and the winter season [vireos (Barlow 1980), tyrant flycatchers (Fitzpatrick 1980), paruline warblers (Keast 1980; Hutto 1985a), and Neotropical migrants in general (Rappole et al. 1983)]. At a continent-wide scale the explanations for these migration patterns are varied and may relate to the location of breeding and wintering areas, major topographical features, availability of suitable resources on the migration route, peculiarities of life history, and prevailing direction of winds during the migration seasons (Rappole et al. 1979, Gauthreaux 1980a, Hutto 1985a). Clearly the mild climate of the Pacific Coast, the north–south

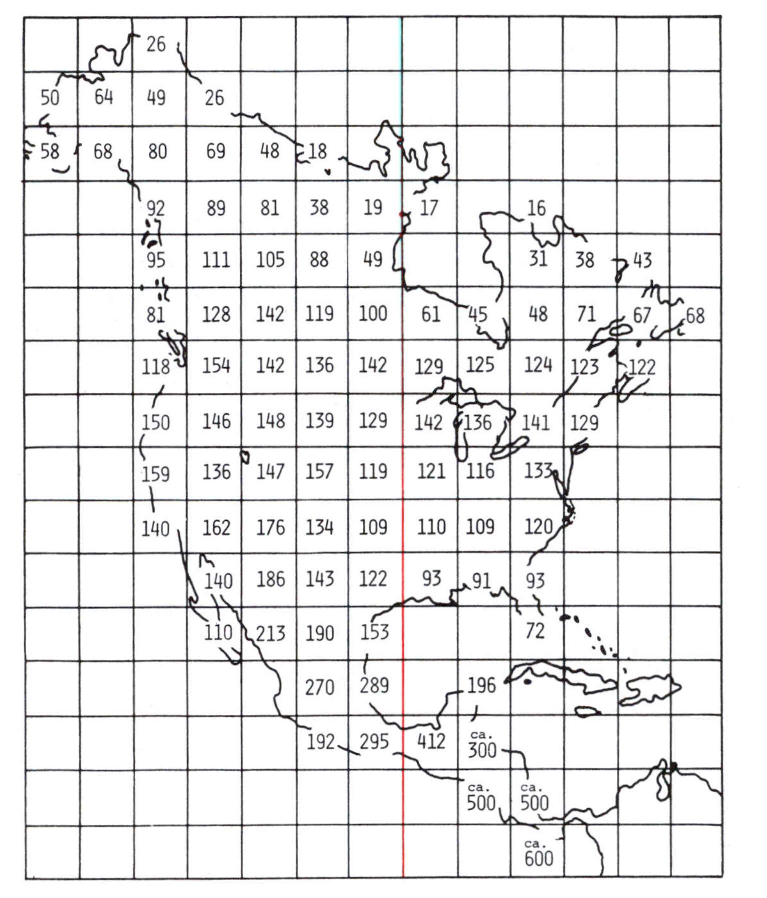

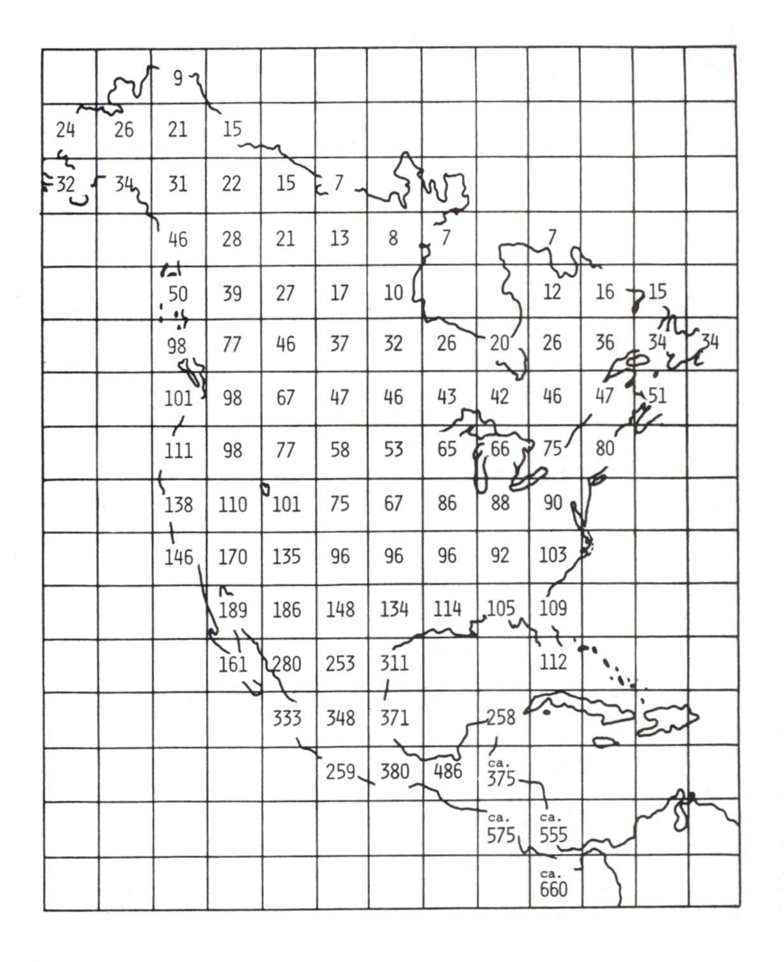

Figure 5-1. Gamma diversity [the total number of landbird species breeding (left) or wintering (right) in quadrats of 500 km per side] in different parts of North and Central America.

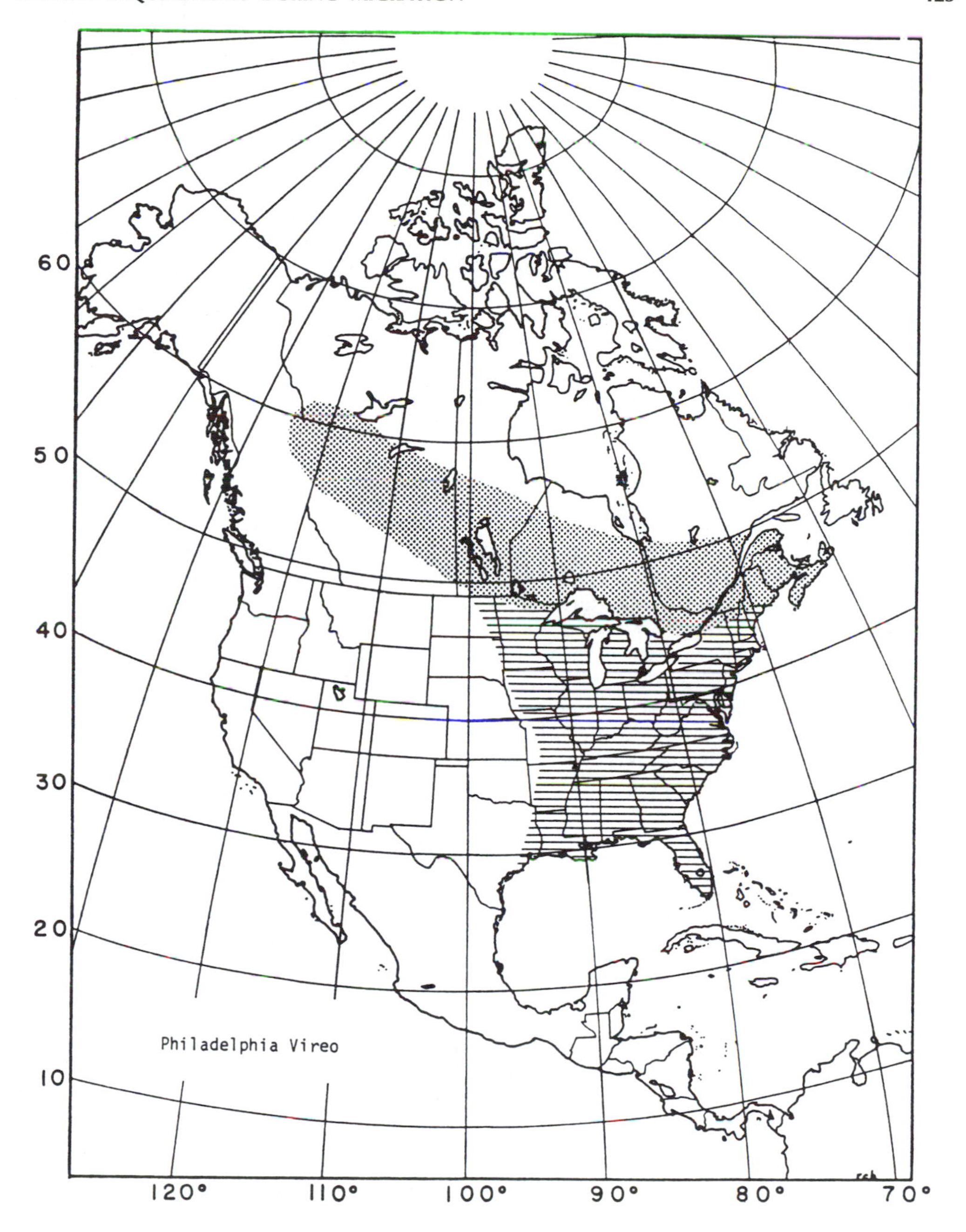

Figure 5-2. The breading (stippled) and migration (hatched) range of the Philadelphia Vireo.

mountain ranges of western North America, and the grasslands to the east of the mountains play an important role in maintaining the integrity of western landbird migration patterns. Likewise the western mountains, extensive grasslands, and prevailing westerly winds help to maintain the eastern bias to the seasonal movements of "eastern" Neotropical migrants during migration.

Weather and the Pattern of Migration

Continent-wide, seasonal differences in migration pathways can be related to prevailing wind patterns at different latitudes such that, in spring, Neotropical migrants are biased westward by prevailing easterlies at low latitudes and are biased eastward at higher latitudes by prevailing westerlies (Bellrose and Graber 1963, Gauthreaux 1980a). These prevailing wind patterns (Fig. 5-3a) produce a clockwise pattern of migratory pathways (Fig. 5-3b) that account for many Neotropical migrants being more abundant in the fall on the East Coast and over the western Atlantic Ocean (Williams et al. 1977) as they move toward their tropical wintering grounds. In spring the clockwise flow (Fig. 5-4a) biases many migrants departing from the tropics toward the northern and northwestern coast of the Gulf of Mexico, and the lower Mississippi Valley (Fig. 5-4b) with reduced numbers in most of Florida and extreme southeastern United States (e.g., Blackburnian Warbler; see Crawford 1981, Crawford and Stevenson 1984).

Migration in relation to the Gulf of Mexico further illustrates how an extrinsic factor such as weather and prevailing winds constrains habitat use at different spatial scales. The likelihood of a successful flight across the Gulf of Mexico is tied to the occurrence of favorable flight conditions for this long, nonstop flight (Gauthreaux 1971, Buskirk 1980). In spring, the peak of spring trans-Gulf migration is in the latter half of April through early May and corresponds to a period of predictable southerly airflow. Although migrants are observed crossing the Gulf of Mexico in fall, favorable conditions for a trans-Gulf flight (passage of a cold front with northerly winds) occur irregularly during the height of fall passage (Able 1972, Buskirk 1980). Prevailing weather conditions during the peak of fall migration along the northern coast of the Gulf of Mexico facilitate movements parallel to the coast rather than across this barrier.

Prevailing wind patterns, in concert with geographical differences in overwintering areas, also influence the relative magnitude of trans-Gulf and circum-Gulf spring migration such that several species that winter primarily in the Greater Antilles (e.g., Cape May Warbler and Black-throated Blue Warbler are abundant in Florida in the spring and become quite rare westward along the northern Gulf Coast (Robertson and Woolfenden 1995). Depending on the winds aloft during spring trans-Gulf flights, migrants may be "transported" anywhere from the coast of Mexico to the coast of Florida.

Weather surveillance radar (WSR-57) has been used to delimit the geographical pattern of landing areas of trans-Gulf migrants as they arrive on the Louisiana coast in spring (Gauthreaux 1975). Virtually every day between the beginning of April and the middle of May, trans-Gulf flights consisting of a variety of species of Neotropical migrants arrive on the northern Gulf coast when winds across the Gulf are favorable. With fair weather (about 80% of the time) the majority of these birds overfly the 25–30 mile (40–48 km) width of the coastal marshes and alight in inland forested areas (Fig. 5-5). When migrants encounter head winds or rain over the northern Gulf of Mexico they are likely to stop immediately along the coast regardless of their energetic condition (e.g., Moore and Kerlinger 1987).

INTRINSIC SUITABILITY AND SELECTION AMONG HABITATS

Although we might expect migrants to settle in habitats on the basis of relative suitability (*sensu* Fretwell and Lucas 1970; see Moore and Simons 1992), that outcome is not assured. Over the course of a season's migration a songbird like a Philadelphia Vireo encounters a variety of habitats, most of them new habitats with associated new food, new competitors, and new predators. After a night's migration a songbird often finds itself in a habitat that is very different from the one it occupied the previous day, let alone the previous year. Moreover, favorable habitat, where migrants can rapidly and safely accumulate energy stores, is probably limited in an absolute sense during

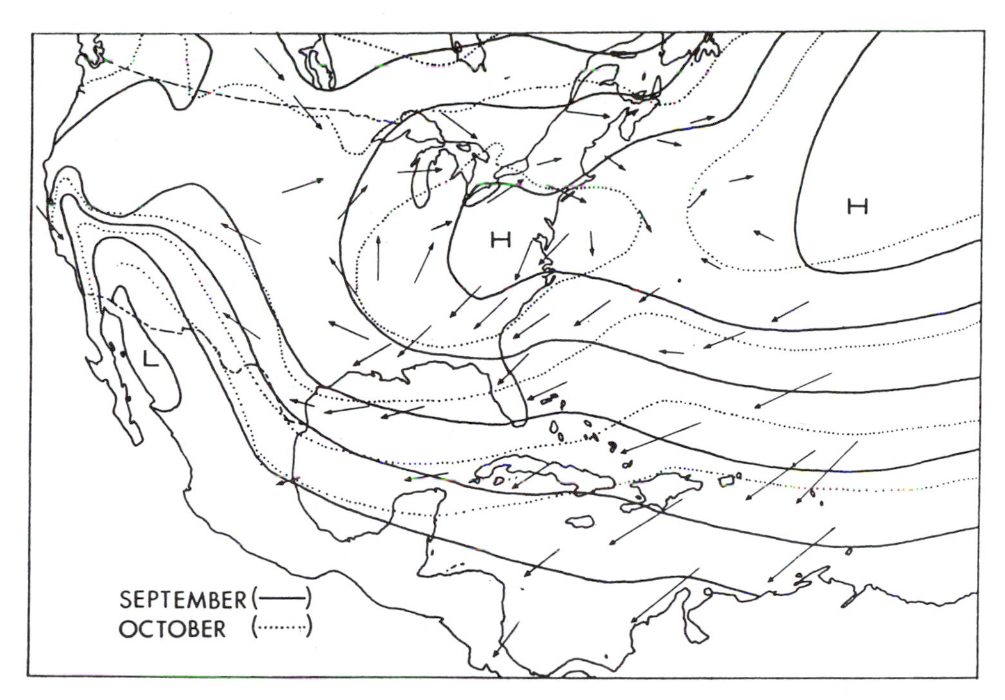

(a)

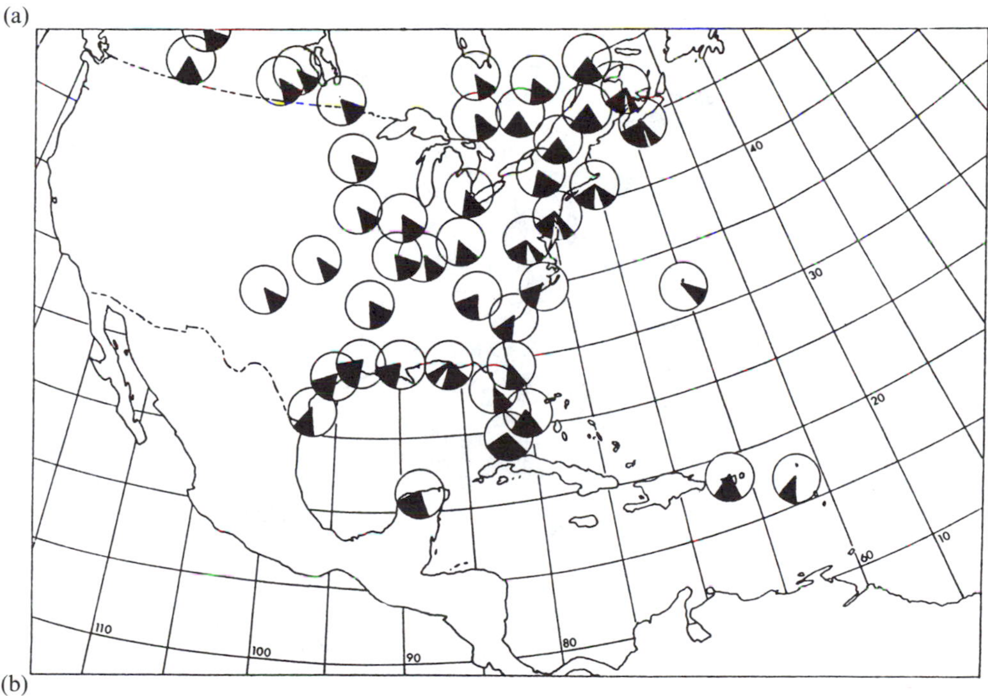

(b)

Figure 5-3. (a) The distribution of sea-level barometric pressure patterns in fall. Continuous and dotted lines connect points of equal pressure for the months of September and October, respectively. The arrows indicate the resultant direction of surface winds. (b) The directional tendencies of nocturnal passerine migration in fall. The circular plots show the predominant direction for a given area. The thickness of the wedge approximates the usual variability in direction. When two major directional tendencies exist for an area, they are both indicated.

(a)

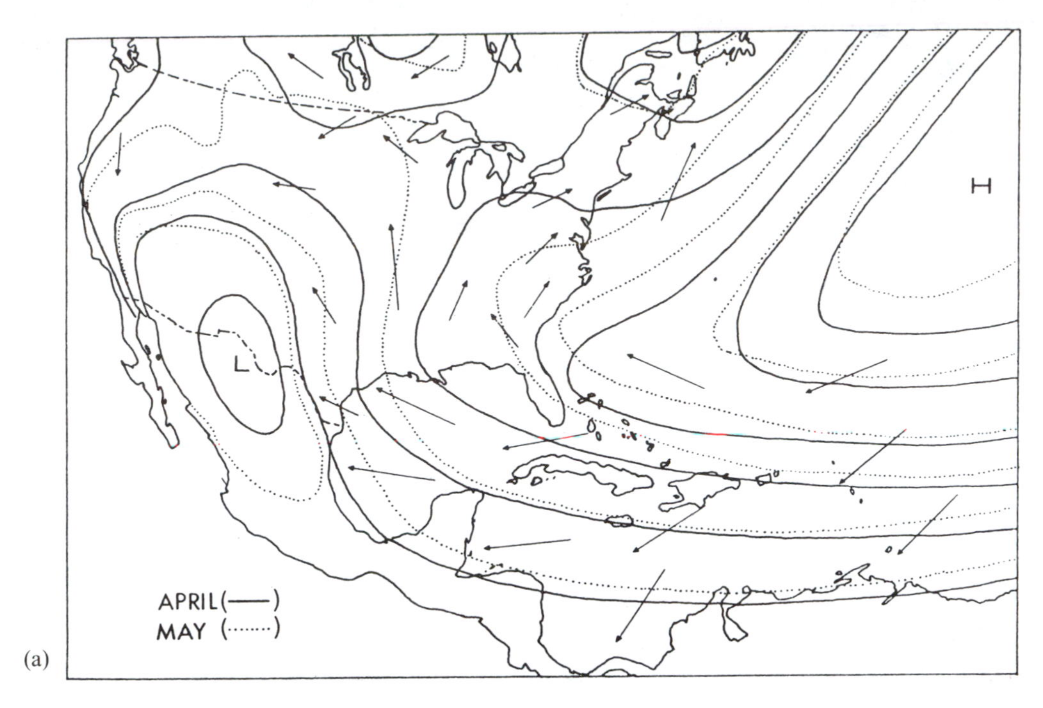

(b)

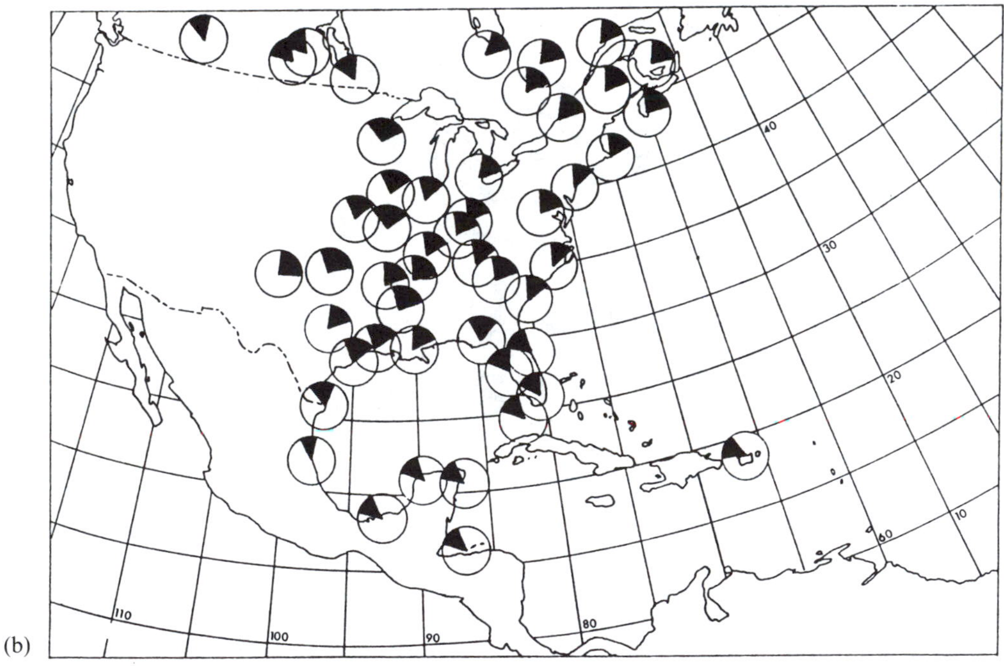

Figure 5-4. (a) The distribution of sea-level barometric pressure patterns in spring. Continuous and dotted lines connect points of equal pressure for the months of September and October, respectively. The arrows indicate the resultant direction of surface winds. (b) The directional tendencies of nocturnal passerine migration in spring. The circular plots show the predominant direction for a given area. The thickness of the wedge approximates the usual variability in direction.

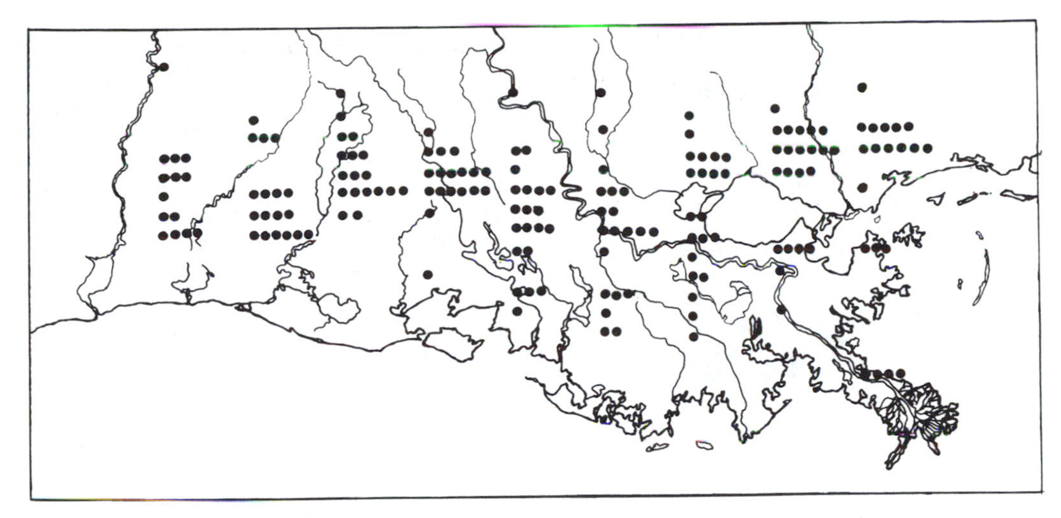

Figure 5-5. Distribution of major landing areas of spring trans-Gulf migrants in southern Louisiana. Each dot represents approximately 5% occurrence of landing in a particular area.

migration (Sprunt 1975, Hutto 1985a, Martin and Karr 1986), or effectively because migrants are constrained to minimize time in passage (Alerstam and Lindström 1990).

Despite the diversity of habitat types encountered during passage, several lines of evidence indicate that migrants prefer certain habitats and select among alternatives during stopover:

1. Species-specific patterns of distribution among different habitats are consistent with habitat selection during migration (Berthold 1988a). For example, transient through southwestern Germany show year-to-year constancy in their pattern of habitat use during fall stopover (Bairlein 1983).
2. The distribution of migrants among habitat types is correlated with changes in food availability (e.g., Martin 1985, Martin and Karr 1986). For example, shifts in habitat use by foliage-gleaning, insectivorous migrants from one migratory season to the next during passage through southeastern Arizona were tied to changes in the availability of insect prey (Hutto 1985b). Eastern Kingbirds occupy open, marsh–meadow habitat on Horn Island in spring (see Moore et al. 1990) and hawk insect prey items, but spend much of their time in areas of Slash Pine (*Pinus elliotti*) with a well-developed shrub layer during fall passage where they consume fruit (Moore and Woodrey 1995). Studies in the St Crois River Valley, Minnesota, revealed seasonal shifts in the distribution of Northern Waterthrush among swamp, floodplain and willow habitats, while Swainson's Thrush shifted from drier habitats in spring toward wetter sites in autumn in the same study area (Weisbrod et al. 1993; see also Winker et al. 1992b).
3. Use of habitat out of proportion to its availability is indicative of habitat selection (cf. Johnson 1980). For example, when the use of five habitat types was examined on Horn Island, a barrier island off the northern coast of the Gulf of Mexico (Moore et al. 1990), the distribution of spring trans-Gulf migrants deviated from that expected based on the availability of habitats (Fig. 5-6). Whereas Scrub/Shrub comprised 14% of available habitat, it was characterized by the greatest number of species, the highest species diversity, and the largest number of individuals. Bird-habitat associations among fall migrating landbirds have also been examined within the coastal region of the Cape May and Delmarva peninsulas

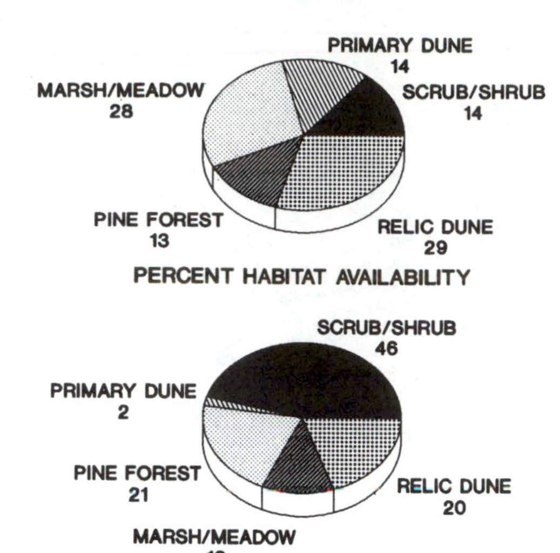

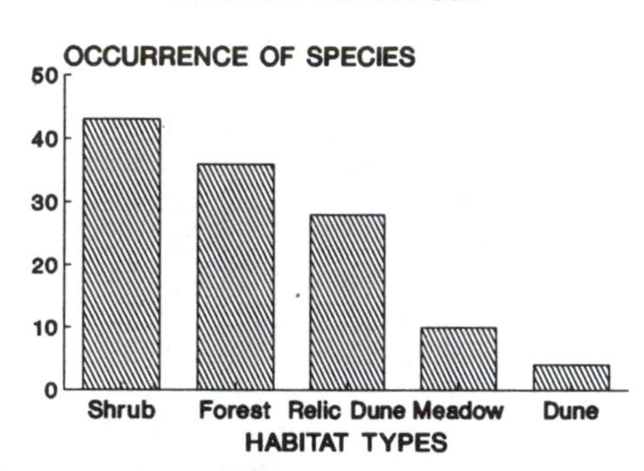

Figure 5-6. Differential use of five habitat types by landbird migrants on Horn Island, a barrier island off the northern coast of the Gulf of Mexico (after Moore et al. 1990).

(Mabey et al. 1992). Species richness varied with habitat type, being greatest for "mixed" forest habitat followed closely by deciduous forests and scrub/shrub, and individuals of certain species associated with specific habitat types (Fig. 5-7) A closer look at certain habitat categories reveals variation among vegetation communities within a habitat type. For example, three scrub/shrub habitats were recognized: old field scrub, young pine scrub, and coastal dune scrub. The latter was characterized by the highest mean bird abundance and species richness, whereas the fewest birds and fewest species were associated with pine scrub.

Determinants of Habitat Suitability

Habitat selection during migration occurs because the probability a migrant will meet its energetic requirements and achieve safe passage depends on the intrinsic suitability of stopover habitat. Possibly the single most important constraint during migration is to acquire enough food to meet energetic requirements, especially for long-distance migrants which must overcome geographic barriers (e.g., Biebach 1990, Moore 1991). Several studies conducted at disparate locations and with different species stress the importance of food availability in relation to the use of different habitats (Table 5-1).

Because migrants with similar food requirements and heightened energy demands are often concentrated in a small area, competition may occur during stopover (Moore and Yong 1991). Observations of territoriality among transients (e.g., Kale 1967, Rappole and Warner 1976, Kodric-Brown and Brown 1978, Bibby and Green 1980, Sutherland and Brooks 1981, Sutherland et al. 1982, Hixon et al. 1983, Carpenter et al. 1983, 1993a,b, Sealy 1988, 1989) and density-dependent patterns of settlement during migration (Viega 1986) are consistent with the occurrence of competition. If competition does occur during stopover, we would expect a decrease in the rate at which migrants replenish energy reserves (gain mass) either because the availability of food is depressed (Moore and Yong 1991) or because migrants have a direct effect on each other's intake rates (e.g., Carpenter et al. 1993a). Even when food availability is not depressed sufficiently to affect rates of mass gain, resource depression increases search time, which could conflict with migration timing.

Water economy might constrain migratory range and could explain why some individuals stop despite sufficient reserves for continued migration (Fogden 1972, Nachtigall 1990; but see Haas and Beck 1979, Biebach 1990). Moreover, lean migrants that have mobilized carbohydrate or protein sources in response to increased energy demand might experience a more serious water-balance problem than birds that have relied solely on their lipid reserves.

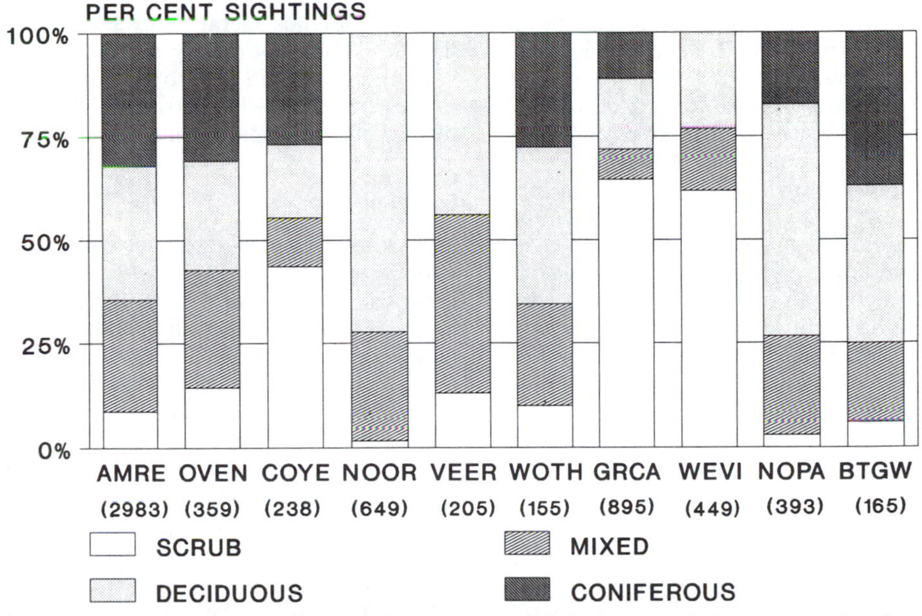

Figure 5-7. En-route habitat associations for several landbird migrants within the coastal region of the Cape May and Delmarva peninsulas August through October 1991: AMRE (American Redstart), OVEN (Ovenbird), COYE (Common Yellowthroat), NOOR (Northern Oriole), VEER (Veery), WOTH (Wood Thrush), GRCA (Gray catbird), WEVI (White-eyed Vireo), NOPA (Northern Parula), BTGW (Black-throated Green Warbler). (After Mabey et al. 1992).

Table 5-1. Evidence that migrants select habitat in relation to food availability.

Species	Context	Reference
Chaffinch Brambling	Reorientation in autumn along southern coast of Sweden to more "suitable" habitats	Lindström and Alerstam (1986)
Neotropical landbird migrants	Distribution of spring migrant among Great Plain shelterbelts tied to food abundance	Martin (1980)
Frugivorous landbird migrants	Attraction to second-growth woodlands in Panama	Martin (1985)
Insectivorous landbird migrants	Seasonal changes in density over different habitats, SE Arizona, matched availability	Hutto (1985b)
Sedge Warbler Reed Warbler	Fall rates of mass gain at two sites in France related to aphid abundance	Bibby and Green (1983)
Sedge Warbler	Year-to-year variation in rates of gain in French marsh related to aphid abundance	Bibby et al. (1976)
European Robin	Year-to-year variation in mass at island site on Skagerrak coast during fall migration related to food availability	Mehlum (1983a,b)
Wood-warblers (Parulinae)	Food consumption varied with the distribution and abundance of larvae during spring migration in Illinois	Graber and Graber (1983)
Pied flycatcher	Different rates of mass gain between sites in central Spain (fall passage)	Viega (1986)
Veery	"Track" fruit crops on different species as availability changes	Skeate (1987)
Migrant frugivores	Fall movements oriented toward forest gaps with abundant fruit	Hoppes (1987)

Predation may constitute a significant hazard to migrants e.g., Rudebeck 1950, 1951, Kerlinger 1989, Lindström 1989, Moore et al. 1990), and stopover habitats may vary in terms of predation risks. In Cape May, New Jersey, where as many as 80,000 migrating hawks and falcons have been counted in a single autumn migration season, predation on songbirds is widespread (Wiedner et al. 1992). Kerlinger (1989) speculated that some hawks migrate along coasts because of the seasonal concentration of potential prey, notably energetically stressed birds, which might be easy prey (see also Aborn 1994). When habitat offers refuge from predators and food in sufficient quantity to satisfy energy demand, little question exists about suitability. When the best areas for depositing fat are also the most dangerous, the migrant must trade off energy gain against mortality risks. For example, migrating Bramblings often shift from rape fields, where both energy gain and predation pressure are high, to beech forests, where energy gain and predation risk are low, in years of beech mast (Lindström 1990). In general, we expect fat-depleted migrants to be more willing to trade off the risk of predation to meet energetic requirements than are birds that arrive with unmobilized fat stores (cf. Ydenberg and Dill 1986, Lima and Dill 1990, Moore 1994).

The physical structure of habitat, including plant species composition and foliage structure, influences habitat suitability by affecting how birds move through the habitat and how they see and capture prey (Holmes and Robinson 1981, Robinson and Holmes 1982, 1984). Such constraints could affect not only the rate at which migrants replenish energy stores (Whelan 1987), but also a migrant's susceptibility to predator attack.

Habitat extent or "patchiness" also contributes to habitat suitability. At one level of analysis, bird species require different threshold levels of habitat area below which they find habitat unsuitable (Martin 1980, Blake 1983; see also Robbins et al. 1989). Sensitivity to area might affect habitat use during migration and the rate at which migrants replenish energy stores. At another level of analysis, suitable en-route habitat may be fragmented (e.g., coastal areas where many woodlands average only a few hectares in area). The opportunity to gain access to conditions wherein fat stores can be safely replenished would be restricted if fragments of suitable habitat are widely dispersed.

Length of Stopover

How long migrants stay at a stopover site varies from a few hours to several days and depends on several factors. In the absence of adverse weather such as rain and head winds, many birds depart the day or night of their arrival (Gauthreaux 1971). Generally lean migrants stay longer than birds that have not mobilized fat stores (e.g., Bairlein 1985, Pettersson and Hasselquist 1985, Biebach et al. 1986, Moore and Kerlinger 1987, Kuenzi et al. 1991). Even among lean birds the probability of staying is dependent on habitat suitability (e.g., Rappole and Warner 1976, Graber and Graber 1983, Kuenzi et al. 1991), and subject to time constraints (e.g., Safriel and Lavee 1988, Alerstam and Lindström 1990, Winker et al. 1992a). The combination of low fat stores coupled with a high probability of rebuilding stores should induce a migrant to stay at a stopover site. Low fat combined with a low probability of replenishment should favor departure and the search for more suitable stopover habitat (Rappole and Warner 1976, Terrill 1988).

When do Migrants Select Habitat During Passage?

Most Neotropical landbird migrants fly at night. Nocturnal migration commences shortly after sunset, peaks prior to 22:00 h, and is virtually complete by midnight or shortly thereafter (Kerlinger and Moore 1989). Exceptions occur, especially when night migrants must cross water barriers (see Gauthreaux 1971, 1972, Moore and Kerlinger 1991) or deserts (Moreau 1972; but see Bairlein 1987, Biebach 1990). In these cases, migrants have little choice but to continue migration until "suitable" habitat is found.

Most nocturnally migrating songbirds end their migratory flight well before dawn (Kerlinger and Moore 1989), so selection of a location to make a migratory stopover probably occurs during daylight hours, and most likely early in the morning. Nocturnal migrants have been observed making "morning flights" at several locations in North America (Gauthreaux 1978, Bingman 1980, Hall and Bell 1981, Wiedner et al. 1992) and Europe (Alerstam 1978 Lindström and Alerstam 1986). "Morning flight" differs from normal nocturnal migration in that: (1) it occurs during daylight, usually within the first two hours after dawn; (2) it occurs at low altitudes (sometimes from treetop to treetop); (3) flights are of short duration; and (4) migrants are often in flocks. In addition, the direction of "morning flight" is rarely the same as the previous night's migration (but see Bingman 1980).

On the Cape May peninsula, New Jersey, thousands of fall migrants can be observed in "morning flight" to the north, away from the end of the peninsula, toward the forested areas up the Delaware Bayshore (Wiedner et al. 1992) . At other sites in the New Jersey coastal plain, "morning flight" is to the west or northwest, again toward forested areas (Gauthreaux, pers and communication). Once the birds reach forested areas they diffuse, presumably into their preferred habitats. Although some of these birds may be compensating for drift experienced during nocturnal migration (e.g., Gauthreaux 1978, Moore 1990), evidence suggests that migrants are seeking more suitable habitat in which to rest and forage (see Lindström and Alerstam 1986, Wiedner et al. 1992). For example, landbirds tend to be "attracted" to riparian areas following a night's migration in the southwestern United States (Terrill and Ohmart 1984).

Mechanism(s) of Habitat Selection

The habitat "decisions" migrants make during passage are undoubtedly affected by innate (programmed) preferences (see Hilden 1965, Berthold 1990). It is not unusual for migrants to occupy habitat during the nonbreeding season that resembles their breeding habitat (e.g., Parnell 1969, Power 1971, Lack and Lack 1973). Such behavior, which is consistent with the existence of innate preferences, may be especially beneficial for hatching-year birds given their lack of experience with different habitat types.

Observations of migrants arriving along the northern coast of the Gulf of Mexico following a trans-Gulf crossing suggest that migrants assess alternative habitats during an initial exploratory phase shortly after arrival (Moore et al. 1990). An "exploratory phase" to habitat selection would be adaptive when migrants encounter a variety of habitat types and the availability of suitable habitat is unpredictable (see Hutto 1985a). Migrants could arrive at a stopover site with prior information of the distribution of resources in the environment, which might increase foraging efficiency (e.g., Valone 1992). However, migrating birds experience a variety of unfamiliar habitats and often do not spend much time in one location—circumstances likely to preclude the use of prior information.

What cues migrants use to select among alternative habitats are not known. Selection might be based on structural features of the habitat (Sherry and Holmes 1985, Morse 1989; but see Whelan 1989). Migrants may "cue" on the feeding rate of other individuals, their level of activity (cf. Krause 1992), or simply the number of other migrants present in a habitat. Presumably a more suitable habitat would attract more individuals, although more migrants would more rapidly deplete resources, potentially increasing competition. Information could be gathered from sampling a habitat, and might include the number of food items harvested, the time spent in a habitat, or the time since the last food item was consumed. If migrants behave so as to minimize time spent in passage, time becomes an important component of the habitat selection. When time for habitat assessment is brief, information must be obtained on the quality of different habitats using cues that are virtually instantly assessible (cf. Sullivan 1994).

Some Constraints on Habitat Selection

Migrating birds do not enjoy the luxury of selecting habitat solely on the basis of intrinsic factors. For example, when adverse weather conditions are encountered while aloft, a migrant might be forced to land in habitat it would otherwise bypass. If energy stores are depleted, stopover "options" are more narrowly circumscribed. For a fat-depleted migrant unfamiliar with the availability of a favorable stopover habitat, the benefits of rejecting suboptimal habitat may be outweighed by the cost of finding a better site.

Consider the stopover of migrants within a landscape of habitat types which vary in suitability (Fig. 5-8). As long as favorable habitat is readily available, search time is reduced, more time is available for sampling among habitat types, and birds are able to locate better habitat. As better habitat patches are lost and accessibility declines, migrants may be constrained to settle in a less suitable habitat. Fragmentation and loss of a more suitable stopover habitat will be especially problematic for fat-depleted migrants, not to mention for young, hatching-year migrants inexperienced at finding a patch of better habitat (e.g., Bruderer and Jenni 1988).

"Problems" en-route are undoubtedly magnified for hatching-year birds on their first migration (e.g. Murry 1966, Alerstam 1978, Ralph 1978, Gauthreaux 1982, DeSante 1983, Moore 1984, Lindström and Alerstam 1986, Terrill 1987), and individuals with different levels of migratory experience can be expected to respond differently to the exigencies of migration (e.g., Ketterson and Nolan 1985, 1988, Terrill 1988; see also Greenberg 1984b, Metcalfe and Furness 1984). Younger birds are often less efficient foragers (e.g., Burger 1988). They are also often behaviorally subordinate to adults (Terrill 1987), which could be a serious handicap should status affect the opportunity to deposit essential energy stores. For example, immature American Redstarts are behaviorally subordinate to adult birds during migration (Woodrey personal communication), which may explain why immature birds have deposited less fat when captured at a stopover site along the northern coast of the Gulf of Mexico in fall (see

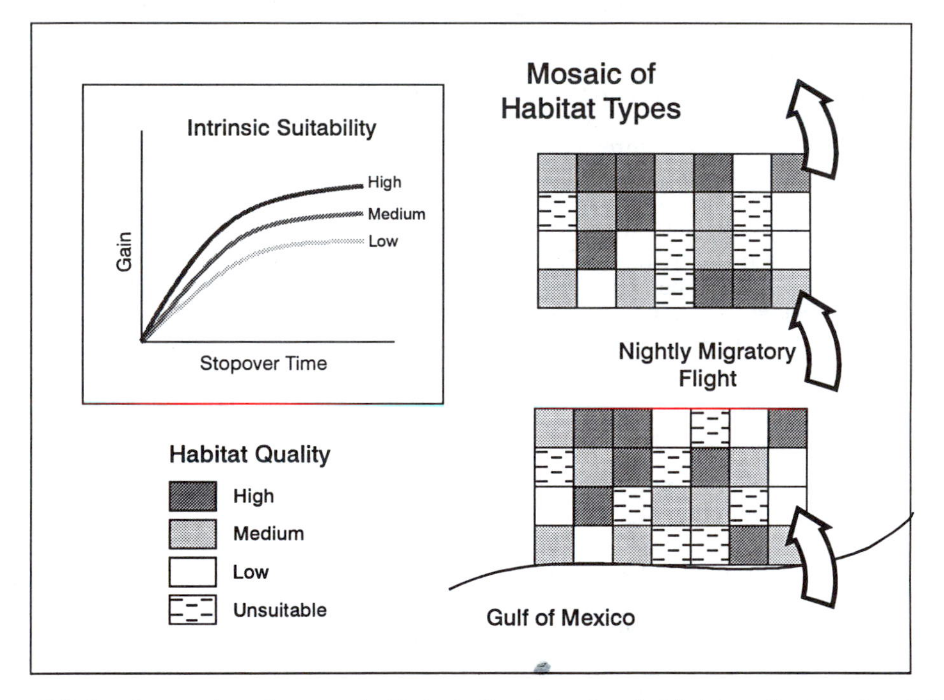

Figure 5-8. Landscape view of en-route habitat selection given three habitat types that vary in suitability (i.e., rate of energy gain). Hypothetical landscapes consist of a mosaic of habitats that differ in availability of the most suitable habitat.

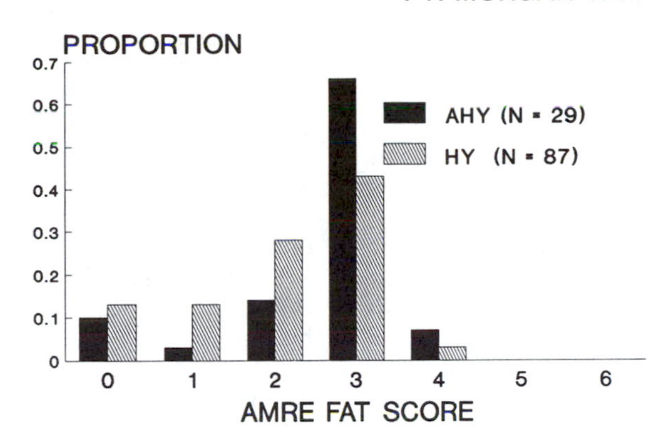

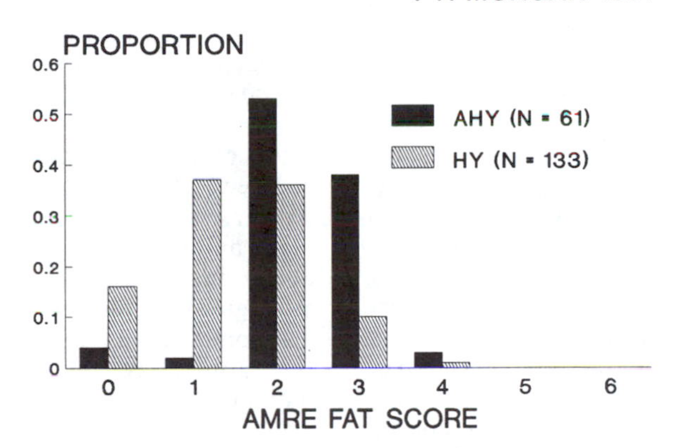

Figure 5-9. Distribution of fat scores (after Helms and Drury 1960) of immature (HY) and adult (AHY) American Redstarts (AMRE) captured on Fort Morgan peninsula, Alabama, fall 1990 and 1991.

Fig. 5-9; Woodrey and Moore, unpublished data). Inefficiency and lack of experience in young migrants could become important if food becomes scarce *or* when migrants experience heightened energy demand in anticipation of the possibility of energetic "short fall" (Moore and Simm 1985, 1986, Loria and Moore 1990). The "price" of inadequate energy stores is high for migrants about to cross a physical barrier or inhospitable habitat (Wood 1982, Moore and Kerlinger 1991). For example, American Redstarts that carry less than 3 g of stored fat (fat score ≤ 2, Fig. 5-9) will not succeed in crossing the Gulf of Mexico (Woodrey and Moore, unpublished data).

Behavioral Plasticity During Migration

When foraging ecology of landbird migrants is examined in changing environments (e.g., Hutto 1981, Greenberg 1984a, Rabenold 1980, Bennett 1980), across geographically distant breeding habitats (Petit et al. 1990), and during migration (e.g., Parnell 1969, Graber and Graber 1983, Hutto 1985b, Loria and Moore 1990, Martin and Karr 1990) considerable variation is revealed. Behavioral variability during migration should come as little surprise given the different vegetation structures, wide variations in resource quality and quantity, and changes in competitive pressures and community composition experienced at that time.

Morse (1971) suggested that such variability may represent adaptive plasticity which permits the migrant to occupy a diverse array of habitat types successfully as well as respond to novel circumstances (see Herrera 1978, Lack 1986, Greenberg 1990, Petit et al. 1990). For example, flexibility in foraging behavior during stopover permits an effective response to the energy demands of migration (Loria and Moore 1990; see also Martin and Karr 1990). Lean, fat-depleted Red-eyed Vireos broadened their use of microhabitat and increased their repertoire of foraging maneuvers following trans-Gulf migration—changes that could be viewed as a response to an anticipated decrease in expected gain. As a consequence, a lean bird was more likely to gain mass during stopover than birds carrying unmobilized reserves because they adjusted their behavior to compensate for increased energy demand. Migrating *Catharus* thrushes responded similarly in the face of heightened energy demand following trans-Gulf migration (Yong and Moore, unpublished data).

Studies of migrants following spring trans-Gulf flight (Kuenzi et al. 1991, Moore 1991) suggest that rates of mass change are inversely related to arrival mass (i.e., presumptive energy demand)—a pattern consistent with the notion that fat-depleted migrants compensate for heightened energy demand. If passage migrants are selected to minimize time spent migrating (Alerstam and Lindström 1990), it is reasonable to expect the evolution of compensatory mechanisms to meet nutritional demands and to prevent delays in the migratory schedule. If energetically stressed migrants alter their foraging behavior to increase their rate of energy acquisition, a favorable energy budget is achieved more quickly, length of stopover decreases, and the speed of migration increases.

On the other hand, a migrant's ability to respond to changing circumstances is limited and probably a matter of degree. Evidence indicates that migrants are able to adjust certain aspects of their foraging behavior, while other components remain essentially unchanged (e.g., Hutto 1981, Loria and Moore 1990, Martin and Karr 1990). Given the close association between morphology and habitat use (Winkler and Leisler 1985, Leisler 1990), morphology undoubtedly circumscribes a migrant's ability to change habitat opportunistically (e.g., Bairlein 1992). Moreover, the degree of plasiticity a migrant displays may be determined by the intensity of its neophobia (Greenberg 1983, 1984b).

Migrants are faced today with a rapidly changing landscape. Although migratory bird populations display the evolutionary potential to respond to environmental change (Berthold 1988b, 1990), flexibility inherent in the behavior of migratory birds may not be sufficient to accommodate large scale, rapid shifts in habitat structure (Wiedenfeld 1992).

MANAGEMENT IMPLICATIONS AND RECOMMENDATIONS

The ecological diversity of migratory species, coupled with the often variable weather patterns that steer migratory movements, make an assessment of habitat requirements and the development of management strategies for migrants particularly difficult. The complexity of this issue, and the fact that the abundance of migrants found at individual stopover sites can vary dramatically from year to year, makes it tempting to devalue the migratory period when developing conservation programs for these birds. Moreover, because Neotropical migrants spend more of their lives in breeding and wintering habitats they would seem a natural target for conservation efforts. Certainly, the characteristics and distribution of breeding and wintering habitats are somewhat easier to define.

It is generally assumed that higher energetic costs and mortality rates experienced by birds during migration are offset by higher productivity and/or survival on the breeding and wintering grounds (e.g., Lack 1946, Greenberg 1980). If mortality is concentrated in the migratory period, then we must assume that factors that increase the cost of migration. could have a disproportionate influence on overall populations levels. Thus, while individual fragmented woodlots may represent local population sinks on the breeding grounds, birds that find themselves in these habitats can often select alternative or more productive habitats. In contrast, the rigors of migration often place birds close to their physiological limits in unfamiliar landscapes, where they simply do not have the luxury of selecting alternative habitats. Therefore, a lack of suitable stopover habitat will result in death and contribute substantially to future population declines.

Geographic Scale

First, it is clear that migratory pathways are only loosely defined and are shaped by seasonal weather patterns. Second, radar and field studies confirm that the importance of an individual patch of habitat varies from year to year; a function of the number of migrants stopping and their energetic condition. The conservation implications of these observations argue for the protection of suitable stopover sites across the breadth of known migratory pathways. When managing for landbird migrants, a matrix of smaller and more widely distributed habitats may be more effective than a small number of large habitat areas.

Once migration pathways are recognized and the geographical variances in the route and timing are documented, one can delimit important stopover habitat. Landbird migrants require extensive stopover habitat, particularly along the Atlantic Coast in fall, and along the Gulf Coast in spring and fall. Stopover habitat in the interior is also critically important in North America wherever extensive deforestation and fragmentation have occurred (intensive and extensive agricultural areas) and, of course, in expansive areas of grassland and desert where stopover habitat for migrants that prefer woodlands is almost nonexistent except for narrow riparian woodlands. Such areas must be maintained and even increased (e.g., Platte River in Nebraska).

The continent-wide pattern of migration concentrates migrants in relation to physical and ecological barriers. Crossing barriers such as water bodies, mountain ranges, deserts, and more recently, agricultural and urban landscapes, can place extreme energetic demands on migrants (Alerstam 1981, Moore et al. 1990, Kuenzi et al. 1991). Furthermore, the problems associated with crossing barriers are magnified for hatching-year birds on their first migration. Protection and management of habitats used by migrants before and after they cross ecological barriers should be a prominent conservation priority. Examples include the narrow woodlands along the northern coast of the Gulf of Mexico, riparian habitats in the southwestern United States and maritime forests and shrub communities along the eastern seaboard.

The challenge of identifying and protecting coastal habitats will be heightened by the explosive population growth taking place in these areas. The concentration of the United States population along our coasts is projected to continue well into the next century (Cullitan et al. 1990; see Moore et al. technical report). About half of our total population now live in coastal areas. By the year 2010, coastal populations will have grown from 80 million to more than 127 million people, an increase of 60%. The northern Gulf coast, possibly the most important migratory stopover area in North America, is expected to see significant population increases. The southern migration of industry coupled with changing demographics will increase development pressure of stopover habitats in the decades ahead. Some coastal habitats spared from development are threatened by accelerating rates of coastal erosion (Dolan et al. 1989). The combined effects of coastal subsidence, the disruption of sediment supplies, and sea-level rise will add further to the loss of important stopover habitats. The creation of new habitats to replace those lost to coastal erosion will be a major conservation challenge in the next century.

Landscape Scale

Recent studies along the northern coast of the Gulf of Mexico (e.g., Moore et al. 1990, Kuenzi et al. 1991), in the Upper Mississippi Valley (Winker et al. 1992b, Weisbrod et al. 1993), and the Delmarva Peninsula (Mabey et al. 1992) have begun to identify some of the local and landscape scale features that are important to migrants. For example, spring migrants clearly preferred habitats with greater structural diversity when they arrived on the northern Gulf coast following a trans-Gulf flight (Moore et al. 1990). Structurally complex habitats, comprised of forest with a mixed shrub layer contained the greatest diversity and abundance of migrants. The maintenance of shrub communities in urban and agricultural landscapes as well as managed forests should improve habitat quality for migrants. For example, city parks can host dozens of species and many individuals during migration. Without these habitat "islands" many of these birds would not have any place in which to stop during migration. Although these "migrant traps" are important habitat, they should not be construed as alternatives to larger, undeveloped areas. In general, given the ecological diversity of this group of birds, maintenance of floristic and structural diversity at stopover sites should be a habitat management goal.

Within-Habitat Scale

A pressing need exists for information on the ecology of migrants during passage and the habitat requirements of landbird migrants. Unfortunately, we know little about the fine-scale habitat characteristics that influence prey availability or other aspects of habitat quality. For example, satisfying the energy demand of migration is not simply a matter of hyperphagia. The availability of nutrients that specifically enhance migratory fattening has as strong an effect on the course of migration as the abundance of food at a particular stopover site (Bairlein 1991).

A variety of foods, including fruits and insects, is important in both spring and fall migration. Fruit facilitates fat deposition and probably provides a rapid (short-term) solution to nutrient deficiencies which result from prolonged migratory activity (Jordano 1982). Consequently, management practices that reduce food abundance should be scrutinized (e.g., pesticide application), and prescribed burn schedules within managed landscapes should take into account the passage of migrants.

Migrating birds use habitat en-route in different ways for different reasons (Kuenzi et al. 1991, Winker et al. 1992a): some birds try to deposit lipid stores, others use the site as a molting ground (e.g., Winker et al. 1992a), and others simply rest until nightfall (e.g., Biebach 1990). As we refine our understanding of the determinants of habitat suitability, we must combine this knowledge with an analysis of habitat status and trends to develop future conservation priorities.

ACKNOWLEDGMENTS

We thank D. Aborn, W. Barrow, J. Clark, W. Hunter, T. Martin and M. Woodrey for their comments on the manuscript, S. Mabey for sharing ideas and unpublished data, and P. Simons and C. Belser for help with figures and manuscript preparation. Our research with Neotropical landbird migrants has been supported by the National Park Service, National Wetlands Research Center (USFWS), USDA Forest Service, National Science Foundation (BSR-9020530 and BSR-9100054 to FRM), and the National Geographic Society.

LITERATURE CITED

Able, K. P. 1972. Fall migration in coastal Louisiana and the evolution of migration patterns in the Gulf region. Wilson Bull. 84:231–242.

Aborn, D. 1994. Correlation between raptor and songbird numbers at a migratory stopover site. Wilson Bull. 106:160–154.

Alerstam, T. 1978. Reoriented bird migration in coastal areas: dispersal to suitable resting grounds? Oikos 30:405–408.

Alerstam, T. 1981. The course and timing of bird migration. Pp. 9–54 *in* Animal migration (D. J. Aidley, ed.). Cambridge University Press, Cambridge, England.

Alerstam, T. 1990. Ecological causes and consequences of bird orientation. Experientia 46:405–415.

Alerstam, T., and Å. Lindström. 1990. Optimal bird migration: the relative importance of time, energy, and safety. Pp. 331–351 *in* Bird migration (E. Gwinner, ed.). Springer-Verlag, Berlin.

Askins, R. A., J. F. Lynch, and R. Greenberg. 1990. Population declines in migratory birds in eastern North America. Curr. Ornithol. 7:1–57.

Bairlein, F. 1983. Habitat selection and associations of species in European passerine birds during southward, post-breeding migrations. Ornis Scand. 14:239–245.

Bairlein, F. 1985. Body weights and fat deposition of Palearctic passerine migrants in the central Sahara. Oecologia 66:141–146.

Bairlein, F. 1987. The migratory strategy of the Garden Warbler: A survey of laboratory and field data. Ringing Migration 8: 59–72.

Bairlein, F. 1991. Nutritional adaptations to fat deposition in the long-distance migratory Garden Warbler Sylvia borin. Acta XX Congr. Internat. Ornithol. 2149–2158.

Bairlein, F. 1992. Morphology–habitat relationships in migrating songbirds. Pp. 356–369 *in* Ecology and conservation of Neotropical migrant landbirds (J. M. Hagan, III and D. W. Johnston, eds). Smithsonian Institution Press, Washington, DC.

Barlow, J. C. 1980. Patterns of ecological interactions among migrant and resident vireos on the wintering grounds. Pp. 79–107 *in* Migrant birds in the Neotropics: ecology, behavior, distribution, and conservation (A. Keast and E. S. Morton, eds). Smithsonian Institution Press, Washington, DC.

Bellrose, F. C. and R. R. Graber. 1963. A radar study the flight directions of nocturnal migrants. Proc. XIII Internat. Ornith. Congr. 362–389.

Bennett, S. E. 1980. Interspecific competition and the niche of the American Redstart (*Setophaga ruticilla*) in wintering and breeding communities. Pp. 319–332 *in* Migrant birds in the Neotropics: ecology, behavior, distribution, and conservation (A. Keast and E. S. Morton, eds). Smithsonian Institution Press, Washington, DC.

Berthold, P. 1975. Migration: control and metabolic physiology. Pp. 77–128 *in* Avian biology, Vol. 5 (D. S. Farner and J. R. King, eds). Academic Press, New York.

Berthold, P. 1988a. The control of migration in European warblers. Acta XIX Ornithol. Congr. 215–249.

Berthold, P. 1988b. Evolutionary aspects of migratory behavior in European warblers. J. Evol. Biol. 1:195–209.

Berthold, P. 1990. Genetics of migration. Pp. 269–280 *in* Bird migration (E. Gwinner, ed.). Springer-Verlag, Berlin.

Berthold, P. and S. Terrill. 1991. Recent advances in studies of bird migration. Annu. Rev. Ecol. Syst. 22:357–378.

Bibby, C. J. and R. E. Green. 1980. Foraging behaviour of migrant Pied Flycatchers, *Ficedula hypoleuca*, on temporary territories. J. Anim. Ecol. 49:507–521.

Bibby, C. F., and R. E. Green. 1983. Food and fattening of migrating warblers in some French marshlands. Ringing Migration 4:175–184.

Bibby, C. F., R. E. Green, G. R. M. Pepler, and P. A. Pepler. 1976. Sedge warbler migration and reed aphids. Br. Birds 69:384–399.

Biebach, H. 1990. Strategies of trans-Sahara migrants. Pp. 352–367 *in* Bird migration (E. Gwinner, ed.). Springer-Verlag, Berlin.

Biebach, H., W. Friedrich, and G. Heine. 1986. Interaction of body mass, fat, foraging and stopover period in trans-Sahara migrating passerine birds. Oecologia 69:370–379.

Bingman, V. 1980. Inland morning flight behavior of nocturnal passerine migrants in eastern New York. Auk 97:465–472.

Blake, J. G. 1983. Trophic structure of bird communities in forest patches in east-central Illinois. Wilson Bull. 95:416–430.

Blem, C. R. 1980. The energetics of migration. Pp. 175–224 *in* Animal migration, orientation, and navigation. (S. A. Gauthreaux, Jr., ed.). Academic Press, New York.

Bruderer, B. and L. Jenni. 1988. Strategies of bird migration in the area of the Alps. Acta XIX Congr. Internat. Ornithol. 2150–2161.

Burger, J. 1988. Effects of age on foraging in birds. Acta XIX Congr. Internat. Ornithol. 1127–1140.

Buskirk, W. H. 1980. Influence of meteorological patterns and trans-Gulf migration on the calendars of latitudinal migrants. Pp. 485–491 *in* Migrant birds in the Neotropics (A. Keast and E. S. Morton, eds). Smithsonian Press, Washington, DC.

Carpenter, F. L., D. C. Paton, and M. A. Hixon. 1983. Weight gain and adjustment of feeding territory size in migrant hummingbirds. Proc. Natl Acad. Sci. USA 80: 7259–7263.

Carpenter, F. L., M. A. Hixon, R. W. Russell, D. C. Paton, and E. J. Temeles. 1993a. Interference asymmetries among age-classes of rufous hummingbirds during migratory stopover. Behav. Ecol. Sociobiol. 33:297–304.

Carpenter, F. L., M. A. Hixon, E. J. Temeles, R. W. Russell, and D. C. Paton. 1993b. Exploitative compensation by subordinate age-classes of migrant rufous hummingbirds. Behav. Ecol. Sociobiol. 33:305–312.

Cox, G. W. 1985. The evolution of avian migration systems between temperate and tropical regions of the New World. Amer. Natur. 126:451–474.

Crawford, R. L. 1981. Bird casualties at a Leon Country, Florida TV tower: a 25-year migration study. Bull. Tall Timbers Res. Stn. 22:1–30.

Crawford, R. L. and H. M. Stevenson. 1984. Patterns of spring and fall migrating in northwest Florida. J. Field Ornithol. 55:196–203.

Culliton, T. J., M. A. Warren, T. R. Goodspeed, D. G. Remer, C. M. Blackwell, and J. J. McDonough, III. 1990. Fifty years of population change along the nation's coasts 1960–2010. Coastal Trends Series, Report No. 2. NOAA, Strategic Assessement Branch, Rockville, MD.

DeSante, D. F. 1983. Annual variability in the abundance of migrant landbirds on southeast Farallon Island, California. Auk 100:826–852.

Dingle, H. 1980. Ecology and evolution of migration. Pp. 1–101 *in* Animal migration, orientation, and navigation (S. A. Gauthreaux, Jr, ed.). Academic Press, New York.

Dolan, R., S. J. Trossbach, and M. K. Buckley. 1989. Patterns of erosion along the Atlantic Coast. Pp. 17–21 *in* Barrier islands: process and management (D. K. Stauble, ed.). ASCE, New York.

Evans, P. R., N. C. Davidson, T. Piersma, and M. W. Pienkowski. 1991. Implications of habitat loss at migration staging posts for shorebird populations. Acta XX Congr. Internat. Ornithol. 2228–2235.

Fitzpatrick, J. W. 1980. Wintering of North American tyrant flycatchers in the Neotropics. Pp. 67–78 *in* Migrant birds in the Neotropics: ecology, behavior, distribution, and conservation (A. Keast and E. S. Morton, eds). Smithsonian Institution Press, Washington, DC.

Fogden, M. P. L. 1972. Premigratory dehydration in the reed warbler, *Acrocephalus scirpaceus* and water as a factor limiting migratory range. Ibis 114:548–552.

Francis, C. M., and F. Cooke. 1986. Differential timing of spring migration in wood warblers (Parulinae). Auk 103:548–556.

Fretwell, S. D. and J. L. Lucas. 1970. On territorial behavior and other factors influencing habitat distribution in birds. I. Theoretical development. Acta Biotheretica 19:16–36.

Gauthreaux, S. A., Jr. 1971. A radar and direct visual study of passerine spring migration in southern Louisiana. Auk 88:343–365.

Gauthreaux, S. A., Jr. 1972. Behavioral responses of migrating birds to daylight and darkness: a radar and direct visual study. Wilson Bull. 84:136–148.

Gauthreaux, S. A., Jr. 1975. Coastal hiatus of spring trans-Gulf bird migration. Pp. 85–91 *in* A rationale for determining Louisiana's coastal zone. Report No. 1, Coastal Zone Management Series (W. G. McIntire, M. J. Hershman, R. D. Adams, K. D. Midboe, and B. B. Barrett, eds). Center for Wetland Resources, Louisiana State University, Baton Rouge, LA.

Gauthreaux, S. A., Jr. 1978. The ecological significance of behavioral dominance. Pp. 17–54 *in* Perspectives in ethology, Vol. 3 (P. P. G. Bateson and P. H. Klopfer, eds). Plenum Press, New York.

Gauthreaux, S. A., Jr. 1980a. The influence of global climatological factors on the evolution of bird migratory pathways. Proc. XVIIth Internat. Ornithol. Congr. 517–525.

Gauthreaux, S. A., Jr. 1980b. The influences of long-tern and short-term climatic changes on the dispersal and migration of organisms. Pp. 103–174 *in* Animal migration, orientation, and navigation (W. Keeton and H. Wallraff, eds). Academic Press, New York.

Gauthreaux, S. A., Jr. 1982. Age-dependent orientation in migratory birds. Pp. 68–74 *in* Avian Navigation (F. Papi and H. Wallraff, eds). Springer-Verlag, Heidelberg.

Graber, J. W., and R. R. Graber. 1983. Feeding rates of warblers in spring. Condor 85:139–150.

Gradwohl, J. and R. Greenberg. 1989. Conserving nongame migratory birds: a strategy for monitoring and research. Pp. 297–328 *in* Audubon Wildlife Report 1989/1990 (W. Chandler et al., ed.). Academic Press, New York.

Greenberg, R. 1980. Demographic aspects of long-distance migration. Pp. 493–516 *in* Migrants in the Neotropics (A. Keast and E. Morton, eds). Smithsonian Institution Press, Washington, DC.

Greenberg, R. 1983. The role of neophobia in determining the degree of foraging specialization in some migrant warblers. Amer. Natur. 122:444–453.

Greenberg, R. 1984a. The winter exploitation system of Bay-breasted and Chestnut-sided Warblers in Panama. Calif. Publ. Zool. 116:1–120.

Greenberg, R. 1984b. Neophobia in the foraging-site selection of a Neotropical migrant bird: an experimental study. Proc. Natl Acad. Sci. USA 81:3778–3780.

Greenberg, R. 1990. Ecological plasticity, neophobia, and resource use in birds. Stud. Avian Biol. 13:431–437.

Haas, W., and P. Beck. 1979. Zum Fruhjahrszug palaarktischer Vogel uber die westliche Sahara. J. Ornithol. 120:237–246.

Hall, G. A. and R. K. Bell. 1981. The diurnal migration of passerines along an Appalachian ridge. Amer. Birds 35:135–138.

Helms, C. W., and W. H. Drury, Jr. 1960. Winter and migratory weight and fat: field studies on some North American buntings. Bird Banding 31:1–40.

Herrera, C. M. 1978. Ecological correlates of residence and nonresidence in a Mediterranean passerine bird community. J. Anim. Ecol. 47:871–890.

Hilden, O. 1945. Habitat selection in birds. Ann. Zool. Fenn. 2:53–75.

Hixon, M. A., F. L. Carpenter, and D. C. Patton. 1983. Territory area, flower density, and time budgeting in hummingbirds: an experimental and theoretical analysis. Amer. Natur. 122:366–391.

Holmes, R. T., and S. K. Robinson. 1981. Tree species preferences of foraging insectivorous birds in a northern hardwoods forest. Oecologia 48:31–35.

Hoppes, W. G. 1987. Pre- and post-foraging movements of frugivorous birds in an eastern deciduous forest woodland, USA. Oikos 49:281–290.

Hutto, R. L. 1981. Seasonal variation in the foraging behavior of some migratory western wood warblers. Auk 98:765–777.

Hutto, R. L. 1985a. Habitat selection by nonbreeding, migratory land birds. Pp. 455–476 *in* Habitat selection in birds (M. Cody, ed.). Academic Press, New York.

Hutto, R. L. 1985b. Seasonal changes in the habitat distribution of transient insectivorous birds in southeastern Arizona: competition mediated? Auk 102:120–132.

Johnson, D. L. 1980. The comparison of usage and availability measurements for evaluating resource preference. Ecology 61:65–71.

Jordano, P. 1982. Migrant birds are the main seed dispersers of blackberries in southern Spain. Oikos 38:183–193.

Kale, H. W. 1967. Aggressive behavior by a migrating Cape May Warbler. Auk 84: 120–121.

Keast, A. 1980. Spatial relationships between migratory parulid warblers and their ecological counterparts in the Neotropics. Pp. 109–130 *in* Migrant birds in the Neotropics: ecology, behavior, distribution, and conservation (A. Keast and E. S. Morton, eds). Smithsonian Institution Press, Washington, DC.

Kerlinger, P. 1989. Flight strategies of migrating hawks. University of Chicago Press, Chicago, IL.

Kerlinger, P. and F. R. Moore. 1989. Atmospheric structure and avian migration. Curr. Ornithol. 6:109–142.

Ketterson, E., and V. Nolan, Jr. 1982. The role of migration and winter mortality in the life history of a temperate-zone migrant, the Dark-eyed Junco as determined from demographic analyses of winter populations. Auk 99:243–259.

Ketterson, E., and V. Nolan, Jr. 1983. The evolution of differential bird migration. Curr. Ornithol. 3:357–402.

Ketterson, E., and V. Nolan, Jr. 1985. Intraspecific variation in avian migration patterns. Pp. 553–579 *in* Migration: mechanisms and adaptive significance (M. A. Rankin, ed.). Contribution Marine Sci. Suppl. 27, University of Texas, Austin, TX.

Ketterson, E., and V. Nolan, Jr. 1988. A possible role for experience in the regulation of the timing of bird migration. Proc. XIX Internat. Ornithol. Congr. 2169–2179.

Kodric-Brown, A., and J. H. Brown. 1978. Influence of economics, interspecific competition, and sexual dimorphism on territoriality of migrant rufous hummingbirds. Ecology 59:285–296.

Krause, J. 1992. Ideal free distribution and the mechanism of patch profitability assessment in three-spined sticklebacks (*Gasterosteus aculeatus*). Behaviour 123: 27–37.

Kuenzi, A. J., F. R. Moore, and T. R. Simons. 1991. Stopover of Neotropical landbird migrants on East Ship Island following trans-Gulf migration. Condor 93:869–883.

Lack, D. 1946. Do juvenile birds survive less well than adults? Br. Birds 39:258–264.

Lack, D., and P. Lack. 1973. Wintering warblers in Jamaica. Living Bird 11:129–153.

Lack, P. C. 1986. Ecological correlates of migrants and residents in a tropical African savannah. Ardea 74:111–119.

Lavee, D., and S. Safriel. 1989. The dilemma of cross-desert migrants—stopover or skip a small oasis? J. Arid Environ. 17:679–681.

Leisler, B. 1990. Selection and use of habitat by wintering migrants. Pp. 156–174 *in* Bird migration (E. Gwinner, ed.). Springer-Verlag, Berlin.

Levery, D. J., and F. G. Stiles. 1992. Evolutionary precursors of long-distance migration: Resource availability and movement patterns in Neotropical landbirds. Amer. Natur. 140:447–476.

Lima, S. L., and L. M. Dill. 1990. Behavioral decisions made under the risk of predation: a review and prospectus. Can. J. Zool. 68:597–600.

Lindström, Å. 1989. Finch flock size and risk of hawk predation at a migratory stopover site. Auk 106:225–232.

Lindström, Å. 1990. The role of predation risk in stopover habitat selection in migrating Bramblings *Fringilla montifringilla*. Behav. Ecol. 1:102–106.

Lindström, Å., and T. Alerstam. 1986. The adaptive significance of reoriented migration of chaffinches *Fringilla coelebs* and bramblings *F. montifringilla* during autumn in southern Sweden. Behav. Ecol. Sociobiol. 19:417–424.

Loria, D. E., and F. R. Moore. 1990. Energy demands of migration on red-eyed vireos, *Vireo olivaceus*. Behav. Ecol. 1:24–35.

Lowery, G. H. 1951. A quantitative study of the nocturnal migration of birds. Univ. Kansas Publ. Mus. Nat. Hist. 3:361–472.

Lowery, G. H., and R. J. Newman. 1955. Direct studies of nocturnal bird migration. Pp. 238–263 *in* Recent studies in avian biology (A. Wolfson, ed.). University of Illinois Press, Urbana, IL.

Lowery, G. H., and R. J. Newman. 1966. A continentwide view of bird migration on four nights in October. Auk 83:547–86.

Mabey, S. E., J. McCann, L. J. Niles, C. Bartlett, and P. Kerlinger. 1992. Neotropical migratory songbirds regional coastal corridor study. Final Report. National Oceanic and Atmospheric Administration, Contract # NA90AA-H-CZ839.

MacArthur, R. H. 1959. On the breeding distribution pattern of North American birds. Auk 76:318–325.

Martin, T. E. 1980. Diversity and abundance of spring migratory birds using habitat islands on the Great Plains. Condor 82:430–439.

Martin, T. E. 1985. Selection of second-growth woodlands by frugivorous migrating birds in Panama: an effect of fruit size and plant density. J. Tropical Ecol. 1: 157–170.

Martin, T. E., and J. R. Karr. 1986. Patch utilization by migrating birds: resource oriented? Ornis Scand. 17:165–174.

Martin, T. E., and J. R. Karr. 1990. Behavioral plasticity of foraging maneuvers of migratory warblers: multiple selection periods for niches? Stud. Avian Biol. 13:353–359.

Mehlum, F. 1983a. Weight changes in migrating Robins (*Erithacus rubecula*) during stopover at the island of Store Faerder, Outer Oslofjord, Norway. Fauna norv. Ser. C, Cinclus 6:57–61.

Mehlum, F. 1983b. Resting time in migrating Robins (*Erithacus rubecula*) at Store Faerder, Outer Oslofjord, Norway. Fauna norv. Ser. C, Cinclus 6:62–72.

Metcalfe, N. B., and R. W. Furness. 1984. Changing priorities: the effect of premigratory fattening on the trade-off between foraging and vigilance. Behav. Ecol. Sociobiol. 15:203–206.

Moore, F. R. 1984. Age-dependent variability in the migratory orientation of the savannah sparrow, *Passerculus sandwichensis*. Z. Tierpsychol. 67:144–153.

Moore, F. R. 1990. Evidence for redetermination of migratory direction following wind displacement. Auk 107:425–428.

Moore, F. R. 1991. Ecophysiological and behavioral response to energy demand during migration. Acta XX Congr. Internat. Ornithol. 753–760.

Moore, F. R. 1994. Resumption of feeding under risk of predation: Effect of migratory condition. Anim. Behav. 48:975–977.

Moore, F. R., and P. Kerlinger. 1987. Stopover and fat deposition by North American wood-warblers (Parulinae) following spring migration over the Gulf of Mexico. Oecologia 74:47–54.

Moore, F. R., and P. Kerlinger. 1991. Nocturnality, long-distance migration, and ecological barriers. Acta XX Congr. Internat. Ornithol. 1122–1129.

Moore, F. R., and P. Simm. 1985. Migratory disposition and choice of diet by the Yellow-rumped Warbler (*Dendroica coronata*). Auk 102:820–826.

Moore, F. R., and P. Simm. 1986. Risk-sensitive foraging by a migratory warbler (*Dendroica coronata*). Experientia 42:1054–1056.

Moore, F. R., and T. R. Simons. 1992. Habitat suitability and stopover ecology of Neotropical landbird migrants. Pp. 345–355 *in* Ecology and conservation of Neotropical migrant landbirds (J. M. Hagan, III and D. W. Johnson, eds). Smithsonian Institution Press, Washington, Dc.

Moore, F. R., and M. S. Woodrey. 1995. Stopover habitat and its importance in the conservation of landbird migrants. Proc. Annu. Conf. Southeast. Assoc. Fish Wildl. Agencies 47 (in press).

Moore, F. R., and W. Yong. 1991. Evidence of food-based competition among passerine migrants during stopover. Behav. Ecol. Sociobiol. 28:85–90.

Moore, F. R., P. Kerlinger, and T. R. Simons. 1990. Stopover on a Gulf coast barrier island by spring trans-Gulf migrants. Wilson Bull. 102:487–500.

Moreau, R. E. 1972. The Palearctic–African bird migration system. Academic Press, New York.

Morrison, M. L., B. G. Marcot, and R. W.

Mannan. 1992. Wildlife–habitat relationships. University of Wisconsin Press, Madison, WI.

Morse, D. H. 1971. The insectivorous bird as an adaptive strategy. Annu. Rev. Ecol. Syst. 2:177–200.

Morse, D. H. 1989. American warblers. Harvard University Press, Cambridge, MA.

Murray, D. G., Jr. 1966. Migration of age and sex classes of passerines on the Atlantic coast in autum. Auk 83:352–360.

Nachtigall, W. 1990. Wind tunnel measurements of long-time flights in relation to the energetics and water economy of migrating birds. Pp. 319–327 *in* Bird migration (E. Gwinner, ed.). Springer-Verlag, Berlin.

National Geographic Society. 1983. Field guide to the birds of North America. National Geographic Society, Washington, DC.

Nisbet, I. C. T., and L. Medway. 1972. Dispersion, population ecology, and migration of Eastern Great Reed Warblers *Acrocephalus orientalis* wintering in Malaysia. Ibis 114:451–494.

Parnell, H. F. 1969. Habitat relations of the Parulidae during spring migration. Auk 86:505–521.

Petit, D. R., K. E. Petit, and L. J. Petit. 1990. Geographic variation in foraging ecology of North American insectivorous birds. Stud. Avian Biol. 13:254–263.

Pettersson, J., and D. Hasselquist. 1985. Fat deposition and migration capacity of Robins *Erithacus rubecula* and Goldcrest *Regulus regulus* at Ottenby, Sweden. Ringing Migration 6:66–75.

Piersma, T. 1990. Pre-migratory "fattening" usually involves more than the deposition of fat alone. Ringing Migration 11:113–115.

Power, D. M. 1971. Warbler ecology: diversity, similarity and seasonal differences in habitat segregation. Ecology 52:434–443.

Rabenold, K. N. 1980. The Black-throated Green Warbler in Panama: geographic and seasonal comparisons of foraging. Pp. 297–307 *in* Migrant birds in the Neotropics: ecology, behavior, distribution, and conservation (A. Keast and E. S. Morton, eds). Smithsonian Institution Press, Washington, DC.

Ralph, C. J. 1978. Disorientation and possible fate of young passerine coastal migrants. Bird Banding 49:237–247.

Ramos, M. A. 1988. Eco-evolutionary aspects of bird movements in the northern Neotropical region. Acta XIX Internat. Ornithol. Congr. 251–293.

Rappole, J. H., and D. W. Warner. 1976. Relationships between behavior, physiology and weather in avian transients at a migration stopover site. Oecologia 26:193–212.

Rappole, J. H., M. A. Ramos, R. J. Oehlenschlager, D. W. Warner, and C. P. Barkan. 1979. Timing of migration and route selection in North American Songbirds. Pp. 199–214 *in* Proc. First Welder Wildl. Found. Symp. (D. L. Drawe, ed). Welder Wildlife Foundation, Sinton, TX.

Rappole, J. H., E. S. Morton, T. E. Lovejoy, III, and J. L. Ruos. 1983. Nearctic avian migrants in the Neotropics. US Dept. Interior, Fish Wildl. Service. US Government Printing Office.

Richardson, W. J. 1978. Timing and amount of bird migration in relation to weather: a review. Oikos 30:224–272.

Robertson, W. B., Jr. and G. E. Woolfenden. 1995. Florida bird species: an annotated list. Spec. Publ. no. 6. Florida Ornithological Society, Gainesville, FL.

Robbins, C. S., B. Bruun, H. Zim, and A. Singer. 1983. A guide to field indentification: birds of North America. Golden Press, New York.

Robbins, C. S., J. R. Sauer, R. S. Greenberg, and S. Droege. 1989. Population declines in North American birds that migrate to the neotropics. Proc. Natl Acad. Sci. USA 86:7658–7662.

Robinson, S. K., and R. T. Holmes. 1982. Foraging behavior of forest birds: the relationships among search tactics, diet, and habitat structure. Ecology 63:1918–1931.

Robinson, S. K., and R. T. Holmes. 1984. Effects of plant species and foliage structure on the foraging behavior of forest birds. Auk 101:672–684.

Rudebeck, G. 1950. The choice of prey and modes of hunting predatory birds with special reference to their selective effects. Oikos 2:65–88.

Rudebeck, G. 1951. The choice of prey and modes of hunting predatory birds with special reference to their selective effects. Oikos 3:200–231.

Safriel, U. N., and D. Lavee. 1988. Weight changes of cross-desert migrants at an oasis—Do energetic considerations alone determine length of stopover? Oecologia 76:611–619.

Schneider, D., and B. A. Harrington. 1981. Timing of shorebird migration in relation to prey depletion. Auk 98:801–811.

Sealy, S. G. 1988. Aggressiveness in migrating Cape May Warblers: defense of an aquatic food source. Condor 90:271–274.

Sealy, S. G. 1989. Defense of nectar resources by migrating Cape May Warblers. J. Field Ornithol. 60:89–93.

Sherry, T. W., and R. T. Holmes. 1985. Dispersion patterns and habitat responses of birds in northern hardwoods forests. Pp. 283–309 *in*

Habitat selection in Birds (M. L. Cody, ed.). Academic Press, New York.

Skeate, S. T. 1987. Interactions between birds and fruits in a northern Florida hammock community. Ecology 68:297–309.

Sprunt, A. 1975. Habitat management implications of migration. Pp. 81–86 *in* Proc. Symp. Management of Forest and Range Habitats for Non-game Birds. USDA Forest Serv. Gen. Tech. Rep. WO-1.

Sullivan, M. S. 1994. Mate choice as an information gathering process under time constraint: implications for behaviour and signal design. Anim. Behav. 47:141–151.

Sutherland, G. D., C. L. Gass, P. A. Thompson, and K. P. Lertzman. 1982. Feeding territoriality in migrant rufous hummingbirds: defense of yellow-bellied sapsucker (*Sphyrapicus varius*) feeding sites. Can. J. Zool. 60:2046–2050.

Sutherland, W. J., and D. J. Brooks. 1981. Territorial behaviour of little stints on spring migration. Br. Birds 74:522–523.

Terrill, S. B. 1987. Social dominance and migratory restlessness in the dark-eyed junco (*Junco hyemalis*). Behav. Ecol. Sociobiol. 21:1–11.

Terrill, S. B. 1988. The relative importance of ecological factors in bird migration. Proc. XIX Internat. Ornithol. Congr. 2180–2190.

Terrill, S. B., and R. D. Ohmart. 1984. Facultative extension of fall migration by Yellow-rumped Warblers (*Dendroica coronata*). Auk 101:427–438.

Valone, T. J. 1992. Information for patch assessement: a field investigation with black-chinned hummingbirds. Behav. Ecol. 3: 211–222.

Viega, J. P. 1986. Settlement and fat accumulation by migrant Pied Flycatchers in Spain. Ringing Migration 7:85–98.

Weisbrod, A. R., C. J. Burnett, J. G. Turner, and D. W. Warner. 1993. Migrating birds at a stopover site in the Saint Croix River Valley. Wilson Bull. 105:265–284.

Whelan, C. J. 1987. Effects of foliage structure on the foraging behavior of insectivorous forest birds. PhD dissertation, Dartmouth College, Hanover, NH.

Whelan, C. J. 1989. An experimental test of prey distribution learning in two paruline warblers. Condor 91:113–119.

Wiedenfeld, D. A. 1992. Foraging in temperate- and tropical-breeding and wintering male Yellow Warblers. Pp. 321–328 *in* Ecology and conservation of Neotropical migrant landbirds (J. M. Hagan, III and D. W. Johnson, eds). Smithsonian Institution Press, Washington, DC.

Wiedner, D. S., P. Kerlinger, D. A. Sibley, P. Holt, J. Hough, and R. Crossley. 1992. Visible morning flights of Neotropical landbird migrants at Cape May, New Jersey. Auk 109:500–510.

Williams, T. C., J. M. Williams, L. C. Ireland, and J. M. Teal. 1977. Autumnal bird migration over the western North Atlantic Ocean. Amer. Birds 31:251–267.

Winker, K., D. W. Warner, and A. R. Weisbrod. 1992a. Daily mass gains among woodland migrants at an inland stopover site. Auk 109:853–862.

Winker, K., D. W. Warner, and A. R. Weisbrod. 1992b. Migration of woodland birds at a fragmented inland stopover site. Wilson Bull. 104:580–598.

Winkler, H., and B. Leisler. 1985. Morphological aspects of habitat selection in birds. Pp. 415–434 *in* Habitat selection in birds (M. Cody, ed.). Academic Press, New York.

Wood, B. 1982. The trans-Saharan spring migration of yellow wagtails (*Motacilla lava*). J. Zool. Lond. 197:267–283.

Ydenberg, R. C., and L. M. Dill. 1986. The economics of fleeing from predators. Adv. Study Behav. 16:229–249.

6

HABITAT USE AND CONSERVATION IN THE NEOTROPICS

DANIEL R. PETIT, JAMES F. LYNCH, RICHARD L. HUTTO, JOHN G. BLAKE, AND ROBERT B. WAIDE

INTRODUCTION

Habitat, the vegetative, physical, and topographic features associated with the location in which an animal lives (Odum 1971), has important ramifications for the persistence of a species because it can influence reproductive and mortality rates. The evolutionary significance of habitat selection is clearly reflected by the close association between habitat use and morphology, behavior, and life-history traits of a species (Cody 1985). Understanding habitat requirements has motivated much of the study of avian ecology during this century and, consequently, habitat studies form the framework for conservation efforts aimed at preserving species diversity (Cody 1985, Probst and Crow 1991). Wildlife ecologists agree that conservation of populations of wild animals depends upon protection of habitats at appropriate spatial scales (Harris and Kangas 1988, Grumbine 1990). This prospect grows increasingly difficult, however, due to encroachment of civilization into even the most remote ecosystems on earth.

Migratory birds that breed in North America and overwinter in the Neotropics have been affected by human alteration of natural landscapes (Robbins et al. 1989a, Sauer and Droege 1992). Declines in populations of many Neotropical migrants have been documented through long-term studies at continental, regional (Breeding Bird Survey of the US Fish and Wildlife Service; Robbins et al. 1986), and local (Ambuel and Temple 1982, Wilcove 1988, Johnston and Hagan 1992) scales during breeding (Briggs and Criswell 1978, Robbins et al. 1989a), migration (Stewart 1987, Gauthreaux 1992, Hussell et al. 1992), and winter periods (Faaborg and Arendt 1992). Two main alternative hypotheses, both habitat based, have been advanced to explain population declines of long-distance migratory species in recent years (see Askins et al. 1990, Wilcove and Robinson 1990): (1) habitat loss and fragmentation on the breeding grounds have led to increased predation and cowbird parasitism due to edge effects, and to decreased opportunities to breed because of shortages of suitable habitat (Briggs and Criswell 1978, Brittingham and Temple 1983, Wilcove 1985); and (2) deforestation in Neotropical wintering areas has forced some individuals to occupy marginal habitats, which causes increased mortality (Rappole and Morton 1985, Rappole et al. 1989, Robbins et al. 1989b). These two alternatives are not mutually exclusive. However, partitioning breeding season effects from those during the boreal winter (or during migration; Moore and Simons 1992, Moore et al., chapter 5, this volume) is essential for understanding population declines and, subsequently, for taking action to reverse those trends. Presently, most evidence implicates events during the breeding season (Gates and Gysel 1978, Wilcove 1985, Yahner and Scott 1988, Gibbs and Faaborg 1990, Wilcove and Robinson 1990, Martin 1992; but see Rappole and McDonald 1994) in population declines of migrants, but this may be misleading because research efforts have been skewed toward North American breeding grounds. Evidence of an effect of tropical deforestation is largely circumstantial, and is based upon concurrent population declines of migrants and conversion of tropical broadleaved forests to other land uses (Hutto 1988a). Nevertheless, the overwintering period could be a time of intense selective pressure on Neotropical migrants because of mortality associated with stress induced from long migratory flights and increased competition for food due to inflated densities of potential competitors (Morse 1980). Because these pressures should be exacerbated in the face of widespread destruction of tropical habitats, a conceptual basis exists for claims that events on wintering areas also contribute substantially to many of the observed population trends (Robbins et al. 1989a).

Knowledge of habitat requirements of species has greatly facilitated the ways in which conservation biologists and wildlife managers attempt to sustain viable populations of long-distance migratory birds. Clearly, documentation of habitat relationships is critical but it is not the only type of information that is necessary to formulate effective conservation plans. To understand why some species are more susceptible to anthropogenic disturbances than others, a better comprehension of the ecological and evolutionary factors governing habitat selection is required; e.g., the relative roles of food resources, predators, and competitors, as well as the spatial scale upon which habitat selection is based. This mechanistic approach facilitates the decoupling of complex relationships and helps isolate causal from spurious associations. Furthermore, behavioral patterns of species may provide an important link between habitat use and habitat suitability (Hutto 1990), and may enhance the predictive capabilities for estimating the effects of certain land use practices on migratory birds.

This chapter provides a general overview of habitat use by long-distance migratory birds in the Neotropics and examines sources of variation influencing those patterns. In addition, it addresses the importance of integrating multiple subdisciplines of conservation biology, for example, behavioral and landscape ecology, into the process by which appropriate conservation and land-use practices are determined. Finally, recommendations are offered of how to best define, conserve, and manage habitats for migratory birds in the Neotropics.

HISTORICAL VIEWS OF HABITAT USE AND GEOGRAPHIC DISTRIBUTION

Behavioral ecology of single species (e.g., Schwartz 1964) or groups of species (e.g., Moynihan 1962, Willis 1966, Leck 1972a) dominated the study of overwintering Neotropical migratory birds during its emergence in the 1960s. As a result, few published quantitative descriptions of habitat use by overwintering migrants (e.g., Willis 1966, Lack and Lack 1972, Tramer 1974) were available when Karr (1976) presented the first overview of migrant habitat use in the New World Tropics. Karr (1976, p. 456) relied upon several published studies, reports from other ornithologists, and his own work, to conclude that migratory birds as a group were found predominantly in "disturbed, transitory, or isolated patches of habitat," and that interiors of mature, moist forests and grassland/savanna vegetation were avoided by migrants (also see Monroe 1970, Russell 1970). In addition, Karr noted that low (<1000 m) and high (>2500 m) elevations supported fewer migratory birds than middle elevations (1000–2500 m), and that mainland areas sustained fewer migrants than islands. Those generalities remained intact through the scrutiny afforded by a host of studies that followed several years later in a Smithsonian Institution publication devoted to the ecology of migrant birds in the Neotropics (Keast and Morton 1980). Based upon those studies, Terborgh (1980) concluded that migrants were most abundant (1) at higher tropical latitudes (see also Keast 1980), and (2) in disturbed vegetation types. The presumption that migrants showed an affinity toward middle elevations was generally supported, but varied somewhat among geographic regions (Terborgh 1980).

Despite publication of numerous studies of winter habitat use by migrants since Terborgh's (1980) compendium, little attempt has been made to summarize available data and reassess the above paradigms. Initial identification of areas most appropriate for designation as protected reserves can be facilitated through application of information on broad patterns of species habitat use and geographic distribution (Margules and Usher 1981, Diamond 1985). Below, a detailed examination of winter habitat use by Neotropical migratory birds is used to identify those habitats and geographic areas that are most critical in sustaining diversity of migratory species.

HABITAT AND GEOGRAPHIC DISTRIBUTION OF MIGRATORY BIRDS IN THE NEOTROPICS

Data were complied from approximately 30 published studies (representing >200 study sites) to identify those habitats and regions in the tropics supporting relatively large numbers of migratory species during winter. Only those studies that met the following criteria were included.

1. The study was conducted largely or entirely during the overwintering period (as opposed to migration). This restriction was necessary because many species are more generalized in habitat use along migratory routes compared to the overwintering (or breeding) period (Petit et al. 1993). Habitat selection during migration needs to be examined separately from other periods (Moore et al. 1993).
2. Study sites were located north of the Tropic of Capricorn (23.5° S) and south of the Tropic of Cancer (23.5° N), with the exception of one study located in Florida. This area represents the geographic limits of the Neotropics and is the region within which most migratory landbirds overwinter.
3. Data on habitat use were gathered through formal visual/auditory surveys (e.g., point counts or transects). Simple species lists usually were excluded from analyses. Estimates of species richness acquired through mistnetting were not included in these analyses because, in some taller habitats, mist-netting may provide a severely biased appraisal of the bird community (Karr 1981, Morton and Greenberg 1989). Alternatively, using only data that were collected by visual/auditory techniques biases against densely vegetated habitats of short stature because of the difficulty of detecting individuals that do not vocalize in those habitats compared with birds in taller, more open forests. For example, in young second growth, Blake and Loiselle (1992) captured 31 migratory species in mist nests, but detected only 17 species via visual/auditory surveys. In nearby primary forest, however, visual (13 species) and mist nest (15 species) surveys produced nearly identical numbers of migratory species. Thus, results presented below should be viewed cautiously because of the probable conservative estimates of species richness in young, densely vegetated habitats. Nevertheless, inclusion of mist-netting results does not substantially change the broad conclusions reached through these analyses (see Petit et al. 1993).
4. Data were collected within a single vegetation type. Surveys conducted over several habitats and presented as a single value were not used because of the need to ascertain habitat-specific abundances of migratory birds.

Several types of information were extracted from each study, including: (1) the percentage of all species surveyed that were Neotropical migrants, and (2) the total number of migratory species detected. Absolute *densities* of migrants were not possible to obtain because most survey techniques used in the tropics are not amenable to estimation of densities. Thus, the number of migratory species detected on study sites was used as an index of the importance of habitat types or regions to migratory birds. The significance of certain habitats or regions to migratory birds compared to those used by resident species was evaluated by examining the relative representation of migratory birds on each study site (i.e., percentage of all species that were Neotropical migrants). Because resident and migratory birds may require distinct conservation and management actions (Petit et al. 1993), this latter analysis can help identify those habitats that merit special attention for one group or the other. The number of migratory species and the percentage of migratory species on a site were significantly related (Pearson's $r = 0.559$, $P < 0.001$, $n = 185$).

The percentage of all *individuals* on a site that were migrants was also derived from each study but this parameter was not included in analyses because: (1) estimates of abundance in field studies are more tenuous than determination of simple presence/absence; (2) a close mathematical relationship exists between probability of detecting a species and its local abundance (Wright 1991); (3) siting of reserves and development of conservation programs are often based upon the types and numbers of *species* that are present (e.g., Burley 1988); and (4) the relationship between percentage of migratory individuals and the percentage of migratory species on a site was robust ($r = 0.779$, $P < 0.001$, $n = 159$).

Habitat and topographic descriptions and geographical information of each site were based upon descriptions provided by the author or from other published sources. That information was used to assess the effects of: (1) habitat type; (2) vegetation height; (3) level of disturbance; (4) elevation; (5) insularization; and (6) latitude on the distribution of migratory species. Study sites were classified in broad categories (Table 6-1) for statistical analyses. Prior to analysis, percentage of species was arcsine transformed and number of species square root transformed to improve normality of sampling distributions. Analysis of variance (ANOVA)

Table 6-1. Habitat, geographic, and environmental classes used in analyses of migratory bird distribution in the Neotropics.

Variable	Category	Description
Vegetation type	Moist forest	Semi-evergreen to evergreen forests >10 m tall in regions generally receiving >150 cm rain/year. Examples include wet, moist, montane, and cloud forests.
	Dry forest	Semi-evergreen to deciduous forests >5–10 m tall in regions generally receiving <150 cm rain/year with a pronounced dry season. Examples include dry, oak, and sclerophyll, as well as arid limestone, forests.
	Pine forest	Habitats dominated by coniferous trees. Examples include pine and pine–oak–fir forests and pine–savanna.
	Scrub	Early successional or naturally occurring broadleaved habitats <5–8 m tall. Examples include early successional dry and moist forests, thornscrub, and savanna–scrub.
	Open	Natural or human-altered habitats low in stature with few woody plants. Examples include open field, pasture, coastal dunes, marsh, and savanna.
	Artificial	Vegetation types heavily altered for agricultural or residential uses, but with vegetation >4 m tall. Examples include urban areas/parks, citrus and coffee plantations.
Vegetation height	≤5 m	
	6–10 m	
	11–20 m	
	>20 m	
Disturbance	Disturbed	Moderate to heavy disturbance. Examples of disturbances include logging, fragmentation, agriculture, residential, and clearing of forest understory.
	Undisturbed	Very slight to nonexistent disturbance.
Elevation	⩽200 m	
	201–500 m	
	501–1000 m	
	1001–2000 m	
	>2000 m	
Insularization	Mainland	
	Island	
Latitude	⩽5° N	
	6°–15° N	
	16°–25° N	
	>25° N	

was used to assess the null hypothesis that no differences in migratory bird abundance exist among categories for each environmental variable. If significant differences were detected, Tukey's test (Zar 1984) was used to identify statistically unique groups. For all hypotheses and in all analyses, a critical probability of 0.10 was set for statistical significance to reduce the chances of committing Type II statistical errors, which often are of greater concern than Type I errors in conservation research (Askins et al. 1990, Petit et al. 1992).

The analyses described above are subject to some biases because differences in plot size, sampling intensity, and sampling technique among studies were largely ignored. These uncontrolled sources of variation would serve to obscure the potentially important ecological effects that were being assessed. Thus, although the broad-scale trends that are documented should be viewed cautiously, these results probably represent conservative estimates of the environmental effects on habitat use by migratory birds.

Vegetation Type

Neotropical migrants were not distributed uniformly across vegetation types ($F = 5.89$; df = 5, 206; $P < 0.001$). Migratory species represented more than one-third of all species in residential/agricultural and pine habitats, 30% in early successional and dry forest habitats, and less than one-quarter of the total in open areas and moist forests; however, definitive habitat groupings were difficult to define (Fig. 6-1B). In contrast, native broadleaved vegetation types (wet forest, dry forest, and early successional growth) contained a greater number of migratory species per site than more artificial, managed habitats (pine, residential/agricultural), and open habitats with little woody vegetation ($F = 4.12$; df = 5, 188; $P = 0.001$; Fig. 6-1A). However, only open grassland/savanna habitats were conspicuously devoid of migrants. On average, approximately 9–11 species of migratory birds per site were identified in all habitat types except open areas, which averaged only five species (Fig. 6-1A).

The general tenet that migratory birds are

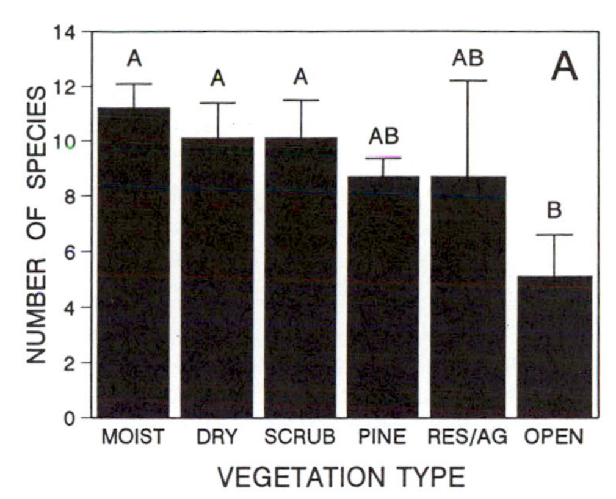

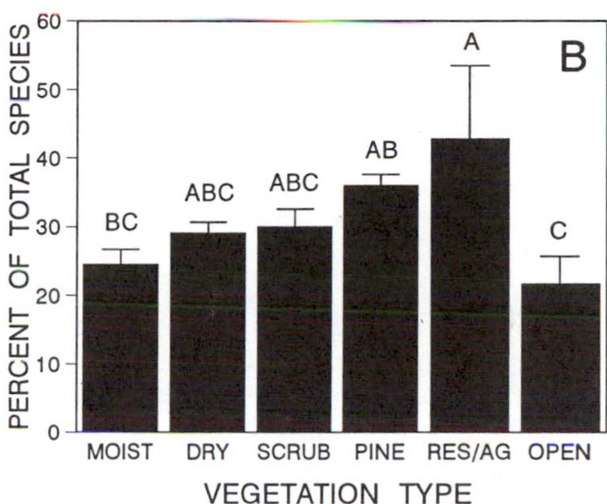

Figure 6-1. (A) Average number of migratory species recorded per study site, and (B) average percentage of total species (migrants and residents) represented by migrants in different vegetation types. MOIST = moist forest; DRY = dry forest; SCRUB = scrub or early successional growth; PINE = pine forest; RES/AG = artificial habitats, such as urban parks, residential areas, and citrus groves; and OPEN = open habitats, such as grassland and pasture. Error bar represents one standard error. Vegetation types with different letters are significantly ($P < 0.10$) different. See text and Table 6-1 for details.

distributed widely across habitat types—natural and artificial, disturbed and undisturbed—is supported by these analyses. In addition, migrants were proportionally underrepresented in moist, evergreen forest and open grassland/savanna (compared to other vegetation types) during the boreal winter (Karr 1976, Terborgh 1980). Comparison of numbers of species found in each vegetation type, however, suggests that moist and dry forests, and early successional/scrub habitats contain the most migratory species. Residential/agricultural habitats and pine forests were not statistically distinguishable from those broadleaved vegetation types but contained, on average, 14–22% fewer species. These analyses provide a general pattern of habitat use by migrants. However, because sites within each vegetation category varied greatly in such characteristics as level of disturbance, age, and geographic location, these trends do not identify specific habitat types in which migrants were most abundant. A more detailed examination of migrant use of major vegetation types will be presented later in this chapter.

Vegetation Height

Migrants represented a greater proportion of the bird community in intermediate vegetation (5–20 m tall) than in relatively short or tall vegetation ($F = 16.66$; df = 3, 208; $P < 0.001$; Fig. 6-2B). Analysis based on numbers of migratory species in each height category showed that, generally, more species occurred on sites with canopies that were > 10 m tall than on sites where canopies were of shorter stature ($F = 7.44$; df = 3, 190; $P < 0.001$; Fig. 6-2A).

This analysis cannot directly equate vegetation heights with particular habitat types. However, typically only moist forest sites exceeded 20 m and open grassland/savanna habitats were < 5 m tall, a further indication that migrants are underrepresented within the bird community occupying tall, moist forests and open grassland habitats. However, migrant species richness clearly increased with canopy height, which supports the theory that availability of additional resources (niches) allows for persistence of a greater number of species (MacArthur and MacArthur 1961, Orians 1969).

Disturbance Level

Disturbed sites (see Table 6-1), irrespective of habitat type, contained a greater proportion of migratory species than more pristine areas ($F = 17.20$; df = 1, 210; $P < 0.001$; Fig. 6-3B). On average, disturbed sites supported 14% more species than undisturbed sites, but this trend was not significant ($F = 2.45$; df = 1, 192; $P = 0.119$; Fig. 6-3A). These results further support the paradigm that migratory birds have a stronger affinity than resident species to disturbed habitats in the Neotropics (Karr 1976, Terborgh 1980). Nevertheless, from a conservation standpoint, little distinction can be made among habitats based on levels of disturbance because undisturbed habitats support nearly as many species on a local scale as do human-modified vegetation types. Furthermore, the importance of disturbance to migratory birds is most effectively assessed by examining different levels of intrusion *within* a

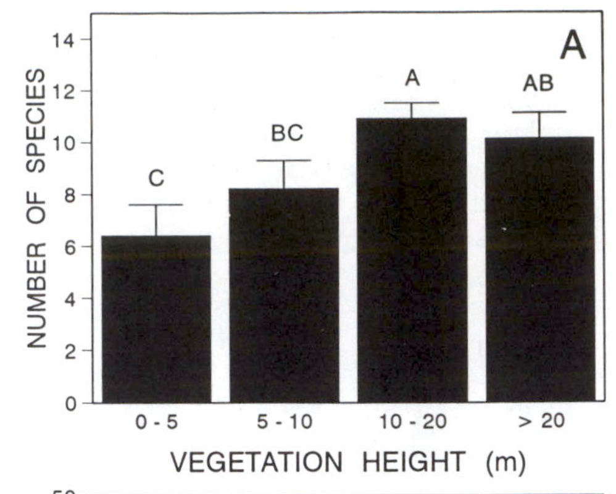

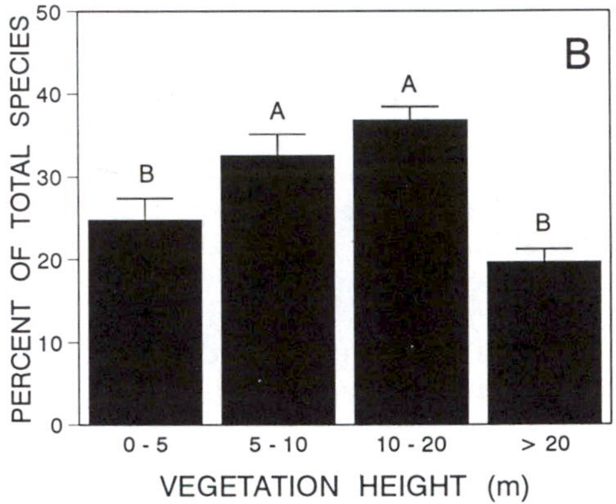

Figure 6-2. (A) Average number of migratory species recorded per study site, and (B) average percentage of total species (migrants and residents) represented by migrants in habitats with different vegetation heights. Error bar represents one standard error. Vegetation heights with different letters are significantly ($P < 0.10$) different. See text and Table 6-1 for details.

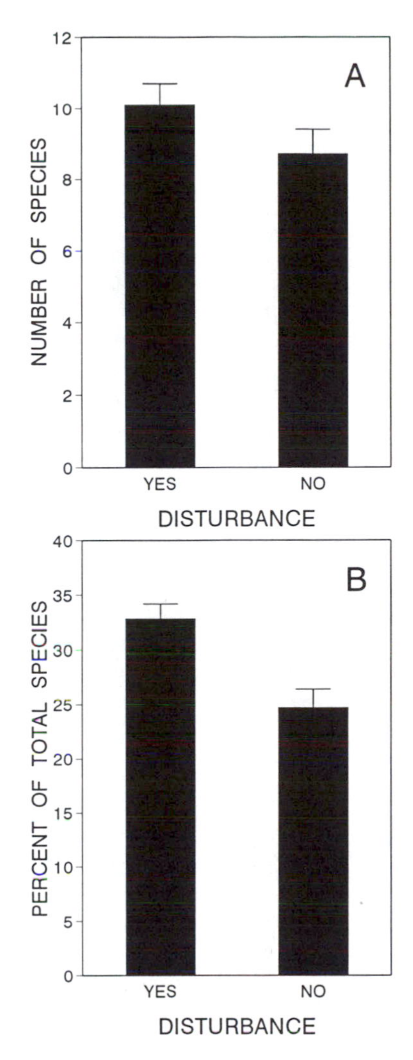

Figure 6-3. (A) Average number of migratory species recorded per study site, and (B) average percentage of total species (migrants and residents) represented by migrants in disturbed and undisturbed habitats. Error bar represents one standard error. See text and Table 6-1 for details.

vegetation type (see below). Although Neotropical migratory birds are slightly more abundant in disturbed habitats, not all disturbed habitats are suitable for migrants. Land-use practices that allow some of the natural canopy and subcanopy trees to remain, such as selective logging, low-density residential areas, or cacao plantations, provide habitat more suitable for migratory birds than habitats where most native woody vegetation has been cleared (e.g., pasture, row crops, sugarcane; Saab and Petit 1992). One exception to this generalization may be *Citrus* groves, where migratory bird abundance (richness and density) can reach high levels (Mills and Rogers 1992, Robbins et al. 1992).

Elevation

Migratory birds were well-represented in lowland (< 1000 m) habitats in both relative and absolute terms. The proportion of migratory species within local communities declined linearly with elevation (Spearman's correlation coefficient; $r = -1.0$, $n = 5$, $P < 0.001$), although Tukey's multiple comparison test showed broadly overlapped elevational groupings ($F = 7.32$; df = 4, 207; $P < 0.001$; Fig. 6-4B). The number of migratory species per site also decreased with an increase in elevation ($r = -0.90$, $n = 5$, $P = 0.037$) but this difference was not significant with analysis of variance ($F = 0.77$; df = 4, 189; $P = 0.548$; Fig. 6-4A). On average, approximately 18% more migratory species were detected per site in habitats below < 1000 m (9.9 species) compared with middle and high elevations (8.1 species). Thus, the often-cited generalization that migrants occur disproportionately at middle elevations (e.g., Leck 1972b, Karr 1976) appears to be unwarranted.

Insularization

Migratory birds comprised a slightly greater proportion of the total avifauna on island (32.4%) compared to mainland (28.4%) sites ($F = 2.75$; df = 1, 210; $P = 0.098$; Fig. 6-5B), but this result was due to a latitudinal effect (see below) associated with the fact that most Caribbean islands used in this study are located at high latitudes (> 15° N) within the tropics. When analyses were restricted to sites between 16° and 25° N, the trend was reversed; mainland sites (34.6%) contained significantly ($F = 3.40$; df = 1, 124; $P = 0.068$) greater proportions of migratory species than island sites (28.1%).

Nearly twice as many migratory species ($F = 13.90$; df = 1, 192; $P < 0.001$) were

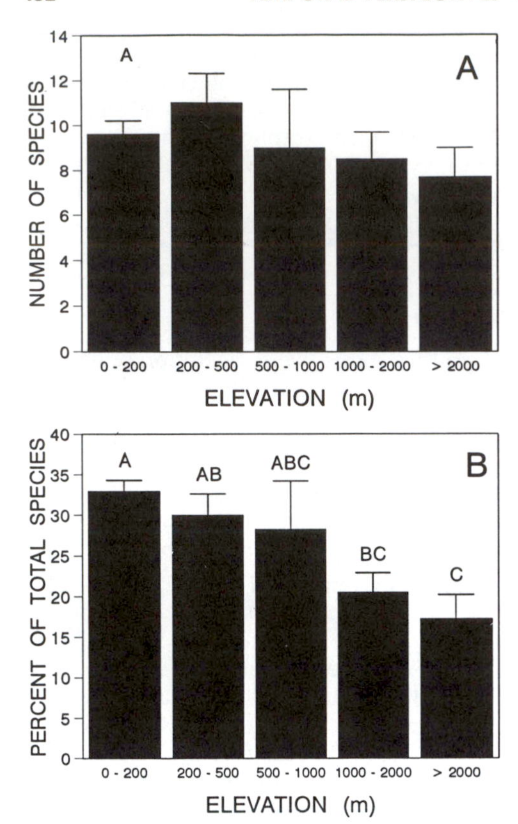

Figure 6-4. (A) Average number of migratory species recorded per study site, and (B) average percentage of total species (migrants and residents) represented by migrants at different elevations. Error bar represents one standard error. Elevations with different letters are significantly ($P < 0.10$) different. See text and Table 6-1 for details.

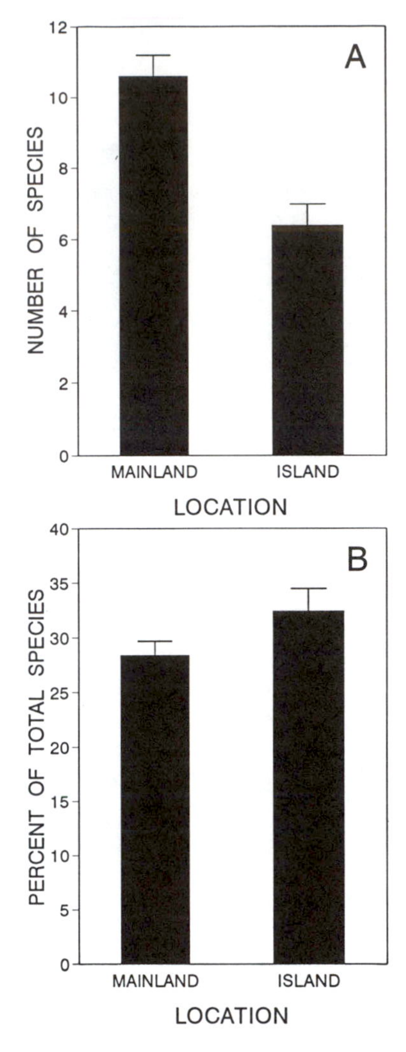

Figure 6-5. (A) Average number of migratory species recorded per study site, and (B) average percentage of total species (migrants and residents) represented by migrants on islands and mainland. Error bar represents one standard error. See text and Table 6-1 for details.

recorded on continental study sites (10.6 species) than on island sites (6.4 species; Fig. 6-5A), contrary to patterns described by Karr (1976) and Leck (1972b). These results suggest that migrants and residents may be influenced in similar ways by ecological factors thought to dictate the composition of insular faunal communities (MacArthur and Wilson 1967), for example, distance from the mainland (see also Terborgh 1980), availability of habitats, predators, and competitors. However, because migrants in moist, dry, and pine forests on continents still were significantly more abundant than their counterparts in the same habitat types on islands (see below), habitat availability probably cannot account for the greater number of migratory species in each of the mainland habitats. Rather, the paucity of migratory species on islands may be related to the relative energetic costs associated with flying long distances over unsuitable (water) habitat in search of relatively small land masses (Moreau 1972, Terborgh and

Faaborg 1980). In support of this, Wunderle and Waide (1993) found that migratory bird abundance on Caribbean islands was negatively related to the distance from the North American mainland. Nevertheless, the importance of Caribbean islands in providing migratory stopover areas and overwintering habitats for migratory birds should not be underrated (e.g., see Lack and Lack 1972, Pashley 1988, Arendt 1992).

Latitude

Migratory birds represented an increasing proportion of the total overwintering avifauna from lower to higher latitudes in the Neotropics ($F = 46.34$; df = 3, 208; $P < 0.001$; Fig. 6-6B). Migrants comprised about 5% of the overwintering species in South America (excluding the northern coastal areas), 15% in southern Central America (south of 15° N), 34% between Belize/Guatemala and central Mexico (25° N), and nearly 40% in northern Mexico and southern Florida (see also Keast 1980). The northern two regions, however, did not differ significantly in their proportions of migrants. The number of migratory species per study site peaked between 16° and 25°, which represented twice the number of species recorded per site in South America ($F = 5.40$; df = 3, 190; $P = 0.001$; Fig. 6-6A). Fewer migratory species at lower latitudes may be related to the increased costs associated with flying longer distances during migration (Greenberg 1980, Terborgh and Faaborg 1980). However, because the centers of winter distribution for many migratory species lie between 10° N and 20° N, and not further north, other considerations such as habitat availability and historical factors may partly define overwintering ranges.

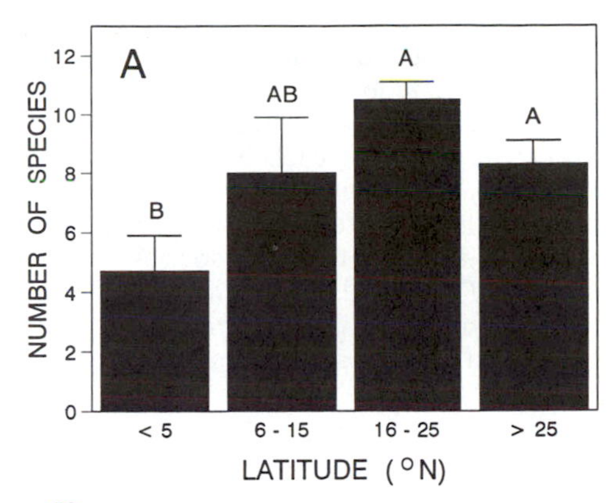

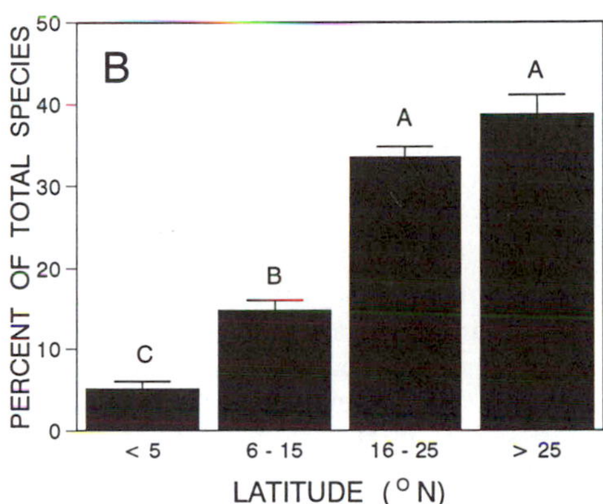

Figure 6-6. (A) Average number of migratory species recorded per study site, and (B) average percentage of total species (migrants and residents) represented by migrants at different latitudes. Error bar represents one standard error. Latitudes with different letters are significantly ($P < 0.10$) different. See text and Table 6-1 for details.

IMPORTANCE OF VARIOUS HABITAT TYPES TO NEOTROPICAL MIGRATORY BIRDS

Knowledge of general vegetation relationships of migratory birds is critical during the initial phases of conservation efforts in the Neotropics. Vegetative associations can be defined by unique combinations of climatic, topographic, edaphic, and biological parameters, such that several dozen naturally occurring life zones (Holdridge et al. 1971) are distinguishable within the Neotropics. Many additional vegetative associations also could be defined if natural combinations were considered in the context of anthropogenic disturbance (Hartshorn 1983, p. 120). Below, a more detailed examination of the relative use of major vegetative associations by overwintering migratory birds is presented. The trends identified are meant to serve as general justification for conservation prioritization of habitat types. However, bird–habitat relationships are far more complex than discussions presented here would indicate, and identification of actual sites that are appropriate for a reserve system need to be evaluated at a finer scale of resolution. Species listed under each habitat type were derived largely from a quantitative summarization of the literature (see "Geographic variation in habitat use by species" below; see also Petit et al. 1993), with additional information taken from several key studies conducted within each vegetation type.

Moist/Wet Forest

Moist forests, popularly referred to as rain forests, occur where the dry season is relatively short and annual precipitation exceeds 150 cm (Whitmore 1991). The mostly evergreen forests supported under these climatic conditions vary greatly in physiognomy and tree species composition (Sutton et al. 1983), thereby making generalizations about the associated avifauna tenuous unless certain distinctions are made among forest associations. The most widely recognized formations of moist tropical forest are based upon Holdridge's (1947) life-zone concept and applied to elevational gradients (also see Whitmore 1991). Here, five broad categories of moist forest are considered: coastal (<200 m), lowland evergreen (200–500 m), lower montane (500–1000 m; includes premontane), montane (1000–2000 m), and upper montane (>2000 m). Coastal forests were considered separately from lowland evergreen forests because of differences in human settlement patterns in these two areas (e.g., see Myers 1980, p. 150; Lynch 1992).

Proportional representation of migrants was inversely related to vegetation height, such that migrants comprised only 16% of the total species in mature forests greater than 20 m tall compared with 40–50% of the avifauna in shorter moist forests ($F = 22.99$; $df = 2, 63$; $P < 0.001$). In contrast, the number of migratory species detected per site was greatest in forests taller than 15 m ($F = 4.05$; $df = 2, 54$; $P = 0.023$).

Disturbed moist forests contained more migratory species, both proportionally (28% vs 21%; $F = 3.09$; $df = 1, 64$; $P = 0.084$) and in absolute terms (13.0 vs 9.9 species; $F = 2.87$; $df = 1, 55$; $P = 0.096$), than undisturbed forests. This trend was evident even when forest canopy height was statistically controlled, suggesting that small-scale forest fragmentation, forest openings, and other minor-to-moderate disturbances enhance the suitability of moist forest habitats for many species of overwintering migratory birds.

Overall, percentage of migrant species ($r_s = -0.90$, $P = 0.037$, $n = 5$) and absolute number of species ($r_s = -0.60$, $P = 0.285$, $n = 5$) inhabiting moist forests declined at higher elevations, although distinct altitudinal preferences were difficult to detect with multiple comparison tests (percent species: $F = 8.82$; $df = 4, 61$; $P < 0.001$; number of species: $F = 2.92$; $df = 4, 52$; $P = 0.029$). In *disturbed* forests these elevational relationships were weak (P values >0.10). Alternatively, migratory birds in *undisturbed* moist forests exhibited a strong decline in proportional representation and absolute numbers up an elevational gradient (P values <0.05). For example, undisturbed coastal and lowland evergreen forest sites contained twice as many migratory species as undisturbed, high elevation (>1000 m) forest plots.

In nearly all moist forest associations, disturbed sites supported significantly more migratory species than relatively pristine sites, although small sample sizes in some cases hindered interpretation of results. In general, migrants as a group show a strong affinity for lowland (<500 m) evergreen forests compared to montane forests, but this relationship is lessened when forests are disturbed. Migrants may increase use of disturbed forests at higher elevations if slight-to-moderate levels of disturbance enhance suitability of moist forests for those species (Petit 1991). Nevertheless, higher species diversity is often obtained at the expense of species that are sensitive to forest disruption—those species currently most in danger of population declines or extinction (Whitcomb et al. 1976, Petit et al. 1993). Thiollay (1992) and Johns (1989) found that selective logging of moist forest decreased the local diversity of *resident* bird species in French Guiana and Malaysia, respectively.

Migratory birds inhabiting moist forests exhibited distributional patterns similar to those found for the entire migratory bird community, as they were best represented on mainland sites and at middle latitudes (6°–25° N), compared to islands and more extreme latitudes within the Neotropics, respectively.

Although the general pattern of wet forest use suggests that disturbed and lowland forests support the greatest number of migratory birds, all types of moist forest provided suitable habitats for overwintering migrants. On average, 11 migratory species were detected within each plot of evergreen forest (Fig. 6-1B), which represented about one-quarter of the total *local* winter avifauna. Moreover, mature evergreen forest types are principal winter habitats for numerous migratory landbirds, such as Olive-sided and Yellow-bellied Flycatchers, Wood Thrush, Yellow-throated Vireo, Scarlet Tanager, Northern Waterthrush, and Blackburnian, Worm-eating, Cerulean, Prothonotary, and Kentucky Warblers. In summary, conservation of many migratory species, in addition to hundreds of species of resident tropical birds, is highly dependent upon preservation of moist evergreen forests. The current emphasis among conservationists on preserving moist forests, therefore, seems appropriate.

Dry Forest

Tropical dry forests occur most frequently in areas of western Mexico and central America, northern and western South America, and on the Yucatan Peninsula and several Caribbean islands. Low precipitation (usually <150 cm/year) and stressful edaphic conditions often produce climax forest types represented by short (<20 m), sclerophyllous trees or thorn forests (Murphy and Lugo 1986). Dry, semi-deciduous forests are usually more open and less structurally complex than moist, evergreen forests (Holdridge et al. 1971).

The number of migratory species within each dry forest study plot (ca. 10) was similar to that found in moist forests (11), and migrants constituted similar proportions of the local avifauna in both forest types (Fig. 6-1). The relative importance of moist forest becomes evident, however, when the effects of the unique geographic distributions of these forest types are removed (that is, dry forests included in this study were mainly found in northern Central America and Mexico, whereas moist forests were distributed throughout Latin America). For example, in northern Latin America, moist forest sites (13 species) supported approximately 40% more migratory species on average than dry forest sites (nine species)

Disturbance appeared to have a weak effect on numbers of species occupying dry forests ($F = 0.94$; df = 1, 23; $P = 0.343$); disturbed sites contained nearly 40% more species, but small sample size in this comparison obscured clear interpretation of results. Numbers of migratory species were greater in relatively tall (>10 m) dry forests compared to forests 5–10 m tall ($F = 3.71$; df = 2, 22; $P = 0.041$). Mainland dry forests contained more migratory species ($F = 3.10$; df = 1, 23; $P < 0.092$), but comprised a similar proportion of the overall avifauna ($F = 0.23$; df = 1, 25; $P = 0.639$), compared to similar island sites.

Dry forests represent an important habitat to migrants during winter, especially at more northerly latitudes (Petit et al. 1993). Species commonly found in these habitats include Wood Thrush, Least Flycatcher, Blue-gray Gnatcatcher, Solitary, White-eyed, and Warbling Vireos, and Orange-crowned, Nashville, Blue-winged, Black-and-white, Black-throated Gray, Magnolia, and Hooded Warblers.

Retention of deciduous and semideciduous forests is especially important for migratory bird conservation because much of Mexico and the Pacific slope of Central America are composed of dry forest formations, and because a large proportion of all migratory birds (especially western North American species) overwinter in that region (Hutto 1980, Terborgh 1980). Little is known about migrant use of dry forest habitats in South America (Bosque and Lentino 1987). Furthermore, some types of dry forest are often the first to be cut for agriculture (Holdridge 1970, Hartshorn 1992; see also Sader and Joyce 1988). This selectivity of tropical dry-forest removal during the past century has resulted in a proportional reduction exceeding that of nearly all other tropical vegetation associations (Janzen 1988).

Pine Forest

Habitats dominated by pine are of two general types in the tropics: (1) pure pine or mixed pine–hardwoods (also fir) at higher elevations (>500 m), and (2) lowland (<200 m) pine–savanna associations. Highland pine forests are often used for timber production. Both types of pine-dominated habitats historically have been maintained largely through natural and human-induced fires because burning inhibits invasion of broadleaved species. Pine habitats are locally and geographically restricted in Mesoamerica and the Caribbean, covering small proportions (usually <20%) of the total land area of Mexico, Belize, Guatemala, Honduras, Nicaragua, Bahamas, and Cuba, Hispaniola, Puerto Rico, Jamaica, and several other smaller Caribbean islands. Introduced pine plantations extend as far south as Panama and into South America, but do not represent significant habitats for migratory birds in those areas (L. J. Petit, unpublished data).

Overall, migratory birds composed a relatively large proportion of the avifauna in pine habitats, second in importance only to residential/agricultural habitats (Fig. 6-1B). In contrast, numbers of migratory species in pine habitats were slightly lower than in broadleaved forest types (Fig. 6-1A). Differences between migrant use of lowland and highland pine forests were minimal, as were effects of disturbance (P values >0.10). Tall (>10 m tall) stands and mainland sites contained significantly more migratory species than younger successional stages ($F = 4.38$; df = 3, 54; $P = 0.008$) and islands ($F = 14.53$; df = 1, 56; $P < 0.001$), respectively. Several paruline warblers show strong affinity for pine forests, such as Hermit, Yellow-throated, Black-throated Gray, Black-throated Green, and Prairie Warblers.

Pine forests are not often recognized as habitats of concern in conservation efforts because of the relatively rapid growth of pine trees and because management practices have traditionally sustained those forests (Howell 1970, Monroe 1970, Russell 1970). Although the total land area of pine has not decreased substantially in the past few decades (Hartshorn 1992), Leonard (1987) suggested that the quality of those forests may be declining due to overharvesting and shorter rotational schedules. Furthermore, management techniques used to increase the vigor of pines frequently involve herbicides or, more commonly, burning to control broadleaved understory plants. However, broadleaved understory and overstory are important components of pine habitats for many Neotropical migrants, as well as for resident species (Cruz 1988, Hutto 1988b). For example, in Belize, middle elevation pine forests that contained patches of broadleaved understory supported an average of nine migratory species, whereas stands with sparse understory (due to cutting and burning) contained only 2–5 species (D. R. Petit, unpublished data). Wise management of pine forests must consider the habitat requirements of associated wildlife, including migratory birds. This includes not only maintenance of some broadleaved understory, but also acceptable regeneration techniques and rotation periods.

Early Successional Growth/Scrub

These habitats represented both naturally occurring climax and early-successional vegetative associations. The only shared attribute was that canopy height rarely exceeded 8 m (5 m for mature vegetation associations). In this section, distinctions between mature associations and early successional growth will be made where appropriate, although emphasis will be placed on early successional habitats because of the large proportion (ca. 80%) of sites in this analysis that represented young (usually <10 years) growth.

Scrub and early-successional habitats supported numbers of migratory species comparable to most other habitats examined here (Fig. 6-1A). However, recently cut forests contained significantly more migratory species (ca. 11) than did natural, low-stature vegetation (ca. four species; $F = 6.77$; df = 1, 27; $P = 0.015$). This trend may result from the fact that many mature vegetative associations of low stature occur under extreme climatic or edaphic conditions (e.g., lack of moisture). Low food (arthropods and fruit) abundance may be a corollary of those climatic regimes, such that few migrants can be supported. Additionally, the relatively high net primary productivity of early successional forests may support a large biomass of herbivorous insects (Janzen 1973), which make those habitats attractive to insectivorous migrants.

Migrant species richness increased at more northerly latitudes (>15° N), and peaked between 15° N and 25° N ($F = 4.22$; df = 2, 26; $P = 0.026$). Approximately four more migratory species were found in the typical mainland scrub habitat ($\bar{x} = 11$ species) compared to island sites (seven species), but this disparity was not significant ($F = 1.12$; df = 1, 27; $P = 0.299$).

Early successional habitats are important for several dozen migratory species, such as Bell's Vireo, Yellow-breasted Chat, Common Yellowthroat, Ovenbird, Gray Catbird, Least Flycatcher, Blue Grosbeak, and Indigo Bunting (e.g., Karr 1976, Waide 1980, Martin and Karr 1986, Hutto 1989, Lynch 1989, Blake and Loiselle 1992, Petit et al. 1992). In fact, often >50% of the local overwintering migratory species are found in early successional habitats, although most species are usually detected within other habitats as well (Waide 1980, Lynch 1989, Petit 1991, Blake and Loiselle 1992, Hutto 1992, Kricher and Davis 1992). Thus, many migrants appear capable of using forested habitats in many stages of succession, especially in northern areas of the Neotropics (e.g., Lynch 1989, Hutto 1992).

Early successional habitats are not the target of conservation efforts in the tropics because of the current plethora of these vegetation types, a result of rapid expansion of shifting cultivation during this century. However, while the current conservation emphasis is on species associated with mature moist forests (Wilson 1988), conservationists must recognize that development of permanent pasture and agricultural plots consumes both mature forest *and* young second growth. Thus, the current conservation status of early-successional birds could be easily reversed as disturbed second-growth sites are converted to permanent agriculture.

Open Habitats

Open habitats represented a wide variety of natural (e.g., marsh, grassland, coastal dune) and modified (e.g., pasture, open agricultural field) vegetation types that were characterized by few broadleaved, woody plants and low canopy closure (usually <10%). As such, these habitats might be expected to support few bird species. Sample size (19 sites) was small for this vegetation type, so trends should be viewed with caution.

Migrants were relatively uncommon in open habitats, although they did comprise one-fifth of all species detected there (Fig. 6-1B). The average number of migratory species recorded within disturbed, open plots (seven species) was more than twice that found in natural, open habitats (three species; $F = 2.95$; df = 1, 17; $P = 0.104$). These data indicate that, whereas open pastures and agricultural fields support relatively few species, marshes, dunes, and natural grasslands may be used even to a lesser extent by migratory landbirds (see also Karr 1976). Although not a critical habitat for migratory land birds as a group, natural open habitats should not be characterized as "expendable" because marshes and grasslands are important habitats for overwintering shorebirds (Myers 1980) and many resident species. Migratory landbirds often found in these low stature habitats include Scissor-tailed Flycatcher, Yellow-rumped Warbler, Palm Warbler, Common Yellowthroat, Indigo Bunting, Painted Bunting, Savanna Sparrow, Grasshopper Sparrow, Bobolink, and Dickcissel.

Wildlife management practices used in the highly agricultural areas of North America and Europe also may be applicable to agricultural landscapes in the Neotropics. For example, the maintenance of hedgerows, woodlots, stream-side (wooded) buffer zones, and scattered overstory trees could provide some migratory species with suitable habitats, even when most of the original forest in an area had been converted to pasture or cropland (Lynch 1989, Gradwohl and Greenberg 1991, Saab and Petit 1992). Gradwohl and Greenberg (1991) suggested that, for conservationists, these practices represent "making the best of a bad situation". Indigenous farmers in Latin

America have incorporated these principles into their land-use practices for centuries, but with mounting population pressures and a switch from family gardens to large-scale monocultures, these conventions are likely to be reduced or eliminated in many areas (Alcorn 1990).

Artificial Habitats

Artificial habitats in this study were predominantly residential yards and agricultural plantations dominated by tree crops (e.g., *Citrus*, *Coffea*), thereby representing highly disturbed areas. The unnatural vegetation structure (e.g., hedges, widely spaced trees) and sometimes exotic tree species result in artificial habitats, which may or may not resemble native vegetative associations. The discussion that follows is supplemented by published accounts based upon mist-netting because few studies based upon visual surveys were available. Trends identified here should be considered preliminary and are not based upon statistical analyses.

Residential and agricultural habitats supported numbers of migratory species comparable to the other habitat types considered in this paper (Fig. 6-1A). Within these habitats, *Citrus* groves supported approximately 25% more species, on average, than did residential or agricultural areas with overstory trees. That *Citrus* plantations seem to harbor disproportionate numbers of migratory species has been noted by others (e.g., Mills and Rogers 1992, Robbins et al. 1992), but reasons for this pattern remain obscure. In the present study, relative migrant abundance in residential/agricultural habitats may be misleading because most of those studies were conducted at higher latitudes (e.g., Mexico, Belize), where geographic location could have accounted for the inflated numbers (see above). However, comparison of migrant species richness at sites located between 16° and 25° N confirmed that residential/agricultural habitats supported numbers of migrants similar to those found within native broadleaved vegetation types. Thus, certain residential and agricultural habitats do not appear to be incompatible with habitat requirements of some migratory species. Twice as many migratory species were found on mainland sites compared to island sites.

Overall, migrants represented a high proportion of the total species (43%) in residential and agricultural habitats (Fig. 6-1B), which suggests that these habitats may be poor for resident birds. As an example, Mills and Rogers (1992) captured 25 migratory and 37 resident species during 2 years of mist-netting in a *Citrus* grove in Belize. In comparison, Petit et al. (1992; D.R. Petit, unpublished data) detected 20–25 migratory species, but approximately 2.5 times as many resident species within each of three broadleaved habitat types <50 km away from Mills and Rogers' *Citrus* grove. Mills and Rogers (1992) suggested that migrants were more abundant than residents in *Citrus* groves because migrants were attracted by the abundance of insects associated with rotting fruit, and that nearly all insects found there were too small (nearly all insects were <5 mm) for large-billed residents to exploit effectively. Likewise, Champe (1993) recorded similar numbers of long-distance migratory species in *Citrus* groves and native, broadleaved forests in south Florida during winter. However, she found that *densities* of most of these species were lowest in *Citrus*, highlighting one of the limitations of using only species lists in conservation planning.

Coffee plantations cover extensive areas of tropical America (Leonard 1987) and represent potentially important winter habitats for migratory birds (Wunderle and Waide 1993). Coffee can be grown under management regimes that vary in their suitability for some migratory species. For example, "shade coffee" is cultivated as an understory crop with an overstory of canopy trees. "Sun coffee," on the other hand, is grown without the benefit of shade trees. Shade coffee often supports more migratory species because the presence of additional vegetative layers provides habitat for canopy and subcanopy foragers (Robbins et al. 1992, Wunderle and Waide 1993; L. J. Petit, unpublished data). Canopy trees that flower or fruit during the boreal winter also provide resources for nectarivorous and frugivorous migrants in coffee plantations (L. J. Petit, unpublished data).

Other crops, such as corn (*Zea*), rice (*Oryza*), and sugarcane (*Saccharum*), appear to provide few migratory species with appropriate habitat (Robbins et al. 1992; L. J. Petit, unpublished data). The lack of vegetative structure and the often heavy use of pesticides in these fields result in low-quality habitat for most species. Several species, however, can occur in high abundance in rice (Indigo Bunting; Robbins et al. 1992) and sugarcane (Barn Swallow; L. J. Petit, unpublished data).

A broad spectrum of migratory birds use agricultural and residential habitats (e.g. Tramer 1974, Mills and Rogers 1992, Robbins et al. 1992, Baltz 1993), including species more typical of forests (e.g., Wood Thrush, Hooded Warbler, Black-and-white Warbler, Northern Waterthrush), edges and second growth (White-eyed Vireo, Magnolia Warbler, American Redstart, Orchard Oriole) and grasslands (Common Yellowthroat, Indigo Bunting). Because many areas in the tropics are being converted to *Citrus* groves, coffee plantations, agricultural fields, and residential areas, determination of the suitability and importance of these habitats for migrants will be an important component of future conservation research.

Summary of Habitat Use

Analyses presented here provide an indication of the relative use of different tropical habitats by migratory birds. However, caution must be exercised when these results are applied to specific situations because the relative importance of each environmental variable (e.g., vegetation type, vegetation height, latitude) in the context of all other variables could not be fully assessed in this study (small sample sizes prohibited use of more sophisticated statistical analyses). Clearly, all tropical habitats contained some migratory birds. Disturbed habitats, whether natural (e.g., regenerating or selectively logged wet forest) or unnatural (*Citrus* plantation), typically supported a greater number of migratory species than similar undisturbed habitats. Some arguments caution against using relative abundance of organisms as an index of suitability (e.g., Van Horne 1983, Gibbs and Faaborg 1990) because high local densities could reflect rigid dominance relationships (i.e., ideal despotic distribution; Brown 1969, Fretwell and Lucas 1970) and not distributions based upon relative habitat suitabilities (i.e., ideal free; Fretwell and Lucas 1970). However, in some tropical habitats, such as undisturbed, moist forest, density of migrants can be especially low (Hutto 1980, Willis 1980, Mabey and Morton 1992) and winter territories of individuals of some species often are not contiguous (Bennett 1980, Chipley 1980), which suggests that some species are not at their carrying capacities in these habitats. In other areas, however, migrants do occur at high densities (Rappole and Warner 1980, Holmes et al. 1989), and dominance (i.e., territorial) relationships among individuals could force some birds into less suitable habitats. For example, young birds or females may be forced into suboptimal, disturbed habitats, where they overwinter in high densities relative to undisturbed forest (see below). If so, the mere distributions of individuals across habitat types does not offer an accurate reflection of the relative suitabilities of those habitats, or indeed of habitat "preference" patterns. Rather, over-winter survival and other potential indicators of habitat quality, such as physical condition of individuals and age and sex ratios, need to be integrated with estimates of relative abundance into an assessment of the value of each habitat to migratory birds (Holmes et al. 1989).

On the other hand, distributions of individuals among habitats may indeed provide a reliable indication of the relative suitabilities of those vegetation types (Orians and Wittenberger 1991). For example, physical condition (Greenberg 1992a) and survival rates (estimated indirectly through recapture rates; Rappole and Warner 1980, Robbins et al. 1987, Blake and Loiselle 1992, Conway et al. 1995) are not necessarily improved for "forest-dwelling" species in undisturbed forests relative to disturbed habitats (but, see Rappole et al. 1989). If this classical interpretation of habitat use by populations is correct (that is, density reflects suitability), then results presented here and those of others suggest that many migratory species are not adversely affected by at least some current land-use practices in Latin America and the Caribbean. Indeed, migrants often reach their greatest abundance in certain disturbed natural and artificial habitats. The task for students of migratory bird ecology is to ascertain the level (and form) of disturbance tropical forests can withstand and still be suitable as winter habitat for Neotropical migrants (also see Lynch 1992). Clearly, however, wildlife conservation plans also must consider resident tropical birds (and other taxa), which often are more adversely affected by alteration of natural forests (e.g., Hutto 1989, Kricher and Davis 1992).

At present, only a few general types of land use appear to be particularly incompatible with habitat use by all but a few species of migratory landbirds: (1) actively grazed pastures that contain few woody plants (Karr 1976, Saab and Petit 1992; also see Lynch 1992, p. 192; Robbins et al. 1992, p. 216; Champe 1993); and (2) large monocultures of monocotyledonous crops, such as bananas (Robbins et al. 1992), rice (Robbins et al. 1992; L. J. Petit, unpublished data), and sugarcane (L. J. Petit, unpublished data). Most migrants occupying those agricultural lands are habitat generalists (e.g., Yellow-rumped Warbler) or are characteristic of highly disturbed vegetation types (e.g., Common Yellowthroat, Indigo Bunting) in their wintering areas, such that winter habitats are not presently limited for those species.

Analyses presented above indicate that migratory birds as a group show a stronger affinity toward disturbed habitats than do resident species, which suggests that some forms of tropical forest alteration may affect migratory species to a much lesser extent than residents. Examination of avifaunal response to direct habitat destruction demonstrates that, indeed, migratory birds as a group may be more resilient to natural and human-caused habitat disruption than are resident tropical species. For example, on the Yucatan Peninsula, dry forests that were completely defoliated by Hurricane Gilbert in September 1988 were recolonized by numerous migrants within a few weeks after the storm. Less than 2 years after the hurricane, the abundance, diversity and species composition of the migrant community in unburned hurricane-damaged forest were similar to pre-hurricane values (Lynch 1991, 1992). The recovery of the migrant community to pre-hurricane conditions was slower in hurricane-damaged forest that was further disturbed by wildfires. However, even in tracts where >90% of the trees had been killed and the ground cover of vegetation completely eliminated by fire, a diverse assemblage of migrants was present several months later. Only two previously common forest-associated migrant species (Wood Thrush and Kentucky Warbler) were absent from burned hurricane-damaged forest 17 months after wildfires razed the area, and one of these (Wood Thrush) reappeared during the second winter after the fire.

In contradiction to the relatively minor and transitory effects of the hurricane and fire on migratory species, the effects of the hurricane and wildfire on resident birds were both more severe and longer lasting. Many resident species, particularly frugivores and nectarivores, completely disappeared from the study area when it burned, and still had not recolonized the forest $3\frac{1}{2}$ years later (Lynch 1991; J. F. Lynch, unpublished data). As a result, migrants made up a higher proportion of both species and individuals in burned than in unburned forest for at least several years after the disturbance occurred. Askins and Ewert (1991) also showed that resident birds on a Caribbean island were more adversely affected by Hurricane Hugo than were overwintering migratory species. Likewise, in Los Tuxtlas, Mexico, a greater percentage of the individuals and species that were mist-netted after part of a forest fragment was cleared for cattle pastures were migrants (individuals = 62%, species = 42%) compared with their representation before alteration of the forest (54% and 33%, respectively; Rappole and Morton 1985).

REGIONAL PATTERNS OF HABITAT USE

The above summary of habitat use can be used to formulate general recommendations as to which vegetation types and areas are most important in general as overwintering habitats for migratory birds. However, because vegetative and historical factors unique to certain regions may greatly influence the habitats used by migratory birds, as well as the species found there, these general patterns may not be fully applicable across all regions in the Neotropics. Below, habitat use by migrants in six broad ecopolitical regions in Latin America and the Caribbean is briefly reviewed. These overviews also are meant to identify those regions and habitats most in need of further ornithological investigation.

Eastern Mexico

The ecology of overwintering Neotropical migrants has been studied intensively in two areas of eastern Mexico (defined here as sites east of the continental divide). Rappole and coworkers (Rappole et al. 1989, Rappole and Warner 1980, Rappole and Morton 1985, Winker et al. 1990) have surveyed migrants at lower elevations in Los Tuxtlas, a humid (annual rainfall = 400–500 cm), mountainous area along the Gulf coast of southern Veracruz. Several investigators have studied Neotropical migrants in the northern half of the Yucatan Peninsula (states of Campeche, Quintana Roo, and Yucatan). This area supports natural vegetation from xeric thorn scrub in Yucatan (annual rainfall = 50–75 cm), to semi-evergreen tropical forest in southern Campeche and Quintana Roo (annual precipitation = 110–140 cm). Migrants in the Yucatan have been surveyed in dry and moist forests, various stages of successional growth, pastures and agricultural fields, residential areas, mangrove swamps, savannas and coastal dune scrub (Tramer 1974, Waide 1980, Waide et al. 1980, Lynch et al. 1985, Lynch 1989, 1991, 1992, Lopez Ornat 1990, Lopez Ornat and Lynch 1991, Greenberg 1992a).

Several general patterns have emerged from these and other studies in eastern Mexico. First, migrants occur across essentially the entire available spectrum of vegetation types in eastern Mexico, but species richness is often greatest in early successional growth/scrub. For example, the number of migratory species recorded per site in early successional growth in eastern Mexico (16 species) was substantially higher than that found overall in this habitat type (10 species). In contrast, numbers of species in east Mexican moist (12 species) and dry (10 species) forests were similar to those observed throughout the Neotropics in general (see Fig. 6-1A). Migrants in eastern Mexico represent a slightly more substantial proportion of the local avifauna in broad-leaved (moist and dry) forests (31% of all species) compared with other broad-leaved forest communities throughout the Neotropics (26%; see also Fig. 6-1B). Migratory species that occupy a broad range of habitats during winter in the Yucatan Peninsula include Least Flycatcher, Gray Catbird, White-eyed Vireo, Northern Parula, and Magnolia Warbler (Lynch 1989).

Second, overwintering migrants are both diverse and abundant in eastern Mexico. In the Yucatan, despite the fact that the species pool of residents far exceeds that of migrants (Paynter 1955), migrants commonly constitute 30–50% of the individual birds, and 30–40% of the species surveyed in a given habitat (Lynch 1989, 1992). In Los Tuxtlas, about 30% of the bird species present during winter are Neotropical migrants (Coates-Estrada et al. 1985), and migrants made up >50% of the individuals that were mist-netted in both heavily and lightly disturbed moist forest (Rappole and Morton 1985).

Finally, many migrants are more abundant in forest than in heavily disturbed or open habitats, although few migratory species are totally restricted to mature forest in eastern Mexico. Disturbed vegetation is, indeed, heavily used by some common migratory species in eastern Mexico, notably Common Yellowthroat, Yellow-breasted Chat, Indigo Bunting, and Painted Bunting. However, migrants such as Yellow-bellied Flycatcher, Wood Thrush, Yellow-throated Vireo, Black-and-white Warbler, Kentucky Warbler, Hooded Warbler, American Redstart, Black-throated Green Warbler, and Blue-winged Warbler are more common in mature forest than in second-growth habitats (Rappole and Warner 1980, Lynch 1989, Greenberg 1992a).

Many ornithological investigations have been conducted in eastern Mexico, especially compared to other regions, but the geographic scope is still relatively limited.

More detailed attention is needed, for example, in the dry forest and scrub habitats in northeastern Mexico, and in the moist montane and cloud forest associations of the eastern highlands of Veracruz and Chiapas.

Western Mexico

Western Mexico is characterized by relatively low annual precipitation (typically < 150 cm) and varied topographic relief associated with tall mountain ranges. These factors have contributed to the predominance of four major forest types characteristic of dry climates and/or high elevations in that region: (1) thorn forest, (2) deciduous and semideciduous broadleaved forest, (3) cloud forest, and (4) mixed (pine–oak–fir) forest. The overwintering migratory bird fauna is derived almost exclusively from North American breeding populations west of the 100th meridian, which results in only moderate overlap in winter bird community composition between western Mexico and sites on the Caribbean slope of Mexico (e.g., compare Hutto 1992 and Lynch 1992). The majority of survey work on Neotropical migrants in western Mexico has been conducted during the past 15 years by Hutto (1980, 1989, 1992), who has demonstrated that, although migrants are most abundant in disturbed habitats, many species can be found in undisturbed natural vegetation.

In dry deciduous forests of western Mexico, disturbance does not appear to influence migrant richness greatly, much as in dry forests of the Yucatan, but in contrast to patterns observed in moist forests throughout the Neotropics (see above). The greatest number of migratory species in western Mexico overwinter in deciduous, cloud, and mixed coniferous forests (Hutto 1992); cloud and coniferous forests appear to be particularly important for migrants here compared to those vegetation types in other regions (Petit et al. 1993). Some migratory species exhibit close associations with specific vegetation types in western Mexico: Western Kingbird and Virginia's Warbler in thorn forest; Western Flycatcher Ovenbird, and Summer Tanager in deciduous forest; Black-throated Green Warbler and MacGillivray's Warbler in cloud forest; Broad-tailed and Calliope Hummingbirds, Red-naped Sapsucker, and Red-faced, Townsend's, and Hermit Warblers in mixed coniferous forest; and Bell's Vireo, Lucy's Warbler, and Grasshopper Sparrow in young second growth. However, many other species are habitat generalists (Hutto 1992), such as Warbling Vireo, Orange-crowned Warbler, Nashville Warbler, Black-throated Gray Warbler, Black-and-white Warbler, and Northern Oriole. Because populations of many long-distance migratory species that breed in western North America are concentrated in western Mexico during the winter, maintaining intact forests in this region is particularly critical to conservation efforts. Hutto's (1980, 1992) work has provided a wealth of information on migratory bird use of forested habitats throughout much of western Mexico. However, a gap exists in our knowledge of the use of heavily modified habitats (e.g., residential and agricultural areas, pasture) by those species, limiting predictions of the effects of future land use practices on migrants.

Northern Central America

Northern Central America (here defined as Belize, Guatemala, Honduras, and El Salvador) supports a diverse array of migrants, both because of the great variability in vegetation types present within the region and because of the close proximity to North American breeding grounds. The dry Pacific slope (annual rainfall typically 100–200 cm) of the northern isthmus is partitioned from the moist Caribbean lowlands (rainfall 150–500 cm) by a *cordillera* often exceeding 2500 m. Consequently, vegetation communities vary from cloud, montane and pine forests at higher elevations to thorn, semideciduous and wet-forest associations in lowlands. Disturbed and artificial habitats (e.g., plantations, residential areas, forest fragments) are, as elsewhere in the Neotropics, commonplace in northern Central America. The Pacific slope harbors many migratory species from western North America, thereby further enhancing the already rich regional avifauna originating from eastern North America.

Few studies designed to examine habitat use by migratory birds have been conducted in northern Central America, especially compared to the large-scale surveys in Mexico, for example. Most of the following discussion is based upon quantitative surveys conducted in Belize and from qualitative accounts of habitat use in Guatemala, Honduras, and El Salvador. However, because of the limited data on habitat use by migrants on the dry Pacific coast, this general summary may not be applicable to El Salvador, or certain areas of Honduras and Guatemala.

Habitat use by migratory birds in northern Central America is generally similar to patterns for the Neotropics as a whole (see Fig. 6-1), although migrants in northern Central America do not appear to be as well represented in residential habitats. Migrants often reach their greatest abundance in disturbed moist forests, including early successional growth and forest edges (e.g., Kricher and Davis 1992, Petit et al. 1992), and in *Citrus*, cacao, *Coffea*, and other tree crop plantations (Mills and Rogers 1992, Robbins et al. 1992, Vannini 1994, D. R. Petit, personal observation). Pastures managed by repeated mowing or burning, and active agricultural fields support few migratory species (Robbins et al. 1992, Saab and Petit 1992). Undisturbed moist forests contain about 25% fewer migratory species and individuals than disturbed forests in northern Central America (e.g., Robbins et al. 1992, Petit et al. 1992), although several species (e.g., Wood Thrush, Kentucky Warbler, and Worm-eating Warbler) often reach their greatest local densities in the interior of undisturbed forest (Russell 1964, Monroe 1968, Land 1970, Petit 1991, Whitacre et al. 1993). Migrants comprise a smaller proportion of the total avifauna in undisturbed compared to disturbed forests. For example, in Belize (this study; $n = 18$ sites), migrants represented 38% of species detected within disturbed forest habitats, but only 21% in undisturbed forests. Moreover, mangroves also appear to be underutilized by migrants in this region compared to areas outside of Central America (Petit et al. 1993).

Lowland pine savanna and upland pine forest cover large areas in Northern Central America and represent important winter habitats for at least several migratory species. In the pine–savanna scrub of Belize, the Black-and-white Warbler, Black-throated Green Warbler, Yellow-throated Warbler, and Common Yellowthroat form mixed-species flocks with residents, while narrow bands of broadleaved vegetation along small water courses in pine-dominated habitats often contain American Redstarts, Magnolia Warblers, and Ovenbirds (Petit 1991, D. R. Petit, unpublished data).

Because little research in northern Central America has been conducted outside of Belize, future efforts to quantify migrant use of native and disturbed habitats in Honduras, Guatemala, and El Salvador would improve our understanding of migratory bird–habitat relationships in this region (e.g., see Whitacre et al. 1993). Indeed, the relatively high density of migrants in northern Central America requires that intense research and monitoring activities be integral components of an overall conservation plan for migratory birds.

Southern Central America

Lower Mesoamerica (Nicaragua, Costa Rica, and Panama), like the northern region, exhibits a general dichotomy in rainfall between Pacific and Caribbean slopes. The distribution of moist and dry forests reflects this climatological difference. Although broadleaved forests are well represented in this region, pine–savanna, grasslands, agriculture, pasture, and early second growth also cover extensive areas. Holdridge et al. (1971), Howell (1971), Janzen (1973), and Hartshorn (1983) provide descriptions of some of the ecological life zones and habitats of this region. Extensive research has been conducted on migratory birds in certain areas of Panama and Costa Rica, but not in Nicaragua.

Whereas many migratory species in southern Central America can be found in early second growth, forest edge, and open forest, only a small (<20%) proportion appear to be dependent upon undisturbed mature forest (e.g., Willis 1980, Stiles 1985,

Blake et al. 1990, Blake and Loiselle 1992). For example, based upon 28 moist forest stands used in this study (incorporating both visual and mist net surveys), slightly-to-moderately disturbed sites supported twice as many migratory species ($\bar{x} = 8$) than relatively pristine forests (four species). However, because more migratory species were found in mature moist forest (disturbed and undisturbed; $\bar{x} =$ six species) than in either early successional growth (4) or in natural grasslands and pasture (3), the importance of forested tracts to migrants in southern Central America should not be underestimated. In fact, at least a dozen species use heavy forest cover in this region, including Chuck-will's-widow, Olive-sided Flycatcher, Acadian Flycatcher, Wood Thrush, Northern Waterthrush, and Blackburnian, Worm-eating, and Kentucky Warblers (Ridgely and Gwynne 1989, Stiles and Skutch 1989, L. J. Petit, unpublished data).

Few studies or quantitative surveys have been conducted in mangroves in this region, but the current literature review suggests that fewer species use this habitat type in southern Central America than in regions outside of Central America (Petit et al. 1993). In Panama, Northern Waterthrushes and Prothonotary Warblers are highly dependent upon mangroves, whereas American Redstarts and Black-and-white Warblers occur there in small numbers (L. J. Petit, unpublished data). Natural grassland/savanna and cattle pastures are avoided by migratory birds in Panama (Karr 1976, L. J. Petit, unpublished data) and Nicaragua (Howell 1971). Many intensive studies on migratory and resident birds have been conducted within several sites in this region (e.g., Barro Colorado Island and Pipeline Road, Panama; and Estacion Biologica La Selva, Costa Rica), but few simultaneous surveys of native habitat types have been initiated. Moreover, Nicaragua is in dire need of ornithological investigation, and few quantitative studies have surveyed highly disturbed habitats (e.g., agriculture, pasture, forest edge, fragments) in any country in lower Mesoamerica. Furthermore, the most remote areas in this region, such as Darien and Bocas del Toro in Panama, have received inadequate attention from avian ecologists.

South America

Natural vegetative formations in South America are highly diverse (for example, desert, open pampas, dry thorn forest, deciduous forest, rain forests, cloud forest, and barren paramo above 3000 m), so the low density and richness of migratory birds in this region probably are not due to lack of suitable habitats *per se*. General habitat relationships of many species inhabiting South America are reasonably well known but, because few formal surveys have been conducted, a quantitative summary of habitat use by migratory species is difficult. In an extensive survey of forests along an elevational gradient in Peru, Robinson et al. (1988) found that Neotropical migrants were most abundant in early successional and edge habitats along river systems. Orejuela et al. (1980) and Pearson (1980) also found that migrants were underrepresented in undisturbed moist forests compared to dry forest and early second growth. Even within moist forest associations in South America, significantly more migratory species occurred on average in disturbed (six species) than undisturbed (four species) forests (this study, using both mist-netting and visual survey data).

Most migratory species that overwinter in South America are associated with relatively tall, moist/wet forests, although, as already mentioned, these include moderately disturbed sites, gaps and forest edges. Thus, the importance of maintaining these forests is obvious; conversion to agricultural lands or pasture would remove the critical structural components of these habitats. In fact, a relatively large proportion of the species that overwinter in South America have been identified in a companion paper (Petit et al. 1993) as being particularly susceptible to alteration of broadleaved forests [e.g., Chuck-will's-widow, Olive-sided Flycatcher, Veery, Prothonotary Warbler, (mangroves), Blackburnian Warbler, Bay-breasted Warbler, Cerulean Warbler Blackpoll Warbler, and Scarlet Tanager], highlighting the urgency of maintaining large tracts of such habitats. Despite the large land area of tropical South America, relatively few quantitative investigations of overwintering migratory birds have been conducted there. Studies of the effects of deforestation on migratory birds are urgently needed because many species that overwinter in South America are highly dependent upon undisturbed forest (Petit et al. 1993).

Caribbean Islands

With few exceptions, all major vegetation associations found on the mainland are represented on Caribbean islands (see Lugo et al. 1981). Arendt (1992) provided a synopsis of the major "bird habitats" in the region. High human population densities on many Caribbean islands have led to severe deforestation during the past two centuries and this situation poses special conservation problems for the region, i.e., there are few large tracts of forest remaining and developed areas comprise a large portion of the total land area (e.g., McElroy et al. 1990). Two major reviews of habitats used by migratory birds in the Caribbean have been published (Arendt 1992, Wunderle and Waide 1993). These and other studies are summarized below.

Overall, migrant species richness and abundance were higher in disturbed than in more natural habitats, but this trend varied greatly among islands and sites. For example, on Cuba and the Dominican Republic, moist forest habitats contained the greatest number of species (as estimated by rarefaction), but on Jamaica, dry forest and shade coffee plantations supported the most species (Wunderle and Waide 1993). The most consistent pattern from the Caribbean islands is the high use of coastal mangroves by migrants. A survey of ornithologists by Arendt (1992) revealed that >50% of migratory landbirds on Caribbean islands used mangroves to some extent during winter. In the Greater Antilles, Yellow-bellied Sapsucker, Black-and-white Warbler, American Redstart, Magnolia Warbler, and Northern Waterthrush reached their greatest densities in mangroves (Wunderle and Waide 1993).

Preservation of native habitats in this region is particularly critical for several migratory species whose winter populations are largely or entirely restricted to Caribbean islands; for example, Kirtland's, Prairie, Cape May, and Black-throated Blue Warblers. Because single natural or human-induced catastrophies, such as hurricanes, droughts, or forest conversion, can severely affect species with limited geographic ranges, migrants whose winter populations are limited to the Caribbean (or are similarly restricted on the mainland) may be the most likely to undergo severe population declines or extinction (Wunderle and Waide 1994). For example, Kirtland's Warbler, confined to several islands in the Bahamas and Greater Antilles (Arendt 1992), exhibited major declines in breeding populations in Michigan following severe drought in the Caribbean during the previous winter (Radabaugh 1974). Similarly, the presumed extinction of Bachman's Warbler, also limited to several Caribbean islands during winter, may be linked to relatively local events on the wintering grounds (Hamel 1986). Ornithologists have devoted considerable attention to many of the habitats and islands in the Caribbean, such that ample information is now available on the importance of many habitat types to migrants. Because most forested habitats on many islands have already been extensively logged or converted to other land uses, perhaps conservation efforts and research in the Caribbean should be directed at assessing the effects of agricultural and urban land uses on populations of migratory birds (see also Arendt 1992). Protection of migratory (and resident) birds on densely populated Caribbean islands depends in large part on identifying and implementing land-use practices that are compatible with both economic goals and preservation of natural vegetation types. Arendt (1992), in collaboration with other bird experts from the region, outlined four major requirements necessary to achieve that goal in the Caribbean: establishment of protected reserves; restoration of degraded habitats; legislation and education, and ecological research (also see Wunderle and Waide 1994).

GEOGRAPHIC VARIATION IN COMPOSITION OF MIGRANT ASSEMBLAGES

The similarity of winter migrant assemblages occupying broadleaved forest tends to decrease with increasing distance between sites (Fig. 6-7). However, similarity of assemblages is not always a simple function of geographic proximity. For example, few migratory species are shared between forest sites in western Mexico (Hutto 1980, 1992) and sites in the northern Yucatan Peninsula (Lynch 1989), but the latter area shows a relatively high faunal resemblance with migrant communities in Costa Rica and Panama (Table 6-2, Fig. 6-7). In general, similarities between the composition of winter migrant communities in *northern Central America and Mexico*, where most migratory species overwinter, are much greater from NW–SE along the Gulf–Caribbean coastal plain than they are from E–W between the Atlantic and Pacific coastal plains (Table 6-2). In these regions, distinct migratory bird communities may arise because of restrictions placed by tall mountain ranges on movement of individuals. Alternatively, differences in bird community composition among these and other regions in the Neotropics may reflect variation in vegetation composition associated with climatic changes due to major montane barriers or to latitudinal gradients. For example, migratory species that breed in the dry and montane portions of western North America tend to overwinter in the dry and montane regions of western Mexico and Guatemala. In contrast, the humid Atlantic coastal slope of Mexico and Central America tends to be inhabited by overwintering migrants that breed in mesic habitats of eastern North America (Hutto 1985). Thus, species may tend to occupy similar vegetation types throughout the year, and availability of those habitats may

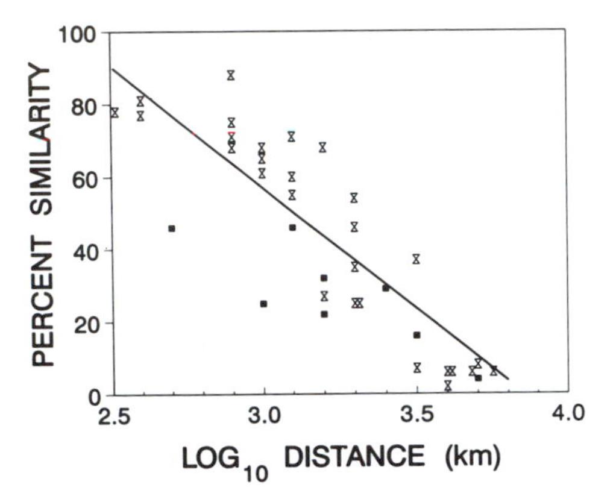

Figure 6-7. Relationship between the similarity (coefficient of community; Pielou 1975) in the composition of migratory bird assemblages in forest and the distance (km) between sites. Equation for regression line: percent similarity = 265.7 − 70.1 ($\log_{10}$ distance); $R^2 = 0.74$. Squares represent relationships involving western Mexico, and illustrate that migratory bird assemblages in western Mexico differ from other locations in Latin America and the Caribbean more than would be expected simply based on their geographic proximity.

Table 6-2. Coefficients of community (Pielou 1975) for pairwise comparisons among nine forested sites in the Neotropics. Possible values range from 0 (no species in common at the two sites) to 100 (all species occur at both sites).

Site	Site Number								
	1	2	3	4	5	6	7	8	9
St John's, Virgin Islands (1)	—								
Western Mexico (2)	16	—							
Yucatan, Mexico (3)	35	46	—						
Los Tuxtlas, Mexico (4)	37	46	88	—					
Belize (5)	46	25	78	75	—				
La Selva, Costa Rica (6)	25	32	65	68	68	—			
Santa Rosa, Costa Rica (7)	25	22	61	60	68	77	—		
Panama (8)	27	29	55	54	71	81	71	—	
Peru (9)	0	8	6	5	6	6	6	7	—

Data sources: Stiles (1980), Willis (1980), Coates-Estrada et al. (1985), Lynch (1989, 1992), Terborgh et al. (1990), Hutto (1992), Askins et al. (1992), and D. R. Petit (unpublished data).

restrict the areas where a species can spend the winter.

Factors associated with the breeding range of a species may also constrain where it overwinters in the Neotropics. For example, most species that breed in western North America tend to overwinter in the dry habitats of western Mexico and the Pacific versant of northern Central America despite the great variability of habitats used by those species during the breeding season. This observation argues against vegetation being the sole determinant of winter ranges. Variation in bird communities among regions may also relate to historical factors, and not to present-day characteristics of these landscapes. For example, Kirtland's Warbler overwinters exclusively on Caribbean islands, often in coastal scrub and dry forest habitats similar to those found on the Yucatan Peninsula of Mexico. Furthermore, overwintering Kirtland's Warblers are restricted to several islands in the Bahamas and Greater Antilles (Arendt 1992), although dry scrub and forest are found on many nearby islands.

Composition of winter migrant assemblages in South America appears to be particularly distinct from that in other regions of the Neotropics (Table 6-2). The similarity between the migratory avifauna occupying forested sites in Central America/Caribbean and the migrant assemblage at Cocha Cashu (Peru) was independent of geographic proximity. In contrast, migratory bird composition on St John's in the Caribbean Sea was more similar to that found on the mainland, varying as a function of distance between sites (Table 6-2).

These analyses suggest that geographic proximity is a good predictor of similarity in the composition of migratory birds inhabiting tropical broadleaved forests, especially within Central America and eastern Mexico. As such, it may provide justification for a region-wide approach to conservation and management of forest-dwelling birds in the Neotropics. Most current nongame management efforts in North America are aimed at large assemblages of species, often indirectly through indicator species (Salwasser et al. 1983) or ecological guilds (Severinghaus 1981). Recently, a direct ecosystem-level approach (e.g., Hutto et al. 1987, Grumbine 1990) has been proposed. Despite the widely divergent practical application of these three concepts, however, all are based upon the idea of minimization of the number of management prescriptions (hence monetary cost and labor) necessary to protect habitats and species. Typically, as the ecological (or taxonomic) similarity increases among species assemblages, so too does the specificity and effectiveness of management plans. Hence, development of conservation plans on a region-by-habitat basis may be a cost-effective and ecologically sound approach to the protection of overwintering migratory birds. Although formation of a single plan to cover each habitat type might be more efficient than multiple regional strategies, disparities in bird species composition among regions could seriously reduce the utility of such an approach because of the disparate ecological requirements of species comprising those assemblages. An argument could be made that a single plan might be effective for all moist forests if a "conservative" approach were adopted; i.e., one that was implemented for the most "disturbance-sensitive" species. However, conservation biologists must justify their recommendations convincingly, especially in countries where there is great pressure on the land base. A conservative plan that is not fully applicable in some regions, although it may be in others, is not constructive in merging conservation efforts with long-term economic and social needs of developing nations (Geerling et al. 1986, Gow 1992).

HABITAT USE AT THE SPECIES LEVEL

Documentation of geographic and regional patterns of habitat use for migratory birds as a group provides critical information on those habitats and areas that are most important for overwintering migrants, but it offers little insight into which habitats are used by individual species. Clearly, species exhibit unique responses to available habitats during winter, much as they do during the breeding season. Recognition of habitat associations of individual species allows identification of those species not adequately protected under general conservation and management plans. While many researchers have noted the urgent need to document habitat needs of migratory species during winter, surprisingly little *quantitative* data have been collected in this area (Greenberg 1986, Hutto 1989, Lynch 1989). The lack of an objective, quantitative approach to classification of winter habitats by avian ecologists has resulted in confusion over the habitat associations of migratory species, thereby making conservation recommendations dubious (see, for example, Martin 1992). Future research on habitat associations of migrants should concentrate on quantitative assessments of habitat requirements, following, for example, the approach of Blake and Loiselle (1992), Petit (1991), Powell et al. (1992), and Wunderle and Waide (1993).

A review of habitat use by each species is beyond the scope of this chapter. However, many avian ecologists believe that the long-distance migratory species currently most susceptible to severe population declines are those that depend upon undisturbed forests during the boreal winter, as well as during the summer (e.g., Terborgh 1989). In light of the need to identify those species that may require special conservation attention, a list of those migrants that are generally thought to be in the greatest danger of population declines due to destruction of forests on the wintering grounds is presented below; the quantitative methods used to derive this list are presented in Petit et al. (1993). The 45 species identified in Table 6-3 represent 42% of the 107 Neotropical migratory species that overwinter primarily in mature phases of tropical forests (Rappole et al. 1983). This is a conservative list, however, because not all taxa were equally evaluated because of lack of information about winter habitat requirements (e.g. swifts, swallows). Members of the families Tyrannidae, Muscicapidae, Vireonidae, and Emberizidae (Parulinae) comprised nearly the entire register, the same taxa (and species) that typically are included on similar lists derived from habitat studies during the *breeding* season. Evident, then, is one of the major problems confronting ecologists attempting to unravel the cause(s) of population declines of migratory birds: many species tend to breed and overwinter in the same general habitat type, thereby hampering partitioning of habitat-based events (e.g., mortality) between seasons. Petit et al. (1993) showed that migratory species highly restricted to undisturbed broadleaved forests during winter had average breeding season population declines more than eight times that exhibited by forest-dwelling migrants that were more tolerant of winter habitat disturbance. Robbins et al. (1989a), also using Breeding Bird Survey data, suggested that events on the wintering grounds were the main cause of population declines. However, Hussell et al. (1992), using an independent data set collected during migration, found no firm evidence to support that conclusion.

Table 6-3. Forty-five long-distance migratory landbirds most likely to be negatively affected by destruction of tropical forests. Information was taken from Rappole et al. (1983), Diamond (1991), Reed (1992), and Petit et al. (1993). Only species listed by more than one author are included (but see footnote). The list of species is incomplete because little information on winter habitat use exists for certain families (e.g., Apodidae, Trochilidae, Hirundinidae) and because all authors did not necessarily consider all species of migratory land birds in their summaries (e.g., Reed 1992).

Black-billed Cuckoo	Black-throated Blue Warbler
Chuck-will's-widow	Townsend's Warbler
Whip-poor-will	Black-throated Green Warbler
Yellow-bellied Sapsucker	Golden-cheeked Warbler
Olive-sided Flycatcher	Blackburnian Warbler
Western Wood-pewee	Kirtland's Warbler
Yellow-bellied Flycatcher	Bay-breasted Warbler
Acadian Flycatcher	Blackpoll Warbler[a]
Great Crested Flycatcher	Cerulean Warbler
Veery	American Redstart
Gray-cheeked Thrush	Prothonotary Warbler
Swainson's Thrush	Worm-eating Warbler
Wood Thrush	Swainson's Warbler
Black-capped Vireo	Ovenbird
Solitary Vireo	Northern Waterthrush[a]
Yellow-throated Vireo	Louisiana Waterthrush
Philadelphia Vireo	Kentucky Warbler
Blue-winged Warbler	Connecticut Warbler
Golden-Winged Warbler	Hooded Warbler
Colima Warbler	Canada Warbler
Lucy's Warbler	Scarlet Tanager
Chestnut-sided Warbler	Black-headed Grosbeak
Magnolia Warbler	

[a] Included only on the "highly vulnerable" list of Petit et al. (1993).

Continuation or initiation of large-scale studies designed to monitor bird populations and, especially, habitat conditions in breeding, migration, and wintering areas are needed to understand better the relative effects of seasonal events on populations of migratory birds (e.g., Powell et al. 1992, Sherry and Holmes, chapter 4, this volume, Moore et al., chapter 5, this volume). Equally important are intensive, small-scale studies that examine the relative seasonal effects on local population dynamics (e.g., Holmes et al. 1986, 1989). Large-scale studies provide broad, long-term trends that are important in assessing the scope of the problem, whereas local studies may provide better insight into the potential causes of those trends.

Even at the species level, habitat use can vary considerably among locations and individuals, thereby increasing the difficulty in managing habitats for species that have been recognized as susceptible to habitat alteration. Variation in habitat use and ways in which that variability can be incorporated into conservation efforts are discussed below.

Geographic Variation in Habitat Use by Species

Habitat use by most migratory species is believed to be consistent across winter ranges. For example, a species that occurs in moist, mature forest in eastern Mexico will tend to occupy structurally and climatically similar (though floristically different) habitats in Belize, Costa Rica, or Panama. Verifying geographic consistency in habitat use is difficult however, because of the lack of quantitative data within different regions for most species.

Understanding geographic variation in habitat use has obvious conservation implications. Most importantly, it allows for identification of appropriate spatial scales at which conservation and management plans can be enacted (see above; also see Van Horne and Wiens 1991). Martin (1992) provided one example of the danger associated with assuming geographic consistency in habitat use.

To address the issue of geographic variation in habitat use, information was summarized from approximately 50 publications that examined habitat affiliations of individual migratory species (see Petit et al. 1993). Reports derived from study of single vegetation types were not used, as those provided no information on habitat "selection," which is defined operationally as differential abundance of a species in one or more habitats relative to others. Data from some studies were presented qualitatively (e.g., Land 1970, Stiles and Skutch 1989) and others quantitatively (e.g., Waide 1980, Lynch 1989, Blake and Loiselle 1992, Hutto 1992). Habitats used outside the wintering period (e.g., during migration) were excluded when possible. Thus, for each species included in a given study, two types of information were extracted: (1) all "possible" habitats where the species could have occurred in the study area (defined by the author and including only those that were actually surveyed; and (2) the habitat(s) that contained a disproportionate number of individuals or that were stated by the author to be "preferred" by the species. A total of 11 habitats were identified for this analysis: moist forest, dry forest, cloud forest, mangroves, advanced second growth, forest edge, open woodlands/forest, coniferous/mixed coniferous forest, early second growth, residential, and grassland/pasture. Thus, for each species the number of both "available" and "occupied" (preferred) sites for each habitat type were derived. The proportion of available sites that were occupied (occupied/available) provided an indication of the extent of use of each habitat type while controlling for the differences in habitats available across studies and geographic regions. For each habitat type for each species, proportions ranged from 0 (species never occurred there or was less common than expected) to 1.0 (species always occurred there with relatively high abundance). Six regions (identified above) were used to examine geographic variation in habitat use.

Some species showed little variability in habitat use across regions. For example, Common Yellowthroat and Summer Tanager were relatively consistent in their distributions across habitats throughout their ranges (Fig. 6-8). Other species, however, exhibited extensive variation, which suggests that disagreement among avian ecologists about

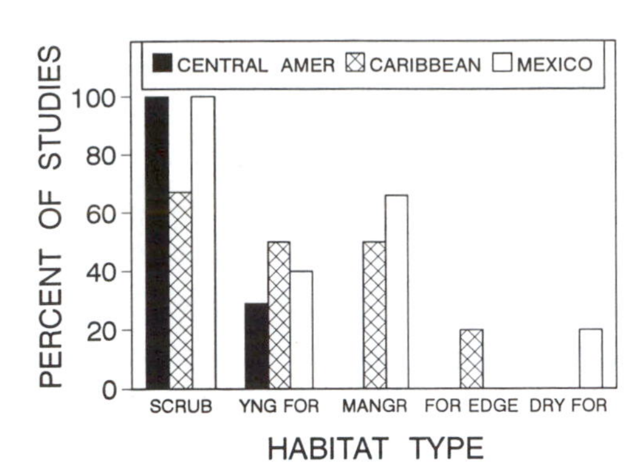

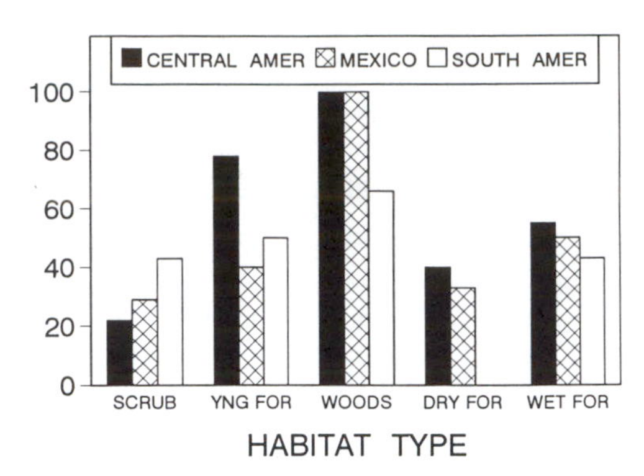

Figure 6-8. Geographic variation in habitat use by Common Yellowthroats and Summer Tanagers. The ordinate represents the number of studies in which a species was found to occur in a given habitat type divided by the number of studies in which each species *could* have occurred in that habitat. See text for details.

habitat requirements of certain migrant species in the tropics (e.g., Greenberg 1986, Martin 1992) probably has been to a large extent a result of geographic variation in their habitat use and not necessarily because of incompatibilities of various surveying methods (e.g., see Morton 1992). For example, Gray Catbirds and Northern Waterthrushes both exhibited pronounced regional differences in habitat use (Fig. 6-9). Potential reasons for geographic variation in habitat use include different availabilities of habitats, and regional differences in preference for certain habitat types. The Black-throated Green Warbler, for instance, is commonly found in pine stands in northern Central America and on Caribbean islands, but occupies broadleaved forest associations in eastern Mexico and Southern Central America, where pines are scarce (this study).

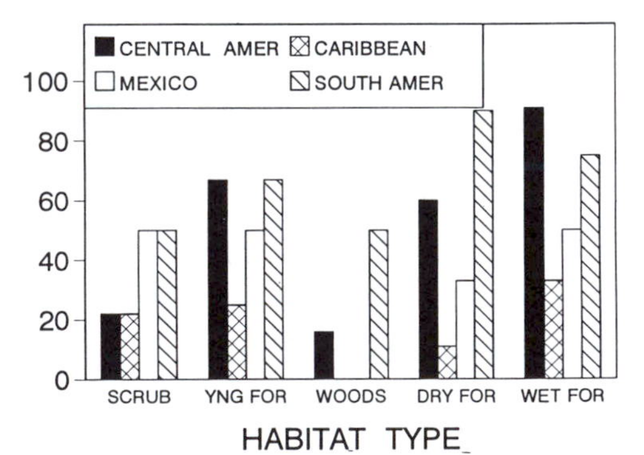

Figure 6-9. Geographic variation in habitat use by Gray Catbirds and Northern Waterthrushes. The ordinate represents the number of studies in which a species was found to occur in a given habitat type divided by the number of studies in which each species *could* have occurred in that habitat. See text for details.

Black-throated Green Warblers at the edge of their winter range in western Mexico, however, make extensive use of both broadleaved cloud forest and mixed pine forest (Hutto 1992, R. L. Hutto, personal observation). Louisiana and Northern Waterthrushes provide other examples of how availability of a given habitat type in a region might influence habitat use. In continental areas, waterthrushes generally do not occupy residential areas during winter but, on densely populated and developed Caribbean islands [compare land-use and population data in Leonard (1987, Table A.2) with that in McElroy et al. (1990, Table 21.1)], this species uses residential gardens (this study; see also Arendt 1992). Furthermore, for the migratory bird community as a whole, certain habitats are of high regional importance, e.g., mangroves in the Caribbean and cloud forests in western Mexico and South America (Fig. 6-10), suggesting that regional availability may have influenced the evolution, or at least patterns, of bird–habitat relationships. These data corroborate the previous recommendation that conservation and management plans for migratory bird habitat should incorporate a regional (or local) perspective.

Geographic variability in habitat use may also indicate localization of individuals of different subspecies (see Ramos and Warner 1980). For instance, the Yellow-rumped Warbler occurs in two rather different situations within its Middle American wintering range. In the interior of Mexico and Central America, this species is typically encountered in high-elevation pine–oak–fir forests, although it occurs less commonly at lower elevation in thorn-scrub and disturbed urban and agricultural areas (Binford 1989, Hutto 1992, R. L. Hutto, personal observation). In the Mexican Yucatan, which lacks tall mountains, Yellow-rumped Warblers are characteristic of grassy coastal dunes and extremely open disturbed habitats (including recently burned sites) at or near sea level (Russell 1964, Lynch 1989, 1991, 1992, Lopez Ornat and Lynch 1991). The species also occurs sporadically in low-elevation grasslands in Panama (Karr 1976). Virtually all

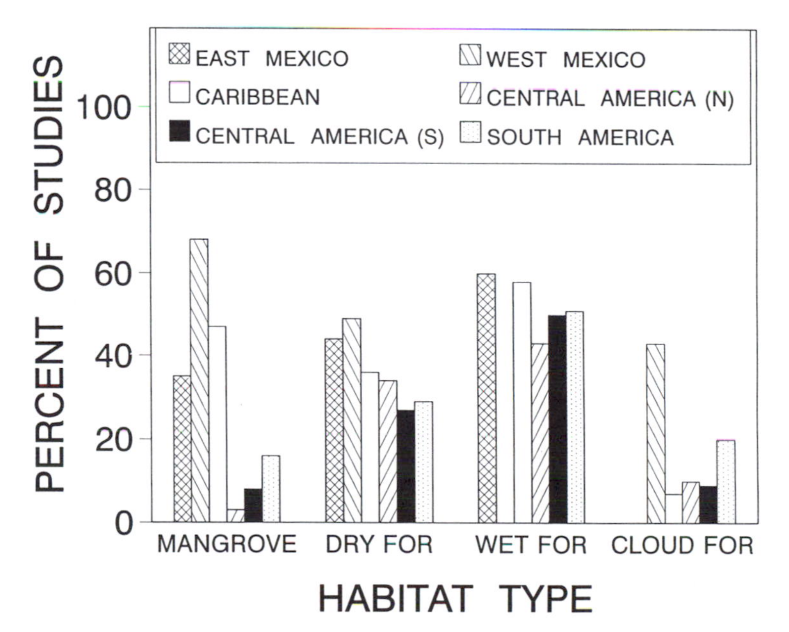

Figure 6-10. Example of geographic variation in habitat use by migratory birds as a group. The ordinate represents the number of studies in which a species was found to occur in a given habitat type divided by the number of studies in which each species *could* have occurred in that habitat, summed over all migratory species. See text for details.

of the Yellow-rumped Warblers that overwinter in the interior mountains of Oaxaca, Mexico, are members of the western subspecies, (*D. c. auduboni*), formerly know as "Audubon's Warbler" (Binford 1989). In contrast, Yellow-rumped Warblers that occur in the Yucatan belong to the race *D. c. coronata*, formerly known as the "Myrtle Warbler". In Honduras, the two races of Yellow-rumped Warbler also tend to segregate geographically, with *D. c. auduboni* occurring mainly in montane pine–oak habitats, and *D. c. coronata* overwintering along the Caribbean coast and on offshore islands, with occasional inland occurrences (Monroe 1968). Separation of subspecies along geographic/habitat lines breaks down in the highlands of western Guatemala, where an isolated breeding population of *D. c. auduboni* coexists with migratory *D. c. coronata* during winter. However, the Atlantic coastal plain of Guatemala is inhabited exclusively by *D. c. coronata* (Land 1970). Thus, in some parts of its winter range, the Yellow-rumped Warbler seems to offer one of the few known examples of geographic differences in habitat use that correlate with genetic differentiation. Regional variation in winter habitat use by migratory birds may reflect segregation of individuals from distinct breeding populations, whose winter habitat preferences may be related to breeding habitat.

Geographic variation in habitat use could result as an outcome of selective pressures that vary regionally. For example, geographic differences in habitat use could be influenced by regional differences in assemblages of competing species (e.g., Bennett 1980, Keast 1980), characteristics of the food base supported by vegetative associations (Janzen 1973), or predation pressure (Buskirk 1976). Whether geographic differences in habitat use reflect variation in these factors or others (e.g., population density, climate) is unknown. However, at this stage of conservation efforts, knowledge of geographic variation in habitat use at least allows conservation planners to recognize that variability does exist so that it can be incorporated into management and conservation prescriptions.

Sex- and Age-Specific Habitat Use

Habitat use by some migrants varies by sex or age of the individual, such that overall habitat use by a species reflects the combined distribution of all age and sex classes. Lynch et al. (1985) showed that male and female Hooded Warblers in the Yucatan region differed in selection of winter habitat: males tended to occur in relatively tall, closed-canopy forest, while females were most often found in old fields, native coastal scrub, treefall gaps, and other disturbed, low-stature vegetation. Furthermore, laboratory and field experiments suggested that this difference reflects active choice, not merely exclusion of females from forest by the dominant males (Morton et al. 1987) and that sexual differences in habitat selection in Hooded Warblers is innate (Morton 1990). In other species, however, behavioral dominance by males appears to drive sex-related patterns of habitat use (Wunderle 1992, Marra et al. 1993). Sex-specific habitat use has been documented for about ten other migratory parulines (Lopez Ornat and Greenberg 1990, Wunderle 1992, Wunderle and Waide 1993) and, in most cases, males occur in more mature habitats. Despite its potential importance, the evolutionary significance of habitat partitioning by sexes has not received much attention and little attempt has been made to incorporate it into conservation recommendations. For example, if habitat use reflects p*references* of the sexes (Morton 1990), then all habitats used are important to the species and should be maintained. However, if sex-specific habitat patterns result from dominance relationships, then habitats used by the subordinate sex might be suboptimal, and conservation efforts should focus on increasing or improving those habitats used by the dominant sex.

Age-related exclusion of some individuals from certain habitats (or actual preferences for those habitats) also may occur on the wintering grounds, although evidence is sparse. In Belize, L. J. Petit et al. (unpublished data) found that young ($\leqslant 2$ years) male American Redstarts occupied habitats that were intermediate between those of females and older (>2 years) males (Fig. 6-11).

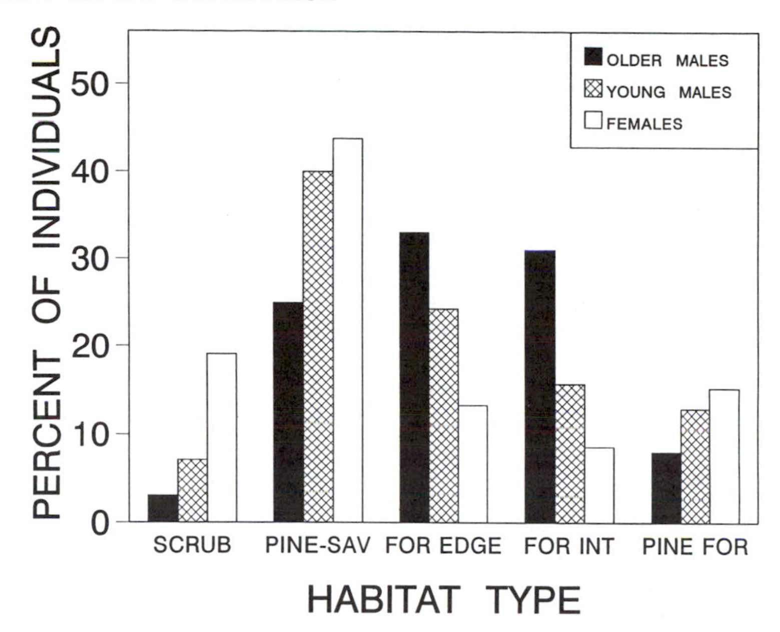

Figure 6-11. Sex- and age-specific habitat use by American Redstarts overwintering in five habitat types in Belize, Central America. Age classes for males represent individuals ⩾2 years old (older males) and <2 years old (young males). Ages of females were not determined in most cases. Data are based upon visual/auditory surveys conducted during winters 1988–1989 and 1989–1990 (L. J. Petit et al., unpublished data).

Behavioral observations suggested that age-related habitat use may reflect the outcome of behavioral dominance of older individuals, as observed during the breeding season (Sherry and Holmes 1989, Marra et al. 1993).

These findings raise two important questions. First, do sexual and age differences in habitat use translate into differences in the geographic distributions of males and females, and age groups? Koplin (1971) demonstrated that habitat differences between wintering male and female American Kestrels were correlated with a strong geographic gradient in the sex ratio. However, whether sexual habitat differences were driving sexual differences in geographic distribution, or vice versa, was not established. Pearson (1980) also described a possible latitudinal gradient of age classes for overwintering Summer Tanagers. For Hooded Warblers, Lynch et al. (1985) found a male : female ratio of 2.0 : 1 in the northwestern Yucatan Peninsula, which is characterized by lower, semideciduous second growth, as compared with a ratio of 3.4 : 1 in the eastern portion of the Peninsula, where the predominant vegetation is closed-canopy semi-evergreen forest. However, because the sampling scheme in that study was not random across all habitats and regions, more data are needed to confirm this possible geographic gradient in sex ratio.

The second question is whether numbers of males or females are differentially limited by the availability of habitats on the wintering grounds. This possibility would most likely arise in situations where a species exhibits a strong degree of sexual habitat segregation (such as the Hooded Warbler) and human-related habitat destruction greatly reduced the preferred habitat of one or the other sex. Recent observations of effects of hurricane and fire on overwintering Hooded Warblers suggest that sexual habitat differences are variable over space and time (Lynch 1992, personal observation), but the generality of these observations is unknown.

Sex- and age-specific habitat use has important ramifications for conservation of

Neotropical migrants. First, partitioning of winter habitats could lead to differential, habitat-based mortality rates either due to intrinsic factors or to more extensive loss of certain habitats than others. The resulting skewed sex ratios would reduce the effective size of breeding populations and could dramatically enhance the probability of local extirpation (Shaffer 1981). For example, Sherry, Holmes and coworkers (Holmes et al. 1989, Sherry and Holmes 1989, Marra et al. 1993, Parrish and Sherry 1994) have articulately documented the territorial system, habitat use, and survival of American Redstarts overwintering in Jamaica. They found that male dominance was the basis for sex-specific habitat distribution, with males occupying the more seasonally stable, food-rich mangroves and females residing in scrub habitats that decreased markedly in quality throughout the winter period. Marra et al. (1993) suggested that differences in winter habitat quality, combined with sex-specific habitat use, could account for the male-biased sex ratio in breeding populations of redstarts (Sherry and Holmes 1991). Clearly, more investigation of this type is needed to identify those factors that limit populations of Neotropical migrants on their wintering (and breeding) grounds (see Holmes and Sherry 1992, Sherry and Holmes, chapter 4, this volume).

Second, placement of reserves must incorporate potential sex-based dichotomies in habitat use by assuring that appropriate habitat for both sexes is adequately represented. Finally, if innate or dominance-based habitat use is manifested in distinct geographic winter ranges for sex or age classes (e.g., Ketterson and Nolan 1983), then development of protected reserves or application of appropriate land-use practices must be implemented throughout *entire* winter ranges of species.

Temporal Variation in Habitat Use

Many migratory species are territorial during winter and, therefore, temporal stability in habitat use would be expected. However, in apparent response to food availability, several species are known to range over extensive areas during the overwintering period (Rappole et al. 1989, Loiselle and Blake 1991), such that patterns of habitat use may also covary. For example, migrants have been shown to respond to changing fruit abundance throughout the winter in the tropics (Leck 1972a, Greenberg 1981). The result is a shift in habitat use as the winter progresses, with migrants sometimes concentrating in early second growth and other disturbed habitats where fruit can be abundant (Martin 1985, Loiselle and Blake 1991; also see Blake and Loiselle 1992). Knowledge of the importance of such "landscape complementation" (Dunning et al. 1992), where multiple habitats within a landscape are used by an individual (or population) for different purposes, is critical to the understanding of how best to manage and conserve wildlife populations. For example, species such as Bay-breasted Warbler and Wood Thrush may spend most of the winter in mature forest, but shift to younger second growth during late winter to fulfill their energetic needs for migration (Martin 1985). Thus, both habitats and their proximity to each other may be important for overwinter survival and for returning to north temperate breeding grounds. In this light, a landscape-level approach to managing tropical habitats for wildlife would seem warranted. Furthermore, these observations also may have consequences for the way in which researchers gather and examine habitat data: information collected over only a portion of the winter, or collected over the entire winter and then lumped together, may overlook important temporal variation in habitat requirements of species (Greenberg 1992b). Clearly, the ecological and evolutionary correlates of habitat use by migrants is central to development of long-term conservation plans. Incorporation of principles of behavioral ecology into research programs may heighten understanding of the ultimate and proximate bases for patterns of habitat use by migrants in the Neotropics.

BEHAVIORAL ASPECTS OF HABITAT USE AND MIGRATORY BIRD CONSERVATION

Species–habitat relationships are manifestations of evolutionary processes that have shaped the ecology of species to exploit their environments effectively. Thus, general relationships between the behavior (Holmes et al. 1979) and morphology (e.g., Miles and Ricklefs 1984) of species and the *micro*-habitats that they occupy are not unexpected. However, close association between the behavior or morphology of species and *broad* habitat types are much less evident due to the diverse assemblages of species occupying habitats at that scale; i.e., a relationship between the "average" behavior or morphology of local assemblages and habitat type are not well pronounced. Thus, basing conservation plans simply on the habitats species use may not adequately protect all species within a given area because of the unique ecological needs of each species found there. Firm understanding of how to conserve and manage habitats for Neotropical migrants during winter requires knowledge of the ways in which individuals search for and secure insects and fruit, the types of food taken, and the significance of different social systems. Below, these issues are reviewed and ways in which such information can be applied to conservation efforts are identified.

Most Neotropical migrants are highly insectivorous. The evolution of long-distance migration traditionally has been explained by year-round reliance on insects (Griffin 1964; see Levey and Stiles 1992 for alternative view). More recently, however, research on behavioral ecology of migrants has revealed that up to one-third of all long-distance migratory landbirds consume some fruit during winter (e.g., Leck 1972b, Morton 1980, Greenberg 1981, Rappole et al. 1983, Blake and Loiselle 1992) and that fruit-eating species, in general, display winter social systems different from those of many highly insectivorous species. Unlike many insectivores, total or partial frugivores often do not defend territories over extended periods of time, but instead range widely in search of fruit (Levey and Stiles 1992). Thus, abundances of frugivorous migrants (and residents) in a given habitat can vary greatly over time in response to fruit availability (Loiselle and Blake 1991). Furthermore, because those movements often are not limited to a single habitat patch, habitat type, or even watershed (Karr et al. 1982, Karr and Freemark 1983, Loiselle and Blake 1991), the presence of corridors (*sensu* MacClintock et al. 1977) and adjacent suitable habitats may greatly facilitate movements among areas. Rappole et al. (1989) radiotracked Wood Thrushes in eastern Mexico and showed that individuals that wandered suffered greater mortality from predators than those individuals that held stable territories. They predicted that mortality of wintering Wood Thrushes would increase as tropical deforestation continued because it would force more individuals to wander in search of food (territories). Forest destruction reduces the connectivity of landscape elements (Forman and Godron 1986), further inflating the hazards faced by Wood Thrushes moving from one patch to another. Additionally, several migratory species are believed to move from mature forest into younger second growth during late spring in order to exploit the abundant fruit there (e.g., Martin and Karr 1986). Thus, appropriate composition of *landscapes* may be necessary for some migratory and many resident frugivores and nectarivores to complete their annual cycles (Loiselle and Blake 1991).

The manner in which insectivorous migrants search for and capture arthropod prey varies widely among species, including gleaning from foliage (e.g., Northern Parula, Chestnut-sided Warbler), hovering at leaves (American Redstart, Yellow-rumped Warbler), gleaning from trunks and branches (Yellow-bellied Sapsucker, Black-and-white Warbler), flycatching (Acadian Flycatcher, Olive-sided Flycatcher), aerial pursuit (Chuck-will's-widow, Chimney Swift), searching ground substrates (Wood Thrush, Ovenbird) and aerial leaf litter (Worm-eating Warbler, Blue-winged Warbler), probing flower parts (Cape May Warbler, Orange-crowned Warbler), gleaning from open ground (Palm Warbler, Louisiana Waterthrush) or water (Northern Waterthrush), and gleaning from herbaceous vegetation (Common Yellowthroat). Examination of avian foraging behavior can provide insight into the ecology of overwintering migrants and might allow extension of conservation efforts beyond simple habitat-based plans to dissection of the reasons *why* some species or more vulnerable than others to conversion of native tropical landscapes. For example, certain forestry practices in North America have a differential effect on forest-dwelling bird species, and much of that variation has been associated with foraging behavior or other aspects of life history (Franzreb 1983, Webb et al. 1977). Similarly, modifications of habitat structure in the Neotropics may result in unique responses by foraging guilds of both migrants and residents. Thiollay (1992) showed that selective logging in French Guiana resulted in a decrease in the number of resident ground foragers and flycatchers, but an increase in density of small frugivores. Johns (1989) found a similar effect of logging on avian foraging guilds in a Malaysian rain forest (but see Wong 1986).

These observations suggest that overwintering migratory birds could be *indirectly* affected by forest disturbance through alteration of food resources. Anthropogenic disturbances may differentially affect abundances of insects and fruit in a habitat, and thereby impact populations of insectivorous birds differently than frugivorous birds. For example, isolation of a moist tropical forest may lead to drier microclimates within the patch (Kapos 1989, Williams-Linera 1990), which, in turn, has been shown to be related to lower abundances of arthropods in foliage (Janzen and Schoener 1968) and especially, leaf litter (Strickland 1947, Plowman 1979). Alternatively, fragmentation enhances penetration of sunlight (because of a larger edge: interior ratio), which has been related to increased fruit density (e.g., Blake and Hoppes 1986, Levey 1988). Thus, some feeding guilds may be better able to withstand disturbances than others. (Forest reserves cannot be so small, however, as to inhibit normal production and dispersal of fruit.) Likewise, foraging generalists (*sensu* Morse 1971) are thought to be more resilient to disturbances and environmental unpredictability than are specialists (Glasser 1982, 1984).

These examples illustrate the complex relationship among behavioral traits, habitat use, and disturbance (Gordon 1991), and highlight the need to consider the behavioral ecology of migrants in projecting the effects of land-use practices on migrant winter population dynamics. For instance, a reserve system comprised of many small forest patches connected by corridors may be more suitable to frugivorous and leaf-gleaning insectivorous migrants than to ground-foraging insectivores because a small forest patch may not adequately buffer ground leaf litter from the desiccating effects of increased wind and temperature; see results of Johns (1989) and Thiollay (1992) above. Interestingly, of Neotropical migrants typically considered to be forest dwellers during winter, ground foragers tend to be more closely associated with undisturbed forests and less with other habitats than species in other foraging guilds (e.g., Robbins et al. 1987, Greenberg 1992a, Lynch 1992, Petit et al. 1993, Wunderle and Waide 1993).

As stated above, the degree of territoriality exhibited by migratory species is often related to their diet; insectivores tend to hold more stable territories (Sliwa and Sherry 1992). Most insectivorous migrants defend small territories throughout the winter. But many insectivores (and some frugivores/omnivores) also join mixed-species flocks as either obligate or, especially, facultative members (join flocks only when they pass through an individual's territory). Examples of migrants that are often found in mixed-species flocks include Solitary Vireo, Magnolia Warbler, Black-and-white Warbler, Worm-eating Warbler, Northern Parula, Townsend's Warbler, Hermit Warbler, Olive Warbler, American Redstart, and Blue-gray Gnatcatcher. Many migratory parulines in western Mexico appear to be obligate flock members (Hutto 1987, 1988b). Flocks in the Neotropics typically have both "core" (or "nuclear") species (Moynihan 1962) and "attendant" (or "satellite") species (Rand 1954). Core species are required for a flock to form and contribute to its cohesiveness, while attendant species tend to follow core individuals. Flocks occupying forests situated on the mainland almost always have resident core species, whereas migrants are typically attendants (Buskirk et al. 1972, Powell 1979, Tramer and Kemp 1980). However, in lowland deciduous forest of western Mexico, migratory Nashville Warblers occupy the role of nuclear species (R. L. Hutto, unpublished manuscript). In the Caribbean, flocks comprised of both migrants and residents are uncommon (Eaton 1953, Lack and Lack 1972, Post 1978, Ewert and Askins 1991).

Proposed advantages of flocking center around foraging efficiency and predator avoidance (Powell 1988). The presence of flocks may be important for overwinter survival of certain species of migrants in some regions because of the benefits accrued by flock members. Thus, understanding flocking behavior could provide supplementary information for conservation of migratory birds. Participation by migratory birds in mixed-species flocks is dependent upon the presence of core species, which may be more sensitive to forest alteration and fragmentation than are migrants because core species are typically residents. For example, forest canopy flocks in Peru preferred areas of forest with unbroken canopies (Munn 1985). In addition, some resident core species have relatively large home ranges (often ≥5–10 ha; Munn 1985, Powell 1985, Terborgh et al. 1990). Therefore, alteration and fragmentation of forest tracts may restrict occupancy of those tracts by core species, such that migratory species also may not settle there because of absence of other flock members (Rappole and Morton 1985). Experimental fragmentation of an Amazonian forest resulted in the virtually complete disappearance of species that forage in mixed species flocks (Bierregaard 1990). The flexibility of migrant winter social systems may be important in determining population responses to habitat alteration and fragmentation (Ewert and Askins 1991, Staicer 1992).

CONSERVATION OF HABITATS IN THE NEOTROPICS

Conserving species usually requires protection of habitats (Grumbine 1990). Indeed, by documenting bird–habitat associations, as well as understanding the sources of variation, and the behavioral and ecological correlates of these relationships, sound conservation plans can be developed for migratory birds in the Neotropics. However, despite the best intentions of ecologists, conservation designs are futile unless ample suitable habitat exists or can be restored within landscapes. In this chapter and elsewhere (Petit et al. 1993), a general overview of the types of land-use practices that are compatible with migratory bird ecology in the Neotropics has been presented. These arguments have stressed that, if global populations of migratory birds are to be sustained, habitats in wintering (as well as breeding and migration) areas must be protected.

Much of the focus of tropical conservation has been directed at two broad forest associations: wet/moist and dry broadleaved forests. Together, these two forest types originally comprised more than two-thirds of all land area in the Neotropics (Murphy and Lugo 1986, Myers 1991). Clearly, preservation of these forest types should remain at the forefront of conservation efforts, if only because of their extensive geographic coverage. However, within these major forest types, and in other forest and nonforest vegetative associations, certain habitats that are important to migratory and resident bird species are highly restricted in area or geographically. Such habitats present special challenges for conservation biologists because of the fragile existence of such localized plant communities (Goodman 1987, Reznicek 1987).

Fragmentation and isolation of Neotropical forests is an ominous threat to the integrity of both resident and migratory bird populations. Fragmentation of forests has forced ecologists to incorporate the issues of reserve size and placement into conservation plans. This leads to the concept of conserving landscapes, the "greater ecosystem" (Grumbine 1990) approach to managing biological diversity. Below, three issues pertinent to the conservation of habitats for migratory birds in the Neotropics are discussed. First, migrant use of restricted natural habitats and the unique problems faced by conservation biologists attempting to protect these areas are reviewed. Next, how forest-dwelling migratory birds might respond to severe reductions in size of tropical forest tracts are evaluated. Finally, the first two topics and others are incorporated into a discussion on landscape-level approaches to conservation of migratory bird habitats within the Neotropics.

Conservation of Naturally Restricted Habitats

Most migratory species occupy a range of broadleaved forest habitats during winter, which generally are widespread. As a consequence, many migratory species can be protected on their wintering grounds by preserving various seral stages of major tropical vegetative associations, such as lowland or montane moist forest or dry semideciduous forest (see Petit et al. 1993). However, there are other, less publicized, natural vegetation types that are potentially important for certain migratory (and resident) bird species. These habitats are often locally restricted, such that wildlife species that rely upon them are likely to be at greater risk of extinction (Terborgh and Winter 1980). Several examples are given below of migrant use of naturally restricted vegetation types. These examples are not exhaustive, but are illustrative of the potentially tenuous existence of these types of habitats, as well as the social, economic, and ecological factors that must be incorporated into plans for their protection.

Mangrove swamps fringe many salt and brackish water ecosystems throughout the world, mostly in the tropics. In the New World tropics, these forests are disappearing mainly due to development along coastlines (Christensen 1983, Leonard 1987, World Resources Institute 1992). Mangroves act as a principal habitat for at least three long-distance migrants (Prothonotary Warbler, Swainson's Warbler, and Northern Waterthrush; possibly migratory Yellow Warblers) throughout much of those species' wintering ranges. Mangroves are not a primary habitat for migrants in general, although in certain regions, such as the Caribbean, mangroves provide important wintering areas for migratory birds (see above). For example, several parulines, such as American Redstart, Ovenbird, Black-and-white Warbler, Hooded Warbler, and Magnolia Warbler, use mangroves extensively on Caribbean islands, but are much less commonly found in those swamps in Central and South America (see above). Thus, mangroves may provide an important alternative *forested* habitat in areas where deforestation has already claimed much of the surrounding lowland forests.

Mangroves occupy areas that are highly valued for development projects. Intense pressure for continued destruction of mangrove swamps exists because of the foreign revenue generated from coastal development (i.e., resort areas). However, economic incentives also exist to *preserve* mangrove ecosystems; these systems are critical for local economies because of the benefits they provide to commercial fisheries (Christensen 1983, Hartshorn et al. 1984). Mangroves also help control shoreline erosion and are important in nutrient flow between marine and terrestrial ecosystems (Kennish 1990). These observations highlight the need to base conservation plans largely upon both the long-term economic and ecological benefits derived from preservation of mangroves; arguments based solely on wildlife conservation may be insufficient to curtail conversion of important habitats (Freese and Saavedra 1991). Protection of mangrove swamps will require concerted legislation and subsequent enforcement by national governments.

Pine forests are widely dispersed, yet patchy in their distribution, throughout Mexico, northern Central America, and the Caribbean. The importance of pine forests to migratory birds was discussed above and in a companion paper (Petit et al. 1993). Pine associations have received little conservation attention, probably because of their low species diversity compared to tropical broadleaved forests. However, because more than a dozen species of migratory landbirds rely on pine-dominated forests as a principal habitat during winter, these habitats cannot be considered trivial in attempts to sustain diverse migratory bird populations. Moreover, many migratory species using coniferous forests are habitat specialists, being relatively restricted to those habitats. For example, Hutto (1992) calculated the habitat breadth (Levins 1968) of migratory species overwintering in western Mexico. His data show that species that make extensive use of coniferous forests (detected more often in pine than in any other habitat type; only species that were found on ⩾3 sites were included) had a lower average habitat breadth (mean = 1.22 ± 0.38 SD) than species that use deciduous forest (1.33 ± 0.57), early second growth (1.49 ± 0.62), thorn forest (1.49 ± 0.52), or cloud forest (1.88 ± 0.67).

Thus, migrants that inhabit pine forests not only occupy geographically restricted habitats, but they are also limited in their distribution across habitat types in western Mexico. Habitat specialization and habitat restriction are two ecological factors that have been closely linked to a species' susceptibility to population declines (Terborgh and Winter 1980, Rabinowitz et al. 1986). Most continental pine forests are found in mountainous highlands, and are often exploited and managed for timber production (Hartshorn et al. 1984, Leonard 1987). Although total area of pine forest has not declined appreciably in recent years, mature and old-growth stands are contracting at the expense of younger, more intensively managed plantations (Leonard 1987). Implications of this shift in management practices for overwintering migratory birds are unknown. Given the patchy nature of pine forests and the potential value of the pine resources to national economies, special attention needs to be devoted to this vegetation type to ensure that forests remain not only sustainable, but also ecologically viable for wildlife, including migratory birds. For example, loss of broadleaved understory in pine forests causes a severe reduction in quality of these sites for most migratory and resident birds (see above). Relatively little pressure has been exerted on Neotropical pine forests in the form of agriculture or cattle grazing because pines frequently grow on soils that are too poor to support those land-use practices. At the same time, however, pine forests do not provide major natural history attractions to foreign tourists because of the relatively depauperate fauna and flora found there; consequently, one generally cannot cite foreign currency derived from ecotourism as a benefit of preserving pine forests. Thus, the fate of these forests for wildlife may rest with the degree of responsibility exercised by natural resource agencies and timber harvesters. Wildlife management principles need to be incorporated into the rotation and management schemes of tropical pine forests. Here lies a situation in which international cooperation could substantially help ensure an ecologically and economically viable solution to a relatively simple problem. Exchanges of technical information on management of forests for the dual purpose of wildlife habitat and timber production, and initial training of Latin Americans at foreign institutions, are thought to be important requisites for the long-term preservation of tropical forests (Thelen 1990, Cornelius 1991).

Effects of Forest Fragmentation on Overwintering Migrants

Extensive research in North America has shown that the presence of migratory birds breeding in temperate forests is intricately related to tract size (Whitcomb et al. 1981, Lynch and Whigham 1984, Blake and Karr 1987, Robbins et al. 1989b). Similar information on the effects of insularization in Latin America, however, is meager. Fragmentation and isolation of tropical broadleaved forest may have pronounced biological ramifications not only for plants and animals that are directly displaced, but also for individuals that remain after isolation. For example, the microclimate of forest fragments can be substantially altered within several hundred meters of edges, such that a concomitant change in plant species often results; this, in turn, is partly responsible for the redistribution or local extinction of the fauna inhabiting forest islands (e.g., Lovejoy et al. 1986, Williams-Linera 1990, Laurance 1991, Laurance and Yensen 1991, Saunders et al. 1991). Tropical forest fragments generally do not contain as many resident bird species as contiguous tracts during either the breeding or nonbreeding season (Terborgh 1974, Willis 1979, Robbins et al. 1987, Karr 1990). Most long-distance migrants, however, do not appear to exhibit such pronounced relationships with forest area during the boreal winter (Robbins et al. 1987). In fact, many migrants that are area sensitive on the breeding grounds show no such relationship, or are found more often in forest remnants than in larger tracts, during winter (Robbins et al. 1987, 1992). Askins et al. (1992) recorded no significant relationship between forest fragment size and migratory bird abundance on St John and St Thomas Islands in the Caribbean. In addition, Ewert and Askins (1991) found that fragmentation of forest tracts on Caribbean islands did not alter the social system (flocking behavior) of migratory warblers. These few studies support the previous conclusions that migrants as a group may not be greatly affected by slightly disturbed and fragmented habitats.

Despite the apparent lack of detrimental effects of forest fragmentation on most migratory species in the Neotropics, few data have been collected to examine fragmentation issues, and results must be considered tentative. Indeed, several migratory species have exhibited disproportionate use of large forested stands (Robbins et al. 1987) and some of the migratory species identified as highly susceptible to forest alteration (Table 6-3) may be equally selective in use of large tracts. Askins et al. (1992) attributed relatively low migrant abundance in forests on St Thomas to urbanization and fragmentation. Furthermore, Rappole et al. (1989) showed that fragmentation of forest can force individuals to wander in search of suitable habitat, behavior that in turn was correlated with increased mortality. Most research on fragmentation, however, has not been based upon a rigid experimental design. Thus, future research must address both direct and indirect consequences of fragmentation (see Terborgh 1992) on migratory bird populations through well-designed studies.

Landscape-Level Approach to Conservation of Migrants

Wildlife-habitat relationships historically have been examined at the microhabitat or stand level (e.g., MacArthur and MacArthur 1961, James 1971, Holmes et al. 1979). Recently, however, emphasis in North America and Europe has been placed on incorporating higher-level spatial scales into the study of wildlife populations (Forman and Godron 1981, O'Neill et al. 1986). Indeed, conservation biologists have realized that because habitat patches are not closed systems and that individuals of a population can move freely among habitat patches, landscape-level patterns and processes (Turner 1989) can greatly influence the composition of local faunal assemblages (Lynch and Whigham 1984, Saunders et al. 1991). Therefore, maintenance of ecosystem integrity and function is fundamental to any long-term effort to preserve biological diversity.

From a theoretical standpoint, few can question the value of a landscape-level approach to conservation. Practical application of theory is difficult to implement, however, because of the accumulation of problems associated with increases in land area required for conservation plans. Imagine, for example, the increased number of political, social and cultural boundaries that a land-use planner would have to cross in establishing a 100,000 ha biosphere reserve compared to, say, a 200 ha reserve. No easy solutions exist for proper management of entire watersheds or landscapes. A system of large reserves is often offered as the best hope of preserving the natural diversity of tropical communities (Rubinoff 1983, Cornelius 1991). In most situations, human population pressure continues to place escalating demands on tropical resources, such that landscapes managed for both native species and sustained economic development is the most practical approach to maintaining natural diversity on a long-term basis (Sayer 1991, Woodwell 1991). A pivotal issue is the development of a protocol that incorporates both wildlife and socioeconomic considerations into long-term maintenance of landscapes. Many conservationists believe that such plans are technically, economically, and politically feasible in the Caribbean and Latin America, albeit with prodigious effort (e.g., Fearnside 1989, Vaughan 1990).

Several key components of a landscape-level approach to the conservation of migratory birds in the Neotropics have already been examined in this chapter: fragmentation effects (patch size), corridors, and landscape composition and structure (Turner 1989). While detailed examination of the theory and application of landscape ecology is beyond the scope of this chapter, below are several general considerations for landscape-level conservation of migratory bird habitats in the Neotropics.

Landscape-level designs for conservation of wildlife during nonbreeding periods must provide the quantity, diversity, and quality of habitat to support local populations of naturally occurring species and to maximize their survival rates. The goal of maintaining high integrity (*sensu* Karr and Dudley 1981) of wildlife communities in tropical landscapes requires that at least three ecological concepts be integrated into long-term conservation blueprints: (1) native plant species and landscape composition; (2) minimum area requirements; and (3) landscape connectivity.

The highly coevolved interactions between tropical plants and animals is well documented (Gilbert 1980, Terborgh 1986). Those tightly coupled systems make existence precarious for species intricately dependent upon others, such that removal of a keystone species (Paine 1969, Terborgh 1992) can lead to cascading losses of dozens of other species. In the context of this chapter, loss of keystone plants (Mills et al. 1993) could cause degeneration of entire ecosystems upon which migratory birds are reliant (Terborgh 1986). More generally, alteration of native plant species composition through anthropomorphic disturbance, such as fragmentation (Laurance and Yensen 1991) or fire (Kellman and Tackaberry 1993), could have serious consequences for migratory birds and other animals dependent upon them for overwinter survival (Terborgh 1989). Furthermore, introduction of exotic tree species of economic value can render "forested" areas virtually useless to migratory birds. For example, pine plantations in Panama do not offer many migratory or resident bird species suitable habitat (L. J. Petit, unpublished data). By maintaining accurate representation of native plant species, conservationists increase the probability of maximizing diversity of native animal species, including migratory birds (Scott et al. 1987).

The composition of landscapes refers to the representation of different native and artificial cover types (Dunning et al. 1992). Within the winter geographic range of migratory species, critical habitats must be present in proportions sufficient to maintain overwinter survival. Above, an example of "landscape complementation" was provided that highlighted the relevance of maintaining an accurate representation of natural landscapes. Native plant species in relatively natural landscapes are the "building blocks" of any long-term conservation strategy.

Research is needed critically not only on the types of landscape elements (*sensu* Forman and Godron 1986) or habitats that are used by migratory birds, but also on the minimum *sizes* of those elements. Minimum area requirements are know only for a handful of forest-dwelling migratory species (see above). However, available information on these requisites is limited and, as more data are gathered (especially in South America), additional species undoubtedly will be identified as requiring large forested tracts. Nevertheless, sizable areas of unbroken forest are imperative for many resident species and for maintaining the integrity of tropical forest ecosystems (Terborgh 1992); that integrity will ultimately determine the long-term persistence of many migratory and resident species.

Migratory birds must be able to travel among habitat patches during overwintering and migration periods. Conservation biologists have recognized that, by connecting patches of similar habitat via corridors (Noss and Harris 1986), animal movement is facilitated, which may result in easier access to critical food resources (Loiselle and Blake 1991) and, ultimately, in reduced mortality rates (Rappole et al. 1989). Maximization of landscape connectivity is believed to optimize the ecological benefits accrued through protection of critical habitat patches (Noss 1987, Hansson 1991), although certain risks also are inherent with this scheme (Simberloff and Cox 1987). An essential first step for development of landscape and regional conservation plans is to quantify the distribution and extent of different land use and cover types within national boundaries. Although lack of information on the extent of forest conversion and other land uses has been a major problem plaguing Latin American governments, recent progress in this area (mainly through application of remote sensing) has been encouraging (e.g., Sader and Joyce 1988, Dirzo and Garcia 1992, Powell et al. 1992). Moreover recently developed conservation plans that emphasize landscape connectivity for migrating birds are being implemented in the Neotropics (e.g., Powell and Bjork 1994).

One application of a landscape or regional approach to migratory bird conservation could be identification and conservation of those areas used as major stopover sites by migrating birds (Rappole 1991). Although birds may not remain at stopover sites for more than several days, these areas are critical for completion of the migratory trek (Moore et al., chapter 5, this volume). Protection of large areas of native habitat along coastlines and on Caribbean islands seems especially significant for trans-Gulf migrants (Moore et al. 1993).

FUTURE RESEARCH ON HABITAT USE AND CONSERVATION

Elsewhere, guidelines for the conservation of migratory birds in the Neotropics have been presented (Petit et al. 1993). However, much additional research needs to be completed on the Neotropical wintering grounds before comprehensive conservation plans that cover all species and regions can be devised. Several important directions for future research on the distribution and habitat use by migratory birds are summarized below.

Geographic Distribution of Migrants

Knowledge of geographic distribution of species is a basic element in development of conservation plans (Terborgh and Winter 1983, Margules et al. 1988). Historically, the wintering ranges of migratory birds, as depicted in field guides or by text in the American Ornithologists' Union (AOU) check-list (AOU 1983), have been determined through a synthesis of information available from published regional accounts and from museum specimen tags (C. S. Robbins, personal communication). The National Biological Service (Laurel, Maryland) has much of this information mapped for species that overwinter in North America, but data from south of the United States border are limited. In addition, locations of individuals captured in the Neotropics and banded with US Fish and Wildlife Service (USFWS) aluminum bands are stored on computer at the Bird Banding Laboratory in Laurel.

In general, the distributional data available for migrants are imprecise because the fine-scale details from local accounts are lost when information is summarized at relatively broad geographic scales (e.g., states or countries). Furthermore, individuals are not distributed equally throughout a species' range, and centers of abundance are not revealed in range maps. Improvement of winter range maps for migratory birds is dependent upon efforts aimed at: (1) collection of more information on the local distribution of species, and (2) a better method of information storage and retrieval. Additional data on the abundances of species in different areas could be gathered either through a systematic survey of habitats and regions (see below), or through compilation of information from field notes of naturalists. In this latter case, if bird lists were available from as few as ten people who visited a given area, a reasonable index of abundance could be estimated (because frequency of occurrence is related to abundance; e.g., Hutto et al. 1986, Wunderle and Waide 1993). Comparisons using these data would have to be restricted by habitat, but it would be reasonable to assume that a bird species detected by 80 of 100 people in one place is more common there than in another location where the species was detected by only 30 of 100 people. Many professional and nonprofessional field biologists have notebooks full of valuable information that may never be used. Perhaps the National Biological Service or a nongovernmental organization could serve as a repository for such information.

These data, along with those from museum specimens, banding records, and systematic surveys (see below), could be entered into a geographic information system (GIS), such that distributional maps for various types of data (sex or age distribution) or for different spatial scales (range-wide or regional) could be produced. Within a few years, generation of winter distribution maps for tropical latitudes that are as accurate as those produced for temperate North America from Christmas Bird Counts (Bock and Root 1981) may be possible. The importance of accurate information on the winter geographic distribution of migrants cannot be overstated. Without such data, identification of areas most appropriate for conservation of migratory birds will remain speculative.

Defining Habitat Types Used by Migratory Birds

A major obstacle to summarizing information on habitat use by migrants is the inconsistent fashion in which avian ecologists identify habitat types; there is a genuine need for standardization. Also, many authors provide vague descriptions of study sites. Researchers should include the following information in site descriptions: (1) elevation; (2) latitude and longitude; (3) type of vegetative association, as defined by a recognized regional botanical guide; (4) canopy height; (5) major plant taxa; and (6) extent and types of disturbance. This information could be used to classify habitat types by means of a systematic, multivariable procedure. For example, in western Mexico, R. L. Hutto classifies habitats by two main criteria: (1) broad habitat cover type [as defined by Rzedowski (1983)], such as marsh/wetland, streamside riparian, mangrove, tropical evergreen forest, tropical deciduous forest, cloud forest, oak woodland, pine–oak woodland, and fir forest; and (2) the type and extent of disturbance, such as undisturbed, selectively cut, understory removal, severely cut and understory replaced by crops (e.g., coffee), clear-cut presently in early stages of regrowth (<1 m), clear-cut presently replaced by tall-stature agriculture (e.g., banana, citrus), and clear-cut largely replaced by man-made structures (e.g., towns). Thus, if there are ten habitat types and eight kinds of disturbance, there would be 80 different categories possible. Some of these combinations probably would not occur and some categories could be lumped together to reduce the number of categories. An advantage of this system is that it allows analyses to be performed for many different combinations of habitat groupings, which enhances the types of questions that can be addressed.

Habitat Associations of Migratory Species

A broad overview of winter habitat use by migratory birds has been presented in this chapter. However, because of relatively scant data and the great geographic variability in habitat use exhibited by many species, this preliminary assessment must be interpreted cautiously. Indeed, the validity of applying the general information presented here to specific local field situations is known. A comprehensive, quantitative survey of habitat use by resident and migratory birds in each of the major physiographic regions (or life zones) of each country is sorely needed. These data are already available for certain regions of some countries (e.g., Yucatan Peninsula and high-elevation life zones of western Mexico). The product of these efforts would be an abundance-weighted habitat and geographic analysis of the winter distributions of migratory (and resident) birds, a central component necessary to build a conservation strategy for migratory birds in the Neotropics (Terborgh 1974, Greenberg 1986).

To assess habitat use and changes in long-term abundances of migratory birds, surveys based upon point counts are recommended (Hutto et al. 1986). The exact method needs to be determined (see Ralph et al. 1994), but would require distinguishing individuals that are both within and beyond some threshold distance (e.g., 25 m or 50 m) from the observer (see Hutto et al. 1986). Mist-netting could be used to supplement visual/auditory surveys and may provide important information about habitat-based survival rates. However, comprehensive visual/auditory surveys should also be conducted in all regions because of the biases associated with mist-netting.

Density estimates of each species within habitats and regions would provide the most useful data for assessing habitat relationships. *Indices* of abundance, as are typically generated through point-count techniques, may be misleading in comparisons among habitats because of inconsistent probabilities of detection of species in different vegetation types (Petit et al. 1995). Estimates of density based upon point counts or transects, however, are subject to a host of similar criticisms (Ralph and Scott 1981) and, therefore, may not provide information any more accurate than indices of abundance. Nevertheless, in theory, estimates of density offer the best opportunity for survey data to identify habitats and areas of greatest importance to migratory birds. Those survey and analytical techniques need to be pursued.

Long-term monitoring of populations also would provide information on the temporal suitability of habitat associations during winter and the relative importance of certain habitats as refugia for migrants during "bottleneck" (*sensu* Wiens 1977) years. For example, during drought years migrants may be unable to find sufficient food in some habitats, such as dry forests, and be forced to relocate to moister habitats (Faaborg et al. 1984). In this case, the benefit of access to moist forest in sustaining long-term populations is clear, although in most years moist forests may contain relatively few individuals.

Behavioral, Ecological, and Evolutionary Correlates of Habitat Use

Several hypotheses have been used to explain the general pattern of habitat use by overwintering migratory birds in the Neotropics: (1) migrants track sources of superabundant food (Willis 1966, Karr 1976); (2) migratory birds spend only 6 months on their wintering grounds and, therefore, are forced out of certain habitats by dominant resident species (Slud 1960, Leck 1972a); and (3) migrants may select wintering habitats that are similar to those used during the breeding season (Morse 1971, Hutto 1980). Conflicting evidence exists for all three of these nonmutually exclusive hypotheses because of the complex environmental factors influencing migrant distributions. The food hypothesis seems to be favored by many avian ecologists (e.g., Levey and Stiles 1992), but definitive evidence has yet to be collected (Petit 1991). While resolution of these and other questions relating to the proximate and ultimate correlates of habitat selection may at first seem to be a purely "academic" exercise, answers to these inquiries can provide important insight into conservation issues. For example, if habitat use by insectivorous migratory species is closely linked to insect density, then development of thorough conservation plans for migrants requires an understanding of the effects of disturbance on insect populations. Future research in the Neotropics needs to be directed at determining the biotic factors influencing habitat use by migrants, and how anthropogenic disruption of these factors might affect overwintering populations of migratory birds.

Effects of Land-Use Practices on Habitat Use and Overwinter Survival

Economists, ecologists, and conservationists need to collaborate to evaluate the potential importance and compatibility of certain sustainable land uses and wildlife conservation (Freese and Saavedra 1991). However, few detailed research projects have focused on the relationship between sustainable forestry or agroforestry and bird populations, and few of those projects have actually been implemented across multiple sites (Cornelius 1991). In principle, sustainable land uses offer the most promising alternatives for long-term stability of tropical economies and wildlife populations (Petit et al. 1993). An important area of future tropical research for avian ecologists is to determine the relative benefits and costs of different modes of sustainable forestry (e.g., Brokaw 1995), and of different land-use practices in general, to migratory and resident bird species. Furthermore, as agricultural lands expand throughout Latin America, traditional wildlife management practices (e.g., maintenance of hedgerows and woodlots; Gradwohl and Greenberg 1991) need to be incorporated into landscape planning. A valuable type of information is the survival rates of species occupying natural and anthropogenically disturbed habitats (both survival rates and reproductive success are necessary for resident tropical species). Without those data, development of conservation plans that incorporate sustainable land-use practices will remain tenuous.

SUMMARY AND CONCLUSIONS

The downward trends in populations of many species of Neotropical migratory birds have been documented through long-term monitoring of populations in the United States. Those alarming patterns, in light of the widespread alteration and loss of habitats in the Western Hemisphere, suggest that humans have played a fundamental role in the declines of these intercontinental migrants. The first step in developing effective conservation plans for these species is to identify habitats, areas, and species in need of greatest protection. This review provided a general synthesis of the mass of ecological information on Neotropical migratory birds available to conservation planners, as well as potential approaches that may be effective in conserving this natural resource in tropical wintering areas.

Migratory birds were found to occupy many different vegetation types during the boreal winter and only highly managed pasture and large monocultures of monocotyledonous crops, such as sugar cane, banana, and rice, seemed to be overtly incompatible with the goal of protecting migratory birds. Native broadleaved forests, in all stages of succession, supported more migrants than highly managed and/or artificial habitats, such as pine plantations, residential areas, and pastures. Generally, slightly disturbed native forests, such as forest fragments and selectively logged tracts, harbored a greater number of migratory species than pristine forests. However, at least one dozen migratory species, and many resident tropical species, are highly dependent upon large tracts of undisturbed forest. Because of this habitat specialization, these are the species most in danger of severe population declines. Thus, the current focus among environmentalists on preservation of tropical forests seems warranted for migratory birds as well.

Contrary to conventional wisdom, migratory birds did not concentrate at middle elevations during winter. Rather, migrants represented a greater proportion of the entire avifauna at lower elevations, this being a result of a greater number of migratory species occupying lowland sites compared with higher altitudinal locations. This elevational relationship disappeared, however, in disturbed forests. Neotropical migrants also were underrepresented on Caribbean islands and in habitats in lower Mesoamerica and South America, which may be associated with the energetic and evolutionary costs of travelling long distances or over inhospitable areas (water).

Patterns of habitat use varied geographically for several species, and for the migratory bird community as a whole. Furthermore, different geographic regions supported distinct assemblages of migratory birds. Thus, a regional approach based upon, for example, physiographic boundaries, may provide the most effective means of identifying, implementing, and coordinating conservation practices necessary to maintain winter populations of migratory birds. In addition, ecologists must incorporate into conservation directives the variation in habitat use associated with age, sex, and time of season.

Knowledge of the natural history and behavioral ecology of species may assist in predicting the potential impacts of local land-use practices on migratory birds. For example, forest fragmentation may have greater repercussions for species that search for insects in ground leaf litter or forage within multispecies flocks because of the relative influence of tract size on leaf litter invertebrates and presence of core-flocking species, respectively. Furthermore, that information can assist land-use planners in developing optimal strategies for landscape-level conservation of tropical wildlife. Large-scale approaches to conservation also must incorporate state-of-the-art knowledge of habitat and minimum area requirements of target species, as well as principles of landscape ecology.

Further research on the ecology of Neotropical migrants needs to be conducted before reasonable confidence can be placed in conservation plans. That research must focus on quantifying the geographic distributions, habitat requirements, and behavioral ecology of migrants, as well as their survival rates in native habitats and under various agricultural, agroforestry, and wildlife management regimes. Those goals further mandate that ecologists devise more effective ways to quantify, store, and distribute information on migrant abundance in different habitats and geographic regions.

The fate of Neotropical migratory birds depends upon the active participation of an international coalition of ecologists, economists, sociologists, politicians, private landowners, national governments, nongovernmental organizations, academic institutions, business leaders and private citizens. Equally important to understanding the science behind sustaining native populations of wildlife, is the urgency to identify sustainable land-use practices that allow coexistence of wildlife and expanding human populations in Latin America. Neotropical migratory birds appear to be more resilient to small-scale alteration of native vegetation than are tropical resident species. Thus, efforts aimed at improving the welfare of rural human populations, while protecting habitats of resident wildlife species, may offer the most effective means of sustaining long-term viability of all native species, including Neotropical migratory birds.

ACKNOWLEDGMENTS

Bob Askins, Christine Champe, Russ Greenberg, Tom Martin, Charlie Paine, Ken Petit, Lisa Petit, Chan Robbins, and Tom Sherry provided many insightful suggestions for the improvement of our work. Bob Askins, Christine Champe, Courtney Conway, Lisa Petit, Tom Sherry, Jay Vaninni, and Joe Wunderle generously offered access to unpublished manuscripts and data.

LITERATURE CITED

Alcorn, J. B. 1990. Indigenous agroforestry strategies meeting farmers' needs. Pp. 141–151 *in* Alternatives to deforestation: steps toward sustainable use of the Amazon rain forest (A. B. Anderson, ed.). Columbia University Press, New York.

Ambuel, B., and S. A. Temple. 1982. Songbird populations in southern Wisconsin forests: 1954 and 1979. J. Field Ornithol. 53:149–158.

American Ornithologists' Union. 1983. Check-list of North American birds, 6th ed. American Ornithologists' Union, Lawrence, KS.

Arendt, W. J. 1992. Status of North American migrant landbirds in the Caribbean region: a summary. Pp. 143–174 *in* Ecology and conservation of Neotropical migrant landbirds (J. M. Hagan, III and D. W. Johnston, eds). Smithsonian Institution Press, Washington, DC.

Askins, R. A., J. F. Lynch, and R. Greenberg. 1990. Population declines in migratory birds in eastern North America. Curr. Ornithol. 7:1–57.

Askins, R. A., and D. N. Ewert. 1991. Impact of Hurricane Hugo on bird populations on St John's US Virgin Islands. Biotropica 23:481–487.

Askins, R. A., D. N. Ewert, and R. L. Norton. 1992. Abundance of wintering migrants in fragmented and continuous forests in the US Virgin Islands. Pp. 197–206 *in* Ecology and conservation of Neotropical migrant landbirds (J. M. Hagan, III and D. W. Johnston, eds). Smithsonian Institution Press, Washington, DC.

Baltz, M. E. 1993. Abundance of Neotropical migrant songbirds on North Andros Island, Bahamas. Florida Field Natur. 21: 115–117.

Bennett, S. E. 1980. Interspecific competition and the niche of the American Redstart (*Setophaga ruticilla*) in wintering and breeding communities. Pp. 319–335 *in* Migrant birds in the Neotropics: ecology, behavior, distribution, and conservation (A. Keast and E. S. Morton, eds). Smithsonian Institution Press, Washington, DC.

Bierregaard, R. O., Jr. 1990. Avian communities in the understory of Amazonian forest fragments. Pp. 333–343 *in* Biogeography and ecology of forest bird communities (A. Keast, ed.). SPB Academic Publishing, The Hague, The Netherlands.

Binford, L. C. 1989. A distributional survey of the birds of Oaxaca. Ornithol. Monogr. no. 43:1–418.

Blake, J. G., and W. G. Hoppes. 1986. Influence of resource abundance on use of tree-fall gaps by birds in an isolated woodlot. Auk 103:328–340.

Blake, J. G., and J. R. Karr. 1987. Breeding birds of isolated woodlots: area and habitat relationships. Ecology 68:1724–1734.

Blake, J. G., and B. A. Loiselle. 1992. Habitat use by Neotropical migrants at La Selva Biological Station and Braulio Carrillo

National Park, Costa Rica. Pp. 257–272 *in* Ecology and conservation of Neotropical migrant landbirds (J. M. Hagan, III and D. W. Johnston, eds). Smithsonian Institution Press, Washington, DC.

Blake, J. G., G. F. Stiles, and B. A. Loiselle. 1990. Birds of La Selva: habitat use, trophic composition, and migrants. Pp. 161–182 *in* Four Neotropical forests (A. Gentry, ed.). Yale University Press, New Haven, CT.

Bock, C. E., and T. L. Root. 1981. The Christmas Bird Count and avian ecology. Pp. 17–23 *in* Estimating numbers of terrestrial birds (C. J. Ralph and J. M. Scott, eds). Stud. Avian Biol. no. 6. Cooper Ornithological Society, Lawrence, KS.

Bosque, C., and M. Lentino. 1987. The passage of North American migratory land birds through xerophytic habitats of the western coast of Venezuela. Biotropica 19:267–273.

Briggs, S. A., and J. H. Criswell. 1978. Gradual silencing of spring in Washington. Atlantic Natur. 32:19–26.

Brittingham, M. C., and S. A. Temple. 1983. Have cowbirds caused forest songbirds to decline? BioScience 33:31–35.

Brokaw, N. 1995. Mahogany silviculture and migrant bird conservation in Belize. Partners in Flight Newsletter 4:25–26.

Brown, J. L. 1969. Territorial behavior and population regulation in birds. Wilson Bull. 81:293–329.

Burley, F. W. 1988. Monitoring biological diversity for setting priorities in conservation. Pp. 227–230 *in* Biodiversity (E. O. Wilson, ed.). National Academy Press, Washington, DC.

Buskirk, W. H. 1976. Social systems in a tropical forest avifauna. Amer. Natur. 110:293–310.

Buskirk, W. H., G. V. N. Powell, J. R. Wittenberger, R. E. Buskirk, and T. U. Powell. 1972. Interspecific bird flocks in tropical highland Panama. Auk 89:612–624.

Champe, C. M. 1993. Bird communities in native and agricultural habitats in south-central Florida. MS thesis, University Florida, Gainesville, FL, 73 pp.

Chipley, R. M. 1980. Nonbreeding ecology of the Blackburnian Warbler. Pp. 309–317 *in* Migrant birds in the Neotropics: ecology, behavior, distribution, and conservation (A. Keast and E. S. Morton, eds). Smithsonian Institution Press, Washington, DC.

Coates-Estrada, R., A. Estrada, D. Pashley, and W. Barrow. 1985. List of the birds of the Los Tuxtlas Biological Station. Instituto de Biologia, Universidad Nacional Autonoma de Mexico, Mexico, DF.

Christensen, B. 1983. Mangroves—what are they worth? Unasylva 34:2–15.

Cody, M. L. (ed.). 1985. Habitat selection in birds. Academic Press, San Diego, CA.

Conway, C. J., G. V. N. Powell, and J. D. Nichols. 1995. Overwinter survival of Neotropical migratory birds in early-successional and mature tropical forests. Conserv. Biol. (in press).

Cornelius, S. E. 1991. Wildlife conservation in Central America: will it survive the 90's? Trans. North Amer. Wildl. Nat. Resour. Conf. 56:40–49.

Cruz, A. 1988. Avian resource use in a Caribbean pine plantation. J. Wildl. Manag. 52:274–279.

Diamond, A. W. 1985. The selection of critical areas and current conservation efforts in tropical forest birds. Pp. 33–48 *in* Conservation of tropical forest birds (A. W. Diamond and T. E. Lovejoy, eds). ICBP Tech. Publ. no. 4.

Diamond, A. W. 1991. Assessment of the risks from tropical deforestation to Canadian songbirds. Trans. North Amer. Wildl. Nat. Resour. Conf. 56:177–194.

Dirzo, R., and M. A. Garcia. 1992. Rates of deforestation in Los Tuxtlas, a Neotropical area in southeast Mexico. Conserv. Biol. 6:84–90.

Dunning, J. B., B. J. Danielson, and H. R. Pulliam. 1992. Ecological processes that affect populations in complex landscapes. Oikos 65:169–175.

Eaton, S. W. 1953. Wood warblers wintering in Cuba. Wilson Bull. 65:169–174.

Ewert, D. N., and R. A. Askins. 1991. Flocking behavior of migratory warblers in winter in the Virgin Islands. Condor 93:864–868.

Faaborg, J., and W. J. Arendt. 1992. Long-term declines of winter resident warblers in a Puerto Rican dry forest: which species are in trouble? Pp. 57–63 *in* Ecology and conservation of Neotropical migrant landbirds (J. M. Hagan, III and D. W. Johnston, eds). Smithsonian Institution Press, Washington, DC.

Faaborg, J., W. J. Arendt, and M. S. Kaiser. 1984. Rainfall correlates of bird population fluctuations in a Puerto Rican dry forest: a nine year study. Wilson Bull. 96:575–593.

Fearnside, P. M. 1989. Extractive reserves in Brazilian Amazonia. BioScience 39:387–393.

Forman, R. T. T., and M. Godron. 1981. Patches and structural components for a landscape ecology. BioScience 31:733–740.

Forman, R. T. T., and M. Godron. 1986. Landscape ecology. John Wiley and Sons, New York.

Franzreb, K. E. 1983. A comparison of avian foraging behavior in unlogged and logged mixed-coniferous forest. Wilson Bull. 95:60–76.

Freese, C. H., and C. J. Saavedra. 1991. Prospects for wildlife management in Latin America and the Caribbean. Pp. 430–444 *in* Neotropical wildlife use and conservation (J. G. Robinson and K. H. Redford, eds). University of Chicago Press, Chicago, IL.

Fretwell, S. D., and H. L. Lucas. 1970. On territorial behavior and other factors affecting habitat distribution in birds. I. Theoretical development. Acta Biotheoretica 19:16–36.

Gates, J. E., and L. W. Gysel. 1978. Avian nest dispersion and fledgling success in field-forest ecotones. Ecology 59:871–883.

Gauthreaux, S. A., Jr. 1992. The use of weather radar to monitor long-term patterns of trans-Gulf migration in spring. Pp. 96–100 *in* Ecology and conservation of Neotropical migrant landbirds (J. M. Hagan, III and D. W. Johnston, eds). Smithsonian Institution Press, Washington, DC.

Geerling, C., H. Breman, and E. T. Berczy. 1986. Ecology and development: an attempt to synthesize. Environ. Conserv. 13:211–214.

Gibbs, J. P., and J. Faaborg. 1990. Estimating the viability of Ovenbird and Kentucky Warbler populations in forest fragments. Conserv. Biol. 4:193–196.

Gilbert, L. E. 1980. Food web organization and the conservation of Neotropical diversity. Pp. 11–33 *in* Conservation biology: an evolutionary–ecological perspective (M. E. Soule and B. A. Wilcox, eds). Sinauer Associates, Sunderland, MA.

Glasser, J. W. 1982. A theory of trophic strategies: the evolution of facultative specialists. Amer. Natur. 119:250–262.

Glasser, J. W. 1984. Evolution of efficiencies and strategies of resource exploitation. Ecology 65:1570–1578.

Goodman, D. 1987. The demography of chance extinction. Pp. 11–34 *in* Viable populations for conservation (M. E. Soule, ed.). Cambridge University Press, Cambridge, England.

Gordon, D. M. 1991. Variation and change in behavioral ecology. Ecology 72:1196–1203.

Gow, D. D. 1992. Forestry for sustainable development: the social dimension. Unasylva 43:41–45.

Gradwohl, J., and R. Greenberg. 1991. Small forest reserves: making the best of a bad situation. Clim. Change 18:253–256.

Greenberg, R. 1980. Demographic aspects of long-distance migration. Pp. 493–504 *in* Migrant birds in the Neotropics: ecology, behavior, distribution, and conservation (A. Keast and E. S. Morton, eds). Smithsonian Institution Press, Washington, DC.

Greenberg, R. 1981. Frugivory in some migrant tropical forest wood warblers. Biotropica 13:215–223.

Greenberg, R. 1986. Competition in migrant birds in the nonbreeding season. Curr. Ornithol. 3:281–307.

Greenberg, R. 1992a. Forest migrants in non-forest habitats on the Yucatan Peninsula. Pp. 13–22 *in* Ecology and conservation of Neotropical migrant landbirds (J. M. Hagan, III and D. W. Johnston, eds). Smithsonian Institution Press, Washington, DC.

Greenberg, R. 1992b. The nonbreeding season: introduction. Pp. 175–177 *in* Migrant birds in the Neotropics: ecology, behavior, distribution, and conservation (A. Keast and E. S. Morton, eds). Smithsonian Institution Press, Washington, DC.

Griffin, D. R. 1964. Bird migration. Natural History Press, Garden City, New York.

Grumbine, E. 1990. Protecting biological diversity through the greater ecosystem concept. Nat. Areas J. 10:114–120.

Hamel P. B. 1986. Bachman's Warbler: a species in peril. Smithsonian Institution Press, Washington, DC.

Hansson, L. 1991. Dispersal and connectivity in metapopulations. Biol. J. Linnean Soc. 42:89–103.

Harris, L. D., and P. Kangas. 1988. Reconsideration of the habitat concept. Trans. North Amer. Wildl. Nat. Resources Conf. 53:137–144.

Hartshorn, G. S. 1983. Plants: introduction. Pp. 118–157 *in* Costa Rican natural history (D. H. Janzen, ed.). University of Chicago Press, Chicago, IL.

Hartshorn, G. S. 1992. Forest loss and future options in Central America. Pp. 13–22 *in* Ecology and conservation of Neotropical migrant landbirds (J. M. Hagan, III and D. W. Johnston, eds). Smithsonian Institution Press, Washington, DC.

Hartshorn, G. S., L. Nicolait, L. Hartshorn, G. Bevier, R. Brightman, J. Cal, A. Cawich, W. Davidson, R. Dubois, C. Dyer, G. Gibson, W. Hawley, J. Leonard, R. Nicolait, D. Weyer, H. White, and C. Wright. 1984. Belize country environmental profile. Trejos. Hnos Sucs. SA, San Jose, Costa Rica.

Holdridge, L. R. 1947. Determination of world plant formations from simple climatic data. Science 105:367–368.

Holdridge, L. R. 1970. Natural vegetation and reservation prospects in northern Latin

America. Pp. 27–33 *in* The avifauna of northern Latin America (H. K. Buechner and J. H. Buechner, eds). Smithsonian Contr. Zool. no. 26. Smithsonian Institution Press, Washington, DC.

Holdridge, L. R., W. Grenke, W. Hatheway, T. Liang, and J. Tosi. 1971. Forest environments in tropical life zones: a pilot study. Pergamon, Elmsford, NY.

Holmes, R. T., R. E. Bonney, Jr., and S. W. Pacala. 1979. Guild structure of the Hubbard Brook bird community: a multivariate approach. Ecology 60:512–520.

Holmes, R. T., and T. W. Sherry. 1992. Site fidelity of migratory warblers in temperate breeding and Neotropical wintering areas: implications for population dynamics, habitat selection, and conservation. Pp. 563–575 *in* Ecology and conservation of Neotropical migrant landbirds (J. M. Hagan, III and D. W. Johnston, eds). Smithsonian Institution Press, Washington, DC.

Holmes, R. T., T. W. Sherry, and F. W. Sturges. 1986. Bird community dynamics in a temperate deciduous forest: long-term trends at Hubbard Brook. Ecol. Monogr. 50: 201–220.

Holmes, R. T., T. W. Sherry, and L. Reitsma. 1989. Population structure, territoriality and over-winter survival of two migrant warbler species in Jamaica. Condor 91:545–561.

Howell, T. R. 1970. Avifauna in Nicaragua. Pp. 58–62 *in* The avifauna of northern Latin America (H. K. Buechner and J. H. Buechner, eds). Smithsonian Contr. Zool. no. 26. Smithsonian Institution Press, Washington, DC.

Howell, T. R. 1971. An ecological study of the birds of the lowland pine savanna and adjacent rainforest in northeastern Nicaragua. Living Bird 10:185–242.

Hussell, D. J. T., M. H. Mather, and P. M. Sinclair. 1992. Trends in numbers of tropical- and temperate-wintering migrant landbirds in migration at Long Point, Ontario, 1961–1988. Pp. 101–114 *in* Ecology and conservation of Neotropical migrant landbirds (J. M. Hagan, III and D. W. Johnston, eds). Smithsonian Institution Press, Washington, DC.

Hutto, R. L. 1980. Winter habitat distribution of migratory land birds in western Mexico with special reference to small, foliage-gleaning insectivores. Pp. 181–204 *in* Migrant birds in the Neotropics: ecology, behavior, distribution, and conservation (A. Keast and E. S. Morton, eds). Smithsonian Institution Press, Washington, DC.

Hutto, R. L. 1985. Habitat selection by nonbreeding, migratory land birds. Pp. 455–476 *in* Habitat selection in birds (M. L. Cody, ed.). Academic Press, New York.

Hutto, R. L. 1987. A description of mixed-species insectivorous bird flocks in western Mexico. Condor 89:282–292.

Hutto, R. L. 1988a. Is tropical deforestation responsible for the reported declines in Neotropical migrant populations? Amer. Birds 42:375–379.

Hutto, R. L. 1988b. Foraging behavior patterns suggest a possible cost associated with participation in mixed-species bird flocks. Oikos 51:79–83.

Hutto, R. L. 1989. The effect of habitat alteration on migratory land birds in a west Mexican tropical deciduous forest: a conservation perspective. Conserv. Biol. 3:138–148.

Hutto, R. L. 1990. Measuring the availability of food resources. Pp. 20–28 *in* Avian foraging: theory, methodology, and applications (M. L. Morrison, C. J. Ralph, J. Verner, and J. R. Jehl, Jr, eds). Stud. Avian Biol. no. 13. Cooper Ornithological Society, Lawrence, KS.

Hutto, R. L. 1992. Habitat distributions of migratory landbird species in western Mexico. Pp. 211–239 *in* Ecology and conservation of Neotropical migrant landbirds (J. M. Hagan, III and D. W. Johnston, eds). Smithsonian Institution Press, Washington, DC.

Hutto, R. L., S. M. Pletschet, and P. Hendricks. 1986. A fixed-radius point count method for nonbreeding and breeding season use. Auk 103:593–602.

Hutto, R. L., S. Reel, and P. B. Landres. 1987. A critical evaluation of the species approach to biological conservation. Endangered Species Update 4:1–4.

James, F. C. 1971. Ordinations of habitat relationships among breeding birds. Wilson Bull. 83:215–236.

Janzen, D. H. 1973. Sweep samples of tropical foliage insects: effects of seasons, vegetation types, elevation, time of day, and insularity. Ecology 54:687–708.

Janzen, D. H. 1988. Tropical dry forests: the most endangered major tropical ecosystem. Pp. 130–137 *in* Biodiversity (E. O. Wilson, ed.). National Academy Press, Washington, DC.

Janzen, D. H., and T. W. Schoener. 1968. Differences in insect abundance and diversity between wetter and drier sites during a tropical dry season. Ecology 49:96–110.

Johns, A. D. 1989. Recovery of a peninsular Malaysian rain forest avifauna following selective timber logging: the first twelve years. Forktail 4:89–105.

Johnston, D. W., and J. M. Hagan, III. 1992. An analysis of long-term breeding bird censuses from eastern deciduous forests. Pp. 75–84 *in* Ecology and conservation of Neotropical migrant landbirds (J. M. Hagan, III and D. W. Johnston, eds). Smithsonian Institution Press, Washington, DC.

Kapos, V. 1989. Effects of isolation on the water status of forest patches in the Brazilian Amazon. J. Tropical Ecol. 5:173–185.

Karr, J. R. 1976. On the relative abundance of migrants from the north temperate zone in tropical habitats. Wilson Bull. 88:433–458.

Karr, J. R. 1981. Surveying birds in the tropics. Pp. 548–553 *in* Estimating numbers of terrestrial birds (C. J. Ralph and J. M. Scott, eds). Stud. Avian Biol. no. 6. Cooper Ornithological Society, Lawrence, KS.

Karr, J. R. 1990. Avian extinction rates and the extinction process on Barro Colorado Island, Panama. Conserv. Biol. 4:391–397.

Karr, J. R., and D. R. Dudley. 1981. Ecological perspective on water quality goals. Environ. Manag. 5:55–68.

Karr, J. R., and K. E. Freemark. 1983. Habitat selection and environmental gradients: dynamics in the "stable" tropics. Ecology 64:1481–1494.

Karr, J. R., D. W. Schemske, and N. Brokaw. 1982. Temporal variation in the undergrowth bird community of a tropical forest. Pp. 441–453 *in* Ecology of a tropical forest: seasonal rhythms and long-term changes (E. G. Leigh, A. S. Rand, and D. Windsor, eds). Smithsonian Institution Press, Washington, DC.

Keast, A. 1980. Spatial relationships between migratory parulid warblers and their ecological counterparts in the Neotropics. Pp. 109–130 *in* Migrant birds in the Neotropics: ecology, behavior, distribution, and conservation (A. Keast and E. S. Morton, eds). Smithsonian Institution Press, Washington, DC.

Keast, A., and E. S. Morton (eds). 1980. Migrant birds in the Neotropics: ecology, behavior, distribution, and conservation. Smithsonian Institution Press, Washington, DC.

Kellman, M., and R. Tackaberry. 1993. Disturbance and tree species coexistence in tropical riparian forest fragments. Global Ecol. Biogeog. Lett. 3:1–9.

Kennish, M. J. 1990. Ecology of estuaries. Vol. II, Biological aspects. CRC Press, Boca Raton, FL.

Ketterson, E. D., and V. Nolan, Jr. 1983. The evolution of differential bird migration. Curr. Ornithol. 1:357–402.

Koplin, J. R. 1973. Differential habitat use by sexes of American Kestrels wintering in northern California. J. Raptor Res. 7:39–42.

Kricher, J. C., and W. E. Davis, Jr. 1992. Patterns of avian species richness in disturbed and undisturbed habitats in Belize. Pp. 240–246 *in* Ecology and conservation of Neotropical migrant landbirds (J. M. Hagan, III and D. W. Johnston, eds). Smithsonian Institution Press, Washington, DC.

Lack, D., and P. Lack. 1972. Wintering warblers in Jamaica. Living Bird 11:129–153.

Land, H. C. 1970. Birds of Guatemala. Livingston, Wynnewood, PA.

Laurance, W. F. 1991. Edge effects in tropical forest fragments: application of a model for the design of nature reserves. Biol. Conserv. 57:205–219.

Laurance, W. F., and E. Yensen. 1991. Predicting the impacts of edge effects in fragmented habitats. Biol. Conserv. 55:77–92.

Leck, C. F. 1972a. The impact of some North American migrants at fruiting trees in Panama. Auk 89:842–850.

Leck, C. F. 1972b. Observations of birds at *Cecropia* trees in Puerto Rico. Wilson Bull. 84:498–500.

Leonard, H. J. 1987. Natural resources and economic development in Central America: a regional environmental profile. Transaction Books, New Brunswick, NJ.

Levey, D. J. 1988. Spatial and temporal variation in Costa Rican fruit and fruit-eating bird abundance. Ecol. Monogr. 58:251–269.

Levey, D. J., and F. G. Stiles. 1992. Evolutionary precursors of long-distance migration: resource availability and movement patterns in Neotropical landbirds. Amer. Natur. 140:447–476.

Levins, R. 1968. Evolution in changing environments. Princeton University Press, Princeton, NJ.

Loiselle, B. A., and J. G. Blake. 1991. Temporal variation in birds and fruits along an elevational gradient in Costa Rica. Ecology 72:180–193.

Lopez Ornat, A. 1990. Ecologia de las passeriformes en la reserva de la biosfera de Sian Kaan, Mexico. PhD thesis, University Complutense de Madrid, Facultad de Ciencias Biologicas, Madrid.

Lopez Ornat, A., and R. Greenberg. 1990. Sexual segregation by habitat in migratory warblers in Quintana Roo, Mexico. Auk 107:539–543.

Lopez Ornat, A., and J. F. Lynch. 1991. Landbird communities in the coastal dune scrub of the Yucatan Peninsula: species composition, ecology, and zoogeographic affinities. Fauna Silv. Neotrop. 2:21–31.

Lovejoy, T. E., R. O. Bierregaard, Jr, A. B. Rylands, J. R. Malcolm, C. E. Quintela, L. H. Harper,

K. S. Brown, Jr, A. H. Powell, G. V. N. Powell, H. O. R. Schubart, and M. B. Hayes. 1986. Edge and other effects of isolation on Amazon forest fragments. Pp. 257–285 *in* Conservation biology: the science of scarcity and diversity (M. E. Soule, ed.). Sinauer Associates, Sunderland, MA.

Lugo, A. E., R. Schmidt, and S. Brown. 1981. Tropical forests in the Caribbean. Ambio 10:318–324.

Lynch, J. F. 1989. Distribution of overwintering nearctic migrants in the Yucatan Peninsula, I: general patterns of occurrence. Condor 91:515–544.

Lynch, J. F. 1991. Effects of Hurricane Gilbert on birds in a dry tropical forest in the Yucatan Peninsula. Biotropica 23:488–496.

Lynch, J. F. 1992. Distribution of overwintering nearctic migrants in the Yucatan Peninsula, II: use of native and human-modified vegetation. Pp. 178–196 *in* Ecology and conservation of Neotropical migrant landbirds (J. M. Hagan, III and D. W. Johnston, eds). Smithsonian Institution Press, Washington, DC.

Lynch, J. F., and D. F. Whigham. 1984. Effects of forest fragmentation on breeding bird communities in Maryland, USA. Biol. Conserv. 28:287–324.

Lynch, J. F., E. S. Morton, and M. E. van der Voort. 1985. Habitat segregation between the sexes of wintering Hooded Warblers (*Wilsonia citrina*). Auk 102:714–721.

Mabey, S. E., and E. S. Morton. 1992. Demography and territorial behavior of wintering Kentucky Warblers in Panama. Pp. 329–336 *in* Ecology and conservation of Neotropical migrant landbirds (J. M. Hagan, III and D. W. Johnston, eds). Smithsonian Institution Press, Washington, DC.

MacArthur, R. H., and J. MacArthur. 1961. On bird species diversity. Ecology 42:594–598.

MacAthur, R. H., and E. O. Wilson. 1967. The theory of island biogeography. Princeton University Press, Princeton, NJ.

MacClintock, L., R. F. Whitcomb, and B. L. Whitcomb. 1977. Island biogeography and "habitat islands" of eastern forest. II. Evidence of the value of corridors and minimization of isolation in preservation of biotic diversity. Amer. Birds 31:6–16.

Margules, C., and M. B. Usher. 1981. Criteria used in assessing wildlife conservation potential: a review. Biol. Conserv. 21:79–109.

Margules, C., A. O. Nicholls, and R. L. Pressey. 1988. Selecting networks of reserves to maximise biological diversity. Biol. Conserv. 43:63–76.

Marra, P. P., R. T. Holmes, and T. W. Sherry. 1993. Territorial exclusion by a temperate–tropical migrant warbler in Jamaica: a removal experiment with American Redstarts (*Setophaga ruticilla*). Auk 110:565–572.

Martin, T. E. 1985. Selection of second-growth woodlands by frugivorous migrating birds in Panama: an effect of fruit size and plant density? J. Tropical Ecol. 1:157–170.

Martin, T. E. 1992. Breeding productivity considerations: what are the appropriate habitat features for management? Pp. 455–473 *in* Ecology and conservation of Neotropical migrant landbirds (J. M. Hagan, III and D. W. Johnston, eds). Smithsonian Institution Press, Washington, DC.

Martin, T. E., and J. R. Karr. 1986. Temporal dynamics of Neotropical birds with special reference to frugivores in second-growth woods. Wilson Bull. 98:38–60.

McElroy, J. L., B. Potter, and E. Towle. 1990. Challenges for sustainable development in small Caribbean islands. Pp. 299–316 *in* Sustainable development and environmental management of small islands (W. Beller, P. d'Ayala, and P. Hein, eds). Man and the Biosphere Series, Vol. 5. UNESCO, Paris.

Miles, D. B., and R. E. Ricklefs. 1984. The correlation between ecology and morphology in deciduous forest passerine birds. Ecology 65:1629–1640.

Mills, E. D., and D. T. Rogers, Jr. 1992. Ratios of Neotropical migrants and Neotropical resident birds in winter in a citrus plantation in central Belize. J. Field Ornithol. 63:109–116.

Mills, L. S., M. E. Soule, and D. F. Doak. 1993. The keystone-species concept in ecology and conservation. BioScience 43:219–224.

Monroe, B. L. 1968. A distributional survey of the birds of Honduras. Ornithol. Monogr. no. 7:1–458.

Monroe, B. L. 1970. Effects of habitat change on population levels of the avifauna in Honduras. Pp. 38–44 *in* The avifauna of northern Latin America (H. K. Buechner and J. H. Buechner, eds). Smithsonian Contr. Zool. no. 26. Smithsonian Institution Press, Washington, DC.

Moore, F. R., and T. R. Simons. 1992. Habitat suitability and stopover ecology of Neotropical landbird migrants. Pp. 345–355 *in* Migrant birds in the Neotropics: ecology, behavior, distribution, and conservation (A. Keast and E. S. Morton, eds). Smithsonian Institution Press, Washington, DC.

Moore, F. R., S. A. Gauthreaux, Jr, P. Kerlinger, and T. R. Simons. 1993. Stopover habitat: management implications and guidelines. Pp. 58–69 *in* Status and management of Neotropical migratory birds (D. M. Finch and P. W. Stangel, eds). Gen. Tech. Rep. RM-229. USDA Forest Serv., Rocky Mt. Forest Range Exp. Sta., Fort Collins, CO.

Moreau, R. E. 1972. The Palearctic–African bird migration systems. Academic Press, New York.

Morse, D. H. 1971. The insectivorous bird as an adaptive strategy. Annu. Rev. Ecol. Syst. 2:177–200.

Morse, D. H. 1980. Population limitation: breeding or wintering grounds? Pp. 505–516 *in* Migrant birds in the Neotropics: ecology, behavior, distribution, and conservation (A. Keast and E. S. Morton, eds). Smithsonian Institution Press, Washington, DC.

Morton, E. S. 1980. Adaptations to seasonal changes by migrant land birds in the Panama Canal Zone. Pp. 437–453 *in* Migrant birds in the Neotropics: ecology, behavior, distribution, and conservation (A. Keast and E. S. Morton, eds). Smithsonian Institution Press, Washington, DC.

Morton, E. S. 1990. Habitat segregation by sex in the Hooded Warbler: experiments on proximate causation and discussion of its evolution. Amer. Natur. 135:319–333.

Morton, E. S. 1992. What do we know about the future of migrant landbirds? Pp. 579–589 *in* Ecology and conservation of Neotropical migrant landbirds (J. M. Hagan, III and D. W. Johnston, eds). Smithsonian Institution Press, Washington, DC.

Morton, E. S., and R. Greenberg. 1989. The outlook for migratory songbirds: "Future shock" for birders. Amer. Birds 43:178–183.

Morton, E. S., J. F. Lynch, K. Young, and P. Mehlhop. 1987. Do male Hooded Warblers exclude females from nonbreeding territories in tropical forest? Auk 104:133–135.

Moynihan, M. 1962. The organization and probable evolution of some mixed species flocks of Neotropical birds. Smithsonian Misc. Coll. 143:1–140.

Munn, C. A. 1985. Permanent canopy and understory flocks in Amazonia: species composition and population density. Pp. 683–712 *in* Neotropical ornithology (P. A. Buckley, M. S. Foster, E. S. Morton, R. S. Ridgely, and F. G. Buckley, eds). Ornithol. Monogr. no. 36. American Ornithologists' Union, Washington, DC.

Murphy, P. G., and A. E. Lugo. 1986. Ecology of tropical dry forest. Annu. Rev. Ecol. Syst. 17:67–88.

Myers, N. 1980. Conversion of tropical moist forests. National Academy of Sciences, Washington, DC.

Myers, N. 1991. Tropical forests: present status and future outlook. Clim. Change 19:3–32.

Noss, R. F. 1987. From plant communities to landscapes in conservation inventories: a look at the Nature Conservancy (USA). Biol. Conserv. 41:11–37.

Noss, R. F., and L. D. Harris. 1986. Nodes, networks and MUMs: preserving diversity at all scales. Environ. Manag. 10: 299–309.

Odum, E. P. 1971. Fundamentals of ecology. Saunders, Philadelphia, PA.

O'Neill, R. V., D. L. DeAngelis, J. B. Wade, and T. F. H. Allen. 1986. A hierarchy concept of ecosystems. Princeton University Press, Princeton, NJ.

Orejuela, J. E., R. J. Raitt, and H. Alvarez. 1980. Differential use by North American migrants of three types of Colombian forests. Pp. 253–264 *in* Migrant birds in the Neotropics: ecology, behavior, distribution, and conservation (A. Keast and E. S. Morton, eds). Smithsonian Institution Press, Washington, DC.

Orians, G. H. 1969. The number of bird species in some tropical forests. Ecology 50:783–801.

Orians, G. H., and J. F. Wittenberger. 1991. Spatial and temporal scales in habitat selection. Amer. Natur. 137:S29–S49.

Paine, R. T. 1969. The *Pisaster–Tegula* interaction: prey patches, predator food perference, and intertidal community structure. Ecology 50:950–961.

Parrish, J. D., and T. W. Sherry. 1994. Sexual habitat segregation by American Redstarts wintering in Jamaica: importance of resource seasonality. Auk 111:38–49.

Pashley, D. N. 1988. Warblers of the West Indies. II. The Lesser Antilles. Carib. J. Sci. 24:112–126.

Paynter, R. A. 1955. The ornithogeography of the Yucatan Peninsula. Peabody Mus. Nat. Hist., Yale Univ. Bull. 9:1–347.

Pearson, D. L. 1980. Bird migration in Amazonian Ecuador, Peru, and Bolivia. Pp. 273–283 *in* Migrant birds in the Neotropics: ecology, behavior, distribution, and conservation (A. Keast and E. S. Morton, eds). Smithsonian Institution Press, Washington, DC.

Petit, D. R. 1991. Habitat associations of migratory birds wintering in Belize, Central America: implications for theory and conservation. PhD dissertation, Univeristy of Arkansas, Fayetteville, AR.

Petit, D. R., L. J. Petit, and K. G. Smith. 1992. Habitat associations of migratory birds overwintering in Belize, Central America. Pp. 246–256 *in* Ecology and conservation of Neotropical migrant landbirds (J. M. Hagan, III and D. W. Johnston, eds). Smithsonian Institution Press, Washington, DC.

Petit, D. R., J. F. Lynch, R. L. Hutto, J. G. Blake, and R. B. Waide. 1993. Management and conservation of migratory landbirds overwintering in the Neotropics. Pp. 70–92 *in* Status and management of Neotropical migratory birds (D. M. Finch and P. W. Stangel, eds). Gen. Tech. Rep. RM-229. USDA Forest Serv., Rocky Mt. Forest Range Exp. Sta., Fort Collins, CO.

Petit, D. R., L. J. Petit, V. A. Saab, and T. E. Martin. 1995. Fixed-radius point counts in forests: factors influencing effectiveness and efficiency. Pp. 54–61 *in* Proceedings of the symposium on monitoring bird population trends by point counts (C. J. Ralph, J. R. Sauer, and S. Droege, eds). Gen. Tech. Rep. PSW-GTR. USDA Forest Serv., Pacific Southwest Res. Sta., Albany, CA.

Pielou, E. C. 1975. Ecological diversity. John Wiley and Sons, New York.

Plowman, K. P. 1979. Litter and soil fauna of two Australian subtropical forests. Austral. J. Ecol. 4:87–104.

Post, W. 1978. Social and foraging behavior of warblers wintering in Puerto Rican coastal scrub. Wilson Bull. 90:197–214.

Powell, G. V. N. 1979. Structure and dynamics of interspecific flocks in a Neotropical mid-elevation forest. Auk 96:375–390.

Powell, G. V. N. 1985. Sociobiology and adaptive significance of interspecific foraging flocks in the Neotropics. Pp. 713–732 *in* Neotropical ornithology (P. A. Buckley, M. S. Foster, E. S. Morton, R. S. Ridgely, and F. G. Buckley, eds). Ornithological Monogr. no. 36. American Ornithologists' Union, Washington, DC.

Powell, G. V. N. 1988. Mixed species flocking as a strategy for Neotropical residents. Pp. 813–819 *in* Acta XIX Congressus Internationalis Ornithologici (H. Ouellet, ed.). University of Ottawa Press, Ottawa, Canada.

Powell, G. V. N., and R. D. Bjork. 1994. Implications of altitudinal migration for conservation strategies to protect tropical biodiversity: a case study of the Resplendent Quetzal *Pharomacrus mocinno* at Monteverde, Costa Rica. Bird Conserv. Int. 4:161–174.

Powell, G. V. N., J. H. Rappole, and S. A. Sader. 1992. Neotropical migrant landbird use of lowland Atlantic habitats in Costa Rica: a test of remote sensing for identification of habitat. Pp. 287–298 *in* Ecology and conservation of Neotropical migrant landbirds (J. M. Hagan, III and D. W. Johnston, eds). Smithsonian Institution Press, Washington, DC.

Probst, J. R., and T. R. Crow. 1991. Integrating biological diversity and resource management. J. Forestry 89:12–17.

Rabinowitz, D., S. Cairns, and T. Dillon. 1986. Seven forms of rarity and their frequency in the flora of the British Isles. Pp. 182–204 *in* Conservation biology: the science of scarcity and diversity (M. E. Soule, ed.). Sinauer Associates, Sunderland, MA.

Radabaugh, B. E. 1974. Kirtland's Warbler and its Bahama wintering grounds. Wilson Bull. 86:374–383.

Ralph, C. J., and J. M. Scott (eds). 1981. Estimating numbers of terrestrial birds. Stud. Avian Biol. no. 6, Cooper Ornithological Society, Lawrence, KS.

Ralph, C. J., J. R. Sauer, and S. Droege (eds). 1995. Proceedings of the symposium on monitoring bird population trends by point counts. Gen. Tech. Rep. PSW-GTR. USDA Forest Serv., Pacific Southwest Res. Sta., Albany, CA.

Ramos, M. A., and D. W. Warner. 1980. Analysis of North American subspecies of migrant birds wintering in Los Tuxtlas, southern Veracruz, Mexico. Pp. 173–180 *in* Migrant birds in the Neotropics: ecology, behavior, distribution, and conservation (A. Keast and E. S. Morton, eds). Smithsonian Institution Press, Washington, DC.

Rand, A. L. 1954. Social feeding behavior of birds. Fieldiana Zool. 36:5–71.

Rappole, J. H. 1991. Migrant birds in Neotropical forest: a review from a conservation perspective. Pp. 259–277 *in* Conserving migratory birds (T. Salathe, ed.). ICBP Tech. Publ. no. 12, Cambridge, England.

Rappole, J. H., and M. V. McDonald 1994. Cause and effect in population declines of migratory birds. Auk 111:652–660.

Rappole, J. H., and D. W. Warner. 1980. Ecological aspects of avian migrant behavior in Veracruz, Mexico. Pp. 353–393 *in* Migrant birds in the Neotropics: ecology, behavior, distribution, and conservation (A. Keast and E. S. Morton, eds). Smithsonian Institution Press, Washington, DC.

Rappole, J. H., and E. S. Morton. 1985. Effects of habitat alteration on a tropical avian forest community. Pp. 1013–1021 *in* Neotropical ornithology (P. A. Buckley, M. S. Foster, E. S. Morton, R. S. Ridgely, and F. G. Buckley, eds). Ornithol. Monogr. no. 36. American Ornithologists' Union, Washington, DC.

Rappole, J. H., Morton, E. S., T. E. Lovejoy, III, and J. L. Ruos. 1983. Nearctic avian migrants in the Neotropics. US Fish Wildl. Service, Washington, DC.

Rappole, J. H., M. A. Ramos, and K. Winker. 1989. Wintering Wood Thrush movements and mortality in southern Veracruz. Auk 106:402–410.

Reed, J. M. 1992. A system for ranking conservation priorities for Neotropical migrant birds based on relative susceptibility to extinction. Pp. 524–536 *in* Ecology and conservation of Neotropical migrant landbirds (J. M. Hagan, III and D. W. Johnston, eds). Smithsonian Institution Press, Washington, DC.

Reznicek, A. A. 1987. Are small reserves worthwhile for plants? Endangered Species Update 5:1–3.

Ridgely, R. S., and J. A. Gwynne. 1989. A guide to the birds of Panama. Princeton University Press, Princeton, NJ.

Robbins, C. S., D. Bystrak, and P. H. Geissler. 1986. The breeding bird survey: its first fifteen years, 1965–1979. Resource Publ. no. 157. US Fish Wildl. Service, Washington, DC.

Robbins, C. S., B. A. Dowell, D. K. Dawson, J. Colon, F. Espinoza, J. Rodriguez, R. Sutton, and T. Vargas. 1987. Comparison of Neotropical winter bird populations in isolated patches versus extensive forest. Acta Oecol./Oecol. Gen. 8:282–292.

Robbins, C. S., J. R. Sauer, R. S. Greenberg, and S. Droege. 1989a. Popoulation declines in North American birds that migrate to the Neotropics. Proc. Natl. Acad. Sci. 86:7658–7662.

Robbins, C. S., D. K. Dawson, and B. A. Dowell. 1989b. Habitat area requirements of breeding forest birds of the middle Atlantic States. Wildl. Monogr. no. 103. The Wildlife Society, Washington, DC.

Robbins, C. S., B. A. Dowell, D. K. Dawson, J. A. Colon, R. Estrada, A. Sutton, R. Sutton, and D. Weyer. 1992. Comparison of Neotropical migrant landbird populations wintering in tropical forest, isolated forest fragments, and agricultural habitats. Pp. 207–220 *in* Ecology and conservation of Neotropical migrant landbirds (J. M. Hagan, III and D. W. Johnston, eds). Smithsonian Institution Press, Washington, DC.

Robinson, S. K., J. Terborgh, and J. W. Fitzpatrick. 1988. Habitat selection and relative abundance of migrants in southeastern Peru. Pp. 2298–2307 *in* Acta XIX Congressus Internationalis Ornithologici (H. Ouellet, ed.). University of Ottawa Press, Ottawa, Canada.

Rubinoff, I. 1983. A strategy for preserving tropical forests. Pp. 465–476 *in* Tropical rain forest: ecology and management (S. L. Sutton, T. C. Whitmore, and A. C. Chadwick, eds). Blackwell Scientific Publications, Oxford, England.

Russell, S. M. 1964. A distributional study of the birds of British Honduras. Ornithological Monogr. no. 1. American Ornithologists' Union, Washington, DC.

Russell, S. M. 1970. Avifauna in British Honduras. Pp. 45–49 *in* The avifauna of northern Latin America (H. K. Buechner and J. H. Buechner, eds). Smithsonian Contr. Zool. no. 26. Smithsonian Institution Press, Washington, DC.

Rzedowski, J. 1983. Vegetacion de Mexico. Editorial Limusa, Mexico, DF.

Saab, V. A., and D. R. Petit. 1992. Impact of pasture development on winter bird communities in Belize, Central America. Condor 94:66–71.

Sader, S. A., and A. T. Joyce. 1988. Deforestation rates and trends in Costa Rica. Biotropica 20:11–19.

Salwasser, H., C. K. Hamilton, W. B. Krohn, J. F. Lipscomb, and C. H. Thomas. 1983. Monitoring wildlife and fish: mandates and their implications. Trans. North Amer. Wildl. Nat. Resour. Conf. 48:297–307.

Sauer, J. R., and S. Droege. 1992. Geographic patterns in population trends of Neotropical migrants in North America. Pp. 26–42 *in* Ecology and conservation of Neotropical migrant landbirds (J. M. Hagan, III and D. W. Johnston, eds). Smithsonian Institution Press, Washington, DC.

Saunders, D. A., R. J. Hobbs, and C. R. Margules. 1991. Biological consequences of ecosystem fragmentation: a review. Conserv. Biol. 5:18–32.

Sayer, J. 1991. Conservation and protection of tropical rain forests: the perspectives of the World Conservation Union. Unasylva 42:40–45.

Schwartz, P. 1964. The Northern Waterthrush in Venezuela. Living Bird 3:169–184.

Scott, J. M., B. Csuti, J. D. Jacobi, and J. E. Estes. 1987. Species richness: a geographic approach to protecting future biological diversity. BioScience 37:782–788.

Severinghaus, W. D. 1981. Guild theory development as a mechanism for assessing environmental impact. Environ. Manag. 5:187–190.

Shaffer, M. 1981. Minimum population sizes for species conservation. BioScience 31:131–134.

Sherry, T. W., and R. T. Holmes. 1989. Age-specific social dominance affects habitat use by breeding American Redstarts (*Setophaga ruticilla*): a removal experiment. Behav. Ecol. Sociobiol. 25:327–333.

Sherry, T. W., and R. T. Holmes. 1991. Population age structure of long-distance migratory passerine birds: variation in space and time. Pp. 1542–1556 *in* Acta XX Congressus Internationalis Ornithologici. New Zealand Ornithological Congress Trust Board, Wellington.

Simberloff, D., and J. Cox. 1987. Consequences and costs of conservation corridors. Conserv. Biol. 1:63–71.

Sliwa, A., and T. W. Sherry. 1992. Surveying winter Warbler populations in Jamaica: point counts with and without broadcast vocalizations. Condor 94:924–936.

Slud, P. 1960. The birds of Finca "La Selva", Costa Rica: a tropical wet forest locality. Bull. Amer. Mus. Natural Hist. 121:1–148.

Staicer, C. A. 1992. Social behavior of the Northern Parula, Cape May Warbler, and Prairie Warbler wintering in second-growth forest in southwestern Puerto Rico. Pp. 308–320 *in* Ecology and conservation of Neotropical migrant landbirds (J. M. Hagan, III and D. W. Johnston, eds). Smithsonian Institution Press, Washington, DC.

Stewart, P. A. 1987. Decline in numbers of wood warblers in spring and autumn migrations through Ohio. North Amer. Bird Bander 12:58–60.

Stiles, F. G. 1980. Evolutionary implications of habitat relations between permanent and winter resident landbirds in Costa Rica. Pp. 421–435 *in* Migrant birds in the Neotropics: ecology, behavior, distribution, and conservation (A. Keast and E. S. Morton, eds). Smithsonian Institution Press, Washington, DC.

Stiles, F. G., 1985. Conservation of forest birds in Costa Rica: problems and perspectives. Pp. 141–168 *in* Conservation of tropical forest birds (A. W. Diamond and T. E. Lovejoy, eds). ICBP Tech. Publ. no. 4.

Stiles, F. G., and A. F. Skutch. 1989. A guide to the birds of Costa Rica. Cornell University Press, Ithaca, NY.

Strickland, A. H. 1947. The soil fauna of two contrasted plots of land in Trinidad, British West Indies. J. Anim. Ecol. 16:1–10.

Sutton, S. L., T. C. Whitmore, and A. C. Chadwick (eds). 1983. Tropical rain forest ecology and management. Blackwell Scientific Publications, Oxford, England.

Terborgh, J. 1974. Preservation of natural diversity: the problem of extinction prone species. BioScience 24:715–722.

Terborgh, J. 1980. The conservation status of Neotropical migrants: present and future. Pp. 21–30 *in* Migrant birds in the Neotropics: ecology, behavior, distribution, and conservation (A. Keast and E. S. Morton, eds). Smithsonian Institution Press, Washington, DC.

Terborgh, J. 1986. Community aspects of frugivory in tropical forests. Pp. 371–384 *in* Frugivores and seed dispersal (A. Estrada and T. H. Flemming, eds). Dr. W. Junk, The Hague, The Netherlands.

Terborgh, J. 1989. Where have all the birds gone? Princeton University Press, Princeton, NJ.

Terborgh, J. 1992. Maintenance of diversity in tropical forests. Biotropica 24:283–292.

Terborgh, J., and J. R. Faaborg. 1980. Factors affecting the distribution and abundance of North American migrants in the eastern Caribbean region. Pp. 145–156 *in* Migrant birds in the Neotropics: ecology, behavior, distribution, and conservation (A. Keast and E. S. Morton, eds). Smithsonian Institution Press, Washington, DC.

Terborgh, J., and B. Winter. 1980. Some causes of extinction. Pp. 119–134 *in* Conservation biology: an evolutionary–ecological perspective (M. E. Soule, ed.). Sinauer Associates, Sunderland, MA.

Terborgh, J., and B. Winter. 1983. A method for siting parks and reserves with special reference to Colombia and Ecuador. Biol. Conserv. 27:45–58.

Terborgh, J., S. K. Robinson, T. A. Parker, III, C. A. Munn, and N. Pierpont. 1990. Structure and organization of an Amazonian forest bird community. Ecol. Monogr. 60:213–238.

Thelen, K. D. 1990. Networking to share technical knowledge on wildlife management in Latin America. Unasylva 41:33–38.

Thiollay, J.-M. 1992. Influence of selective logging on bird species diversity in a Guianan rain forest. Conserv. Biol. 6:47–63.

Tramer, E. J. 1974. Proportions of wintering North American birds in disturbed and undisturbed dry tropical habitats. Condor 76:460–464.

Tramer, E. J., and R. Kemp. 1980. Foraging ecology of migrant and resident warblers and vireos in the highlands of Costa Rica. Pp. 285–296 *in* Migrant birds in the Neotropics: ecology, behavior, distribution, and conservation (A. Keast and E. S. Morton, eds). Smithsonian Institution Press, Washington, DC.

Turner, M. G. 1989. Landscape ecology: the effect of pattern on process. Ann. Rev. Ecol. Syst. 20:171–197.

Van Horne, B. 1983. Density as a misleading indicator of habitat quality. J. Wildl. Manag. 47:893–901.

Van Horne, B., and J. A. Wiens. 1991. Forest bird habitat suitability models and the development of general habitat models. Fish Wildl. Res. no. 8. US Fish Wildl. Service, Washington, DC.

Vannini, J. P. 1994. Nearctic avian migrants in coffee plantations and forest fragments of south-western Guatemala. Bird Conserv. Int. 4:209–232.

Vaughan, C. 1990. Patterns in natural resource destruction and conservation in Central America: a case for optimism? Trans. North Amer. Wildl. Nat. Resour. Conf. 55:407–422.

Waide, R. B. 1980. Resource partitioning between migrant and resident birds: the use of irregular resources. Pp. 337–352 *in* Migrant birds in the Neotropics: ecology, behavior, distribution, and conservation (A. Keast and E. S. Morton, eds). Smithsonian Institution Press, Washington, DC.

Waide, R. B., J. T. Emlen, and E. J. Tramer. 1980. Distribution of migrant birds in the Yucatan Peninsula: a survey. Pp. 165–171 *in* Migrant birds in the Neotropics: ecology, behavior, distribution, and conservation (A. Keast and E. S. Morton, eds). Smithsonian Institution Press, Washington, DC.

Webb, W. L., D. F. Behrend, and B. Saisorn. 1977. Effects of logging on songbird populations in a northern hardwood forest. Wildl. Monogr. no. 55. The Wildlife Society, Washington, DC.

Whitacre, D. F., J. M. Madrid, C. Marroquin, M. Schulze, L. Jones, J. Sutter, and A. J. Baker. 1993. Migrant songbirds, habitat change, and conservation prospects in northern Peten, Guatemala: some initial results. Pp. 339–345 *in* Status and management of Neotropical migratory birds (D. M. Finch and P. W. Stangel, eds). Gen. Tech. Rep. RM-229. USDA Forest Serv., Rocky Mt. Forest Range Exp. Sta., Fort Collins, CO.

Whitcomb, R. F., J. F. Lynch, P. A. Opler, and C. S. Robbins. 1976. Island biogeography and conservation: strategy and limitations. Science 193:1030–1032.

Whitcomb, R. F., C. S. Robbins, J. F. Lynch, B. L. Whitcomb, M. K. Klimkiewicz, and D. Bystrak. 1981. Effects of forest fragmentation on avifauna of eastern deciduous forest. Pp. 125–205 *in* Forest island dynamics in man-dominated landscapes (R. L. Burgess and D. M. Sharpe, eds). Springer-Verlag, New York.

Whitmore, T. C. 1991. An introduction to tropical rain forests. Clarendon Press, Oxford, England.

Wiens, J. A. 1977. On competition and variable environments. Amer. Sci. 65:590–597.

Wilcove, D. S. 1985. Nest predation in forest tracts and the decline of migratory songbirds. Ecology 66:1211–1214.

Wilcove, D. S. 1988. Changes in the avifauna of the Great Smokey Mountains: 1947–1983. Wilson Bull. 100:256–271.

Wilcove, D. S., and S. K. Robinson. 1990. The impact of forest fragmentation on bird communities in eastern North America. Pp. 319–331 *in* Biogeography and ecology of forest bird communities (A. Keast, ed.). SPB Academic Publishing, The Hague, The Netherlands.

Williams-Linera, G. 1990. Vegetation structure and environmental conditions of forest edges in Panama. J. Ecol. 78:356–373.

Willis, E. O. 1966. The role of migrant birds at swarms of army ants. Living Bird 5:187–231.

Willis, E. O. 1979. The composition of avian communities in remanescent woodlots in southern Brazil. Papeis Avulsos Zool. 33:1–25.

Willis, E. O. 1980. Ecological roles of migratory and resident birds on Barro Colorado Island, Panama. Pp. 205–225 *in* Migrant birds in the Neotropics: ecology, behavior, distribution, and conservation (A. Keast and E. S. Morton, eds). Smithsonian Institution Press, Washington, DC.

Wilson, E. O. 1988. The current state of biological diversity. Pp. 3–18 *in* Biodiversity (E. O. Wilson, ed.). National Academy Press, Washington, DC.

Winker, K., J. H. Rappole, and M. A. Ramos. 1990. Population dynamics of the Wood Thrush in southern Veracruz, Mexico. Condor 92:444–460.

Wong, M. 1986. Trophic organization of understory birds in a Malaysian dipterocarp forest. Auk 103:100–116.

Woodwell, G. M. 1991. Forests in a warming world: a time for new policies. Clim. Change 19:245–251.

World Resources Institute. 1992. World resources 1992–93. Oxford University Press, Oxford, England.

Wright, D. H. 1991. Correlations between incidence and abundance are expected by chance. J. Biogeog. 18:463–466.

Wunderle, J. M., Jr. 1992. Sexual habitat segregation in wintering Black-throated Blue Warblers in Puerto Rico. Pp. 299–307 *in* Ecology and conservation of Neotropical migrant landbirds (J. M. Hagan, III and D. W. Johnston, eds). Smithsonian Institution Press, Washington, DC.

Wunderle, J. M., Jr., and R. B. Waide. 1993. Distribution of overwintering Nearctic migrants in the Bahamas and Greater Antilles. Condor 95:904–933.

Wunderle, J. M., Jr. and R. B. Waide. 1994. Future

prospects for Nearctic migrants wintering in Caribbean forests. Bird Conserv. Int. 4:191–207.

Yahner, R. H., and D. Scott. 1988. Effects of forest fragmentation on depredation of artificial nests. J. Wildl. Manag. 52:158–161.

Zar, J. H. 1984. Biostatistical analysis. Prentice Hall, Englewood Cliffs, NJ.

PART III

FOREST MANAGEMENT

7

IMPACTS OF SILVICULTURE: OVERVIEW AND MANAGEMENT RECOMMENDATIONS

FRANK R. THOMPSON, III, JOHN R. PROBST, AND MARTIN G. RAPHAEL

INTRODUCTION

Recent declines in population sizes of Neotropical migratory birds (NTMBs) have been attributed to problems on the breeding grounds as well as nonbreeding areas (Hutto 1988, Robbins et al. 1989a, Askins et al. 1990, Wilcove and Robinson 1990). Habitat loss and fragmentation, and the resultant area and edge-related decreases in reproductive success, are at least partially responsible for some local declines or extirpations (see Faaborg et al., Chapter 13, this volume, Freemark et al., Chapter 14, this volume, for reviews). Silvicultural practices alter landscape structure, forest age and structure, and create edges, causing concern for the impacts of these practices on NTMBs. This concern is often greater when timber is harvested on public forest lands because they are some of the least fragmented forests remaining in parts of North America (Wilcove 1988). Much research and management effort have been directed at the impacts of silvicultural practices on forest songbirds, partly evident by a series of regional workshops on management of forest birds from 1975 to 1980 (Smith 1975, DeGraaf 1978, 1979, 1980). However, most research on silviculture and its impact on birds has occurred at the stand or habitat level, and only occasionally are large-scale inferences made. Researchers and managers now are realizing they not only must be concerned with the impacts of silvicultural practices at the forest-stand level, but also with the regional and landscape context of a stand, the cumulative and landscape level effects of management practices, and species interactions such as brood parasitism and predation.

We review common silvicultural systems used in North America and their impacts on forest-dwelling NTMBs. We begin with a brief review of landscape and habitat factors that affect breeding forest birds, then review some basic concepts of silviculture and the potential impacts of these systems with emphasis on harvest and regeneration methods. Other forest practices and regional practices are addressed in this volume by Rotenberry et al. (Chapter 3), Dickson et al. (Chapter 9), Freemark et al. (Chapter 14), and Hejl et al. (Chapter 8). We approach this topic from a regional-landscape scale to a stand-habitat scale, rather than the traditional stand-level approach. We do not provide a complete review of lilerature on this topic, but identify what we believe are the major impacts and processes impacting Neotropical migrant birds in managed forests, and document these with representative citations. Our review in part reflects the availability of information. There is a great deal of information, for example, on changes in NTMB abundance following clearcutting, but very little information on species abundances in selectively cut stands, or reproductive success of any species in forests managed by any regeneration method. We also caution managers in their interpretation of results. In our review of silvicultural impacts, we treat all species equally, and report positive and negative impacts on early and late successional species, and forest interior and forest edge species. To apply this information, land managers will need established priorities and goals. Therefore, we conclude by outlining a hierarchical conservation approach in which local management decisions are made to complement regional- and landscape-level patterns and goals. This "top-down" approach allows for special emphasis on regionally rare or critically threatened species and ecosystems without compromising habitat for more common species.

REGIONAL, LANDSCAPE AND LOCAL FACTORS AFFECTING POPULATIONS

Population sizes and viability (long-term sustainability) are determined by interactions between local habitat factors, and regional or landscape features, such as total habitat area, habitat context and biogeography. Habitat selection can be viewed at successive levels of resolution (Wiens et al. 1987), and habitat selection and distribution may shift dramatically in response to changes in landscape structure and composition, as well as imbalances between regional population size and habitat quantity and quality (eg., Probst and Weinrich 1993).

Generally, a large area of suitable habitat will support larger populations with lower local extinction rates and greater potential to produce excess individuals for dispersal, than will a small patch of habitat (see reviews in Faaborg et al., Chapter 13 this volume, and Freemark et al., Chapter 14, this volume). Smaller habitat patches not only have higher local extinction rates, but are less likely to be colonized or recolonized. Species requiring large patches of fairly homogeneous habitat are said to be "area sensitive." Many birds in the eastern United States are considered area sensitive because they are often absent from small habitat fragments (Whitcomb et al. 1981, Ambuel and Temple 1983, Blake and Karr 1984, Hayden et al. 1985, Robbins et al. 1989b, Faaborg et al., Chapter 13, this volume). In addition to biogeographic factors, many birds may be area sensitive because they have lower reproductive success near forest edges and in edge-dominated forest fragments due to increased predation and brood parasitism (Gates and Gysel 1978, Brittingham and Temple 1983, Temple and Cary 1988, Robinson 1992). Although species may be present in these landscapes, their populations may be population sinks where reproduction is insufficient to compensate for adult mortality (Pulliam 1988). While edge-related declines in reproductive success in forests fragmented by non-forest habitats are a likely cause of area sensitivity (Temple and Cary 1988), the effects of edges created by timber harvest in predominantly forested landscapes is unclear.

Large-scale (regional, landscape) factors may impose important "top-down constraints" on the way NTMBs respond locally to silviculture. For example, regional population size will affect habitat distribution of species within landscapes. The effects of edge and openings created by timber harvest on levels of nest predation and parasitism may depend on the landscape context (Martin 1992). Examples of important context considerations for NTMB are landscape composition (relative proportions of ecosystems, such as amount of forest versus agricultural land), degree of forest fragmentation, and stand age-class distribution. In some fragmented forests, brood parasitism and predation are extremely high but unrelated to distance to edge. Predator and cowbird numbers may be so high in these landscapes that they saturate forest habitats (Robinson et al. 1993). In extensively forested landscapes, however, cowbird and predator numbers may be so low that their influence is limited to forest edges.

At a local or habitat level, birds appear to select nesting and foraging habitats based on an array of factors including vegetation structure, life forms, presence or volume of vegetative strata, plant- or tree-species composition, and special features such as snags, streams, or cliffs (MacArthur et al. 1962, James 1971, Karr and Roth 1971, Willson 1974, Balda 1975, and others). Stand- or habitat-level factors affecting these local features include forest type, history of disturbance, forest age, and site quality. Forest type and disturbance history (natural or man-induced) affect successional pathways and rate of succession, plant composition and vegetation structure. Forest age affects attributes such as tree size, foliage volume, foliage stratification, horizontal patchiness, bark surface area, cavity formation, coarse woody debris, and other special features. Birds use forests of all ages, but the

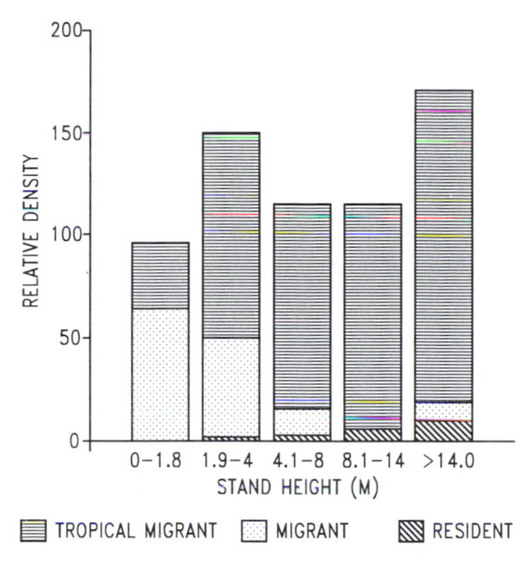

Figure 7-1. The community composition by migration status in different age (height) aspen forests. Adapted from Probst et al. (1992).

importance of different-aged forests to NTMBs varies at least in some forest types (Fig. 7-1). Site quality affects forest-type composition, successional pathways, rate of succession, and vegetation structure, especially stature or tree height. Silvicultural practices directly or indirectly affect local habitat structure and vegetation composition through practices that regenerate stands, or directly manipulate stand structure and composition.

SILVICULTURE

Silviculture is the theory and practice of controlling forest establishment, composition, structure, and growth to achieve management objectives (Smith 1986). Silviculture usually is thought of in the context of timber production, though it should be interpreted more broadly to include other possible objectives such as conservation of biological diversity or NTMBs. Indeed, the Forest Management Act of 1976 (P. L. 94-588) directs the US Forest Service to maintain animal and plant diversity in its management activities.

Silvicultural treatments are applied at the stand level. A stand is a contiguous group of trees sufficiently uniform in species composition and structure to serve as a management unit. Stands are usually identified by the composition and structure of vegetation currently occupying a site, but sometimes are based on ecological classification systems as well. Stands often are equated to animal habitats, communities, or even ecosystems (Hunter 1990). This may be appropriate in some instances but often is an oversimplification or inappropriate structuring of animal habitat requirements or ecosystems. Management is usually planned at a larger scale, referred to as the forest, which is a collection of stands administered as an integrated unit (Smith 1986). A forest is often subdivided into management compartments consisting of groups of stands, usually for uniform distribution of harvest and stand age classes.

Silvicultural Systems

A silvicultural system is a program of forest management for an entire rotation of a stand. It includes harvest cutting, regeneration of the stand, and intermediate treatments. Silvicultural systems often are named by the regeneration method used because these practices have such a large impact on the future of a stand. Regeneration methods usually harvest timber and establish tree reproduction simultaneously. Timber is harvested according to a forest plan that provides for regulation of the harvest to meet certain goals; in the absence of regulation harvest may be exploitative.

The four silvicultural systems commonly used in North America are selection, shelterwood, seed tree, and clearcutting (USDA Forest Service 1973). An important distinction among silvicultural systems is whether they maintain even-aged or uneven-aged stands. In even-aged stands, trees are of one or two age classes, although they may vary somewhat in diameter. An uneven-aged stand contains at least three age classes. Often the height profile of a stand is characteristic of its age-class distribution; an even-aged stand tends to have a level canopy while an uneven-aged stand is distinctly irregular in height. The selection method is used to maintain uneven-aged forests; the clearcut, seed tree, and shelter-

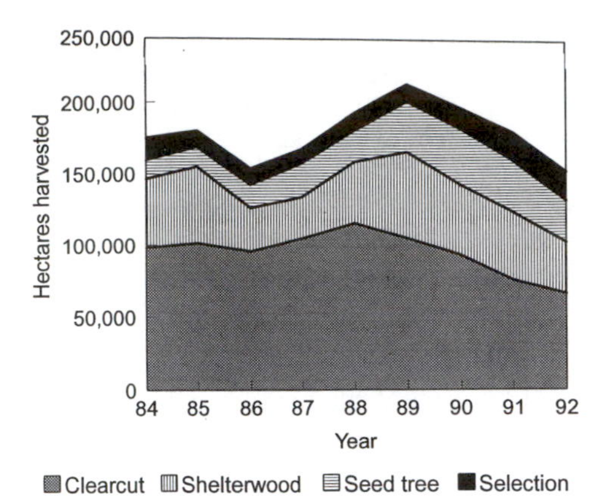

Figure 7-2. The area regenerated by different regeneration methods on US National Forests 1984–1992 (USDA Forest Service, Washington DC).

wood methods maintain even-aged stands. Some modifications of these traditional even-aged practices maintain two-aged stands. The use of clearcutting has been declining on US National Forest lands, and alternative regeneration and harvest methods are now used on more than half the area treated (Fig. 7-2).

Even-aged Systems

Under even-aged management, harvest and regeneration are planned by area (hectares or acres) and are a function of rotation age, which is the age at which a stand is regenerated. Rotation age is based on management objectives that could be economic, esthetic, structural, or ecological. The goal of harvest scheduling (sometimes called regulation) is usually to provide a sustained yield of products, or other uses and values over time. It is important to realize that sustained yield is a forest objective and not a stand objective. Three even-aged methods have traditionally been used to harvest or regenerate stands under even-aged systems (Smith 1986):

- *Clearcutting method.* Removal of most trees in one cutting. Size of the stand varies from small patches (<1 ha) to extensive (>100 ha).
- *Seed tree method.* Removal as in clearcutting except a small number of trees (commonly 4–8) are left, singly or in groups, to reseed the harvested area.
- *Shelterwood method.* Gradual removal of most or all trees in a series of partial cuttings, which extend over a fraction of the rotation. Regeneration is established under the protection of a partial overstory before the final removal cut.

A number of alternative regeneration methods have recently been tried in an attempt to meet public opposition to clearcutting and to address ecological concerns such as structural and compositional diversity. Patch cutting essentially involves creating small clearcuts (<1 ha). It differs from selection cutting because cutting is regulated by area, as with other even-aged practices, and not stand structure as in selection cutting. "Esthetic" shelterwoods are similar to traditional shelterwoods except the removal cut is done over widely spaced entries, or a final removal is never made and a portion of the original stand is left. Two-age silviculture does not fit neatly into uneven-aged or even-aged systems, though it most closely resembles even-aged systems in its application. It is accomplished by removing half the trees every half rotation, which results in two distinct age classes present throughout the rotation (Marquis 1989).

Uneven-aged Systems

In uneven-aged systems, single trees or small groups of trees are periodically harvested. Trees are selected on the basis of age, diameter, vigor, form, and species with the objective of maintaining a relatively consistent stand structure. Sustained yield can be accomplished within a stand if a balanced size-class distribution is maintained within the stand. The desired size class distribution for a balanced stand is defined by the largest desired tree size and the ratio of the number of trees in successive diameter classes (q-value). Thus harvest is planned by volume and diameter, rather than by area as with even-aged management. There is tremendous variation in the implementation of the selection system though harvest is classified as one of two methods:

- *Single-tree selection.* Individual trees are removed throughout the stand.
- *Group selection.* Trees are removed in small groups.

Often single-tree selection and group selection are performed together; this is sometimes referred to as selection with groups (Law and Lorimer 1989). Groups may be harvested to establish regeneration of less tolerant species and single trees removed to balance larger diameter classes or regenerate tolerant species.

Silvicultural Practices

Silvicultural practices can be divided into two broad categories: regeneration practices and intermediate treatments. The objective of regeneration practices is to establish a new stand, whereas the objective of intermediate treatments is to regulate stand composition, structure, and growth, as well as provide some early forest products (Smith 1986). Many other practices associated with silviculture and forest management may affect NTMBs, such as pest control, salvage, fire management, and road building, but these are beyond the scope of this paper.

Regeneration Practices

Following or during harvest, a stand is treated to create conditions favorable for regeneration of desired species. Site preparation may dispose of slash (debris left from harvest cuts), reduce competition from unharvested vegetation, or prepare the soil for new trees. Slash may be removed to reduce potential fuel for wildfires, or because it creates too much shade or physically impedes the regeneration of the stand. Slash disposal commonly occurs in the western forests in combination with planting. Slash is disposed of by broadcast burning, piling and burning, lopping and scattering, windrowing, or chopping on site. Seedbed preparation usually consists of removing organic matter to expose mineral soil. Predominant methods are prescribed burning and scarification, that is, the mechanical removal or mixing of the organic matter with mineral soil. Competing vegetation may be controlled by prescribed burning, mechanical treatment, or herbicides. Prescribed burning also may be used to promote desirable species that are adapted to or dependent on fire.

Artificial regeneration occurs by planting young trees or seeding before or after removing the old stand. Artificial regeneration is most commonly used for conifers because of low natural regeneration, and the probability of success and high financial yield are often greater than for deciduous trees. Natural regeneration occurs from natural seeding, or from stump and root sprouts. The essential step in natural regeneration is to ensure that there is an adequate seed source, advanced reproduction, or potential for sprouting. Advance reproduction is young trees that are present before a stand is regenerated.

Intermediate Treatments

Intermediate treatments are those done between regeneration periods, and are sometimes called timber stand improvement (TSI). Release cuttings are applied to stands that are still in the sapling stage to free desirable trees from competing vegetation. Three types of release cuttings are: weeding, which removes all competitors; cleaning, which removes overtopping competitors of the same age; and liberation, which removes overtopping competitors that are older. Because competing vegetation often resprouts if simply cut or girdled, herbicides are sometimes used alone, or in combination with cutting or girdling.

Thinnings are the selective removal of trees in stands past the sapling stage. They utilize some trees that normally would die from competition in immature stands and, perhaps more importantly, they accelerate growth of selected trees that are released. There are two general types of thinnings. Low thinning removes trees from lower crown classes, salvaging trees that would normally die, and possibly reducing root competition. Crown thinning removes trees from middle and upper portions of the canopy (dominant or codominant crown classes) to favor development of selected trees.

Salvage and sanitation cuttings are other important intermediate practices. Salvage is the harvest of dead, dying, damaged, or deteriorating trees to derive economic benefits before decay processes reduce such values. Salvage cutting also reduces the risk of wildfire. Salvage cutting is a widespread practice often employed after insect outbreaks, fire, windstorms, and other natural disturbances. Sanitation cutting removes trees to inhibit actual or anticipated spread of insects and diseases.

Related Practices

Forest operations involve a number of additional practices not included in the major types of activities mentioned above. Nonetheless, these practices impact forest structure and processes, and merit discussion.

Road building and maintenance is a major activity. National Forests in the United States contain about 570,000 km of roads designated as "local" (75%), single-lane dirt or gravel; "collector" (20%), single-lane gravel providing all-weather access; or "arterial" (5%), double-lane, paved. The Forest Service constructs about 11,000 km of new roads per year, but this is expected to decline as the total system nears completion (USDA Forest Service 1990) and in response to environmental concerns.

Fire suppression is an activity that can result in long-term changes in forest structure and composition. The policy of the US Forest Service is to "suppress all wildfires in a timely, energetic, and thorough manner" (USDA Forest Service 1989, p. 49). This policy is applied in both wilderness and nonwilderness areas, although different techniques may be employed. To implement this policy, one major program involves fire prevention through fuels management. Fuel reduction is accomplished through prescribed fire, salvage harvest, and piling and burning of debris. Underburning is used in pine forests of the South to reduce surface and understory fuels at 3–5 year intervals (approximately one million ha/year; Wenger 1984, p. 238). About 140,000 ha of National Forest lands are treated by prescribed burning annually in the United States.

Pesticides and herbicides are used by forest managers for insect and disease prevention and suppression, vegetation management, and for animal damage control. From 1984 to 1988, the area of National Forest lands treated with pesticide varied from 194,000 to 446,000 ha. Most pesticide use is directed toward suppression of insect pests, primarily gypsy moth (*Porthetria dispar*), southern pine beetle (*Dendroctomus frontalis*), western spruce bud-worm (*Choristoneura fumifera*), and mountain pine beetle (*D. ponderosae*).

Fuelwood harvest occurs along roadways and in designated areas over much forested land. This practice is widespread close to towns where home woodburning is practiced, and is difficult to regulate. The amount of fuel wood harvested, usually as snags or downed logs, is tied to fossil fuel prices and air quality standards.

IMPACTS ON HABITAT AND BIRDS

Even-aged Systems

Landscape Composition

Even-aged management creates a specific age-class distribution of forest habitats that usually differs from forests with no timber harvest. Assuming timber harvest is regulated to provide sustained yield over time, rotation age will determine the amount of forest in any given age class. For instance, an oak–hickory forest managed by regulated clearcutting on a 100-year rotation would comprise approximately 10% regeneration (stands 1–10 years old). Forests managed by even-aged management could have more or less early successional forest than natural landscapes, depending on rotation age and frequency of natural disturbances. Timber harvest in excess of levels that provide sustained yield may severely threaten some NTMB populations.

Forests managed by even-aged management often contain more early successional forests and NTMBs than unmanaged forests, or than before the onset of large-scale logging (Raphael et al. 1988), or compared to present-day forests without logging (Thompson et al. 1992) (Fig. 7-3). For example, Raphael et al. (1988) modeled large-scale changes in bird populations in Douglas-fir forests of northwestern California

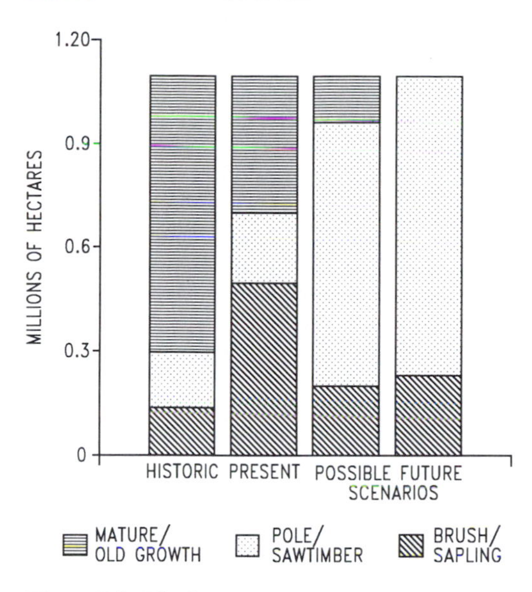

Figure 7-3. The forest area occupied by three seral stages of Douglas-fir forest in northwestern California in historic times, at present, and under two projected trends (Raphael et al. 1988).

based on the impacts of forest management on landscape composition. They compared presettlement, present-day, and future bird populations given current management trends. They predicted that early seral species were currently at a peak compared to historic levels, and that mature forest species had declined and would continue to do so. Among 34 species of NTMBs, 47% were estimated to have increased over historic populations, whereas 26% had decreased. Raphael et al. projected that 65% of NTMB species will have increased in the future and that 24% will have decreased. Thus, stand age distribution caused by logging in northwestern California forests may increase populations of many NTMB species, and reduce populations of resident species and some NTMBs. These trends should be evaluated within the regional context in which they would occur, which would include the status and distribution of the species that are projected to increase or decrease.

Thompson et al. (1992) compared NTMBs in central hardwood forest landscapes managed by the clearcutting method to those in wilderness areas with no timber harvest. Total density of early successional NTMBs was much greater and forest interior NTMBs slightly lower in landscapes managed by clearcutting (Fig. 7-4). Hof and Raphael (1993) predicted the optimum mix of forest age classes under an objective to maintain the highest possible abundance of all bird species based on the relative abundance of birds in several seral stages of forest of northwestern California. In their study, the optimum mix was a nearly even proportion of all age classes because each stage in succession is important for some species. The optimum mix of age classes likely would be different if they had maximized numbers of early or late successional species, as opposed to all species. Similarly, Raphael et al. (1988) predicted that as the mix of seral stages changes due to intensity of timber harvest, some species of birds will increase in abundance whereas others will decrease. Thus, any change in the mix of seral stages will benefit some species and will be detrimental to others. The relative proportion of the forest in each seral stage should be evaluated relative to species status, species habitat requirements, and the silvicultural options available across ownerships.

Spatial Distribution and Edge Effects

The spatial distribution of different aged stands also may impact NTMBs. Stand size determines the size of habitat patches created by regeneration cuts and is usually in the range of 5–20 ha on public lands. Natural disturbances and openings occur much more frequently at small scales than at large scales, but have a wide range of sizes (Hunter 1990). Without special considerations, even-aged management can result in an unnatural uniformity of habitat patch size and distribution, excluding small and very large patches.

The juxtaposition of different aged stands may result in increased amounts of edge in the forest, which may affect the reproductive success of NTMBs (Wilcove 1988). It is less clear, however, how edges created by timber harvest in forested landscapes affect NTMBs (Martin 1992). Several studies have found higher nest parasitism or predation near openings created by timber harvest (Brittingham and Temple 1983, Yahner and Scott

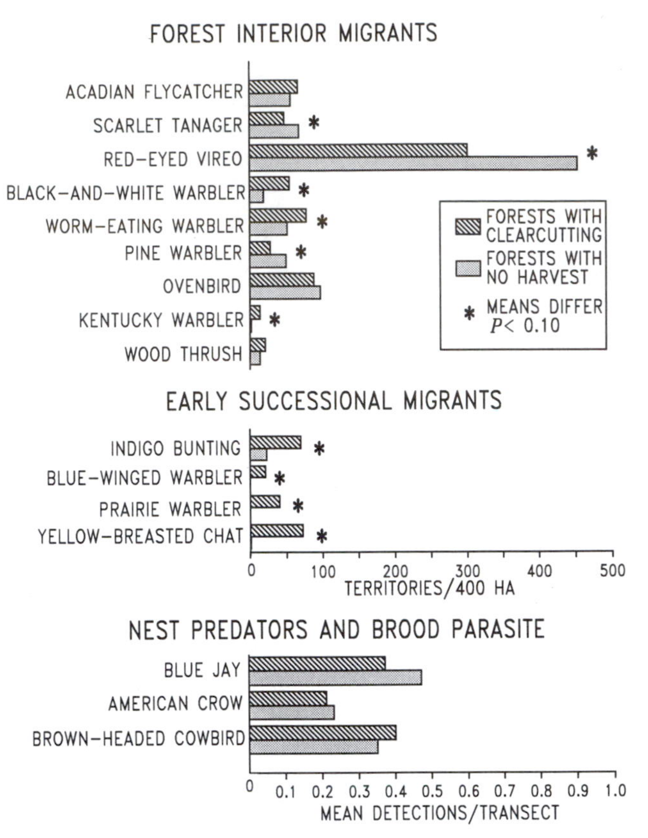

Figure 7-4. Numbers of forest interior Neotropical migrants and early successional Neotropical migrants in forested landscapes managed by clearcutting and landscapes with no timber harvest (Thompson et al. 1992).

1988, D. Whitehead unpublished data), while others have not (Ratti and Reese 1988, Rudnicky and Hunter 1993). Reproductive success may depend greatly on characteristics of the surrounding landscape. In highly fragmented forests in agricultural landscapes, cowbirds and predators may be so abundant they saturate the forest, and parasitism and predation rates may be high throughout the forest with no relation to edges of clearcuts or wildlife openings (Robinson et al. 1993). While many forest interior species remain abundant in managed forests (Thompson et al. 1992) the reproductive success and ability of these populations for self-maintenance is largely unknown.

While the presence of edge effects has been documented, its impact on populations has only been inferred or demonstrated by models. The impact of edge effects on a NTMB population may depend on the extent of edge in the landscape. Temple and Cary (1988) used simulation modeling to demonstrate that edge effects could explain the decline or loss of a habitat interior bird from a highly fragmented landscape. Thompson (1993) used a similar model to investigate the impacts of different timber management practices in a forested landscape on a forest interior bird that nests in mature forest. Simulated population size declined up to 40% in a forest managed by clearcutting on a 100-year rotation, but projected populations were only slightly different in models with and without edge effects (Fig. 7-5). This suggests that most of the decline was due to the conversion of older stands to younger stands (a reduction in habitat suitability or

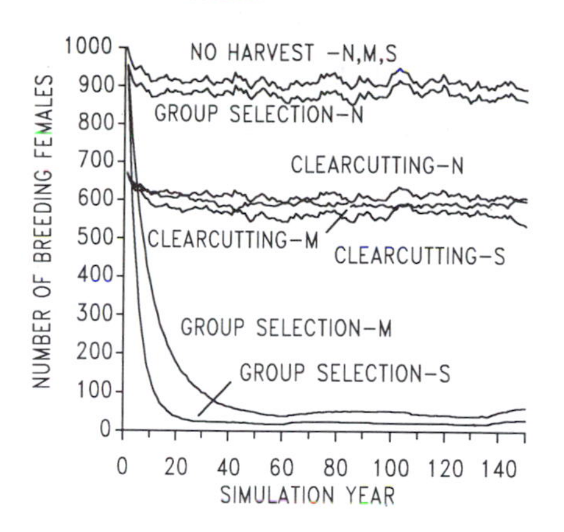

Figure 7-5. Simulated population levels of a forest interior bird population in a 1000 ha forest with no timber harvest, clearcutting, and group selection harvest. The clearcutting method is based on a 100 year rotation and 10 ha clearcuts; under group selection 300 0.14 ha groups are harvested every 15 years (Thompson 1993). N = no effect; M = moderate effect; S = strong edge effect.

carrying capacity) because even-aged management did not create enough edge to affect production greatly, and that forest interior bird populations may persist (at reduced levels) in managed forests even with deleterious edge effects (Thompson 1993).

The extent to which even-aged management exacerbates cowbird parasitism and nest predation also depends on the ultimate factors or mechanisms causing edge effects. For instance, in forested landscapes cowbirds are likely limited by potential feeding sites such as pasture and feedlots (Robinson et al., Chapter 15, this volume). Unlike fragmentation resulting from the loss of forest habitat to agricultural or suburban land uses, regulated (sustained yield) timber harvest may do little to increase the number of cowbirds in the landscape because cowbirds are dependent on agricultural or suburban habitats for feeding. Cowbirds that breed and incidentally feed in clearcuts in the morning still commute daily to agricultural habitats to feed in flocks (Thompson 1994). Thompson et al. (1992) observed similar numbers of cowbirds and avian nest predators in landscapes with and without clearcutting, but cowbirds were more abundant in clearcuts. So, while overall numbers of cowbirds in the landscape may not differ, cowbirds may be attracted to clearcuts. Potential reasons for this attraction could be higher host densities or forest structure more suitable to locating nests. If host species are attracted away from more productive interior habitats to edge habitats, then these edges are functioning as true ecological traps (Gates and Gysel 1978) and may impact host populations. If hosts are not attracted to edges associated with timber harvest, then timber harvest may simply redistribute parasitism but not result in a net increase in parasitism of forest NTMBs. Studies are needed to assess the effects of cowbirds on early successional NTMBs in and adjacent to different sizes, distributions, and densities of regeneration cuts.

Some species of NTMB seem to benefit from the presence of edges. For example, Rosenberg and Raphael (1986) studied the impacts of forest fragmentation on both NTMB and residents in Douglas-fir forests of northwestern California. In that study, 11 species of NTMBs were found to occur frequently on or near edges between forest and cutover areas. In addition, the relative abundance of six species (Olive-sided Flycatcher, Western Wood-pewee, Western Flycatcher, House Wren, Warbling Vireo, and Wilson's Warbler) increased as the amount of edge around the sample stand increased. Abundance of only one species, Western Tanager, decreased with increasing edge. They did not, however, determine the reproductive success of these species.

Temporal Distribution of Forest Age-Classes

Management for sustained and constant yield of timber requires the maintenance of a balanced stand age-class distribution. This type of regulated harvest scheduling also provides a relatively constant availability of habitats. On lands with an unbalanced age-class distribution the area of each successional stage may vary greatly through time.

Species Turnover Within Stands

Regeneration or harvest cuts remove a mature forest community and replace it with a young forest community. Numerous studies have documented bird species turnover associated with regeneration practices (e.g., Conner and Adkisson 1975, Webb et al. 1977, Conner et al. 1979, Crawford et al. 1981, Franzreb and Ohmart 1978, Thompson and Fritzell 1990). In their review of silvicultural impacts on forest birds in the Rocky Mountains, Hejl et al. (Chapter 8, this volume) concluded that about half the NTMBs increased after timber harvest and half decreased. Similarly, in their review of silvicultural impacts in southeastern forests, Dickson et al. (Chapter 9, this volume) concluded that about 50% of NTMBs in central hardwood forests preferred early successional stands created by harvest, but only 10–20% preferred these habitats in southern pine forests. These changes are largely due to changes in vegetation structure resulting from regeneration of the stand.

Tree-species composition also may change when a stand is regenerated. The most obvious example is the use of artificial regeneration because the dominant species in the future forest is largely determined by the selection of planting stock. Planting stock can potentially be anything a site can support, including exotics. Conversion of deciduous stands to pine, and the use of exotic tree species, are decreasing on some public lands. However, artificially regenerated stands are still usually planted with one or a few species. Changing the forest type or reducing tree species diversity in a stand may change the NTMB community or reduce NTMB species richness. Birds may prefer specific tree species because of their structure and availability of foods (Robinson and Holmes 1984, Morrison et al. 1985). Many NTMBs are associated with hardwoods in conifer plantations, so control of competing hardwood vegetation may further limit NTMB diversity. Conifer plantations often have dense, closed canopies at certain ages or if they are not thinned. Closed-canopy plantations often have limited vertical and horizontal vegetative-structural diversity, which also may result in low NTMB diversity. For example, pine plantations in Michigan may have only two NTMB species (Probst and Rakstad, unpublished data). Selection of a regeneration method also can affect the species composition of natural regeneration. Regeneration methods range from clearcutting, which favors shade intolerant trees, to single tree selection, which favors shade-tolerant species. Small changes in tree-species composition in eastern deciduous forests probably have little effect on breeding birds because of high tree-species diversity and because similar vegetative structure or life forms are maintained.

Rotation Age

Rotation age greatly affects stand structure. Rotation ages have usually been defined to maximize economic returns from a stand and typically range from 30 to 100 years, which is often shorter than the average frequency of natural disturbances. As a result, even-aged management may truncate succession and prevent development of structural characteristics associated with old stands (Edgerton and Thomas 1978, Bunnell and Kemsater 1990). This includes development of large trees, accumulation of downed and standing dead wood, and development of high vertical foliage density due to canopy layering. Commercial rotations could result in fewer cavity-nesting, bark-foraging, foliage-gleaning, or canopy-nesting species resulting in lower within-stand species diversity (Probst 1979). In Douglas-fir forests of the Pacific northwest, nine species of NTMBs reach peak abundance in the most mature stands (Raphael et al. 1988). Long rotations in managed forests would provide more mature stands and benefit these species.

Residual Structure

Regeneration practices could result in felling of all trees (including snags) and disposal of slash. This can result in a stand (and forest over a rotation) deficient in downed dead woody material and snags, and with little variation in tree age and structure. Retaining snags, woody debris, and some live trees from the previous stand will result in a more structurally diverse stand, provide habitat features needed by certain species, and result in greater diversity and abundance of species in deciduous and coniferous stands (Balda 1975, Scott 1979, Dickson et al. 1983, Niemi and Hanowski 1984). Retention of snags primarily benefits NTMBs that are secondary cavity nesters. Retention of live trees in regeneration cuts provides structure for open canopy species such as Blue-gray Gnatcatchers, Northern Orioles, and Yellow-billed Cuckoos, or species from surrounding forests such as Red-eyed Vireos and some flycatcher species (Pearson and Probst 1979, Rakstad and Probst 1981, Probst et al. 1992). A number of alternative silvicultural systems and revised practices are currently under development in response to increased public concern over the ecological effects (whether real or presumed) of traditional forestry practices (Kessler et al. 1992). This includes the use of shelterwoods without final removal cuts and management of two-aged stands, which retain more residual structure than clearcutting.

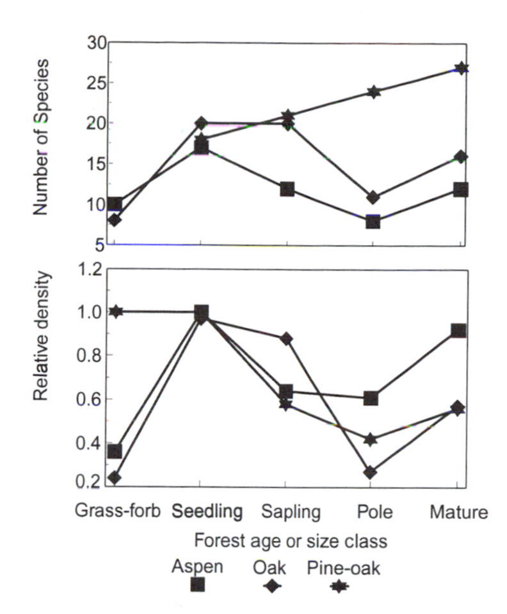

Figure 7-6. Numbers of bird species and bird density in different size (age)-class even-aged stands. Results are from studies in aspen forests (Probst et al. 1992), mixed oak forests (Conner and Adkisson 1975), and pine–oak forests (Conner et al. 1979). Density values from original papers were rescaled to 0–1.0 to be comparable.

Stand Succession

Avian density and diversity generally increase with succession following land abandonment (Johnston and Odum 1956, Karr 1971, Shugart and James 1973, Shugart et al. 1975). However, birds often respond differently to stand regeneration than secondary succession. Breeding bird densities in regenerating forests often are similar to or much greater than those in mature stands, with densities often lowest in mid-successional pole-sized stands (Conner and Adkisson 1975, Conner et al. 1979, Dickson and Selquist 1979, Probst 1979, Horn 1984, Yahner 1986. Thompson and Fritzell 1990, Probst et al. 1992). Species richness and density also may show an early peak in regenerating stands (Conner and Adkisson 1970, Conner et al. 1979, Dickson and Segelquist 1979, Probst 1979, Horn 1984, Yahner 1986, Thompson and Fritzell 1990) (Fig. 7-6). Bird-species richness did not differ, however, among young, mature, and old-growth forest stands in Douglas-fir forests of northwestern California (Raphael 1991). Early peaks in NTMB density and diversity in regeneration stands may be due to dense tree foliage in seedling–sapling stands and horizontal patchiness resulting from small patches of failed regeneration that create small herbaceous openings, as well as intruding bird species from adjacent older stands.

Bird community structure changes as a result of timber harvest and throughout the life of a stand. In general, young forests are dominated by foliage gleaning NTMBs and ground-foraging short-distance migrants. Young trees provide fruit, seeds, foliage insects, and nest sites for ground foragers and shrub–sapling gleaners. (Similar successional trends occur in old-field reforestation). Aerial foragers and a few "pursuers," such as flycatchers, are also present in early successional stages. Most pursuers and aerial foragers are NTMBs.

As canopy closure progresses, understory vegetation diminishes. As a result, ground foragers decrease, but reappear in older forests when the tree canopy opens and the understory is released. Thus, ground foraging birds decrease with forest age (Maurer and Whitmore 1981) or have a bimodal distribution with forest age (Probst 1979, Probst et al. 1992). In broadleaf stands with vigorous coppice formation, canopy closure can occur quite rapidly. In most conifer stands and during old-field succession, however, the ratio of ground to foliage gleaners changes more slowly because trees must recolonize the site rather than sprout.

In northern broadleaf forests, one-quarter to half the species and individuals may be associated with shrubs and understory (Thompson and Capen 1988, Probst et al. 1992). Understory gleaners usually decrease with forest succession, however, they may be more numerous in disturbed floodplain forests, older forests, or uneven-aged forests. Pursuers also may show a bimodal stand–age distribution, but they are much more common in open forests than in younger stages (Flack 1976, Probst et al. 1992) as well as Western riparian zones. Bark-foraging birds (usually residents) tend to increase with forest age and tree diameter (Flack 1976, Probst 1979, Maurer and Whitmore 1981, Probst et al. 1992), but are present in young forests if sufficient residual snags and trees were left when the stands was regenerated.

Uneven-aged Systems

Comparatively little information exists on responses of forest birds to uneven-aged management or selection cutting, but some effects can be inferred from what is known about avian habitat associations. Even-aged regeneration methods result in the near complete removal of the previous stand and, as a result, a near complete turnover in breeding birds. Selection cutting maintains a specific tree-diameter distribution in the stand through periodic removal of selected trees. Hence, there is less change in vegetation structure and bird communities than under even-aged management. Selectively cut stands typically retain much of the mature forest-bird community, and provide habitat for some species that use the ground–shrub–sapling layer. Whereas changes in density of NTMBs are typically small with such practices, they may be significant when summed across the landscape.

Landscape-level Impacts

Unlike even-aged management, selection cutting maintains a mature tree component at all times and does not create a mosaic of different-aged stands (unless mixed with even-aged practices). Selection cutting does not provide the landscape-level temporal and spatial diversity that even-aged management does. This may benefit area-sensitive or forest-interior species that prefer mature forests but will not provide habitat for species that require larger openings, even-aged stands, early seral conditions, or a diversity of even-aged stands. In many forests, selection cutting occurs in combination with even-aged systems so the landscape will be a mosaic of even-aged and uneven-aged stands.

Single and Multi-tree Gaps

Canopy gaps resulting from harvest of single trees or groups of trees provide habitat for a variety of migrant birds associated with young second-growth forests or gaps. In the Midwest, for example, the Hooded Warbler, Kentucky Warbler, and Indigo Bunting breed in small gaps created by single-tree and group selection, whereas other species such as Yellow-breasted Chats, Blue-winged Warblers, and Prairie Warblers, only breed in larger openings such as clearcuts and shelterwoods (S. Robinson, unpublished data, E. Annand and F. Thompson, unpublished data). Species such as the Kentucky Warbler and Hooded Warbler are generally considered forest-interior, area-sensitive species that are adapted to internal forest disturbances such as tree-fall gaps. There is a dearth of information on the area sensitivity of species requiring early successional forest or gaps. Kirtland's Warbler is one well-studied counter example that is concentrated in <6 large early successional stands and is rarely found in stands of less than 35 ha (Probst 1986).

Canopy gaps resulting from selection cutting also may attract cowbirds and result in higher levels of brood parasitism. Cowbirds occur in greater numbers in selectively cut stands in Illinois and Missouri (S. Robinson unpublished data, Ziehmer 1992) than in uncut mature forest. Brittingham and Temple (1983) found increased levels of brood parasitism near edges of forest openings as small as 0.2 ha, which is comparable to small group-selection openings. Brood parasitism and nest depredation were higher for a few species in selectively cut stands than uncut stands in Illinois (S. Robinson, unpublished data). As previously stated, cowbird and predator numbers may be largely determined by characteristics of the surrounding landscape but, if cowbirds were attracted to selection cuts, they would have to be more effective at parasitizing nests near selection openings, or hosts attracted to edges for parasitism to have an additive effect (the ecological trap hypothesis). Thompson (1993) addressed this issue by simulating population responses of a forest interior bird to clearcutting and selection cutting. Simulations suggest that, if edges around group selection openings function as ecological traps, they could result in local extirpations or population sinks because, while small, these openings could be much more numerous and widely dispersed than those created by clearcutting. If openings created by selection cutting do not function as ecological traps, then selectively cut forests were predicted to maintain higher populations of a mature-forest species than forests managed by clearcutting (Fig. 7-5) (Thompson 1993).

Change in Stand Structure

Uneven-aged stands usually have a well-developed understory and subcanopy because of frequent canopy gaps. The presence of several well-developed vegetation levels and complex habitat structure results in higher within-stand bird species diversity than in even-aged stands. For instance, in southwestern Ponderosa pine forests, bird density and diversity was higher in uneven-aged managed stands and strip-cut stands than uncut controls, "severely thinned stands," and clearcut stands (Szaro and Balda 1979). Mature uneven-aged stands of mixed conifers in the western Sierra Nevada had a slightly more diverse bird community than mature even-aged stands, in part due to greater canopy development (Morrison 1992). Maintenance of a mature tree component at all times should provide habitat for canopy-dwelling species at all times. However, the loss of some large trees and potential large snags is likely to result in lower densities of bark foragers (mostly resident species), canopy-foliage gleaners, and cavity nesting species (e.g., Raphael et al. 1987). The few studies that have compared selection cutting or partial cuts to unlogged stands have in fact found that some bark foragers and foliage gleaners decrease, and some ground and shrub foragers or nesters increase (Medin 1985, Medin and Booth 1989, S. Robinson, unpublished data). In Missouri oak–pine forests, some species that typically breed in mature even-aged forests occur in comparable or greater numbers in selectively cut stands (Eastern Wood-pewee, Red-eyed Vireo, Summer Tanager, Worm-eating Warbler); while others occur in lower numbers in selectively cut stands (Pine Warbler, Ovenbird, Scarlet Tanager, Yellow-billed Cuckoo, Wood Thrush) (E. Annand and F. Thompson, unpublished data).

Single-tree selection will primarily maintain shade-tolerant trees and group selection shade-tolerant and intermediate shade-tolerant trees. As previously discussed, tree-species composition may impact bird communities.

RECOMMENDATIONS

Individuals select habitats based on local characteristics. Populations, however, are determined by an interaction between local habitat factors, the landscape context of habitats, and regional or continental context of habitat biogeography and population levels. We believe that the only way to incorporate the diverse needs of Neotropical migratory birds with other resource considerations, such as timber, is a hierarchical approach that begins at a continental scale, identifies opportunities at regional scales, sets composition and structure goals at landscape- and management-unit scales, and matches management prescriptions to goals at a habitat-stand level. Single-resource or single-species approaches (including indicators) originating at stand- or management-unit level scales will not yield a holistic, comprehensive management strategy. This approach ensures that finer scale management decisions are made with the knowledge of large scale patterns and goals. We present this approach in the context of incorporating

NTMB needs into forest management but it is valid for all resources including biodiversity in general.

Step 1: Establish Regional Context (Scales: multistate, province, ecoregion)

An understanding of ecoregion differences, species geographic distributions, and habitat distribution is basic to viability assessments, habitat conservation plans, and landscape or local management plans. Landscape-local decisions and options should be developed in the context of land potentials, species distributions, biogeographic factors, and landscape-demographic interactions. Establish the management area in a regional context by identifying the spatial patterns of ecosystems and NTMB ranges in the region. Locate or prepare a complete list of NTMBs for the region with information on their status, habitat associations, and geographic location. Determine desired regional ecosystem–vegetation patterns. Consider historical and current vegetation patterns, trends in vegetation types, and habitat needs of NTMBs on the regional list with consideration of their status. Finer scale, local-level management should occur with knowledge of the species or ecosystems status in the region, and should complement regional goals.

Step 2: Determine Desired Landscape Composition and Structure (Scales: landform, watershed, mountain range, national forest or refuge)

Determine the desired amounts and distribution of forest types, forest age-classes, and nonforest habitats from a multiresource perspective. Set objectives based on the relative abundance of ecosystems, forest types, and forest age-classes and habitats in a landscape, and the unique or complementary contribution that a landscape can provide relative to species ranges and neighboring landscape patterns. These objectives should complement regional goals within the potential of ecosystems and landscapes. Consider natural tendencies such as site capability, natural disturbance frequency and pattern, and successional pathways. Next consider NTMB needs in terms of habitats including spatial relationships (size, shape, juxtaposition).

Because of diverse habitat needs, and edge or area sensitivity of NTMBs, landscape-level forest planning is extremely important to NTMB conservation. At this level, the simplest approach is a coarse filter approach that assumes a representative variety of ecosystems will contain the vast majority of species in a region (Hunter et al. 1988, Hunter 1990). For instance, management and restoration efforts might be directed toward regionally rare ecosystems such as bottomland hardwoods, lowland conifers, old-growth, and savannas. This will address needs of regionally rare species, including NTMBs. However, concerns for impacts of forest fragmentation and edge on NTMBs, population size and viability, as well as source–sink relationships require careful spatial planning for even common habitats such as upland forest. In extensively forested regions or landscapes, cowbirds and predator numbers may be sufficiently low that edge effects are not a concern. In these areas, silviculturally sound, regulated harvest, which maintains the naturally occurring diversity of forest types, should be compatible with NTMB conservation. In highly fragmented, edge-dominated landscapes, forest habitats may already be saturated with cowbirds and predators, and any additional edge effects resulting from timber harvest may be inconsequential. However, in the wide range of landscapes between these extremes, careful spatial planning may be required. For instance, NTMBs that breed in mature forest would benefit from large blocks of unfragmented forest reserved from timber harvest and other anthropogenic disturbances. These areas would support productive source populations of forest interior NTMBs. Harvest and other activities could be concentrated in the more fragmented parts of these landscapes. This type of planning would produce a diversity of landscapes, some with undisturbed, mature-contiguous forest and others with more successional diversity.

In areas where timber is harvested, a balance of selection cutting and even-aged systems should be used to create small openings for gap species, large openings for early successional forest migrants, and a balanced age-class distribution to maintain sufficient mature forest habitats. Where late successional or edge-sensitive species are featured, single-tree selection or even-aged systems with long rotations should be used. Larger regeneration cuts and longer rotations will increase the amount of late successional forest, and decrease the amount of early successional forest and their edges in the landscape.

Step 3: Establish Management Unit Goals (Scales: An administrative unit, compartment, group of habitats).

Set vegetation composition and structural goals across the management unit including forest age-classes, vertical stratification, horizontal pattern and special features. These may be uniform or varied based on species needs and management unit context. Maximizing diversity at this scale could compromise landscape and regional diversity by fragmenting mature forest or homogeneous forest habitats. Instead, manage to meet landscape and regional diversity goals for forest types and age-classes, and to complement management in other management units. Emphasize landscape-local opportunities not found elsewhere. Consider natural forest type diversity. Examine spatial relationships of stands based on elements of structure and composition. Group regeneration cuts to minimize impacts on area and edge-sensitive NTMBs where appropriate. In coniferous forest, maintain deciduous components where these are declining. Conversely, in deciduous forests, allow some habitats to develop a coniferous component—typically through succession. Mix silvicultural options across units, unless specific concerns dictate otherwise.

Step 4: Develop Stand-Habitat Level Management Prescriptions (Scale: habitat or stand)

At this level, the manager needs to use the best practices to complement goals established for the management unit and landscape, and to match site capabilities and natural tendencies. Prescriptions should address NTMB diversity and habitat requirements of priority species. Priority species will vary depending on landscape and region as established in steps 1–3. For instance, in many Midwest landscapes, forest-interior and prairie species will be a priority, while in some New England landscapes, early successional migrants may be more important. We offer the following suggestions for enhancing specific components of the NTMB community as well as diversity at the stand level. However, we reiterate the need to manage stands to address goals set at larger scales.

NTMB Diversity

Do not maximize within-stand diversity at the expense of landscape or regional diversity. For example, selection cutting may produce high within-stand diversity but an entire landscape of selectively cut uneven-aged forest would be lacking some NTMBs. A better approach might be to use a mix of silvicultural practices (even and uneven-aged), and to reserve some areas from harvest, if there is a particular concern about the viability of species that are edge sensitive and that use mature forests. Maintain deciduous and coniferous components in mixed stands. Limit control of hardwoods in regenerating conifer plantations, and allow conifers to develop under deciduous forests by extending rotations. Use variable and wider spacing in conifer plantations.

Area- or Edge-sensitive NTMB

Increase stand size and regeneration cuts to benefit early and late-successional species. Cluster regeneration cuts when possible. Emphasize even-aged systems and single-tree selection cuts. Reserve some of the least fragmented areas from timber harvest.

Cavity Nesting and Bark-foraging NTMBs

Lengthen rotation ages in even-aged systems and increase the proportion of larger trees (decrease q-values) in uneven-aged systems. Retain some snags and live residual trees in regeneration cuts.

Canopy Gleaners

Lengthen rotation ages in even-aged systems and increase the proportion of larger trees (decrease q-values) in uneven-aged systems.

Ground Foragers and Understory Gleaners

Mix silvicultural systems to provide regenerating even-aged stands and selectively cut uneven-aged stands. Use wide and variable spacing in plantations with minimum hardwood control.

Early Successional Species

Use even-aged systems. Shorten rotations.

ACKNOWLEDGMENTS

We thank D. E. Capen, W. D. Dijak, P. S. Johnson, S. R. Shifley, and M. J. Schwalbach for their review of this manuscript.

LITERATURE CITED

Ambuel, B., and S. A. Temple. 1983. Area-dependent changes in the bird communities and vegetation of southern Wisconsin forests. Ecology 64:1057–1068.

Askins, R. A., J. F. Lynch, and R. Greenberg. 1990. Population declines in migratory birds in eastern North America. Curr. Ornithol. 7:1–57.

Balda, R. P. 1975. Vegetation structure and breeding bird diversity. Pp. 59–80 *in* Proc. Symp. Manag. Forest Range Habitats Nongame Birds (D. R. Smith, tech. coord.) USDA For. Serv. Gen. Tech. Rep. WO-1, Washington, DC.

Blake, J. G., and J. R. Karr. 1984. Species composition of bird communities and the conservation benefit of large versus small forests. Biol. Conserv. 30:173–187.

Brittingham, M. C., and S. A. Temple. 1983. Have cowbirds caused forest songbirds to decline? BioScience 33:31–35.

Bunnell, F. L., and L. L. Kremsater. 1990. Sustaining wildlife in managed forests. Northwest Environ. J. 6:243–269.

Conner, R. N., and C. S. Adkisson. 1975. Effects of clearcutting on the diversity of breeding birds. J. For. 73:781–785.

Conner, R. N., J. W. Via, and I. D. Prather. 1979. Effects of pine–oak clearcutting on winter and breeding birds in southwestern Virginia. Wilson Bull. 91(2):301–316.

Crawford, H. S., R. G. Hooper, and R. W. Titterington. 1981. Songbird response to silvicultural practices in central Appalachian hardwoods. J. Wildl. Manag. 45:680–692.

DeGraaf, R. M. (tech coord.). 1978. Proc. Workshop Nongame Bird Habitat Manag. Conif. Forests Western United States. USDA Forest Serv. Pacific Northwest Forest Range Exp. Sta. Gen. Tech. Rep. PNW-64, Portland, OR.

DeGraaf, R. M. (tech. coord.). 1979. Proc. Workshop Manag. North Central Northeastern Forests Nongame birds. USDA For. Serv. North Central Forest Exp. Sta. Gen. Tech. Rep. NC-51, St Paul, MN.

DeGraaf, R. M. (tech. coord.). 1980. Workshop Proc. Manag. Western Forests Grasslands Nongame Birds. USDA For. Serv. Intermountain For. Range Exp. Sta. Gen. Tech. Rep. INT-86, Ogden, UT.

Dickson, J. G., and C. A. Segelquist. 1979. Breeding bird populations in pine and pine–hardwood forest in Texas. J. Wildl. Manag. 43(2):549–555.

Dickson, J. G., R. N. Conner, and J. H. Williamson. 1983. Snag retention increases use of clear-cut. J. Wildl. Manag. 47:799–804.

Edgerton, P. J., and J. W. Thomas. 1978. Silvicultural options and habitat values in coniferous forests. Pp. 56–65 *in* Proc. Workshop Nongame Bird Habitat Manag. Conif. Forests Western United States (R. M. DeGraaf, tech. coord.). USDA Forest Serv. Pacific Northwest Forest Range Exp. Sta. Gen. Tech. Rep. PNW-64, Portland, OR.

Flack, J. A. D. 1976. Bird populations of aspen forests in western North America. Ornithol. Monogr. no. 19. 97 pp.

Franzreb, K. E., and R. D. Ohmart. 1978. The effects of timber harvesting on breeding birds in a mixed-coniferous forest. Condor 80: 431–441.

Gates, J. E., and L. W. Gysel. 1978. Avian nest dispersion and fledgling success in field forest ecotones. Ecology 58:871–883.

Hayden, T. J., J. Faaborg, and R. L. Clawson. 1985. Estimates of minimum area requirements for Missouri forest birds. Trans. Missouri Acad. Sci. 19:11–22.

Hof, J. G., and M. G. Raphael. 1993. Some mathematical programming approaches for optimizing timber age class distributions to meet multi-species wildlife population objectives. Can. J. For. Res. 23:828–834.

Horn, J. C. 1984. Short-term changes in bird

communities after clearcutting in western North Carolina. Wilson Bull. 96:684–689.

Hunter, M. L., Jr. 1990. Wildlife, forests, and forestry: principles of managing forests for biological diversity. Prentice-Hall, Englewood Cliffs, NJ.

Hunter, M. L., G. L. Jacobson, Jr., and T. Webb, III. 1988. Paleocology and the coarse-filter approach to maintaining biological diversity. Conserv. Biol. 2:375–385.

Hutto, R. L. 1988. Is tropical deforestation responsible for the reported declines in Neotropical migrant populations? Amer. Birds 42:375–379.

James, F. C. 1971. Ordinations of habitat relationships among breeding birds. Wilson Bull. 83:215–236.

Johnston, D. W., and E. P. Odum. 1956. Breeding bird populations in relation to plant succession on the Piedmont of Georgia. Ecology 37: 50–62.

Karr, J. R. 1971. Structure of avian communities in selected Panama and Illinois habitats. Ecol. Monogr. 41:207–233.

Karr, J. R., and R. R. Roth. 1971. Vegetation structure and avian diversity in several new world areas. Amer. Natur. 105:423–435.

Kessler, W. B., H. Salwasser, C. W. Cartwright, Jr, and J. A. Caplan. 1992. New perspectives for sustainable natural resources management. Ecol. Applic. 2(3):221–225.

Law, J. R., and C. G. Lorimer. 1989. Managing uneven-aged stands. Pp. 6.08–6.086 *in* Central Hardwood Notes (F. B. Clark and J. G. Hutchinson, eds). USDA Forest Serv. North Central For. Exp. Sta., St Paul, MN.

MacArthur, R. H., J. W. MacArthur, and J. Preer. 1962. On bird species diversity. II. Predictions of bird census from habitat measurements. Amer. Natur. 96:167–174.

Marquis, D. A. 1989. Alternative silvicultural systems—East. Pp. 29–35 *in* Proc. Nat. Silviculture Workshop. USDA Forest Serv. Timber Manag., Washington, DC.

Martin, T. E. 1992. Landscape considerations for viable populations and biological diversity. Trans. 57th North Amer. Wildl. Nat. Res. Conf. 283–291.

Maurer, B. A., and R. C. Whitmore. 1981. Foraging of five bird species in two forests with different vegetation structure. Wilson Bull. 93:478–490.

Medin, D. E. 1985. Breeding bird responses to diameter-cut logging in west central Idaho. USDA Forest Serv. Intermountain Res. Sta. Res. Pap. INT-355, Ogden, UT, 6 pp.

Medin, D. E., and G. D. Booth. 1989. Responses of birds and small mammals to single-tree selection logging in Idaho. USDA Forest Serv. Intermountain Res. Sta. Res. Pap. INT-408, Ogden, UT, 11 pp.

Morrison, M. L. 1992. Bird abundance in forests managed for timber and wildlife resources. Biol. Conserv. 60:127–134.

Morrison, M. L., I. C. Timossi, K. A. With, and P. N. Manley. 1985. Use of tree species by forest birds during winter and summer. J. Wildl. Manag. 49:1098–1102.

Niemi, G. J., and J. M. Hanowski. 1984. Relationships of breeding birds to habitat characteristics in logged areas. J. Wildl. Manag. 48:438–443.

Pearson, C. J., and J. R. Probst. 1979. Aspen clearcut (2nd year). Amer. Birds 33(1):77.

Probst, J. R. 1979. Oak forest bird communities. Pp. 80–88 *in* Proc. Workshop Manag. North Central Northeastern Forests Nongame Birds (R. M. Degraaf, tech. coord.). USDA Forest Serv. North Central Forest Exp. Sta. Gen. Tech. Rep. NC-51, St Paul, MN.

Probst, J. R. 1986. A review of factors limiting the Kirtland's warbler on its breeding grounds. Am. Midl. Nat. 116:87–100.

Probst, J. R., and J. Weinrich. 1993. Relating Kirtland's warbler population to changing landscape composition and structure. Landscape Ecol. 8:257–271.

Probst, J. R., D. S. Rakstad, and D. J. Rugg. 1992. Breeding bird communities in regenerating and mature broadleaf forests in the USA Lake States. Forest Ecol. Manag. 49:43–60.

Pulliam, H. R. 1988. Sources, sinks, and population regulation. Amer. Natur. 132: 652–661.

Rakstad, D., and J. R. Probst. 1981. Aspen clearcut (4th year). Amer. Birds 35(1):59.

Raphael, M. G. 1991. Vertebrate species richness within and among seral stages of Douglas-fir forests in Northwestern California. Pp. 415–423 *in* Wildlife and vegetation of unmanaged Douglas-fir forests (L. F. Ruggiero, K. B. Aubry, A. B. Carey, and M. H. Huff, tech. coords). Pacific Northwest Res. Sta. Gen. Tech. Rep. PNW-285, Portland, OR.

Raphael, M. G., M. L. Morrison, and M. P. Yoder-Williams. 1987. Breeding bird populations during 25 years of postfire succession. Condor 89(3):614–626.

Raphael, M. G., K. V. Rosenberg, and B. G. Marcot. 1988. Large-scale changes in bird populations of Douglas-fir forests, Northwest California. Pp. 63–83 *in* Bird conservation 3 (J. A. Jackson, ed.). University of Wisconsin Press, Madison, WI.

Ratti, J. T., and K. P. Reese. 1988. Preliminary test of the ecological trap hypothesis. J. Wildl. Manag. 52(3):484–491.

Robbins, C. S., J. R. Sauer, R. S. Greenberg, and

S. Droege. 1989a. Population declines in North American birds that migrate to the neotropics. Proc. Natl Acad. Sci. USA 86:7658–7662.

Robbins, C. S., D. K. Dawson, and B. A. Dowell. 1989b. Habitat area requirements of breeding forest birds of the Middle Atlantic States. Wildl. Monogr. 103:1–34.

Robinson, S. K. 1992. Population dynamics of breeding Neotropical migrants in a fragmented Illinois landscape. Pp. 408–418 *in* Ecology and conservation of Neotropical migrant landbirds (J. M. Hagan, III and D. W. Johnston, eds). Smithsonian Institution Press, Washington, DC.

Robinson, S. K., and R. T. Holmes. 1984. Effects of plant species and foliage structure on the foraging behavior of forest birds. Auk 101:672–684.

Robinson, S. K., J. A. Grzybowski, S. I. Rothstein, M. C. Brittingham, L. J. Petit, and F. R. Thompson, III. 1993. Management implications of cowbird parasitism for Neotropical migrant songbirds. Pp. 93–102 *in* Status and management of Neotropical migrant birds (D. M. Finch and P. W. Stangel, eds). USDA Forest Serv. Rocky Mt. Exp. Sta. Gen. Tech. Rep. RM-229, Fort Collins, CO.

Rosenberg, K. V., and M. G. Raphael. 1986. Effects of forest fragmentation on vertebrates in Douglas-fir forests. Pp. 263–272 *in* Wildlife 2000: modeling habitat relationships of terrestrial vertebrates (J. Verner, M. L. Morrison, and C. J. Ralph, eds). University of Wisconsin Press, Madison, WI.

Rudnicky, T. C., and M. L. Hunter. 1993. Avian nest depredation in clearcuts, forests, and edges in a forest-dominated landscape. J. Wildl. Manag. 57:358–364.

Scott, V. E. 1979. Bird responses to snag removal in ponderosa pine. J. For. 77:26–28.

Shugart, H. H., Jr, and D. James. 1973. Ecological succession of breeding bird populations in northwestern Arkansas. Auk 90:62–77.

Shugart, H. H., S. H. Anderson, and R. H. Strand. 1975. Dominant patterns in bird populations of the eastern deciduous forest biome. Pp. 90–95 *in* Proc. Symp. Manag. Forest Range Habitats Nongame Birds (D. R. Smith, tech. coord.). USDA Forest Serv. Gen. Tech. Rep. WO-1, Washington, DC.

Smith, D. M. 1986. The practice of silviculture. John Wiley and Sons, Inc., New York.

Smith, D. R. (tech. coord.). 1975. Proc. Symp. Manag. Forest Range Habitats Nongame Birds. USDA Forest Serv. Gen. Tech. Rep. WO-1, Washington, DC.

Szaro, R. C., and R. P. Balda. 1979. Effects of harvesting ponderosa pine on nongame bird populations. USDA Forest Serv. Rocky Mt. Forest Range Exp. Sta. Res. Pap. RM-212, Fort Collins, CO, 8 pp.

Temple, S. A., and J. R. Cary. 1988. Modeling dynamics of habitat-interior bird populations in fragmented landscapes. Conserv. Biol. 2:340–347.

Thompson, F. R., III. 1993. Simulated responses of a forest interior bird population to forest management options in central hardwood forests of the United States. Conserv. Biol. 7:325–333.

Thompson, F. R., III. 1994. Spatial and temporal patterns of brown-headed cowbirds in the midwestern United States. Auk 111:977–988.

Thompson, F. R., III, and D. E. Capen. 1988. Avian assemblages in seral stages of a Vermont forest. J. Wildl. Manag. 52:771–777.

Thompson, F. R., III., and E. K. Fritzell. 1990. Bird densities and diversity in clearcut and mature oak–hickory forest. USDA Forest Serv. North Central Forest Exp. Sta. Res. Pap. NC-293. St Paul, MN, 7 pp.

Thompson, F. R., III, W. D. Dijak, T. G. Kulowiec, and D. A. Hamilton. 1992. Breeding bird populations in Missouri Ozark forests with and without clearcutting. J. Wildl. Manag. 56(1):23–30.

USDA Forest Service. 1973. Silvicultural systems for the major forest types of the United States. Agric. Handbook 445, 114 pp.

USDA Forest Service. 1989. Report of the Forest Service, Fiscal Year 1988. USDA, Washington, DC.

USDA Forest Service. 1990. The Forest Service plan for forest and rangeland resources: a long-term strategic plan. USDA, Washington, DC.

Webb, W. L., D. F. Behrend, and B. Saisorn. 1977. Effect of logging on songbird populations in a northern hardwood forest. Wildl. Monogr. no. 55, 35 pp.

Wenger, K. F. 1984. Forestry handbook, 2nd ed. Edited for the Society of American Foresters. John Wiley and Sons, New York.

Wiens, J. A., J. T. Rotenberry, and B. Van Horne. 1987. Habitat occupancy patterns of North American shrubsteppe birds: the effects of spatial scale. Oikos 48:132–147.

Whitcomb, R. F., C. S. Robbins, J. F. Lynch, B. L. Whitcomb, M. K. Klimkiewicz, and D. Bystrak. 1981. Effects of forest fragmentation on avifauna of eastern deciduous forest. Pp. 125–205 *in* Forest island dynamics in man dominated landscapes (R. L. Burgess and D. M. Sharpe, eds). Springer-Verlag, New York.

Wilcove, D. S. 1988. Forest fragmentation as a

wildlife management issue in the eastern United States. Pp. 146–150 *in* Healthy forests, healthy world: Proc. 1988 Soc. Amer. Foresters Natl Convention, Rochester, NY.

Wilcove, D. S., and S. K. Robinson. 1990. The impact of forest fragmentation on bird communities in Eastern North America. Pp. 319–331 *in* Biogeography and ecology of forest bird communities (A. Keast, ed.). SPB Academic Publishing, The Hague, The Netherlands.

Willson, M. F. 1974. Avian community organization and habitat structure. Ecology 55:1017–1029.

Yahner, R. H. 1986. Structure, seasonal dynamics and habitat relationships of avian communities in small even-aged forest stands. Wilson Bull. 98:61–82.

Yahner, R. H., and D. P. Scott. 1988. Effects of forest fragmentation on depredation of artificial nests. J. Wildl. Manag. 49:508–513.

Ziehmer, R. L. 1992. Effects of uneven-aged timber management of forest bird communities. MS thesis, University of Missouri, Columbia, MO.

8

EFFECTS OF SILVICULTURAL TREATMENTS IN THE ROCKY MOUNTAINS

SALLIE J. HEJL, RICHARD L. HUTTO, CHARLES R. PRESTON, AND DEBORAH M. FINCH

INTRODUCTION

Neotropical migrants have been affected by the loss and fragmentation of forests in the eastern United States (Askins et al. 1990). Changes in western forests and the effects of these changes on birds may be different from those in the East. While timber harvesting is widespread in the western United States, the purpose of silvicultural systems on public land is to perpetuate forests, not to convert forests to agricultural land or residential areas. Western logging has resulted in landscapes that are primarily composed of forests of different ages and treatments, rarely isolated forests. The silvicultural systems used to manage these forests, however, include timber harvesting, intermediate treatments, and stand-regeneration practices that usually result in forests different from presettlement ones (Thompson et al., Chapter 7, this volume). Managers interested in maintaining western birds need not be as concerned with deforestation as with the loss of old growth and whether managed forests successfully substitute for unmanipulated forests. While baseline studies on birds in most forest types in the Rocky Mountains exist (e.g., Marshall 1957, Salt 1957, Flack 1976, Winternitz 1976, Balda and Masters 1980, Finch and Reynolds 1987, Raphael 1987a,b, Scott and Crouch 1988, Block et al. 1992, Morrison et al. 1993), studies on the effects of silvicultural practices on songbird populations in the Rocky Mountains are relatively rare. Most studies consider only the effects of timber harvesting.

The purpose of this synthesis is to summarize knowledge about the effects of silvicultural practices on birds in Rocky Mountain forests, to suggest future research needs, and to suggest how managers can make decisions based on what we currently know. We offer a description of current forest structure and basic information on bird distribution across forest habitats in the Rockies as background knowledge for understanding the effects of silviculture on birds in these habitats. We also offer a comparison between the effects of silviculture, and the effects of fire and fire suppression on forest birds in an attempt to give a holistic perspective on the health of forest birds in the Rockies.

CURRENT FOREST STRUCTURE

Three major factors contribute to the current forest structure of the Rocky Mountains. These three factors are: (1) floristic composition; (2) natural disturbances, especially fire; and (3) human-induced disturbances, including logging, fire exclusion, and other forest management activities.

Floristic Composition of Rocky Mountain Forests

Plant composition will be discussed in the context of Southern, Central, and Northern Rocky Mountains, as defined by Daubenmire (1943). The Southern Rockies extend from Mexico to the northern borders of Arizona and New Mexico, and the Central Rockies extend from that border to the middle of Wyoming. The Northern Rockies encompass the region from northern Wyoming to central

Alberta and British Columbia. All three provinces share many tree species, but both the Southern and Northern Rockies exhibit some species specific to their region. Many distinctive pine and oak species grow in the southern region. Some of the species unique to the northern region are typical of forests on the western slope of the Cascades in the Pacific Northwest.

The Rocky Mountains are dominated by coniferous forests (Gleason and Cronquist 1964, Peet 1988) with the only widespread and abundant broadleaved tree being quaking aspen (*Populus tremuloides*). Excluding the riparian zone, Peet (1988) recognized nine forest types: (1) madrean pine–oak woodland; (2) pygmy conifer woodland; (3) ponderosa pine woodland; (4) Cascadian forests; (5) Douglas-fir forest; (6) spruce-fir forest; (7) subalpine white pine forests; (8) treeline vegetation; and (9) montane seral forests.

A diverse assemblage of pines and oaks characterize the madrean pine–oak woodlands of the Southern Rocky Mountains. Prominent woodland species include Arizona pine (*Pinus ponderosa* var. *arizonica*), Chihuahua pine (*P. leiophylla*), Mexican pinyon (*P. cembroides*), Gambel oak (*Quercus gambelii*), Arizona white oak (*Q. arizonica*), and Emory oak (*Q. emoryi*).

The *pygmy conifer woodland* of the Central and Southern Rockies often forms the transition between grasslands of the plains and montane forests. Dominant trees in this zone on the eastern slope of the Rockies are pinyon pine (*P. edulis*) and one-seed juniper (*Juniperus monosperma*). On the western slope, pinyon pine or singleleaf pinyon (*P. monophylla*) and Utah juniper (*J. osteosperma*) characterize much of this zone.

Ponderosa pine woodland is widespread in the Rocky Mountains. West of the continental divide in the northern region, the primary tree is Pacific ponderosa pine (*P. ponderosa* var. *ponderosa*). East of the divide and south through the Rockies, Rocky Mountain ponderosa pine (*P. Ponderosa* var. *scopulorum*) is generally dominant. Rocky Mountain ponderosa pine, Arizona pine, Chihuahua pine, or Apache pine (*P. engelmannii*) are characteristic trees in the Southern Rockies.

The *Cascadian forests* of the Northern Rockies, the most mesic forests in the Rockies, are restricted to the Pacific maritime-influenced climate of northern Idaho and adjacent Montana, Washington, and British Columbia. Species typical of the Cascade Mountains in the Pacific Northwest comprise these forests and include western hemlock (*Tsuga heterophylla*), western red cedar (*Thuja plicata*), grand fir (*Abies grandis*), and Pacific yew (*Taxus brevifolia*).

Douglas fir (*Pseudotsuga menziesii*) is characteristic of the *Douglas-fir forest* throughout the Rocky Mountains. In the Southern and Central Rockies, white fir (*Abies concolor*), blue spruce (*Picea pungens*), ponderosa pine, limber pine (*Pinus flexilis*), and quaking aspen are associated species in the Douglas-fir forests. Associated species in the Northern Rockies include grand fir, ponderosa pine, and western larch (*Larix occidentalis*). Technically, "mixed coniferous" forests are found in several of these forest types and, where the types abut, but we use the term in reference to mixed Douglas-fir forests.

The subalpine *spruce-fir forest*, dominated by subalpine fir (*Abies lasiocarpa*) and Engelmann spruce (*Picea engelmannii*), occurs throughout the Rocky Mountains. In the northernmost Rockies, white spruce (*P. glauca*) replaces Engelmann spruce. Spruce-fir forests are poorly developed in the mountains of Mexico.

Subalpine white pine forests are found on dry ridges and exposed southern slopes of the subalpine zone. Whitebark pine (*Pinus albicaulis*) is the dominant white pine in the Northern Rockies, bristlecone pine (*P. aristata*) in the Central Rockies, and intermountain bristlecone pine (*P. longaeva*) on peaks in the Great Basin. Limber pine ranges across much of the Northern and Central Rockies. Mexican white pine (*P. strobiformis*) replaces limber pine in the Southern Rockies.

Treeline vegetation typically is subalpine fir and Engelmann spruce throughout much of the Rockies. Whitebark pine, bristlecone pine, and limber pine are also important treeline species, especially on dry or exposed ridges. Other pines dominate at treeline in the high mountains of Mexico. Subalpine larch (*Larix lyallii*) and mountain hemlock

(*Tsuga mertensiana*) are characteristic of treeline habitats in the northern Rockies.

Montane seral forests dominate a large portion of the landscape in the Rocky Mountains, since disturbances are common, especially fire. Quaking aspen and lodgepole pine (*P. contorta*) are two species that are important postfire seral tree species. Aspen and lodgepole are often replaced by stands of more shade-tolerant species, such as Douglas fir or subalpine fir in the Douglas fir and spruce-fir zones, respectively. In the absence of other conifer species, aspen can form stable, self-maintained stands (Johnston 1984, Mueggler 1985). Western larch and western white pine (*P. monticola*) are both seral species of the Northern Rockies. Other species (e.g., Douglas fir, ponderosa pine, and limber pine) all act as successional species on sites more mesic than those on which they are typically climax.

Natural Disturbances

Fire, wind, insects, ungulate browsing, avalanches, landslides, extreme weather, and disease are sources of natural disturbance in the Rocky Mountains (Peet 1988). Historically, fire has been the most important and extensive disturbance to vegetation, influencing the development of landscape patterns (Habeck and Mutch 1973, Gruell 1983, Peet 1988). Historic fire regimes ranged from low-intensity, high-frequency fires in lower elevation forests to high-intensity, low-frequency fires in upper elevation forests (Peet 1988). As a consequence of fire suppression, fire frequency has decreased and intensity has increased in many forests since early in the 20th century. Fire suppression has altered the natural fire regimen with the result that the structure of many forests has changed from open to closed stands.

Logging History and Current Silvicultural Recommendations

Some of the logging activities throughout the Rockies in the past 100 years have stemmed from silvicultural prescriptions. Silvicultural suggestions have changed as our knowledge of the ecologies of these forests has increased. Public opinion, political expediency, and individual personalities have also affected how the land has been managed, often irrespective of silvicultural requirements, site conditions, and conflicting objectives (Mustian 1977). Therefore, a great diversity of logging practices has occurred in the Rockies (S. Arno, personal communication), including overstory removal, selection, seed-tree, shelterwood, and clearcut logging (Hejl 1992).

A general rule of historic logging was that the most accessible and commercially valuable trees were logged before less accessible and less valuable trees. The result was that, in general, low elevations and preferred species such as western white pine, ponderosa pine, and western larch were logged before high elevations and Douglas fir, western red cedar, western hemlock, lodgepole pine, subalpine fir, and Engelmann spruce. Some upper-elevation forests were logged early on, because they were accessible. Both even-aged and, to a lesser extent, uneven-aged systems have been used in most forest types.

Cutting regimes in the Rocky Mountains have varied from large clearcuts that are often densely spaced in moist forests (e.g., spruce/fir and cedar/hemlock forests) to repeated entries of selective cutting in drier forests (e.g., ponderosa pine, western larch, or Douglas fir forests) (Arno, personal communication). These logging activities along with fire suppression have resulted in changes in overstory and understory species composition, forest structure, and landscape heterogeneity (Thompson et al., Chapter 7, this volume).

A major use of silvicultural treatments is to harvest and regenerate trees to obtain desired species of trees in a stand of suitable structure (Burns 1983, Thompson et al., Chapter 7, this volume). The precise silvicultural practice that is recommended for a particular habitat is constantly changing (Mustian 1977). Current silvicultural recommendations vary for each floristic zone in the Rocky Mountains depending on tree species, site conditions, and management objectives. Recent silvicultural recommendations, primarily for wood production, include: (1) single-tree selection or group selection for pinyon-juniper (Meeuwig and Bassett 1983), and for climax ponderosa pine in the Northern Rockies (Ryker and Losensky 1983); (2) a combination of uneven-aged and even-aged management for ponderosa pine in the Southern Rockies (Ronco and Ready 1983), for Cascadian forests (Graham et al. 1983), for mixed-conifer forests containing inland Douglas fir (Ryker and Losensky 1983), and for spruce-fir forests (Alexander and Engelby 1983); and (3) even-aged management for larch forests (Schmidt et al. 1983), for lodgepole pine (Alexander et al. 1983), and for aspen (Shepperd and Engelby 1983). We have not found any silvicultural suggestions for pine–oak woodlands or whitebark pine forests.

GENERAL BIRD–HABITAT RELATIONS IN NATURAL FORESTS

Methods for Estimating Bird Distribution across Forest Habitats

We created a species list for eight forest habitats (trying to emphasize mature or older stands) with a relative abundance rating for each species during the breeding season. The lists were based on a variety of bird community studies from each habitat (e.g., Marshall 1957, Salt 1957, Flack 1976, Winternitz 1976, Balda and Masters 1980, Raphael 1987a,b, Finch and Reynolds 1987, Scott and Crouch 1988, Block et al. 1992), on the 19 studies examining birds in logged and unlogged habitats, our field experience, and the opinions of other field naturalists. The abundance ratings were subjectively derived from reading the literature and from field experience, since combining information across studies with very different methods precluded our ability to give exactly comparable, objective ratings.

Our ratings also reflect the geographic locations where most studies have been conducted or where we had experience. Many of these habitats span a large geographic area especially from north to south, and we were only able to express some of the differences within habitat use in some forest types (indicated by a superscript c in Table 8-1). In general, pine–oak, pygmy conifer, and ponderosa pine woodlands reflect southern distributions of species, and mixed conifer, primarily Douglas-fir, forests reflect northern distributions. Information was sparse and summarized references spanning 50 years. The purpose of Table 8-1 is to give a general idea of some species commonly found in each of these habitats, not a definitive description of what should be in any one specific locale.

General Bird–Habitat Associations

Of 215 species present in Rocky Mountain forests, 72 (34%) are permanent residents, 69 (32%) are long-distance migrants, 50 (23%) are short-distance migrants, and 24 (11%) are migrants that breed primarily south of the United States/Mexico border (Table 8-1). Twenty-three species (nine residents, four long-distance migrants, and ten short-distance migrants) are found in all eight forest types. No species is common in all eight types, but eight species—Hairy Woodpecker, Northern Flicker, Red-breasted Nuthatch, American Robin, Yellow-rumped Warbler, Chipping Sparrow, Dark-eyed Junco, and Pine Siskin—are common or abundant in all but one type. Forty-six species (13 residents, 11 long-distance migrants, four short-distance migrants, and 18 migrants that breed primarily south of the United States/ Mexico border) are found only in pine–oak and/or pinyon–juniper woodlands; 44 of these species were uncommon and/or rare in one or both of these habitats.

Methods for Evaluating Differences in Birds among Natural Stands of Different Ages

We summarized the results of two studies in natural forests of different ages in the Rocky Mountains to find possible indications of birds associated with particular stand ages. In the Northern Rockies, birds were compared in natural pole-sapling, mature, and old-growth forests in lodgepole–spruce–fir, and in mature and old-growth Douglas fir (Catt 1991, Moore 1992). We used the results of statistical tests if available.

Table 8-1. Relative abundance of species (A = abundant; C = common; U = uncommon; R = rare) in the breeding season in eight forest types [pine–oak woodland = PO; pygmy conifer (pinyon–juniper) = PJ; Cascadian = CA; ponderosa pine = PP; mixed conifer (primarily dominated by Douglas fir) = MC; lodgepole pine = LP; spruce fir = SF; aspen = AS] in the Rockies.

Species	NTMB Status[a]	PO	PJ	CA	PP	MC	LP	SF	AS
Turkey Vulture	B	C	C	—	U	C	—	U	U
Bald Eagle	R	R	R	—	—	—	—	—	R
Sharp-shinned Hawk	B	C	R	—	C	C	U	C	C
Cooper's Hawk	B	C	U	C	C	C	C	U	C
Northern Goshawk	B	U	R	C	C	C	C	C	C
Common Black-Hawk[b]	C	R	R	—	—	—	—	—	—
Swainson's Hawk	A	R	R	—	—	—	—	—	U
Zone-tailed Hawk	C	R	U	—	—	—	—	—	—
Red-tailed Hawk	B	C	C	—	C	U	U	C	C
Ferruginous Hawk	B	R	U	—	U[c]	—	—	—	R
Golden Eagle	B	U	U	—	U	R	U	U	U
American Kestrel	B	U	C	—	C	U	—	C	C
Merlin	A	R	R	—	R	R	—	R	R
Peregrine Falcon	A	R	R	—	R	R	R	—	R
Prairie Falcon	B	—	U	—	U	—	—	—	—
Ring-necked Pheasant	R	R	R	—	—	—	—	—	R
Spruce Grouse	R	—	—	—	—	—	—	C[c]	—
Blue Grouse	R	U	R	—	U	—	—	C	C
Ruffed Grouse	R	—	—	—	—	U[c]	—	U	C
Wild Turkey	R	U	U	—	R	—	—	U	C
Montezuma Quail	R	U	U	—	—	—	—	—	—
Gambel's Quail	R	U	U	—	U	—	—	—	—
Killdeer	B	U	U	—	R	—	—	—	U
Band-tailed Pigeon	A	U	R	—	U	U[c]	—	U	U
White-winged Dove	C	R	U	—	—	—	—	—	—
Mourning Dove	B	C	C	—	U	U	—	U	U
Thick-billed Parrot	R	R	—	—	—	—	—	—	—
Greater Roadrunner	R	R	U	—	R	—	—	—	—
Barn Owl	R	R	R	—	—	—	—	—	—
Flammulated Owl	A	A	U	R	A	C	—	R	R
Western Screech-Owl[b]	R	U	C	U	R	U	—	—	C
Whiskered Screech-Owl	R	C	R	—	—	—	—	—	—
Great Horned Owl	R	C	C	C	C	C	R	U	C
Northern Hawk Owl	R	—	—	—	—	—	—	R	R
Northern Pygmy-Owl	R	U	U	U	C	C	R	C	U
Elf Owl	C	R	R	—	—	—	—	—	—
Spotted Owl (Mexican)	R	C	R	—	U[c]	C[c]	—	R[c]	R[c]
Barred Owl	R	—	—	C	U[c]	C[c]	—	—	—
Great Gray Owl	R	—	—	—	—	U	R	U	R
Long-eared Owl	B	R	U	—	U	U	R	U	U
Boreal Owl	R	—	—	—	—	—	—	C	U
Northern Saw-whet Owl	R	U	R	U	C	C	R	U	C
Lesser Nighthawk	A	—	U	—	—	—	—	—	—
Common Nighthawk	A	U	U	—	C[c]	U	U	C	C
Common Poorwill	B	C	C	—	U[c]	—	—	—	U
Whip-poor-will	A	R	R	—	U[c]	—	—	—	—
Black Swift	A	R	—	—	—	—	—	U	U
Vaux's Swift	A	R	R	R	—	R	—	—	—
White-throated Swift	A	C	U	—	U[c]	R[c]	—	C	C
Broad-billed Hummingbird	C	R	R	—	—	—	—	—	—
White-eared Hummingbird	C	R	—	—	—	—	—	—	—
Berylline Hummingbird	C	R	—	—	—	—	—	—	—
Blue-throated Hummingbord	C	U	R	—	—	—	—	—	—
Magnificent Hummingbird	C	U	U	—	—	—	—	—	—
Lucifer Hummingbird	C	R	R	—	—	—	—	—	—
Black-chinned Hummingbird	A	U	C	—	—	—	—	—	—
Anna's Hummingbird	B	U	U	—	—	—	—	—	—
Costa's Hummingbird	A	—	R	—	—	—	—	—	—

Table 8-1 (*cont.*)

Species	NTMB Status[a]	PO	PJ	CA	PP	MC	LP	SF	AS
Calliope Hummingbird	A	—	—	—	—	U	R	C	C
Broad-tailed Hummingbird	A	C	U	—	C	U	C	C	C
Rufous Hummingbird	A	—	—	U	U[c]	U	—	C	C
Elegant Trogon	C	U	R	—	—	—	—	—	—
Eared Trogon	C	R	R	—	—	—	—	—	—
Lewis' Woodpecker	B	U	R	—	U[c]	—	—	—	R
Acorn Woodpecker	R	U	U	—	U[c]	—	—	—	—
Yellow-bellied Sapsucker	B	R	—	—	—	—	—	—	R
Red-naped Sapsucker	B	C	U	C	—	U	R	C	C
Williamson's Sapsucker	B	R	—	—	C[c]	U	U	U	U
Ladder-backed Woodpecker	R	—	R	—	—	—	—	—	—
Downy Woodpecker	R	U	R	—	U[c]	R	—	C	C
Hairy Woodpecker	R	C	C	C	C	C	U	C	C
Strickland's Woodpecker	R	C[c]	U[c]	—	—	—	—	—	—
White-headed Woodpecker	R	—	—	—	U[c]	U[c]	—	—	—
Three-toed Woodpecker	R	R	—	R	U[c]	R	U	U	U
Black-backed Woodpecker	R	R	—	—	—	R	R	R	R
Northern Flicker	B	C	C	U	C	C	C	C	C
Pileated Woodpecker	R	—	—	C	—	U	—	U	U
Olive-sided Flycatcher	A	U	R	U	U	U	R	C	C
Greater Pewee	C	C	R	—	C[c]	C[c]	—	—	—
Western Wood-Pewee	A	C	R	U	C	U	—	C	C
Willow Flycatcher[b]	A	R	R	C	—	R	—	U	U
Hammond's Flycatcher	A	U	R	A	—	C	—	C	C
Dusky Flycatcher	A	U	U	R	—	C	—	C	C
Gray Flycatcher	A	U	U	—	U[c]	—	—	—	—
Cordilleran Flycatcher	A	U	R	U	U	C	C	U	U
Buff-breasted Flycatcher	C	U	R	—	—	—	—	—	—
Black Phoebe[b]	R	U	U	—	—	—	—	—	—
Say's Phoebe	B	R	U	—	—	R	—	—	—
Vermilion Flycatcher[b]	A	U	U	—	—	—	—	—	R
Dusky-capped Flycatcher	C	U	R	—	R	U[c]	—	—	—
Ash-throated Flycatcher	A	C	A	—	U	R[c]	—	—	—
Brown-creasted Flycatcher[b]	C	U	U	—	R	—	—	—	—
Sulphur-bellied Flycatcher	A	C	R	—	—	R[c]	—	—	—
Cassin's Kingbird	A	U	C	—	U	—	—	—	—
Thick-billed Kingbird[b]	C	R	R	—	—	—	—	—	—
Western Kingbird	A	U	U	—	R	—	—	—	R
Rose-throated Becard	C	U	R	—	—	—	—	—	—
Horned Lark	B	—	R	—	—	—	—	—	—
Purple Martin	A	R	—	—	R	—	—	R	R
Tree Swallow[b]	B	U	U	U	U	R	—	C	C
Violet-green Swallow	A	C	U	—	C	R	—	U	U
Cliff Swallow	A	—	R	—	—	—	—	—	—
Gray Jay	R	R	R	C	—	C	A	C	C
Steller's Jay	R	C	U	U	C	U	U	C	C
Blue Jay	R	—	—	—	U[c]	—	—	—	—
Scrub Jay	R	U	C	—	R	—	—	—	—
Gray-breasted Jay	R	C[c]	R[c]	—	—	—	—	—	—
Pinyon Jay	R	C	C	—	U	—	—	—	—
Clark's Nutcracker	R	—	—	—	U[c]	U	C	A	C
Black-billed Magpie	R	U	U	—	U	—	—	R	—
American Crow	R	U	U	—	U	R	—	R	U
Common Raven	R	C	C	C	U[c]	U	U	C	C
Black-capped Chickadee	R	—	C	C	C[c]	C	—	C	C
Mexican Chickadee	R	C[c]	R[c]	—	U[c]	—	—	—	—
Mountain Chickadee	R	U	R	A	A	A	A	C	A
Boreal Chickadee	R	—	—	—	—	—	—	A[c]	C[c]
Chestnut-backed Chickadee	R	—	—	A	—	R	—	—	—
Bridled Titmouse	R	C[c]	U[c]	—	R[c]	—	—	—	—

(continued)

Table 8-1 *(cont.)*

Species	NTMB Status[a]	PO	PJ	CA	PP	MC	LP	SF	AS
Plain Titmouse	R	U	C	—	—	—	—	—	—
Bushtit	R	R	C	—	—	—	—	—	—
Red-breasted Nuthatch	R	C	R	A	C[c]	C	C	C	C
White-breasted Nuthatch	R	C	U	—	C	U	—	C	C
Pygmy Nuthatch	R	C	U	—	A	U	—	U	R
Brown Creeper	B	U	R	C	C	C	C	C	C
Rock Wren	B	U	C	—	R	—	—	R	U
Canyon Wren	R	R	U	—	U[c]	—	—	—	R
Bewick's Wren	R	C	C	—	U	—	—	—	—
House Wren	A	C	R	—	C	U	—	C	A
Winter Wren	R	R	—	C	—	U	—	C	—
American Dipper[b]	R	—	—	—	—	—	—	U	U
Golden-crowned Kinglet	R	—	R	A	—	C	U	A	C
Ruby-crowned Kinglet	B	C	U	A	—	A	C	A	C
Blue-gray Gnatcatcher	A	U	C	—	—	R[c]	—	—	—
Western Bluebird	B	U	R	—	A[c]	R[c]	—	—	R
Mountain Bluebird	B	U	C	—	U	U	—	U	U
Townsend's Solitaire	B	U	U	C	U	C	C	C	C
Swainson's Thrush	A	R	R	A	—	C	C	C	C
Hermit Thrush	B	C	—	U	C	C	A	C	C
American Robin	B	C	U	C	C	A	A	C	C
Varied Thrush	R	R	—	A	—	U	—	U	U
Northern Mockingbird	B	R	C	—	R	—	—	—	R
Curve-billed Thrasher	R	—	C	—	—	—	—	—	—
Bohemian Waxwing	R	—	—	—	—	—	—	C	U
Cedar Waxwing	B	U	R	—	—	R	—	R	R
Loggerhead Shrike	B	R	R	—	R	—	—	—	—
European Starling	R	—	—	—	—	R	—	—	—
Gray Vireo	A	R	C	—	—	—	—	—	—
Solitary Vireo	A	C	R	C	C	C	U	—	—
Hutton's Vireo	R	U[c]	—	—	—	—	—	—	—
Warbling Vireo	A	U	—	U	C	C	—	C	A
Red-eyed Vireo	A	—	—	U	U[c]	—	—	—	—
Tennessee Warbler	A	—	—	—	—	—	—	R[c]	—
Orange-crowned Warbler	A	U	R	U	—	C	—	U	U
Nashville Warbler	A	—	—	R	—	U	—	—	—
Virginia's Warbler	A	C	C	—	U[c]	U[c]	—	—	—
Lucy's Warbler[b]	C	R	R	—	—	—	—	—	—
Northern Parula	A	R	—	—	—	—	—	—	—
Yellow Warbler[b]	A	R	R	U	—	U	—	C	C
Yellow-rumped Warbler	B	C	U	A	C	A	A	A	A
Black-throated Gray Warbler	A	C	C	—	U	C[c]	—	U	U
Townsend's Warbler	A	—	—	A	—	C	—	C	C
Hermit Warbler	A	—	—	—	—	R[c]	—	—	—
Grace's Warbler	A	C[c]	—	—	A[c]	C[c]	—	—	—
American Redstart[b]	A	—	—	—	—	U	—	—	R
Ovenbird	A	—	—	—	C[c]	R	—	R	U
Northern Waterthrush[b]	A	—	—	R	—	U	—	U	—
MacGillivray's Warbler	A	R	—	C	U	C	—	U	U
Common Yellowthroat[b]	A	—	—	—	—	—	—	—	R
Wilson's Warbler	A	—	—	U	—	U	—	C	U
Red-faced Warbler	C	C	—	—	U[c]	C[c]	—	—	—
Painted Redstart	C	C	U	—	—	C[c]	—	—	—
Olive Warbler	C	R	—	—	A[c]	U[c]	—	—	—
Hepatic Tanager	A	C	R	—	U[c]	R[c]	—	—	—
Summer Tanager	A	U	—	—	—	—	—	—	—
Western Tanager	A	C	R	C	C	A	U	C	C
Northern Cardinal	R	R	U	—	R	—	—	—	—
Yellow Grosbeak	C	R	R	—	—	—	—	—	—
Black-headed Grosbeak	A	C	U	U	U[c]	C	—	R	U
Lazuli Bunting	A	U	U	—	—	U	—	U	U

Table 8-1 (*cont.*)

Species	NTMB Status[a]	PO	PJ	CA	PP	MC	LP	SF	AS
Green-tailed Towhee	A	U	U	—	—	R	—	C	C
Rufous-sided Towhee	B	C	C	—	R	U[c]	—	—	R
Canyon Towhee	R	R[c]	U[c]	—	—	—	—	—	—
Chipping Sparrow	A	C	C	C	C	A	U	C	C
Brewer's Sparrow	A	—	R	—	—	—	—	—	—
Black-chinned Sparrow	A	—	U	—	—	—	—	—	—
Vesper Sparrow	B	R	U	—	—	R	—	R	R
Lark Sparrow	A	—	C	—	—	—	—	—	—
Black-throated Sparrow	B	—	U	—	—	—	—	—	—
Sage Sparrow	B	—	R	—	—	—	—	—	—
Fox Sparrow	B	—	—	C	—	U	—	U[c]	R
Song Sparrow[b]	B	R	—	C	U[c]	U	—	U	U
Lincoln's Sparrow	A	—	—	—	—	R	C	U	U
White-crowned Sparrow	B	R	—	—	—	R	—	C	U
Dark-eyed Junco	B	C	U	A	A	A	A	A	A
Yellow-eyed Junco	R	C[c]	U[c]	—	A[c]	—	—	—	—
Western Meadowlark	B	R	U	—	—	—	—	—	R
Yellow-headed Blackbird[b]	A	R	U	—	—	—	—	—	R
Brewer's Blackbird	B	U	C	—	U[c]	R	—	—	U
Common Grackle	R	—	U	—	—	R	—	—	R
Bronzed Cowbird	C	R[c]	U[c]	—	—	—	—	—	—
Brown-headed Cowbird	B	C	C	—	R	C	R	R	C
Hooded Oriole	A	U	U	—	—	—	—	—	—
Northern Oriole	A	R	U	—	—	—	—	—	R
Scott's Oriole	A	C	U	—	—	R[c]	—	—	—
Pine Grosbeak	R	—	—	—	U[c]	U	C	C	U
Purple Finch	B	R	—	—	—	—	—	C[c]	—
Cassin's Finch	B	U	R	U	C[c]	C	U	U	U
House Finch	R	U	U	—	—	—	—	—	R
Red Crossbill	R	R	R	C	C	C	C	C	U
White-winged Crossbill	R	—	—	—	—	—	R	R	—
Pine Siskin	B	U	C	C	A	A	A	C	C
Lesser Goldfinch	B	U	U	—	R	—	—	—	—
American Goldfinch	B	U	U	—	U[c]	R	—	—	R
Evening Grosbeak	R	U	U	U	U	C	—	C	U
House Sparrow	R	U	U	—	R	—	—	—	R

[a] As designated by the *Partners in Flight* preliminary list: A = long-distance migrant species, those that breed in North America and spend their nonbreeding period primarily south of the United States; B = short-distance migrant species, those that breed and winter extensively in North America; C = migrants whose breeding range is primarily south of the United States/Mexican border, and enter the United States along the Rio Grande Valley or where the Mexican highlands extend across the United States border. These populations largely vacate the United States during the winter months. R = permanent resident species that primarily have overlapping breeding and nonbreeding areas.
[b] Species associated with wet areas in these habitats.
[c] Species at least locally found in that habitat type. While there is probably a north/south difference in bird species in most habitats, it is most notable in ponderosa pine and mixed conifer.

Differences in Birds among Natural Stands of Different Ages

No common results for any one species nor obvious trends for any particular migrant group were found in the two studies comparing natural stands of different ages (Catt 1991, Moore 1992). From his study of forest succession in spruce-fir forests, Catt (1991) found 12 species in all four forested successional stages (pole-sapling, young, mature, and old-growth forests), while six species were associated with the three oldest stages. Two of these species (Golden-crowned Kinglet and Townsend's Warbler) were more abundant in mature and old-growth forests. Three-toed Woodpeckers, Winter Wrens, and White-winged Crossbill were only found in mature and old-growth forests, with Winter Wrens being clearly more abundant in old-growth stands. A few species were found only in one stand age. Lincoln's Sparrow and Evening Grosbeaks were only present in mature forests. Dusky

Flycatcher and Townsend's Solitaire were only found in pole-sapling stands.

Of 24 common species in mature and old-growth Douglas-fir forests in Montana, White-crowned Sparrow was only present in, and American Robin and Chipping Sparrow were more abundant in, old-growth stands (Moore 1992). No common species was significantly more abundant in the mature stands. Of 19 uncommon species, three were found only in mature and six in old-growth forests. Three-toed Woodpecker was the only uncommon species seen on all four old-growth sites.

COMPARISONS OF BIRDS AMONG LOGGING TREATMENTS

Methods for Evaluating Effects of Silvicultural Treatments on Forest Birds

Since most of the ornithological literature describes the effects of silvicultural practices on birds in Cascadian, ponderosa pine, mixed-conifer (primarily dominated by Douglas fir), lodgepole pine, spruce-fir, and aspen forests, we concentrated our efforts on these habitats. We searched through federal publications, university dissertations and theses, and the major ornithological and ecological journals for studies on effects of timber harvesting on birds. We also included unpublished data. We classified data from a given study site in community-wide studies into one of the following forest cover types: (1) Cascadian; (2) ponderosa pine; (3) mixed conifer; (4) lodgepole pine; (5) spruce fir; or (6) aspen. The cover type was also classified into one of seven disturbance categories: (1) uncut; (2) group selection; (3) overstory removal; (4) shelterwood cut (before re-entry to remove the remaining overstory trees); and (5) three ages of clearcuts. Because few studies (most with very few replicates in any one treatment) had been conducted in any one forest type (those from conifer forests are listed in Table 8-2) or for any particular silvicultural method, we combined data from all of the studies in conifer forests and made comparisons between birds in uncut forests with those in four developmental classes: low shrub clearcuts (from grass-forb to small shrub stage, in general, 0–10 years old), tall shrub clearcuts (including tall shrubs and seedlings, in general from 11–20 years old), pole-sapling clearcuts (in general 21–40 years old), and partial cuts (any cutting treatment besides clearcut; categories 2–4 above). We preferentially used descriptions of the vegetation to determine how to categorize each site. We do not know if the "uncut"

Table 8-2. Distribution of study sites by habitat and logging treatment from 19 studies[a] on the effects of logging treatments on birds in conifer forests throughout the Rocky Mountains. A study was required to have a control as well as a treated area to be included in our analyses. Several studies compared several treatments with one control. The five forest types include Cascadian (CA), ponderosa pine (PP), mixed conifer (dominated primarily by Douglas fir; MC), lodgepole pine (LP), and spruce fir (SF).

Logging Treatment	Forest Types					
	CA	PP	MC	LP	SF	Total
Group selection	0	0	2	0	2	4
Overstory removal	0	3	4	0	0	7
Shelterwood cut	0	0	2	0	0	2
Low-shrub clearcut	3	1	4	2	3	13
Tall-shrub clearcut	3	0	1	2	0	6
Pole-sapling clearcut	2	0	0	1	0	3
Total	8	4	13	5	5	35

[a] Studies include: Austin and Perry (1979), Brawn and Balda (1988), Case and Hutto (1980, unpublished field notes), Catt (1991), Davis (1976), Franzreb and Ohmart (1978), Hallock (1989–1990), Holmes et al. (1991, unpublished field notes), Keller and Anderson (1992), McClelland (1980), Medin (1985), Medin and Booth (1989), Mitchell and Bratkovich (1992), Peterson (1982), Scott and Gottfried (1983), Scott et al. (1982), Siegel (1989), Tobalske et al. (1991), and Wetmore et al. (1985).

sites or "control" sites from most studies were truly never cut. We assumed that, if anything, they were lightly cut. We also do not know the age of all of these uncut stands but we assume that they were mature or old-growth forests. Studies were conducted in British Columbia, Idaho, Montana, Wyoming, Utah, Colorado, and Arizona.

To evaluate the effect of timber harvesting on birds in conifer forests, we analyzed only those community-wide studies specifically designed for that purpose. We scored each bird species as one that was less abundant (−1), similarly abundant (0), or more abundant (+1) in each logged site compared to an unlogged site from the same study. We used the results of significance tests when they were available but most studies had not used statistical tests. One tally was used for each treatment from studies that included multiple treatments. We assumed that differences resulted from the timber harvesting activity. Our methods were subjective, but we minimized the effect of analyzer bias by having just one of us (RLH) make the decision as to whether a species was more, less or similarly abundant in treated than in untreated areas. The potential effect of each timber harvest activity on each species was determined by calculating the average score over all such studies, resulting in an index. Thus, an index of 1.0 indicates that every study reported more birds in the treated than in the untreated areas, and an index of −1.0 indicates that every study reported more birds in the untreated areas. An index of 0.0 indicates either that a species had similar abundances in treated and untreated areas in every study, or that no obvious trend held across studies. Species that were encountered in at least three studies are emphasized in results. The original sampling effort varied among the studies—from one site to many sites per treatment, from one to many years, and from one to several observers. A few studies had data from study sites before and after a treatment, but most compared data from sites that had been treated to those that had not been. These differences among studies add sources of variation to the analyses. The power of our analyses comes from comparing many studies.

We summarized the effects of silviculture on birds in aspen and the effects of chaining on birds in pinyon–juniper separately. Information was sparse on the effects of silviculture in aspen forests, so we simply noted a few potentially relevant facts. There was no information on the effects of silvicultural practices from madrean pine–oak woodlands, pygmy-conifer woodlands, or whitebark pine. Some information, however, was available on the effects of chaining (knocking down trees) on birds in pygmy-conifer woodlands. While chaining is not a silvicultural method, chaining initially affects the landscape in some similar ways to logging and is an important source of human-induced change in pinyon–juniper forests. We therefore included a brief summary of the effects of chaining on birds in pinyon-juniper. While we could not find any studies examining silvicultural effects in whitebark pine or associated communities, it is important to realize that whitebark pine is sometimes logged (more extensively in the past; Losensky 1990) and is an important but endangered resource for some bird species (Kendall and Arno 1990).

To find possible indications of old-growth associates, we summarized the results of four community-wide studies in the Rocky Mountains (Peterson 1982, Mannan and Meslow 1984, Mannan and Siegel 1988, Hejl and Woods 1991, Hejl, unpublished data). These studies compared birds in uncut or lightly cut "old-growth" forests with those in immature or mature second-growth stands. Three were conducted in Cascadian or mixed-conifer forests in the Northern Rockies, and one in ponderosa pine in the southern Rockies. We used the results of statistical tests if available, but most categorizations were subjective.

Finally, we used additional methods for evaluating the effects of silvicultural treatments on raptors. Most community-wide studies are not useful for evaluating raptor abundance or occurrence. We include raptors in our syntheses of community studies, if raptors were mentioned in the bird list, but we realize that they are more inadequately sampled than are the other species by these methods. The results from community studies may even be misleading, since many raptors are secretive birds and are more likely to be seen from an edge, opening, or clearcut, even if they rarely frequent these habitats. Therefore, we also searched the literature separately for specific studies addressing raptors and silvicultural treatments, and we describe those results separately.

Effects of Silvicultural Treatments on Birds in Conifer Forests

From community-wide studies, 26 species were less abundant in treated areas as compared to unlogged areas in general (Table 8-3). In contrast, 15 species were generally more abundant in treated areas than in unlogged ones.

Comparing recent, low-shrub clearcuts to unlogged forests, we found that 17 species were less abundant in the clearcuts in every case and 25 species in some cases (Table 8-3). All resident species were less abundant in these recent clearcuts than in uncut forests. Sixty-eight per cent of long-distance migrants and 52% of short-distance migrants were less abundant in recent clearcuts in most studies. In contrast, 13 species were generally more abundant in recent clearcuts. No species was more abundant in low-shrub clearcuts in all studies, but 21% of long-distance migrants and 43% of short-distance migrants were more abundant in most cases.

In a comparison of partially logged and unlogged areas, three species were less abundant in partially logged areas in all cases and 26 species in some cases (Table 8-3). Ninety-four per cent of the resident species were less abundant in partially logged forests in most studies. Thirty-three per cent of long-distance migrants and 42% of the short-distance migrants were usually less abundant in partially logged forests. In contrast, 19 species were sometimes more abundant in partially logged areas and two species always more abundant. Sixty-one per cent of long-distance migrants and 50% of the short-distance migrants were sometimes to always more abundant in partially logged forests.

Each species responded uniquely to the harvesting treatments. Brown Creeper exhibited the clearest difference between harvested and unharvested treatments; creepers were always less abundant in clearcuts or partially logged forests than in uncut areas (Table 8-3). Red-breasted Nuthatch was always less abundant in any age of clearcut than in uncut forest. Seven other species (e.g., Golden-crowned Kinglet and Swainson's Thrush) were always less abundant in low- and tall-shrub clearcuts than in uncut forests, but not always so in partially cut forests. Five other species were less abundant in low-shrub clearcuts in all studies in which they were present. Pygmy Nuthatch and Pine Grosbeak were always less abundant in partially logged areas, but Pine Grosbeak was more abundant in clearcuts in some studies.

While many species were noticeably more abundant in one or two of the categories of harvested areas, no species was always more abundant in all classes (Table 8-3). Mountain Bluebirds were more abundant in low-shrub clearcuts in almost all studies. Nine species (e.g., Warbling Vireo, MacGillivray's Warbler and Rufous Hummingbird) were more abundant in tall-shrub clearcuts in all cases. Four species were always more abundant in pole-sapling clearcuts. Calliope Hummingbird and Rock Wren were always more abundant in partially logged areas than in uncut forest. All of the species that were consistently more abundant in logged areas were migrant species. For example, 69% of the species more abundant in recent clearcuts were short-distance migrants; the rest were long-distance migrants. Forty-three per cent of the species more abundant in partially cut areas were short-distant migrants and 52% were long-distant migrants. Hairy Woodpecker, Steller's Jay, and Clark's Nutcracker were the only resident species that were sometimes more abundant in treated than in untreated areas.

Some species did not seem to be negatively or positively affected by a particular silvicultural treatment. Rufous Hummingbird, Cassin's Finch, and Lincoln's Sparro were equally abundant in recent clearcuts and uncut areas (Table 8-3). Williamson's Sapsucker and Cordilleran Flycatcher were equally abundant in partially logged and uncut areas.

There are a few species for which sample size (number of studies) was too low to include in the table (we chose three studies

Table 8-3. Indices of the tendency for a bird species to be more or less abundant in clearcut or partially cut forest than in uncut forest. A species was scored as being more abundant (+1), less abundant (−1), or similarly abundant (0) in treated vs. untreated areas. Values in the table are averages of these scores over all studies on which the species was recorded. Species are ranked in ascending order from −1.00 based on low-shrub clearcut column. Sample sizes in parentheses (we only included sample sizes ≥ 3).

Species	NTMB Status[a]	Clearcuts			Partially Cut
		Low Shrub	Tall Shrub	Pole-sapling	
Mountain Chickadee	R	−1.00 (10)	−1.00 (5)	0.00 (3)	−0.77 (13)
Red-breasted Nuthatch	R	−1.00 (10)	−1.00 (5)	−1.00 (3)	−0.70 (10)
Brown Creeper	B	−1.00 (10)	−1.00 (4)	—	−1.00 (12)
Golden-crowned Kinglet	R	−1.00 (9)	−1.00 (3)	—	−0.60 (10)
Ruby-crowned Kinglet	B	−1.00 (9)	−1.00 (4)	—	−0.40 (10)
Winter Wren	R	−1.00 (7)	—	—	−0.20 (5)
Swainson's Thrush	A	−1.00 (7)	−1.00 (3)	—	−0.50 (6)
Varied Thrush	R	−1.00 (7)	−1.00 (3)	—	−0.75 (4)
Townsend's Warbler	A	−1.00 (7)	−1.00 (3)	—	−0.40 (5)
Three-toed Woodpecker	R	−1.00 (6)	—	—	−0.50 (6)
Black-capped Chickadee	R	−1.00 (6)	−0.67 (3)	—	−0.67 (3)
Solitary Vireo	A	−1.00 (5)	0.33 (3)	—	0.33 (9)
Hammond's Flycatcher	A	−1.00 (4)	−1.00 (4)	—	—
Evening Grosbeak	R	−1.00 (4)	—	—	—
Pileated Woodpecker	R	−1.00 (3)	−0.67 (3)	—	—
Chestnut-backed Chickadee	R	−1.00 (3)	—	—	—
White-breasted Nuthatch	R	−1.00 (3)	—	—	−0.14 (7)
Pygmy Nuthatch	R	—	—	—	−1.00 (5)
Western Tanager	A	−0.86 (7)	−1.00 (4)	—	0.09 (11)
Hermit Thrush	B	−0.71 (7)	—	—	−0.80 (10)
Steller's Jay	R	−0.67 (6)	0.33 (3)	—	−0.29 (7)
Clark's Nutcracker	R	−0.67 (6)	—	—	0.33 (3)
Warbling Vireo	A	−0.67 (6)	1.00 (4)	—	0.33 (9)
Yellow-rumped Warbler	B	−0.67 (12)	−0.50 (6)	1.00 (3)	−0.46 (13)
Gray Jay	R	−0.60 (10)	−0.50 (4)	0.00 (3)	−0.25 (4)
Black-headed Grosbeak	A	−0.62 (8)	0.40 (5)	—	0.22 (9)
Orange-crowned Warbler	A	−0.60 (5)	—	—	−0.50 (4)
Violet-green Swallow	A	—	—	—	−0.60 (5)
Pine Grosbeak	R	−0.50 (4)	—	—	−1.00 (3)
Pine Siskin	B	−0.45 (11)	0.00 (6)	0.00 (3)	−0.08 (12)
Western Wood-pewee	A	−0.43 (7)	—	—	−0.50 (4)
House Wren	A	−0.40 (5)	0.00 (3)	—	0.86 (7)
Hairy Woodpecker	R	−0.36 (11)	−0.33 (6)	0.33 (3)	−0.25 (12)
Cooper's Hawk	B	−0.33 (3)	—	—	−0.67 (3)
Common Raven	R	−0.33 (9)	−0.25 (4)	—	−0.17 (6)
Brown-headed Cowbird	B	−0.33 (3)	—	—	—
Red Crossbill	R	−0.33 (3)	−0.25 (4)	—	−0.33 (3)
Common Nighthawk	A	−0.25 (4)	−0.33 (3)	—	−0.50 (4)
Northern Flicker	B	−0.18 (11)	0.67 (6)	0.67 (3)	−0.17 (12)
Wilson's Warbler	A	−0.17 (6)	0.67 (3)	—	—
Fox Sparrow	B	−0.17 (6)	0.67 (3)	—	—
Red-naped Sapsucker	B	−0.14 (7)	0.00 (5)	0.67 (3)	0.17 (6)
MacGillivray's Warbler	A	−0.12 (8)	1.00 (4)	—	0.17 (6)
American Robin	B	−0.08 (13)	0.50 (6)	1.00 (3)	0.15 (13)
Rufous Hummingbird	A	0.00 (6)	1.00 (3)	—	0.33 (3)
Cassin's Finch	B	0.00 (5)	−0.20 (5)	0.67 (3)	0.60 (5)
Lincoln's Sparrow	A	0.00 (3)	0.67 (3)	—	—
Cordilleran Flycatcher	A	—	—	—	0.00 (6)
Williamson's Sapsucker	B	—	—	—	0.00 (5)
Chipping Sparrow	A	0.18 (11)	0.67 (6)	1.00 (3)	0.60 (10)
Western Bluebird	B	—	—	—	0.20 (5)
Olive-sided Flycatcher	A	0.25 (12)	0.25 (4)	—	0.67 (9)
Broad-tailed Hummingbird	A	0.33 (3)	1.00 (3)	—	0.25 (4)
Tree Swallow	B	0.40 (5)	—	—	—

(continued)

Table 8-3 (*cont.*)

Species	NTMB Status[a]	Clearcuts			Partially Cut
		Low Shrub	Tall Shrub	Pole-sapling	
Dark-eyed Junco	B	0.46 (13)	1.00 (6)	1.00 (3)	0.38 (13)
Northern Goshawk	B	0.50 (4)	−0.60 (5)	0.00 (3)	—
Red-tailed Hawk	B	0.50 (4)	0.33 (3)	—	0.33 (3)
Mourning Dove	B	0.50 (4)	—	—	0.67 (3)
White-crowned Sparrow	B	0.50 (6)	—	—	—
Townsend's Solitaire	B	0.57 (7)	0.25 (4)	—	−0.25 (8)
American Kestrel	B	0.67 (3)	1.00 (4)	0.67 (3)	—
Dusky Flycatcher	A	0.67 (3)	1.00 (3)	—	—
Mountain Bluebird	B	0.90 (10)	1.00 (5)	0.33 (3)	0.67 (6)
Song Sparrow	B	—	1.00 (3)	—	—
Calliope Hummingbird	A	—	—	—	1.00 (3)
Rock Wren	B	—	—	—	1.00 (3)

[a] See footnote a of Table 8-1.

as a cutoff) but for which we guess that there are likely to be real differences among treatments. For example, Boreal Chickadee and Pygmy Nuthatch were always less abundant in recent clearcuts in the literature we searched. Flammulated Owl, Pileated Woodpecker, and Grace's Warbler were always less abundant in partially logged forests. Also, Virginia's Warbler, Grace's Warbler, Red-faced Warbler, and Olive Warbler were not present in clearcuts in a multiyear study but were present in treated or uncut forests (Brawn and Balda 1988).

Differences in Birds between Cut and Uncut Aspen Forests

Only a few avian species are more closely associated with aspen than other forest habitats in the Rockies (Table 8-1). They include Red-naped Sapsucker, Black-capped Chickadee, House Wren, Warbling Vireo (Finch and Reynolds 1987, Scott and Crouch 1988), and perhaps the Northern Saw-whet Owl in some areas. Results of two studies on the effects of logging treatments on birds in aspen forests (DeByle 1981, Scott and Crouch 1987) serve to underscore the need for more specific, practical information for managers. The two studies were conducted in different areas (Utah, Colorado), and involved treatments on vastly different scales (50% of a 4 ha site clearcut in Utah, 25% of a 930 ha site clearcut in Colorado). The combined results are equivocal; therefore, no assessments can be made as to the effects of cutting aspen on any particular migrant group. For example, the House Wren declined in abundance after clearcutting in Utah, but increased in Colorado. The Warbling Vireo and Dark-eyed Junco declined in Utah, but showed essentially no response in Colorado. Cordilleran Flycatchers declined in the Colorado study, but were not present in the Utah study. Estimated bird density increased in the Colorado site, but decreased in Utah after clearcutting. Both studies indicate that bird species richness increased after clearcutting and that Hermit Thrushes were adversely affected whereas Song Sparrows and Mountain Bluebirds benefited from clearcutting, but the evidence is more anecdotal than analytic.

Differences in Birds between Chained and Unchained Pinyon–Juniper Woodland

The effects of chaining on birds in pinyon–juniper woodland are similar to the effects of clearcutting on birds in other Rocky Mountain forests (O'Meara et al. 1981, Sedgwick and Ryder 1987). In one study in Colorado, the 8- and 15-year-old chained areas had no breeding species in common with unchained areas (O'Meara et al. 1981). Only ground- and shrub-nesters were found on the chained areas; the 10 species found only in the unchained area typically require trees for nesting and foraging. In the other study in

Colorado, more similar ties in bird presence were noted between the recently chained and adjacent, unchained plots (Sedgwick and Ryder 1987). Some ground-searchers or ground-nesters used the chained plot, but this use varied by year. Foliage-and-timber searchers, aerial foragers, foliage nesters, and cavity nesters mainly used the unchained plot. Gray Flycatcher, White-breasted Nuthatch, Mountain Bluebird, Hermit Thrush, Solitary Vireo, and Black-throated Gray Warbler were more abundant in unchained areas and Rock Wren was more abundant in chained areas in both studies. Differences in abundances between chained and unchained areas did not obviously correlate with the migratory status of the birds.

Old-growth and Second-growth Associates

Although 15 species were more abundant in old growth in at least one study, no species was consistently more abundant in old growth in all four studies that compared old-growth with old second-growth stands (Peterson 1982, Mannan and Meslow 1984, Mannan and Siegel 1988, Hejl and Woods 1991, Hejl, unpublished data) (Table 8-4). Woodpeckers, nuthatches, and thrushes, however, were more abundant in old growth in general, and six species (Brown Creeper, Golden-crowned Kinglet, Varied Thrush, Swainson's Thrush, Hermit Thrush, and Townsend's Warbler) were relatively more abundant in old-growth stands in two studies. Of 13 species that were more abundant in second-growth forests, Chipping Sparrow was most abundant in mature second growth in three studies that compared such stands to old growth. Two species (Dusky Flycatcher, Brown-headed Cowbird) were relatively more abundant in mature, second-growth stands in two studies. Contrary to these trends, Brown Creepers were more abundant in mature second-growth in Idaho. Six other species also had conflicting trends. No trends were obvious for any particular migrant group.

Peterson's (1982) study of ten different stand ages (spanning from recent clearcuts to old-growth forests) gives additional information not found in the other three studies that compared just older-aged stands (forests older than 60 years). In Table 8-4, we compared bird abundances from Peterson's mature and old-growth stands to provide a comparison with other studies that had compared only older stand ages. In this comparison, some species proved to be more abundant in mature or old-growth stands when only those two stand ages were compared. Upon comparing abundances among all ten ages of logged stands as Peterson did, however, no bird species was obviously more abundant in old-growth or mature stands than in all other stand ages in Idaho. For example, while Golden-crowned Kinglets would be considered old-growth associates in a comparison with just mature stands, they were as abundant in pole stands as they were in old growth. This fact reminds us to realize the limitation of data from only two types of stands (i.e., the other three studies in Table 8-4).

Peterson's (1982) results, however, may simply underscore the uniqueness of each species' relation to stand age, depending on forest type or geographic area. Whereas Swainson's Thrushes were old-growth associates in mixed-coniferous forests in Oregon and Montana, Peterson found them in somewhat similar abundances in all forested stages of Cascadian forests. Townsend's Warblers, an old-growth associate in mixed-coniferous forests in Oregon and Montana, were most abundant in tall shrubs mixed with some conifers and in sapling conifers in Peterson's Cascadian forests in Idaho.

The Effects of Silvicultural Treatments on Raptors

Only four raptor species were sampled adequately enough in community-wide studies to be listed in our assessment of the presence of birds in various logging treatments across forests in the Rocky Mountains (Table 8-3). Cooper's Hawks were less abundant in low-shrub clearcuts and partially cut forests than in uncut forests in most studies. Northern Goshawks appeared to be negatively affected by clearcuts in some studies but not others; they were less abundant in tall-shrub clearcuts but more abundant in low-shrub clearcuts than in uncut areas. Red-tailed Hawks and American

Table 8-4. Locations in the Rocky Mountains where individual species were found to be old-growth (O) or second-growth (M = mature; I = immature) associates in comparisons of birds in old-growth and second-growth stands. Some species were present but not clearly associated with any habitat (P) and other species were not recorded in that location (—).

Species	NTMB Status[a]	Locations[b]			
		ID	OR	MT/ID	AZ
Common Nighthawk	A	—	P	P	I
Williamson's Sapsucker	B	—	P	P	O
Hairy Woodpecker	R	P	P	P	O
Three-toed Woodpecker	R	—	—	P	O
Pileated Woodpecker	R	P	P	O	—
Western Wood-pewee	A	O	—	M	P
Hammond's Flycatcher	A	P	P	O	—
Dusky Flycatcher	A	—	M	M	—
Clark's Nutcracker	R	—	—	M	P
Black-capped Chickadee	R	P	—	M	—
Chestnut-backed Chickadee	R	M	—	—	—
Mountain Chickadee	R	P	P	P	I
Red-breasted Nuthatch	R	M	O	P	—
White-breasted Nuthatch	R	—	P	P	O
Pygmy Nuthatch	R	—	—	—	O
Brown Creeper	B	M	O	P	O
Golden-crowned Kinglet	R	O	O	P	—
Ruby-crowned Kinglet	B	O	M	P	—
Western Bluebird	B	—	—	P	O
Townsend's Solitaire	B	P	P	P	I
Varied Thrush	R	O	P	O	—
Swainson's Thrush	A	P	O	O	—
Hermit Thrush	B	—	O	P	O
American Robin	B	P	P	P	I
Solitary Vireo	A	P	P	M	O
Warbling Vireo	A	—	—	P	O
Yellow-rumped Warbler	B	M	P	P	P
Townsend's Warbler	A	P	O	O	—
Grace's Warbler	A	—	—	—	O
Chipping Sparrow	A	M	M	M	P
Dark-eyed Junco	B	P	M	P	I
Brown-headed Cowbird	B	—	M	M	—
Cassin's Finch	B	—	M	P	O
Red Crossbill	R	—	P	M	P
Evening Grosbeak	R	—	P	M	O

[a] See footnote a of Table 8-1.
[b] Idaho (ID) study from Peterson (1982), Oregon (OR) study from Mannan and Meslow (1984), Montana and Idaho (MT/ID) study reported in Hejl and Woods (1991) and unpublished data from Hejl, and Arizona (AZ) study from Mannan and Siegel (1988).

Kestrels were more abundant in treated areas than in uncut forests in most studies. Flammulated Owls were not mentioned in our summary of old-growth associates, since they were present in very low numbers, but they were only found in old-growth stands in community studies from mixed-conifer forests in the Northern Rockies (Mannan and Meslow 1984, Hejl and Woods 1991).

General literature suggests that accipiters may be more affected than other hawks by intensive silvicultural activity in the short term. Northern Goshawks seem to prefer a 20–30 acre stand of large trees and high canopy closure surrounding their nest sites (Reynolds et al. 1982, Moore and Henny 1983, Reynolds 1983, Crocker-Bedford 1990, Reynolds et al. 1992). All three accipiters forage in the forest canopy (Reynolds and Meslow 1984) or for a limited distance into openings (R. Reynolds, personal communication), so foraging habitat would probably be reduced by clearcutting large openings, but not necessarily by partial cutting. Cooper's Hawks may be more capable of nesting in fragmented forests than the other accipiters (Beebe 1974, Evans 1982). Often, American Kestrels (Palmer 1988, Johnsgard 1990) and Red-tailed Hawks (Schmutz et al. 1980) are found in association with forest openings.

General literature suggests that at least four owl species may be associated with old-growth components and habitats in the Rocky Mountains. Male Flammulated Owls tend to establish territories in mature to old-growth stands of ponderosa pine, aspen, Douglas fir, or ponderosa pine mixed with Douglas fir (Richmond et al. 1980, Webb 1982, Howie and Ritcey 1987, Reynolds and Linkhart 1987, 1992, Jones 1987, 1991, Bull et al. 1990). Some of these forests have been selectively harvested (Howie and Ritcey 1987, Bull et al. 1990). In a study in ponderosa pine and mixed-conifer forests in northern Arizona (Ganey and Balda 1989), all Mexican Spotted Owls had activity centers located in old-growth forests and visited other portions of the home range infrequently. Great Gray Owls nest most often in mature and older stands (Bull et al. 1988), and they sometimes forage along edges of openings and clearcuts (G. Hayward, personal communication). Boreal Owls seem to be associated with mature and old-growth forests in spruce fir in central Idaho (Hayward 1989, Hayward et al. 1993), western Montana (Holt and Hillis 1987), and Colorado (Palmer 1986, Ryder et al. 1987). Mature and old-growth spruce-fir may provide optimum nesting habitat for Boreal Owls (G. Hayward, personal communication), but nests have also been found in old mixed-conifer, old Douglas-fir, and aspen forests. In contrast, the Great Horned owl and Barred Owl may successfully use fragmented areas in the Rockies, as has been suggested, but not substantiated, for these owls in the Pacific Northwest (Thomas et al. 1990). No relationships were found between any particular migratory group of raptors and distribution in certain habitats.

EFFECTS OF FIRE AND FIRE SUPPRESSION ON FOREST BIRDS

Methods for Evaluating Effects of Fire and Fire Suppression on Birds

Since fire is the most important natural disturbance in the Rocky Mountains (Peet 1988), we briefly summarize the literature on the importance of fire to birds in these forests (for additional information, see Rotenberry et al., Chapter 3, this volume).

Effects of Fire and Fire Suppression on Forest Birds

Teasing apart the effects of fires on forest birds is difficult, since fires vary in intensity, duration, frequency, location, shape, and extent (Rotenberry et al., Chapter 3, this volume). In spite of this fact, we attempted to make a few generalizations. These generalizations are based on limited and often anecdotal data.

Fire seems to affect birds in the Rocky Mountains differently depending on its intensity (Hejl 1994). High-intensity fires often create habitat for primary and secondary cavity nesters (Taylor and Barmore 1980), and one species (Black-backed Woodpecker) seems to be nearly restricted in distribution to recently burned forests (Hutto 1995). Primary cavity nesters often dramatically increase for the first few years following an intense burn, with secondary cavity nesters increasing in following years. The benefits may be short term, as snags fall down and are not replaced (for an example in the Sierra Nevada, see Raphael et al. 1987).

Moderate and low-intensity burns show less dramatic immediate effects than high-intensity burns. For the first few years after a moderate burn, birds characteristic of severely burned and unburned forests were present (Taylor and Barmore 1980). Low-intensity fires may have their greatest effect on forest birds in the long term. The cumulative effect of low intensity fires is the maintenance of park-like forests, resulting in habitats for birds that prefer open forests (Marshall 1963). Open forest species may be lost with fire suppression (Marshall 1963). Burns and fire exclusion may even have the opposite effects on many forest birds (Hejl 1994).

COMPARISON OF THE EFFECTS OF LOGGING AND FIRE ON FOREST BIRDS

Landbird communities associated with the standing dead "forests" remaining after high-intensity fires are unique and distinctly different from those associated with clearcuts (Hutto 1995). The distinction is largely due to the relative abundance of species that are nearly restricted in their habitat distribution within the Rocky Mountains to early postfire conditions (e.g., Black-backed Woodpecker), and to species that are not restricted to, but are relatively abundant in, early postfire habitats (e.g., Olive-sided Flycatcher). These patterns have been well documented in the western United States, if anecdotally (Blackford 1955, Bock and Lynch 1970, Bock and Bock 1974, Davis 1976, Pfister 1980, Taylor and Barmore 1980, Harris 1982, Raphael et al. 1987, Skinner 1989; but see Blake 1982).

Logging the trees and snags remaining in a burn after a fire affects the quality of the habitat for many species. Cavity-nesting bird density is likely to decline after snag removal in burns, as has been shown in the Sierra Nevada (Raphael and White 1984, Raphael et al. 1987). Logging burns may also decrease the quality of the habitat for tree nesters (e.g., Western Wood-Pewee, Overturf 1979).

DISCUSSION

We cannot offer managers as complete a synthesis as we would like. Too few studies have been conducted on the effects of silvicultural practices on birds in forests in the Rocky Mountains to make robust conclusions. Our results are limited in that they focus on short-term distributional changes as the result of two broad categories of timber harvesting (clearcutting and partial logging) lumped across conifer forests. The data indicate that many forest birds were less abundant in clearcuts than in uncut forests, and species that frequent open forests or open habitats were more abundant in clearcuts than in uncut forests. Most permanent residents were less abundant after either kind of harvesting treatment, whereas about half of the migrant species were less abundant and half more abundant in harvested areas. The effects of partial cutting were less dramatic than those of clearcutting; these results may be partly due to the fact that partial cutting included many different kinds of harvesting treatments.

Our information was limited to how distribution and abundance during the breeding season in general may be affected by clearcutting and partial cutting. We do not know what the actual effects of these harvest practices are on individual species (e.g., if nesting success for any individual species is lower in clearcuts than in uncut forests), what the effects of other silvicultural treatments are, how effects vary among forest-cover types and regions, and what landscape effects these treatments are creating. Because we examined studies of breeding habitat, studies on reproductive success in these habitats would help us interpret distributional patterns. We could not summarize results from other seasons because we found too little information on birds in other seasons. Understanding the effects of silvicultural practices on the distribution of individual species during nonbreeding seasons may be as critical in the maintenance of viable populations of these species.

Our clearest results were for common species. Whereas cumulative effects on each common species may be important in the long-term viability of that species, short-term effects may be greatest on uncommon species whose declines go unnoticed for lack of an adequate sample size. To demonstrate the effects of forest management on raptors, woodpeckers, and other species that are difficult to detect and/or have large home ranges will require intensive, individual species' studies of their density and demographics in treated and untreated areas (Hejl 1994). Therefore, our statements about those species should be viewed with caution. The effects of timber harvesting on rare species might be of even graver concern than we now know.

Neither long- nor short-distant migrants can be treated as a guild for managing forests.

Each species responded individually to silvicultural treatments. For example, some long-distance migrants (Swainson's Thrush and Townsend's Warbler) are found more often in unlogged forests and others (Broad-tailed Hummingbird and Dusky Flycatcher) in clearcuts when these areas are compared. Emphasizing any one silvicultural practice would favor some birds at the expense of others. Our results suggest that proportionally more resident than migrant species will be deleteriously affected by the loss of uncut forests.

The greatest effect of silvicultural practices may come through changes in landscape pattern. Unfortunately, too few studies have examined the effects on forest birds of silvicultural changes in Rocky Mountain landscapes for any conclusions to be made (Freemark et al., Chapter 14, this volume). Indeed, it is difficult to isolate the relative contribution of stand-level and landscape-level factors to a particular species' distribution in a logged area (Dobkin 1994). Some of the patterns that we noted for silvicultural treatments might be attributable to changes in the landscapes caused by those treatments.

Our dissatisfaction with this synthesis stems from the inadequate number of studies on the effects of silvicultural treatments on bird populations in the Rocky Mountains. We could only find studies that examined timber harvesting, not other silvicultural treatments. Within those studies, we did not have enough information to make specific conclusions for any logging treatment or stand age in a particular forest type or region. Not only are there too few studies in any one habitat or about a particular silvicultural treatment, but individual studies are often based on few replicates. Most authors did not deal with potential interpretive problems associated with observer differences. A large amount of variation or "noise" enters into our analysis as a result of having to lump across conifer habitats, harvesting methods, and studies based on different sampling methods and sample sizes. We could not examine increasing or decreasing trends that may be evident with gradations in severity of treatment.

We had enormous problems summarizing the literature, because the authors of studies often did not describe their control or treated areas very well. Patterns are undoubtedly confounded not only because of having to group survey data from all conifer forests into only two broad classes of harvesting methods, but because of the variety of postharvest treatments that may have been applied. Unfortunately, most studies failed to describe postharvest treatments. Indeed, we could not determine the preharvest age of the stand in many instances. To tease apart effects of all these variables over a broad range of vegetation cover types will take a lot more thoroughly described data than have been collected to this point in time.

Long-term study sites with many replicates in all habitats throughout the Rockies would help us assess the short-term and long-term effects of various silvicultural practices on as many individual species as possible. We need studies that are designed to distinguish between effects due to timber harvesting at the stand level versus the landscape level. Because most studies on western forest birds have shown great yearly fluctuations in bird numbers (Raphael and White 1984, Szaro and Balda 1986, Hejl et al. 1988), long-term studies at various locations are necessary to identify avifaunal changes due to timber management practices independent of weather and other factors. Basic autecological studies are needed to determine why a species responds as it does to habitat alteration. Of particular concern are species' responses to truncated succession, loss of early-successional and old-growth forests, loss of snags, especially in burns, and loss of all types of burned habitats [similar concerns to those in Thomas et al. (1975)].

Our judgment of an "effect" of timber harvesting is colored by the fact that we are comparing limited data from only one uncut and two cut vegetation types across the Rocky Mountains. Knowledge of the complete distribution of a species among habitats (more rigorously derived than our Table 8-1) and the distribution of habitats is required before we could say whether or not a local population decline in response to timber harvesting translates into a serious population problem. A certain level of decline presumably is much less serious a concern for a species that occurs over a broad range of additional habitat types than it is for a species confined to a single uncut forest type. We have presented preliminary information on species' distributions based on at least a few known, reported studies in each habitat, but more information is needed on bird distribution across habitats and the distribution of habitats themselves to assess which species are truly of concern.

Philosophically, determining the "effects of timber harvesting" is very complicated. The "effect" can be measured as either a short-term or a long-term consequence of the harvesting activity and on small or large spatial scales. Our review deals with short-term, small-scale consequences but the managers' goal should be one of placing these results in a long-term, broad-scale perspective (Bartlett and Jones 1992, Kessler et al. 1992) with a focus on managing the land in an ecological manner that will serve to sustain natural populations, abiotic and biotic interactions, patterns, and processes. We agree with the recommendation of Thompson et al. (Chapter 7, this volume) to make management decisions first at the large scale and secondarily at the small scale. From such a perspective, a manager might want to consider that a timber-harvesting practice that might immediately cause a relatively great amount of change from preharvest conditions may be one component of a strategy for maintaining populations of all wildlife species for the long term. In other words, we suggest that any one individual piece of a landscape might be managed to the detriment of some species and benefit of others, with the goal of maintaining enough variety within the different pieces of the landscape (i.e., in the constantly shifting mosaic of logging treatments and successional stages) that all native species are being managed simultaneously over a broad landscape [for an example from the Pacific Northwest, see Hof and Raphael (1993)]. While it is clearly important to emphasize the maintenance and restoration of old-growth forests, it is similarly important to consider the maintenance of early successional and other ages of forests. This concept may be especially important for areas that experience frequent and widespread disturbance, but such judgments require more knowledge of the way birds are affected by natural processes. Rather than simply asking what the short-term effect of a given harvest method is, we should also be asking which methods best operate to mimic natural patterns and processes, and how we can manage for those species that do not benefit from this approach; for a review of some information for the Northern Rockies, see Hejl (1992).

To illustrate the point of mimicking natural patterns and processes, consider that the Northern Rocky Mountain conifer forests are part of fire-maintained systems (Hejl 1992). Much less vegetation cover in early successional stages exists now than prior to fire control in some cover types (Gruell 1983). If, of all timber-harvesting practices, clearcuts come closest to matching the pattern of a naturally intense fire regime, then perhaps the method affecting the greatest change from preharvest conditions in an immediate sense (i.e., clearcuts) is the best practice in a long-term sense. We caution that we need hard data to answer this question, but to many bird species, clearcuts are *not* the same as intense canopy burns (Hutto 1995). "Sloppy" clearcuts (some snags and trees remaining) or selection cutting may come the closest to mimicking intense burns, depending upon forest cover type. Nonetheless, current thinking and current research efforts need to be directed along these lines if we are to make progress in managing the land for the maintenance of migratory landbirds, resident landbirds, all other plant and animal species, and their interactions (i.e., biological diversity).

We think that fire is so important as a creator of variety in landscapes that the conservation of native diversity may only be accomplished through the maintenance of fire as a process. Some bird species may simply need the maintenance of open forests as occurs with low-intensity fires (Marshall 1963). Frequent, low-intensity understory fires, however, do not satisfy the needs of all fire-dependent species. Some of these species probably rely on the presence of large, high-intensity crown fires that characterize the historical fire regime of many conifer forest types (Loope and Gruell 1973; Heinselman 1981, 1985).

Finally, we suggest the following goal for managing forests: to maintain natural bird populations, ecological patterns, and ecological processes over broad landscapes. Suggested steps to work toward that goal include: (1) maintain all habitats (e.g., forest-cover types and successional stages) and important habitat components (e.g., snags); (2) strive to mimic (either retain or restore within the range of variation of) presettlement ("natural") proportions and distribution of forest types, successional stages, and habitat components; (3) allow or reintroduce natural disturbance patterns (e.g., let fires burn or use prescribed fire); and (4) constantly monitor birds to see how this plan is working and redirect efforts if need be (with special emphasis for species that seem to be declining).

We emphasize sustaining species and ecosystems within a flexible framework (i.e., use adaptive management; Holling 1978), while acknowledging the constraints imposed by current landscape patterns. Current landscape patterns are the result of continual habitat modification. Burns have been salvage logged. Fire suppression has led to the change in forest structure and composition in many habitats in the Rocky Mountains and, in addition, a great proportion of old-growth forests have been logged. We merely suggest the above steps as goals. In future research efforts, we need to determine whether or not these steps will lead to the maintenance of forest bird populations in the Rocky Mountains.

ACKNOWLEDGMENTS

We are grateful for comments on floristic composition by Peter Stickney, for information on logging history from Steve Arno and Clint Carlson, for critical evaluation of the effects of silvicultural treatments on raptors by Joe Ganey, Denver Holt, and especially Greg Hayward and Richard Reynolds, for critical reviews by Kay Franzreb, Tom Martin, and two anonymous reviewers, and for assistance editing the final draft by Beth Beringer.

LITERATURE CITED

Alexander, R. R., and O. Engelby. 1983. Engelmann spruce/subalpine fir. Pp. 59–62 *in* Silvicultural systems for the major forest types of the United States (R. M. Burns, tech. comp.). USDA Forest Serv., Agricultural Handbook no. 445.

Alexander, R. R., J. E. Lotan, M. J. Larson, and L. A. Volland. 1983. Lodgepole pine. Pp. 63–66 *in* Silvicultural systems for the major forest types of the United States (R. M. Burns, tech. comp.). USDA Forest Serv., Agricultural Handbook no. 445.

Askins, R. A., J. F. Lynch, and R. Greenberg. 1990. Population declines in migratory birds in eastern North America. Curr. Ornithol. 7:1–57.

Austin, D. D., and M. L. Perry. 1979. Birds in six communities within a lodgepole pine forest. J. For. 77:584–586.

Balda, R. P., and N. Masters. 1980. Avian communities in the pinyon–juniper woodland: a descriptive analysis. Pp. 146–167 *in* Workshop proceedings: management of western forests and grasslands for nongame birds (R. M. DeGraff, tech. coord.). Gen. Tech. Rep. INT-86. USDA Forest Serv., Intermountain Forest Range Exp. Sta., Ogden, UT.

Bartlett, E. T., and J. R. Jones (eds). 1992. Rocky Mountain new perspectives, proceedings of a workshop. Gen. Tech. Rep. RM-200. USDA Forest Serv., Rocky Mt. Forest Range Exp. Sta., Fort Collins, CO.

Beebe, F. J. 1974. Falconiformes of British Columbia. Br. Col. Proc. Mus. Occ. Papers 17:1–163.

Blackford, J. L. 1955. Woodpecker concentration in burned forest. Condor 57:28–30.

Blake, J. G. 1982. Influence of fire and logging on nonbreeding bird communities of ponderosa pine forests. J. Wildl. Manag. 46:404–415.

Block, W. M., J. L. Ganey, K. E. Severson, and M. L. Morrison. 1992. Use of oaks by neotropical migratory birds in the Southwest. Pp. 65–70 *in* Ecology and management of oak and associated woodlands: perspectives in the southwestern United States and northern Mexico (P. F. Ffolliott, G. J. Gootfried, D. A. Bennett, C. Hernandez, V. Manuel, A. Ortega-Rubio, and R. H. Hamre, tech. coords). Gen. Tech. Rep. RM-218. USDA Forest Serv., Rocky Mt. Forest Range Exp. Sta., Fort Collins, CO.

Bock, C. E., and J. H. Bock. 1974. On the geographical ecology and evolution of the three-toed woodpeckers, *Picoides tridactylis* and *P. arcticus*. Amer. Midl. Natur. 92:397–405.

Bock, C. E., and J. F. Lynch. 1970. Breeding bird populations of burned and unburned conifer forest in the Sierra Nevada. Condor 72:182–189.

Brawn, J. D., and R. P. Balda. 1988. The influence of silvicultural activity on ponderosa pine forest bird communities in the southwestern United States. Bird Conserv. 3:3–21.

Bull, E. L., M. G. Henjum, and R. S. Rohweder. 1988. Nesting and foraging habitat of great gray owls. J. Raptor Res. 22:107–115.

Bull, E. L., A. L. Wright, and M. G. Henjum. 1990. Nesting habitat of flammulated owls in Oregon. J. Raptor Res. 24:52–55.

Burns, R. M. 1983. Introduction. Pp. 1–2 *in* Silvicultural systems for the major forest types of the United States (R. M. Burns, tech. comp.). USDA Forest Serv., Agricultural Handbook no. 445.

Catt, D. J. 1991. Bird communities and forest succession in the subalpine zone of Kootenay National Park, British Columbia. MS thesis, Simon Fraser University, British Columbia.

Crocker-Bedford, D. C. 1990. Goshawk reproduction and forest management. Wild. Soc. Bull. 18:262–269.

Daubenmire, R. F. 1943. Vegetational zonation in the Rocky Mountains. Botan. Rev. 9:325–393.

Davis, P. R. 1976. Response of vertebrate fauna to forestfire and clearcutting in south central Wyoming. PhD thesis, University of Wyoming, Laramie, WY.

Debyle, N. V. 1981. Songbird populations and clearcut harvesting of aspen in northern Utah. Res. Note INT-32. USDA Forest Serv., Intermountain Forest Range Exp. Sta., Ogden, UT.

Dobkin, D. S. 1994. Conservation and management of Neotropical migrant landbirds in the Northern Rockies and Great Plains. University of Idaho Press, Moscow, ID.

Evans, D. L. 1982. Status reports on twelve raptors. US Dept. Interior, Fish Wildl. Serv. Wildl. 238.

Finch, D. M., and R. T. Reynolds. 1987. Bird response to understory variation and conifer succession in aspen forests. Pp. 87–96 *in* Proceedings of a national symposium: issues and technology in the management of impacted wildlife (J. Emerick, S. Q. Foster, L. Hayden-Wing, J. Hodgson, J. W. Monarch, A. Smith, O. Thorne, II, and J. Todd, eds). Thorne Ecological Institute, Boulder, CO.

Flack, J. A. D. 1976. Bird populations of aspen forests in western North America. Ornithol. Monogr. no. 19.

Franzreb, K. E., and R. D. Ohmart. 1978. The effects of timber harvesting on breeding birds in a mixed-coniferous forest. Condor 80:431–441.

Ganey, J. L., and R. P. Balda. 1989. Home-range characteristics of spotted owls in northern Arizona. J. Wildl. Manag. 53:1159–1165.

Gleason, H. A., and A. Cronquist. 1964. The natural geography of plants. Columbia University Press, New York.

Graham, R. T., C. A. Wellner, and R. Ward. 1983. Mixed conifers, western white pine, and western redcedar. Pp. 67–69 *in* Silvicultural systems for the major forest types of the United States (R. M. Burns, tech. comp.). USDA Forest Serv., Agricultural Handbook no. 445.

Gruell, G. E. 1983. Fire and vegetative trends in the Northern Rockies: interpretations from 1871–1982 photographs. Gen. Tech. Rep. INT-158. USDA Forest Serv., Intermountain Forest Range Exp. Sta., Ogden, UT.

Habeck, J. R., and R. W. Mutch. 1973. Fire dependent forests in the northern Rocky Mountains. Quatern. Res. 3:408–424.

Hallock, D. 1989–1990. A study of breeding and winter birds in different age-classed lodgepole pine forests. Colorado Field Ornithol. J. 24:2–16.

Harris, M. A. 1982. Habitat use among woodpeckers in forest burns. MS thesis, University of Montana, Missoula, MT.

Hayward, G. D. 1989. Habitat use and population biology of boreal owls in the Northern Rocky Mountains, USA. PhD thesis, University of Idaho, Moscow, ID.

Hayward, G. D., P. H. Hayward, and E. O. Garton. 1993. Ecology of boreal owls in the Northern Rocky Mountains, USA. Wildl. Monogr. 124:1–59.

Heinselman, M. L. 1981. Fire and succession in the conifer forests of northern North America. Pp. 374–405 *in* Forest succession: concepts and applications (D. C. West, H. H. Shugart, and D. B. Botkin, eds). Springer-Verlag, New York.

Heinselman, M. L. 1985. Fire regimes and management options in ecosystems with large high-intensity fires. Pp. 101–109 *in* Proceeding—Symposium and workshop on wilderness fire (J. E. Lotan, B. M. Kilgore, W. C. Fischer, and R. W. Mutch, eds). Gen. Tech. Rep. INT-82. USDA Forest Serv., Intermountain Forest Range Exp. Sta., Ogden, UT.

Hejl, S. J. 1992. The importance of landscape patterns to bird diversity: a perspective from the Northern Rocky Mountains. Northwest Environ. J. 8:119–137.

Hejl, S. J. 1994. Human-induced changes in bird populations in coniferous forests in Western North America during the past 100 years. Stud. Avian Biol. 15:232–246.

Hejl, S. J., and R. E. Woods. 1991. Bird assemblages in old-growth and rotation-aged Douglas-fir/ponderosa pine stands in the northern Rocky Mountains: a preliminary assessment. Pp. 93–100 *in* Symposium Proceedings: Interior Douglas-fir: the species and its management (D. M. Baumgartner and J. E. Lotan, eds). Washington State University, Pullman, WA.

Hejl, S. J., J. Verner, and R. P. Balda. 1988. Weather and bird populations in true fir forests of the Sierra Nevada, California. Condor 90:561–574.

Hof, J. G., and M. G. Raphael. 1993. Some mathematical programming approaches for optimizing timber age-class distributions to meet multispecies wildlife population objectives. Can. J. Forest Resources 23: 828–834.

Holling, C. S. (ed.). 1978. Adaptive environmental assessment and management. John Wiley and Sons, London, England.

Holt, D. W., and M. Hillis. 1987. Current status and habitat associations of forest owls in western Montana. Pp. 281–286 *in* Symposium Proceedings: Biology and conservation of northern forest owls (R. W. Nero, R. J. Clark, R. J. Knapton, and R. H. Hamre, eds). Gen. Tech. Rep. RM-142. USDA Forest Serv., Rocky Mt. Forest Range Exp. Sta., Fort Collins, CO.

Howie, R. R., and R. Ritcey. 1987. Distribution, habitat selection, and densities of flammulated owls in British Columbia. Pp. 249–254 *in* Symposium Proceedings: Biology and conservation of northern forest owls (R. W. Nero, R. J. Clark, R. J. Knapton, and R. H. Hamre, eds). Gen. Tech. Rep. RM-142. USDA Forest Serv., Rocky Mt. Forest Range Exp. Sta., Fort Collins, CO.

Hutto, R. L. 1995. The composition of bird communities following stand-replacement fires in northern Rocky Mountain conifer forests. Conserv. Biol. (in press).

Johnsgard, P. A. 1990. Hawks, eagles, and falcons of North America: biology and natural history. Smithsonian Institution Press, Washington, DC.

Johnston, B. C. 1984. Aspen site classification in the Rocky Mountains. Pp. 14–24 *in* Proceedings of the aspen symposium, May 22–24, 1984 (J. Capp and L. Gadt, symp. coords). Colorado Springs, CO.

Jones, S. 1987. Breeding status of small owls in Boulder County, Colorado. Colorado Field Ornithol. J. 21:35.

Jones, S. 1991. Distribution of small forest owls in Boulder County, Colorado. Colorado Field Ornithol. J. 25:55–70.

Keller, M. E., and S. H. Anderson. 1992. Avian use of habitat configurations created by forest cutting in southeastern Wyoming. Condor 94:55–65.

Kendall, K. C., and S. F. Arno. 1990. Whitebark pine—an important but endangered wildlife resource. Pp. 264–273 *in* Proceedings of a symposium on whitebark pine ecosystems: ecology and management of a high-mountain resource (W. C. Schmidt and K. J. McDonald, comps). Gen. Tech. Rep. INT-270. USDA Forest Serv., Intermountain Forest Range Exp. Sta., Ogden, UT.

Kessler, W. B., H. Salwasser, C. W. Cartwright, Jr, and J. A. Caplan. 1992. New perspectives for sustainable natural resources management. Ecol. Applic. 2:221–225.

Loope, L. L., and G. E. Gruell. 1973. The ecological role of fire in the Jackson Hole area, northwestern Wyoming. Quatern. Res. 3:425–443.

Losensky, B. J. 1990. Historical uses of whitebark pine. Pp. 191–197 *in* Proceedings of a symposium on whitebark pine ecosystems: ecology and management of a high-mountain resource (W. C. Schmidt and K. J. McDonald, comps). Gen. Tech. Rep. INT-270. USDA Forest Serv., Intermountain Forest Range Exp. Sta., Ogden, UT.

Mannan, R. W., and E. C. Meslow. 1984. Bird populations and vegetation characteristics in managed and old-growth forests, northeastern Oregon. J. Wildl. Manag. 48:1219–1238.

Mannan, R. W., and J. J. Siegel. 1988. Bird populations and vegetation characteristics in immature and old-growth ponderosa pine forests, northern Arizona. Final report. School of Renewable Natural Resources, University of Arizona, Tucson, AZ.

Marshall, J. T., Jr. 1957. Birds of pine–oak woodland in southern Arizona and adjacent Mexico. Pacific Coast Avifauna 32.

Marshall, J. T., Jr. 1963. Fire and birds in the mountains of southern Arizona. Pp. 135–141 *in* Proceedings of the tall timbers fire ecology conference, Tall Timbers Research Station, Tallahassee, FL.

McClelland, B. R. 1980. Influences of harvesting and residue management on cavity-nesting birds. Pp. 469–496 *in* Proceedings of the symposium on environmental consequences of timber harvesting in Rocky Mountain coniferous forests. Gen. Tech. Rep. INT-90. USDA Forest Serv., Intermountain Forest Range Exp. Sta., Ogden, UT.

Medin, D. E. 1985. Breeding bird responses to diameter-cut logging in west-central Idaho. Res. Paper INT-355. USDA. Forest Serv., Intermountain Forest Range Exp. Sta., Ogden, UT.

Medin, D. E., and G. D. Booth. 1989. Responses of birds and small mammals to single-tree selection logging in Idaho. Res. Paper INT-408. USDA Forest Serv., Intermountain Forest Range Exp. Sta., Ogden, UT.

Meeuwig, R. O., and R. L. Bassett. 1983. Pinyon–juniper. Pp. 84–86 *in* Silvicultural systems for the major forest types of the United States (R. M. Burns, tech. comp.). USDA Forest Serv., Agricultural Handbook no. 445.

Mitchell, M. C., and A. Bratkovich. 1992. Bird abundance in old-growth forest fragments and harvested stands in the Kootenai National Forest. Internal report for USDA Forest Serv., Kootenai National Forest, Libby Ranger District.

Moore, K. R., and C. J. Henny. 1983. Nest site characteristics of three coexisting accipiter hawks in northeastern Oregon. Raptor Res. 17:65–76.

Moore, R. L. 1992. Breeding birds in old-growth forests and snag management for birds. Forest Plan monitoring project report. USDA Forest Serv., Gallatin National Forest, Bozeman, MT.

Morrison, M. L., L. S. Hall, J. J. Keane, A. J. Kuenzi, and J. Verner. 1993. Distribution and abundance of birds in the White Mountains, California. Great Basin Natur. 53:246–258.

Mueggler, W. F. 1985. Vegetation associations. Pp. 45–55 *in* Aspen: ecology and management in the western United States (N. V. DeByle and R. P. Winokur, eds). Gen. Tech. Rep. RM-119. USDA Forest Serv.

Mustian, A. P. 1977. History and philosophy of silviculture and management systems in use today. Pp. 3–11 *in* Proceedings of a workshop on uneven-aged silviculture and management in the western United States. Timber Management Research, USDA Forest Serv., Washington, DC.

O'Meara, T. E., J. B. Haufler, L. H. Stelter, and J. G. Nagy. 1981. Nongame wildlife responses to chaining of pinyon–juniper woodlands. J. Wildl. Manag. 45:381–389.

Overturf, J. H. 1979. The effects of forest fire on breeding bird populations of ponderosa pine forests of northern Arizona. MS thesis, Northern Arizona University, Flagstaff, AZ.

Palmer, D. A. 1986. Habitat selection, movements and activity of boreal and saw-whet owls. MS thesis, Colorado State University, Fort Collins, CO.

Palmer, R. 1988. Handbook of North America birds, Vols 4–5. Yale University Press, New Haven, CT.

Peet, R. K. 1988. Forests of the Rocky Mountains. Pp. 63–101 *in* North American terrestrial vegetation (M. G. Barbour and W. D. Billings, eds). Cambridge University Press, Cambridge, England.

Peterson, S. R. 1982. A preliminary survey of forest bird communities in northern Idaho. Northwest Sci. 56:287–298.

Pfister, A. R. 1980. Postfire avian ecology in Yellowstone National Park. MS thesis, Washington State University, Pullman, WA.

Raphael, M. G. 1987a. Nongame wildlife research in subalpine forests of the central Rocky Mountains. Pp. 113–122 *in* Management of subalpine forests: building on 50 years of research. Gen. Tech. Rep. RM-149. USDA Forest Serv., Rocky Mountain Forest Range Exp. Sta., Fort Collins, CO.

Raphael, M. G. 1987b. The Coon Creek wildlife project: effects of water yield augmentation on wildlife. Pp. 173–179 *in* Management of subalpine forests: building on 50 years of research. Gen. Tech. Rep. RM-149. USDA Forest Serv., Rocky Mt. Forest Range Exp. Sta., Fort Collins, CO.

Raphael, M. G., and M. White. 1984. Use of snags by cavity-nesting birds in the Sierra Nevada. Wildl. Monogr. 86:1–66.

Raphael, M. G., M. L. Morrison, and M. P. Yoder-Williams. 1987. Breeding bird populations during twenty-five years of postfire succession in the Sierra Nevada. Condor 89:614–626.

Reynolds, R. T. 1983. Management of western coniferous forest habitat for nesting accipiter hawks. Gen. Tech. Rep. RM-102. USDA Forest Serv., Rocky Mt. Forest Range Exp. Sta., Fort Collins, CO.

Reynolds, R. T., and B. D. Linkhart. 1987. The nesting biology of Flammulated owls in Colorado. Pp. 239–248 *in* Symposium proceedings: biology and conservation of northern forest owls (R. W. Nero, R. J. Clark, R. J. Knapton, and R. H. Hamre, eds). Gen. Tech. Rep. RM-142. USDA Forest Serv., Rocky Mt. Forest Range Exp. Sta., Fort Collins, CO.

Reynolds, R. T., and B. D. Linkhart. 1992. Flammulated owls in ponderosa pine: evidence of preference for old growth. Pp. 166–169 *in* Workshop on old-growth forests in the southwest and rocky mountain regions (M. R. Kaufmann, W. H. Moir, and R. L. Bassett, eds). Gen. Tech. Rep. RM-213.

USDA Forest Serv., Rocky Mt. Forest Range Exp. Sta., Fort Collins, CO.

Reynolds, R. T., and E. C. Meslow. 1984. Partitioning of food and niche characteristics of coexisting *Accipiter* during breeding. Auk 101:761–779.

Reynolds, R. T., E. C. Meslow, and H. M. Wight. 1982. Nesting habitat of coexisting *Accipiter* in Oregon. J. Wildl. Manag. 46:124–138.

Reynolds, R. T., R. T. Graham, M. H. Reiser, R. L. Bassett, P. L. Kennedy, D. A. Boyce, Jr, G. Goodwin, R. Smith, and E. L. Fisher. 1992. Management recommendations for the Northern Goshawk in the southwestern United States. Gen. Tech. Rep. RM-217. USDA Forest Serv., Rocky Mt. Forest Range Exp. Sta., Fort Collins, CO.

Richmond, M. L., L. R. Deweese, and R. E. Pillmore. 1980. Brief observations on the breeding biology of the Flammulated Owl in Colorado. Western Birds 11:35–46.

Ronco, F., Jr, and K. L. Ready. 1983. Southwestern ponderosa pine. Pp. 70–72 *in* Silvicultural systems for the major forest types of the United States (R. M. Burns, tech. comp.). USDA Forest Serv., Agricultural Handbook no. 445.

Ryder, R. A., D. A. Palmer, and J. J. Rawinski. 1987. Distribution and status of the Boreal Owl in Colorado. Pp. 169–174 *in* Symposium proceedings: biology and conservation of northern forest owls (R. W. Nero, R. J. Clark, R. J. Knapton, and R. H. Hamre, eds). Gen. Tech. Rep. RM-142. USDA Forest Serv., Rocky Mt. Forest Range Exp. Sta., Fort Collins, CO.

Ryker, R. A., and J. Losensky. 1983. Ponderosa pine and Rocky Mountain Douglas-fir. Pp. 53–55 *in* Silvicultural systems for the major forest types of the United States (R. M. Burns, tech. comp.). USDA Forest Serv., Agricultural Handbook no. 445.

Salt, G. W. 1957. An analysis of avifaunas in the Teton Mountains and Jackson Hole, Wyoming. Condor 59:373–393.

Schmidt, W. C., R. C. Shearer, and J. R. Naumann. 1983. Western larch. Pp. 56–58 *in* Silvicultural systems for the major forest types of the United States (R. M. Burns, tech. comp.). USDA Forest Serv., Agricultural Handbook no. 445.

Schmutz, J. K., S. M. Schmutz, and D. A. Boag. 1980. Coexistence of three species of hawks (*Buteo* spp.) in the prairie-parkland ecotone. Can. J. Zool. 58:1075–1089.

Scott, V. E., and G. L. Crouch. 1987. Response of breeding birds to commercial clearcutting of aspen in southwestern Colorado. Res. Note RM-475. USDA Forest Serv., Rocky Mt. Forest Range Exp. Sta., Fort Collins, CO.

Scott, V. E., and G. L. Crouch. 1988. Summer birds and mammals of aspen-conifer forests in west-central Colorado. Res. Paper RM-280. USDA Forest Serv., Rocky Mt. Forest Range Exp. Sta., Fort Collins, CO.

Scott, V. E., and G. J. Gottfried. 1983. Bird response to timber harvest in a mixed conifer forest in Arizona. Res. Paper RM-245. USDA Forest Serv., Rocky Mt. Forest Range Exp. Sta., Fort Collins, CO.

Scott, V. E., G. L. Crouch, and J. A. Whelan. 1982. Responses of birds and small mammals to clearcutting in a subalpine forest in central Colorado. Res. Note RM-422. USDA Forest Serv., Rocky Mt. Forest Range Exp. Sta., Fort Collins, CO.

Sedgwick, J. A., and R. A. Ryder. 1987. Effects of chaining pinyon–juniper on nongame wildlife. Pp. 541–551 *in* Proceedings—pinyon–juniper conference (R. L. Everett, comp.). Gen. Tech. Rep. INT-215. USDA Forest Serv., Intermountain Forest Range Exp. Sta., Ogden, UT.

Shepperd, W. D., and O. Engelby. 1983. Rocky Mountain aspen. Pp. 77–79 *in* Silvicultural systems for the major forest types of the United States (R. M. Burns, tech. comp.). USDA Forest Serv., Agricultural Handbook no. 445.

Skinner, N. G. 1989. Seasonal avifauna use of burned and unburned lodgepole pine forest ecotones. MS thesis, University of Montana, Missoula, MT.

Siegel, J. J. 1989. An evaluation of the minimum habitat quality standards for birds in old-growth ponderosa pine forests, northern Arizona. MS thesis, University of Arizona, Tuscon, AZ.

Szaro, R. C., and R. P. Balda. 1986. Relationships among weather, habitat structure, and ponderosa pine forest birds. J. Wildl. Manag. 50:253–260.

Taylor, D. L., and W. J. Barmore, Jr. 1980. Post-fire succession of avifauna in coniferous forests of Yellowstone and Grand Teton National Parks, Wyoming. Pp. 130–145 *in* Workshop proceedings: management of western forests and grasslands for nongame birds (R. M. DeGraff, tech. coord.). Gen. Tech. Rep. INT-86. USDA Forest Serv., Intermountain Forest Range Exp. Sta., Ogden, UT.

Thomas, J. W., G. L. Crouch, R. S. Bumstead, and L. D. Bryant. 1975. Silvicultural options and habitat values in coniferous forests. Pp. 272–287 *in* Proceedings of the symposium on management of forest and range habitats for nongame birds (D. Smith, tech. coord.). Gen. Tech. Rep. WO-1. USDA Forest Serv.

Thomas, J. W., E. D. Forsman, J. B. Lint, E. C. Meslow, B. R. Noon, and J. Verner. 1990. A conservation strategy for the Northern Spotted Owl: Report of the Interagency Scientific Committee to address the conservation of the Northern Spotted Owl. US Government Printing Office, Washington, DC.

Tobalske, B. W., R. C. Shearer, and R. L. Hutto. 1991. Bird populations in logged and unlogged western larch/Douglas-fir forest in northwestern Montana. Res. Paper INT-442. USDA Forest Serv., Intermountain Forest Range Exp. Sta., Ogden, UT.

Webb, B. 1982. Distribution and nesting requirements of montane forest owls in Colorado. Part III: Flammulated Owl (*Otus flammeolus*). Colorado Field Ornithol. J. 16:76–82.

Wetmore, S. P., R. A. Keller, and G. E. J. Smith. 1985. Effects of logging on bird populations in British Columbia as determined by a modified point-count method. Can. Field Natur. 99:224–233.

Winternitz, B. L. 1976. Temporal change and habitat preference of some montane breeding birds. Condor 78:383–393.

9

SILVICULTURE IN CENTRAL AND SOUTHEASTERN OAK–PINE FORESTS

JAMES G. DICKSON, FRANK R. THOMPSON, III, RICHARD N. CONNER, AND KATHLEEN E. FRANZREB

INTRODUCTION

The objective of this chapter is to document the Neotropical migratory bird NTMB communities of eastern pine–oak and oak–hickory forests and provide information on how silviculture practices affect those bird communities. The area covered is the southeastern pine–oak forests from Virginia to eastern Oklahoma and eastern Texas and the oak-dominated forests northerly, particularly the midwestern United States.

Eastern pine–oak and oak–hickory forest are characterized by conifers and hardwoods, particularly pines and oaks in the south, oaks and hickories northerly from the east to midwest, and bottomland hardwoods along rivers and streams. Climate is generally mild and annual precipitation averages about 100–150 cm. The moderate climate and diverse vegetative systems support abundant and diverse wildlife communities. The current Neotropical migrants found in this region represent a group of robust species that were able to survive the drastic land use practices of the 1800s and early 1900s. Species associated with old-growth forests have declined. Exotics, and species associated with man-altered habitats and younger forests have generally increased over the long term (Smith and Petit 1988). Silviculture, or the culture of trees, is the manipulation of forest stand establishment, composition, and growth. Even-aged stands are regenerated by clearcuts, seedtree, or shelterwood cuts, uneven-aged stands usually are regenerated by individual tree or group selection. Details are covered in Thompson et al. (Chapter 7, this volume). Particular techniques employed in each major ecosystem will be addressed in each appropriate following section.

Studies have shown bird communities are associated with physiography and vegetative structure, such as moisture gradient (Smith 1977), vertical foliage layers (MacArthur and MacArthur 1961), foliage volume (Willson 1974), habitat patchiness (Roth 1976) or edge (Strelke and Dickson 1980), and successional stage of stand (Shugart and James 1973, Dickson and Segelquist 1979). In general, the number of bird species and the density are positively related to the stand foliage volume and diversity. In southern pine and hardwood stands, bird density and diversity are usually high in young brushy stands, decrease in dense pole stands as canopies close and shade out understories, and are highest in older stands with diverse vegetation. Bird species in forest stands have specific habitat requirements and any changes to stand characteristics influence the bird communities. The hardwood component of pine stands is the main determinant of the bird community composition and abundance (Johnston and Odum 1956, Dickson and Segelquist 1979). Tree harvesting drastically alters bird habitat and bird communities (Webb et al. 1977, Crawford et al. 1981, McComb et al. 1989, Thompson et al. 1992). Generally, partial removal of a forest overstory results in decreases in some species, increases in others, and little change in relative abundance of other species. Landscape–bird community relationships are treated in several other chapters of this publication.

CENTRAL HARDWOOD FORESTS

Forest Composition

The central hardwood forest is one of the largest forest areas in the United States. Oak–hickory forests dominate the area but give way to mixed hardwoods in the east and oak–pine forest to the south (Eyre 1980, Sander and Fischer 1989). The oak–hickory type is the most widespread in the region. The dominant species are white, black, scarlet, and northern red oak (scientific names of plants in Appendix) (Braun 1950, Eyre 1980, Sander and Fischer 1989). The oak–pine forest type is very similar to the oak–hickory type except shortleaf, loblolly, pitch, and Virginia pine make up 25–50% of the forest. Mixed hardwoods are found on moister, more productive sites primarily east of the Mississippi River. Principal species are yellow poplar, white oak, northern red oak, and sugar maple (Braun 1950, Eyre 1980, Sander and Fischer 1989).

History

The present central hardwood forest has been influenced strongly by human disturbance. Fire and logging were important perturbations in the region. There is evidence of repeated burning of moister oak forests of the east as well as the drier forest prairie interface along the western boundary of the region. Native Americans and early European settlers practiced slash and burn agriculture. There is some debate concerning to what extent fire affected oak forests (Steyermark 1959, Beilman and Brenner 1951). However, there is evidence of frequent fires in some parts of the region. In the Missouri Ozarks, fires occurred most frequently between 1785 and 1810 during a period of influx of Native Americans and European explorers. This period was followed by a period of decreased fire frequency beginning with an exodus of Native Americans around 1815 (Guyette and Cutter 1991).

Logging had a substantial impact on the area through the 1800s as it accompanied human expansion through the area. For instance, with the influx of settlers into the Ozarks from 1850 to 1874 came the railroad and a period of accelerated logging (Smith and Petit 1988). The lumber boom came to an end in the Ozarks in the early 1900s and much of the land was sold by lumber companies sight unseen through the mail as farmland. Widespread burning and open-range grazing of the forest occurred up until the 1930s in parts of the region. During the 1930s most of the land became tax delinquent and some was purchased by the US Forest Service. These past practices have resulted in widespread oak-dominated stands (Parker and Weaver 1989).

The amount of forest in the region declined due to clearing for agriculture and urban expansion through the first half of this century, but now this decline has leveled off and is reversing in some parts of the region. Recent forest inventories of central hardwood states show 7.5–31% increases in the amount of forest land from the 1960s to 1980s (Smith and Golitz 1988, Raile and Leatherberry 1988, Brand and Walkowiak 1991), primarily the result of farmland and pasture reverting to forest. There are also trends in forest composition indicating central hardwood forests are becoming older and shifting to shade tolerant species such as maple and beech (Raile and Leatherberry 1988, Smith and Golitz 1988).

Current Silvicultural Practices

Both even-aged and uneven-aged silvicultural systems are used in central hardwood forests. Central hardwood forests are almost exclusively regenerated by natural regeneration, of which the primary source is advance (existing) reproduction instead of stump and root sprouts or seed (Johnson 1989, Smith and Sander 1989). In the 1960s, even-aged management, usually with clearcutting, was widely adopted throughout the central hardwood region because it met the ecological requirements of the commercially valuable, shade intolerant species (oaks, ash, cherry, poplar), it was economical, and was simple to implement. Clearcutting is most successful for regenerating oaks in drier ecosystems that accumulate advance oak reproduction, and less successful in more mesic ecosystems, which may accelerate succession of oak-dominated forests to mixed mesophytic (Johnson 1993). The use of clearcutting has declined recently, particularly on public lands, in favor of other methods that focus less on commodities and more on other values (Salwasser 1990). The shelterwood system is being used as one alternative to clearcutting because it may be useful in regenerating oaks in more mesic ecosystems and may be more esthetic than clearcutting. The seedtree method is not recommended for regenerating oaks because it provides too little reproduction, but may be useful in sustaining acorn production for wildlife and creating structural diversity in forest stands.

Group selection is increasing in use as an alternative to even-aged systems and clearcutting. It is potentially suitable for commercially valuable, shade-intolerant species because it creates regeneration openings in the main canopy, while maintaining some characteristics of mature forest. Single-tree selection has generally been considered inappropriate for managing oak forests (Sander and Clark 1971) because it encourages the development of more shade-tolerant species. However, it may be used in oak forests together with group selection (Law and Lorimer 1989). On private, nonindustrial forest lands, diameter limit cutting or "high grading" is still commonly practiced.

Intermediate treatments in central hardwood forests are usually only economically justified on high quality sites with special species. Precommercial treatments are usually limited to crop tree release, and commercial treatments include thinning.

Table 9-1. Abundance[a] of Neotropical migratory birds in central hardwood forests[b] (reproduced from Dickson et al. 1993).

Species	Stand Age[c]					
	R	S	P	M	G	T
Whip-poor-will	U	U	U	U	U	U
Ruby-throated Hummingbird	C	N	N	N	?	N
Acadian Flycatcher	N	N	C	A	N	A
Eastern Wood-Pewee	N	N	U	A	N	A
Eastern Phoebe[d]	N	N	U	U	N	U
Great Crested Flycatcher	C	C	C	C	C	C
Blue-gray Gnatcatcher	A	C	C	C	C	C
Eastern Bluebird[d]	C	N	N	N	N	N
Wood Thrush	U	C	C	C	U	C
Gray Catbird	C	C	N	N	?	N
White-eyed Vireo	C	C	N	N	?	N
Yellow-throated Vireo	N	N	N	U	N	U
Red-eyed Vireo	U	U	A	A	U	A
Blue-winged Warbler	A	C	N	N	?	N
Golden-winged Warbler	C	U	N	N	?	N
Northern Parula	N	N	U	C	N	C
Chestnut-sided Warbler	C	C	N	N	?	N
Yellow-throated Warbler	N	N	U	U	N	U
Prairie Warbler	A	C	N	N	?	N
Black-and-white Warbler	C	C	C	C	C	C
Worm-eating Warbler	U	C	C	C	C	C
Ovenbird	U	C	C	C	U	U
Louisiana Waterthrush	N	U	C	C	C	C
Common Yellowthroat	A	U	N	N	?	N
Kentucky Warbler	A	C	U	U	A	C
Hooded Warbler	C	C	U	U	C	C
Yellow-breasted Chat	A	C	N	N	?	N
Orchard Oriole	U	N	N	N	N	N
Summer Tanager	C	C	C	A	C	C
Scarlet Tanager	U	U	C	A	U	A
Indigo Bunting	A	C	U	U	A	C
Rufous-sided Towhee[d]	A	U	N	N	C	N
Brown-headed Cowbird[d]	A	C	C	C	C	C
American Goldfinch[d]	U	N	N	N	N	N

[a] A = abundant; C = common or regular; P = present; U = uncommon; N = not present.
[b] Includes oak–hickory, mixed hardwood, and oak–pine forest types. Habitat associations based on Conner and Adkisson (1975), Conner et al. (1979), Dickson and Segelquist (1979), Dickson et al. (1980), Evans and Kirkman (1981), Yahner (1986), Thompson and Fritzell (1990), Thompson et al. (1992), Whitehead unpublished data, Robinson unpublished data.
[c] R = regeneration; S = sapling; P = poletimber; M = mature; G = group selection; T = single-tree selection.
[d] Undenoted species breed in North America and winter primarily south of the United States, denoted species (*d*) breed and winter extensively in North America, although some populations winter south of the United States.

Effects of Even-aged Management

Regenerating Stands

After harvesting, herbaceous ground cover develops quickly, but the harvested areas become dominated by tree regeneration from sprouts and advance regeneration resulting in as many as 25,000 stems/ha (Gingrich 1971). The first year after clearcutting there is usually a drastic reduction in total bird numbers and a nearly complete turnover in species (Table 9-1). American Goldfinch and Field Sparrow often prefer stands of dominant grass/forb vegetation. As tree regeneration dominates the site, Yellow-breasted Chat, Indigo Bunting, Prairie Warbler, Blue-winged Warbler, Kentucky Warbler, Common Yellowthroat, White-eyed Vireo, Gray Catbird, and Rufous-sided Towhee occupy the stands (Thompson et al. 1992). To the east, Chestnut-sided and Hooded warblers may also be common. In shelterwood and seedtree cuts, and clearcuts with residual live trees and snags, some mature-forest, canopy-dwelling species may continue to use the stand.

Sapling Stands

From age 10–20 years the stands are dominated by tree saplings with a closed canopy. At around age 20, the number of stems has been reduced to 3400–6200/ha and the larger trees on good sites have reached 18 cm diameter breast height (dbh)(Gingrich 1971). Many birds typical of regenerating stands persist in these stands at lower densities (Thompson et al. 1992). Black-and-white, Worm-eating, and Kentucky warblers seem to prefer the high stem densities and closed canopies of this age class. Ovenbird, Wood Thrush, and Red-eyed Vireo may begin using stands at this age.

Poletimber Stands

From age 20–60 years, some 90% of the trees usually die due to competition for space and light. The canopy remains closed and there is little understory development. As a result, common species tend to be canopy nesters such as Red-eyed Vireo, Scarlet Tanager, Eastern Wood-Pewee, and Wood Thrush, or ground nesters such as Ovenbird and Black-and-white Warbler (Thompson et al. 1992).

Mature Stands

Mature forest structure and composition varies widely throughout the region. Depending on soils, geology, climate, and geography, mature stands may have sparse to dense groundcover and understory. Decay and deaths of large trees result in cavities, snags, and tree fall gaps not present in short rotation stands. Rotations for central hardwood stands managed for sawtimber are usually 60–120 years. At age 80 dominant trees will range from 30 to 46 cm dbh. If left undisturbed these stands will slowly become uneven-aged as they age and individual trees die. However, because of widespread logging, burning, and grazing of this region in the late 1800s and early 1900s, much of the mature forest in the region is even-aged ranging from 60 to 100 years old. There are no known obligate old-growth NTMBs in these forests. The most abundant species throughout the region in mature forests is the Red-eyed Vireo. Other abundant or common species in this age class include Eastern Wood-Pewee. Acadian Flycatcher, Blue-gray Gnatcatcher, Ovenbird, Worm-eating Warbler, Scarlet and Summer tanagers, and Bluejay (Thompson et al. 1992). Pine and Yellow-throated warblers are common in oak–pine stands.

Changes in Stand Composition

Tree species composition usually remains consistent after harvest in central hardwood stands because of advance reproduction and stump sprouts. Small changes in tree-species composition have little effect on breeding birds because it is generally believed most birds select breeding habitat by vegetation structure. Past practices of converting low-value hardwood stands to pine have been largely abandoned on public lands but may persist on some private lands. Pine plantations generally support a lower density and diversity of breeding birds than stands of mixed species because of their structural simplicity.

Effects of Uneven-aged Management

Single and Multi-tree Gaps

Canopy gaps resulting from the harvest of single trees or groups of trees provide habitat for a variety of NTMBs. Species such as the Hooded Warbler and Indigo Bunting appear to make use of small gaps typical of single-tree selection whereas other species such as Yellow-breasted Chat, Blue-winged Warbler, and Prairie Warbler require large openings more typical of clearcuts (F. Thompson, personal observation). There is a dearth of information on the area sensitivity of species requiring early successional forest or gaps. These canopy gaps also may be attractive to cowbirds and NTMB predators.

Change in Stand Structure

Uneven-aged stands have a well-developed understory and subcanopy because of frequent canopy gaps. The presence of several well-developed vegetation levels and more complex habitat structure than similar-aged even-aged stands could result in higher within-stand bird species diversity. For instance, selectively cut stands in Illinois contained NTMBs associated with mature forest habitats as well as some species associated with young second growth.

There are indications that uneven-aged stands may be a poor habitat for some mature forest species. There were fewer Ovenbirds in mature uneven-aged than in even-aged forests, and all Red-eyed Vireo males in a selectively cut stand in Missouri were unpaired, compared to 80% paired in a nearby uncut mature forest (Ziehmer 1992).

Management Recommendations

General recommendations are provided at the end of this chapter. Central hardwood forests range from large forested areas with low levels of cowbird parasitism and nest predation (i.e., the Missouri Ozarks; J. Faaborg and R. Clawson, unpublished data) to highly fragmented forest with some of the worst reported levels of parasitism and predation (i.e., southern Illinois; Robinson, unpublished data). In fragmented landscapes with high brood parasitism or nest predation, there are three basic management options: (1) reforestation to reduce cowbirds and predators associated with agricultural and urban–suburban land uses; (2) direct control of cowbirds or predators; or (3) management of habitats as population sinks for some species in some years.

LOBLOLLY–SHORTLEAF PINE FORESTS

Forest Composition

This ecosystem is characterized by a species composition of at least 50% pines (either loblolly, shortleaf, or a mix). Associated species include oaks, hickories, sweetgum, blackgum, winged elm, and red maple (Garrison et al. 1977). The degree of hardwood inclusion in these forests is largely determined by past frequency and intensity of natural and prescribed fire. More than 60% of the area of this ecosystem is forested and about 20% is in cropland.

History

The loblolly–shortleaf pine ecosystem has been severely impacted by human activities. By 1860 more than 40% of the lands in the South had been harvested and converted to croplands (USDA Forest Service 1988). The harvesting continued throughout the South until it peaked around 1920, and soon after the old-growth timber of the uncut forest was gone (Maxwell and Baker 1983). Only a few small tracts scattered over the South remained as representatives of this ecosystem. Some lands that were cleared for agriculture, but found unsuitable, were abandoned during the late 1800s and early 1900s. The boll weevil and agricultural depression of the early 1920s caused many other hectares of crop and pasture lands to become idle (USDA 1988) and reforested through natural regeneration or planting.

Another major impact on the loblolly–shortleaf ecosystem was the reduction of the frequency and intensity of fire normally associated with the ecosystem. Croplands, roads, and pastures acted as fire breaks. Longleaf pine stands, which carried fires well, was declining, and their decline diminished the spread of fire into the loblolly–shortleaf stands. Many fires were fought and extinguished before they covered much area. With this reduction of fire came an increase in hardwoods within the loblolly–shortleaf pine ecosystem. Later, in the 1970s and 1980s, the value of fire in maintaining the ecosystem was beginning to be realized and use of fire became more accepted.

Current Silvicultural Practices

Harvesting of loblolly and shortleaf pines and subsequent stand regeneration can be successfully accomplished with both even- and uneven-aged silvicultural techniques (Baker and Balmer 1983, Lawson and Kitchens 1983, Society of American Foresters 1981,

Reynolds et al. 1984, Murphy and Farrar 1985). Throughout the South, clearcutting followed by planting has been the most widely used harvesting technique in the loblolly-shortleaf ecosystem. Seedtree and shelterwood are other even-aged harvesting methods that regenerate stands through natural seeding. These methods are used much less often than clearcutting but recently have increased on National Forests. However, residual pines left standing during the initial harvest as seed sources are subsequently removed 1–5 years after adequate seedling establishment. Thus, from a forest structure point of view, both seedtree and shelterwood harvesting provide a savannah-like forest for Neotropical migrants for only a few years, then take on the structural characteristics of a regenerating clearcut. Clearcutting and planting remains the regeneration method of choice on private commercial forest lands.

Rotation ages for even-aged management vary between public and private industrial forest lands. Loblolly pine currently is harvested after reaching about 70 years of age on National Forest lands, whereas shortleaf pine rotation age is 80 years. Private commercial pine forests are harvested at about 35–40 years for saw logs and lumber, and between 20 and 30 years for pulp wood. Cut sizes on National Forest lands are typically limited to less than 32 ha but cut sizes on private industrial lands are more variable in size.

Even-aged pine stands are often thinned prior to reaching final commercial maturity. Thinnings can be precommercial, where some trees are cut and left on the ground, or commercial thinnings where trees are large enough to be of economic value. Thinning opens the stand up, releases uncut trees for faster diameter and crown growth, and stimulates growth of understory vegetation because more sunlight reaches the ground.

During the past 30 years uneven-aged timber management has rarely been implemented on both public and private industrial forest lands because of the economic advantages of clearcutting. Both group selection and single-tree selection are feasible harvest methods for both loblolly and shortleaf pines (Baker and Balmer 1983, Society of American Foresters 1981, Reynolds et al. 1984, Murphy and Farrar 1985). Structurally, group selection can create small clearcuts up to 0.8 ha in size and result in a mosaic of small forest patches of different ages. Single-tree selection creates a more homogeneous stand but has greater structural diversity at any one location in the stand.

There are difficulties in maintaining pines with implementation of uneven-aged management regeneration techniques. Natural plant succession favors hardwoods. Fire kills many hardwoods but it can also kill younger loblolly and shortleaf pines, destroying pine regeneration and younger age classes of pines. In a managed forest using single-tree selection, prescribed fire would have to be withheld from stands until pines had grown large enough with thick enough bark to withstand the heat. This could create major gaps between age classes of subsequent generations of pines and permit considerable hardwood invasion of stands between fires. Many hardwoods would grow too large for fire to kill and could result in hardwood dominance. Logistically, it would be very difficult to implement prescribed fire in group selection harvesting because of the patchwork of young stands. Currently, hardwoods growing in the few uneven-aged pine stands are controlled with applications of herbicides.

Bird Communities

The loblolly–shortleaf pine ecosystem provides habitat for a diverse array of birds, many of which are Neotropical migrants. In general, two major characteristics of the forest are important to Neotropical migrants. Forest structure includes both vertical and horizontal presence or absence of foliage, and the relative patchiness of the distribution of foliage in these dimensions. Tree species composition and distribution are also important determinants of avian species occurrence. Most Neotropical migrants that breed in the loblolly–shortleaf ecosystem are associated with deciduous foliage (e.g., Dickson and Segelquist 1979, Conner et al. 1983), but some migrants find coniferous foliage attractive. Thus, the effect of forest management on tree species, and the abundance and distribution/patchiness of foliage will determine NTMB species.

Effects of Even-aged Management

Early successional stands are characterized by little vertical foliage diversity. As forest stands age, height and additional vertical layers of foliage are added until the forest again reaches maturity. Different stages of stand development are attractive to different species of birds.

Seedling and Sapling Stands

Breeding season. Clearcutting today and over the past several decades has produced habitat conditions similar to those created by large natural landscape-level disturbances. A diverse group of Neotropical migrants (e.g., Prairie Warblers, Field Sparrows, Blue Grosbeaks) are attracted to the youngest (0–3-year-old) stands during the breeding season, many responding to the presence of hardwoods that grow among the young pines (Table 9-2) (Noble and Hamilton 1976, Meyers and Johnson 1978, Conner et al. 1979, 1983; Dickson and Segelquist 1979, Darden 1980, Dickson et al. 1980, 1984, Childers et al. 1986). When the stands become 3 years old, the number of neotropical migrants using them for breeding sites increases (Dickson et al. in press). Indigo Buntings, Painted Buntings, White-eyed Vireos, Yellow-breasted Chats, and Common Yellowthroats are some of the more common species found in the well-developed shrubby vegetation of young clearcuts.

As even-aged stands develop, foliage patchiness in the 0–3 m layer increases and faster growing trees reach heights of 4–5 m. A few additional species of Neotropical migrants begin to occupy the older sapling loblolly–shortleaf stands. In the northern portion of the ecosystem, Ovenbird, Rufous-sided Towhee, Black-and-white Warbler, and American Redstart begin to appear. Further south, only Black-and-white Warblers are added in any numbers. During this later sapling stage of succession, some species, such as Field Sparrow and Blue Grosbeak, begin to disappear as foliage dominates old-field type spots of bare ground and grasses.

Table 9-2. Abundance[a] of Neotropical migratory birds in loblolly–shortleaf pine forests[b] (reproduced from Dickson et al. 1993).

Species	Stand Age[c]				
	R	S	P	M	O
Whip-poor-will	N	N	N	N	?
Ruby-throated Hummingbird	U	U	U	U	U
Acadian Flycatcher	N	N	U	C	C
Eastern Wood-Pewee	N	U	P	C	A
Eastern Phoebe[d]	N	N	N	N	?
Great Crested Flycatcher	N	N	U	P	C
Blue-gray Gnatcatcher	N	N	U	C	A
Eastern Bluebird[d]	U	U	N	N	P
Wood Thrush	N	N	U	C	A
American Robin[d]	N	U	U	U	U
Gray Catbird	U	U	N	U	U
White-eyed Vireo	U	A	P	U	C
Yellow-throated Vireo	N	N	U	A	C
Red-eyed Vireo	N	U	C	A	A
Blue-winged Warbler	N	N	N	N	N
Golden-winged Warbler	N	N	N	N	N
Northern Parula	N	N	U	U	C
Chestnut-sided Warbler	N	N	N	N	N
Prairie Warbler	C	A	N	N	U
Black-and-white Warbler	N	U	C	C	C
Worm-eating Warbler	N	N	C	C	C
Chuck-will's-widow	U	U	U	U	U
Ovenbird	N	N	U	C	C
Louisiana Waterthrush	N	N	N	P	P
Kentucky Warbler	N	U	C	P	P
Hooded Warbler	N	U	C	A	C
Yellow-breasted Chat	C	A	U	N	P
Summer Tanager	N	N	U	C	C
Scarlet Tanager	N	N	U	U	U
Indigo Bunting	N	A	P	U	P
Rufous-sided Towhee[d]	N	P	C	C	P
Brown-headed Cowbird[d]	P	C	P	P	N
American Goldfinch[d]	U	U	U	U	U
Blue Grosbeak	C	U	N	N	N

[a] See footnote *a* of Table 9-1.
[b] Includes oak–hickory, mixed hardwood, and oak–pine forest types. Habitat associations based on Conner and Adkisson (1975), Conner et al. (1979), Crawford et al. (1981), Yahner (1986), Thompson and Fritzell (1990), Thompson et al. (1992), Hamel (1992), Robinson (unpublished data), Whitehead (unpublished data).
[c] R = regeneration; S = sapling; P = poletimber; M = mature; O = oldgrowth.
[d] See footnote *d* of Table 9-1.

Winter. Many migrants use sapling loblolly–shortleaf pine stands as wintering habitat (Noble and Hamilton 1976, Dickson and Segelquist 1977). Winter Wren, Brown Thrasher, American Robin, Hermit Thrush, Eastern Bluebird, Ruby-crowned and Golden-crowned kinglets, Pine and

Yellow-rumped warblers, Dark-eyed Junco, and Field, Song, Lincoln's, and White-throated sparrows all use young pine plantations during winter.

Pole Stands

Breeding season. When young pines and hardwoods reach the pole stage (approximately, 12–25 years old), most early successional migrants are no longer found. The taller foliage gradually forms a canopy that reduces light penetration to the understory. A few White-eyed Vireos and Yellow-breasted Chats remain in some areas, such as wind rows, where some open patches remain. Kentucky and Black-and-white warblers are fairly common in pole stands. Red-eyed Vireo, Worm-eating, Pine, and Hooded warblers, Yellow-billed Cuckoo, and Summer Tanager begin to appear in the pole stands and sing as if on territory (Noble and Hamilton 1976, Meyers and Johnson 1978, Conner et al. 1979, 1983, Dickson and Segelquist 1979, Darden 1980, Dickson et al. 1980, 1984, Childers et al. 1986). Their actual productivity in this and other successional stages is unknown. Occasionally, Wood Thrushes and Brown Thrashers are detected in such stands.

Winter. Both pine and pine–hardwood pole stands are used by a variety of migrant birds during winter. Brown Creeper, Winter Wren, Hermit Thrush, Eastern Bluebird, Golden-crowned and Ruby-crowned kinglets, Black-and-white, Pine, and Yellow-rumped warblers, and White-throated Sparrow find winter cover and food in pole stands of loblolly and shortleaf pine forests (Noble and Hamilton 1976, Dickson and Segelquist 1977).

Mature Loblolly–Shortleaf Pine Stands

Breeding season. After 35–50 years the developing stand begins to achieve some characteristics of maturity. An overstory canopy is present, and midstory and understory foliage are present in varying degrees depending on how much light filters through the canopy. White-eyed Vireo and Indigo Bunting are the only early successional species that may persist in low numbers into the more mature stages of forest growth. These species depend on the open patches where sufficient light has penetrated to stimulate growth of understory and midstory foliage for nesting sites. Wood Thrush, Red-eyed Vireo, Black-and-white Warbler, Eastern Wood-Pewee, Great Crested and Acadian flycatchers, Pine, Hooded, and Kentucky warblers, Summer Tanager, and Blue-gray Gnatcatcher are now abundant (Noble and Hamilton 1976, Meyers and Johnson 1978, Conner et al. 1979, 1983, Dickson and Segelquist 1979, Darden 1980, Dickson et al. 1980, 1984, Childers et al. 1986). Yellow-throated Warbler, Northern Parula, and Yellow-billed Cuckoo are also present. Pine Warblers are attracted to the pine foliage, whereas most of the other species are primarily attracted to the deciduous foliage that has regenerated along with the pine trees.

Winter. Mature loblolly–shortleaf pine and pine–hardwood forests provide winter cover and food for American Robin, Hermit Thrush, Golden-crowned and Ruby-crowned kinglets, Black-and-white Warbler, and White-throated Sparrow (Noble and Hamilton 1976, Dickson and Segelquist 1977).

Management Activities Affecting Bird Communities

Variations in the stand development scenario can affect bird communities. Generally, procedures to eliminate hardwoods decrease NTMB diversity.

Site preparation. When loblolly–shortleaf pine forest stands are harvested, cutting is often followed by mechanical activities that prepare the area for planting of pine seedlings. The intensity of mechanical site preparation affects the amount of hardwood vegetation that regenerates with the pine seedlings. Intensive K-G blading (bulldozer blade) and chopping of roots will reduce the amount of hardwood regeneration substantially. Less intensive site preparation, such as prescribed burning, permits more hardwood vegetation to survive and grow along with the pines. NTMB diversity will be reduced as the amount of hardwoods within all ages of pine stands decreases, as will the abundance of fruits that serve as food for many songbirds (Stransky and Richardson 1977).

Herbicides to control hardwoods. Herbicides sometimes are used to control hardwoods in young pine plantations. Presumably, reduction of hardwood vegetation negatively affects avian diversity.

Thinning. Thinning opens the canopy, permits more light into the understory, and promotes understory vegetation growth, and increased foliage layers (Blair and Enghardt 1976, Blair 1982). Thinning is beneficial to some neotropical migrants and has a significant positive influence on bird abundance and species richness during the breeding season (Chritton 1988). Indigo Bunting, Pine Warbler, and Brown-headed Cowbird increased in abundance following thinning of a loblolly plantation in Texas. Other Neotropical migrants were negatively affected, including White-eyed Vireo, and Worm-eating and Hooded warblers. The Black-and-white Warbler appeared to be unaffected by thinning.

Thinning also affects migrants that winter in the loblolly-shortleaf pine ecosystem in the South (Chritton 1988). During winter, bird abundance and species richness was higher in thinned pine stands than in unthinned stands. Pine Warbler, Ruby-crowned and Golden-crowned Kinglets, and Dark-eyed Junco increased in numbers following thinning in Texas.

Rotation ages. Rotation age affects stand development and maturity. Saw-log rotations are generally 35–50 years on private industrial lands and 70–80 years on federal lands. The longer rotations of most public lands permit some old-growth attributes to develop and provide habitat for many species of neotropical migrants that prefer mature pine forest for breeding or wintering habitat. Red-eyed Vireo, Northern Parula, and Hooded, Pine, and Yellow-throated warblers are some species that would be favored by the more mature habitat.

Pulp-wood rotations on private industrial lands have become shorter during the past decade. Pines can be cut after only 20–30 years and, in some cases, 15 years. These rotation ages will greatly affect the ability of the lands to provide habitat for Neotropical migrants requiring stands beyond the pole timber stage.

Seedtree and Shelterwood Harvesting

Trees left in shelterwood harvesting serve as shelter for the developing stands and can be of value to Neotropical migrant birds (Hall 1987).

Breeding Season. Neotropical migrants using seedtree and shelterwood cuts during the breeding season include: Eastern Kingbird, Acadian Flycatcher, Pine Warbler, Prairie Warbler, Yellow-breasted Chat, Wilson's Warbler, Hooded Warbler, Orchard Oriole, Indigo Bunting, and Chipping and Field sparrows. Thus, these cuts are attractive to some species that normally are associated with either early successional or late successional seral stages, and result in higher bird species diversity.

Winter. During winter, seedtree and shelterwood harvesting provide habitat for both early and late successional bird species. Migrants using these cuts included American Kestrel, Ruby-crowned Kinglet, Yellow-rumped Warbler, Common Yellowthroat, Red-winged Blackbird, Purple Finch, Pine Siskin, American Goldfinch, and Grasshopper, Henslow's, Lark, Field, White-crowned, White-throated, Swamp, and Song sparrows (Hall 1987).

Unfortunately, the benefits for bird species richness gained by the presence of the residual pines is lost when the residuals are removed following the establishment of the new pine stand.

Effects of Uneven-aged Timber Management

There is little published information on the bird communities that inhabit stands managed under single-tree selection and group selection harvesting methods. Information on bird communities in uneven-aged pine forests is being collected by the US Forest Services Southern Forest Experiment

Station's Laboratory in Nacogdoches, but currently we can only speculate on how different bird species might respond to selection harvesting based on what we know about their habitat characteristics and their use of seedtree and shelterwood cuts, and heavily thinned stands.

Single-tree selection. Bird communities found in single-tree selection harvested stands will probably be similar to those that Chritton (1988) found in thinned pine stands in eastern Texas (see section on thinning above). Both harvesting techniques remove some overstory pines and open up the forest canopy. Species that use early successional stands, small gaps within forests, and "edge" would probably be favored by this harvest method. However, we do not know if the open canopy will have a negative effect on some of the mature forest species such as the Red-eyed Vireo that seem to prefer contiguous forest.

Group selection. We expect that the bird community would respond to group-selection harvesting in a fashion similar to single-tree selection. However, there will be larger gaps in the forest canopy, sometimes up to 1 ha. This could have positive and negative impacts. Early successional species that require larger gaps or patches of low shrubby vegetation, (e.g,. Yellow-breasted Chat, Indigo Bunting, and Prairie Warbler) will probably be added to the overall bird community. However, mature forest species, which prefer a continuous canopy may be negatively impacted and perhaps the Brown-headed Cowbird increased.

LONGLEAF- AND SLASH-PINE FORESTS

Forest Composition

Longleaf Pine

Longleaf pine occurs in monotypic stands, with other pine species, or as an associate of slash pine where their distributions overlap. The natural distribution of longleaf pine includes nine southern states (Virginia, North Carolina, South Carolina, Georgia, Florida, Alabama, Mississippi, Louisiana, and Texas). Maintenance of longleaf pine depends on fire as this forest type is a fire subclimax. Typical hardwood associates include southern red oak, flowering dogwood, blackjack oak, water oak, sassafras, blackgum, persimmon, and sweetgum.

Slash Pine

The flatwoods of the slash-pine ecosystem comprise more than 40% of the Lower Coastal Plain and are among the most intensively managed forests in the United States (Walker 1962). Slash pine can grow wherever longleaf pine does, but its natural range is more restricted (South Carolina to central Florida and southeast Louisiana). It has been planted as far north as North Carolina and west to eastern Texas. When found in longleaf forests, slash pine often occurs in the creek drainages where fires have been excluded (Walker 1962). Slash pine can reproduce under its own canopy even though it is not considered shade tolerant. It is basically a stable type where it occurs without hardwoods.

Slash-pine stands usually have more hardwood associates than stands of other southern pine species, and moist areas may contain the slash pine–hardwood type (Walker 1962). Associates on wetter areas include a wide variety of moist-site hardwoods, such as swamp tupelo, blackgum, red maple, pond pine and pond cypress. On dry sites, a wide variety of dry-site hardwoods are also encountered as well as longleaf, loblolly, and sand pines.

Longleaf–Slash Pine

Most of the typical longleaf–slash pine cover type is found in the flatwoods of Florida and Georgia. Slash pine seedlings are more shade tolerant than those of longleaf pine; hence, slash pine may begin as an understory and remain subordinate to the longleaf pine overstory.

On well-drained upland soils, common associated trees include flowering dogwood, post oak, blackjack oak, southern red oak, hickories, yaupon, persimmon, and hawthorn. Wetter sites on the flatwoods of the Coastal Plains or adjacent to creeks in the uplands often contain red maple, sweetgum, blackgum, water oak, and laurel oak. On sites that are flooded periodically, bald cypress, pond cypress, blackgum, and water tupelo may occur (Grelan 1980).

Fire is important in the determination of these types. Longleaf- and slash-pine seedlings <5 m in height are vulnerable to fire, but longleaf-pine seedlings in the grass stage can benefit from fire (Walker 1962). Frequent ground fires prevent species other than longleaf pine from growing into the overstory within forest openings (Platt et al. 1988). Fire suppression for as little as 10 years can result in a substantial increase in the size and abundance of hardwood stems in longleaf pine forests (Heyward 1939, Edmisten 1963).

History

The former longleaf pine ecosystems are among the most disturbed landscapes in the United States. Estimates of the amount of presettlement longleaf-pine habitat include 37 million ha (Frost 1993), 24.3 million ha (Croker 1979), and 12–24 million ha (Boyer 1980). Land has been used for the last 100–400 years, primarily for agriculture, livestock grazing, and logging, often followed with regeneration with faster growing slash or loblolly pine. These uses, coupled with fire suppression, have reduced the longleaf pine ecosystem to less than 3% (<0.5 million ha) of what was historically present (Frost 1993). Approximately 0.7 million ha remained in longleaf pine and slash pine in 1990 (Frost).

Historically, fire has played a major role in these forests. Means and Grow (1985) suggest that the original longleaf pine forest may have burned every 3–5 years. The extensive fire suppression activities that developed in the longleaf pine region were largely responsible for an intrusion of hardwoods and less fire-tolerant pine species.

Silviculture and Bird Communities

Of the 106 summer-resident bird species within the geographic range of the longleaf pine/slash pine forests, 49 are associated with pine stands and 54 of the 114 winter species are pine associates (Wood and Niles 1978). Bird–habitat relationships in truly old-growth forests has yet to be adequately measured (Jackson 1988).

Pine stands normally contain some hardwoods and hardwood removal influences bird-community composition (Johnston and Odum 1956, Dickson and Segelquist 1979, Dickson 1981). Methods to reduce hardwoods, such as burning or herbiciding can have deleterious effects on birds using the hardwood component for cover, nesting, or foraging. However, the effects of burning may be beneficial for ground foragers as fire reduces leaf litter, exposes seeds, and enhances forbs, which produce seed. Slash-pine sites in southeastern Georgia that contained hardwoods supported 17 species that either were absent or occurred in low densities in the pure slash pine (Johnson and Landers 1982). Avian response to hardwood intrusion in longleaf pine stands is probably similar to that observed in the above studies of other pine species, but currently studies have not been conducted in longleaf pine forests.

Several studies have examined longleaf and slash-pine stands of various ages to determine how avian community structure changes as stands mature (Table 9-3). In slash-pine stands in southeastern Georgia, Johnson and Landers (1982) found that total bird numbers tended to be lowest in 1-year-old pine plantations, increased in the 2–6-year-old stands as understories developed, and then declined again until approximately mid-rotation (16 years). Once stands passed mid-rotation age (16–28 years), the initial stand treatments, such as site preparation and regeneration method, had no discernible effect on the avifauna. Ground foragers, primarily insectivores, were the most common species (47%) in the fallow areas. Stands 11–15 years old (pole stage) showed significant changes in the habitat structure resulting in an increase in ground foraging insectivores and a decline in shrub users. In age classes 16–20 and 21–28 years, the avifauna was mainly canopy (43%) and ground foraging insectivores (28%).

Of the 14 breeding NTMB species, five were found in the fallow areas with only the Blue-gray Gnatcatcher being abundant (Johnson and Landers 1982). Five species

Table 9-3. Longleaf- and slash-pine stand suitability[a] for Neotropical migratory birds.[b]

Species	Stand Age[c]			
	1	2	3	4
Longleaf pine				
Common Nighthawk	S	S	M	M
Chuck-will's-widow			M	M
Eastern Wood-Pewee			M	O
Great Crested Flycatcher			M	S
Purple Martin	M			
Barn Swallow	M			
Prairie Warbler		M		
Summer Tanager			M	S
Longleaf pine–slash pine				
Osprey				M
American Swallow-tailed Kite				M
Common Nighthawk	M	M	M	M
Chuck-will's-widow			M	M
Ruby-throated Hummingbird		M		
Eastern Wood-Pewee			M	O
Great Crested Flycatcher			M	S
Eastern Kingbird	M	M	M	M
Purple Martin	M			
Barn Swallow	M			
Blue-gray Gnatcatcher			M	M
White-eyed Vireo				S
Yellow-throated Warbler			M	M
Prairie Warbler		M	M	M
Common Yellowthroat	M	S	M	M
Yellow-breasted Chat		M		
Summer Tanager			M	S
Indigo Bunting		S	M	M

[a] O = optimum; S = suitable; and M = marginal habitat.
[b] Adopted from Hamel (1992). See also Johnson and Landers (1982), O'Meara et al. (1985), Repenning and Labisky (1985), and Dickson (1991). Note that this list may be incomplete as limited data are available on birds inhabiting these vegetation types and stand ages.
[c] 1 = grass/forb; 2 = shrub/seedling; 3 = sapling/poletimber; and 4 = sawtimber.

were observed regularly in the regenerating areas. Of the seven species found in the seedling–sapling stage, the Indigo Bunting, Common Yellowthroat, and Ruby-throated Hummingbird were either abundant or common. The Common Yellowthroat, Blue-gray Gnatcatcher, and Eastern Wood-Pewee were the most commonly observed of the 10 NTMB species in the pole stage.

The effects of prescribed burning on a 20-year-old slash-pine stand resulted in a drastic decline in ground cover and shrub foliage in Everglades National Park, Florida (Emlen 1970). No significant difference was detected in NTMB numbers or foraging guilds after the burn. Emlen speculated that food or shelter needs were not drastically affected by the fire, perhaps because of individual attachments to home range and familiar foraging sites.

Harris et al. (1974) compared effects of site-preparation techniques on birds in three 9-year-old slash-pine stands with a naturally regenerated mature slash-pine stand and a mature longleaf-pine stand, which had been prescribed burned the previous year. Stands with high-intensity site preparation, where most hardwood shrubs were eliminated, had somewhat fewer individuals than other stands and the lowest number of species. Bird abundance was higher in areas that had undergone low-intensity site preparation than in mature slash pine; however, the number of species was significantly lower. There were nine times more birds in mature longleaf pine stands than in the low-intensity site-prepared slash-pine stands and 60 times more than in the high-intensity treated site-prepared stands.

Repenning and Labisky (1985) compared the avian community in three naturally regenerated longleaf pine (>50 years) to slash-pine plantations of four different ages (1, 10, 24, and 40 years old). Bird densities, biomass, and species richness were strongly positively correlated with stand age. Breeding-bird density was lowest in the 1-year-old stand (45 birds/km^2), increased in the 40-year-old stand (149/km^2), and was highest in the longleaf stands (288/km^2). Of the seven species of Neotropical migrant breeding birds, two species were found in 1-year-old stands (13 birds/km^2), four in 10-year-old stands (19 birds/km^2), two in 24-year-old stands (22 birds/km^2), four in 40-year-old stands (33 birds/km^2), and five in mature longleaf-pine forest (33 birds/km^2). The foraging guilds were somewhat different between stands of different ages. Ground, foliage, and cavity nesters were lowest in the youngest stand, increased with stand age, and were highest in older longleaf-pine stands.

Conversely, wintering birds were most numerous in the youngest slash-pine stand, perhaps the result of the abundance of seed-producing grasses and forbs (Repenning and Labisky 1985). The bird community, with four species dominating (Chipping and Vesper Sparrows, Eastern Meadowlark, and

American Goldfinch), bore little resemblance to that of longleaf-pine stands.

Repenning and Labisky (1985) concluded that pine plantations managed on a rotation of 30 years or less will not provide winter or breeding habitat for birds primarily associated with mature natural pinelands, but some young slash-pine stands, because of their shrub layer, provide habitat for birds not normally found in mature longleaf pine.

O'Meara et al. (1985) and Rowse and Marion (1980) compared birds in unharvested 35-year-old slash-pine stands, regenerating slash-pine (recently clearcut) stands, cypress and edge areas. Only three Neotropical migrant bird species, the Great Crested Flycatcher, White-eyed Vireo, and Common Yellowthroat, were observed in spring or summer in the 35-year-old stands. In the young slash-pine stand only the Common Nighthawk, Common Yellowthroat, and Blue Grosbeak occurred at low breeding densities.

Winter densities in clearcut and older slash-pine habitat were higher than in the breeding season. Both the edge and cypress habitats supported large numbers of breeding and wintering birds. Overall the results of this study are similar to that of Repenning and Labisky (1985) in that an immediate result of clearcutting slash pine was a replacement by species adapted to early successional stages, a lower density of breeding birds, reduced bird species diversity, and a much larger wintering than breeding population.

The previous studies assessed even-aged management on the avifauna including NTMBs. However, little work has been conducted on the response of birds to uneven-aged harvesting techniques. Single-tree and group selection harvesting would probably benefit species that use edge habitats or small gaps, but it is not clear if the habitat fragmentation created by this approach would be detrimental to forest interior birds.

Recommendations

General recommendations appropriate for the different ecosystems are presented at the end of this chapter. Because of the extensive conversion of longleaf pine to other pine types, longleaf stands should be restored and older stands maintained. Apparently, historic longleaf stands were resistant to early mortality agents and were long lived (sometimes >200 years old). To maintain longleaf- and slash-pine forests requires a regular burning regime (3–5 years) to control the hardwood understory, and to allow sufficient open areas for pine regeneration. Prescribed fire should not be used in seedling slash-pine stands or in small sapling stages of either longleaf or slash pine. Hardwoods should be allowed to develop in some stands to accommodate those species that depend on hardwoods. Low-intensity site-preparation measures tend to be less drastic to the avifauna than high intensity site preparation and are recommended.

OAK–GUM–CYPRESS FORESTS

Forest Composition

Oak–gum–cypress forests (also called bottomland hardwoods) occur on mesic to hydric sites along streams or rivers throughout the southeastern United States. These bottomland forests covered about 11 million ha (in 1977) from Virginia to East Texas and along the Mississippi river drainage to Indiana (USDA Forest Service 1982). Of the previous total bottomland area only about 10% percent remains in forest, with the remainder roughly evenly divided between pasture and crops primarily cotton, soybeans, and corn (Garrison et al. 1977).

Bottomland soils were formed by alluvial deposits. Dominant vegetative communities of this complex are closely associated with sites, which are determined mainly by soils, elevation, and hydroperiod. There are eight major general forest types and many variations within the oak–gum–cypress complex (Putnam 1951). The major types include: sweetgum–water oak, white oak–red oak–other hardwoods, hackberry–elm–ash, overcup oak–bitter pecan, cottonwood, willow and associates, riverfront hardwoods, and cypress–tupelo gum. These different types are associated with different sites.

History

Oak–gum–cypress forest area has declined but varies regionally. There were an estimated 10 million ha in the entire Mississippi River delta system originally (Dunaway 1980). By the early 1970s, delta hardwood area in the primary states of Louisiana, Mississippi, and Arkansas had dropped to 2.9 million ha (Sternizke 1976), mainly from conversion to soybeans, cotton, and pasture. Bottomland hardwoods also have been lost to reservoirs in many areas. For example in eastern Texas, Toledo Bend and Sam Rayburn reservoirs occupy more than 100,000 ha, which once were vegetated primarily by bottomland hardwoods (Dickson 1978a). Considerable bottomland hardwood habitat also has been lost to stream channelization and clearing of associated mature vegetation (e.g., Barclay 1978).

However, since the heavy losses in the 1960s, bottomland habitat loss has declined (Birdsey and McWilliams 1986). Most of the land suitable for agriculture has already been converted and only marginal sites remain. Apparently, bottomland hardwoods in Florida, Georgia, South Carolina, North Carolina, and Virginia have not been subjected to such losses, and area occupied has remained relatively stable over the last four decades (Langdon et al. 1981). Recently there has been some effort to restore bottomland forests that have been converted to agricultural uses and to protect the remaining bottomland forests.

Cutting

Bottomland hardwood forests have been altered substantially by past cutting. In the late 1800s and early 1900s as colonists settled new land and cut trees for hardwood products, most of the oak–gum–cypress forests were harvested. What was cut and what left was quite variable depending on site, accessibility, availability, and the desired commodity. For example, cypress heartwood is naturally decay resistant, was in demand for wood products, such as houses and boats, and was virtually eliminated. In Louisiana, the last virgin bald-cypress stands were cut from 1890 to 1925. The area of cypress stands in the state has dwindled from estimates of 1–3 million ha originally to 0.14 million ha now (Conner and Toliver 1990). Also, stands of mixed species often were "high graded;" the highest quality trees were taken and trees of lesser value or culls were left. Virtually all stands have been selectively harvested in the past, and present stand composition reflects harvest decisions.

Silviculture

Most harvest regeneration methods practiced today involve clearcuts or group-selection cuts, which favor commercially valuable, shade-intolerant species such as eastern cottonwood, yellow poplar, ash, and sweetgum; or mid-tolerant oaks. Some mixed stands have been replaced by plantations of the fast-growing species, such as cottonwood, sycamore, and sweetgum, but this practice has largely been abandoned today and area in plantations remains small. Most clearcuts today are regenerated by seedlings which were present when the cuts were made or suckers from roots when the stands are opened and sunlight penetrates. Size of clearcuts or group-selection cuts in bottomland hardwood forests is usually small (i.e., from 0.3 to 20 ha), but sometimes can be sizable (e.g., 200 ha). Minimum size is that required to allow sufficient sunlight to permit seedling growth and the harvest cut radius is usually at least equal to the height of dominant trees of the surrounding stand.

Rotation is quite variable in oak–gum–cypress forests, depending on a number of factors, including tree species, size, quality, and wood product. Some government land is maintained as old-growth with no harvesting. Long rotations (e.g., from 80 to over 100 years) are used when large trees and knot-free wood are desired for quality wood products, or when mast production is emphasized. Variable rotations long enough to allow trees to grow to sufficient size are used to produce wood products of lesser value, such as pallets and cross ties. Short rotations (e.g., 40 years) are used to maximize growth in trees destined for fragmented wood products such as hardwood chips for quality paper or reconstituted products, such as fiber board.

Improvement cuts or thinnings (a portion of the trees in a stand harvested) are conducted for wood products from harvested trees and to improve stand composition and/or concentrate growth on remaining trees. The remaining overstory and understory vegetation growth is enhanced by the increased sunlight after a partial cut.

Bird Habitat

The avifauna of oak–gum–cypress forests is abundant and diverse. Hydroperiod influences stand structure as bird habitat and bird communities (Stauffer and Best 1980, Swift et al. 1984). Flooding provides supplemental nutrients and soil accretion into riparian systems, and influences stand and foliage structure. Diverse foliage layers on higher sites should promote high bird diversity in stands with limited foliage layers. However, extended flooding during the growing season can kill trees and eliminate most foliage near the ground, which would be detrimental for most birds. Ground-nesting birds such as Kentucky Warblers (Dickson and Noble 1978) could be excluded by seasonal flooding. Extended flooding could create habitat for aquatic birds, afford protection from mammalian predators to colonial nesters, and create dead trees suitable for cavity nests and foraging sites in the short term.

Breeding Birds

There are a variety and abundance of breeding birds in the mature bottomland hardwood forests of the South. In eastern Texas, Anderson (1975) found estimated bird densities were much higher in a bottomland hardwood stand (1050/km^2) than in a pine (835/km^2) or pine–hardwood stand (422/km^2). The number of bird species and species diversity were similar in the structurally diverse bottomland hardwood and pine–hardwood stands, but higher than the number and diversity in the pine stand. In the Louisiana–East Texas area, Dickson (1978a) compared birds in six pine stands, four pine–hardwood stands, and three bottomland stands. Bird densities in the mature bottomland hardwood stands were 2–4 times greater than densities in the upland pine and pine–hardwood stands of different ages. Bird species diversities in the mature bottomland hardwoods were similar to those in mature upland pine and pine–hardwood stands, but higher than diversities in young pine and pine hardwood stands. The bottomland hardwood stands had high tree densities (29–45 m^2/ha basal area), which limited light penetration and shrubby vegetation; otherwise, bird diversities would probably have been higher. Studies have shown flooded forests also supported more birds than upland forests in the Midwest (Stauffer and Best 1980) and in New England (Swift et al. 1984).

In a Louisiana floodplain forest about half of the number and species of breeding season birds were Neotropical migrants and about half were permanent residents (Dickson 1978b). The proportion of neotropical migrant breeders was lower than in more northerly and seasonally harsher climates.

Oak–gum–cypress forests are special habitat for many species of birds (Dickson 1978a, 1988, Hamel et al. 1982). A survey of breeding bird censuses from seven mature stands (Dickson et al. 1980) showed that the Yellow-billed Cuckoo, Acadian Flycatcher, and Red-eyed Vireo were consistently abundant in oak–gum–cypress habitat. Other species regularly inhabit these stands, and some have special affinities for this habitat (Table 9-4). Many long-legged waders nest and forage in aquatic woodlands. The Wood Stork, which is now endangered, nests in tall cypress and hardwoods and feed in associated aquatic systems.

Several migratory raptors inhabit bottomland hardwoods. Mississippi and Swallow-tailed kites, Cooper's Hawk, and Osprey frequent this habitat. Also, Purple Gallinule and Common Moorhen are found in appropriate aquatic habitat. The Acadian Flycatcher is strongly associated with forested wetlands (Shugart and James 1973, Smith 1977), and the Wood Thrush is a common breeding bird in the mesic sites.

White-eyed Vireos are common in low, shrubby foliage and Red-eyed Vireos in

Table 9-4. Neotropical migrant breeding bird species present in southeastern oak–gum–cypress forests (reproduced from Dickson et al. 1993).

Species
American Anhinga[a]
Green-backed Heron
Little Blue Heron
Cattle Egret[b]
Great Egret[a]
Snowy Egret
Tricolored Heron
Black-crowned Night-Heron[a]
Yellow-crowned Night-heron[a]
Wood Stork
Glossy Ibis[a]
White Ibis
Hooded Merganser[a]
American Swallow-tailed Kite
Mississippi Kite
Cooper's Hawk[a]
Osprey
Purple Gallinule
Common Moorhen[a]
Mourning dove[a,b]
Yellow-billed Cuckoo
Chimney swift[b]
Ruby-throated Hummingbird
Belted Kingfisher[a]
Great Crested Flycatcher
Eastern Phoebe[b]
Acadian Flycatcher
Eastern Wood-Pewee
Barn Swallow[b]
Purple Martin[b]
Wood Thrush
Blue-gray Gnatcatcher
White-eyed Vireo
Yellow-throated Vireo
Red-eyed Vireo
Black-and-white Warbler
Prothonotary Warbler
Swainson's Warbler
Worm-eating Warbler
Northern Parula
Bachman's Warbler
Black-throated Green Warbler
Yellow-throated Warbler
Prairie Warbler[c]
Ovenbird
Louisiana Waterthrush
Kentucky Warbler
Common Yellowthroat[c]
Yellow-breasted Chat[c]
Hooded Warbler
American Redstart
Eastern Meadowlark[a,b]
Red-winged Blackbird[b]
Brown-headed Cowbird[a,b]
Orchard Oriole
Northern Oriole
Summer Tanager
Blue Grosbeak[c]
Indigo Bunting[c]
Painted Bunting[c]
Rufous-sided Towhee[a,c]

[a] Undenoted species breed in North America and winter primarily south of the United States, denoted species (*a*) breed and winter extensively in North America, although some populations winter south of the United States.
[b] Associated with human-altered nonforest habitat.
[c] Associated with early successional stands.

canopy foliage (Dickson and Noble 1978). There are many warblers in this habitat, some with special affinities. Prothonotary, Swainson's, Northern Parula, Kentucky, and Hooded Warblers are strongly associated with this habitat. The Prothonotary Warbler nests in cavities, often in small, flood-killed trees. The Northern Parula constructs its nest with Spanish moss. The Swainson's Warbler is primarily associated with understory thickets of southern river floodplains and the southern Appalachian Mountains (Meanly 1971). The habitat of the Bachman's Warbler, which is probably extinct, is bottomlands and headwater swamps subject to disturbances (Hooper and Hamel 1977). Both Swainson's and Bachman's warblers are associated with cane thickets, which were once extensive in southern bottomland forests (Meanly 1971, Remsen 1986). Kentucky and Hooded warblers are usually found in the moist understory of bottomland hardwoods (Dickson and Noble 1978). Other warblers often found in mature stands include Black-and-white, Worm-eating, Yellow-throated, Ovenbird, American Redstart, and Louisiana Waterthrush (Hamel et al. 1982).

Bird communities are dependent on the age and development of forest stands (Shugart and James 1973, Dickson and Segelquist 1979). Breeding birds discussed previously have been those associated with mature stands of mixed species, and those species are usually present in middle-aged stands. Young stands would have mostly a different bird composition. In the earliest stages of hardwood stand development the Dickcissel and Red-winged Blackbird would be characteristic species (Weinell 1989). Neotropical migrant birds typifying the avian community in young brushy stands include the Yellow-breasted Chat, Indigo Bunting, Painted Bunting, Prairie Warbler, Common Yellowthroat, and White-eyed Vireo.

Winter Birds

Mature oak–gum–cypress provide critical habitat for wintering birds, and these forests support very high densities of winter birds (1400–2000 individuals/km^2, Dickson 1978b). Many winter resident species, such as Common Grackle and White-throated Sparrow, do not winter in the tropics. However, these mature bottomland forests are regular habitat for several species that winter from southern forests into the tropics, including Yellow-bellied Sapsucker, American Robin, Hermit Thrush, Ruby-crowned Kinglet, and Orange-crowned Warbler.

Silviculture and Bird Communities

There is little specific information from studies of bird community changes related to silvicultural practices in oak–gum–cypress forests. However, some information from general habitat relationships of bird species might provide insight into how silvicultural practices affect bird communities in bottomland hardwoods. Tree harvesting has a drastic effect on bird communities. The replacement of mature stands of mixed hardwoods by hardwood plantations alters bird communities. These plantations and natural stands of pure black willow or cottonwood lack vegetative diversity and support fewer birds and a less diverse bird community than natural mixed stands (Wesley et al. 1976). However, in areas where most of the land is in mature mixed stands and the plantations represent a small land commitment, the overall bird diversity on a landscape scale could be increased because of the birds associated with early successional stands that inhabit the plantations. In Mississippi, Red-winged Blackbird, Common Yellowthroat, Yellow-breasted Chat, Northern and Orchard oriole, Rufous-sided Towhee, and Warbling Vireo were common in plantations, but not in natural stands (Wesley et al. 1976).

Clearcuts with natural regeneration generally would favor edge species such as the Wood Pewee, and early successional species, such as the Indigo Bunting, Prairie Warbler, White-Eyed Vireo, and Yellow-breasted Chat (Dickson and Segelquist 1979, McComb et al. 1989, Thompson et al. 1992).

Harvest regimes in which some trees are harvested and some left, such as improvement cuts or thinnings, would have a less drastic effect on bird communities and would favor early successional and edge species (e.g., Webb et al. 1977). Partial cuts or small clearcuts usually result in higher bird diversity and most species associated with mature forests remain in forested stands where some mature trees or stands remain. Understory vegetation growth in the opened stands would probably favor understory associated species such as Kentucky and Hooded warblers (Dickson and Noble 1978, McComb et al. 1989, Thompson et al. 1992).

A few forest interior species associated with closed-canopy forest would probably dwindle with tree harvest and stand opening. Studies have shown that Ovenbird and Wood Thrush abundance were negatively correlated with stand harvest (Crawford et al. 1981, Webb et al. 1977).

Rotation age also affects bird-community composition in forest stands. Short rotations would favor early successional species, whereas long rotations should favor cavity using species such as the Great Crested Flycatcher and canopy associated species such as Red-eyed and Yellow-throated Vireo, Northern Parula, and Summer Tanager (Dickson and Noble 1978).

Recommendations

General recommendations for this ecosystem are common with the other ecosystems and are included in overall recommendations. The primary effort in behalf of bottomland hardwood systems and their associated avifauna would be to keep current forests and prevent further conversion to other uses, and reclaim some previously converted land to forest. Also, broad-scale research should be directed to provide information on how the water regime, nutrients, vegetation, and associated wildlife interact.

OVERALL RECOMMENDATIONS

Local habitat factors as well as landscape composition determine NTMB communities. At a landscape level, the single most important consideration is to maintain large areas in breeding and wintering forest habitats to provide for large NTMB populations, and minimize numbers of cowbirds and predators associated with agricultural, suburban, and urban land uses. At the habitat level the most basic management step is to maintain natural ecosystems. Rare ecosystems and habitats required by threatened or endangered species, and regional species of high concern should be emphasized. A priority in southeastern forests is to protect existing old-growth stands and corridors, and to allow new old-growth stands to develop. Restoration and maintenance of natural ecosystems that have been substantially reduced or altered, such as longleaf and oak–gum–cypress forests, should be accelerated.

Species of concern breed in all stages of forest succession (Thompson et al. 1992), so a diversity of successional stages should be provided. Unless specific concerns dictate otherwise, both selection cutting and even-aged management should be used to create small openings for gap species, large openings for early successional forest migrants, and a balanced age–class distribution to maintain sufficient mature forest habitats. This range of opening sizes more closely imitates the range in size of natural openings or disturbances in forests than the use of any one regeneration practice. Where late successional, area or edge-sensitive NTMBs are a concern (e.g., Red-eyed Vireo, Ovenbird, Wood Thrush) some blocks of unfragmented forest should be set aside from timber harvest, larger regeneration cuts on longer rotations used in even-aged systems, and single-tree selection favored over group selection. Even-aged systems should be used to provide young forest habitats for early successional species (e.g., Prairie Warbler, Yellow-breasted Chat, Blue-winged Warbler, etc.) because openings created by selection cutting may be too small for many of these species. Other stand-level practices that will maintain NTMB-community viability include retaining live cavity trees and snags when stands are regenerated, and maintaining both coniferous and deciduous components of mixed stands. An extensive monitoring program should be implemented that tracks bird species abundance and viability over the long term.

Research into NTMBs and their forest habitat at the stand and landscape levels should be expanded to develop more complete information for management of NTMBs. The components and function of different forested ecosystems should be explored more fully. Species density may not always be a suitable measure of habitat quality (Van Horne 1983). A better understanding of species demographics, productivity, cowbird parasitism, and nest predation is essential. Moreover, additional information relating avian communities to forest composition, distribution, fragmentation and various silviculture practices is needed to ensure the future of sensitive NTMBs.

LITERATURE CITED

Anderson, R. M. 1975. Bird populations in three kinds of eastern Texas forests. MS thesis, Stephen F. Austin State University, Nacogdoches, TX.

Baker, J. B., and W. E. Balmer. 1983. Loblolly pine. Pp. 149–152 *in* Silvicultural systems for the major forest types of the United States (R. M. Burns, tech. comp.). USDA Forest Serv., Agricultural Handbook no. 445.

Barclay, J. S. 1978. The effects of channelization on riparian vegetation and wildlife in south central Oklahoma. Pp. 129–138 *in* Strategies for protection and management of flood-plain wetlands and other riparian ecosystems symposium (R. R. Johnson and J. F. McCormick, tech. coord.). USDA Forest Serv. Gen. Tech. Rep. WO-12.

Belilman, A. P., and L. G. Brenner. 1951. The recent intrusion of forest in the Ozarks. Ann. Missouri Bot. Garden 38:261–282.

Birdsey, R. A., and W. H. McWilliams. 1986. Midsouth forest area trends. USDA Forest Serv. Res. Bull. SO-107.

Blair, R. M. 1982. Growth and nonstructural carbohydrate content of southern browse species as influenced by light intensity. J. Range Manag. 35:756–760.

Blair, R. M., and H. G. Enghardt. 1976. Deer forage and overstory dynamics in a loblolly pine plantation. J. Range Manag. 29:104–108.

Boyer, W. D. 1980. Longleaf pine. Pp. 51–52 *in* Forest cover types of the United States and Canada (F. H. Eyre, ed.). Society of American Foresters, Washington DC.

Brand, G. J., and J. T. Walkowiak. 1991. Forest statistics for Iowa, 1990. USDA Forest Serv. North Central Forest Exp. Sta. Resource Bull.

Braun, E. L. 1950. Deciduous forests of eastern North America. The Blakiston Company, Philadelphia, PA.

Childers, E. L., T. L. Sharik, and C. S. Adkisson. 1986. Effects of loblolly pine plantations on songbird dynamics in the Virginia Piedmont. J. Wildl. Manag. 50:406–413.

Chritton, C. A. 1988. Effects of thinning a loblolly pine plantation on nongame bird populations in east Texas. MS thesis, Stephen F. Austin State University, Nacogdoches, TX.

Conner, R. N., and C. S. Adkisson. 1975. Effects of clearcutting on the diversity of breeding birds. J. For. 73:781–785.

Conner, R. N., J. G. Dickson, B. A. Locke, and C. A. Segelquist. 1983. Vegetation characteristics important to common songbirds in east Texas. Wilson Bull. 95:349–361.

Conner, R. N., J. W. Via, and I. D. Prather. 1979. Effects of pine–oak clearcutting on winter and breeding birds in southwestern Virginia. Wilson Bull. 91:301–316.

Conner, W. H., and J. R. Toliver. 1990. Long-term trends in bald-cypress (*Taxodium distichum*) resource in Louisiana (USA). Forest Ecol. Manag. 33/34:543–557.

Crawford, H. S., R. G. Hooper, and R. W. Titterington. 1981. Songbird population response to silvicultural practices in central Appalachian hardwoods. J. Wildl. Manag. 45:680–692.

Darden, T. L., Jr. 1980. Bird communities in managed loblolly–shortleaf pine stands in east central Mississippi. MS thesis, Mississippi State University. Miss. State, MS.

Dickson, J. G. 1978. Forest bird communities of the bottomland hardwoods. Pp. 66–73 *in* Proceedings of the workshop: management of southern forests for nongame birds (R. M. DeGraff, tech. coord.). USDA Forest Serv. Gen. Tech. Rep. SE-14.

Dickson, J. G. 1978b. Seasonal bird populations in a south central Louisiana bottomland hardwood forest. J. Wildl. Manag. 42:875–883.

Dickson, J. G. 1981. Impact of forestry practices on wildlife in southern pine forests. Pp. 224–230 *in* Proceedings: increasing forest productivity. SAF Publ. 82-01. Society of American Foresters, Bethesda, MD.

Dickson, J. G. 1988. Bird communities in oak–gum–cypress forests. Pp. 51–62 *in* Bird conservation 3. International Council for Bird Preservation (J. A. Jackson, ed.). University of Wisconsin Press, Madison, WI.

Dickson, J. G. 1991. Birds and mammals of pre-colonial southern old-growth forests. Natural Areas J. 11:26–33.

Dickson, J. G., and R. E. Noble. 1978. Vertical distribution of birds in a Louisiana bottomland hardwood forest. Wilson Bull. 90:19–30.

Dickson, J. G., and C. A. Segelquist. 1977. Winter bird populations in pine and pine–hardwood stands in east Texas. Proc. Annu. Conf. Southeast Assoc. Fish Wildl. Agencies 31:134–137.

Dickson, J. G., and C. A. Segelquist. 1979. Breeding bird populations in pine and pine–hardwood forests in Texas. J. Wildl. Manag. 43:549–555.

Dickson, J. G., R. N. Conner, and J. H. Williamson. 1980. Relative abundance of breeding birds in forest stands in the southeast. South. J. Appl. For. 4:174–179.

Dickson, J. G., R. N. Conner, and J. H. Williamson. 1984. Bird community changes in a young pine plantation in east Texas. South. J. Appl. For. 8:47–51.

Dickson, J. G., F. R. Thompson, III R. N. Conner, and K. E. Franzreb. 1993. Effects of silviculture on Neotropical migratory birds in central and southeastern oak pine forests. Pp. 374–385 *in* Status and management of Neotropical migratory birds (D. M. Finch and P. W. Stangel, eds). USDA Forest Serv., Gen. Tech. Rep. RM-229.

Dickson, J. G., R. N. Conner, and J. W. Williamson. In press. Neotropical migratory birds in a developing pine plantation. Proc. Ann. Conf. Southeast Assoc. Fish Wildl. Agencies 47.

Dunaway, A. 1980. The delta's diminishing lands. Mississippi Outdoors 43(3):16–17.

Edmisten, J. A. 1963. The ecology of the Florida pine flatwoods. PhD dissertation, University of Florida, Gainesville, FL.

Emlen, J. T. 1970. Habitat selection by birds following a forest fire. Ecology 51:343–345.

Evans, K. E., and R. A. Kirkman. 1981. Guide to bird habitat of the Ozark Plateau. USDA Forest Serv. Gen. Tech. Rep. NC-68.

Eyre, F. H. (ed.). 1980. Forest cover types of the United States and Canada. Society of American Foresters, Washington, DC.

Frost, C. C. 1993. Four centuries of changing landscape patterns in the longleaf pine ecosystem. Tall Timbers Fire Ecol. Conf., 30 May–2 June 1991. Tall Timbers Res. Stat. and the Nature Conservancy, Tallahassee, FL.

Garrison, G. A., A. J. Bjungstad, D. A. Duncan, M. E. Lewis, and D. R. Smith. 1977. Vegetation and environmental features of forest and range ecosystems. USDA Agricultural Handbook no. 475.

Gingrich, S. F. 1971. Management of young and intermediate stands of upland hardwoods. USDA Forest Serv. Res. Pap. NE-195.

Grelan, H. E. 1980. Longleaf pine–slash pine. Pp. 52–53 *in* Forest cover types of the United States and Canada. Society of American Foresters, Washington, DC.

Guyette, R. P., and B. E. Cutter. 1991. Tree ring analysis of fire history of a post oak savanna in the Missouri Ozarks. Natural Area J. 11(2):93–99.

Hall, S. B. 1987. Habitat structure and bird species diversity in seedtree and clearcut regeneration areas in east Texas. MS thesis, Stephen F. Austin State University, Nacogdoches, TX.

Hamel, P. B. 1992. The land manager's guide to the birds of the South. The Nature Conservancy and the US Forest Service, Southern Region, Chapel Hill, NC.

Hamel, P. B., H. E. LeGrand, Jr, M. R. Lennartz, and S. A. Gauthreaux, Jr. 1982. Bird–habitat relationships on southeastern forest land. USDA Forest Serv. Gen. Tech. Rep. SE-22.

Harris, L. D., L. D. White, J. E. Johnston, and

D. G. Milchunas. 1974. Impact of forest plantations on north Florida wildlife and habitat. Proc. Southern Assoc. Game Fish Comm. 28:659–667.

Heyward, F. 1939. The relation of fire to stand composition of longleaf pine forests. Ecology 20:287–304.

Hooper, R. G., and P. B. Hamel. 1977. Nesting habitat of Bachman's Warbler—a review. Wilson Bull. 89:373–379.

Jackson, J. A. 1988. The southern pine ecosystem and its birds: past, present, and future. Pp. 119–159 *in* Bird conservation 3 (J. A. Jackson, ed.). International Council for Bird Preservation, University of Wisconsin Press, Madison, WI.

Johnson, A. S., and J. L. Landers. 1982. Habitat relationships of summer resident birds in slash pine flatwoods. J. Wildl. Manag. 46:416–428.

Johnson, P. S. 1989. Principles of natural regeneration. Pp. 3.010–3.015 *in* Central hardwood notes (F. B. Clark and J. G. Hutchinson, eds). North Central Forest Exp. Sta., St Paul, MN.

Johnson, P. S. 1993. Perspectives on the ecology and the silviculture of oak-dominated forests in the central and eastern states. USDA Forest Serv., Gen. Tech. Rep. NC-153.

Johnston, D. W., and E. P. Odum. 1956. Breeding bird populations in relation to plant succession on the Piedmont of Georgia. Ecology 37:50–62.

Langdon, O. G., J. P. McClure, D. D. Hook, J. M. Crockett, and R. Hunt. 1981. Extent, condition, management, and research needs of bottomland hardwood cypress forests in the southeastern United States. Pp. 71–85 *in* Wetlands of bottomland hardwood forests. Elsevier Scientific Publishing Co., New York.

Law, J. R., and C. G. Lorimer. 1989. Managing uneven-aged stands. Pp. 6.080–6.086 *in* Central hardwood notes (F. B. Clark and J. G. Hutchinson, eds). North Central Forest Exp. Sta., St Paul, MN.

Lawson, E. R., and R. N. Kitchens. 1983. Shortleaf pine. Pp. 157–161 *in* Silvicultural systems for the major forest types of the United States (R. M. Burns, tech. comp.). USDA Forest Serv. Agricultural Handbook no. 445.

MacArthur, R. H., and J. W. MacArthur. 1961. On bird species diversity. Ecology 42:594–598.

Maxwell, R. S., and R. D. Baker. 1983. Sawdust empire, the Texas lumber industry, 1830–1940. Texas A&M University Press, College Station, TX.

McComb, W. C., P. L. Groetsch, G. E. Jacoby, and G. A. McPeek. 1989. Response of forest birds to an improvement cut in Kentucky. Proc. Southeast Assoc. Fish Wildl. Agencies 43:313–325.

Meanly, B. 1971. Natural history of the Swainson's Warbler. North Am. Fauna no. 69.

Means, D. B., and G. Grow. 1985. The endangered longleaf pine community. ENFO, pp. 1–8. Florida Conservation Foundation, Winter Park, FL.

Meyers, J. M., and A. S. Johnson. 1978. Bird communities associated with succession and management of loblolly–shortleaf pine forests. Pp. 50–65 *in* Proceedings of the workshop management of southern forests for nongame birds. USDA Forest Serv. Gen. Tech. Rep. SE-14.

Murphy, P. A., and R. M. Farrar, Jr. 1985. Growth and yield of uneven-aged shortleaf pine stands in the interior highlands. USDA Forest Serv. Res. Pap. SO-218.

Noble, R. E., and R. B. Hamilton. 1976. Bird populations in even-aged loblolly pine forests of southeastern Louisiana. Proc. Southeast. Assoc. Game Fish Comm. 29:441–450.

O'Meara, T. E., L. A. Rowse, W. R. Marion, and L. D. Harris. 1985. Numerical responses of flatwoods avifauna to clearcutting. Florida Sci. 48:208–219.

Parker, G. R., and G. T. Weaver. 1989. Ecological principles: climate, physiography, soil, and vegetation. Pp. 2.010–2.014 *in* Central hardwood notes (F. B. Clark and J. G. Hutchinson, eds). North Central Forest Exp. Sta., St. Paul, MN.

Platt, W. J., G. W. Evans, and S. L. Rathbun. 1988. The population dynamics of a long-lived conifer (*Pinus plaustris*). Amer. Natur. 131: 491–525.

Putnam, J. A. 1951. Management of bottomland hardwoods. USDA Forest Serv. Sov. For. Exp. Stn. Occ. Pap. 116.

Raile, G. K., and E. C. Leatherberry. 1988. Illinois' forest resource. USDA Forest Serv. North Central For. Exp. Sta. Resour. Bull. NC-105.

Remsen, J. V., Jr. 1986. Was Bachman's Warbler a bamboo specialist? Auk 103:216–219.

Repenning, R. W., and R. F. Labisky. 1985. Effects of even-age timber management on bird communities of the longleaf pine forest in northern Florida. J. Wildl. Manag. 49: 1088–1098.

Reynolds, R. R., J. B. Baker, and T. T. Ku. 1984. Four decades of selection management on the Crossett farm forestry forties. Agric. Exp. Sta. Univ. Arkansas Bull. 872.

Roth, R. R. 1976. Spatial heterogeneity and bird species diversity. Ecology 57:773–782.

Rowse, L. A., and W. R. Marion. 1980. Effects of silvicultural practices on birds in a north Florida flatwoods. Pp. 349–357 *in* First

Biennial Southern Silvicultural Res. Conf. (J. P. Barnett, ed.). USDA Forest Serv. Gen. Tech. Rep. SO-34.

Salwasser, H. 1990. Gaining perspective: forestry for the future. J. For. 88(11):32–38.

Sander, I. L., and F. B. Clark. 1971. Reproduction of upland hardwood forests in the central states. USDA Agriculture Handbook no. 405.

Sander, I. L., and B. C. Fischer. 1989. Central Hardwood forest types. Pp. 1.021–1.022 *in* Central hardwood notes (F. B. Clark and J. G. Hutchinson eds). North Central Forest Exp. Sta., St Paul, MN.

Shugart, H. H., Jr, and D. James. 1973. Ecological succession of breeding birds populations in northwestern Arkansas. Auk 90:62–77.

Smith, K. G. 1977. Distribution of summer birds along a forest moisture gradient in an Ozark watershed. Ecology 58:810–819.

Smith, H. C., and I. L. Sander. 1989. Silvicultural systems for harvesting mixed hardwood stands. Pp. 2.070–2.076 *in* Central Hardwood Notes (F. B. Clark and J. G. Hutchinson, eds). North Central Forest Exp. Sta., St Paul, MN.

Smith, K. G., and D. R. Petit. 1988. Breeding bird and forestry practices in the Ozarks: past, present, and future relationships. Pp. 23–49 *in* Bird Conservation 3 (J. A. Jackson, ed.). University of Wisconsin Press, Madison, WI.

Smith, W. B., and M. F. Golitz. 1988. Indiana forest statistics, 1986. USDA Forest Serv. North Central For. Exp. Sta. Resour. Bull. NC-108.

Society of American Foresters. 1981. Choices in silviculture for American Forests. Soc. Amer. For., Washington, DC.

Stauffer, D. F., and L. B. Best. 1980. Habitat selection by birds of riparian communities: evaluating effects of habitat alterations. J. Wildl. Manag. 44:1–15.

Sternitzke, H. S. 1976. Impact of changing land use on delta hardwood forests. J. For. 74:25–27.

Steyermark, J. A. 1959. Vegetational history of the Ozark forest. Univ. Mo. Stud. 31:1–38.

Stransky, J. J., and D. Richardson. 1977. Fruiting of browse plants affected by pine site preparation in east Texas. Proc. Annu. Conf. Southeast Assoc. Fish Wildl. Agencies 31:5–7.

Strelke, W. K., and J. G. Dickson. 1980. Effect of forest clear-cut "edge" on breeding birds in east Texas. J. Wildl. Manag. 44:559–567.

Swift, B. L., J. S. Larson, and R. M. DeGraff. 1984. Relationship of breeding bird density and diversity to habitat variables in forested wetlands. Wilson Bull. 96:48–59.

Thompson, F. R., III, and E. K. Fritzell. 1990. Bird densities and diversity in clearcut and mature oak–hickory forest. USDA Forest Serv. Res. Pap. NC-293.

Thompson, F. R., III, W. D. Dijak, T. G. Kulowiec, and D. A. Hamilton. 1992. Breeding bird populations in Missouri Ozark forests with and without clearcutting. J. Wildl. Manag. 56:23–30.

USDA Forest Service. 1982. An analysis of the timber situation in the United States 1952–2030. Forest Res. Rep. no. 23.

USDA Forest Service. 1988. The south's fourth forest: alternatives for the future. Forest. Resour. Rep. no. 24.

Van Horne, B. 1983. Density as a misleading indicator of habitat quality. J. Wildl. Manag. 47:893–901.

Walker, L. C. 1962. The Coastal Plain's southern pine region. Pp. 246–295 *in* Regional Silviculture of the United States (J. W. Barrett, ed.). Ronald Press Co., New York.

Webb, W. L., D. F. Behrend, and B. Saisorr. 1977. Effect of logging on songbird populations in a northern hardwood forest. Wildl. Monogrs. 55:6–35.

Weinell, D. E. 1989. Initial effects of hardwood reforestation and plant succession upon avian richness at the Ouachita Wildlife Management Area, Louisiana. MS thesis, Northeast Louisiana University, Monroe, LA.

Wesley, D. E., C. J. Perkins, and A. D. Sullivan. 1976. Preliminary observations of cottonwood plantations as wildlife habitat. Pp. 460–476 *in* Proc. Symp. East. Cottonwood and Related Species. Greenville, MS.

Willson, M. F. 1974. Avian community organization and habitat structure. Ecology 55:1017–1029.

Wood, G. W., and L. J. Niles. 1978. Effects of management practices on nongame bird habitat in longleaf–slash pine forests. Pp. 40–49 *in* Proceedings of the workshop on management of southern forests for nongame birds (R. M. DeGraaf, tech. coord.). USDA Forest Serv. Gen. Tech. Rep. SE-14.

Yahner, R. H. 1986. Structure, seasonal dynamics and habitat relationships of avian communities in small even-aged forest stands. Wilson Bull. 98:61–82.

Ziehmer, R. L. 1992. Effects of uneven-aged timber management of forest bird communities. MS thesis, University of Missouri, Columbia, MO.

APPENDIX

Scientific names of plants

Ash, white	(*Fraxinus americana*)
Aster	(*Aster* spp.)
Bald cypress	(*Taxodium distichum*)
Basswood, white	(*Tilia heterophylla*)
Beautyberry, American	(*Callicarpa americana*)
Beech, American	(*Fagus grandifolia*)
Blackberry	(*Rubus* spp.)
Blackgum	(*Nyssa sylvatica*)
Black bayberry	(*Myrica heterophylla*)
Black Cherry	(*Prunus serotina*)
Black Walnut	(*Juglans nigra*)
Blueberry	(*Vaccinium* spp.)
Bluestem	(*Andropogon* spp.)
Buckeye	(*Aesculus* spp.)
Buckwheat-tree	(*Cliftonia monophylla*)
Cane	(*Arundinaria* spp.)
Cottonwood, eastern	(*Populus deltoides*)
Curtis dropseed	(*Sporobolus curtissii*)
Deers-tongue	(*Trilisa odoratissima*)
Elm, winged	(*Ulmus alata*)
Flowering dogwood	(*Cornus florida*)
Gallberry	(*Ilex glabra*)
Gopher-apples	(*Chrysobalanus oblongifolius*)
Greenbriar	(*Smilax* spp.)
Hawthorn	(*Crataegus* spp.)
Hazel, beaked	(*Corylus cornuta*)
Hickory	(*Carya* spp.)
Bitternut	(*C. cordiformis*)
Mockernut	(*C. tomentosa*)
Pignut	(*C. glabra*)
Shagbark	(*C. ovata*)
Water	(*C. aquatica*)
Hophornbeam, eastern	(*Ostrya virginiana*)
Hornbeam, American	(*Carpinus caroliniana*)
Huckleberry	(*Gaylussacia* spp.)
Large (or sweet) gallberry	(*Ilex coriacea*)
Laurel, mountain	(*Kalmia latifolia*)
Lyonias	(*Lyonia* spp.)
Maple	(*Acer* spp.)
Red	(*A. rubrum*)
Sugar	(*A. saccharum*)
Oak	(*Quercus* spp.)
Black	(*Q. velutina*)
Blackjack	(*Q. marilandica*)
Bur	(*Q. macrocarpa*)
Chestnut	(*Q. prinus*)
Chinkapin	(*Q. muhlenbergii*)
Laurel	(*Q. laurifolia*)
Northern pin	(*Q. palustris*)
Northern red	(*Q. rubra*)
Overcup	(*Q. lyrata*)
Post	(*Q. stellata*)
Scarlet	(*Q. coccinea*)
Southern red	(*Q. falcata*)
Water	(*Q. nigra*)
Panicum	(*Panicum* spp.)
Partridge pea	(*Cassia fasciculata*)
Paspalums	(*Passpalum* spp.)
Persimmon	(*Diospyros virginiana*)
Pine	(*Pinus* spp.)
Loblolly	(*P. taeda*)
Longleaf	(*P. palustris*)
Pitch	(*P. rigida*)
Pond	(*P. serotina*)
Sand	(*P. clausa*)
Shortleaf	(*P. echinata*)
Slash	(*P. elliotti*)
Virginia	(*P. virginiana*)
Pond cypress	(*Taxodium ascendens*)
Poplar, yellow	(*Liriodendron tulipifera*)
Redbud, eastern	(*Cercis canadensis*)
St. Andrews cross	(*Hypericum hypericoides*)
Sassafras	(*Sassafras albidum*)
Saw-palmetto	(*Serenoa* spp.)
Serviceberry	(*Amelanchier* spp.)
Shining sumac	(*Rhus copallina*)
Sourwood	(*Oxydendrum arboreum*)
Spanish moss	(*Tillandsia usneoides*)
Swamp cyrilla	(*Cyrilla racemiflora*)
Swamp tupelo	(*Nyssa biflora*)
Sweet pepperbrush	(*Clethra* spp.)
Sweetbay	(*Magnolia virginiana*)
Sweetgum	(*Liquidambar styraciflua*)
Sycamore, American	(*Platanus occidentalis*)
Tickclovers	(*Desmodium* spp.)
Viburnum	(*Viburnum* spp.)
Water tupelo	(*Nyssa aquatica*)
Willow, black	(*Salix nigra*)
Wiregrass	(*Aristida stricta*)
Witchhazel	(*Hamamelis virginiana*)
Yaupon	(*Ilex vomitoria*)

PART IV

GENERAL HUMAN EFFECTS

10

EFFECTS OF AGRICULTURAL PRACTICES AND FARMLAND STRUCTURES

NICHOLAS L. RODENHOUSE, LOUIS B. BEST, RAYMOND J. O'CONNOR, AND ERIC K. BOLLINGER

INTRODUCTION

Agriculture probably has had greater impact on populations of Neotropical migratory birds than any other human activity. Although crop types differ greatly among regions, most counties in the United States include agricultural land (Sommer and Hines 1991), about 52% of the land area of the contiguous 48 states (USDA 1992) and 11% of Canada (Statistics Canada 1986) is in farms. Conversion of native habitats to agricultural land has resulted in local extirpation of migratory bird species (e.g., Owens and Myres 1973, Bowles et al. 1980, Blankespoor and Krause 1982), shifts in species composition and relative abundances (e.g., Graber and Graber 1963, 1983, Dinsmore 1981, Herkert 1991), and alteration of geographic ranges (e.g., Hurley and Franks 1976, Brittingham and Temple 1983, Dolbeer 1990). However, agriculture has also created habitat features used by Neotropical migratory birds (Freemark et al. 1991).

Field and edge areas created by agriculture are used by Neotropical migratory birds when breeding or migrating. Fields are the areas worked for crop production (including forage crops that may be grazed), and birds found in these fields often were formerly grassland species. Edges include field borders (e.g., fencerows, roadsides) but also uncultivated areas within fields, such as grassed waterways or terrace berms. Edges support many migratory bird species not found in fields (e.g., Best et al. 1990). The distinction between fields and edges is important not only because these landscape elements support largely different bird communities, but also because they are managed differently.

Most landbirds using crop fields and edges are Neotropical migratory birds, and agriculture has been implicated in the decline of some species associated with each (Rodenhouse et al. 1993). How agriculture may have contributed to these declines is often not clear. Nevertheless, changes in agricultural practices and farmland structure have coincided with the ongoing declines of Neotropical migratory birds in farmland (Rodenhouse et al. 1993). Agricultural practices are the methods used in crop production, including the choice of crop grown; and farmland structure comprises the types, relative coverage, and spatial distribution of habitat features in agricultural landscapes, including cultivated and uncultivated areas. Consideration of how changes in agricultural practices and farmland structure affect migratory birds will be an important part of developing conservation programs for these species.

The purpose of this chapter is: (1) to review how crop production practices and farmland structure influence the use of agricultural habitats by Neotropical migratory birds and; (2) to suggest management considerations relevant to migratory birds. Because studies of these species in North American agriculture are few, we draw on studies from agriculture in the temperate zone around the world and apply them to Neotropical migratory birds. Different crops and regions are disproportionately represented in this review because the effects of agriculture on birds have been studied in relatively few crops and regions. We focus on migratory birds during the breeding season and migration because research on the impacts of agriculture on migratory birds in winter, although potentially of great importance to their conservation (Greenberg 1992, Petit et al., Chapter 6, this volume), is very limited at present.

The results of this review suggest that the primary focus of efforts to conserve Neotropical migratory birds in farmland should be placed on creating landscape structures suitable for these species. Declines of migratory birds have probably occurred primarily because of loss of suitable habitat rather than declining habitat quality. Thus encouraging agricultural practices that benefit migratory birds will probably be effective only when suitable agricultural landscapes are created. Because the impact of both farmland structure and agricultural practices on Neotropical migratory birds has been little studied, we conclude by identifying research that is urgently needed to guide management decisions.

AGRICULTURAL PRACTICES

Neotropical migratory birds on farmland use fields, edge areas, or both for many types of activities: foraging, nesting, singing, preening, etc. Field and/or edge areas, therefore, must provide all the requisites for these activities to be suitable for bird use. For example, crop fields that provide food for Vesper Sparrows but that lack elevated song perches may be unacceptable or less preferred than fields that provide both (Best and Rodenhouse 1984). Furthermore, "suitable" habitat may result in harm to nesting migratory birds if it is frequently disturbed by agricultural operations, or if it also favors the abundance or activity of the predators or brood parasites of migratory birds. Thus, agricultural practices can affect migratory birds either directly or indirectly. Direct effects include harm to adults, disturbance of nests, and exposure to chemicals toxic to birds. Nest disturbance can be caused by tillage, planting, cultivation, burning of crop residues or uncropped vegetation, purposeful flooding of crop fields (e.g., for rice production), mowing, or harvesting. Juvenile and adult birds, particularly incubating females, can be injured by machinery. Indirect effects occur via changes in food supplies, vegetation, or natural enemies of Neotropical migratory birds. Such direct and indirect impacts of agricultural practices on migratory birds using croplands during breeding or migration are addressed in this section.

Agricultural Chemicals

Chemicals potentially harmful to birds include various insecticides, fungicides, nematicides, herbicides, and fertilizers. The use of these chemicals in agriculture has increased dramatically since the 1940s (Pimentel et al. 1991). Agricultural chemicals probably affect Neotropical migratory birds both directly via toxic effects, and indirectly by changes in bird food supplies or habitat. Because Gard and Hooper (Chapter 11, this volume) review the effects of agricultural chemicals on these species, we only address this topic briefly to add emphasis and to address points not covered by their review.

Pesticides continue to be a widespread threat to Neotropical migratory birds but the magnitude of this threat is largely unknown (Gard and Hooper, Chapter 11, this volume). Persistent pesticides that biomagnify in food webs have largely been eliminated from temperate agriculture in industrialized countries, although migratory birds are still exposed to these chemicals in the tropics (Risebrough 1986). These pesticides have been replaced by highly toxic but nonpersistent pesticides that have been associated with local "bird kills." Although there is little evidence that such local events have long-term impact on populations in farmland (Gard and Hooper, Chapter 11, this volume), effects vary greatly among bird species (Hart 1990). Furthermore, effects of pesticides on Neotropical migratory birds using farmland seldom have been studied (Gard and Hooper, Chapter 11, this volume). Despite this lack of study, some effects are obvious. Some intensively sprayed crops such as orchard crops, cotton, fruits (e.g., strawberries, blueberries), and vegetables are largely devoid of beneficial predators of crop pests, including birds, because of toxic sprays and repeated disturbance (some crops, e.g., apples and cotton, are sprayed more than 20 times per growing season; Pimentel et al.

1991). Birds are purposefully excluded from some crops because of the damage that they cause to the crop (see below).

Concern about the present reliance on chemical control of crop pests is growing because of the problems of pesticide resistance and environmental contamination (e.g., National Research Council 1989, Higley et al. 1992). This concern is leading to reductions in the rate of pesticide application by some farmers and to increases in the precision of pesticide targeting (Andow and Rosset 1990, Luna and House 1990), each of which reduces the exposure of migratory birds to pesticides. However, indirect effects of pesticides, particularly herbicides, continue to be a major concern. Herbicides are of concern because they simplify the vegetation structure and species composition of crop fields and uncropped areas that are purposefully or inadvertently sprayed (Risebrough 1986, Freemark and Boutin 1995). This simplification potentially reduces the cover and food resources available to migratory birds (Shelton and Edwards 1983, Chiverton and Sotherton 1991, Dover 1991).

The effects of inorganic fertilizers on the suitability of farmland for Neotropical migratory birds are seldom addressed. It is important, however, to consider these effects because inorganic fertilizers are used extensively in temperate agriculture and are required for most high-yielding crop varieties. Furthermore, intensive use of inorganic fertilizer has contributed to at least half of the 2–3-fold increases in crop productivity during the last half century (Boyer 1982). However, these chemicals probably have altered bird food supplies and the vegetation of uncropped areas. Use of inorganic fertilizers occurs with cultural practices that lead to declines in soil organic matter (Reganold et al. 1987, Doran and Werner 1990). Soil organic matter is essential to the maintenance of the soil fauna (Tate 1987, Hendrix et al. 1990). Fertilization also tends to sustain plant communities in early stages of plant succession (Carson and Barrett 1988). Thus fertilized fields are usually bordered by plant communities dominated by one or a few plant species. Both the loss of organic matter and the simplification of plant communities have potentially reduced food supplies for birds. Birds preferentially forage in fields with high organic matter content such as those fertilized with manure (e.g., Tucker 1992), and numerous studies demonstrate the positive relationship between plant species diversity, and the diversity and abundance of arthropods (e.g., Altieri and Whitcomb 1980, Shelton and Edwards 1983, Kirchner 1977, Chiverton and Sotherton 1991). Whether the reductions in arthropod abundance or the changes in vegetation structure and composition caused by fertilization and herbicide use have significantly reduced populations of Neotropical migratory birds in farmland is unknown, but it seems likely that neither inorganic fertilizers nor pesticides enhance the favorability of cropland for most of these species.

Tillage, Planting, and Cultivation

These disturbances to crop fields affect Neotropical migratory birds both directly and indirectly. Direct effects are predominantly via the destruction of nests, and the proportion of nests lost to such disturbances varies widely (reviewed by Rodenhouse et al. 1993). Although nest destruction can be reduced by using subsurface tillage methods (e.g., Rodgers 1983) or no-tillage farming (Basore et al. 1986), the impact of nest losses to agricultural operations on populations of migratory birds is largely unknown. Nests lost to field operations early in the season probably are replaced and whether such losses ultimately reduce annual breeding productivity enough to lower recruitment has not been studied. Beintema and Muskens (1987), who studied the nesting success of birds breeding in very productive Dutch agricultural grasslands, suggest that such losses may not threaten the persistence of local populations.

Indirect effects of tillage occur via reductions in sheltering cover and food abundance for birds. Tillage usually buries 75% or more of crop residues (Slonecker and Moldenhauer 1977). These residues include waste grains or weed seeds that are foods for numerous migratory species (Rodenhouse et al. 1993), and decomposing organic residues are the habitat or food of the detritivorous arthropods (Hendrix et al. 1986) that are frequently consumed by omnivorous and insectivorous migratory birds. Burial of crop residues kills some arthropods associated with litter (Hendrix et al. 1986, Doran and Werner 1990) and speeds litter decomposition (Buchanan and King 1993). However, turning the soil can provide a temporary abundance of food for birds by exposing soil dwelling arthopods and worms (O'Connor and Shrubb 1986).

Reduced tillage and no-tillage methods of farming that leave significant quantities of crop residues on the soil surface have frequently been associated with greater abundance and diversity of birds than conventionally tilled fields (Warbuton and Klimstra 1984, Castrale 1985). Numerous studies indicate that reduced tillage fields have greater abundance of litter arthropods (reviewed by Rodenhouse et al. 1993) but such fields may (House and Stinner 1983) or may not (Basore et al. 1987) support greater abundances of foliage dwelling arthropods. The concealing cover of crop residues may also make reduced tillage fields more attractive to nesting and foraging birds (Castrale 1985, Rodenhouse and Best 1994). However, the frequency of nesting losses to agricultural operations may be no less in reduced than in conventional tillage fields (Basore et al. 1986). Consequently, whether reduced tillage is a net benefit or detriment for bird populations in cropland is unknown (Wooley et al. 1985, Basore et al. 1986, Best 1986). The growing acceptance of reduced tillage for agroeconomic reasons (Weersink et al. 1992) provides opportunities to test whether these practices will be of long-term benefit to breeding or migrating birds.

Burning of Crop Residues

Residues remaining after harvest of a variety of crops are sometimes burned for pest control or to speed the release of nutrients (Martin et al. 1976). Crop residues burned include alfalfa, wheat, sugar cane, asparagus, soybeans, rice, blue grass (for seed), and prairie grasses (for seed or pasture). Such burning is usually conducted after the nesting season of most migratory birds (with the exception of some relay cropping systems that are harvested in June or July). Burning of residues may affect breeding or migrating birds by reducing arthropod abundances in fields, or by reducing the amount of cover available to foraging or nesting birds. Whether such alterations signficantly affect migratory bird populations is unknown. To our knowledge, the effects of burning crop residues on breeding or migrating birds have not been reported.

Burning is sometimes used to "renovate" pastures. The effect of burning on migratory birds using these areas is affected by the timing of the burn. If burning occurs late in the spring, nests can be destroyed and birds displaced from territories. Fall burning is less likely to have adverse effects on migratory birds. Shortly after burning, vegetation is often short and sparse, favoring species such as Horned Larks and Vesper Sparrows. As the density and biomass of vegetation increases, species such as Savannah Sparrows and Bobolinks are favored (Cody 1985).

Uncropped areas such as fencerows and road verges may also be burned by farm operators, primarily to control noxious weeds or to maintain a farm that looks "clean." Burning of uncropped areas usually occurs in spring (but may occur at any time of the season), and this practice may, therefore, destroy active nests. However, no information on the prevalence or impacts of this practice on migratory birds is available. Studies of prairie burns indicate that, after burning, regrowth of vegetation and arthropod communities may provide more food for birds than in unburned areas (Warren et al. 1987, Collins and Wallace 1990). Burning, however, also eliminates woody vegetation and some forbs from field borders, simplifying vegetation structure and composition. Such changes in vegetation structure may benefit grassland birds while reducing the habitat available to species that use woody vegetation. Burning, therefore, is potentially an effective tool for managing uncultivated habitats but its effects on the requisites of migratory birds are poorly known at present.

Flooding of Fields

Flooding is typical of the production of only one major crop, rice, which is grown primarily in Arkansas, Louisiana, Mississippi, Texas, and California (Martin et al. 1976). Rice fields are usually plowed or the residues burned soon after fall harvest, and the resulting bare fields in spring are attractive to few Neotropical migratory birds. Hence rice flooding, normally carried out in spring, probably affects few migratory birds. Rice fields remain flooded during the growing season. Studies of Neotropical migratory birds using rice fields are few, including studies of the effects of pesticides on birds (e.g., Flickinger et al. 1980) and studies of crop damage (e.g., Brugger et al. 1992). Red-winged Blackbirds cause the most damage to rice crops.

Mowing or Harvesting

The direct and indirect effects of mowing on Neotropical migratory birds of grasslands are well documented. Mowing often results in the destruction of nests and in the abandonment of territories (e.g., Beintema and Muskens 1987, Bollinger et al. 1990, Frawley and Best 1991, Bryan and Best 1994). Mowing at night can also cause high mortality of adult birds that are attending nests or roosting (Frawley, personal communication). Indirect effects of mowing also occur via altered habitat structure and plant species composition. Consequently, harvesting of hay or mowing of uncropped areas such as roadsides is associated with reduced diversity and density of some Neotropical migratory birds [e.g., Bobolink (Bollinger et al. 1990), Savannah Sparrow (Warner 1992)] and may actually favor individuals of other species (Laursen 1981), particularly those that nest on the ground only in areas of sparse vegetation (e.g., Killdeer, Horned Lark, Vesper Sparrow) or those that prefer to forage in the short vegetation of mowed areas [e.g., American Robin (Eiserer 1980), Northern Mockingbird (Roth 1979)].

The frequency and timing of mowing determine the impact of this practice on Neotropical migratory birds. Hayfields are typically mowed at least once per year to harvest the forage crop, with two to four cuts per field per season being more common. Dates of the initial harvest are typically in late May or early June in midwestern and northeastern United States (Warner and Etter 1989, Bollinger et al. 1990, Frawley and Best 1991) and coincide with the peak of the nesting season for Neotropical migratory birds such as Dickcissels, Bobolinks, meadowlarks, and Grasshopper Sparrows (earlier dates for first mowing occur farther south). Mowing in late May or early June is especially detrimental to long-distance migrants (e.g., Bobolink, Dickcissel) that have later nesting seasons because mowing at this time typically results in 0% nesting success. Because of weather and logistical constraints, however, many fields may not be cut until later in June or July, allowing some successful nests in these habitats. Nevertheless, Bollinger et al. (1990) estimated that production of fledgling Bobolinks was reduced by 40% in their New York study area by mowing of hayfields. They concluded (as did Frawley and Best 1991) that mowing is probably contributing to declining populations of grassland-nesting species. Besides nest destruction, mowing greatly reduces vegetation density and complexity, making mowed areas unsuitable for species such as Bobolinks, Red-winged Blackbirds, Dickcissels, Sedge Wrens, and Common Yellowthroats, but acceptable (and perhaps preferable) for species such as Savannah Sparrows, meadowlarks, Horned Larks, Vesper Sparrows, and Grasshopper Sparrows (Bollinger 1988, Frawley and Best 1991). Even these species, however, may not be able to complete nesting before subsequent mowings occur.

Current trends in forage crop production toward earlier first-cutting and more rapid rotation of hayfields may result in harm to grassland species (Bollinger and Gavin 1992). Dates of first-cutting of hayfields are currently 1–3 weeks earlier than they were 50 years ago (Warner and Etter 1989), largely because of the development of faster growing (or regrowing) forage crop hybrids and the shift to alfalfa hays that mature earlier than the timothy–clover hays that were formerly grown. Earlier first-cutting probably has resulted in increased nest mortality of migratory birds (Bollinger et al. 1990) and the trend is likely to continue. Rapid rotation of hayfields may be detrimental because uniform alfalfa hays are less preferred as breeding habitat by many grassland bird

Table 10-1. Crops were tested for significant differences in the strength of the species–crop associations between Neotropical migratory birds that eat the crop versus those that do not eat it. Association values (percentage of variance explained) were from O'Connor and Boone (1992) and Lauber (1991) who used classification and regression tree methods (Breiman et al. 1984, Clark and Pregibon 1992) to determine associations. The statistical significance of the small percentages was due to the large samples involved. The primary source of information for bird diets was Martin et al. (1961).

Crop	Eat		Do not Eat		Mann–Whitney U-test[a]	
	N[b]	% Variance	N[b]	% Variance	z	P
Sorghum	8	1.7	20	0.8	−1.53	0.06
Oats	10	1.3	8	1.0	−1.52	0.06
Corn for grain	6	3.4	20	1.1	−2.38	0.01
Winter wheat	10	1.5	10	1.2	−0.95	0.17
Spring wheat[c]	5	3.8	2	0.1		
Durum wheat	4	1.1	5	0.2	−2.46	0.01
All wheat	4	3.8	8	1.3	−2.21	0.01
Sunflower[c]	6	1.5	1	0.6		

[a] One-tailed test of the hypothesis that birds eating the crop would have a stronger association with the crop than those not eating the crop.
[b] Number of bird species significantly associated with the crop that either eat or do not eat the crop.
[c] Sample of bird species was too small for the statistical test.

species than the more diverse vegetation of timothy–clover mixtures (e.g., Bobolinks; Bollinger 1988). Alfalfa fields become a more suitable habitat in time but current practices are to rotate hayfields to other crops before this vegetational succession occurs (Bollinger and Gavin 1992).

Summer harvesting of crops, usually small grains, is a potential threat to bird nests and adults. Summer harvesting occurs in multiple cropping systems (e.g., the barley–cotton–rice rotation of the southern United States or the small grain–soybeans rotation of the mid latitudes). Multiple-cropping systems are increasingly used (Stinner and Blair 1990), and 10% of all soybeans are now grown in multiple-cropping systems (Rice and Vandermeer 1990). Multiple cropping is used more in mid and low latitudes of North America than at high latitudes where the short growing season precludes two crops. The timing of summer harvests, therefore, strongly depends on latitude (i.e., the length of the growing season) as well as on the particular crop harvested. The impact of summer harvesting on nesting birds depends on the timing of field operations relative to avian nesting cycles. We are unaware of any studies of the ecology of Neotropical migratory birds in multiple-cropping systems.

Summer or fall harvesting may result in waste grain remaining in fields after harvest. These seeds are foods for many Neotropical migratory birds [24 of the 52 Neotropical migratory species examined by Rodenhouse et al. (1993)], and some probably have benefited from the presence of waste grains. Using data from O'Connor and Boone (1992), we found that the migratory species consuming grains had stronger positive associations with small grain crops than nonconsumers (Table 10-1). However, the benefit of waste grains even for consumers may be offset by brood parasitism. Brittingham and Temple (1983) noted that the dramatic increase in cowbird abundance during this century coincided with the expansion of rice production in southern United States. Waste rice is a major food of wintering cowbirds (Meanley 1971).

Effects of Agricultural Practices on Brown-headed Cowbirds and Nest Predators

The effects of specific agricultural practices on the abundance of and parasitism by Brown-headed Cowbirds in breeding areas have not been investigated (Robinson et al. 1993). However, it is unlikely that any relationships would be detected because cowbirds search for nests and forage for food over broad areas (Rothstein et al. 1984). Practices that minimize waste grains in croplands, however, may reduce feeding opportunities for this parasitic species (Robinson et al. 1993).

Predation is a major cause of nesting losses of Neotropical migratory birds using farmland (Table 10-2), but little is known about the impacts of agricultural practices on nest predators. In general, farmland provides the food and habitat supporting a wide variety of predators (e.g., Roseberry and Klimstra 1970) that prey on the nest contents, fledglings, and adults of Neotropical migratory birds. Many of these predators are omnivorous, including: corvids (Corvidae: crows, jays, magpies), larids (Laridae: gulls), canids (Canidae: domestic and feral dogs, foxes, coyotes), and various other mammals [racoons, opossums, skunks, and rodents (Rodentia: squirrels, chipmunks, mice)]. Population sizes of these species in farmland may be enhanced by the crops, waste grains, and arthropods available in fields as well as by the fruits available in uncultivated areas [e.g., mulberry (*Morus* spp.) and raspberry (*Rubus* spp.)]. Carnivorous predators of Neotropical migratory birds include domestic and feral cats, snakes (primarily Colubridae), and avian predators, some of which are themselves Neotropical migratory birds (e.g., Loggerhead Shrike, American Kestrel). Avian carnivores have been studied in farmland (e.g., Newton and Marquiss 1986, Smith and Kruse 1992), but the low densities of these predators in farmland probably obviates significant impact on the distribution or nesting success of their avian prey.

Unfortunately, the population dynamics and predatory activities of the predators of Neotropical migratory birds have seldom been studied in North American farmland. In European farmland, predation rates can be equal to or greater than those in uncropped habitats (Angelstam 1986, O'Connor and Shrubb 1986), suggesting that predator abundances may be enhanced in farmland or that nests there are more vulnerable to predation. The nearly ubiquitous presence in farmland of predators that are human commensals suggests that some forms of agriculture may enhance predator densities. Furthermore, commensals can be efficient predators (Liberg 1984). Cats killed a wide variety of birds nesting in and near a rural English village, and cats were a major predator of certain species (Churcher and Lawton 1987). Specific agricultural practices, however, may have little effect on predator abundances where croplands are only used for foraging by predators. Indeed, nearly all predators of Neotropical migratory birds in farmland require uncropped areas for reproduction and wintering, and hence, are only sustained in landscapes that have these elements.

FARMLAND STRUCTURE

The structure of farmland can be characterized by the types, relative abundances, and spatial distribution of landscape elements such as crops, pasture, and uncropped areas (Freemark et al. 1993). The latter include fencerows, grassed waterways, road verges, woodlots, riparian strips adjacent to streams or ponds, as well as farm buildings and yards. Characteristics of these elements that influence their use by Neotropical migratory birds include vegetation structure, disturbance regimes, the presence of specialized nesting sites (e.g., buildings for swallows), and the size and spatial juxtaposition of these elements. Consequently, landscape elements and landscapes differ greatly in the resources available to migratory birds. In this section, we address what is known about how different types of crops, uncropped areas, and spatial structures of landscapes influence Neotropical migratory birds in farmland. Consideration of crop types necessarily includes assessment of field sizes and cropping systems. Regarding uncropped areas, we address the ongoing loss and fragmentation of these elements, and we note how the Conservation Reserve Program (CRP) and agroforestry may influence present trends. Last, we address how the spatial structure of agricultural landscapes influences migratory birds, their natural enemies, and the crop damage caused by migratory birds.

Crops, Field Sizes, and Cropping Systems

Associations between Crops and Migratory Birds

Rodenhouse et al. (1993) reported that 50 of 52 migratory bird species examined were

Table 10-2. The productivity (fledglings per breeding pair per season) of Neotropical migratory birds breeding in farmland. The level needed to balance mortality of adults and juveniles is about ≥3 fledglings/pair/year (e.g., Rodenhouse and Best 1983, Probst 1986, Sullivan 1989) except for Loggerhead Shrikes which is 5.5 (Brooks and Temple 1990). (From Rodenhouse et al. 1993.)

Migrant Species	Number of Nests	Fledglings/ Breeding Pair/Year	Nesting losses (%)[a]			Nesting Habitat	Reference
			Predation	Agriculture	Parasitism		
Vesper Sparrow	45	2.8	29	27	11	Corn/soybeans	Rodenhouse and Best (1983)
	74	2.9[b]				Corn/soybeans	Perritt and Best (1989)
	10	2.4	50	10	0	Alfalfa	Frawley (1989)
	35	1.4[c]	54	0	9	No-till/strip cover[d]	Basore and Best (unpublished data)
Grasshopper Sparrow	41	0.8[c]	80	2	2	No-till/strip cover	Basore and Best (unpublished data)
Loggerhead Shrike	222	2.2	86	0	0	Roadside[e]	DeGeus (1990)
	100	3.3				Pasture	Tyler (1992)
Bobolink	33	0.3	9	85	0	Hayfield	Bollinger et al. (1990)
Dickcissel	34	0.2	18	50	21	Alfalfa	Frawley (1989)
	69	1.7[c]	28	23	3	Oat field	Frawley and Best (unpublished data)
	27	2.2	44	8	19	Waterway[f]	Bryan (1990), Bryan and Best (1994)
Red-winged Blackbird	41	0.4	29	41	10	Alfalfa	Frawley (1989)
	133	1.2[c]	50	1	20	No-till/strip cover	Basore and Best (unpublished data)
	65	0.9[c]	20	42	5	Oat field	Frawley and Best (unpublished data)
	63	1.0	27	33	16	Waterway	Bryan (1990), Bryan and Best (1994)
	73		55	1	4	Roadside[e]	Camp and Best (1994)
Western Meadowlark	9	0.1	56	1	0	Alfalfa	Frawley (1989)
	15	0.7[c]	47	20	0	No-till/strip cover	Basore and Best (unpublished data)
Killdeer	12	6.9[c]	8	0	0	No-till/strip cover	Basore and Best (unpublished data)
Mourning Dove	13	0.4[c]	31	8	0	Oat field	Frawley and Best (unpublished data)
	12	1.5[c]	33	17	0	No-till/strip cover	Basore and Best (unpublished data)

[a] Losses (as a percentage of all nests) to predation, agricultural activity or brood parasitism. Nests considered lost to parasitism were deserted due to this cause or fledged only cowbird young.
[b] Calculated as the mean number of successful nests per female for 1984 and 1985 (mean = 0.77) times mean clutch size (3.8) from Rodenhouse and Best (unpublished data). Missing values indicate data either not gathered or reported.
[c] Calculated as (number of fledgings per successful nest) × (nesting success) × (two nesting attempts). Each female was assumed to make two nesting attempts.
[d] Includes nests in no-till corn and soybean fields and adjacent strip cover.
[e] Roadsides adjacent to corn and soybean fields.
[f] Grassed waterways within corn and soybean fields.

Table 10-3. Number of positive or negative associations between Neotropical migratory bird species ($N = 52$ species) and major crop types. Association was determined using Breeding Bird Survey abundances and USDA agricultural statistics by county for 1973–1989, and are taken from Lauber (1991) and O'Connor and Boone (1992).

Crop category	Number of Significant Associations		Chi-square[a]	Probability
	Positive	Negative		
CRP[b]	19	2	13.76	<0.01
Soybeans	17	13	0.53	0.47
Corn for grain	21	8	5.83	0.02
Corn for silage	4	5	0.11	0.74
Sorghum	16	15	0.03	0.86
Barley	14	3	7.12	<0.01
Oats	18	6	6.00	0.01
Spring wheat	11	4	3.27	0.07
Durum wheat	8	5	0.69	0.41
Winter wheat	20	5	9.00	<0.01
All wheat[c]	14	2	9.00	<0.01
Alfalfa	7	2	2.78	0.10
Other hay[d]	5	3	0.50	0.48
All hay[e]	9	4	1.92	0.17
Sunflower seed	9	1	6.40	0.01
Peanuts	0	5		
Cotton	3	3		
Flax seed	5	0		
Rice	4	1		
Sugar beets	3	3		
Dry beans	3	0		
Tobacco	5	0		
Potatoes	1	2		
Total	216	91		

[a] Chi-square value for a test of equal frequencies of migrants and residents associated with each crop category. No values are given where sample sizes were too small for a valid test.
[b] Conservation Reserve Program.
[c] Includes winter wheat, spring wheat and durum wheat.
[d] Includes all types of hay excluding alfalfa.
[e] Includes the categories alfalfa and other hay.

significantly correlated with the coverage of one or more of 22 crops (percentage coverage within counties) or the amount of land in the CRP (see Appendix). Because both bird abundances and CRP or crop coverages were measured at the county level, these correlations indicated bird responses to the types of farmland dominated by these crops or CRP land, including their coverage, spatial arrangement, and associated uncropped areas. Crop types differed greatly in the number of bird species with which they were associated, ranging from the 31 species correlated with sorghum coverage down to three species each for beans and potatoes (Table 10-3). In fact, all of the crops with significantly more positive than negative associations were grains potentially consumed by migratory birds. Less expected was the prevalence of positive correlations, both overall (216 of 307 significant associations or 70%) and within some individual crops (Table 10-3). However, even this favorable bias (for migratory birds) was dwarfed by the positive effects of the CRP acreage. No less than 19 of 21 species (90%) were positively correlated with the CRP, i.e., displayed greater abundance in counties with a larger proportion of CRP lands. The greater prevalence of positive correlations for conservation acreage compared to those for alfalfa (seven of nine correlations or 78% were positive) and hay (nine of 13 or 69%) indicated that uncropped land in the CRP was highly valuable for migratory birds. What was less certain was the extent to which CRP management was responsible for these benefits. Lauber (1991) found that some bird species had long been more abundant in areas where subsequent enrollment in the CRP was heavy, though for other species, density increases accorded with the institution of the CRP.

Benefits of certain crop types or CRP for some species may be offset by detriments for other species. This was indicated by the significant positive correlation between the number of positive and negative associations for the 22 crop categories plus CRP tested ($r_s = 0.44$, df $= 21$, $P = 0.04$). Such detrimental effects might be expected, where increasing crop coverage is associated with increasing field sizes and hence decreases in the amount of edge.

Field Size

Field sizes have increased throughout central North America (Best et al. 1990) as has reportedly occurred elsewhere (O'Connor and Shrubb 1986). The trend toward larger field size is driven by the increasing size of farm machinery and by the specialization of farms on producing one or a few commodities. Increasing field size results in the removal of field-edge vegetation. Data on the loss of edge areas are rarely available for most regions or crops in North America, but in central United States, from 30 to 80% of this habitat has been removed since the 1930s (reviewed by Rodenhouse et al. 1993). Of course, field sizes and edge vegetation may have changed little for some labor intensive crops such as orchards, vineyards, or tobacco, and edge areas of benefit to migratory birds have probably been created by irrigation in semi-arid areas (e.g., southwestern United States and California). We are unaware of avian studies focusing on field size for any of these crops or cropping systems. Yet, the loss of edge areas as a result of increasing field sizes probably significantly reduces the use of croplands by migratory birds because many migratory birds are associated only with edge areas during migration or breeding (e.g., Graber and Graber 1963, Best et al. 1990, Warner 1994).

The trend toward larger field sizes might benefit grassland migratory birds if their nesting requirements are met. For example, both the Kildeer and Horned Lark probably have increased in abundance in the croplands of central North America because they nest in bare fields and begin nesting early (Beason and Franks 1974, Brunton 1988). They are sometimes able to raise a brood before seedbed preparation begins. Most migratory birds of grasslands, however, require perennial vegetation for nesting cover, and they begin nesting later than Killdeer and Horned Larks (when nests in fields are destroyed by field operations). Large fields, therefore, will be used by grassland birds only if they include areas with perennial herbaceous vegetation (e.g., grassed waterways; Bryan and Best 1991), and such fields will be beneficial to migratory birds only if nesting is not disrupted by field operations.

Cropping Systems

In most cropping systems of temperate North America, one crop is raised on each field in each year. With the soil largely bare of living vegetation for 8–9 months of each year, soil erosion is enhanced and essential nutrients are lost (Papendick et al. 1986). Widespread recognition of the soil-fertility and off-farm costs of these systems (e.g., ground-water pollution, Hallberg 1987; siltation of surface waters, National Research Council 1989) are resulting in farm operators increasingly moving to relay cropping (where two crops are raised sequentially in one field each year) (Stinner and Blair 1990) and/or the use of cover crops or green manures that retain soil and soil nutrients from fall to spring (e.g., Sullivan et al. 1991). No studies of migratory birds have compared the effects of these alternative cropping systems on bird populations. It is likely, however, that the impacts of these cropping systems on migratory species nesting in fields will depend on the number and timing of field operations, and hence, on the number of nests destroyed. Relay cropping may increase the number of field operations and thereby raise the probability of nest destruction. For migratory species that only forage in crop fields and nest elsewhere, alternative cropping systems may have the largest impacts via food availability. Whether food availability is increased or decreased will depend on the foraging capabilities of species. For example, rye cover crops or winter wheat may make croplands unsuitable for foraging by edge or grassland species that do not forage in tall grass habitats. Leguminous cover crops, however, may benefit migratory birds that forage in dense vegetation because these plant species typically support more dense arthropod populations than grasses, e.g., Curry (1986).

Uncropped Landscape Elements

Fragmentation and Loss of Uncropped Areas

Two key landscape features are changed by the fragmentation of native habitats to create cropland—the diversity of landscape elements and the sizes of these elements. Landscape diversity is typically decreased because a variety of native habitat types are converted to the few types that compose farmland, including crop fields, fencerows, woodlots, riparian strips, and farmsteads. This decrease in diversity continues as cropping intensity increases, as it has in central North America during the past 50 years (Barrett et al. 1990, Warner 1994). Even where significant areas of cropland are abandoned (e.g., northeastern United States), habitat diversity may not return because most of the abandoned land follows a common pattern of secondary succession (see Confer and Knapp 1981). The size structure of landscape elements is changed by conversion of native habitats to farmland because large homogeneous crop fields are created from a continuously variable habitat mosaic. This shift occurs even in seemingly uniform grassland habitats because patterns of habitat disturbance are altered, e.g., Collins (1992), and patchily distributed habitat types such as wetlands are eliminated.

Loss or degradation of permanent or temporary wetlands (e.g., ponds, sloughs, riparian strips, marshes) have probably affected populations of Neotropical migratory birds negatively in farmland. These highly productive areas can support a high diversity and density of species, e.g., Stauffer and Best (1980), and often the migratory species associated with wetlands (e.g., Say's Phoebe, Bell's Vireo, White-eyed Vireo) are found nowhere else in agricultural landscapes (Dinsmore 1981). Species that use both wetlands and upland areas may have greater nesting success when breeding in or near wetlands, e.g., Robertson (1972) and Wittenberger (1978). However, even seemingly minor alterations in riparian vegetation, e.g., Stauffer and Best (1980), or the surrounding landscape, e.g., Goldwasser et al. (1980), can significantly reduce the quality of riparian habitat for migratory birds. Despite the apparent importance of wetlands to migratory birds, effects of the loss or degradation of these habitats on populations of Neotropical migratory birds have seldom been studied.

The increasing homogeneity of agricultural landscapes during the past 50 years (O'Connor and Shrubb 1986, Warner 1994) is associated with significant declines in both the abundance and diversity of Neotropical migratory birds that use either edge vegetation or crop fields (Rodenhouse et al. 1993). Migratory species that nest in edge areas have declined as edge vegetation has been converted to grasses or eliminated entirely (Best 1986, Warner 1994). Grassland species have declined as the amount of land in hay or pasture has decreased dramatically during past decades (USDA 1990, Bollinger and Gavin 1992). Grassland birds require areas (of various sizes) without trees or shrubs for breeding (Herkert 1994), and many areas in farmland are made unsuitable for breeding by a lack of perennial grassland or by woody vegetation.

Effect of CRP on Landscape Structure and Migratory Bird Populations

The decline in habitat suitable for grassland birds may be reversed by the CRP. This federal program reimburses landowners for setting aside highly erodable land for 10-year periods. Perennial vegetative cover must be established on this land, and it cannot be harvested or grazed when enrolled. Over 14 million ha of land have been enrolled in this program since its inception in 1985, and this land is widely distributed throughout the United States (Lauber 1991). By far the largest amount of land in the CRP has been planted to grasses (over 85%), with less than 10% in woody vegetation and less than 1%

in wetlands of various types (Dunn et al. 1993). The largest areas of land in the program are concentrated in north central United States where small grains are the primary crops (USDA 1990).

The CRP has reversed landscape fragmentation in some intensively cultivated areas (Dunn et al. 1993), but largely by the creation of one type of landscape element—grassland. Whether the CRP land is having positive effects on the abundances of Neotropical migratory birds is still poorly known. However, all studies of CRP impacts on bird populations indicate that migratory birds are using these areas (Lauber 1991, Johnson and Schwartz 1993). Johnson and Schwartz (1993), in particular, found that those grassland birds whose populations declined most in intensively cultivated areas were often those that bred at highest densities on CRP land. They suggested that, if the CRP is continued, breeding by grassland birds in this habitat might reverse the declines in their abundances. Robinson et al. (1993) noted, however, that for grasslands to be of maximum benefit to bird populations, the amount of edge associated with these grasslands should be minimized. This recommendation rests on the finding that grassland birds nesting adjacent to edge areas are more likely to be parasitized by Brown-headed Cowbirds (Johnson and Temple 1990). Benefits of CRP land for Neotropical migratory birds also may be reduced by present management practices. About 50% of the grassland in this program is mowed during the avian breeding season, largely for weed control (Miller and Bromley 1989), and Hays et al. (1989) reported that mowing of CRP land negatively affected at least one nesting migratory species, Eastern Meadowlarks.

Agroforestry

Agroforestry provides opportunities to enhance the amount of edge with trees and shrubs in farmland, to increase landscape diversity greatly (Pimentel et al. 1992), and to contribute to crop production and farm profit (Wilson and Diver 1991). Unfortunately, studies of Neotropical migratory birds in agroforestry systems are lacking. However, the extensive use by migratory birds of shelterbelts (e.g., Cassel and Wiehe 1980, Yahner 1983, Johnson and Beck 1988), wooded fencerows (e.g., Best 1983, Shalaway 1985), and woodlots (e.g., Best et al. 1990) suggests that agroforestry has the potential to increase habitat for species that use edge areas during breeding or migration. Most raptors, in particular, prefer forested areas for roosting and nesting, but forage in adjacent open areas (Kingsley and Nicholls 1993), and hence, may be favored by expansion of agroforestry. Insectivorous Neotropical migratory birds are likely to benefit from the enhanced abundance of arthropod prey associated with trees in agroforestry systems (Pasek 1988, Peng et al. 1993).

Spatial Structure of Agricultural Landscapes

Effects of Spatial Structure on Neotropical Migratory Birds

Diverse habitat types in close proximity have long been recognized as essential for the maintenance of game bird species within farmland (reviewed by O'Connor and Boone 1992). It is likely that some Neotropical migratory species also benefit from diverse landscape structure (see Renken and Dinsmore 1987, Jenny 1990), including both pest (Red-winged Blackbirds; Clark et al. 1986) and threatened species (Loggerhead Shrike; Gawlik and Bildstein 1993). For example, when one crop type is disturbed (e.g., by cultivation, mowing, or harvesting), nesting or foraging birds might find refuge in an adjacent undisturbed area (e.g., Rodenhouse and Best 1983). Similarly, rapid vegetation development in one cropped or uncropped area that makes that area unsuitable for foraging or nesting might have little effect on birds if vegetation in adjacent areas developed at a different rate (Rodenhouse and Best 1994).

Effects of Farmland Structure on Predators and Brown-headed Cowbirds

Landscape structure seems to have a strong impact on the distribution of nest predation, and predation usually causes most nesting losses in crop fields and edge areas (reviewed by Rodenhouse et al. 1993). Most mammalian

(Fritzell 1978, Glueck et al. 1988), avian (Johnson and Adkisson 1985), and reptilian (Durner and Gates 1993) nest predators are highly mobile and travel along corridors created by uncropped land. Thus, farmland structure may imperil migratory birds by increasing the spatial concentration of predators that feed on adults, juveniles, or nest contents (Gates and Gysel 1978). Nests of many migratory bird species are concentrated in or near edge habitat (e.g., Rodenhouse and Best 1983), and nests in or near these travel lanes often have a low probability of success (Gates and Gysel 1978, Basore et al. 1986, DeGeus 1990, Johnson and Temple 1990, Bryan and Best 1994, Camp and Best 1994). Furthermore, nest predation in farmland seems to be equal to or higher than that in uncropped habitats (O'Connor and Shrubb 1986, Angelstam 1986), except in intensively cultivated landscapes largely lacking edge areas and probably nest predators (Warner 1994), or in uncropped habitats lacking an important nest predator. Shalaway (1985) reported high nesting success (58% of nests fledged at least one young) in a wooded fencerow that lacked small mammalian predators which often destroy bird nests. The causes of high rates of nest predation in farmland have not been clearly identified, largely because the activities and densities of predators in farmland seldom have been quantified.

Farmland structures including fields of small grains, livestock paddocks, and uncropped areas with shrubs or trees that provide elevated perches probably enhance parasitism of Neotropical migratory birds by Brown-headed Cowbirds. Most of the hosts parasitized by cowbirds are migratory birds (Robinson 1992a), and brood parasitism often severely reduces reproductive success of these species in farmland (Table 10-2). Small grains and livestock provide foraging opportunities for cowbirds, and cowbirds often search for host nests by using elevated perches (reviewed by Robinson et al. 1993). In fact, brood parasitism is more frequent near field edges with elevated perches than away from such edges (Best 1978, Johnson and Temple 1990). Thus the conversion of native vegetation to farmland creates the food resources and edge that promotes both cowbird abundances and their effectiveness in finding host nests. Only farmland structures that more closely resemble native habitat than present structures are likely to result in reduced rates of cowbird parasitism.

Damage to Crops by Neotropical Migratory Birds

Fewer than 10 of the 215 migratory landbird species are currently reported to cause significant damage to agricultural crops over wide geographic areas (Table 10-4). Both grain and fruit crops are damaged. The

Table 10-4. Examples of damage to agricultural crops in the United States caused by Neotropical migratory birds (after DeGrazio 1978).

Crop	Species Causing Damage	Location	Source
Corn	Red-winged Blackbird	Nationwide	Besser and Brady (1986)
Sorghum	Red-winged Blackbird	Nationwide	Knittle and Guarino (1976)
Oats	Red-winged Blackbird	AR[a]	Agricultural Extension Service (1964)
Rice	Red-winged Blackbird, Brown-headed Cowbird, Bobolink, Dickcissel	AR, LA, CA, TX	Meanley (1971)
Sunflowers	Red-winged Blackbird,	MN, NE	Stone (1973b)
	Yellow-headed Blackbird	ND, SD, MN	Twedt et al. (1991)
Cherries	American Robin, Northern Oriole	MI, NY	Stone (1973a) Tobin et al. (1991)
Grapes	American Robin, orioles, warblers	Nationwide	Crase et al. (1976) Seamans and Caslick (1983)
Blueberries	American Robin, Gray Catbird	Nationwide	Mott and Stone (1973) Tobin et al. (1988)

[a] AR = Arizona; LA = Louisiana; CA = California; TX = Texas; MN = Minnesota; ND = North Dakota; SD = South Dakota; MI = Michigan; NY = New York.

principal migratory species involved in damage to grains are the Red-winged Blackbird, Brown-headed Cowbird, Bobolink, Dickcissel, and Yellow-headed Blackbird. Of these, Red-winged Blackbirds cause by far the most economic damage. Although bird damage typically reduces total yield on a state or nationwide basis by less than 1–2%, e.g., Dolbeer (1990), economic losses to individual farmers may be severe. The migratory birds most frequently mentioned as damaging fruit crops are American Robins, Gray Catbirds, and Northern Orioles. Crops eaten include cherries, grapes, blueberries, and strawberries (e.g., Seamans and Caslick 1983, Tobin et al. 1988, 1991). Bobolinks and Dickcissels are considered pests on their wintering grounds in South America because of their rice-eating habits (Dyer and Ward 1977). Migratory birds visiting feedlots (primarily Brown-headed Cowbirds) may also reduce "yield" of livestock by consuming livestock feed, and by fouling food or transmitting diseases (Glahn 1983).

The amount of crop damage by Neotropical migratory birds is probably influenced by farmland structure. Damage to corn by Red-winged Blackbirds increased in North America from the late 1960s to the early 1980s due perhaps to increases in the area producing grain (White et al. 1985, Clark et al. 1986) and corresponding decreases in alternative foods (Besser and Brady 1986). During this period, the coverages of small grains, hayfields, and uncultivated lands declined, in turn, leading to increased reliance on corn for food by redwings. When crop damage occurs, it is usually concentrated in space and time. Most fields receive little or no damage, but those located near roosting concentrations of birds can be heavily damaged (e.g., Bollinger and Caslick 1985, Brugger et al. 1992). Crop damage also tends to be highest where crop and landscape diversity is lowest (Stone and Danner 1980). Typically, crops are vulnerable only for a short period of time (often just before harvest), and fields developing markedly earlier (or maturing later) than most are damaged more severely (Bollinger and Caslick 1985, Wilson et al. 1989).

INTEGRATION OF AGRICULTURAL PRACTICES AND FARMLAND STRUCTURE

Our review of the effects of agricultural practices and farmland structure on Neotropical migratory birds indicates that management for the conservation of migratory birds in farmland should emphasize farmland structure. Landscape structure largely determines whether birds choose farmland for breeding, foraging, or other activities. The large areas of cropland that do not provide breeding requisites, even at a minimal level, are simply not used by most migratory bird species. The most extreme examples of this situation are found in monocultures with large field sizes, particularly where soil surface litter is buried by tillage and where uncropped areas with perennial vegetation are lacking (Warner 1994). The spatial extent of such monocultures has been increasing in recent decades.

Once cropland is chosen by individuals, agricultural practices then influence the suitability of the habitat, i.e., the ability of the habitat to support migratory bird populations. For example, destruction of nests by agricultural operations may also result in seemingly low "survivorship" of adults because individuals whose nests are destroyed tend not to return to former territories in subsequent years (Bollinger and Gavin 1989). Low recruitment because of nest destruction and reduced site fidelity may lead to population declines. However, it is important to emphasize that farmland is not invariably a suboptimal habitat for migratory birds. Gawlik and Bildstein (1990) reported high reproductive success for Loggerhead Shrikes nesting in farmland edge areas. They attributed this species' decline in abundance to habitat loss, not to declines in habitat quality. Other studies also report the potential for adequate to high reproductive success for Neotropical migratory birds breeding in farmland. Rodenhouse and Best (1983) and Perritt and Best (1989) reported average clutch sizes for Vesper Sparrows breeding in rowcrops, and annual breeding productivity was only below replacement level (about three fledglings per pair per season) because of losses due to agricultural operations. Similarly, Bollinger and Gavin (1992) and

Wittenberger (1978) found potentially high breeding productivity of Bobolinks nesting in hayfields and pastures, but breeding productivity of this species could be severely reduced by losses of nests and adults due to mowing. Clearly, reductions in the number of nests lost to agricultural operations is key to improving the quality of farmland for Neotropical migratory birds, but management to reduce such losses will only benefit migratory birds in landscapes that provide breeding and foraging requisites.

Farmland structures that attract Neotropical migratory birds, but that are combined with harmful agricultural practices, may create ecological traps for these species (Best 1986). Ecological traps attract breeding densities of migratory birds similar to those of source areas (*sensu* Wiens and Rotenberry 1981) but support only below replacement-level reproduction because of high rates of nest failure from agricultural field operations, predation, and/or brood parasitism. Some rowcrop fields, such as no-till corn and soybeans (Wooley et al. 1985, Best 1986), or intensively managed hayfields (Bollinger et al. 1990) may act as ecological traps, but whether migratory birds breeding in such fields could have bred elsewhere with greater success has not been established. The absence of alternatives to breeding in crop fields in extensively cultivated areas [e.g., Iowa is about 62% corn and soybean fields (Skow and Halley 1981) and some regions of Illinois are up to 90% rowcrops (Warner 1994)] and the tendency for some migratory species to breed almost exclusively in crop fields [e.g., Dickcissels (Igl 1991), Bobolinks (Bollinger and Gavin 1992)] create a strong possibility for some crop fields to act as ecological traps.

No evidence to date indicates that edge areas in farmland are acting as ecological traps for Neotropical migratory birds. High nest densities are achieved in many edge areas whether adjacent to crop fields or not (e.g., Graber and Graber 1963, Gates and Gysel 1978, Bryan and Best 1994, Camp and Best 1994), and edges may support above or below replacement-level reproduction largely depending on the abundance and activity of nest predators and brood parasites. Edge areas may be suboptimal for reproduction of migratory birds because of high nest losses or low adult survival, but alternative habitat for most edge breeders may be available in woodlands and shrublands.

MANAGEMENT CONSIDERATIONS

The limited amount of information presently available about Neotropical migratory birds in farmland indicates that management efforts should focus on farmland structure to provide the greatest benefit for the most species. Optimal farmland structures, however, differ for grassland and edge species. Grassland species require areas without trees and shrubs for breeding, and most grassland species need perennial grass cover for nesting sites. Edge species benefit from shrubs and trees, and from a diversity of landscape elements in close proximity. Unfortunately, edge conditions create an enigma for the conservation of Neotropical migratory birds in agricultural landscapes. The wooded edge that supports the greatest diversity and density of Neotropical migratory birds also seems to expose them to the highest levels of brood parasitism and nest predation. No solution to this problem is readily apparent. We suggest, however, that farmland structures that more closely resemble native habitats are likely to be more favorable to Neotropical migratory birds that use either edge or fields. Such structures could be achieved by promoting wooded edge (including agroforestry) in formerly forested landscapes. In prairie landscapes, the reverse is needed; woody vegetation (that provides elevated perches for Brown-headed Cowbirds or avian predators) could be eliminated to make the habitat more suitable for grassland species.

It is important to note that prairie landscapes originally included wetlands with woody vegetation (e.g., sloughs bordered by shrubs, riparian strips). Because the occurrence of these elements has been severely reduced, efforts to protect the remaining wetland habitat or (where possible) to restore wetlands and surrounding buffer strips should receive high priority. Management for grassland birds is primarily relevant to formerly prairie regions, but because of the near complete conversion of prairie to croplands, use of formerly forested areas will be essential to maintaining the migratory birds of grasslands (Bollinger and Gavin 1992).

Agricultural practices influence the local reproduction and survival of Neotropical migratory birds that obtain food, shelter, and nesting sites within farmland. Management considerations, therefore, emphasize these requisites, but specific recommendations should be designed for each crop, landscape, and bird community of concern (Hansen and Urban 1992). In general, crop residues should be retained on the soil surface because these residues contain waste grain, sustain populations of arthropods that are food for migratory birds, and provide cover for foraging or nesting individuals. Chemicals potentially toxic to migratory birds or that indirectly affected them via changes in vegetation structure or food resources (i.e., inorganic fertilizers and pesticides) should be used with restraint. For example, inorganic fertilizers should be applied only based on measured soil requirements because their excessive use can harm soil organisms that are food for some migratory species. Migratory birds might benefit from curtailing pesticide use in strips bordering field edges, as gamebirds have in the United Kingdom (Sotherton 1991). Where pesticides are essential, their use should be a part of integrated pest management (IPM) systems. IPM involves monitoring pest populations (plants and animals) closely, and using pesticides only when and where pests are likely to cause economically important damage (Luna and House 1990). Thus, IPM minimizes exposure of migratory birds to harmful chemicals and reduces the destruction of nontarget arthropods that are food for many migratory bird species.

To protect both arthopod abundances and complex vegetation structure, uncultivated areas being managed for edge bird species (e.g., fencerows) should, wherever possible, neither be sprayed with herbicide nor mowed. Mowing or burning should be used to maintain herbaceous vegetation within areas being managed for grassland species. When necessary, these field operations should be conducted outside the breeding season to prevent destruction of nests. Nest destruction in fields can be minimized by choosing the tillage method that destroys fewest nests (e.g., subsurface tillage or no-till). In hayfields, delaying spring mowing as long as possible, avoiding nighttime mowing, and spacing mowings as widely as possible in time will allow the greatest probability of successful nesting. Establishment of native prairie grasses as forage crops would likely increase the value of pasture and hayfields for wildlife, because of the nesting cover provided and because these warm-season forage crops are mowed or grazed later in the avian breeding season than cool-season forage crops such as alfalfa (e.g., George et al. 1979).

Combinations of landscape structures and practices that benefit both migratory birds and agriculture are possible, and agricultural policies such as CRP could be used to create such landscapes (Dunn et al. 1993). For example, CRP fields could be clustered to create large areas of grassland devoid of trees and shrubs. Retaining these areas as grassland might be made more likely by allowing moderate levels of grazing or late-season haying. However, the benefits of CRP land for migratory birds probably would be altered by permitting grazing (e.g., Taylor and Littlefield 1986) or mowing (Bollinger et al. 1990). Both local extinctions and shifts in relative abundances of migratory bird species might occur as the vegetation structure and composition as well as the disturbance regimes of these grasslands were altered (Wilson and Belcher 1989). Presently, most CRP grassland is mowed during the avian breeding season for weed control (Hays et al. 1989, Miller and Bromley 1989). The extent to which this practice harms nesting migratory birds is unknown, but it is currently being studied (Patterson and Best, unpublished data). Such mowing, however, is a pointed example of how the potential for CRP land to benefit Neotropical migratory birds is largely undeveloped (Miller and Bromley 1989).

Coordinated management for both Neotropical migratory birds and gamebirds is sometimes possible. Such multispecies management merits consideration because landowners obtain social and economic benefit from creating and preserving habitat for game species (e.g., Hanley 1993) that they may not gain from preserving habitat for

Neotropical migratory birds. Gamebirds of farmland edge such as Ring-necked Pheasants or Northern Bobwhite benefit from the same landscapes and practices recommended for migratory birds that use edge areas. Management for these species, however, would probably be detrimental to populations of area-sensitive grassland birds (e.g., Bobolink and Savannah Sparrow; Herkert 1994). Thus gamebird species should not be used as indicators of habitat quality for all migratory bird species. Furthermore, the agricultural practices and farmland structures that best foster gamebirds and Neotropical migratory birds have not been fully identified. Hence, both field-scale and landscape-scale studies of Neotropical migratory birds in farmland are urgently needed.

RESEARCH NEEDS

Studies of Neotropical migratory birds in cropland are few, and most published studies have been conducted in a few crop types grown in central North America. Consequently, studies of migratory birds using crops in the southern and western parts of the continent are urgently needed for better assessment of the impacts of agriculture on Neotropical migratory birds. In future studies, high research priority should be placed on determining the dynamics of migratory birds in farmland (Robinson 1992b, Warner 1994). This will include identifying the farmland structures and agricultural practices that create and sustain populations. Studies of the annual breeding productivity and survival of migratory birds nesting in fields and edges are few, and potential source areas (*sensu* Wiens and Rotenberry 1981) that should be protected or expanded have not yet been identified. Nest predators and brood parasites probably are key determinants of whether a site is a source or sink; thus, assessments of the abundance and activities of these natural enemies of migratory birds should receive high research priority.

Special research emphasis should be placed on determining optimal farmland structures. Studies of the values of specific landscape elements for migratory birds, and of combinations of elements, are largely lacking and are urgently needed. These studies will be most useful to policymakers and farmers when field-scale research (focusing on territory selection, breeding productivity, site fidelity, etc.) is integrated with complementary landscape-scale studies (examining, for example, relationships between field sizes, crop types or diversity, and bird abundances). Such integration of field-scale and landscape-scale research is essential for determining the impacts of farmland structures on Neotropical migratory birds. Landscape-scale studies can be carried out, in part, by using existing data bases including agricultural statistics and bird abundances (see O'Connor and Boone 1992).

Because agriculture is changing rapidly, the impacts of emerging practices such as relay cropping and agroforestry on Neotropical migratory birds should be examined as soon as possible. Results can then be used to guide landscape-scale changes such as those associated with set-aside programs (e.g., the CRP) or agroforestry. A possible method of assessing the impacts of landscape structure on Neotropical migratory birds is outlined by Freemark et al. (1993).

CONCLUSIONS

Agriculture is changing rapidly. This change is driven by high costs of inputs (e.g., fertilizer and pesticides), environmental problems caused by these inputs (e.g., contamination of ground water), and cropland degradation via soil erosion and loss of soil organic matter. Both farmers and policymakers are interested in developing a profitable and sustainable agriculture (National Research Council 1989, Edwards et al. 1990), but farm profitability is the primary concern of the farm operator (Batie and Taylor 1989, Dobbs and Cole 1992). Thus, research and education regarding migratory birds should include the impact of proposed management practices on farm profitability. However, present evidence indicates that farmland structured for Neotropical migratory birds is also of long-term benefit to crop production in numerous ways (Robinson 1991, Rodenhouse et al. 1993). Reduced soil erosion, reduced nonpoint source pollution of freshwater, reduced costs of chemical inputs, reduced energy costs from fewer tillage operations, promotion of soil fertility, and the creation of favorable microclimates for crop growth all are consistent with farm profitability and with the conservation of migratory birds. Furthermore, uncropped areas within farmland also can benefit crop production (Kort 1988, Rodenhouse et al. 1993). Thus we are optimistic that with enlightened management farmland can satisfy both the needs of farm operators and of Neotropical migratory birds.

ACKNOWLEDGMENTS

We thank B. Van Horne, E. Littrell and D. H. Johnson for providing helpful comments on an earlier draft of the manuscript.

LITERATURE CITED

Agricultural Extension Service. 1964. Arkansas state summary on blackbird damage (1963). USDA 1, and University of Arkansas Fayetteville, AR.

Altieri, M. A., and W. H. Whitcomb. 1980. Weed manipulation for insect pest management in corn. Environ. Manag. 4:483–489.

Andow, D. A., and P. M. Rosset. 1990. Integrated pest management. Pp. 413–439 *in* Agroecology. (C. R. Carroll, J. H. Vandermeer, and P. M. Rosset, eds). McGraw-Hill, New York.

Angelstam, P. 1986. Predation on ground-nesting birds' nests in relation to predator densities and habitat edge. Oikos 47:365–373.

Barrett, G. W., N. L. Rodenhouse, and P. J. Bohlen. 1990. Role of sustainable agriculture in rural landscapes. Pp. 624–636 *in* Sustainable agricultural systems (C. A. Edwards, R. Lal, P. Madden, R. H. Miller, and G. House, eds). Soil and Water Conservation Society, Ankeny, IA.

Basore, N. S., L. B. Best, and J. B. Wooley, Jr. 1986. Bird nesting in Iowa no-tillage and tilled cropland. J. Wildl. Manag. 50:19–28.

Basore, N. S., L. B. Best, and J. B. Wooley, Jr. 1987. Arthropod availability to pheasant broods in no-tillage fields. Wildl. Soc. Bull. 15:229–233.

Batie, S. S., and D. B. Taylor. 1989. Widespread adoption of non-conventional agriculture: profitability and impacts. Amer. J. Altern. Agric. 4:128–134.

Beason, R. C., and E. C. Franks. 1974. Breeding behavior of the Horned Lark. Auk 91:65–74.

Beintema, A. J., and G. J. D. M. Muskens. 1987. Nesting success of birds breeding in Dutch agricultural grasslands. J. Appl. Ecol. 24: 743–758.

Besser, J. F., and D. J. Brady. 1986. Bird damage to ripening field corn increase in the United States from 1971 to 1981. US Dept Interior, Fish Wildlife Serv., Fish Wildlife Leaflet 7, Washington, DC.

Best, L. B. 1978. Field sparrow reproductive success and nesting ecology. Auk 95:9–22.

Best, L. B. 1983. Bird use of fencerows: implications for contemporary fencerow management practices. Wildl. Soc. Bull. 11:343–347.

Best, L. B. 1986. Conservation tillage: ecological traps for nesting birds? Wildl. Soc. Bull. 14:308–317.

Best, L. B., and N. L. Rodenhouse. 1984. Territory preference of Vesper Sparrows in cropland. Wilson Bull. 96:72–82.

Best, L. B., R. C. Whitmore, and G. M. Booth. 1990. Bird use of cornfields during the breeding season: the importance of edge habitat. Amer. Midland Natur. 123:84–99.

Blankespoor, G., and H. Krause. 1982. The breeding birds of Minnehaha County, South Dakota: then (1907–1916) and now (1971–1975). Amer. Birds 36:22–27.

Bollinger, E. K. 1988. Breeding dispersion and reproductive success of Bobolinks in an agricultural landscape. PhD dissertation, Cornell University, Ithaca, NY.

Bollinger, E. K., and J. W. Caslick. 1985. Factors influencing blackbird damage to field corn. J. Wildl. Manag. 49:1109–1115.

Bollinger, E. K., and T. A. Gavin. 1989. Effects of site quality on breeding site fidelity in Bobolinks. Auk 106:584–594.

Bollinger, E. K., and T. A. Gavin. 1992. Eastern Bobolink populations: ecology and conservation in an agricultural landscape. Pp. 497–506 *in* (J. M. Hagan, and D. W. Johnston, eds). Ecology and conservation of Neotropical migrant landbirds. Smithsonian Institution Press, Washington, DC.

Bollinger, E. K., P. B. Bollinger, and T. A. Gavin. 1990. Effects of hay-cropping on eastern populations of the Bobolink. Wildl. Soc. Bull. 18:142–150.

Bowles, M. L., K. Kerr. R. H. Thom, and D. E. Birkenholz. 1980. Threatened, endangered and extirpated birds of Illinois prairies. Illinois Audubon Bull. 193:2–11.

Boyer, J. S. 1982. Plant productivity and environment. Science 218:443–448.

Breiman, L., J. H. Friedman, R. A. Olshen, and C. J. Stone. 1984. Classification and regression trees. Wadsworth, Belmont, England.

Brittingham, M. C., and S. A. Temple. 1983. Have cowbirds caused forest songbirds to decline? Bioscience 33:31–35.

Brooks, B. L. and S. A. Temple. 1990. Dynamics of a Loggerhead Shrike population in Minnesota. Wilson Bull. 102:441–450.

Brugger, K. E., R. F. Labinsky, and D. E. Daneke. 1992. Blackbird roost dynamics at Millers Lake, Louisiana: implications for damage control in rice. J. Wildl. Manag. 56:393–398.

Brunton, D. H. 1988. Sexual differences in reproductive effort: time–activity budgets of monogamous killdeer, *Charadrius vociferous*. Anim. Behav. 36:705–717.

Bryan, G. G. 1990. Species abundance patterns and productivity of birds using grassed waterways in Iowa rowcrop fields. MS thesis, Iowa State University, Ames, IA.

Bryan, G. G., and L. B. Best. 1991. Bird abundance and species richness in grassed waterways in Iowa rowcrop fields. Amer. Midland Natur. 126:90–102.

Bryan, G. G., and L. B. Best. 1994. Avian nest density and success in grassed waterways in Iowa rowcrop fields. Wildl. Soc. Bull. 22:583–592.

Buchanan, M., and L. D. King. 1993. Carbon and phosphorus losses from decomposing crop residues in no-till and conventional till agroecosystems. Agron. J. 85:631–638.

Camp, M., and L. B. Best. 1994. Nest density and nesting success of birds in roadsides adjacent to rowcrop fields. Amer. Midland Natur. 131:347–358.

Carson, W. P., and G. W. Barrett. 1988. Succession in old field plant communities: effects of contrasting types of nutrient enrichment. Ecology 69:984–994.

Cassel, J. F., and J. M. Wiehe. 1980. Uses of shelterbelts by birds. Pp. 78–87 *in* Management of western forests and grasslands for nongame birds (R. M. DeGraff, ed.). USDA Forest Serv. Gen. Techn. Rep. INT-86.

Castrale, J. S. 1985. Responses of wildlife to various tillage conditions. Trans. North Amer. Wildl. Natural Resources Conf. 50: 142–156.

Chiverton, P. A., and N. W. Sotherton. 1991. Effects on beneficial arthropods of the exclusion of herbicides from cereal crop edges. J. Appl. Ecol. 28:1027–1039.

Churcher, P. B., and Lawton, J. H. 1987. Predation by domestic cats in an English village. J. Zool. Soc. Lond. 212:439–455.

Clark, L. A., and D. Pregibon. 1992. Tree-based models. Pp. 377–419 *in* Statistical models in S. Pacific. (J. M. Chambers, and T. J. Hastie, eds). Wadsworth and Brooks, Grove, England.

Clark, R. G., P. J. Weatherhead, H. Greenwood, and R. D. Titman. 1986. Numerical responses of Red-winged Blackbird populations to changes in regional land-use patterns. Can. J. Zool. 64:1944–1950.

Cody, M. L. 1985. Habitat selection in grassland and open-country birds. Pp. 191–226 *in* Habitat selection in birds (M. L. Cody, ed.). Academic Press, Orlando, FL.

Collins, S. L. 1992. Fire frequency and community heterogeneity in tallgrass prairie vegetation. Ecology 73:2001–2006.

Collins, S. L., and L. L. Wallace. 1990. Fire in North American tallgrass prairie. University of Oklahoma Press, Norman, OK.

Confer, J. L., and K. Knapp. 1981. Golden-winged Warblers and Blue-winged Warblers: the relative success of a habitat specialist and generalist. Auk 98:108–114.

Crase, F. T., C. P. Stone, R. W. DeHaven, and D. F. Mott. 1976. Bird damage to grapes in the United States with emphasis on California. US Fish Wildl. Serv., Special Sci. Rep.—Wildl. 197.

Curry, J. P. 1986. Above-ground arthropod fauna of four Swedish cropping systems and its role in carbon and nitrogen cycling. J. Appl. Ecol. 23:853–870.

DeGeus, D. W. 1990. Productivity and habitat preferences of Loggerhead Shrikes inhabiting roadsides in a midwestern agroenvironment. MS thesis. Iowa State University, Ames, IA.

DeGrazio, J. W. 1978. World bird damage problems. Pp. 9–24 *in* Proc. 8th Vertebrate Pest Conf. Sacramento, CA.

Dinsmore, J. J. 1981. Iowa's avifauna: changes in the past and prospects for the future. Proc. Iowa Acad. Sci. 88:28–37.

Dobbs, T. L., and J. D. Cole. 1992. Potential effects on rural economies of conversion to sustainable farming systems. J. Altern. Agric. 7:70–80.

Dolbeer, R. A. 1990. Ornithology and integrated pest management: Red-winged Blackbirds *Agelaius phoeniceus* and corn. Ibis 132:309–322.

Doran, J. W., and M. R. Werner. 1990. Management and soil biology. Pp. 205–230 *in* Sustainable agriculture in temperate zones. (C. A. Francis, C. Butler Flora, L. D. King, eds). John Wiley & Sons, New York.

Dover, J. W. 1991. The conservation of insects on arable farmland. Pp. 294–318 *in* The conservation of insects and their habitats (N. M. Collins, and J. A. Thomas, eds). Academic Press, London.

Dunn, C. P., F. Stearns, G. R. Gunmtenspergen, and D. M. Sharpe. 1993. Ecological benefits of the Conservation Reserve Program. Conserv. Biol. 7:132–139.

Durner, G. M., and J. E. Gates. 1993. Spatial ecology of black rat snakes on Remington Farms, Maryland. J. Wildl. Manag. 57:812–826.

Dyer, M. I., and P. Ward. 1977. Management of pest situations. Pp. 267–300 *in* Granivorous birds in ecosystems (J. Pinowski, and S. C. Kendeigh, eds). Cambridge University Press, New York.

Edwards, C. A., R. Lal, P. Maden, R. H. Miller, and G. House. 1990. Sustainable agricultural systems. Soil and Water Conservation Society, Ankeny, IA.

Eiserer, L. A. 1980. Effects of grass length and mowing on foraging behavior of the American Robin. Auk 97:576–580.

Flickinger, E. L., K. A. King, W. F. Stout, and M. M. Mohn. 1980. Wildlife hazards from Furadan 3G applications to rice in Texas. J. Wildl. Manag. 44:190–197.

Frawley, B. J. 1989. The dynamics of nongame bird breeding ecology in Iowa alfalfa fields. MS thesis, Iowa State University, Ames, IA.

Frawley, B. J., and L. B. Best. 1991. Effects of mowing on breeding bird abundance and species composition in alfalfa fields. Wildl. Soc. Bull. 19:135–142.

Freemark, K., and C. Boutin. 1995. Impacts of agricultural herbicide use on terrestrial wildlife in temperate landscapes: a review with special reference to North America. Agric. Ecosys. Environ. 52:67–91.

Freemark, K., H. Dewar, and J. Saltman. 1991. A literature review of bird use of farmland habitats in the Great Lakes–St. Lawrence Region. Can. Wildl. Serv. Tech. Rep. Ser. no. 114.

Freemark, K. E., J. R. Probst., J. B. Dunning, and S. J. Hejl. 1993. Adding a landscape perspective to conservation and management planning. Pp. 346–352 *in* Status and management of Neotropical migratory birds (D. M. Finch, P. W. Stangel, eds). USDA Forest Serv. Rocky Mt. Forest Range Exp. Sta., Gen. Tech. Rep. RM-229:346–352.

Fritzell, E. K. 1978. Habitat use by prairie raccoons during the water fowl breeding season. J. Wildl. Manag. 42:118–127.

Gates, J. E., and L. W. Gysel. 1978. Avian nest dispersion and fledging success in field-forest ecotones. Ecology 59:871–883.

Gauthreaux, S. A. 1992. Preliminary list of migrants for Partners in Flight Neotropical Migratory Bird Program. Partners Flight 2:30.

Gawlik, D. E., and K. L. Bildstein. 1990. Reproductive success and nesting habitat of Loggerhead Shrikes in north-central South Carolina. Wilson Bull. 102:37–48.

Gawlik, D. E., and K. L. Bildstein. 1993. Seasonal habitat use and abundance of Loggerhead Shrikes in South Carolina. J. Wildl. Manag. 57:352–357.

George, R. R., A. L. Farris, C. C. Schwartz, D. D. Humberg, and J. C. Coffey. 1979. Native prairie grass pastures as nest cover for upland birds. Wildl. Soc. Bull. 7:4–9.

Glahn, J. F. 1983. Blackbird and starling depredations of Tennessee livestock farms. Proc. Bird Control Semin. 9:125–134.

Glueck, T. F., W. R. Clark, and R. D. Andrews. 1988. Raccoon movement and habitat use during the fur harvest season. Wildl. Soc. Bull. 16:6–11.

Goldwasser, S., D. Goines, and S. R. Wilbur. 1980. The Least Bell's Vireo in California: a defacto endangered race. Amer. Birds 34:742–745.

Graber, R., and J. Graber. 1963. A comparative study of bird populations in Illinois, 1906–1909 and 1956–1958. Illinois Natural Hist. Serv. Bull. 28:383–528.

Graber, R., and J. Graber. 1983. The declining grassland birds. Illinois Natural History Surv. Rep. no. 227.

Greenberg, R. 1992. Forest migrants in nonforest habitats on the Yucatan Peninsula. Pp. 273–286 *in* Ecology and conservation of Neotropical migrant landbirds (J. M. Hagan, and D. W. Johnston, eds). Smithsonian Institution Press, Washington, DC.

Hallberg, G. R. 1987. Agricultural chemicals in ground water: extent and implications. J. Altern. Agric. 2:3–15.

Hanley, T. A. 1993. Balancing economic development, biological conservation, and human culture: the Sitka black-tailed deer *Odocoileus hemionus sitkensis* as an ecological indicator. Conserv. Biol. 66:61–67.

Hansen, A. J., and D. L. Urban. 1992. Avian response to landscape pattern: the role of species' life histories. Landscape Ecol. 7: 163–180.

Hart, A. D. M. 1990. The assessment of pesticide hazards to birds: the problem of variable effects. Ibis 132:192–204.

Hays, R. L., R. P. Webb, and A. H. Farmer. 1989. Effects of the Conservation Reserve Program on wildlife habitat: results of 1988 monitoring. Trans. North Amer. Wildl. Natural Resources Conf. 54:365–376.

Hendrix, P. F., R. W. Parmelee, D. A. Crossley, Jr, D. C. Coleman, E. P. Odum, and P. M. Groffman. 1986. Detritus food webs in conventional and no-tillage agroecosystems. BioScience 36:374–380.

Hendrix, P. F., D. A. Crossley, Jr, J. M. Blair, and D. C. Coleman. 1990. Soil biota as components of sustainable agroecosystems. Pp. 637–654 *in* Sustainable agricultural systems (C. A. Edwards, R. Lal, P. Madden, R. H. Miller, and G. House, eds). Soil and Water Conservation Society, Ankeny, IA.

Herkert, J. R. 1991. Prairie birds of Illinois: population response to two centuries of habitat change. Illinois Natural Hist. Surv. Bull. 34:393–399.

Herkert, R. 1994. The effects of habitat fragmentation on midwestern grassland bird communities. Ecol. Appl. 4:461–471.

Higley, L. G., M. R. Zeiss, W. K. Wintersteen, and L. P. Pedigo. 1992. National pesticide policy: a call for action. Amer. Entomol. 38:139–146.

House, G. J., and B. R. Stinner. 1983. Arthropods in no-tillage soybeans agroecosystems: community composition and ecosystem interactions. Environ. Manag. 7:23–28.

Hurley, R. J., and E. C. Franks. 1976. Changes in the breeding ranges of two grassland birds. Auk 93:108–115.

Igl, L. D. 1991. The role of climate and mowing on Dickcissel (*Spiza americana*) movements, distribution and abundance. MS thesis, Iowa State University, Ames, IA.

Jenny, M. 1990. Territoriality and breeding biology of Skylark (*Alauda arvensis*) in an intensively farmed area in Switzerland. J. Ornithol. 131:241–265.

Johnson, D. H., and M. D. Schwartz. 1993. The Conservation Reserve Program and grassland birds. Conserv. Biol. 7:934–937.

Johnson, R. G., and S. A. Temple. 1990. Nest predation and brood parasitism of tallgrass prairie birds. J. Wildl. Manag. 54:106–111.

Johnson, R. J., and M. M. Beck. 1988. Influences of shelterbelts on wildlife management and biology. Agricult. Ecosyst. Environ. 22/23: 301–335.

Johnson, W. C., and C. S. Adkisson. 1985. Dispersal of beech nuts by Blue Jays in fragmented landscapes. Amer. Midland Natur. 113:319–324.

Kingsley, N. P., and T. H. Nicholls. 1993. Raptor habitat in the Midwest. Pp. 185–194 *in* Midwest Raptor Management Symposium and Workshop (B. A. Pendleton, ed.). National Wildlife Federation, Washington, DC.

Kirchner, T. B. 1977. The effects of resource enrichment on the diversity of plants and arthropods in a shortgrass prairie. Ecology 58:1334–1344.

Knittle, C. E., and J. L. Guarino. 1976. A 1974 questionnaire survey of bird damage to ripening grain sorghum in the United States. Sorghum Newslett. 19:93–94.

Kort, J. 1988. Benefits of windbreaks to field and forage crops. Agricult. Ecosyst. Environ. 22/23:165–190.

Lauber, T. B. 1991. Birds and the Conservation Reserve Program: a retrospective study. MS thesis, University of Maine, Orono, ME.

Laursen, K. 1981. Birds on roadside verges and the effect of mowing on frequency and distribution. Biol. Conserv. 20:59–68.

Liberg, O. 1984. Food habits and prey impact by feral and house-based domestic cats in a rural area in southern Sweden. J. Mammol. 65:424–432.

Luna, J. M., and G. J. House. 1990. Pest management in sustainable agricultural systems. Pp. 157–173 *in* Sustainable agricultural systems (C. A. Edwards, R. Lal, P. Madden, R. H. Miller, and G. House, eds). Soil and Water Conservation Society, Ankeny, IA.

Martin, A. C., H. S. Zim, and A. L. Nelson. 1961. American wildlife and plants. Dover Publications, Inc., New York.

Martin, J. H., W. H. Leonard, and D. L. Stamp. 1976. Principles of field crop production, 3rd ed. Macmillan Publishing Company, Inc., New York.

Meanley, B. 1971. Blackbirds and the southern rice crop. US Dep. Interior, Fish Wildl. Serv. Resource Publ. 100.

Miller, E. J., and P. T. Bromley. 1989. Wildlife management on Conservation Reserve Program land: the farmer's view. Trans. 54th North Amer. Wildl. Natural Resources Conf. 54:377–381.

Mott, D. F., and C. P. Stone. 1973. Bird damage to blueberries in the United States. US Fish Wildl. Serv. Special Sci. Rep.—Wildl. 172.

National Resource Council. 1989. Alternative agriculture. National Academy Press, Washington, DC.

Newton, I., and M. Marquiss. 1986. Population regulation in sparrowhawks. J. Anim. Ecol. 55:463–480.

O'Connor, R. J., and R. B. Boone. 1992. A retrospective study of agricultural bird populations in North America. Pp. 1165–1184 *in* Ecological indicators, Vol. 2. (D. M. McKenzie, D. E. Hyatt, and V. J. McDonald, eds). Elsevier, New York.

O'Connor, R. J., and M. Shrubb. 1986. Farming and birds. Cambridge University Press, Cambridge, England.

Owens, R. A., and M. T. Myres. 1973. Effects of agriculture upon populations of native passerine birds of an Alberta grassland. Can. J. Zool. 51:697–713.

Papendick, R. I., L. F. Elliot, and R. B. Dahlgren.

1986. Environmental consequences of modern production agriculture: how can alternative agriculture address these issues and concerns? Amer. J. Altern. Agricult. 1:3–10.

Pasek, J. E. 1988. Influence of wind and windbreaks on local insect dispersal. Agricult. Ecosyst. Environ. 22/23:539–554.

Peng, R. K., L. D. Incoll, S. L. Sutton, C. Wright, and A. Chadwick. 1993. Diversity of airborne arthropods in a silvoarable agroforestry system. J. Appl. Ecol. 30:551–562.

Perritt, J. E., and L. B. Best. 1989. Effects of weather on the breeding ecology of Vesper Sparrows in Iowa crop fields. Amer. Midland Natur. 121:355–360.

Pimentel, D., L. McLaughlin, A. Zepp, B. Lakitan, T. Kraus, P. Kleinman, F. Vancini, W. J. Roach, E. Graap, W. S. Keeton, and G. Selig. 1991. Environmental and economic effects of reducing pesticide use. BioScience 41: 402–409.

Pimentel, D., U. Stachow, D. A. Takacs, H. W. Brubaker, A. R. Dumas, J. J. Meaney, J. A. S. O'Neil, D. E. Onsi, and D. B. Corzilius. 1992. Conserving biological diversity in agricultural/forestry systems. BioScience 42:354–362.

Probst, J. R. 1986. A review of factors limiting the Kirtland's Warbler on its breeding grounds. Amer. Midland Natur. 116:87–100.

Reganold, J. P., L. F. Elliot, and Y. L. Unger. 1987. Long-term effects of organic and conventional farming on soil erosion. Nature 330:370–372.

Renken, R. B., and Dinsmore, J. J. 1987. Nongame bird communities on managed grasslands in North Dakota. Can. Field Natur. 101:551–557.

Rice, R. A., and J. H. Vandermeer. 1990. Climate and geography of agriculture. Pp. 21–63 *in* Agroecology. (C. R. Carroll, J. H. Vandermeer, and P. Rosset, eds). McGraw-Hill Publishing Company, New York.

Risebrough, R. W. 1986. Pesticides and bird populations. Curr. Ornithol. 3:397–427.

Robertson, R. J. 1972. Optimal niche space of the Red-winged Blackbird (*Agelaius phoeniceus*). I. Nesting success in marsh and upland habitat. Can. J. Zool. 50:247–263.

Robinson, A. Y. 1991. Sustainable agriculture: the wildlife connection. Amer. J. Altern. Agricult. 6:161–167.

Robinson, S. K. 1992a. Population dynamics of breeding Neotropical migrants in a fragmented Illinois landscape. Pp. 408–418 *in* Ecology and conservation of Neotropical migrant landbirds. (J. M. Hagan, and D. W. Johnston, eds). Smithsonian Institution Press, Washington, DC.

Robinson, S. K. 1992b. The breeding season: introduction. Pp. 405–407 *in* Ecology and conservation of Neotropical migrant landbirds (J. M. Hagan, III, and D. W. Johnston, eds). Smithsonian Institution Press, Washington, DC.

Robinson, S. K., J. A. Grzybowski, S. I. Rothstein, M. C. Brittingham, L. J. Petit, and F. R. Thompson. 1993. Management implications of cowbird parasitism on Neotropical migrant songbirds. Pp. 93–102 *in* Status and management of Neotropical migratory birds (D. M. Finch, and P. W. Stangel, eds). USDA Forest Serv. Rocky Mt. Forest Range Exp. Sta., Gen. Tech. Rep. RM-229.

Rodenhouse, N. L., and L. B. Best. Breeding ecology of Vesper Sparrows in corn and soybeans fields. Amer. Midland Natur. 110: 265–275.

Rodenhouse, N. L., and L. B. Best. 1994. Foraging patterns of Vester Sparrows (*Pooecetes gramineus*) breeding in cropland. Amer. Midland Natur. 131:196–206.

Rodenhouse, N. L., L. B. Best, R. J. O'Connor, and E. K. Bollinger. 1993. Effects of temperate agriculture on Neotropical migrant landbirds. Pp. 280–295 *in* Status and management of Neotropical migratory birds (D. M. Finch, and P. Stangel, eds). USDA Forest Serv. Rocky Mt. Forest Range Exp. Sta., Gen. Tech. Rep. RM-229.

Rodgers, R. D. 1983. Reducing wildlife losses to tillage in fallow wheat fields. Wildl. Soc. Bull. 11:31–38.

Roseberry, J. L., and W. D. Klimstra. 1970. The nesting ecology and reproductive performance of the Eastern Meadowlark. Wilson Bull. 82:243–267.

Roth, R. R. 1979. Foraging behavior of Mockingbirds: the effect of too much grass. Auk 96:421–422.

Rothstein, S. I., J. Verner, and E. Stevens. 1984. Radio-tracking confirms a unique diurnal pattern of spatial occurrence in the parasitic Brown-headed Cowbird. Ecology 65:77–88.

Seamons, T. W. and J. W. Caslick. 1983. An assessment of bird damage and bird control measures in New York vineyards. Cornell University, Agricultural Experiment Station, Natural Resources Research and Extension Series no. 19.

Shalaway, S. D. 1985. Fencerow management for nesting birds in Michigan. Wildl. Soc. Bull. 13:302–306.

Shelton, M. D., and C. R. Edwards. 1983. Effect of weeds on the diversity and abundance of insects in soybeans. Environ. Entomol. 12:296–298.

Skow, D. M., and C. R. Halley. 1981. Iowa agricultural statistics, 1981. Iowa Crop and Livestock Reporting Service, Iowa

Department of Agriculture, Des Moines, IA.
Sloneker, L. L., and W. C. Moldenhauer. 1977. Measuring the amounts of crop residue remaining after tillage. J. Soil Water Conserv. 32:231–236.
Smith, E. L., and K. C. Kruse. 1992. The relationship between land-use and the distribution and abundance of Loggerhead Shrikes in south-central Illinois. J. Field Ornithol. 63:420–427.
Sommer, J. E., and F. K. Hines. 1991. Diversity in US agriculture. USDA Econ. Res. Serv. Agricult. Econ. Rep. no. 646.
Sotherton, N. W. 1991. Conservation headlands: a practical combination of intensive cereal farming and conservation. Pp. 373–397 *in* The ecology of temperate cereal fields (L. G. Firbank, N. Cater, J. F. Darbyshire, and G. R. Potts, eds). Blackwell Scientific Publications, Oxford, England.
Statistics Canada. 1986. Census of Canada, Agriculture. Statistics Canada, Ottawa, Ontario.
Stauffer, D. F., and L. B. Best. 1980. Habitat selection by birds of riparian communities: evaluating effects of habitat alterations. J. Wildl. Manag. 44:1–15.
Stinner, B. R., and J. M. Blair. 1990. Ecological and agronomic characteristics of innovative cropping systems. Pp. 123–140 *in* Sustainable agricultural systems (C. A. Edwards, R. Lal, P. Madden, R. H. Miller, and G. House, eds). Soil and Water Conservation Society, Ankeny, IA.
Stone, C. P. 1973a. Bird damage to tart cherries in Michigan, 1972. Proc. 6th Bird Control Semin. Bowling Green, pp. 19–23.
Stone, C. P. 1973b. Bird damage to agricultural crops in the United States—a current summary. Proc. 6th Bird Control Semin. Bowling Green, pp. 264–267.
Stone, C. P., and C. R. Danner. 1980. Autumn flocking of Red-winged Blackbirds in relation to agricultural variables. Amer. Midland Natur. 103:196–199.
Sullivan, K. A. 1989. Predation and starvation: age-specific mortality in juvenile juncos (*Junco phaenotus*). J. Anim. Ecol. 58:275–286.
Sullivan, P. G., D. J. Parrish, and J. M. Luna. 1991. Cover crop contributions to N supply and water conservation in corn production. Amer. J. Altern. Agricult. 6:106–113.
Tate, R. L., III. 1987. Soil organic matter: biological and ecological effects. Wiley, New York.
Taylor, D. M., and C. D. Littlefield. 1986. Willow Flycatcher and Yellow Warbler response to cattle grazing. Amer. Birds 40:1169–1173.
Tobin, M. E., R. A. Dolbeer, C. M. Webster, and T. W. Seamons. 1991. Cultivar differences in bird damage to cherries. Wildl. Soc. Bull. 19:190–194.
Tobin, M. E., P. P. Woronecki, R. A. Dolbeer, and R. L. Bruggers. 1988. Reflecting tape fails to protect ripening blueberries from bird damage. Wildl. Soc. Bull. 16:300–303.
Tucker, G. M. 1992. Effects of agricultural practices on field use by invertebrate-feeding birds in winter. J. Appl. Ecol. 29:779–790.
Twedt, D. J., W. J. Bleier, and G. L. Linz. 1991. Geographic and temporal variation in the diet of Yellow-headed Blackbirds. Condor 93:975–986.
Tyler, J. D. 1992. Nesting ecology of the Loggerhead Shrike in southwestern Oklahoma. Wilson Bull. 104:95–104.
USDA 1990. 1990 agricultural chartbook. USDA Agriculture Handbook no. 689.
USDA 1992. Agricultural statistics, 1991. USDA, Washington, DC.
Warburton, D. B., and W. D. Klimstra. 1984. Wildlife use of no-tilled and conventional tilled corn fields. J. Soil Water Conserv. 39:327–330.
Warner, R. E. 1992. Nest ecology of grassland passerines on road rights-of-way in central Illinois. Biol. Conserv. 59:1–7.
Warner, R. E. 1994. Agricultural land use and grassland habitat in Illinois: future shock for midwestern birds? Conserv. Biol. 8:147–156.
Warner, R. E., and S. L. Etter. 1989. Hay cutting and the survival of pheasants: a long-term perspective. J. Wildl. Manag. 53:455–461.
Warren, S. D., C. J. Scifres, and P. D. Teel. 1987. Response of grassland arthropods to burning: a review. Agric. Ecosys. Environ. 19:105–130.
Weersink, A., M. Walker, C. Swanton, J. E. Shaw. 1992. Costs of conventional and conservation tillage systems. J. Soil Water Conserv. 47:328–334.
White, S. B., R. A. Dolbeer, and T. A. Bookhout. 1985. Ecology, bioenergetics, and agricultural impacts of a winter-roosting population of blackbirds and starlings. Wildl. Monogr. 93:1–42.
Wilson, E. A., E. A. LeBouef, K. M. Weaver, and D. J. LeBlanc. 1989. Delayed seeding for reducing blackbird damage to sprouting rice in southwestern Louisiana. Wildl. Soc. Bull. 17:165–171.
Wilson, R. J., and S. G. Diver. 1991. The role of birds in agroforestry systems. Pp. 256–273 *in* Proc. 2nd conf. on agroforestry in North America (H. E. Garrett, ed.). University of Columbia, Columbia, MO.
Wilson, S. D., and J. W. Belcher. 1989. Plant and bird communities of native prairie and introduced Eurasian vegetation in Manitoba, Canada. Conserv. Biol. 3:39–44.

Wiens, J. A., and J. T. Rotenberry. 1981. Censusing and the evaluation of avian habitat occupancy. Stud. Avian Biol. 6:522–532.

Wittenberger, J. F. 1978. The breeding biology of an isolated Bobolink population in Oregon. Condor 80:355–371.

Wooley, J. B., Jr, L. B. Best, and W. R. Clark. 1985. Impacts of no-till row cropping on upland wildlife. Trans. North Amer. Wildl. Natural Resources Conf. 50:157–168.

Yahner, R. H. 1983. Seasonal dynamics, habitat relationships, and management of avifauna in farmstead shelterbelts. J. Wildl. Manag. 47:85–104.

APPENDIX

Common name	Migratory status[a]	Number of significant crop associations[b]	Consumes grains[c]
American Goldfinch	B	5	Y
American Kestrel	B	7	N
American Robin	B	3	N
Baird's Sparrow	A	3	Y
Barn Swallow	A	10	N
Bobolink	A	8	Y
Brewer's Blackbird	B	8	Y
Bronzed Cowbird	C	1	Y
Brown-headed Cowbird	B	9	Y
Chestnut-collared Longspur	B	7	Y
Chipping Sparrow	A	6	Y
Common Nighthawk	A	14	N
Common Yellowthroat	A	6	N
Dickcissel	A	10	Y
Eastern Bluebird	B	10	N
Eastern Kingbird	A	12	N
Eastern Meadowlark	B	17	Y
Eastern Phoebe	B	7	N
Ferruginous Hawk	B	3	N
Grasshopper Sparrow	A	16	Y
Gray Catbird	A	2	N
Horned Lark	B	12	Y
House Wren	A	9	N
Indigo Bunting	A	9	Y
Killdeer	B	14	N
Lark Bunting	A	8	Y
Lazuli Bunting	A	6	Y
Lesser Nighthawk	A	0	N
Loggerhead Shrike	B	8	N
Long-billed Curlew	A	5	N
McCown's Longspur	B	1	Y
Mississippi Kite	A	3	N
Mourning Dove	B	10	Y
Northern Mockingbird	B	5	N
Prairie Falcon	B	1	N
Purple Finch	B	9	Y
Red-tailed Hawk	B	4	N
Red-winged Blackbird	B	13	Y
Savannah Sparrow	B	8	Y
Say's Phoebe	B	7	N
Short-eared Owl	B	2	N
Song Sparrow	B	8	Y
Swainson's Hawk	A	9	N
Turkey Vulture	B	6	N
Vesper Sparrow	B	12	Y
Water Pipit	B	0	N
Western Bluebird	B	4	N
Western Kingbird	A	10	N
Western Meadowlark	B	9	Y
White-eyed Vireo	A	8	N
Yellow-headed Blackbird	A	11	Y
Yellow Warbler	A	5	N

[a] Migratory status: A includes species that breed in North America and spend their nonbreeding period south of the United States; B includes species that breed and winter in North America, but some populations winter south of the United States; C includes species that breed primarily south of the United States but their ranges extend north of the United States border (Gauthreaux 1992).
[b] Number of statistically significant associations at the county level between Breeding Bird Survey abundances and the proportion of the county area planted to one of 22 major crop types or in the Conservation Reserve Program; see text for details.
[c] Indicates species that consume crop grains or weed seeds; crop grains may be consumed as waste grain. The primary source of information for bird diets was Martin et al. (1961). Y = yes; N = no.

11

AN ASSESSEMENT OF POTENTIAL HAZARDS OF PESTICIDES AND ENVIRONMENTAL CONTAMINANTS

NICHOLAS W. GARD AND MICHAEL J. HOOPER

INTRODUCTION

Since the discovery in the 1940s of the pesticidal properties of synthetic chlorinated hydrocarbon and organophosphorus compounds, these chemicals have played an increasingly important role in pest control and public health programs worldwide. Large quantities of pesticides and industrial pollutants are released into the environment each year, and are frequently disseminated from point of release by biotic and abiotic processes, resulting in extensive global contamination. Soon after use of organochlorine pesticides became prevalent, adverse effects on nontarget species were detected. Population declines or extensive mortality in several species of birds provided the earliest indications of the hazards posed by environmentally persistent organochlorine compounds such as 1,1'-(2,2,2-trichloroethylidene)bis[4-chlorobenzene] (DDT) (Rudd and Genelly 1955, Hickey and Hunt 1960, Wurster et al. 1965). Other classes of pesticides including organophosphorus compounds, though generally less persistent, were potentially as hazardous to wildlife because of their greater acute toxicity. Monitoring of avian populations has proved to be an important method of assessing effects of chemical contaminants in the environment. In many respects, birds are ideal indicators of environmental pollution. Besides being ubiquitous, conspicuous and intensively studied, birds are often more sensitive to contaminants than other vertebrates (Grue et al. 1983). Migratory bird species are exposed to potentially harmful contaminants on breeding, migration and wintering habitats, and consequently can indicate the magnitude of contamination over vast geographic regions. Despite this, it has often been difficult to evaluate the impact of environmental contaminants on birds accurately due to difficulties in detecting and quantifying the extent of exposure, and ascertaining its contribution to population fluctuations.

This chapter reviews the possible lethal and sublethal effects of several classes of environmental contaminants on Neotropical migrants, and discusses methods that can accurately assess these effects. Unfortunately, the majority of the research directed at elucidating mechanisms of toxicity in avian species has focused on nonmigratory species and this review draws largely from that literature. However, comparable mechanisms of toxicity are likely to occur in migratory species under analogous exposure situations. The intent of the chapter is to highlight gaps in our understanding of contaminant effects on migrant species, and to propose research needed to determine the potential contribution of exposure to pesticides and contaminants to population declines of Neotropical migratory birds.

ORGANOCHLORINE INSECTICIDES AND CONTAMINANTS

The organochlorine (OC) insecticides (e.g. DDT and analogs, dicofol, dieldrin, aldrin, endrin, hexachlorocyclohexanes, mirex, heptachlor) are a diverse family of compounds

Table 11-1. Potential effects of organochlorine pesticides on Neotropical migrants.

Effect	Representative Chemical	Reference
Acute mortality	DDT	Hickey and Hunt (1960)
	Aldrin	Presst and Ratcliffe (1972)
	PCBs	Koeman et al. (1973)
Eggshell thinning	DDT, dicofol	Hickey and Anderson (1968)
Endocrine impairment	PCBs, DDT, dieldrin, etc.	Colborn et al. (1993)
Aberrant reproductive behavior	PCBs, DDT	McArthur et al. (1983)
Feminization of reproductive organs	DDT	Fox (1992)

belonging to three distinct chemical classes, but which exhibit similar biological and environmental effects. These compounds were extensively used in North America from the mid-1940s to the mid-1970s in a wide range of agricultural, forestry, and human health applications. Although use of these chemicals has been virtually eliminated in North America, they are believed to be widely used in many developing regions, including Latin America (Risebrough 1986), although usage rates are rarely given. Other halogenated compounds with similar chemical properties and biological effects are the polychlorinated biphenyls (PCBs), dioxins and furans. These OCs are industrial chemicals or byproducts, not insecticides, and are released into the environment as a result of manufacturing processes. Organochlorine compounds exhibit high chemical stability and slow rates of degradation. The nonpolar chemical structure of OCs causes them to be highly soluble in lipids, and as a result they accumulate in animal tissue and biomagnify in food chains (Peterle 1991). These compounds are widely distributed in the environment, including regions distant from agricultural or industrial activities.

Acute Effects

Organochlorine compounds differ greatly in toxicity to birds. Some OCs, such as DDT and PCBs, have relatively high median lethal dose (LD50) values and are not acutely toxic to birds, but others including endrin, aldrin and dioxins are highly toxic (Peterle 1991). While compounds can have various mechanisms of toxicity, most chlorinated hydrocarbons are neurotoxicants, and acute effects frequently result from interference with transmission of nerve impulses (Ecobichon 1991).

Extensive mortality of birds has been associated with the use of many OCs including DDT (Hickey and Hunt 1960, Rudd and Genelly 1955, Wurster et al. 1965, Smies and Koeman 1980), toxaphene (Keith 1966), aldrin, dieldrin and heptachlor (Koeman et al. 1971, Presst and Ratcliffe 1972), and PCBs (Koeman et al. 1973, Parslow and Jefferies 1973). Despite such poisoning incidents, effects of OCs on avian populations appear to result most frequently from chronic exposure, not acute mortality. Some of the potential chronic effects of OCs are summarized in Table 11-1. Only a few of these effects have been conclusively documented to occur in Neotropical migrant species, although all have been detected in other avian species.

Sublethal Effects

Observations of population declines in birds contributed to the initial discovery of the detrimental environmental effects of chlorinated hydrocarbons (Stickel 1975). Reproductive impairment in many bird species, primarily in the orders Pelecaniformes and Falconiformes, was attributed to effects of DDE, the major decomposition product of DDT, on eggshell thinning (Anderson and Hickey 1972). Eggshell thinning was never shown to be a serious problem in migrant passerine species. Other compounds such as dieldrin, heptachlor epoxide and PCBs do not seem to have induced eggshell thinning, but probably contributed to population declines in some species through direct mortality (Stickel 1975, Nisbet 1988, Risebrough and Peakall 1988). Aberrant behavior (abnormal courtship, nest defense and incubation), decreased fertility and hatching success, and increased nestling mortality resulting from OC- or PCB-induced disruptions of the endocrine system are also major causes of decreased reproductive success in some species (Colborn et al. 1993).

Evidence of adverse impacts of DDT on wildlife populations led to its cancellation for all uses in the USA in 1972. Since then, most chlorinated insecticides have been banned in the USA except in certain restricted-use applications (Szmedra 1991). Following the bans, population recoveries have been seen in many of the most seriously affected species such as Peregrine Falcons (Barclay and Cade 1983) and Brown Pelicans (Anderson and Gress 1983). Residue levels in birds declined rapidly following production bans. Multiyear data on DDT residues in starlings collected by the US Fish and Wildlife Service as part of the National Contaminant Biomonitoring Program showed that levels declined in many areas of North America (White 1979, Cain and Bunck 1983). Johnston (1974) found that DDT and DDE levels in the fatty tissue of ten species of warblers and vireos taken in Florida during fall migration were substantially lower in birds sampled in 1973 as compared with pre-1973 individuals. However, "hot spots" of contamination still remain (White and Krynitsky 1986), and OCs and PCBs continue to be prevalent environmental contaminants (Fleming et al. 1983, Eisler 1986).

Some of the residues still found in migrants are probably accumulated on breeding grounds where OCs persist in spite of no recent usage. However, continuing use of DDT and other chlorinated compounds in Latin America is frequently cited as the major contributor to the elevated contaminant burdens being detected in Neotropical migratory species relative to Nearctic resident species (Johnston 1975, White et al. 1981, Henny et al. 1982, DeWeese et al. 1986). Among migratory passerines, insectivores accumulate OCs to a greater extent than granivores or omnivores (DeWeese et al. 1986). Mora and Anderson (1991) indicated that accumulation of some OCs by birds was greater in winter than summer, possibly magnifying the degree of pesticide accumulation in wintering migrants. Quantitative data on the extent of OC pesticide use in Latin American countries is limited, but these compounds remain important for agricultural and public health purposes (Forget 1991). Only a few Latin American countries such as Costa Rica have enacted regulations banning or governing sale and application of these compounds (von Düszeln 1991). Differences in regional-use patterns are indicated by the different pesticide burdens found in resident and migrant species in various countries (Fyfe et al. 1990). Without accurate information on the magnitude and geographic scope of pesticide use throughout Central and South America, it is difficult to predict which migrant species or wintering populations are most likely to accumulate substantial pesticide burdens. Species at greatest risk are probably predators and insectivores because of biomagnification of OCs at higher trophic levels.

The contribution of OCs to population declines in migrant passerines is difficult to assess, since residues have been measured in only a few species, and burdens needed to cause population declines are unknown. DeWeese et al. (1986) found that some individuals of several migrant species (Tree Swallows and Killdeer) had sufficient carcass concentrations of OCs (DDE, dieldrin, heptachlor epoxide) that rapid utilization of fat reserves, as might occur during migration, could mobilize the amount of contaminant estimated to be lethal in the brain. No evidence of regional declines in abundance was found for any of the species studied but DeWeese et al. (1986) noted that regional population estimates may be insensitive to localized fluctuations in abundance. White and Krynitsky (1986) reported that hot spots of DDE contamination in New Mexico and Texas appeared to be impairing reproduction in several species including Western Kingbirds. DeWeese et al. (1985) found a trend toward higher DDE residues in dead female Tree Swallows and in unattended eggs than in controls, suggesting a possible interference with reproduction due to DDE contamination.

Monitoring Techniques

A variety of methods are available to document OC exposure in birds. Monitoring schemes such as the US Fish and Wildlife Service's Biomonitoring of Environmental Status and Trends (BEST) Program are useful for assessing temporal and geographic trends in pesticide contamination. Analysis of contaminant residues in blood plasma may be a useful nonlethal method of monitoring population pesticide burdens, because residue levels in blood are in equilibrium with levels in other tissues (Henny and Meeker 1981). The use of contaminant-induced biochemical and physiological changes as biological markers (biomarkers) of exposure and environmental quality are becoming increasingly popular (Peakall 1992). Rattner et al. (1989) have indicated that measuring induction of mixed-function oxygenases (enzymes which participate in detoxification of foreign substances) is useful in detecting exposure to a variety of pesticides and industrial contaminants. Fox et al. (1988) used porphyrin induction as a biomarker of exposure to PCBs in Herring Gulls. Newer techniques such as assays to detect suppression of the immune system may be particularly useful as biomarkers of exposure and effect since these techniques are sensitive to very low levels of contaminants (Dickerson et al. 1994).

The value of any biochemical or physiological biomarker as an ecological indicator will ultimately depend on being able to correlate observed changes in an individual with population-level effects. However, predicting population-level effects of OCs or PCBs from LD50 toxicity tests, carcass residue levels or biomarker responses is not a straightforward process. In order for biomarker end points to be useful data, quantitative computer models need to be developed that permit extrapolation to population-level responses (Emlen 1989). Additionally, since wild animals are rarely exposed to only one type of pollutant, models that quantify risk must take into account the possible additive or synergistic effects of contaminants occurring in combination. As Newton (1988) points out, death of some individuals due to chemical contamination may not alter the population size if the mortality rate of survivors is not concomitantly reduced. Above some threshold contaminant level, mortality from contaminant-related causes will be additive, resulting in population declines. If critical threshold levels for various contaminants or combinations of contaminants could be determined, the relevance of residue burdens detected in individual birds to population level responses could be quantified. Otherwise, residue levels in birds or eggs fail to provide any meaningful ecological information.

ORGANOPHOSPHORUS AND CARBAMATE INSECTICIDES

Agricultural, rangeland and silvicultural control practices have increasingly relied on organophosphorus (OP) and carbamate insecticides since the decline in the use of OCs insecticides due to the development of pest resistance and regulatory actions restricting their use. In 1964, OPs and carbamates comprised 28% of insecticide use on major agricultural crops in the USA; by 1982, that figure had risen to 85% (Szmedra 1991). More than 100 OP and carbamate compounds are registered for use as active ingredients in pesticide products, and over 65 million hectares of agricultural crops and forests are treated with these pesticides annually in the USA (Smith 1987). In North America pesticide use is heaviest during times of peak insect abundance, which typically corresponds with the avian breeding season. Therefore population-level effects on migratory birds from exposure to OPs could potentially result either from increased adult mortality or decreased reproductive success. The extent of use of OPs and carbamates in Latin American countries is not well known but is apparently increasing (Forget 1991), suggesting that Neotropical migrants may also be exposed on their wintering grounds.

Birds can be exposed to these pesticides via contact with residues on treated vegetation or by secondary poisoning following consumption of OP-poisoned prey (insects, vertebrates). Compounds applied in granular form may be mistaken for grit or seeds by granivorous birds, or may be unintentionally ingested with food (Best and Gionfriddo

Table 11-2. Potential effects of organophosphorus and carbamate insecticides on Neotropical migrants.

Effect	Representative Chemical	Reference
Acute mortality	Various	Grue et al. (1983)
Decreased body weight	Dicrotophos	Grue and Shipley (1984)
Lethargic behavior	Chlorfenvinphos	Hart (1993)
Reduced territorial maintenance	Fenitrothion	Busby et al. (1990)
Reduced parental attentiveness	Fenitrothion	Busby et al. (1990)
Decreased nestling growth rates	Multiple OPs and carbamates	Patnode and White (1991)
Increased postfledging mortality	Methyl parathion	Hooper et al. (1990)
Reduced return rates in subsequent years	Fenitrothion	Millikin and Smith (1990)
Fluctuations in prey abundance or diversity	Carbaryl	Hunter and Witham (1985)

1991). Dermal absorption and inhalation of insecticides may be important routes of exposure for birds encountering spray clouds during aerial insecticide applications or coming into contact with residue-bearing foliage following such applications.

Acute Effects

Organophosphorus and carbamate insecticides exert their toxic effects by binding to and inhibiting acetylcholinesterase (AChE) enzyme at nerve synapses. Inhibition of AChE creates an accumulation of the acetylcholine neurotransmitter, which leads to excessive nerve stimulation. Common physical symptoms of such poisoning in wildlife include ataxia, convulsions and paralysis. Death usually results from respiratory failure (O'Brien 1967). Diagnosis of OP or carbamate poisoning is usually done by measuring AChE activity in plasma and brain tissue. Symptoms typically appear within several hours of acute exposure, although some OPs are capable of producing delayed long-term neurotoxic effects in birds (Johnson and Barnes 1970). Plasma cholinesterase (ChE) inhibition generally occurs more rapidly and extensively than brain ChE inhibition (Ludke et al. 1975, Westlake et al. 1981) and may be a more sensitive indicator of low-level exposure (Fairbrother et al. 1989, Hooper et al. 1990, Fossi et al. 1992). Recovery to 50% of normal brain ChE activity occurs within several days of OP exposure (Fleming 1981), although up to a month may be required for complete return to normal levels (Fleming and Grue 1981, Busby et al. 1983). Recovery rates depend on the OP or carbamate involved, and the initial extent of brain ChE inhibition (Fleming and Bradbury 1981). Plasma ChE activity returns to normal levels more rapidly than brain ChE (Fleming 1981).

Variations in sensitivity to ChE-inhibiting compounds are influenced by many factors including species and sex (Hudson et al. 1972, Schafer 1972, Smith 1987), age (Grue and Shipley 1984, Hooper et al. 1990), route of exposure (Hudson et al. 1972), differences in rates of enzymatic activation and degradation of OPs or carbamates, and seasonal variations in body fat reserves (Grue 1982). Young altricial birds are more sensitive than adults (Hooper et al. 1990). OPs and carbamates generally have low environmental persistence, but can remain and accumulate in food chains under certain conditions leading to secondary poisoning (Peterle 1991).

Sublethal Effects

Many sublethal effects of acute or chronic exposure to OPs have been recorded (Table 11-2), although their impact on survival and population size of wild birds is not fully understood. Pesticide-induced anorexia and subsequent weight losses of up to 40% are common effects recorded in adult and young birds following subacute dietary exposure to OPs (Pope and Ward 1972, Grue 1982, Grue and Shipley 1984, Holmes and Boag 1990). The extent of weight loss and its duration are affected by the length of time birds remain exposed to a pesticide. Adult European Starlings exposed to a single oral dose of dicrotophos, an OP insecticide, experienced a 14% decrease in body mass, but regained normal mass within 24 h (Grue and Shipley

1984). Chronic exposure resulting in prolonged periods of OP-induced anorexia may deplete energy reserves in small passerines, potentially resulting in greater mortality.

Decreased tolerance to cold is often noted following pesticide exposure, presumably due to the inhibition of cholinergic nerve synapses involved in heat regulation (Grue et al. 1991). Several researchers have suggested that depressions in body temperature indicate an inability to thermoregulate, which may contribute to mortality in severely affected individuals (Rattner and Franson 1984, MacGuire and Williams 1987, Martin and Solomon 1991).

Lethargy in captive and wild birds, including decreases in time spent foraging, flying, singing or displaying is associated with sublethal exposure to OPs (Grue and Shipley 1981, Peakall and Bart 1983, Hart 1993). Exposure to OPs can also lead to lapses in parental attentiveness of eggs or nestlings (White et al. 1983, Busby et al. 1990, Meyers et al. 1990). Female European Starlings dosed with dicrotophos made significantly fewer feeding trips to their nest than undosed controls, and young of dosed parents lost significantly more weight than control nestlings (Grue et al. 1982). In all studies that have noted lapses in parental attentiveness, alterations in normal behaviors were transitory but can persist longer than 24 h following exposure (Busby et al. 1990). Chronic exposure to pesticides may lower reproductive success if disruption of parental attentiveness results in prolonged periods when eggs are not incubated or young are not fed. Nestlings exposed to ChE-inhibiting compounds either through dermal absorption or consumption of contaminated prey items also display altered behavior. Decreased begging behavior and ensuing losses in body weight were observed in starling nestlings after oral dosing (Grue and Shipley 1984).

The effect of sublethal exposure of nestlings to ChE-inhibitors on postfledging survival is not fully understood. Four-day-old White-throated Sparrows given a single oral dose of the OP fenthion suffered perturbations in growth and survivors fledged at lower weights than controls (Pearce and Busby 1980). When nestlings were administered two doses over a 4 day period, two drops in body weight were seen and fledging weights were also depressed relative to controls (Pearce and Busby 1980). Because fledging weight may be positively correlated with future survival (Perrins 1965, Garnett 1981), behavioral modifications of adults or nestlings that result in decreased nestling fledging weight may lower postfledging survival. Stromborg et al. (1988) dosed 16-day-old starlings with dicrotophos at levels expected to result in 13–27% mortality. Nestlings that survived the dose had an average of 46% depression in brain ChE activity 2 days postdose but fledged at near normal weights. Age at fledging, survival to 6 weeks post-fledging, flocking behavior, and habitat use in dosed birds did not differ significantly from that of undosed siblings. However, Hooper et al. (1990) found that nestling starlings that had survived earlier exposure to methyl parathion were much more susceptible than control nestlings to postfledging mortality occurring as a result of predation. Patnode and White (1991) found that daily survival rates of nestling Northern Mockingbirds, Northern Cardinals, and Brown Thrashers, varied inversely with pesticide application frequency in pecan groves. Also, rates of weight gain of mockingbird and thrasher nestlings were reduced with increasing exposure. Other studies (Powell 1984, Spray et al. 1987) have also detected trends toward reduced growth rates of nestlings in sprayed areas. These studies suggest that evaluation of nestling growth rates and postfledging survival may be sensitive techniques for monitoring the consequences of sublethal exposure.

Effects on Reproduction

Numerous cases of avian mortalities resulting from accidental misuse or intentional use of OPs and carbamates have been reported (McLeod 1967, Hill and Fleming 1982, Grue et al. 1983). Grue et al. (1983) suggested that while these mortalities may have temporarily reduced local populations, there was no evidence that insecticides caused long-term population declines in any species. However, the effects of sublethal pesticide exposure on population-level reproduction have not been well quantified. Busby et al. (1990) noted decreased reproductive success for White-throated Sparrows breeding in a New Brunswick forest following two aerial fenitrothion applications for spruce budworm (*Choristoneura fumiferana*) control. Birds in sprayed areas had higher rates of territory abandonment, decreased territorial defense, abnormal incubation, and higher incidences of clutch desertion than birds in control plots. Mortality and territory abandonment reduced the adult breeding population by one-third. Clutch sizes and fledging success of birds remaining to breed in the sprayed areas were not significantly different than for the control population. However, the overall impact of the spraying was a 75% reduction in the number of young fledging relative to the control site. Millikin and Smith (1990) found that return rates of marked individuals of migrant songbirds (flycatchers, vireos, warblers) in boreal forests were lower on treated plots than on control plots in the year after fenitrothion spraying. Breeding populations on treated plots remained stable, however, due to increased recruitment of new birds. Spray et al. (1987) found that fenitrothion spraying in Scottish pine forests did not alter population sizes of the five most common passerine species. Based on their results, Spray et al. (1987) and Millikin and Smith (1990) both concluded that long-term effects of insecticide spraying on the songbird populations were minor. However, because both studies used small experimental spray blocks (49–78 ha), immigration of birds into the study plot from adjacent unsprayed habitat probably obscured spray-related mortality. Busby et al. (1990) performed their study on a 50 ha plot in the center of a 10,000 ha operational spray block, and movement of birds into the sprayed areas from adjacent unsprayed forest was probably negligible. This study is more representative of the potential consequences of widespread spraying, which typically occurs in forest and rangeland control operations.

Organophosphate and carbamate insecticides may also affect avian survival and reproduction indirectly. Indirect effects on individuals result from pesticide-related changes in other components of the environment, including the species' food resources, competitors, predators and cover/habitat. Shifts in foraging sites from sprayed areas into adjacent unsprayed areas by insectivorous birds have been frequently noted following insecticide-induced declines in food abundance (Moulding 1976, DeWeese et al. 1979, Hunter and Witham 1985). Habitat shifts are only likely when localized or patchy spray operations leave unsprayed swaths. Hunter and Witham (1985) found that, following uniform spray coverage, warblers did not alter foraging locations, since unsprayed adjacent habitat was not available. Birds may also alter foraging locations, within a tree, moving away from areas of the canopy with high spray residues, although time–activity budgets may remain unchanged (Hunter and Witham 1985, Millikin and Smith 1990).

Other indirect responses to insecticide applications include shifts in the proportion of insects in the diet, or enlargement of foraging territories due to elimination of prey (Cooper et al. 1990). If insecticide-induced declines in prey abundance greatly increase the travel time and energetic costs of foraging for adults, nestling growth rates and reproductive success of nests in sprayed areas could be adversely affected. Even with the presence of nearby unsprayed areas, the energetic costs to irds of moving to those areas may be too great. Hunter et al. (1984) found that carbaryl-induced reductions in invertebrate biomass resulted in a decrease in the growth rates of Black Ducks and Mallards. Ducklings on treated ponds spent more time searching for food than ducks on untreated ponds, and the rate of movement around ponds was greater. Use of the insect growth regulator diflubenzuron for gypsy moth (*Lymantria dispar*) control led to significantly reduced fat reserves for seven insectivorous Neotropical migrant species, likely resulting from a reduction in available food (Whitmore et al. 1993). No reduction in fat stores was detected in the two resident species studied. Effects of reduced fat reserves on adult or nestling survival were not determined.

Monitoring Techniques

Diagnosing OP or carbamate exposure involves analysis of brain or plasma tissue for detection of ChE inhibition. Analytical methods commonly follow the procedure of Ellman et al. (1961). Modifications of the assay, which enhance its use in field situations, are described in detail by Hill and Fleming (1982) and Fairbrother et al. (1991). Based on studies using a range of OPs, inhibition of brain ChE greater than 20% below mean control activities is considered indicative of exposure to anti-ChE compounds, while inhibition of 50% or greater is sufficient for diagnosis of death due to ChE inhibition (Ludke et al. 1975, Zinkl et al. 1980). Some birds can survive brain ChE inhibition greater than 50% (Grue et al. 1991), and the level of ChE inhibition at the time of sample collection is not likely to equal the maximum level of inhibition (Busby and White 1991). Confirmation of cause of death in birds with depressed ChE activity usually requires chemical analysis of stomach or crop contents for the presence of OP or carbamate residues (Greig-Smith 1991), or tissue reactivation analysis to detect the presence of inhibited enzyme (Fairbrother et al. 1991).

Use of plasma ChE has several advantages over brain ChE. This method allows for repeated sampling over time from an individual, thereby allowing detection of chronic ChE inhibition (Fairbrother et al. 1989, Hooper et al. 1989, 1990). Because low-level exposure can inhibit plasma ChE but not brain ChE, plasma ChE monitoring provides a more sensitive indicator of toxicity than brain ChE, which only evaluates acute poisonings. Use of plasma ChE for diagnostic purposes is hindered by a lack of data on normal plasma ChE activity in different species and by large intraspecific variability in activity (Fairbrother et al. 1991). Some researchers (Fleming 1981, Hill and Fleming 1982) have questioned the value of using plasma ChE for monitoring exposure because its relevance to brain ChE inhibition is not clear. Fairbrother et al. (1989) have indicated that the two enzyme activities do not correlate well in intoxicated birds, but Fossi et al. (1992) found significant correlations at lethal and sublethal doses. Another romising nonlethal technique is fecal–urate analysis to detect alkyl phosphate catabolites of OPs. Hooper et al. (1989) found a good correlation between fecal–urate catabolite concentrations and plasma ChE depression in OP-exposed Red-tailed Hawks.

The reliability of using brain or plasma ChE inhibition as biomarkers of exposure requires baseline enzyme activity data for unexposed individuals against which the ChE activity of birds suspected of exposure can be compared (Hill 1988). Brain and plasma enzyme activity are influenced by many factors including species, age, sex, and reproductive condition (Grue and Shipley 1984, Bennett and Bennett 1991, Rattner and Fairbrother 1991, Gard and Hooper 1993). Different analytical methods also produce different results. Data bases of ChE activity in healthy individuals (e.g., Hill 1988) can provide control data, but concurrent controls permit more accurate diagnosis of pesticide exposure.

Mineau and Peakall (1987) have suggested that population level effects of ChE inhibitors have not been well documented due to inappropriate census methods, including small study plots and use of auditory or visual cues to estimate abundance. Censuses of small plots are subject to bias due to immigration of floaters from adjacent habitat or to emigration of birds which forage in unsprayed areas. Also, pesticide deposition in small plots is often not uniform, and plots may be over- or undersprayed, or not sprayed at all (Peakall and Bart 1983). Therefore, many small plots may be needed to assess the impact of an insecticide application accurately. Further, studies using small, isolated spray plots may not be representative of large-scale spray operations in forests or rangelands. Mineau and Peakall (1987) have suggested that these problems can be overcome in silvicultural operations by censusing along transects in the center of large operational spray blocks. The bias may be more difficult to control in agricultural situations where pesticides are often applied to individual fields while adjacent agricultural or nonagricultural habitats remain untreated or are treated with other pesticides.

Census techniques that rely on auditory (singing males) or visual cues to estimate population size are often used to detect changes in bird abundance after a spray operation. However, sublethal exposure to OPs typically causes reduced activity and vocalizations for as long as 3 days following application (Grue and Shipley 1981). Therefore, decreased detection rates cannot be taken as evidence of mortality or emigration. Further bias is introduced if collection of birds for analysis of ChE inhibition is based on the most visible or audible birds detected immediately after a spray application because these individuals were probably least exposed. Collection of birds over several days following exposure also underestimates the average inhibition because of recovery of inhibited enzyme, particularly following exposure to carbamate insecticides.

Confirmation that declines in census counts are attributable to pesticide-induced mortality requires intensive carcass searches to locate dead or incapacitated birds. Given the difficulty in discovering even a few carcasses due to low searching efficiency and the rapidity with which carcasses disappear (Balcomb 1986), the presence of a small number of carcasses may indicate that substantial mortality has occurred. Failure to detect carcasses cannot be used as evidence of no effect, however, because such census methods are imprecise and unable to evaluate subacute effects such as decreased reproductive success.

The diagnostic procedures discussed above can be applied to several types of monitoring schemes suitable for Neotropical migratory birds. The simplest form of monitoring involves incident sampling in which reports of wildlife mortality from pesticide use are investigated, and carcasses tested for the presence of pesticides or inhibited enzyme. Various federal and state agencies use incident-monitoring schemes, which vary widely in effectiveness due to differences in the thoroughness to which mortality incidents are investigated (Avian Effects Dialogue Group 1989). Incident reports can create useful databases of pesticide-related bird mortality, but a lack of reports does not indicate an absence of adverse effects because geographic coverage is not uniform and sublethal effects are rarely studied. Standardized data collection methods are required in order for incident-monitoring programs to document the frequency of mortality incidents and their potential impact on avian populations reliably.

Targeted monitoring schemes are more detailed than incident reports because they involve design and implementation of planned field studies to evaluate potential acute and subacute effects of pesticides on bird populations under conditions of actual use. The Environmental Protection Agency (EPA) has issued guidelines for designing such field studies to obtain pesticide registration or reregistration data (Fite et al. 1988, Avian Effects Dialogue Group 1991). Because of the time and effort involved in conducting these studies, selection of focal species and study sites considered to represent "worst-case" scenarios is recommended. Although appropriately chosen focal species may represent an entire avian community, such field studies need to be supplemented with census methods that accurately estimate abundance and reproductive success of other migrant species.

At present, the contribution of ChE-inhibiting pesticides to declines in abundance of Neotropical migrants cannot be adequately assessed because of insufficient or inappropriate research. Given the widespread use of OPs and carbamates in a variety of habitats of North America, and their increasing use in Latin American agroecosystems, the potential exists for many migratory species to be exposed to these compounds. Even when exposure does not cause mortality, impaired reproductive success and/or indirect effects through decreased prey abundance probably occur. Evidence that OP and carbamate insecticides are not contributing to long-term changes in population abundance will not be provided by any single assessment technique or monitoring scheme, but will require a combination of techniques sensitive to detection of both acute and sublethal effects.

OTHER ENVIRONMENTAL CONTAMINANTS

Herbicides

Herbicides account for over 60% of all pesticides applied on major field and forage crops in the United States (Szmedra 1991), and are also widely used in tropical agricultural applications (Forget 1991). Herbicides are generally nontoxic to birds (Morrison and Meslow 1983), and the severest impact on avian populations is their potential to produce extensive habitat modification (Morrison and Meslow 1984). Large-scale herbicide applications in silvicultural or agricultural operations can alter avian community structure or reproductive success through loss of suitable habitat or declines in prey abundance of modified habitats (O'Connor and Shrubb 1986).

Best (1972) found that the number of shrub-nesting Brewer's Sparrows declined by 54% in the first year in which sagebrush was treated with the herbicide, 2,4-dichlorophenoxyacetic acid (2,4-D), but the number of ground-nesting Vesper Sparrows was unchanged. The major plant and animal foods of these sparrows were reduced in amount. Two years after 2,4-D treatment to sagebrush, nesting Brewer's Sparrows were reduced by 99% (Schroeder and Sturges 1975). Remaining Brewer's Sparrows nested in surviving sagebrush plants. Savidge (1978) found that 2,4,5-trichlorophenoxyacetic acid (2,4,5-T), used to control brush, reduced the number of individuals and species of birds, with twice as many birds being present on the untreated site as the treated site. The bird species not present on treated sites were associated with the *Caenothus* and manzanita (*Arctostaphylos patula*) brush being controlled.

Herbicides used to control scrub in Norwegian forests reduced the number of bird territories by 30% due to changes in cover quality and food resources (Slagsvold 1977). Bird species preferring early successional stages increased while those preferring old forest decreased. Bird numbers remained depressed on sprayed areas for 5 years. Herbicides used in Oregon clearcuts to remove red alder (*Alnus rubra*) resulted in a 2–3-fold decrease in abundance of Wilson's Warblers at 1 and 4 years after spraying. Concomitant alterations in patterns of habitat use and foraging behavior were alsoseen (Morrison and Meslow 1983). Abundance of other Neotropical migrants (Willow Flycatcher, Swainson's Thrush, MacGillivray's Warbler, and Orange-crowned Warbler) was not significantly different on sprayed plots relative to control sites. Steele (1992) observed a large decline in Black-throated Blue Warbler density on experimental plots where herbicide use reduced shrub density by 76%. Warbler density declined by 61% in the first year after herbicide application, and no individuals were present 3 years after herbicide application.

Herbicides are intended to change the plant community in which they are applied but they can also indirectly affect birds by changing food availability. Potts (1984) concluded that herbicide use in cereal fields in the United Kingdom was the most significant cause of the decline in the Grey Partridge population. The survival of partridge chicks to 6 weeks of age is determined largely by the availability of plant bugs, leafhoppers, sawfly larvae, weevils, and leaf beetles. The use of herbicides halved the overall density of these insects by removing the plants on which they lived. The distances that partridge broods moved to find food were 3–4 times greater on herbicide-treated fields than on control sites (Southwood and Cross 1969). Rands (1985) experimentally tested whether the herbicides and fungicides used on cereal fields reduced Grey Partridge chick survival. Brood size and insect abundance were lower on completely sprayed fields than on fields with unsprayed borders. Based on radiotelemetry data, fields with unsprayed borders had higher chick survival, shorter distances between roosts and a greater proportion of the home range that included unsprayed headlands. Survival rates on completely sprayed fields were below those necessary to maintain a stable population. Similar research evaluating the impact of herbicides on food abundance, home-range size, and reproductive success for insectivorous Neotropical migrants is required. In summary, although most herbicides are not acutely toxic to birds, they can significantly affect bird populations by altering cover and food resources.

Acidification

Vast expanses of aquatic and forest ecosystems in the United States and Canada are undergoing modification due to contamination from acidic precipitation created by combustion of fossil fuels. Impacts on migrants using these ecosystems during the breeding season are likely, particularly indirect effects due to changes in habitat structure or prey availability (McNicol et al. 1987, Blancher 1991).

Reductions in pH levels of aquatic environments due to acidification may reduce the abundance or diversity of acid-sensitive insect species on which Neotropical migrants prey (McNicol et al. 1987). Acid-sensitive molluscs and insects (Ephemeroptera) comprised a smaller proportion of the diet of nestling Tree Swallows hatched in nest boxes near acidified lakes than for a control population at a wetland of higher pH. However, nestlings at acidified lakes were fed a greater proportion of acid-tolerant insects (Diptera), and their total food intake did not differ from that of control nestlings (Blancher and McNicol 1991). Species that specialize on acid-sensitive prey appear to be most at risk from acidification of aquatic ecosystems (Ormerod and Tyler 1987).

Habitat acidification can increase the bioavailability of several nonessential, potentially toxic metals, particularly mercury, aluminum, cadmium and lead, which can move up food chains to piscivorous and insectivorous migrants (Scheuhammer 1991). High dietary intake of aluminum may have caused eggshell defects, decreased clutch sizes and hatching success and increased nest abandonment in Pied Flycatchers nesting by the shore of an acidified lake in Sweden (Nyholm and Myhrberg 1977, Nyholm 1981). Reproductive impairment may have been due to aluminum-induced secondary disruptions of calcium availability and metabolism rather than direct aluminum toxicity (Scheuhammer 1991). Blancher and McNicol (1988) observed that Tree Swallows breeding near acidified wetlands laid smaller clutches and raised smaller young than a control population. Reduced calcium content of the insect prey at the acidified wetlands was believed to be responsible for the decreased reproductive success. Glooschenko et al. (1986) studied reproductive success of Eastern Kingbirds nesting near acidified lakes in Ontario. Lake chemistry parameters influenced by acidification (pH, alkalinity, iron, manganese, and aluminum concentrations) accounted for small but significant variations in nestling growth. Emergent insect prey abundance was not altered by lake acidity.

The effect of acidification of terrestrial environments on birds is less well understood, but the potential exists for negative impacts due to habitat modification or shifts in prey abundance and quality. Dieback of sugar maple stands in Quebec as a result of acidic precipitation has resulted in shifts in avian community structure, with decreases in canopy-feeding birds and concurrent increases in shrubfeeders and dead-tree specialists (DesGranges 1987, DesGranges et al. 1987). An increased frequency of eggs with no shells or poor shells for forest birds (tits and nuthatches) breeding in the Netherlands was attributed to decreased availability of calcium in their insect prey resulting from soil acidification (Drent and Woldendorp 1989).

Metals

Human activities including fossil-fuel combustion, metal smelting and processing, and industrial emissions, are the primary sources of many heavy-metal pollutants, although in some regions naturally occurring metal levels can also pose threats to wildlife (Goyer 1991). Metals have high environmental persistence because natural cycling in the environment is extremely slow. Many heavy metals have the ability to bioaccumulate through food chains, especially in aquatic ecosystems (Goyer 1991). Only about 30 metals are considered toxic, and, of these, the ones of greatest toxicological concern include mercury, cadmium, aluminum, lead, and selenium (Eisler 1985, Scheuhammer 1987). Chronic dietary exposure to some metals can lead to reproductive dysfunction or increased susceptibility to disease in birds (Scheuhammer 1987). Scheuhammer (1987) evaluated the monitoring strategies available for assessing environmental exposure of birds to these metals, but the impact of heavy metal exposure on bird populations, in general, and migratory species, in particular, is poorly understood.

CONCLUSION

Pesticides and environmental contaminants can adversely affect avian populations via direct or indirect mechanisms. Much of the toxicological research that has established these mechanisms has used nonmigratory passerines or gamebirds. With the exception of a few well-documented cases such as DDT-induced eggshell thinning in migratory raptors, very little toxicological research has explicitly focused on Neotropical migrants. This is reflected by the limited number of examples cited in this review that refer to migratory species. Toxic effects of exposure to pesticides and contaminants in migrants are probably similar to those seen in other species. However, the importance of contaminant-induced alterations in survival and reproductive rates as a contributing factor to population declines observed in many migrant species relative to other anthropogenic stresses cannot be reliably quantified at present.

Several critical-research needs must be addressed to evaluate the role of environmental contaminants in population declines of migrant species. Toxicologists must increase collaboration and communication with ornithologists so that the latter are aware of potential pesticide-induced effects that may occur in field studies. In particular, ornithologists should be alert for sublethal effects on birds or chronic effects, which may only become evident long after the pollution event has occurred. Additionally, planned field studies that specifically address toxicant effects on migrants, particularly passerine species, are needed. Toxicologists and ornithologists also need to correlate biomarkers of contaminant exposure and tissue residue levels with relevant demographic parameters, such as survival and reproduction rates.

Without minimizing the importance of pesticide use and pollution in North America, a greater likelihood of adverse impacts probably occurs on the Central and South American wintering ranges, where pesticide use and industrial emissions are less stringently regulated. A critical lack of knowledge exists regarding exposure of migrants to chemicals in regions where they overwinter. Monitoring programs need to be established to provide baseline data on pesticide use rates and residue levels in abiotic and biotic components of the ecosystems where migrants spend the winter. Risk of exposure of migrants to these chemicals and the potential for exposure to increase overwinter mortality must also be quantified. For temporal and geographic continuity, similar standardized monitoring programs should also be initiated on breeding habitats and migration stopover sites. Gard et al. (1993) discuss several criteria that must be fulfilled in order that monitoring programs provide useful information.

Pesticides and contaminants adversely affect all wildlife, including Neotropical migrants. However, with the formulation of newer generation pesticides and stricter pollution-control measures, effects have largely changed from overt mortality to more subtle processes. The focus of pesticide registration procedures and pollution discharge regulations must be shifted from an emphasis on acute effects on wildlife to greater concern for chronic exposures, which cause long-term alterations in fecundity or survival. Currently, many of the toxicity tests used for regulatory purposes do not adequately assess these sublethal effects, and therefore may seriously underestimate the risk pesticides or contaminants pose to wildlife. Additionally, the immense dependence placed on use of pesticides as the primary method of crop protection must be lessened by incorporating less environmentally harmful practices, including biological control and integrated pest management, which can protect crops without adversely affecting agricultural productivity.

LITERATURE CITED

Anderson, D. W., and F. Gress. 1983. Status of a northern population of California Brown Pelicans. Condor 85:79–88.

Anderson, D. W., and J. J. Hickey. 1972. Eggshell changes in certain North American birds. Prox. XVth Int. Ornithol. Congress 514–540.

Avian Effects Dialogue Group. 1989. Pesticides and birds: improving impact assessment. The Conservation Foundation, Washington DC.

Avian Effects Dialogue Group. 1991. Assessing pesticide impacts on birds: discussions of the Avian Effects Dialogue Group (1989–1991). World Wildlife Fund, Washington DC.

Balcomb, R. 1986. Songbird carcasses disappear rapidly from agricultural fields. Auk 103: 817–820.

Barclay, J. H., and T. J. Cade. 1983. Restoration of the Peregrine Falcon in the eastern United States. Pp. 3–40 *in* Bird conservation, Vol. 1 (S. A. Temple, ed.). University of Wisconsin Press, Madison, WI.

Bennett, R. S., and J. K. Bennett. 1991. Age-dependent changes in activity of Mallard plasma cholinesterases. J. Wildl. Dis. 27: 116–118.

Best, L. B. 1972. First-year effects of sagebrush control on two sparrows. J. Wildl. Manag. 36:534–544.

Best, L. B., and J. P. Gionfriddo. 1991. Characterization of grit use by cornfield birds. Wilson Bull. 103:68–82.

Blancher, P. J. 1991. Acidification: implications for wildlife. Trans. North Amer. Wildl. Natural Resources Conf. 56:195–204.

Blancher, P. J., and D. K. McNicol. 1988. Breeding biology of Tree Swallows in relation to wetland acidity. Can. J. Zool. 66:842–849.

Blancher, P. J., and D. K. McNicol. 1991. Tree Swallow diet in relation to wetland acidity. Can. J. Zool. 69:2629–2637.

Busby, D. G., and L. M. White. 1991. Factors influencing variability in brain acetylcholinesterase activity in songbirds exposed to aerial fenitrothion. Pp. 211–232 *in* Cholinesterase-inhibiting insecticides: their impact on wildlife and the environment (P. Mineau, ed.). Elsevier, Amsterdam.

Busby, D. G., P. A. Pearce, N. R. Garrity, and L. M. Reynoolds. 1983. Effect of an organophosphosphorus insecticide on brain cholinesterase activity in White-throated Sparrows exposed to aerial forest spraying. J. Appl. Ecol. 20:255–263.

Busby, D. G., L. M. White, and P. A. Pearce. 1990. Effects of aerial spraying of fenitrothion on breeding White-throated Sparrows. J. Appl. Ecol. 27:743–755.

Cain, B. W., and C. M. Bunck. 1983. Residues of organochlorine compounds in Starlings (*Sturnus vulgaris*), 1979. Environ. Monit. Assess. 3:161–172.

Colborn, T., F. S. vom Saal, and A. M. Soto. 1993. Developmental effects of endocrine-disrupting chemicals in wildlife and humans. Environ. Health Perspect. 101:378–384.

Cooper, R. J., K. M. Dodge, P. J. Martinat, S. B. Donahoe, and R. C. Whitmore. 1990. Effect of diflubenzuron application on eastern deciduous forest birds. J. Wildl. Manag. 54: 486–493.

DesGranges, J.-L. 1987. Forest birds as indicators of the progression of maple dieback in Quebec. Pp. 249–257 *in* The value of birds (A. W. Diamond, and F. L. Filion, eds). ICBP Tech. Publ. No. 6, Cambridge, England.

DesGranges, J.-L., Y. Mauffette, and G. Gagnon. 1987. Sugar maple forest decline and implications for forest insects and birds. Trans. North Amer. Wildl. Natural Resources Conf. 52: 677–689.

DeWeese, L. R., R. R. Cohen, and C. J. Stafford. 1985. Organochlorine residues and eggshell measurements for Tree Swallows *Tachycineta bicolor* in Colorado. Bull. Environ. Contam. Toxicol. 35:767–775.

DeWeese, L. R., C. J. Henny, R. L. Floyd, K. A. Bobal, and A. W. Schultz. 1979. Response of breeding birds to aerial sprays of trichlorfon (Dylox) and carbaryl (Sevin-4-oil) in Montana forests. US Fish Wildl. Serv., Spec. Sci. Rep. Wildl. 224:1–29.

DeWeese, L. R., L. C. McEwen, G. L. Hensler, and B. E. Petersen. 1986. Organochlorine contamination in passeriforms and other avian prey of the Peregrine Falcon in the western United States. Environ. Toxicol. Chem. 5: 675–693.

Dickerson, R. L., M. J. Hooper, N. W. Gard, G. P. Cobb, and R. J. Kendall. 1994. Toxicological foundations of ecological risk assessment: biomarker development and interpretation based on laboratory and wildlife species. Environ. Health Perspect. 102 (Suppl. 12):65–69.

Drent, P. J., and J. W. Woldendorp. 1989. Acid rain and eggshells. Nature 339:431.

Ecobichon, D. J. 1991. Toxic effects of pesticides. Pp. 565–622 *in* Casarett and Doull's toxicology, 4th ed. (M. O. Amdur, J. Doull, and C. D. Klaassen, eds). Pergamon Press, New York.

Eisler, R. 1985. Selenium hazards to fish, wildlife, and invertebrates: a synoptic review. US Fish Wildl. Serv. Biol. Rep. 85(1.5):1–57.

Eisler, R. 1986. Polychlorinated biphenyl hazards to fish, wildlife, and invertebrates: a synoptic review. US Fish Wildl. Serv. Biol. Rep. 85(1.7):1–72.

Ellman, G. L., K. D. Courtney, V. Andres. Jr, and M. R. Featherstone. 1961. A new and rapid colorimetric determination of acetylcholinesterase activity. Biochem. Pharmacol. 7:88–98.

Emlen, J. M. 1989. Terrestrial population models for ecological risk assessment: a state-of-the-art review. Environ. Toxicol. Chem. 8: 831–842.

Fairbrother, A., R. S. Bennett, and J. K. Bennett. 1989. Sequential sampling of plasma cholinesterase in Mallards (*Anas platyrhynchos*) as an indicator of exposure to cholinesterase inhibitors. Environ. Toxicol. Chem. 8:117–122.

Fairbrother, A., B. T. Marden, J. K. Bennett, and M. J. Hooper. 1991. Methods used in determination of cholinesterase activity. Pp. 35–71 *in* Cholinesterase-inhibiting insecticides: their impact on wildlife and the environment (P. Mineau, ed.). Elsevier, Amsterdam.

Fite, E. C., L. W. Turner, N. J. Cook, and C. Stunkard. 1988. Guidance document for conducting terrestrial field studies. US Environmental Protection Agency, Washington DC.

Fleming, W. J. 1981. Recovery of brain and plasma cholinesterase activities in ducklings exposed to organophosphorus pesticides. Arch. Environ. Contam. Toxicol. 10:215–229.

Fleming, W. J., and S. P. Bradbury. 1981. Recovery of cholinesterase activity in Mallard ducklings administered organophosphorus pesticides. J. Toxicol. Environ. Health 8:885–897.

Fleming, W. J., and C. E. Grue. 1981. Recovery of cholinesterase activity in five avian species exposed to dicrotophos, an organophosphorus pesticide. Pest. Biochem. Physiol. 16:129–135.

Fleming, W. J., D. R. Clark, Jr, and C. J. Henny. 1983. Organochlorine pesticides and PCB's: a continuing problem for the 1980's. Trans. North Amer. Wildl. Natural Resources Conf. 48:186–199.

Forget, G. 1991. Pesticides and the Third World. J. Toxicol. Environ. Health 32:11–31.

Fossi, M. C., C. Leonzio, A. Massi, L. Lari, and S. Casini. 1992. Serum esterase inhibition in birds: a nondestructive biomarker to assess organophosphorus and carbamate contamination. Arch. Environ. Contam. Toxicol. 23:99–104.

Fox, G. A. 1992. Epidemiological and pathobiological evidence of contaminant-induced alterations in sexual development in free-living wildlife. Pp. 147–158 *in* Chemically induced alterations in sexual and functional development: the wildlife/human connection. (T. Colborn, and C. Clement, eds). Princeton Scientific Publishing, Princeton, NJ.

Fox, G. A., S. W. Kennedy, R. J. Norstrom, and D. C. Wigfield. 1988. Porphyria in Herring Gulls: a biochemical response to chemical contamination of Great Lakes food chains. Environ. Toxicol. Chem. 7:831–839.

Fyfe, R. W., U. Banasch, V. Benavides, N. H. De Benavides, A. Luscombe, and J. Sanchez. 1990. Organochlorine residues in potential prey of Peregrine Falcons, *Falco peregrinus*, in Latin America. Can. Field Natur. 104:285–292.

Gard, N. W., and M. J. Hooper. 1993. Age-dependent changes in plasma and brain cholinesterase activities of Eastern Bluebirds and European Starlings. J. Wildl. Dis. 29:1–7.

Gard, N. W., M. J. Hooper, and R. S. Bennett. 1993. Effects of pesticides and contaminants on Neotropical migrants. Pp. 310–314 *in* Status and management of Neotropical migratory birds (D. M. Finch, and P. W. Stangel, eds). Gen. Tech. Rep. RM-229, USDA Forest Serv. Rocky Mt. Forest Range Exp. Sta., Fort Collins, CO.

Garnett, M. C. 1981. Body size, its heritability and influence on juvenile survival among Great Tits *Parus major*. Ibis 123:31–41.

Glooschenko, V., P. Blancher, J. Herskowitz, R. Fulthorpe, and S. Rang. 1986. Association of wetland acidity with reproductive parameters and insect prey abundance of the Eastern Kingbird (*Tyrannus tyrannus*) near Sudbury, Ontario. Water Air Soil Pollut. 30:553–567.

Goyer, R. A. 1991. Toxic effects of metals. Pp. 623–680 *in* Casarett and Doull's toxicology, 4th ed. (M. O. Amdur, J. Doull, and C. D. Klaassen, eds). Pergamon Press, New York.

Greig-Smith, P. W. 1991. Use of cholinesterase measurements in surveillance of wildlife poisoning in farmland. Pp. 127–149 *in* Cholinesterase-inhibiting insecticides: their impact on wildlife and the environment (P. Mineau, ed.). Elsevier, Amsterdam.

Grue, C. E. 1982. Response of Common Grackles to dietary concentrations of four organophosphate pesticides. Arch. Environ. Contam. Toxicol. 11:617–626.

Grue, C. E., and B. K. Shipley. 1981. Interpreting population estimates of birds following pesticide applications—behavior of male Starlings exposed to an organophosphate pesticide. Stud. Avian Biol. 6:292–296.

Grue, C. E., and B. K. Shipley. 1984. Sensitivity of nestling and adult Starlings to dicrotophos, an organophosphate pesticide. Environ. Res. 35:454–465.

Grue, C. E., W. J. Fleming, D. G. Busby, and E. F. Hill. 1983. Assessing hazards of organophosphate pesticides to wildlife. Trans. North Amer. Wildl. Natural Resources Conf. 48:200–220.

Grue, C. E., A. D. M. Hart, and P. Mineau. 1991. Biological consequences of depressed brain cholinesterase activity in wildlife. Pp. 151–209 *in* Cholinesterase-inhibiting insecticides: their impact on wildlife and the environment (P. Mineau, ed.). Elsevier, Amsterdam.

Grue, C. E., G. V. N. Powell, and M. J. McChesney. 1982. Care of nestlings by wild female Starlings exposed to an organophosphate pesticide. J. Appl. Ecol. 19:327–335.

Hart, A. D. M. 1993. Relationships between behavior and the inhibition of acetylcholinesterase in birds exposed to organophosphorus pesticides. Environ. Toxicol. Chem. 12:321–336.

Henny, C. J., and D. L. Meeker. 1981. An evaluation of blood plasma for monitoring DDE in birds of prey. Environ. Pollut. Ser. A. 25:291–304.

Henny, C. J., F. P. Ward, K. E. Riddle, and R. M. Prouty. 1982. Migratory Peregrine Falcons, *Falco peregrinus*, accumulate pesticides in Latin America during winter. Can. Field Natur. 96:333–338.

Hickey, J. J., and D. W. Anderson. 1968. Chlorinated hydrocarbons and eggshell changes in raptorial and fish-eating birds. Science 162:271–273.

Hickey, J. J., and L. B. Hunt. 1960. Initial song bird mortality following a Dutch elm disease control program. J. Wildl. Manag. 24: 259–265.

Hill, E. F. 1988. Brain cholinesterase activity of apparently normal wild birds. J. Wildl. Dis. 24:51–61.

Hill, E. F., and W. J. Fleming. 1982. Anticholinesterase poisoning of birds: field monitoring and diagnosis of acute poisoning. Environ. Toxicol. Chem. 1:27–38.

Holmes, S. B., and P. T. Boag. 1990. Effects of the organophosphorus pesticide fenitrothion on behavior and reproduction in Zebra Finches. Environ. Res. 53:62–75.

Hooper, M. J., P. Detrich, C. Weisskopf, and B. W. Wilson. 1989. Organophosphorus insecticide exposure in hawks inhabiting orchards during winter dormant-spraying. Bull. Environ. Contam. Toxicol. 42:651–659.

Hooper, M. J., L. W. Brewer, G. P. Cobb, and R. J. Kendall. 1990. An integrated laboratory and field approach for assessing hazards of pesticide exposure to wildlife. Pp. 271–283 *in* Pesticide effects on terrestrial wildlife (L. Somerville, and C. H. Walker, eds). Taylor and Francis, London.

Hudson, R. H., R. K. Tucker, and M. A. Haegele. 1972. Effects of age on sensitivity: acute oral toxicity of 14 pesticides to Mallard ducks of several ages. Toxicol. Appl. Pharmacol. 22:556–561.

Hunter, M. L., Jr, and J. W. Witham. 1985. Effects of a carbaryl induced depression of arthropod abundance on the behaviour of Parulinae warblers. Can. J. Zool. 63:2612–2616.

Hunter, M. L., Jr, J. W. Witham, and H. Dow. 1984. Effects of a carbaryl-induced depression in invertebrate abundance on the growth and behavior of American Black Duck and Mallard ducklings. Can. J. Zool. 62:452–456.

Johnson, M. K., and J. Barnes. 1970. Age and sensitivity of chicks to delayed neurotoxic effects of some organophosphorus compounds. Biochem. Pharmacol. 19:3045–3047.

Johnston, D. W. 1974. Decline of DDT residues in migratory songbirds. Science 186:841–842.

Johnston, D. W. 1975. Organochlorine pesticide residues in small migratory birds, 1964–73. Pestic. Monit. J. 9:79–88.

Keith, J. O. 1966. Insecticide contaminations in wetland habitats and their effects on fish-eating birds. J. Appl. Ecol. 3(Suppl.):57–70.

Koeman, J. H., H. D. Rijksen, M. Smies, B. K. Na'Isa, and K. J. R. MacLennan. 1971. Faunal changes in a swamp habitat in Nigeria sprayed with insecticide to exterminate *Glossina*. Neth. J. Zool. 21:443–463.

Koeman, J. H., H. C. W. van Velzen-Blad, R. de Vries, and J. G. Vos. 1973. Effects of PCB and DDT in Cormorants and evaluation of PCB residues from an experimental study. J. Reprod. Fert. (Suppl.) 19:353–364.

Ludke, J. L., E. F. Hill, and M. P. Dieter. 1975. Cholinesterase (ChE) response and related mortality among birds fed ChE inhibitors. Arch. Environ. Contam. Toxicol. 3:1–21.

MacGuire, C. C., and B. A. Williams. 1987. Cold stress and acute organophosphorus exposure: interaction effects on juvenile Northern Bobwhite. Arch. Environ. Contam. Toxicol. 16:477–481.

Martin, P. A., and K. R. Solomon. 1991. Acute carbofuran exposure and cold stress: interactive effects in Mallard ducklings. Pestic. Biochem. Physiol. 40:117–127.

McArthur, M. L. B., G. A. Fox, D. B. Peakall, and B. J. R. Philogene. 1983. Ecological significance of behavioral and hormonal abnormalities in breeding Ring Doves fed an organochlorine chemical mixture. Arch. Environ. Contam. Toxicol. 12:324–353.

McLeod, J. M. 1967. The effect of phosphamidon on bird populations in jack pine stands in Quebec. Can. Field Natur. 81:102–106.

McNicol, D. K., B. E. Bendell, and D. G. McAuley. 1987. Avian trophic relationships and wetland acidity. Trans. North Amer. Wildl. Natural Resources Conf. 52:619–627.

Meyers, S. M., J. L. Cummings, and R. S. Bennett. 1990. Effects of methyl parathion on Red-winged Blackbird (*Agelaius phoeniceus*) incubation behavior and nesting success. Environ. Toxicol. Chem. 9:807–813.

Millikin, R. L., and J. N. M. Smith. 1990. Sublethal effects of fenitrothion on forest passerines. J. Appl. Ecol. 27:983–1000.

Mineau, P., and D. B. Peakall. 1987. An evaluation of avian impact assessment techniques following broad-scale forest insectide sprays. Environ. Toxicol. Chem. 6:781–791.

Mora, M. A., and D. W. Anderson. 1991. Seasonal and geographical variation of organochlorine residues in birds from northwest Mexico. Arch. Environ. Contam. Toxicol. 21:541–548.

Morrison, M. L., and E. C. Meslow. 1983. Impacts of forest herbicides on wildlife: toxicity and habitat alteration. Trans. North Amer. Wildl. Natural Resources Conf. 48:175–185.

Morrison, M. L., and E. C. Meslow. 1984. Response of avian communities to herbicide-induced vegetation changes. J. Wildl. Manag. 48:14–22.

Moulding, J. D. 1976. Effects of a low-persistence insecticide on forest bird populations. Auk 93:692–708.

Newton, I. 1988. Determination of critical pollutant levels in wild populations, with examples from organochlorine insecticides in birds of prey. Environ. Pollut. 55:29–40.

Nisbet, I. C. T. 1988. Relative importance of DDE and dieldrin in the decline of Peregrine Falcon populations. Pp. 351–376 *in* Peregrine Falcon populations: their management and recovery (T. J. Cade, J. H. Enderson, C. G. Thelander, and C. M. White, eds). The Peregrine Fund, Inc., Boise, ID.

Nyholm, N. E. I. 1981. Evidence of involvement of aluminum in causation of defective formation of eggshells and of impaired breeding in wild passerine birds. Environ. Res. 26:363–371.

Nyholm, N. E. I., and H. E. Myhrberg. 1977. Severe eggshell defects and impaired reproductive capacity in small passerines in Swedish Lapland. Oikos 29:336–341.

O'Brien, R. D. 1967. Insecticides: action and metabolism. Academic Press, New York.

O'Connor, R. J., and M. Shrubb. 1986. Farming and birds. Cambridge University Press, Cambridge, England.

Ormerod, S. J., and S. J. Tyler. 1987. Dippers (*Cinclus cinclus*) and Grey Wagtails (*Motacilla cinerea*) as indicators of stream acidity in upland Wales. Pp. 191–208 *in* The value of birds (A. W. Diamond, and F. L. Filion, eds). ICBP Tech. Publ. no. 6, Cambridge, England.

Parslow, J. L. F., and D. J. Jefferies. 1973. Relationship between organochlorine residues in livers and whole bodies of Guillemots. Environ. Pollut. 5:87–101.

Patnode, K. A., and D. H. White. 1991. Effects of pesticides on songbird productivity in conjunction with pecan cultivation in southern Georgia: a multiple-exposure experimental design. Environ. Toxicol. Chem. 10:1479–1486.

Peakall, D. B. 1992. Animal biomarkers as pollution indicators. Chapman and Hall, London.

Peakall, D. B., and J. R. Bart. 1983. Impacts of aerial application of insecticides on forest birds. CRC Crit. Rev. Environ. Control. 13:117–165.

Pearce, P. A., and D. G. Busby. 1980. Research on the effects of fenitrothion on the White-throated Sparrow. Pp. 24–27 *in* Environmental surveillance in New Brunswick, 1978–79: effects of spray operations for forest protection against spruce budworm (I. M. Varty, ed.). Committee for Environmental Monitoring of Forest Insect Control Operations. University of New Brunswick, Fredericton, Canada.

Perrins, C. M. 1965. Population fluctuations and clutch-size in the Great Tit, *Parus major*. J. Anim. Ecol. 34:601–647.

Peterle, T. J. 1991. Wildlife toxicology. Van Nostrand Reinhold, New York.

Pope, G. G., and P. Ward. 1972. The effects of small applications of an organophosphorus poison, fenthion, on the weaver-bird, *Quelea quelea*. J. Pestic. Sci. 3:197–205.

Potts, G. R. 1984. Monitoring changes in the cereal ecosystem. Pp. 128–134 *in* Agriculture and the environment (D. Jenkins, ed.). Institute of Terrestrial Ecology, Natural Environment Research Council, UK.

Powell, G. V. N. 1984. Reproduction by an altricial songbird, the Red-winged Blackbird, in fields treated with the organophosphate insecticide fenthion. J. Appl. Ecol. 21:83–95.

Presst, I., and D. A. Ratcliffe. 1972. Effects of organochlorine insecticides on European birdlife. Proc. XVth Int. Ornithol. Congress 486–513.

Rands, M. R. W. 1985. Pesticide use on cereals and the survival of Grey Partridge chicks: a field experiment. J. Appl. Ecol. 22:49–54.

Rattner, B. A., and A. Fairbrother. 1991. Biological variability and the influence of stress on cholinesterase activity. Pp. 89–107 *in* Cholinesterase-inhibiting insecticides: their impact on wildlife and the environment (P. Mineau, ed.). Elsevier, Amsterdam.

Rattner, B. A., and and J. C. Franson. 1984. Methyl parathion and fenvalerate toxicity in American Kestrels: acute physiological responses and effects of cold. Can. J. Physiol. Pharmacol. 62:787–792.

Rattner, B. A., D. J. Hoffman, and C. M. Marn. 1989. Use of mixed-function oxygenases to monitor contaminant exposure in wildlife. Environ. Toxicol. Chem. 8:1093–1102.

Risebrough, R. W. 1986. Pesticides and bird populations. Curr. Ornithol. 3:397–427.

Risebrough, R. W., and D. B. Peakall. 1988. Commentary—The relative importance of the several organochlorines in the decline of Peregrine Falcon populations. Pp. 449–462 *in* Peregrine Falcon populations: their management and recovery (T. J. Cade, J. H. Enderson, C. G. Thelander, and C. M. White, eds). The Peregrine Fund, Inc., Boise, ID.

Rudd, R. L., and R. E. Genelly. 1955. Avian mortality from DDT in California rice fields. Condor 57:117–118.

Savidge, J. A. 1978. Wildlife in a herbicide-treated Jeffrey pine plantation in eastern California. J. Forest. (August 1978):476–478.

Schafer, E. W. 1972. The acute oral toxicity of 369 pesticidal, pharmaceutical and other chemicals to wild birds. Toxicol. Appl. Pharmacol. 21:315–330.

Scheuhammer, A. M. 1987. The chronic toxicity of aluminum, cadmium, mercury, and lead in birds: a review. Environ. Pollut. 46:263–295.

Scheuhammer, A. M. 1991. Effects of acidification on the availability of toxic metals and calcium to wild birds and mammals. Environ. Pollut. 71:329–375.

Schroeder, M. H., and D. L. Sturges. 1975. The effect on the Brewer's Sparrow of spraying big sagebrush. J. Range Manag. 28:294–297.

Slagsvold, T. 1977. Bird population changes after clearance of deciduous scrub. Biol. Conserv. 12:229–244.

Smies, M., and J. H. Koeman. 1980. The effects of tsetse fly control measures on birds in West Africa. Proc. XVIIth Int. Ornithol. Congress 942–948.

Smith, G. J. 1987. Pesticide use and toxicology in relation to wildlife: organophosphorus and carbamate compounds. US Fish Wildl. Serv. Resource Publ. 170:1–171.

Southwood, T. R. E., and D. J. Cross. 1969. The ecology of the partridge. III. Breeding success and the abundance of insects in natural habitats. J. Anim. Ecol. 38:497–507.

Spray, C. J., H. Q. P. Crick, and A. D. M. Hart. 1987. Effects of aerial applications of fenitrothion on bird populations of a Scottish pine plantation. J. Appl. Ecol. 24:29–47.

Steele, B. B. 1992. Habitat selection by breeding Black-throated Blue Warblers at two spatial scales. Ornis Scand. 23:33–42.

Stickel, W. H. 1975. Some effects of pollutants in terrestrial ecosystems. Pp. 25–74 *in* Ecological toxicology research (A. D. McIntyre, and C. F. Mills, eds). Plenum Press, New York.

Stromborg, K. L., C. E. Grue, J. D. Nichols, G. R. Hepp, J. E. Hines, and H. C. Bourne. 1988. Postfledging survival of European Starlings exposed as nestlings to an organophosphorus insecticide. Ecology 69:590–601.

Szmedra, P. I. 1991. Pesticide use in agriculture. Pp. 649–677 *in* CRC handbook of pest management in agriculture (D. Pimentel, ed.). CRC Press, Boca Raton, FL.

von Düszeln, J. 1991. Pesticide contamination and pesticide control in developing countries: Costa Rica, Central America. Pp. 410–428, *in* Chemistry, agriculture and the environment (M. L. Richardson, ed.). The Royal Society of Chemistry, Cambridge, England.

Westlake, G. E., P. J. Bunyan, A. D. Martin, P. I. Stanley, and L. C. Steed. 1981. Organophosphate poisoning. Effects of selected organophosphate pesticides on plasma enzymes and brain esterases of Japanese Quail (*Coturnix coturnix japonica*). J. Agric. Food Chem. 29:772–778.

White, D. H. 1979. Nationwide residues of organochlorine compounds in wings of adult Mallards and Black Ducks, 1976–77. Pestic. Monit. J. 13:12–16.

White, D. H., and A. J. Krynitsky. 1986. Wildlife in some areas of New Mexico and Texas accumulate elevated DDE residues, 1983. Arch. Environ. Contam. Toxicol. 15:149–157.

White, D. H., K. A. King, C. A. Mitchell, and A. J. Krynitsky. 1981. Body lipids and pesticide burdens of migrant Blue-winged Teal. J. Field. Ornithol. 52:23–28.

White, D. H., C. A. Mitchell, and E. F. Hill. 1983. Parathion alters incubation behavior of Laughing Gulls. Arch. Environ. Contam. Toxicol. 31:93–97.

Whitmore, R. C., R. J. Cooper, and B. E. Sample. 1993. Bird fat reductions in forests treated with Dimilin®. Environ. Toxicol. Chem. 12:2059–2064.

Wurster, D. H., C. F. Wurster, and W. N. Strickland. 1965. Bird mortality following DDT spray for Dutch elm disease. Ecology 46:488–499.

Zinkl, J. G., R. B. Roberts, C. J. Henny, and D. J. Lenhart. 1980. Inhibition of brain cholinesterase activity in forest birds and squirrels exposed to aerially applied acephate. Bull. Environ. Contam. Toxicol. 24:676–683.

12

LIVESTOCK GRAZING EFFECTS IN WESTERN NORTH AMERICA

VICTORIA A. SAAB, CARL E. BOCK, TERRELL D. RICH, AND DAVID S. DOBKIN

INTRODUCTION

Livestock grazing is the most widespread economic use of public lands in western North America (Platts 1991). Approximately 86 million hectares of US Federal land in 17 western states are used for livestock production (Sabadell 1982). In the American West, grazing by domestic ungulates began in the 1840s, increased rapidly in the 1870s, and peaked around 1890 (Young and Sparks 1985). By 1900 much rangeland vegetation had been altered by the combination of extreme drought and high intensity grazing (Yensen 1981, Young and Sparks 1985). Range-management practices, including grazing systems (Appendix) and fenced pastures, were initiated in the early 1900s to help restore damaged rangelands (Behnke and Raleigh 1979). By the mid-1960s, management by allotment (designated areas for a prescribed number of livestock under one plan of management) had become an accepted practice on public lands, and is still in use today (Platts 1991).

Grazing by domestic livestock is probably the most controversial issue facing managers of public lands in the American West. This controversy is due in part to the competing economic, social, and conservation interests involved. A unique factor to grazing, as opposed to other land uses, is the fact that herbivory by native hoofed mammals has been a natural, ecological, and evolutionary force in certain nonforested ecosystems, including many in central and western North America (Stebbins 1981, McNaughton 1986). Domestic livestock has greatly intensified the influence of grazing in most of these ecosystems historically, and this influence has been especially damaging to those ecosystems where native grazing ungulates were scarce or absent (e.g., Mack and Thompson 1982, Milchunas et al. 1988, Schlesinger et al. 1990). Nonetheless, for certain habitats it is arguable that livestock grazing simulates a natural ecological event that some native flora and fauna tolerate, or perhaps require. Therefore, assertions about the effects of grazing on Neotropical migratory birds and other organisms must be habitat and species specific, and based on field data.

Birds generally do not respond to the presence of grazing livestock but to the impacts on vegetation as a result of grazing (Bock and Webb 1984). Cattle compact soil by hoof action, remove plant materials, and indirectly reduce water infiltration, all of which can result in decreased vegetation density (Holechek et al. 1989). In turn, these alterations of the structure and floristics in plant communities are known to affect some breeding bird species negatively, while other species respond positively.

Increased numbers of Brown-headed Cowbirds, created by the presence of cattle, is another indirect impact (i.e., nest parasitism) on many breeding Neotropical migratory landbirds (Robinson et al., Chapter 15, this volume). In presettlement times, cowbirds inhabited the Great Plains of central North America and were associated with giant bison herds of that region. Their range, now encompassing most of North America (American Ornithologists' Union 1983), expanded when Europeans arrived with their livestock and cleared forests (Mayfield 1965).

Cowbirds are now associated with domestic livestock, and are sufficiently numerous to pose major threats to the continued survival of several species that are regularly parasitized (Rothstein et al. 1980, 1984).

Livestock grazing is a primary land-use in four habitats important to Neotropical migrants: (1) grasslands of the Great Plains and Southwest; (2) shrubsteppe communities in the Intermountain region; (3) riparian plant communities of the arid West; and (4) montane coniferous forests. The objectives of this chapter are to evaluate the consequences of grazing by domestic ungulates on migratory landbirds using western habitats, and to provide perspectives on management as it relates to conservation of western Neotropical migrants.

METHODS

We reviewed a variety of federal publications, scientific journals, and unpublished reports for studies regarding effects of livestock grazing on landbird communities in western North America. The synthesized information is presented for Neotropical migrants in the following habitat sections: (1) grasslands, (2) shrubsteppe, (3) riparian, and (4) coniferous forests. We evaluated neotropical migrants as listed by Gauthreaux (1992), which excludes waterbirds and most shorebirds. This list includes landbirds that breed in North America and whose winter ranges predominantly extend into the Neotropics (also known as migratory landbirds).

The results of bird survey data are presented in a tabular format to facilitate comparisons between species and vegetative types. A number of important limitations exist in the information presented in the tables. Sizes and numbers of study sites, and season and intensity of grazing varied among studies. In all of the studies listed, data were obtained on the relative abundances of birds in variously grazed habitats, compared either to ungrazed or to lightly grazed sites. We listed a response as positive or negative only in those cases where the differences between treatments were $>20\%$.

For every study, we recorded each bird species as one that increased (+), decreased (−), or was unaffected in abundance as a result of cattle grazing. In each habitat section, we provided a qualitative assessment on patterns in bird responses to grazing. In some cases where data were available, we evaluated differences in bird responses according to grazing intensity and vegetative type (grasslands), and seasonality of grazing (riparian).

When abundance data for species were recorded by two or more studies in shrubsteppe and riparian vegetation, bird responses were analyzed statistically. Abundance data were standardized to evaluate species and guilds most vulnerable to grazing disturbances. We tested the hypothesis that grazing did not affect abundances of species and ecological guilds. Standardized means were tested for differences using a paired *t*-test and derived in the following manner: $Sg = 2Ng/(Nu + Ng)$ and $Su = 2Nu/(Nu + Ng)$; where Sg = standardized mean number of individuals or pairs in a grazed treatment, Ng = number of individuals or pairs in a grazed treatment, Nu = number of individuals or pairs in an ungrazed treatment, and Su = standardized mean number of individuals or pairs in an ungrazed treatment. These proportional data were transformed with an arcsine to obtain a normal distribution. This statistical approach was not applied to grasslands because of the graded response shown by many bird species depending on grassland type.

For the guild analyses, species were categorized into groups associated with nest type (open nesting or cavity nesting; appropriate only for riparian habitats), nest location (ground, shrub, or sub canopy/canopy), and foraging behavior (insectivore, carnivore, nectarivore, or omnivore), based on characteristics described by Harrison (1979), Ehrlich et al. (1988), and Martin (1993). Finally, species were evaluated by their migratory status (Tables 12-1, 12-2, 12-4; Gauthreaux 1992). In coniferous forest vegetation, we lacked information on any bird responses to livestock grazing. Therefore, we based our conclusions on knowledge about effects of livestock on vegetation, and the known habitat requirements of the birds.

GRASSLANDS OF THE GREAT PLAINS

Characteristics of Grassland Habitats

The Great Plains evolved in the rain shadow of the Rocky Mountains (Daubenmire 1978). Today, grasslands reach from the easternmost escarpment of that range out to the mixed deciduous forests of the Midwest, and from Alberta and Saskatchewan south to the arid and semi-arid grasslands surrounding and mixed with the Chihuahuan Desert in Texas, New Mexico, southwest Arizona, and northern Mexico. We ascribe to the plains this liberal southern extent because the Chihuahuan Desert grasslands have much in common climatically, floristically, and evolutionarily with their northern counterparts (Axelrod 1985). Furthermore, a number of migratory birds that breed in the northern plains also winter in the southern plains, so-defined, and therefore are influenced by climate and land-use patterns across the region as a whole (e.g., Wiens 1973, Pulliam and Parker 1979, Dunning and Brown 1982).

Two striking climatic gradients occur within the Great Plains. The first is a north–south temperature gradient, such that the mean annual number of frost-free days in Canadian grasslands is less than 120, while in the warmest Chihuahuan Desert grasslands it is more than 240 (Visher 1966). Running generally perpendicular to this thermocline is an equally if not more important precipitation gradient, reflective of the diminishing influence of the Rocky Mountain rain shadow eastward on to the plains. Specifically, grasslands to the west and south are increasingly water-stressed, due to declining precipitation and increasing evaporation.

We used Bailey's (1976, 1978) ecoregions of the United States to classify the ecosystems of the Great Plains, which includes three major divisions: prairie, steppe, and Chihuahuan Desert. Prairie grasslands are comparatively tall and moist, and, especially in the south and east, frequently include parklands with scattered trees or tall shrubs. The typical grasses of undisturbed tallgrass prairie include bluestems (*Andropogon* spp.) and switch grass (*Panicum virgatum*). Steppe grasslands are drier, experience more frequent droughts, and are dominated by shorter grasses than the tallgrass prairies. Dominant species include blue grama (*Bouteloua gracilis*) and buffalograss (*Buchloe dactyloides*). Northern steppe ecosystems (such as in the western Dakotas, eastern Montana, and the Canadian prairie provinces) frequently support a variety of mid-height bunchgrasses, especially wheatgrasses (*Agropyron* spp.) and needlegrasses (*Stipa* spp.), and fescues (*Festuca* spp.) that are scarce or missing from the comparatively arid shortgrass plains farther south. The Chihuahuan Desert comprises the most arid and heat-stressed part of the Great Plains. Here, grasslands of any sort are hanging on a climatic brink, where environmental perturbations (such as grazing by domestic livestock) can readily convert them into desert shrublands. Desert grassland, dominated by species such as black grama (*Bouteloua eriopoda*), formerly was widespread in the Southwest, but historic overgrazing has converted most of it to desert scrub (Buffington and Herbel 1965). However, in areas with slightly higher annual precipitation, semidesert grasslands persist in the Southwest, and these are important breeding and wintering habitats for Great Plains grassland birds (Bock and Bock 1988).

The preceding discussion of the subdivisions of the plains is much more than simply a lesson in plant geography, because most Neotropical migrant birds respond differently to livestock grazing in different places. The same amount of grazing that can be used to create ideal habitat for a species in a tallgrass prairie may be equally certain to destroy that same species' habitat in a shortgrass steppe or semidesert grassland. Therefore, management recommendations, derived from data synthesized in this section, should be tailored to the various sorts of grasslands involved.

Historical Perspective and Dynamics of Great Plains Ecosystems

The major forces creating the plains grasslands were, and are, drought, fire and grazing by bison and prairie dogs (Sauer 1950, Stebbins 1981, Anderson 1982). Evidence suggests that many plains grasslands are inherently vulnerable to invasions by woody plants, and that climate alone cannot sustain them as grassland (Sauer 1950). Fire retards or reverses invasions by trees or shrubs, while grazing encourages them, both by reducing fuels for fire and by facilitating the dispersal and establishment of the woody invaders (Risser et al. 1981, Bock and Bock 1987, 1988, Humphrey 1987, Steinauer and Bragg 1987, Archer 1989).

Historically, drought, fire, and grazing were not equally important in all plains grasslands. For example, fire appears to have been the major force sustaining tallgrass prairies and parklands against the relentless invasions of trees and shrubs (Gibson and Hulbert 1987). Farther west, grasses characteristic of shortgrass steppe ecosystems, such as blue grama and buffalograss, were those short-stature species equally tolerant of the frequent droughts and of grazing by native ungulates, especially bison (Milchunas et al. 1988). By contrast, there were few if any native ungulate grazers in the desert and semidesert grasslands of the Southwest, and here fire probably was a major factor keeping certain desert shrubs from invading these fragile ecosystems (Bahre 1991).

Introduction of domestic ungulates to the Great Plains greatly increased the role of grazing, relative to fire and drought, in determining the nature of the grasslands. In the desert region livestock grazing, perhaps coupled with drought, degraded many grasslands into essentially pure and permanent desert scrub (Buffington and Herbel 1965, Neilson 1986, Schlesinger et al. 1990). Many former tallgrass and mixed grass ecosystems were converted to grasslands dominated by shorter, more grazing-tolerant species (e.g., Bock and Bock 1993). However, shortgrass steppe ecosystems on the central plains may have changed relatively little (Milchunas et al. 1988), as millions of native ungulates (bison) were obliterated and then replaced by millions of exotic ungulates (cattle, horses, and sheep).

Grassland Avifauna

Density and diversity of birds is low in grasslands of the Great Plains, compared to wetlands, riparian woodlands, or adjacent forested ecosystems (Johnsgard 1979, Cody 1985, Knopf 1988). This has been variously attributed to the structural simplicity, ecological instability, and recent origin of the grassland communities (Udvardy 1958, Mengel 1970; Wiens 1973, 1974, Rotenberry and Wiens 1980a, 1980b, Cody 1985).

Total bird densities in the Great Plains grasslands usually range from 200–400 birds/km^2 (Wiens 1974, Cody 1985). Most tallgrass prairie songbird (passerine) communities include Eastern Meadowlarks, Bobolinks or Dickcissels, and Grasshopper, Savannah, and/or LeConte's Sparrows, with Red-winged Blackbirds and Common Yellowthroats in wetter areas (Cody 1985). Shortgrass steppe songbird assemblages usually include Horned Larks, Western Meadowlarks, Lark Buntings, and McGown's or Chestnut-collared Longspurs. Passerine faunas of mixed grass prairies include elements of both previous types, and they vary with grassland condition. Two species largely restricted to the mixed grass prairie region are Baird's Sparrow and Sprague's Pipit. Most bird species depart plains grasslands in winter but this is not true of semidesert grasslands (Bock et al. 1984). Typical breeding passerines in this habitat include meadowlarks, Horned Larks, and Grasshopper, Cassin's, Botteri's, and Lark Sparrows. Wintering assemblages are dominated by Cassin's, Brewer's, Vesper, and Grasshopper Sparrows, and meadowlarks.

Avian Responses to Livestock Grazing in Grasslands

We found published data about the effects of livestock grazing on 33 Neotropical migrant bird species that breed and/or winter on the Great Plains (Table 12-1). We found no information about the effects of grazing on birds in arid Chihuahuan grasslands, probably because most of these habitats have been grazed out of existence.

Despite inevitable variation in the nature of individual studies, the data for most species reveal consistent, ecologically interpretable patterns. Importantly, one of the patterns emerging from our synthesis is that many species respond differently to the effects of grazing in different grassland types, and that certain species may require grasslands in

Table 12-1. Responses to cattle grazing by Neotropical migrant landbirds breeding in grasslands of the North American Great Plains and Southwest.

Species	Migrant Status[a]	Region	Grassland Type	Grazing Intensity[b]	Response to Grazing[c]	Reference
Northern Harrier	B	S. Dakota	Mixed grass	Moderate	–	Duebbert and Lokemoen (1977)
		N. plains	Mixed grass	Variable	–	Kantrud and Kologiski (1982)
Ferruginous Hawk	B	N. plains	Mixed grass	Variable	+	Kantrud and Kologiski (1982)
		S. Dakota	Mixed grass		+	Lokemoen and Duebbert (1976)
Killdeer	B	Colorado	Shortgrass	Heavy	+	Ryder (1980)
		N. plains	Mixed grass	Moderate	+	Kantrud and Kologiski (1982)
		N. plains	Mixed grass	Heavy	+	Kantrud and Kologiski (1982)
		N. Dakota	Mixed/tall	Moderate	0	Kantrud (1981)
		N. Dakota	Mixed/tall	Heavy	+	Kantrud (1981)
Mountain Plover	A	Colorado	Shortgrass	Heavy	+	Graul (1975)
		Colorado	Shortgrass	Heavy	+	Ryder (1980)
		N. plains	Mixed grass	Moderate	0	Kantrud and Kologiski (1982)
		N. plains	Mixed grass	Heavy	+	Kantrud and Kologiski (1982)
Upland Sandpiper	A	N. Dakota	Mixed grass		–	Kirsch and Higgins (1976)
		N. plains	Mixed grass	Moderate	0	Kantrud and Kologiski (1982)
		N. plains	Mixed grass	Heavy	0	Kantrud and Kologiski (1982)
		N. plains	Mixed grass	Variable	–	Kantrud and Higgins (1992)
		N. Dakota	Mixed/tall	Moderate	0	Kantrud (1981)
		N. Dakota	Mixed/tall	Heavy	+	Kantrud (1981)
		Missouri	Tallgrass		+	Skinner (1975)
Long-billed Curlew	A	Colorado	Shortgrass	Heavy	+	Ryder (1980)
		N. plains	Mixed grass	Heavy	0	Kantrud and Kologiski (1982)
		N. plains	Mixed grass	Moderate	0	Kantrud and Kologiski (1982)
Mourning Dove	B	S. Arizona	Semidesert	Moderate	+	Bock et al. (1984)
		Colorado	Shortgrass	Heavy	0	Ryder (1980)
		N. plains	Mixed grass	Moderate	0	Kantrud and Kologiski (1982)
		N. plains	Mixed grass	Heavy	Mixed	Kantrud and Kologiski (1982)
		N. Dakota	Mixed/tall	Moderate	0	Kantrud (1981)
		N. Dakota	Mixed/tall	Heavy	0	Kantrud (1981)
		S. Texas	Tallgrass	Heavy	+	Baker and Guthery (1990)
Burrowing Owl	A	N. plains	Mixed grass	Moderate	0	Kantrud and Kologiski (1982)
		N. plains	Mixed grass	Heavy	+	Kantrud and Kologiski (1982)
Short-eared Owl	B	S. Dakota	Mixed grass	Moderate	–	Duebbert and Lokemoen (1977)
		N. plains	Mixed grass	Variable	–	Kantrud and Higgins (1992)
Common Nighthawk	A	Colorado	Shortgrass	Heavy	+	Ryder (1980)
		N. plains	Mixed grass	Moderate	0	Kantrud and Kologiski (1982)
		N. plains	Mixed grass	Heavy	+	Kantrud and Kologiski (1982)
Horned Lark	B	S. Arizona	Semidesert	Moderate	+	Bock et al. (1984)
		Colorado	Shortgrass	Heavy	+	Ryder (1980)

(continued)

Table 12-1 (*cont.*)

Species	Migrant Status[a]	Region	Grassland Type	Grazing Intensity[b]	Response to Grazing[c]	Reference
		Plains	Shortgrass	Heavy	+	Wiens (1973)
		Saskatchewan	Mixed/short		+	Maher (1979)
		Alberta	Mixed grass	Heavy	+	Owens and Myres (1973)
		S. Dakota	Mixed grass	Heavy	+	Wiens (1973)
		N. plains	Mixed grass	Moderate	+	Kantrud and Kologiski (1982)
		N. plains	Mixed grass	Heavy	+	Kantrud and Kologiski (1982)
		N. Dakota	Mixed/tall	Moderate	+	Kantrud (1981)
		N. Dakota	Mixed/tall	Heavy	+	Kantrud (1981)
		Oklahoma	Tallgrass	Moderate	0	Risser et al. (1981)
		Oklahoma	Tallgrass	Heavy	+	Risser et al. (1981)
Northern Mockingbird	B	S. Arizona	Semidesert	Moderate	+	Bock et al. (1984)
Sprague's Pipit	B	N. plains	Mixed grass	Moderate	0	Kantrud and Kologiski (1982)
		N. plains	Mixed grass	Heavy	–	Kantrud and Kologiski (1982)
		Saskatchewan	Mixed/short		–	Maher (1979)
		Alberta	Mixed grass	Heavy	–	Owens and Myres (1973)
		N. Dakota	Mixted/tall	Moderate	+	Kantrud (1981)
		N. Dakota	Mixed/tall	Heavy	+	Kantrud (1981)
Common Yellowthroat	A	N. Dakota	Mixed/tall	Moderate	–	Kantrud (1981)
		N. Dakota	Mixed/tall	Heavy	–	Kantrud (1981)
Dickcissel	A	Oklahoma	Tallgrass	Heavy	–	Risser et al. (1981)
		Oklahoma	Tallgrass	Moderate	+	Risser et al. (1981)
		Missouri	Tallgrass	Moderate	+	Skinner (1975)
Botteri's Sparrow	C	S. Arizona	Semidesert	Moderate	–	Webb and Bock (1990)
Cassin's Sparrow	B	S. Arizona	Semidesert	Moderate	–	Bock and Bock (1988)
Clay-colored Sparrow	A	Alberta	Mixed grass	Heavy	0	Owens and Myres (1973)
		N. plains	Mixed grass	Moderate	0	Kantrud and Kologiski (1982)
		N. plains	Mixed grass	Heavy	–	Kantrud and Kologiski (1982)
		N. Dakota	Mixed/tall	Moderate	0	Kantrud (1981)
		N. Dakota	Mixed/tall	Heavy	–	Kantrud (1981)
Brewer's Sparrow	A	S. Arizona	Semidesert	Moderate	+	Bock et al. (1984)
		Colorado	Shortgrass	Heavy	–	Ryder (1980)
		N. plains	Mixed grass	Moderate	0	Kantrud and Kologiski (1982)
		N. plains	Mixed grass	Heavy	–	Kantrud and Kologiski (1982)
Vesper Sparrow	B	S. Arizona	Semidesert	Moderate	0	Bock et al. (1984)
		Alberta	Mixed grass	Heavy	+	Owens and Myres (1973)
		N. plains	Mixed grass	Moderate	0	Kantrud and Kologiski (1982)
		N. plains	Mixed/tall	Heavy	–	Kantrud and Kologiski (1982)
		N. Dakota	Mixed/tall	Heavy	0	Kantrud (1981)
Lark Sparrow	A	S. Arizona	Semidesert	Moderate	+	Bock et al. (1984)
Black-throated Sparrow	B	S. Arizona	Semidesert	Moderate	+	Bock et al. (1984)
Lark Bunting	A	N. Texas	Shortgrass	Heavy	–	Wiens (1973)
		Colorado	Shortgrass	Heavy	–	Ryder (1980)
		N. plains	Mixed grass	Moderate	0	Kantrud and Kologiski (1982)
		N. plains	Mixed grass	Heavy	0	Kantrud and Kologiski (1982)
		N. Dakota	Mixed/tall	Moderate	+	Kantrud (1981)
		N. Dakota	Mixed/tall	Heavy	0	Kantrud (1981)

Table 12-1 (*cont.*)

Species	Migrant Status[a]	Region	Grassland Type	Grazing Intensity[b]	Response to Grazing[c]	Reference
Savannah Sparrow	B	Saskatchewan	Mixed/short		–	Maher (1979)
		Alberta	Mixed grass	Heavy	–	Owens and Myres (1973)
		N. Dakota	Mixed/tall	Moderate	–	Kantrud (1981)
		N. Dakota	Mixed/tall	Heavy	–	Kantrud (1981)
		Minnesota	Tallgrass	Moderate	–	Tester and Marshall (1961)
Baird's Sparrow	A	Saskatchewan	Mixed/short		–	Maher (1979)
		Alberta	Mixed grass	Heavy	–	Owens and Myres (1973)
		N. plains	Mixed grass	Moderate	–	Kantrud and Kologiski (1982)
		N. plains	Mixed grass	Heavy	–	Kantrud and Kologiski (1982)
		N. Dakota	Mixed/tall	Moderate	0	Kantrud (1981)
		N. Dakota	Mixed/tall	Heavy	–	Kantrud (1981)
Grasshopper Sparrow	A	S. Arizona	Semidesert	Moderate	–	Bock et al. (1984)
		Colorado	Shortgrass	Heavy	–	Ryder (1980)
		N. Texas	Shortgrass	Heavy	–	Wiens (1973)
		S. Dakota	Mixed grass	Heavy	–	Wiens (1973)
		N. plains	Mixed grass	Moderate	–	Kantrud and Kologiski (1982)
		N. plains	Mixed grass	Heavy	–	Kantrud and Kologiski (1982)
		N. Dakota	Mixed/tall	Moderate	0	Kantrud (1981)
		N. Dakota	Mixed/tall	Heavy	–	Kantrud (1981)
		Oklahoma	Tallgrass	Moderate	+	Risser et al. (1981)
		Oklahoma	Tallgrass	Heavy	–	Risser et al. (1981)
		Missouri	Tallgrass	Moderate	+	Skinner (1975)
McCown's Longspur	B	Colorado	Shortgrass	Heavy	+	Ryder (1980)
		Saskatchewan	Mixed/short		+	Maher (1979)
		N. plains	Mixed grass	Moderate	+	Kantrud and Kologiski (1982)
		N. plains	Mixed grass	Heavy	+	Kantrud and Kologiski (1982)
Chestnut-collared Longspur	B	S. Arizona	Semidesert	Moderate	+	Bock and Bock (1988)
		Colorado	Shortgrass	Heavy	0	Ryder (1980)
		Saskatchewan	Mixed/short		+	Maher (1979)
		Alberta	Mixed grass	Heavy	+	Owens and Myres (1973)
		N. plains	Mixed grass	Moderate	0	Kantrud and Kologiski (1981)
		N. plains	Mixed grass	Heavy	0	Kantrud and Kologiski (1982)
		N. Dakota	Mixed/tall	Moderate	+	Kantrud (1981)
		N. Dakota	Mixed/tall	Heavy	+	Kantrud (1981)
Bobolink	A	N. plains	Mixed grass	Moderate	–	Kantrud and Kologiski (1982)
		N. plains	Mixed grass	Heavy	–	Kantrud and Kologiski (1982)
		N. Dakota	Mixed/tall	Moderate	–	Kantrud (1981)
		N. Dakota	Mixed/tall	Heavy	–	Kantrud (1981)
		Minnesota	Tallgrass	Moderate	0	Tester and Marshall (1961)
		Missouri	Tallgrass	Moderate	+	Skinner (1975)
Redwinged Blackbird	B	N. Dakota	Mixed/tall	Moderate	–	Kantrud (1981)
		N. Dakota	Mixed/tall	Heavy	–	Kantrud (1981)
		Missouri	Tallgrass	Moderate	+	Skinner (1975)
Eastern Meadowlark	B	S. Arizona	Semidesert	Moderate	0	Bock et al. (1984)
		Oklahoma	Tallgrass	Moderate	+	Risser et al. (1981)
		Oklahoma	Tallgrass	Heavy	–	Risser et al. (1981)
		Missouri	Tallgrass	Moderate	+	Skinner (1975)
		S. Texas	Tallgrass	Heavy	+	Baker and Guthery (1990)

(*continued*)

Table 12-1 (*cont.*)

Species	Migrant Status[a]	Region	Grassland Type	Grazing Intensity[b]	Response to Grazing[c]	Reference
Western Meadowlark	B	Colorado	Shortgrass	Heavy	–	Ryder (1980)
		N. Texas	Shortgrass	Heavy	–	Wiens (1973)
		Saskatchewan	Mixed/short		–	Maher (1979)
		Alberta	Mixed grass	Heavy	0	Owens and Myres (1973)
		S. Dakota	Mixed grass	Heavy	–	Wiens (1973)
		N. plains	Mixed grass	Moderate	0	Kantrud and Kologiski (1982)
		N. plains	Mixed grass	Heavy	–	Kantrud and Kologiski (1982)
		N. Dakota	Mixed/tall	Moderate	0	Kantrud (1981)
		N. Dakota	Mixed/tall	Heavy	0	Kantrud (1981)
Brown-headed Cowbird	B	N. plains	Mixed grass	Moderate	0	Kantrud and Kologiski (1982)
		N. plains	Mixed grass	Heavy	0	Kantrud and Kologiski (1982)
		N. Dakota	Mixed/tall	Moderate	0	Kantrud (1981)
		N. Dakota	Mixed/tall	Heavy	0	Kantrud (1981)

[a] Status "A" contains long-distance migrants, those species that breed in North America and spend their nonbreeding period primarily south of the United States. Status "B" contains short-distance migrants, those species that breed and winter extensively in North America, although some populations winter south of the United States. Status "C" contains those species whose breeding range is primarily south of the United States/Mexican border, and enter the United States along the Rio Grande Velley and where the Mexican Highlands extend across the United States border.

[b] Grazing intensity as reported by original authors in the references list.

[c] Grazing effects on abundance: + = increase; – = decrease; 0 = no effect, as reported by original authors.

different condition to meet different life requirements (e.g., foraging vs nesting).

We found limited information on the responses of avian predators to livestock grazing, probably because most of these birds have such large home ranges that their densities cannot be compared on plots of the sizes used in typical grazing studies (Bock and Bock 1988). Northern Harriers are characteristic of tall, comparatively lush, grasslands (Johnsgard 1979), and they have been found in higher numbers in lightly grazed mixed grass prairie in the northern plains (Table 12-1). Short-eared Owls also have responded positively to the presence of substantial ground cover in the same region. By contrast, Ferruginous Hawks showed a mixed response to grazing, apparently preferring to hunt open grasslands but to nest in areas with more ground cover or with scattered large trees. Burrowing Owls usually nest in abandoned prairie dog burrows on the Great Plains, and they prefer habitat with much bare ground (e.g., Agnew et al. 1986). Livestock, like bison, facilitate establishment and expansion of prairie dog colonies by grazing down dense grasslands that the prairie dogs otherwise are unable to occupy (Coppock et al. 1983, Uresk 1984, Uresk and Paulson 1988). The response of Burrowing Owls to grassland grazing therefore has been positive probably due to an increase in prey availability (Table 12-1).

Shorebird species respond variously though, in general, these birds prefer to nest in relatively sparse grasslands (Kantrud and Higgins 1992). Killdeer and Mountain Plovers nest in such open areas that only heavy livestock grazing appears sufficient to create or maintain their breeding habitat (Table 12-1; see also Graul and Webster 1978, Parrish et al. 1993). In mixed grassland in north-central Montana, Mountain Plovers nested primarily in areas both grazed by cattle and occupied by prairie dogs (Knowles et al. 1982). Long-billed Curlews may prefer grazed areas in steppe grasslands, but more data are needed (Kantrud and Kologiski 1982). Upland Sandpipers clearly require more grass cover than Killdeer or Mountain Plovers, but evidence exists that this shorebird is flexible in its choice of nesting habitats

(Table 12-1). Response to grazing may be related to grassland type, with effects being neutral or negative in mixed grasslands (e.g., Kantrud and Kologiski 1982, Kantrud and Higgins 1992), and positive in tallgrass prairies (Skinner 1975). Shortgrass steppe appears not to be suitable Upland Sandpiper breeding habitat. Upland Sandpipers also may require heavier ground cover for nesting sites than for feeding sites (Kantrud 1981).

Mourning Doves are another species that has shown mixed responses to grazing, suggesting that other environmental factors may be more important to this species. For example, abundances of doves in the northern mixed grass prairie varied widely across plots on different soils, regardless of grazing intensities, although ground nests usually were placed in idle versus currently grazed pastures (Kantrud and Kologiski 1982, Kantrud and Higgins 1992). However, doves frequently nest in trees across the Great Plains (Johnsgard 1979), and therefore may not be completely dependent upon ground cover for successful nesting. Furthermore, doves have such weak legs and feet that they cannot forage effectively in very heavy ground cover (Leopold 1972). They appear to benefit from grazing, at least in comparatively lush grasslands of southern Texas and southeastern Arizona (Bock et al. 1984, Baker and Guthery 1990).

Songbirds exhibited the full range of responses to grazing in grasslands of the Great Plains (Table 12-1). At one extreme is the Horned Lark, one of the most widespread birds in the region and one that has been universally positive in its response to grazing. Other, more narrowly distributed, songbirds shown to benefit from grazing include Lark and Black-throated Sparrows, McCown's Longspur, and Dickcissel (moderate grazing only, see Table 12-1). At the other extreme is a group of songbirds that appear negatively affected by grazing wherever they have been studied. This group includes the Common Yellowthroat, and Botteri's, Cassin's, Savannah, Baird's, and Henslow's Sparrows. Western Meadowlarks also have been negatively affected by grazing, although in many cases differences between plots under different grazing regimes have been minor. Western Meadowlarks are one of the most widespread plains birds, and they appear to tolerate all except the heaviest levels of livestock grazing.

A fourth, and particularly interesting, group of songbirds includes species that apparently require or prefer intermediate amounts of grass cover. They have benefited from grazing in tallgrass and some mixed grass communities, but declined or disappeared in the presence of grazing in shorter grasslands. Perhaps the best-studied example is the Grasshopper Sparrow (Table 12-1), a species experiencing widespread population declines (Robbins et al. 1993). Grasshopper Sparrows were negatively affected by grazing in shortgrass, semidesert, and certain mixed grass communities (Ryder 1980, Kantrud and Kologiski 1982, Bock et al. 1984), but they have responded positively at least to moderate grazing in tallgrass prairies (Skinner 1975, Risser et al. 1981). Other species that have exhibited a similar pattern include Sprague's Pipit, Lark Bunting, Chestnut-collared Longspur, Bobolink, Red-winged Blackbird, and, perhaps, Eastern Meadowlark. Finally, some songbirds have shown weak, inconsistent responses to grazing in different parts of the plains. These include Clay-colored, Brewer's, and Vesper Sparrows, and Brown-headed Cowbird (Table 12-1). At least for Clay-colored and Brewer's Sparrows, distribution and abundance patterns appear to be more strongly associated with shrub than with grass cover.

No obvious relationship existed between migratory status and responsiveness to grazing among the 35 grassland species (Table 12-1). First, while most are listed as Neotropical migrants (Gauthreaux 1992), the great majority in fact winter across the southern portion of North American continent (Knopf 1994). Only four of the 35 are long-distance migrants that winter entirely outside the region (Upland Sandpiper, Common Nighthawk, Dickcissel, and Bobolink). Furthermore, pairs of species with generally similar breeding and wintering distributions often have shown very different responses to grazing. For example, Chestnut-collared and McCown's Longspurs both breed in the northern Great Plains and winter in the south-central United States and northern Mexico; grazing usually has favored the former, while the latter has shown a graded response. Vesper and Savannah Sparrows both breed widely across North America, and winter across the southern United States from California to Florida; the former appears generally unresponsive to grazing, while the latter usually has been negatively affected (Tables 12-1 to 12-4).

Breeding Bird Survey data suggest that grassland birds as a group are showing greater population declines than any other avian assemblage in North America (Robbins et al. 1993, Knopf 1994a). This probably is attributable to habitat modifications including livestock grazing, in addition to fire suppression, prairie dog control, cultivations, and planting exotic grasses.

SHRUBSTEPPE OF THE INTERMOUNTAIN REGION

Characteristics of Shrubsteppe Habitats

Shrubsteppe habitats in western North America are characterized by woody, mid-height shrubs and perennial bunchgrasses (Fautin 1946, Daubenmire 1978, Dealy et al. 1981, Tisdale and Hironaka 1981, Short 1986). Shrubsteppe typically is arid with annual precipitation over much of the region averaging less than 36 cm (Daubenmire 1956, Richard and Vaughan 1988, Rogers and Rickard 1988). Periodic drought, extreme temperatures, wind, poor soil stability and only fair soil quality (Fautin 1946, Wiens and Dyer 1975, Short 1986) manifest a stressful environment for biotic communities.

The shrubsteppe has been delineated in various ways (Kuchler 1964, Wiens and Dyer 1975, Risser et al. 1981). Major differences depend on the inclusion of salt desert shrublands of the Great Basin, shrubsteppe of the southwestern United States, or pinon–juniper types (Short 1986).

Historical Perspective and Dynamics of Shrubsteppe Habitats

Major changes in native shrubsteppe vegetation, particularly the rapid loss of forbs and grasses, took as little as 10–15 years under severe overgrazing that accompanied early settlement of the West (Kennedy and Doten 1901, Cottam and Stewart 1940, Brougham and Harris 1967, McNaughton 1979, West 1979). Some plant species may have been extirpated from the region or driven to extinction, but we assume that most of the species present today were also important historically (Dealy et al. 1981, Tisdale and Hironaka 1981).

Little doubt exists that sagebrush (*Artemisia* spp.) has always been an important component of the Intermountain landscape (Vale 1975, Braun et al. 1976), with a variety of sagebrush vegetative types dominating large areas (McArthur and Welch 1986). Other important shrubs include saltbush (*Atriplex* spp.), rabbitbrush (*Chrysothamnus* spp.), and bitterbrush (*Purshia tridentata*) (West 1979, Tisdale and Hironaka 1981, Yensen 1981, McArthur and Welch 1986). The region is characterized by perennial bunchgrasses (also known as caespitose grasses) including the genera *Agropyron*, *Poa*, *Stipa*, *Elymus*, and *Festuca* (Fautin 1946, West 1979, Yensen 1981). Few rhizomatous or sod-forming grasses occur and they play only a minor role in the ecosystem, in contrast to the prairies farther east (Mack and Thompson 1982).

Domestic livestock grazing has caused major changes in plant species composition of shrubsteppe habitats including loss of the cryptogam layer from trampling, loss of native seral grasses, reduced perennial grass cover, reduced forb cover, increased shrub cover, and invasion by exotic species, particularly cheatgrass (*Bromus tectorum*) (Yensen 1981).

Prior to European settlement, cryptogams such as the lichen Parmelia chlorochroa (MacCracken et al. 1983), covered all undisturbed soil surfaces not populated by vascular plants. Because of the permanent loss of this layer through trampling by domestic livestock (Poulton 1955, Daubenmire 1970, Mack and Thompson 1982), we do not know what role this stratum played in the original ecosystem. Increased soil temperatures, increased erosion, lower soil moisture, lower productivity and a lower rate of seedling establishment are likely negative consequences of the loss of cryptogams (MacCracken et al. 1983).

Because herbaceous species are more palatable than shrubs during the growing season, grazing tends to increase shrub cover, and decrease palatable forbs and grasses (Pickford 1932, Cottam and Stewart 1940, Smith 1967, Tisdale et al. 1969, Smith and Schmutz 1975, Page et al. 1978, Ryder 1980, Blaisdell et al. 1982). More intense grazing will eliminate even less palatable species and lead to domination by woody, unpalatable and spiny species (Ellison 1960).

Generally, cattle grazing favors shrubs and forbs over grasses while sheep grazing shifts the balance towards grass (Allred 1941, Costello and Turner 1941, Tisdale 1947, Cooper 1953, Robertson 1971, Urness 1979). Season of use is also an important influence in shrubsteppe. For example, heavy spring sheep grazing reduces grasses and increases sagebrush, whereas heavy fall sheep grazing has the opposite effect (Craddock and Forsling 1938, Mueggler 1950, Ellison 1960, Laycock 1967). Livestock grazing can also increase the density of junipers (Cottam and Stewart 1940, Woodbury 1947, Springfield 1976, Little 1977) and reduce vegetation diversity (Wiens and Dyer 1975, Reynolds and Rich 1978).

While exotic annuals are found essentially everywhere in the shrubsteppe, it is clear that their dominance increases with disturbance such as livestock grazing. Piemeisel (1938) and Young et al. (1979) submit that cheatgrass, at least, cannot significantly invade healthy shrubsteppe habitats.

Evaluation of Grazing in Shrubsteppe Habitats

Shrubsteppe habitats did not coevolve with large herds of grazing animals, and plant species are not adapted to withstand severe or continuous grazing (Mack and Thompson 1982). Post-Pleistocene native ungulates in shrubsteppe only included American bison and pronghorn. Bison numbers were estimated at 40 million when Europeans arrived (England and DeVos 1969) but it is unlikely that large herds occurred west of the Rockies (Gustafson 1972, Grayson 1977). Few prehistoric bison records exist from the Columbia Plateau (Schroedl 1973) and records are rare elsewhere in the region (Mack and Thompson 1982).

Caespitose grasses depend on seed production rather than rhizomes or stolons to maintain their populations. The effects of grazing, both removal of vegetation and mechanical damage from trampling, are more serious for caespitose species (Mack and Thompson 1982). Consequently, sagebrush–perennial bunchgrass communities are adapted to small, dispersed groups typified by pronghorn, mule deer and elk. Although these species form groups on winter ranges, they largely rely on woody vegetation at that time of year. This lack of adaptation to concentrations of large herbivores has led to "striking susceptibility" of shrubsteppe vegetation to the impact of domestic ungulates (Larson 1940, Tisdale 1961, Dyer 1979, Mack and Thompson 1982).

Classic approaches to grazing management in shrubsteppe habitats are discussed by Stoddart et al. (1975) and Laycock (1983), with novel strategies infrequent, controversial and slow to be substantiated (Savory and Parsons 1980, Savory 1988). The most noteworthy long-term trend on public land in shrubsteppe has been the reduction of destructive season-long cattle grazing where animals are released in early spring and removed in late fall (Appendix).

Multipasture rest–rotation systems (Appendix, Stoddart et al. 1975) have become popular for cattle and are a significant improvement in cattle management for shrubsteppe habitats. The rest–rotation method typically produces more uniform grazing across the landscape rather than areas of high use and areas of little or no use. The system also requires more fencing, water developments, prescribed burns, seedings or other manipulations. Ultimately, more cattle may be allowed in an allotment.

Shrubsteppe Avifauna

While more than 50 species of neotropical migrants breed in this region, the shrubsteppe bird community typically has 2–7 regular breeding species. Densities vary between 100 and 600 individuals/km^2 with over half the individuals at a site belonging to the most common species. Irregular precipitation patterns in shrubsteppe habitats have resulted in annual redistributions of individual birds, locally and regionally (Wiens and Rotenberry 1981a, Wiens 1985).

Certain associations exist between bird species and particular plant species, perhaps in response to arthropod abundance or availability (Wiens and Rotenberry 1981b). Some shrubsteppe birds show a high degree of selectivity for grass seeds of certain species (Goebel and Berry 1976). Thus, selective removal of particular plant species by livestock could have direct effects on individual bird species.

Avian predators occupying shrubsteppe habitats are influenced by their small-mammal prey. Small-mammal community composition, densities and distribution vary with vegetation structure (Feldhamer 1979, Rogers and Hedlund 1980, Gano and Rickard 1982, McGee 1982) and species diversity declines as grazing intensifies (see Kochert 1989). However, the specific ecological relationships between small mammals and shrubsteppe raptors are essentially unknown.

Avian Responses to Livestock Grazing in Shrubsteppe Habitats

We found information from 15 studies that evaluated grazing effects on 34 Neotropical migrants that breed in shrubsteppe vegetation (Table 12-2). Birds considered in this evaluation range from sagebrush obligates to much more widespread species that are only peripherally associated with shrubsteppe (see references in Table 12-2). In a qualitative assessment of grazing effects on shrubsteppe birds, 12 species responded positively, 12 negatively, and 10 species showed no clear response.

The referenced studies reported abundance information on 31 migrant species and only 14 of those were evaluated by two or more studies (Table 12-3), thus limiting conclusions based on quantitative data. No species or ecological guild showed significant differences ($P < 0.05$) in abundances between grazed and ungrazed treatments. Of the 34 species evaluated, only six are considered long-distance migrants and 28 short-distance migrants (Gauthreaux 1992). Combined in groups, neither long-distance migrants [standardized means 0.81 vs 1.20 (grazed vs ungrazed), $T = 1.01$, $P = 0.33$] nor short-distance migrants [standardized means 1.00 vs 0.99 (grazed vs ungrazed), $T = -0.09$, $P = 0.92$) appeared particularly vulnerable to livestock grazing. These results should be viewed with caution considering the little quantitative data available and a lack of information about pristine shrubsteppe habitats (i.e., no controls from which to judge grazing effects).

Most studies were conducted with cattle on a short-term basis during the growing season (references in Table 12-2). Effects of other kinds of livestock (McKnight 1958, Hanley and Brady 1977) during other times of the year might differ substantially. Nevertheless, we make some tentative conclusions based on the limited published information, and knowledge about the effects of grazing on vegetation and the known habitat requirements of the birds.

Wiens and Dyer (1975) suggested that the ecological plasticity of many shrubsteppe birds would make them unresponsive to moderate levels of livestock grazing. Major avifaunal shifts may occur only after some threshold of habitat change has passed. Such thresholds may have passed historically, when livestock were first introduced into the region. However, virtually no pristine ecosystems exist where this hypothesis might be tested. As a result, our conclusions about the effects of grazing on Neotropical migrants must be largely speculative.

Distinguishing between historical and current livestock impacts is important when categorizing bird responses to grazing. For example, species requiring shrubs as nest sites may have benefitted from early, grazing-related increases in shrubs across the West. They may now be harmed by heavy grazing that removes herbaceous cover. Brewer's Sparrows may be an example, and we consider this species to be negatively affected by grazing (Tables 12-2 and 12-3).

Brewer's Sparrow populations have declined significantly both in the western United States and over their entire range during the last 25 years (Robbins et al. 1993, Peterjohn et al., chapter 1, this volume). As this species is the most typical, widespread

Table 12-2. Responses to cattle grazing by Neotropical migrant landbirds breeding in shrubsteppe habitats of western North America.

Species	Migrant Status[a]	Region	Shrubland Type	Grazing Intensity[b]	Response to Grazing[c]	Reference
Northern Harrier	B	Nevada	Greasewood/Great Basin wild rye	Heavy	–	Page et al. (1978)
		Idaho	Big sage/bluebunch wheatgrass	Moderate	+	Reynolds and Trost (1981)
		Oregon	Various	Variable	–	Kochert (1989)
		Idaho	Various	Variable	+	Martin (1987)
Swainson's Hawk[d]	A	Idaho	Big sage/bluebunch wheatgrass	Moderate	+	Reynolds and Trost (1981)
		Oregon	Various	Heavy	–	Littlefield et al. (1984)
		West	Various	Variable	–	Kochert (1989)
Red-tailed Hawk[d]	B	Idaho	Big sage/bluebunch wheatgrass	Moderate	–	Reynolds and Trost (1981)
		West	Various	Variable	Mixed	Kochert (1989)
Ferruginous Hawk	B	Idaho	Big sage/bluebunch wheatgrass	Moderate	–	Reynolds and Trost (1981)
		West	Various	Variable	Mixed	Kochert (1989)
Golden Eagle	B	Idaho	Various	Heavy	–	Kochert (1989)
		Idaho	Big sage	Variable	+	Nydegger and Smith (1986)
American Kestrel[d]	B	Nevada	Low sage/Idaho fescue	Heavy	–	Page et al. (1978)
		Idaho	Big sage/bluebunch wheatgrass	Moderate	+	Reynolds and Trost (1981)
		West	Various	Variable	–	Kochert (1989)
Prairie Falcon	B	Idaho	Big sage/bluebunch wheatgrass	Moderate	–	Reynolds and Trost (1981)
Long-billed Curlew	A	Idaho	Big sage/bluebunch wheatgrass	Moderate	–	Reynolds and Trost (1981)
Mourning Dove	B	Nevada	Greasewood/Great Basin wild rye	Heavy	+	Page et al. (1978)
		Nevada	Shadscale/Indian ricegrass	Heavy	–	Page et al. (1978)
		Idaho	Big sage/bluebunch wheatgrass	Moderate	–	Reynolds (1980)
Burrowing Owl	A	West	Various	Variable	+	Kochert (1989)
		West	Various	Variable	+	Snyder and Snyder (1975)
		Idaho	Big sage/bluebunch wheatgrass	Variable	Mixed	Rich (1986)
		Idaho	Big sage	Variable	–	Gleason (1978)
Long-eared Owl	B	West	Various	Variable	Mixed	Kochert (1989)
Short-eared Owl	B	Idaho	Big sage/bluebunch wheatgrass	Moderate	–	Reynolds (1980)
Common Nighthawk	A	Idaho	Big sage/bluebunch wheatgrass	Moderate	–	Reynolds and Trost (1981)
Common Poorwill	B	Idaho	Big sage/bluebunch wheatgrass	Moderate	+	Reynolds and Trost (1981)
Northern Flicker[d]	B	Nevada	Low sage/Idaho fescue	Heavy	0	Page et al. (1978)
		Nevada	Big sage/bluebunch wheatgrass	Heavy	+	Page et al. (1978)
Gray Flycatcher	A	Nevada	Shadscale/Indian ricegrass	Heavy	+	Page et al. (1978)
		Nevada	Nevada bluegrass/sedge	Heavy	+	Page et al. (1978)
		Idaho	Big sage/bluebunch wheatgrass	Moderate	–	Reynolds and Trost (1981)
Say's Phoebe	B	Idaho	Big sage/bluebunch wheatgrass	Moderate	+	Reynolds and Trost (1981)
Horned Lark	B	Nevada	Greasewood/Great Basin wild rye	Heavy	–	Page et al. (1978)

(continued)

Table 12-2 *(cont.)*

Species	Migrant Status[a]	Region	Shrubland Type	Grazing Intensity[b]	Response to Grazing[c]	Reference
Horned Lark (*cont.*)		Nevada	Shadscale/Indian ricegrass	Heavy	–	Page et al. (1978)
		Nevada	Low sage/Idaho fescue	Heavy	–	Page et al. (1978)
		Idaho	Big sage/bluebunch wheatgrass	Moderate	+	Reynolds (1980)
		Utah	Shadscale/sand dropseed	Heavy	Mixed	Medin (1986)
		Idaho	Big sage	Not reported	+	Olson (1974)
		Idaho	Various	Variable	+	Rotenberry and Knick (1992)
Tree Swallow[d]	B	Nevada	Nevada bluegrass/ sedge	Heavy	–	Page et al. (1978)
Rock Wren	B	Nevada	Shadscale/Indian Ricegrass	Heavy	+	Page et al. (1978)
		Nevada	Big sage/bluebunch wheatgrass	Heavy	+	Page et al. (1978)
		Idaho	Big sage/bluebunch wheatgrass	Moderate	+	Reynolds and Trost (1981)
Mountain Bluebird	B	Nevada	Low sage/Idaho fescue	Heavy	+	Page et al. (1978)
Sage Thrasher	B	Nevada	Greasewood/Great Basin wild rye	Heavy	+	Page et al. (1978)
		Nevada	Nevada bluegrass/ sedge	Heavy	+	Page et al. (1978)
		Idaho	Big sage/bluebunch wheatgrass	Moderate	+	Reynolds (1980)
		Idaho	Big sage	Not reported	+	Olson (1974)
		Idaho	Big sage/bluebunch wheatgrass	Moderate	–	Reynolds and Rich (1978)
Loggerhead Shrike	B	Nevada	Shadscale/Indian ricegrass	Heavy	+	Page et al. (1978)
		Nevada	Low sage/Idaho fescue	Heavy	+	Page et al. (1978)
		Idaho	Big sage/bluebunch wheatgrass	Moderate	0	Reynolds (1980)
		Utah	Shadscale/sand dropseed	Heavy	0	Medin (1986)
Green-tailed Towhee	A	Nevada	Big sage/bluebunch wheatgrass	Heavy	–	Page et al. (1978)
Vesper Sparrow	B	Nevada	Greasewood/Great Basin wild rye	Heavy	+	Page et al. (1978)
		Nevada	Shadscale/Indian ricegrass	Heavy	+	Page et al. (1978)
		Nevada	Low sage/Idaho fescue	Heavy	–	Page et al. (1978)
		Nevada	Big sage/bluebunch wheatgrass	Heavy	–	Page et al. (1978)
		Nevada	Nevada bluegrass/ sedge	Heavy	–	Page et al. (1978)
		Idaho	Big sage	Not reported	–	Olson (1974)
Black-throated Sparrow	B	Utah	Shadscale/sand dropseed	Heavy	Mixed	Medin (1986)
Sage Sparrow	B	Nevada	Greasewood/Great Basin wild rye	Heavy	–	Page et al. (1978)
		Nevada	Shadscale/Indian ricegrass	Heavy	–	Page et al. (1978)
		Nevada	Nevada bluegrass/ sedge	Heavy	+	Page et al. (1978)
		Idaho	Big sage/bluebunch wheatgrass	Moderate	+	Reynolds (1980)

Table 12-2 *(cont.)*

Species	Migrant Status[a]	Region	Shrubland Type	Grazing Intensity[b]	Response to Grazing[c]	Reference
Sage Sparrow *(cont.)*		Idaho	Big sage	Not reported	+	Olson (1974)
		Great Basin	Big Sage	Variable	+	Wiens and Rotenberry (1981b), Wiens (1985)
Brewer's Sparrow	B	Nevada	Greasewood/Great Basin wild rye	Heavy	–	Page et al. (1978)
		Nevada	Shadscale/Indian ricegrass	Heavy	+	Page et al. (1978)
		Nevada	Low sage/Idaho fescue	Heavy	–	Page et al. (1978)
		Nevada	Big sage/bluebunch wheatgrass	Heavy	+	Page et al. (1978)
		Nevada	Nevada bluegrass/ sedge	Heavy	+	Page et al. (1978)
		Idaho	Big sage/bluebunch wheatgrass	Moderate	–	Reynolds (1980)
		Idaho	Big sage	Not reported	–	Olson (1974)
Savannah Sparrow	B	Nevada	Nevada bluegrass/ sedge	Heavy	–	Page et al. (1978)
White-crowned Sparrow	B	Nevada	Greasewood/Great Basin wild rye	Heavy	–	Page et al. (1978)
		Nevada	Big sage/bluebunch wheatgrass	Heavy	+	Page et al. (1978)
		Idaho	Big sage/bluebunch wheatgrass	Moderate	+	Reynolds and Trost (1981)
Red-winged Blackbird	B	Nevada	Nevada bluegrass/ sedge	Heavy	–	Page et al. (1978)
Western Meadowlark	B	Nevada	Greasewood/Great Basin wild rye	Heavy	+	Page et al. (1978)
		Nevada	Shadscale/Indian ricegrass	Heavy	+	Page et al. (1978)
		Nevada	Low sage/Idaho fescue	Heavy	–	Page et al. (1978)
		Nevada	Big sage/bluebunch wheatgrass	Heavy	+	Page et al. (1978)
		Nevada	Nevada bluegrass/ sedge	Heavy	–	Page et al. (1978)
		Idaho	Big sage/bluebunch wheatgrass	Moderate	–	Reynolds (1980)
		Idaho	Big sage	Not reported	–	Olson (1974)
		Idaho	Various	Variable	0	Rotenberry and Knick (1992)
		Great Basin	Big sage	Variable	–	Wiens and Rotenberry (1981b)
Brewer's Blackbird	B	Nevada	Shadscale/Indian ricegrass	Heavy	+	Page et al. (1978)
		Nevada	Nevada bluegrass/ sedge	Heavy	–	Page et al. (1978)
		Idaho	Big sage/bluebunch wheatgrass	Moderate	0	Reynolds and Trost (1981)
Brown-headed Cowbird	B	Idaho	Big sage/bluebunch wheatgrass	Moderate	+	Reynolds and Trost (1981)
		Great Basin	Various	Variable	+	Rich and Rothstein (1985)

[a] Status "A" contains long-distance migrants, those species that breed in North America and spend their nonbreeding period primarily south of the United States. Status "B" contains short-distance migrants, those species that breed and winter extensively in North America, although some populations winter south of the United States.
[b] Grazing intensity as reported by original authors in the references listed.
[c] Grazing effects on abundance: + = increase; – = decrease; 0 = no effect, as reported by original authors.
[d] Species forages in shrubsteppe vegetation but nests in other adjacent habitat.

Table 12-3. Standardized relative abundance for 31 species of migratory landbirds. Original data were taken from 15 studies conducted in shrubsteppe habitats. Sample size is the number of studies from which the data were derived. Standard errors (SE) and *P* values were calculated by a paired *t*-test. No species' abundances differed significantly ($P < 0.05$) between treatments.

Species	Forage Guild[a]	Nest Layer[b]	Nest Type[c]	Sample Size	Standardized Mean		SE	*P* Value
					Grazed	Ungrazed[d]		
Northern Harrier	CA	GR	O	2	0.60	1.40	1.77	0.58
Swainson's Hawk	CA	CA	O	1	1.67	0.33	—	—
Red-tailed Hawk	CA	CA	O	1	0.67	1.33	—	—
Ferruginous Hawk	CA	GR	O	1	0.67	1.33	—	—
American Kestrel	CA	CA	C	2	0.80	1.20	2.21	0.75
Prairie Falcon	CA	CA	C	1	0.00	2.00	—	—
Long-billed Curlew	GI	GR	O	1	0.00	2.00	—	—
Mourning Dove	GI	SH	O	3	0.78	1.22	1.88	0.82
Short-eared Owl	CA	GR	C	1	0.00	2.00	—	—
Common Nighthawk	AI	GR	O	1	0.67	1.33	—	—
Common Poorwill	AI	GR	O	1	2.00	0.00	—	—
Northern Flicker	BI	CA	C	2	1.50	0.50	1.57	0.50
Gray Flycatcher	AI	SH	O	3	1.33	0.67	2.09	0.67
Say's Phoebe	AI	CA	O	1	2.00	0.00	—	—
Horned Lark	GI	GR	O	5	0.42	1.58	0.71	0.07
Tree Swallow	AI	CA	C	1	0.00	2.00	—	—
Rock Wren	GI	GR	C	3	2.00	0.00	0	0
Mountain Bluebird	AI	CA	C	1	2.00	0.00	—	—
Sage Thrasher	FI	SH	O	3	1.76	0.24	0.86	0.11
Loggerhead Shrike	SA	SH	O	4	1.42	0.58	1.01	0.26
Green-tailed Towhee	GI	GR	O	1	0.00	2.00	—	—
Vesper Sparrow	GI	GR	O	5	0.80	1.20	1.54	0.70
Black-thorated Sparrow	FI	SH	O	1	1.23	0.77	—	—
Sage Sparrow	GI	SH	O	4	0.93	1.07	0.37	0.69
Brewer's Sparrow	FI	SH	O	6	0.83	1.17	0.69	0.45
Savannah Sparrow	GI	GR	O	1	0.00	2.00	—	—
White-crowned Sparrow	OM	GR	O	3	1.12	0.88	1.08	0.87
Red-winged Blackbird	OM	SH	O	1	0.00	2.00	—	—
Western Meadowlark	GI	GR	O	6	1.00	1.00	1.22	1.00
Brewer's Blackbird	OM	SH	O	3	0.54	1.46	0.63	0.24
Brown-headed Cowbird	OM	—	P	1	1.78	0.22	—	—

[a] Foraging-guild abbreviations: AI = aerial insectivore; BI = bark insectivore; FI = foliage insectivore; GI = ground insectivore; CA = carnivore; NE = nectarivore; OM = omnivore.
[b] Nest-layer abbreviations: SH = shrub-nesting species; GR = ground-nesting species; CA = subcanopy/canopy-nesting species.
[c] Nest-type abbreviations: O = open; C = cavity; P = parasite.
[d] Includes lightly grazed and fall-grazed treatments.

and common shrubsteppe bird in many locations, their decline is a major cause for concern in sagebrush ecosystems.

Other shrubsteppe species such as Gray Flycatcher, Rock Wren, Green-tailed Towhee, Sage Thrasher, and Lark and Sage Sparrows have shown no significant population trends over the western United States (Peterjohn et al., Chapter 1, this volume). However, Lark Sparrows and Rock Wrens show significant range-wide declines. Black-throated Sparrows, which inhabit more xeric shrub communities, have also shown significant population declines.

Data for two other species also suggest that shrubsteppe bird communities are changing, whether or not livestock grazing is implicated as a major cause. Long-billed Curlews and Burrowing Owls both breed in habitats characterized by a lack of shrubs and large areas of relatively low vegetation. Both species showed significant population increases between 1966 and 1991 in the western United States (Peterjohn et al., Chapter 1, this volume).

Little information exists on responses to grazing by migratory raptors in shrubsteppe (Kochert 1989). Our designations of raptor species increasing or decreasing in abundance were based on grazing-induced habitat alterations, which affect small mammal populations, nest cover and substrates. Intensive grazing and fire suppression favors encroachment by shrubs and trees (especially *Juniperus* spp.). Ground-nesting raptors (e.g., Northern Harrier, Short-eared Owl, and Ferruginous Hawk) are often negatively affected by grazing practices that reduce nest cover (Duebbert and Lokemoen 1977). Raptors and their rodent prey often decrease under conditions with reduced amounts of herbaceous cover and increased shrub densities (see Kochert 1989). Other prey species (e.g., jackrabbits) respond positively to dense shrub conditions (Nydegger and Smith 1986), potentially benefitting their primary predator, the Golden Eagle. Increases in juniper trees could increase availability of nest sites (e.g., Long-eared Owls and Red-tailed Hawks) (Kochert 1989) and perch sites of some raptor species. Another potentially significant indirect effect of grazing on migrants in shrubsteppe is nest parasitism by Brown-headed Cowbirds. However, almost no data are available for shrubsteppe (Rich 1978, Rich and Rothstein 1985) and the degree of impact caused by cowbirds is unknown.

Shrubsteppe birds generally respond negatively after deliberate conversions of native shrub habitats to exotic vegetation for the foraging benefit of livestock (Best 1972, Schroeder and Sturges 1975, Reynolds and Trost 1980, 1981. Castrale 1982). However, some responses may not be detected due to lack of clear population responses due to time lags, site tenacity by individuals and scale of treatment (Wiens and Rotenberry 1985). Thus, short-term before and after surveys in this avian community may be "dangerously misleading" (Wiens et al. 1986).

WESTERN RIPARIAN HABITATS

Characteristics of Riparian Habitats

Riparian zones include assemblages of plant and animal communities occurring at the interfaces between terrestrial and aquatic ecosystems. In arid portions of western North America, riparian areas create well-defined, narrow zones of vegetation along ephemeral, intermittent, and perennial streams and rivers, and are most conspicuous in steppe, shrubsteppe, and desert regions. The diversity and productivity of these systems compared to surrounding uplands are largely attributable to biotic and nutrient exchanges between aquatic and adjacent upland areas (Gregory et al. 1991).

Western riparian woodlands vary from extensive floodplain forests dominated by cottonwoods (*Populus* spp.) along large rivers to narrow bands of aspen (*Populus tremuloides*) woodlands and willow (*Salix* spp.) thickets along small mountain streams. Plant composition varies geographically and climatically, with higher elevation areas often composed of alder (*Alnus* spp.), birch (*Betula* spp.), and dogwood (*Cornus* spp.). Sycamore (*Platanus* spp.), cherry (*Prunus* spp.), hawthorn (*Crataegus* spp.), and hackberry (*Celtis* spp.) are typically found at lower elevations and in drier climates.

Historical Perspective of Riparian Habitats

The critical and disproportionate value of riparian habitat to wildlife has been recognized only within the last two decades (Johnson et al. 1977, Knopf et al. 1988a). Riparian vegetation is used by wildlife more than any other vegetation type (Thomas et al. 1979). Yet, riparian areas are among the most threatened habitats on the continent because they are favored for many land uses including livestock grazing, agriculture, water management, timber harvest, recreation, and urbanization (e.g., Thomas et al. 1979, Knopf et al. 1988a).

Livestock grazing has caused geographically extensive impacts on western riparian zones (Carothers 1977, Crumpacker 1984, Chaney et al. 1990), and these areas are considered the most modified land type in the West (Chaney et al. 1990). Grazing on riparian bottomlands tends to be more damaging than on uplands (Platts and Nelson 1985), especially in arid regions where water, shade, succulent vegetation, and flatter terrain occur near streams (Behnke 1979, Chaney et al. 1990, Platts 1991). Livestock grazing affects riparian habitats by altering, reducing, or removing vegetation, and by actually eliminating riparian areas through channel widening, channel aggrading, or lowering the water table (see Platts 1991).

Evaluation of Grazing Systems in Riparian Habitats

Rangeland grazing practices have been reviewed and evaluated for riparian ecosystems (Platts 1981, Knopf and Cannon 1982, Kauffman et al. 1983, Kauffman and Krueger 1984, Skovlin 1984, Clary and Webster 1989, Platts 1991, Sedgwick and Knopf 1991, Kovalchik and Elmore 1992). Riparian habitats are known to be detrimentally affected by most grazing practices tested to date. This is not surprising because traditional grazing systems were developed for upland grasses, not for riparian plant species (see Platts 1991 for review).

Grazing systems are evaluated by the intensity and seasonality of use by livestock. Riparian areas generally are grazed most in summer and least in winter (Knopf et al. 1988b, Goodman et al. 1989). The resulting summer concentration of use in riparian zones is particularly damaging due to severe trampling and mechanical damage, soil compaction, and plant consumption by livestock. Thus, year-long and growing-season (spring–summer) grazing are particularly damaging to riparian vegetation (Kauffman and Krueger 1984, Platts 1991), and the associated bird communities (Crouch 1982).

Short-term, early spring grazing may be preferable to summer grazing (Clary and Webster 1989). Early season grazing can result in better distribution of livestock because upland vegetation is succulent at this time and because livestock may avoid the wetter riparian soils (Clary and Webster 1989, Platts 1991). However, impacts of soil compaction may be most severe at this time. Early season grazing, followed by complete removal of livestock, allows regrowth of riparian vegetation before the dormant period in autumn.

As herbaceous cover is depleted or as palatability of alternate forage decreases, livestock will shift to browsing riparian shrubs before leaf drop (Kovalchik and Elmore 1992). Therefore, most browsing damage to willows (*Salix* spp.) occurs in late summer and fall (Kauffman et al. 1983, Clary and Webster 1989, Sedgwick and Knopf 1991, Kovalchik and Elmore 1992). Alternatively, light-to-moderate autumn grazing appears to have the least impact on numbers of migratory birds during the breeding season (Kauffman et al. 1982, 1983, Sedgwick and Knopf 1987, Knopf et al. 1988b, Medin and Clary 1991).

In late fall and winter, water levels typically are low, streambanks are dry, and vegetation is dormant, thus minimizing the effects of trampling, soil compaction, erosion, and browsing (Rauzi and Hanson 1966, Knopf and Cannon 1982, Kauffman and Krueger 1984). However, fall–winter grazing should be carefully controlled to leave residual plant cover needed for streambank maintenance during subsequent high spring flows (Clary and Webster 1989).

Kauffman et al. (1982, 1983) evaluated the effects of late season grazing on ten common riparian communities in eastern Oregon by comparing plant and animal communities in enclosed and grazed areas (late August–mid-September, at 1.3–1.7 ha/animal unit month). Avian populations in all plant communities appeared to have little differential response to grazing treatments with respect to species richness, density, or diversity. Meadows and Douglas hawthorn (*Crataegus douglasii*) communities were more heavily used by cattle than other riparian communities, shrub use was light except on willow-dominated gravel bars, and use of plant communities with dense canopy cover (black cottonwood [*Populus trichocarpa*], Ponderosa pine [*Pinus ponderosa*], and thin-leafed alder [*Alnus incana*]) was light.

Sedgwick and Knopf (1987) evaluated the impact of fall (October–November) grazing on breeding densities of six Neotropical migrants (House Wren, Brown Thrasher, American Robin, Common Yellowthroat, Yellow-breasted Chat, and Rufous-sided Towhee) associated with the lower vegetative layer of a cottonwood (*P. sargentii*) riparian forest. Moderate, late-fall grazing had no apparent impact on densities of any of the species, implying that proper seasonal grazing of cottonwood bottomlands is compatible with migratory bird use of a site during the breeding season. Common Yellowthroats and Yellow-breasted Chats were the most individual in their habitat associations and most likely to respond negatively to higher levels of grazing.

Knopf et al. (1988b) compared plant and bird communities between healthy (historically winter grazed) and decadent (historically summer grazed) willow communities within a year. Population densities of habitat generalists (Yellow Warblers, Savannah Sparrows, and Song Sparrows) differed little between winter-grazed and summer-grazed willow communities. Densities of the species intermediate in habitat specialization (American Robins, Red-winged Blackbirds, and Brown-headed Cowbirds) differed more dramatically, while habitat specialists (Willow Flycatchers, Lincoln's Sparrows, and White-crowned Sparrows) were absent or accidental in decadent willows. Brown-headed Cowbirds showed the greatest tendency to increase in numbers in disturbed, summer-grazed riparian areas. Conversely, high local densities of habitat specialists (and possibly Red-winged Blackbirds) occurred in winter-grazed willow communities.

With prescribed, late-season grazing in a cottonwood floodplain in Colorado, herbaceous and shrub vegetation (excluding willows) appeared to be resilient to cattle grazing, at least during the initial 3 years after grazing began, following 31 years of nonuse (Sedgwick and Knopf 1991). The grazing program for this study was strictly controlled by season and intensity of use within the riparian zone. This is unlike most grazing programs, wherein the riparian zone is included as part of a larger allotment and the use of riparian vegetation often exceeds forage use on the uplands (Platts 1991).

Sedgwick and Knopf (1991) cautioned that even a 4 year study is a relatively brief time to study grazing impacts. Longer-term grazing effects may alter composition, structural diversity, and community succession patterns in riparian systems. For example, they were unable to assess grazing impacts on cottonwood seedling survival because seedlings were so few on their study area. Glinski (1977), however, found that cattle grazing reduced cottonwood-seedling establishment along an Arizona stream, and predicted that the future width of the riparian zone would be significantly reduced. Longer term (more than 3 years) studies of dormant-season grazing may well document alterations of herbaceous communities (Sedgwick and Knopf 1991).

Riparian Avifauna

Western riparian areas are key components of migratory bird habitats during all seasons of the year (Stevens et al. 1977, Henke and Stone 1979, Szaro 1980, Terborgh 1989). Riparian vegetation covers less than 1% of the landscape in the arid West, yet more species of breeding birds are found in this limited habitat than in the more extensive surrounding uplands (Knopf et al. 1988a). Migratory landbirds inhabiting western North America are thought to be particularly vulnerable to disturbance because their riparian habitats are fragmented and limited in distribution, thus probably restricting their total populations below those of their eastern counterparts (Terborgh 1989). Because the contribution of these productive areas to avian diversity is disproportionate to other western habitats, riparian woodlands are critical to overall conservation of the continental avifauna.

The highest densities of breeding birds for North America have been reported from southwestern riparian habitats (Carothers and Johnson 1975, Ohmart and Anderson 1986, Rice et al. 1983). More than two-thirds (127 of 166) of southwestern bird species nest in riparian woodlands, and Neotropical migrants comprise 60% of the 98 landbirds (Johnson et al. 1977). In arid portions of the West, several studies documented that most bird species nest in riparian habitats where Neotropical migrants comprise between 60% and 85% of the landbirds (Knopf 1985, Dobkin and Wilcox 1986, Saab and Groves 1992). Probably most migrant landbirds in the western United States are associated with riparian habitats during the breeding season (cf. Mosconi and Hutto 1982, Ohmart and Anderson 1986).

Avian Responses to Livestock Grazing in Riparian Habitats

We know of nine studies that provide some quantitative comparisons of species abundances in systems that were variously

Table 12-4. Responses to cattle grazing by Neotropical migrant landbirds breeding in western riparian habitats.

Species	Migrant Status[a]	Region	Riparian Type	Grazing Intensity[b]	Response to Grazing[c]	Reference
Killdeer	B	Colorado	Cottonwood/ willow	Variable	+	Crouch (1982)
		Oregon	Willow	Variable	+	Taylor (1986)
		Idaho	Herbaceous	Heavy	+	Medin and Clary (1990)
		Colorado	Willow	Heavy	+	Schulz and Leininger (1991)
Long-billed Curlew	A	Idaho	Herbaceous	Heavy	+	Medin and Clary (1990)
American Kestrel	B	Colorado	Cottonwood/ willow	Variable	+	Crouch (1982)
		Montana	Cottonwood/ pine	Heavy vs light	–*	Mosconi and Hutto (1982)
		Nevada	Aspen/willow	Moderate	–	Medin and Clary (1991)
Mourning Dove	B	California/ Nevada	Aspen	Not reported	+	Page et al. (1978)
		Colorado	Cottonwood/ willow	Variable	0	Crouch (1982)
		Montana	Cottonwood/ pine	Heavy vs light	–	Mosconi and Hutto (1982)
		Oregon	Willow	Variable	–	Taylor (1986)
Yellow-billed Cuckoo	A	Colorado	Cottonwood/ willow	Variable	+	Crouch (1982)
Common Nighthawk	A	Colorado	Cottonwood/ willow	Variable	+	Crouch (1982)
Calliope Hummingbird	A	California/ Nevada	Aspen	Not reported	–	Page et al. (1978)
		Montana	Cottonwood/ pine	Heavy vs light	–	Mosconi and Hutto (1982)
Rufous Hummingbird	A	California/ Nevada	Aspen	Not reported	–	Page et al. (1978)
Broad-tailed Hummingbird	A	Nevada	Aspen/willow	Moderate	+	Medin and Clary (1991)
		Colorado	Willow	Heavy	–	Schulz and Leininger (1991)
		California/ Nevada	Aspen	Not reported	–	Page et al. (1978)
Belted Kingfisher	B	Colorado	Cottonwood/ willow	Variable	0	Crouch (1982)
Northern Flicker	B	California/ Nevada	Aspen	Not reported	+	Page et al. (1978)
		Colorado	Cottonwood/ willow	Variable	–	Crouch (1982)
		Montana	Cottonwood/ pine	Heavy vs light	+	Mosconi and Hutto (1982)
		Nevada	Aspen/willow	Moderate	0	Medin and Clary (1991)
		Colorado	Willow	Heavy	–	Schulz and Leininger (1991)
Lewis' Woodpecker	B	Montana	Cottonwood/ pine	Heavy vs light	+	Mosconi and Hutto (1982)
		Nevada	Aspen/willow	Moderate	+	Medin and Clary (1991)
Red-naped Sapsucker	B	California/ Nevada	Aspen	Not reported	–	Page et al. (1978)
		Montana	Cottonwood/ pine	Heavy vs light	–	Mosconi and Hutto (1982)
		Nevada	Aspen/willow	Moderate	0	Medin and Clary (1991)
		Colorado	Willow	Heavy	+	Schulz and Leininger (1991)
Western Kingbird	A	Colorado	Cottonwood/ willow	Variable	0	Crouch (1982)
Eastern Kingbird	A	Colorado	Cottonwood/ willow	Variable	–	Crouch (1982)
		Montana	Cottonwood/ pine	Heavy vs light	+	Mosconi and Hutto (1982)
		Oregon	Willow	Variable	–	Taylor (1986)

Table 12-4 (*cont.*)

Species	Migrant Status[a]	Region	Riparian Type	Grazing Intensity[b]	Response to Grazing[c]	Reference
Eastern Kingbird (*cont.*)		Colorado	Cottonwood/ willow	Variable	0	Crouch (1982)
Western Wood-Pewee	A	California/ Nevada	Aspen	Not reported	–	Page et al. (1978)
		Montana	Cottonwood/ pine	Heavy vs. light	+*	Mosconi and Hutto (1982)
Say's Phoebe	B	Colorado	Cottonwood/ willow	Variable	+	Crouch (1982)
Least Flycatcher	A	Montana	Cottonwood/ pine	Heavy vs light	+	Mosconi and Hutto (1982)
Willow Flycatcher	A	Montana	Cottonwood/ pine	Heavy vs light	–	Mosconi and Hutto (1982)
		Oregon	Willow	Variable	–	Taylor (1986)
		Colorado	Willow	Variable	–*	Knopf et al. (1988b)
Empidonax sp.	A	Nevada	Aspen/willow	Moderate	–	Medin and Clary (1991)
		Colorado	Willow	Heavy	+	Schulz and Leininger (1991)
Tree Swallow	B	California/ Nevada	Aspen	Not reported	–	Page et al. (1978)
		Nevada	Aspen/willow	Moderate	+	Medin and Clary (1991)
		Colorado	Willow	Heavy	+	Schulz and Leininger (1991)
Barn Swallow	A	Colorado	Cottonwood/ willow	Variable	–	Crouch (1982)
House Wren	A	California/ Nevada	Aspen	Not reported	+	Page et al. (1978)
		Colorado	Cottonwood/ willow	Variable	0	Crouch (1982)
		Montana	Cottonwood/ willow	Heavy vs light	+*	Mosconi and Hutto (1982)
		Oregon	Willow	Variable	–	Taylor (1986)
		Colorado	Cottonwood/ willow	Moderate	+	Sedgwick and Knopf (1987)
		Nevada	Aspen/willow	Moderate	+	Medin and Clary (1991)
		Colorado	Willow	Heavy	+	Schultz and Leininger (1991)
Marsh Wren	B	Oregon	Willow	Variable	+	Taylor (1986)
Ruby-crowned Kinglet	B	Colorado	Willow	Heavy	0	Schulz and Leininger (1991)
Mountain Bluebird	B	California/ Nevada	Aspen	Not reported	+	Page et al. (1978)
		Colorado	Cottonwood/ willow	Variable	0	Crouch (1982)
		Colorado	Willow	Heavy	+	Schulz and Leininger (1991)
Veery	A	Montana	Cottonwood/ pine	Heavy vs light	–*	Mosconi and Hutto (1982)
Hermit Thrush	A	California/ Nevada	Aspen	Not reported	–	Page et al. (1978)
American Robin	B	California/ Nevada	Aspen	Not reported	+	Page et al. (1978)
		Colorado	Cottonwood/ willow	Variable	+	Crouch (1982)
		Montana	Cottonwood/ pine	Heavy vs light	+*	Mosconi and Hutto (1982)
		Oregon	Willow	Variable	0	Taylor (1986)
		Colorado	Cottonwood/ willow	Moderate	+	Sedgwick and Knopf (1987)
		Colorado	Willow	Variable	0	Knopf et al. (1988b)
		Nevada	Aspen/willow	Moderate	–	Medin and Clary (1991)
		Colorado	Willow	Heavy	+	Schulz and Leininger (1991)
Gray Catbird	A	Colorado	Cottonwood/ willow	Variable	–	Crouch (1982)
		Montana	Cottonwood/ pine	Heavy vs light	+	Mosconi and Hutto (1982)

(*continued*)

Table 12-4 *(cont.)*

Species	Migrant Status[a]	Region	Riparian Type	Grazing Intensity[b]	Response to Grazing[c]	Reference
Cedar Waxwing	B	Montana	Cottonwood/ pine	Heavy vs light	–	Mosconi and Hutto (1982)
		Oregon	Willow	Variable	0	Taylor (1986)
Solitary Vireo	A	Montana	Cottonwood/ pine	Heavy vs light	+*	Mosconi and Hutto (1982)
Red-eyed Vireo	A	Montana	Cottonwood/ pine	Heavy vs light	+	Mosconi and Hutto (1982)
Warbling Vireo	A	California/ Nevada	Aspen	Not reported	–	Page et al. (1978)
		Montana	Cottonwood/ pine	Heavy vs light	+	Mosconi and Hutto (1982)
		Nevada	Aspen/willow	Moderate	–	Medin and Clary (1991)
		Colorado	Willow	Heavy	+	Schulz and Leininger (1991)
Orange-crowned Warbler	A	Montana	Cottonwood/ pine	Heavy vs light	–	Mosconi and Hutto (1982)
Nashville Warbler	A	Montana	Cottonwood/ pine	Heavy vs light	–*	Mosconi and Hutto (1982)
Yellow-rumped Warbler	B	California/ Nevada	Aspen	Not reported	–	Page et al. (1978)
		Colorado	Cottonwood/ willow	Variable	–	Crouch (1982)
		Montana	Cottonwood/ pine	Heavy vs light	+	Mosconi and Hutto (1982)
		Colorado	Willow	Heavy	–	Schulz and Leininger (1991)
Yellow Warbler	A	Califonia/ Nevada	Aspen	Not reported	+	Page et al. (1978)
		Montana	Cottonwood/ pine	Heavy vs light	–	Mosconi and Hutto (1982)
		Oregon	Willow	Variable	–	Taylor (1986)
		Colorado	Willow	Variable	0	Knopf et al. (1988b)
		Nevada	Aspen/willow	Moderate	0	Medin and Clary (1991)
MacGillivray's Warbler	A	California/ Nevada	Aspen	Not reported	–	Page et al. (1978)
		Montana	Cottonwood/ pine	Heavy vs light	–*	Mosconi and Hutto (1982)
		Nevada	Aspen/willow	Moderate	–	Medin and Clary (1991)
		Colorado	Cottonwood/ willow	Heavy	+	Schulz and Leininger (1991)
Wilson's Warbler	A	California/ Nevada	Aspen	Not reported	–	Page et al. (1978)
		Colorado	Cottonwood/ willow	Variable	+	Crouch (1982)
		Colorado	Willow	Heavy	–*	Schultz and Leininger (1991)
Northern Waterthrush	A	Montana	Cottonwood/ pine	Heavy vs light	–	Mosconi and Hutto (1982)
Common Yellowthroat	A	Colorado	Cottonwood/ willow	Variable	–	Crouch (1982)
		Montana	Cottonwood/ pine	Heavy vs light	–*	Mosconi and Hutto (1982)
		Oregon	Willow	Variable	–	Taylor (1986)
		Colorado	Cottonwood/ willow	Moderate	–	Sedgwick and Knopf (1987)
Yellow-breasted Chat	A	Oregon	Willow	Variable	0	Taylor (1986)
		Colorado	Cottonwood/ willow	Moderate	–	Sedgwick and Knopf (1987)
American Redstart	A	Colorado	Cottonwood/ willow	Variable	+	Crouch (1982)
		Montana	Cottonwood/ pine	Heavy vs light	–*	Mosconi and Hutto (1982)

Table 12-4 (*cont.*)

Species	Migrant Status[a]	Region	Riparian Type	Grazing Intensity[b]	Response to Grazing[c]	Reference
Black-headed Grosbeak	A	California/ Nevada	Aspen	Not reported	–	Page et al. (1978)
		Montana	Cottonwood/ pine	Heavy vs light	+	Mosconi and Hutto (1982)
		Oregon	Willow	Variable	–	Taylor (1986)
		Montana	Cottonwood/ pine	Heavy vs light	–*	Mosconi and Hutto (1982)
Lazuli Bunting	A	Montana	Cottonwood/ pine	Heavy vs light	–*	Mosconi and Hutto (1982)
Green-tailed Towhee	A	California/ Nevada	Aspen	Not reported	+	Page et al. (1978)
Rufous-sided Towhee	B	California/ Nevada	Aspen	Not reported	+	Page et al. (1978)
		Colorado	Cottonwood/ willow	Variable	0	Crouch (1982)
		Colorado	Cottonwood/ willow	Moderate	–	Sedgwick and Knopf (1987)
		Idaho	Herbaceous	Heavy	–	Medin and Clary (1990)
Vesper Sparrow	B	Idaho	Herbaceous	Heavy	–	Medin and Clary (1990)
Savannah Sparrow	B	Oregon	Willow	Variable	–	Taylor (1986)
		Colorado	Willow	Variable	0	Knopf et al. (1988b)
		Idaho	Herbaceous	Heavy	–	Medin and Clary (1990)
Song Sparrow	B	Montana	Cottonwood/ pine	Heavy vs light	–*	Mosconi and Hutto (1982)
		Colorado	Willow	Variable	+	Knopf et al. (1988b)
		Nevada	Aspen/willow	Moderate	+	Medin and Clary (1991)
Chipping Sparrow	B	Colorado	Cottonwood/ willow	Variable	–	Crouch (1982)
		Colorado	Willow	Heavy	–	Schulz and Leininger (1991)
Dark-eyed Junco	B	California/ Nevada	Aspen	Not reported	–	Page et al. (1978)
		Colorado	Cottonwood/ willow	Variable	–	Crouch (1982)
		Colorado	Willow	Heavy	–	Schulz and Leininger (1991)
White-crowned Sparrow	B	California/ Nevada	Aspen	Not reported	–	Page et al. (1978)
		Colorado	Willow	Variable	–*	Knopf et al. (1988b)
		Nevada	Aspen/willow	Moderate	+	Medin and Clary (1991)
		Colorado	Willow	Heavy	–	Schulz and Leininger (1991)
Fox Sparrow	B	California/ Nevada	Aspen	Not reported	–	Page et al. (1978)
		Colorado	Willow	Variable	–*	Knopf et al. (1988b)
		Colorado	Willow	Heavy	–	Schulz and Leininger (1991)
Lincoln's Sparrow	A	Colorado	Willow	Variable	–*	Knopf et al. (1988b)
		Colorado	Willow	Heavy	–	Schulz and Leininger (1991)
Bobolink	A	Oregon	Willow	Variable	–	Taylor (1986)
Western Meadowlark	B	Colorado	Cottonwood/ willow	Variable	+	Crouch (1982)
		Idaho	Herbaceous	Heavy	–	Medin and Clary (1990)
Yellow-headed Blackbird	A	Oregon	Willow	Variable	+	Taylor (1986)
Red-winged Blackbird	B	Colorado	Cottonwood/ willow	Variable	–	Crouch (1982)
		Oregon	Willow	Variable	–	Taylor (1986)
		Colorado	Willow	Variable	–	Knopf et al. (1988b)
		Idaho	Herbaceous	Heavy	–	Medin and Clary (1990)
Brewer's Blackbird	B	California/ Nevada	Aspen	Not reported	+	Page et al. (1978)

(*continued*)

Table 12-4 (*cont.*)

Species	Migrant Status[a]	Region	Riparian Type	Grazing Intensity[b]	Response to Grazing[c]	Reference
Brewer's Blackbird (*cont.*)		Colorado	Cottonwood/ willow	Variable	−	Crouch (1982)
		Oregon	Willow	Variable	0	Taylor (1986)
		Idaho	Herbaceous	Heavy	+	Medin and Clary (1990)
Brown-headed Cowbird	B	California/ Nevada	Aspen	Not reported	+	Page et al. (1978)
		Montana	Cottonwood/ pine	Heavy vs light	+	Mosconi and Hutto (1982)
		Oregon	Willow	Variable	−	Taylor (1986)
		Colorado	Willow	Variable	+	Knopf et al. (1988b)
		Colorado	Willow	Heavy	+	Schulz and Leininger (1991)
Northern Oriole	A	California/ Nevada	Aspen	Not reported	+	Page et al. (1978)
		Colorado	Cottonwood/ willow	Variable	−	Crouch (1982)
		Montana	Cottonwood/ pine	Heavy vs light	−*	Mosconi and Hutto (1982)
		Oregon	Willow	Variable	−	Taylor (1986)
		Nevada	Aspen/willow	Moderate	0	Medin and Clary (1991)
Hooded Oriole	A	California/ Nevada	Aspen	Not reported	+	Page et al. (1978)
Western Tanager	A	Montana	Cottonwood/ pine	Heavy vs light	+*	Mosconi and Hutto (1982)
		Colorado	Willow	Heavy	−	Schulz and Leininger (1991)
Pine Siskin	B	Montana	Cottonwood/ pine	Heavy vs light	+	Mosconi and Hutto (1982)
		Colorado	Willow	Heavy	+	Schulz and Leininger (1991)
American Goldfinch	B	Colorado	Cottonwood/ willow	Variable	−	Crouch (1982)
		Montana	Cottonwood/ pine	Heavy vs light	−	Mosconi and Hutto (1982)
		Oregon	Willow	Variable	−	Taylor (1986)
Cassin's Finch	B	California/ Nevada	Aspen	Not reported	−	Page et al. (1978)
		Montana	Cottonwood/ pine	Heavy vs light	+	Mosconi and Hutto (1982)
		Oregon	Willow	Variable	−	Taylor (1986)
		Nevada	Aspen/willow	Moderate	0	Medin and Clary (1991)
		Colorado	Willow	Heavy	−	Schulz and Leininger (1991)

[a] Status "A" contains long-distance migrants, those species that breed in North America and spend their nonbreeding period primarily south of the United States. Status "B" contains short-distance migrants, those species that breed and winter extensively in North America, although some populations winter south of the United States.
[b] Grazing intensity as reported by original authors in the references listed.
[c] Grazing effects on abundance: + = increase; − = decrease; 0 = no effect, as reported by original authors. Species whose abundances differed statistically between treatments, as reported by original authors, are indicated by an asterisk.

grazed by cattle (Table 12-4). These studies were conducted in six states and most were in cottonwood and willow riparian communities. The studies described the impacts of grazing by comparing avian populations on adjacent grazed and ungrazed sites (Page et al. 1978, Crouch 1982, Sedgwick and Knopf 1987, Medin and Clary 1990, 1991, Schulz and Leininger 1991), on adjacent sites that were subject to different levels of grazing (Mosconi and Hutto 1982, Taylor 1986), and on adjacent sites historically grazed during different seasons of the year (Knopf et al. 1988b). Season and intensity of grazing were not always well defined, and the results of four of the studies (i.e., Page et al. 1978, Mosconi and Hutto 1982; Medin and Clary 1990, 1991) are compromised by the complete absence of treatment replications to evaluate the effects of grazing. Despite shortcomings, we generally found consistent patterns and biologically interpretable responses by many members of the riparian avifauna.

These studies reported abundance data on 68 species of Neotropical migrant landbirds (Table 12-4). In a qualitative assessment of all studies combined, nearly half (46%) of these species decreased in abundance with cattle grazing, 29% increased with grazing, and 25% showed no clear response.

Forty-three of the 68 species were evaluated by two or more studies, and used in statistical analyses (see Methods section). Among these, few species showed significant differences ($P < 0.05$) in abundance between grazed and ungrazed treatments (Table 12-5). Species with significant or near-significant reductions in grazed treatments included Red-winged Blackbirds, Common Yellowthroats, and Willow Flycatchers. These species were about 1.5 times more abundant in ungrazed treatments (Table 12-5), indicating that these species are sensitive to changes resulting from livestock grazing. They also experience high rates of nest parasitism (Hofslund 1957, Sedgwick and Knopf 1988, Weatherhead 1989). All three species nest within the shrub layer and, in forested habitats, songbirds that nest in shrubs generally experience the highest rates of nest predation (Martin 1993). Cattle may further increase nest losses by exposing concealed nests to predators by reducing foliage densities or opening dense patches of vegetation to allow predator access (Knopf 1995).

At least seven more species were probably also harmed by grazing in riparian ecosystems. Three of these were evaluated in only one of the nine studies, but showed strongly negative responses: Veery, Nashville Warbler, and Fox Sparrow; all are ground or near-ground nesters. Other species showed uncertain or inconsistent responses to grazing, but likely would be negatively affected by grazing, based on knowledge of their habitat requirements. Conspicuous among these are the Yellow Warbler (see Taylor and Littlefield 1986), American Redstart, Gray Catbird, and Yellow-breasted Chat. Of these species with limited data but expected to decrease with grazing, only the Veery is experiencing population declines in the West and significant continental declines (Robbins et al. 1993, Peterjohn et al., chapter 1, this volume). Cattle grazing could be one factor contributing to their population declines.

American Robins, Killdeer, and Pine Siskins, all habitat generalists, showed the strongest trend of increasing in abundance with grazing. American Robins and Killdeer appear well adapted to human-modified landscapes in the West, e.g., both commonly nest in residential areas and prefer relatively open habitats (DeGraff et al. 1991, Dobkin 1993). Although Pine Siskins generally prefer coniferous habitats, some are found nesting in western riparian woodlands adjacent to pine forests (Mosconi and Hutto 1982, Schulz and Leininger 1991). This species nests in tree canopies, generally unaffected by livestock grazing in the short term.

Grouping species by nest type, we found that cavity-nesting species appeared least affected by cattle grazing [standardized mean 1.02 vs 0.97 ± 0.35 (grazed vs ungrazed), $T = 0.021$, $P = 0.98$]. Although not significant, abundance of open-nesting birds was more reduced by grazing practices [standardized mean 0.89 vs 1.11 ± 0.17 (grazed vs ungrazed), $T = -1.55$, $P = 0.12$]. These results support those of individual studies evaluating short-term grazing effects, in concluding that woodpeckers and other cavity-nesting species are relatively unaffected (Good and Dambach 1943, Mosconi and Hutto 1982) and sometimes increase in grazed pastures (Butler 1979, Medin and Clary 1991). Cavity-nesting birds place their nests in snags and dead limbs, and frequently forage in tree locations (bark) that are generally not used by cattle. Open-nesting species generally experience lower rates of nest success than cavity-nesting species (Martin and Li 1992), and cattle could further increase nest losses through physical damage to the herbaceous and shrub layers where open-nesting species often nest and forage.

Evaluating species by nest location, we found that ground-nesting species were most susceptible to disturbances created by livestock grazing (Table 12-6). Dark-eyed Juncos, and White-crowned, Savannah,

Table 12-5. Standardized relative abundance for 68 species of migratory landbirds. Original data were taken from nine studies conducted in western riparian habitats. Sample size is the number of studies from which the data were derived. Standard errors (SE) and *P* values were calculated by a paired *t*-test, and species whose abundances differed significantly ($P < 0.05$) are indicated by an asterisk.

Species	Forage Guild[a]	Nest Layer[b]	Nest Type[c]	Sample Size	Standardized Mean		SE	*P* Value
					Grazed	Ungrazed[d]		
Killdeer	GI	GR	O	4	1.53	0.47	0.62	0.12
Long-billed Curlew	GI	GR	O	1	1.29	0.71	—	—
American Kestrel	CA	CA	C	3	0.41	1.59	0.79	0.18
Mourning Dove	GI	SH	O	4	1.09	0.91	1.00	0.71
Yellow-billed Cuckoo	FI	SH	O	1	1.17	0.83	—	—
Common Nighthawk	AI	GR	O	1	1.41	0.59	—	—
Callipe Hummingbird	NE	CA	O	2	0.14	1.86	0.78	0.20
Broad-tailed Hummingbird	NE	CA	O	2	0.89	1.12	0.44	0.68
Rufous Hummingbird	NE	CA	O	1	0.00	2.00	—	—
Belted Kingfisher	CA	CA	C	1	1.09	0.91	—	—
Northern Flicker	OM	CA	C	5	0.88	1.12	0.68	0.53
Lewis' Woodpecker	AI	CA	C	2	1.30	0.70	0.42	0.38
Red-naped Sapsucker	BI	CA	C	4	0.90	1.10	1.30	0.87
Western Kingbird	AI	CA	O	1	2.00	0.00	—	—
Eastern Kingbird	AI	CA	O	3	0.80	1.20	0.62	0.60
Western Wood-pewee	AI	CA	O	2	0.59	1.41	1.75	0.57
Say's Phoebe	AI	CA	O	1	2.00	0.00	—	—
Least Flycatcher	AI	SH	O	1	2.00	0.00	—	—
Willow Flycatcher	AI	SH	O	3	0.33	1.67	0.72	0.13
Empidonax sp.	AI	CA	O	3	0.87	1.13	0.39	0.56
Tree Swallow	AI	CA	C	3	1.07	0.93	1.82	0.95
Barn Swallow	AI	CA	O	1	0.07	1.93	—	—
House Wren	FI	CA	C	7	1.16	0.84	0.71	0.65
Marsh Wren	FI	SH	O	1	2.00	0.00	—	—
Ruby-crowned Kinglet	FI	CA	O	1	1.00	1.00	—	—
Mountain Bluebird	AI	CA	C	3	1.47	0.53	0.91	0.28
Veery	GI	GR	O	1	0.79	1.21	—	—
Hermit Thrush	GI	SH	O	1	0.00	2.00	—	—
American Robin	GI	CA	O	8	1.31	0.69	0.41	0.09
Gray Catbird	FI	SH	O	2	0.86	1.14	0.28	0.61
Cedar Waxwing	FI	CA	O	2	0.74	1.26	0.55	0.50
Solitary Vireo	FI	CA	O	1	2.00	0.00	—	—
Red-eyed Vireo	FI	CA	O	1	1.05	0.95	—	—
Warbling Vireo	FI	CA	O	4	1.15	0.85	0.93	0.58
Orange-crowned Warbler	FI	GR	O	1	0.00	2.00	—	—
Nashville Warbler	FI	GR	O	1	0.00	2.00	—	—
Yellow-rumped Warbler	FI	CA	O	4	0.68	1.32	1.48	0.58
Yellow Warbler	FI	SH	O	5	1.09	0.91	0.73	0.62
MacGillivray's Warbler	FI	SH	O	4	0.50	1.50	1.57	0.39
Wilson's Warbler	FI	GR	O	2	0.43	1.57	1.24	0.27
Northern Waterthrush	GI	GR	O	1	0.43	1.57	—	—
Common Yellowthroat	FI	SH	O	4	0.45	1.55	0.61	0.10
Yellow-breasted Chat	FI	SH	O	2	1.26	0.74	2.07	0.69
American Redstart	FI	CA	O	2	1.39	0.60	1.78	0.59
Black-headed Grosbeak	OM	CA	O	3	1.07	0.93	1.51	0.93
Lazuli Bunting	OM	SH	O	1	0.00	2.00	—	—
Green-tailed Towhee	GI	SH	O	1	2.00	0.00	—	—
Rufous-sided Towhee	OM	GR	O	3	0.98	1.02	1.81	0.98
Vesper Sparrow	OM	GR	O	1	0.95	1.05	—	—
Savannah Sparrow	OM	GR	O	3	0.71	1.29	0.30	0.18
Song Sparrow	GI	SH	O	3	0.92	1.08	0.63	0.79
Chipping Sparrow	OM	SH	O	2	0.85	1.15	0.19	0.35
Dark-eyed Junco	OM	GR	O	3	0.71	1.29	0.19	0.09
White-crowned Sparrow	OM	GR	O	4	0.36	1.64	0.80	0.10
Fox Sparrow	OM	GR	O	1	0.00	2.00	—	—
Lincoln's Sparrow	OM	GR	O	2	0.23	1.77	1.01	0.28
Bobolink	FI	GR	O	1	0.00	2.00	—	—

Table 12-5 *(cont.)*

Species	Forage Guild[a]	Nest Layer[b]	Nest Type[c]	Sample Size	Standardized Mean		SE	*P* Value
					Grazed	Ungrazed[d]		
Western Meadowlark	GI	GR	O	2	1.01	0.99	0.01	0.20
Yellow-headed Blackbird	OM	SH	O	1	1.18	0.82	—	—
Red-winged Blackbird	OM	SH	O	4	0.46	1.54	0.35	0.4*
Brewer's Blackbird	OM	SH	O	4	1.31	0.70	1.06	0.43
Brown-headed Cowbird	GI	—	P	5	1.39	0.61	0.81	0.20
Northern Oriole	OM	CA	O	5	0.83	1.17	1.01	0.74
Hooded Oriole	OM	CA	O	1	2.00	0.00	—	—
Western Tanager	FI	CA	O	2	1.34	0.66	1.91	0.63
Pine Siskin	OM	CA	O	2	1.96	0.04	0.40	0.09
American Goldfinch	OM	SH	O	3	0.61	1.39	0.30	0.11
Cassin's Finch	OM	CA	O	5	0.75	1.25	1.17	0.57

[a] Foraging-guild abbreviations: AI = aerial insectivore; BI = bark insectivore; FI = foliage insectivore; GI = ground insectivore; CA = carnivore; NE = nectivore; OM = omnivore.
[b] Nest-layer abbreviations: SH = shrub-nesting species; GR = ground-nesting species; CA = subcanopy/canopy-nesting species.
[c] Nest-type abbreviations: O = open; C = cavity; P = parasite.
[d] Includes lightly grazed and fall-grazed treatments.

and Lincoln's Sparrows were ground-nesting species that experienced the greatest reductions in grazed areas. These ground nesters are also dependent on the grass–forb–shrub layer for foraging, making them particularly vulnerable to grazing disturbances (Sedgwick and Knopf 1987).

Canopy-nesting birds were least affected in the short term (Table 12-6). These data support other studies indicating that birds are impacted most by habitat perturbations in the vegetative zone in which they occur (Short and Burnham 1982, Verner 1984). Knopf (1995) noted that in the vertical plain, livestock grazing has little direct impact on birds nesting and foraging in forest canopies. However, cattle trampling and browsing of young trees can limit the number of trees that reach maturity, thus reducing future canopy layers.

Grazing had a differential effect on avian foraging guilds (Table 12-7). Aerial and bark insectivores were probably not greatly affected. Aerial insectivores do not rely upon vegetation for feeding substrates and bark insectivores exploit a substrate generally not used by cattle. In contrast, species dependent upon food resources produced directly (nectarivores) or indirectly (omnivores) by understory plants were less represented in grazed compared to ungrazed treatments. Similar responses to grazing by these guilds were observed on tropical wintering grounds of migratory landbirds (Saab and Petit 1992). Local reductions in nectarivores and omnivores could have widespread ramifications, because these species are important pollinators and seed dispersers, respectively (Feinsinger 1983, Herrera 1984).

Table 12-6. Standardized relative abundances for ground-, shrub-, and canopy-nesting birds in grazed and ungrazed habitats. Sample size is the number of occurrences in which a nest layer was represented in each treatment. Original data were taken from nine studies conducted in western riparian habitats (see Table 12-1). Cavity-nesting species were placed in the canopy layer and Brown-headed Cowbirds were excluded from the analysis. Groups of species in each nest layer whose abundances differ significantly (paired *t*-test, $P < 0.05$) between treatments are indicated by an asterisk.

Nest Layer	Sample Size	Standardized Mean		SE	*P* Value
		Grazed	Ungrazed[a]		
Ground	33	0.71	1.29	0.32	0.01*
Shrub	48	0.86	1.14	0.29	0.33
Subcanopy/canopy	86	1.02	0.98	0.27	0.78

[a] Includes lightly grazed and fall-grazed treatments.

Table 12-7. Standardized relative abundances of seven foraging guilds representing 68 species of Neotropical migrants in grazed and ungrazed habitats. Sample size is the number of occurrences in which a guild was represented in each treatment. Original data were taken from nine studies conducted in western riparian habitats (see Table 12-1). Guilds whose abundances differ significantly (paired *t*-test, $P < 0.05$) between treatments are indicated by an asterisk.

Foraging Guild	Sample Size	Standardized Mean		SE	*P* Value
		Grazed	Ungrazed[a]		
Aerial insectivore	24	1.04	0.96	0.43	0.83
Bark insectivore	4	0.89	1.11	1.30	0.88
Carnivore	4	0.58	1.42	0.71	0.21
Foliage insectivore	49	0.91	1.09	0.31	0.44
Ground insectivore	24	1.21	0.79	0.30	0.03*
Nectarivore	5	0.41	1.59	0.66	0.06
Omnivore	64	0.82	1.18	0.25	0.07

[a] Includes lightly grazed and fall-grazed treatments.

Ground insectivores were better represented in grazed areas. Over half the species in this guild are represented by birds (e.g., American Robin, Killdeer) that are well known for their adoption of human-altered habitats. Ground and aerial insectivores were the most commonly found guilds in grazed riparian habitats at various elevations in the Southwest (Szaro and Rinne 1988).

As a group, long-distance migrants (Status A in Table 12-4) appeared more susceptible (Standardized Means 0.86 vs. 1.13 [grazed vs. ungrazed], $T = -1.64$, $P = 0.10$) to disturbances by livestock grazing than short-distance migrants [status B in Table 12-4; standardized means 0.98 vs 1.01 $\pm$ 0.20 (grazed vs ungrazed), $T = 0.10$, $P = 0.92$]. One explanation for this result could be that many short-distance migrants are cavity nesters, whose nesting sites appear not affected by grazing in the short term. Long-distance migrants also might be more energy stressed upon arrival at the breeding grounds, and thus more vulnerable to human-related disturbances.

MONTANE CONIFEROUS FOREST HABITATS

Characteristics of Forested Habitats

Montane coniferous forests of western North America vary in species composition over broad geographic areas in response to the complex interactions produced by climate, elevation, latitude, soils, and the temporal and spatial pattern of disturbance factors such as fire. A highly simplified characterization for the general pattern of coniferous forest distributions would place juniper and xeric-adapted pine (*Pinus*) woodlands at lower elevations. ponderosa pine (*P. ponderosa*) savannahs at moderate elevations providing slightly less xeric conditions, and Douglas-fir (*Pseudotsuga menziesii*) forests and mixed-conifer associations at higher elevations that typically provide more mesic conditions. Throughout the West, lodgepole pine (*Pinus contorta* or *P. murrayana*) forests occur over a wide range of elevations, typically occupying areas following disturbance.

Engelmann spruce (*Picea engelmannii*) and subalpine fir (*Abies lasiocarpa*) are dominant tree species at high-elevation forests throughout the Rocky Mountains (Peet 1988). Mixed cedar–hemlock–pine (*Thuja–Tsuga–Pinus*) forests and grand-fir (*Abies grandis*) Douglas-fir forests also are common at lower elevations in the northern rockies and interior Northwest.

Jeffrey pine (*Pinus jeffreyi*) dominates mid-montane and lower montane forests in the eastern Sierra. Conifer associations of white fir (*Abies concolor*) incense-cedar (*Libocedrus decurrens*), sugar pine (*P. lambertiana*), ponderosa pine, and Douglas fir comprise the forests along the western slopes of the Sierra (Verner 1980).

Historical Perspective of Coniferous Forests

Many forests of western North America were maintained historically by frequent, low-intensity fires carried by fine herbaceous fuels (Cooper 1960, Vale 1977, Barbour 1988, Peet 1988). This pattern in western landscapes has been disrupted by recent changes in land-use management (Franklin 1988).

Domestic livestock, in conjunction with active fire suppression, produced a widespread transformation of woodlands into denser forests as a result of increased seedling establishment by woody species, and decreased cover of perennial grasses and other herbaceous plants (e.g., Johnsen 1962, Vale 1977, Vankat and Major 1978, West 1988). In at least some areas, livestock grazing apparently was the primary cause of this transformation, with fire suppression of secondary importance (Rummell 1951, Madany and West 1983). One result accompanying this metamorphosis from savannah to forest is the impressive accumulations of fuels in these forests that now produce fires that are often catastrophic and frequently result in widespread tree mortality (Peet 1988).

At low elevations, there has been a strong positive correlation between heavy cattle grazing and extensive expansion of juniper woodlands (Johnsen 1962, West 1988). Intense grazing by domestic cattle reduces cover and vigor of dominant grasses, and facilitates the establishment of juniper while simultaneously reducing the frequency of ground fires that otherwise would eliminate woody vegetation. When excessive grazing virtually removes the grasses, fires can no longer be carried (and those that occurred were actively suppressed) with a resulting dramatic increase in shrub and tree densities (West 1984). The net effect has been an increase in tree density to create woodlands where open savannah previously occurred and an expansion of these denser woodlands into grasslands and shrubsteppe that were degraded by livestock at both higher and lower elevations.

Intense sheep grazing has altered conditions of subalpine and montane meadows (Dunwiddie 1977, Vankat and Major 1978). Lodgepole pine has successfully invaded montane meadows after the removal of sheep grazing (Vankat and Major 1978). Heavy sheep use led to increased runoff, increased erosion, and stream entrenchment, resulting in lowered water tables and drier meadows that facilitated conifer seedling establishment (Dunwiddie 1977, Vankat and Major 1978).

Among the varied altitudinally distributed forest communities dominated by conifers, those found at low and moderate elevations (e.g., juniper–pinyon woodlands and ponderosa pine forests) may have been influenced more by livestock grazing by virtue of their longer, snow-free periods each year. However, this conclusion ignores the extraordinary numbers of sheep that once grazed in the forests and on the rangelands of the western United States. Sheep numbers rose spectacularly between 1865 and 1901, reached a peak in 1910, and have declined steadily since that time (Thilenius 1975). Grazing was year-round with flocks moving upslope as they followed the receding snow line in spring. Although the higher altitude ecosystems were usable for grazing only during the summer or late summer snow-free period, they were subject to extreme grazing pressure for the entire brief growing season or until the forage was depleted, whichever occurred first. The intense levels of livestock grazing that typified the late 1800s resulted in greatly increased tree densities that are still visible today in lodgepole pine and subalpine forests.

Evaluation of Grazing Systems in Coniferous Forests

Little information is available about grazing systems in coniferous forests. Within forested landscapes, the impacts of livestock tend to be concentrated in drainage bottoms, wet meadows, and grassy slopes (Willard 1989), although forested areas are frequently used for bedding and shelter (e.g., Warren and Mysterud 1991). The effects of grazing on these areas varies depending upon climate, elevation, and floristic composition, although generally the result is decreased species diversity and density of herbaceous and shrubby vegetation. Intense grazing pressure sometimes leads to enhanced establishment of conifer seedlings and (in conjunction with fire suppression) consequent conversion of montane shrub, meadow, and grassland areas to forested habitats.

In recent decades, a decrease in sheep numbers and an increase in cattle numbers in the western United States has led to decreased livestock use of many previously used coniferous forest areas at higher elevations. Although their numbers are now much reduced compared to the turn of the century, sheep are still the principal livestock using subalpine ecosystems. Cattle are confined generally to elevations below the subalpine zone, while horses (both recreational and feral) can be significant in wilderness areas (Willard 1989).

Avian Responses to Livestock Grazing in Coniferous Forests

We found no studies that specifically evaluated the influence of livestock grazing on Neotropical migrants using coniferous forests in the western United States. Thus, we can only speculate that the birds most likely to have been negatively affected by livestock grazing were: (1) those species dependent on herbaceous and shrubby ground cover for nesting and/or foraging; and (2) species requiring open savannahs as opposed to closed-canopy forests. Likely examples from the first group include the Hermit Thrush, Nashville Warbler, Fox and Lincoln's Sparrows; examples from the second group include the Lewis' Woodpecker, Violet-green Swallow, and Mountain Bluebird.

A reduction in suitable nesting and foraging habitats for ground-nesting species may be the most likely form of livestock-induced negative impact experienced by migratory landbirds within western coniferous forests. A second, only slightly less direct, impact of livestock grazing is the decreased availability of nesting and roosting cavities in standing snags as a result of the reduction in fire frequency due to diminished fine fuels. The reduction in fire-caused standing snags and in postfire habitat generally must have translated into reduced temporal and spatial availability of potential nest sites for primary and secondary cavity nesters, especially for species such as Mountain Bluebird, which appear to be postfire habitat specialists in parts of their range (Hutto, unpublished MS).

DISCUSSION

Few studies have been conducted that adequately address effects of different grazing systems on Neotropical migratory birds. In only a few cases were we able to discuss the effects of seasonality and intensity of grazing in different vegetative types. Few data exist from outside the breeding season, no quantitative data are available on reproductive or survival success in relation to grazing treatments, and no studies have evaluated the effects of grazing on birds using coniferous forests. In addition, historical conditions of many western ecosystems were altered before studies on effects of land-use activities. Therefore, current habitat conditions rarely reflect true controls from which to evaluate grazing effects on migrant birds and other organisms.

Grazing studies often were difficult to interpret. Some studies included detailed quantitative descriptions of birds and vegetation, but many did not. One worker's concept of moderate grazing at one site might be another's idea of heavy grazing at a different site (Wiens 1973). Thus, there is an urgent need to quantify descriptions of livestock use, bird use, and vegetation.

Nevertheless, introductions of domestic grazers can depress populations of some species while favoring populations of other species. Grazing appeared particularly detrimental to those species dependent on dense ground cover for nesting and/or foraging, and in riparian habitat, species apparently vulnerable to cowbird parasitism. Species occurring in the lower vegetation level that consistently declined with grazing across habitats included Northern Harrier, Short-eared Owl, Common Yellowthroat, and Savannah Sparrow.

Species showing little response in the short term were those that generally forage on open ground and in the air, those nesting and foraging in forest canopies (riparian habitats), and in shrub and riparian habitats, a species directly attracted to livestock (Brown-headed Cowbird). Species that consistently increased with grazing across habitats included Killdeer and Mountain Bluebird. Other species that occurred across habitats showed mixed responses depending on the vegetative type, and timing and duration of grazing.

Because of differential responses by Neotropical migrants to grazing, management programs for single species would be difficult to develop and generally cost prohibitive (Knopf 1995). Management of single species would become necessary only when they become species of special concern and only for the time it takes for species recovery (Hejl et al., Chapter 8, this volume).

An alternative to single species management would be to manage for "ecological guilds." Whereas patterns of bird responses emerged using this approach, at least in riparian habitats, it would be difficult to apply in all situations and would have to be implemented on a habitat specific basis. For example, in grasslands, several species showed a graded response depending on the grassland vegetative type and grazing intensity. Some species (e.g., Upland Sandpiper, Grasshopper Sparrow, and Bobolink) tended to be negatively affected in shorter grasslands (at least with heavy grazing), and responded positively to grazing in taller grasslands (at least with moderate grazing). Thus, we still need a management approach that considers individual species, but one implemented with habitat monitoring on a landscape level (Hejl et al., Chapter 8, this volume). A combination of species and vegetative community monitoring will help in determining whether population declines are caused by local perturbations. If monitoring indicates no changes in vegetation yet populations are declining, that would indicate that local habitat conditions are not responsible for those declines.

Studies of grazing effects on small landbirds have reported exclusively bird abundance data, primarily during the breeding season, and evaluated only localized, short-term consequences of grazing. Land managers' legal mandates require long-term, landscape-level considerations that allow only land-use patterns that maintain natural populations, patterns, and processes. Grazing practices may not cause a great short-term change in some bird populations but we do not know the long-term consequences or whether there are widespread ramifications over the landscape. Data on long-term reproductive success, in a variety of vegetative cover types, over a broad scale would be a better index of the health of bird populations than abundance data in the short term (e.g., Van Horne 1983).

Species habitat requirements may be different from those predicted by information gathered from a limited area and time of the year, where only part of a species' life-history requirements are met (Mosconi and Hutto 1982). The ideal study would include large replicated areas totally protected from grazing for long periods. We need to assess the influence of livestock grazing in areas that are used for migration and overwintering, not merely for breeding habitats. This is particularly important in riparian habitats because they are critical for bird migration corridors (Stevens et al. 1977, Szaro 1980, Knopf et al. 1988a).

Clearly most plant and animal communities in the western United States, excluding some grasslands, have not evolved with widespread grazing repeated annually in the same locations. Thus, heavy grazing is likely to harm many species over the long term. Objectives for public lands must consider many other resources. Recently, the needs of wildlife, recreation, water quality, exemplary natural communities and biodiversity have been incorporated into management plans. For any given management unit, the objectives are apt to be specific to that geographic area, being both more complex and more detailed than was the case historically.

Given the ubiquity of livestock in much of the American West, plants and animals intolerant of activities by livestock grazing have relatively few places left to inhabit. This is undoubtedly true for birds and their habitats, which evolved in the absence of large herbivorous mammals. Protection and restoration of ungrazed habitats resembling their prehistoric counterparts must be fundamental to any conservation plan for Neotropical migrants and all other plant and animal species in western North America.

MANAGEMENT RECOMMENDATIONS

Livestock management decisions about western habitats will affect many Neotropical migrants significantly. Despite limitations to our current knowledge, we offer the following general management recommendations.

Grasslands

First, substantially increase the amount of public rangeland from which all livestock are permanently excluded (Bock et al. 1993). Of particular importance on the Great Plains are the US National Grasslands, which include more than 1.5 million ha presently managed by the US Forest Service largely for livestock production (Lewis 1989, West 1990). Many public rangelands presently are managed by applying some sort of rotational grazing strategy. However, the frequency of rotation is far too high to permit postgrazing ecological succession to proceed to the point where habitat is created for those Neotropical migrant birds (or any other species) generally intolerant of the impacts of grazing mammals. Furthermore, rotational livestock management fails to create heavily grazed habitats that may be required by some species. Therefore, we recommend establishing a system that creates a mosaic of heavily grazed habitats mixed with large (at least 1000 ha), permanent livestock exclosures, which would include a significant portion (perhaps 20%) of public lands presently dedicated to livestock production (Bock et al. 1993).

We are aware of the difficulties involved in designating public land as biological preserves and we recognize the competing interests involved. However, it is also important to recognize the declining agricultural value of many of these lands and their likely increase in value to the public as natural landscape (Popper and Popper 1991). We call only for an effort to restore to these lands something resembling their prehistoric condition. The obvious first step should be to free a portion of these lands from the controlling influence of domestic grazers.

Our second recommendation is to continue a modified version of the Federal Conservation Reserve Program (CRP), to encourage landowners to convert and maintain formerly tilled croplands as grazing lands planted to native vegetation. CRP lands remain vulnerable to recultivation and this decision rightly is in the hands of the landowners. However, it would be ecologically unfortunate if CRP lands were tilled (setting ecological succession back to zero), only to be returned to grassland when crop prices or future government incentives once again make it economically attractive. From the standpoint of indigenous flora and fauna, it would be much better to find ways of making the CRP grasslands valuable to landowners, perhaps by encouraging moderate amounts of livestock grazing. This strategy would be doubly valuable if it could somehow be coupled with creation of permanent livestock exclosures on the public rangelands, including especially the National Grasslands on the Great Plains.

Third, fire should be reintroduced to many grasslands from which it has recently been excluded, and where it is a natural ecological process.

Fourth, caution should be taken in implementing short-duration grazing as a grassland management tool. Short-duration grazing is advocated as a means of increasing livestock production, while improving rangeland condition (Savory 1988). Most field tests of this grazing system have failed to support either claim (e.g., Weltz and Wood 1986, Heitschmidt et al. 1986, Taylor et al. 1993). Furthermore, we found no studies evaluating the impacts of such high-intensity grazing on ground-nesting birds.

Shrubsteppe

First, exclude or significantly reduce livestock grazing. Although we have not documented likely responses of bird populations to this management change, avian communities are expected to respond positively in a landscape that resembles historical conditions. Where livestock are grazed, the short-term goal should be to maintain adequate herbaceous cover to conceal nests through the first incubation period. This could be accomplished by maintaining current season growth through 15 July, and allowing more than 50% (see Pond 1960) of the annual vegetative growth of perennial bunchgrasses to persist through the following nesting season.

Second, restore perennial bunchgrasses,

forbs, shrubs and plant-species diversity to historical levels. Seedings of native species, prescribed burns, and fall–winter grazing must be more carefully controlled to ensure the maintenance of residual plant cover.

Third, avoid fencing and water developments in circumstances where protection is needed for maintenance of plant communities and for population sources of species of special concern. This could result in the concentration of livestock in some areas while creating de facto protected areas in other locations.

Fourth, eliminate the conversion of shrubsteppe habitats to seedings of exotic grasses for the purpose of livestock grazing. Attempts should be made to restore burned areas of shrubsteppe to native vegetation rather than exotic seedings for livestock.

Fifth, determine methods for recovering soil cryptogam layers to increase soil moisture, increase seedling germination, reduce soil erosion, and enhance soil productivity.

Sixth, initiate long-term research that will help us understand the following problems: the direct effects of grazing on shrubsteppe avifauna, the indirect effects on the avifauna mediated through changes in vegetation, the influence of livestock on the distribution of Brown-headed Cowbirds, and the effects of cowbird parasitism on the productivity of breeding birds.

Riparian

First, the condition of riparian areas must be considered critically when implementing grazing systems. Given their scarcity, fragility, and importance to neotropical migrants and other wildlife, western riparian ecosystems should be excluded from livestock grazing wherever possible. Managers should evaluate how local activities alter potential dispersal opportunities for riparian species (Knopf and Samson 1994). Season of use, livestock numbers and livestock distributions must be strictly controlled within riparian zones to implement grazing programs that are compatible with riparian avifauna.

Where livestock must have access to riparian zones for water, restricted-access fencing can localize and minimize their impacts on streambanks and riparian vegetation. Development of alternate water sources also could help reduce concentration of livestock in riparian zones. When the cost of fencing is prohibitive, uplands and riparian zones must be managed together and grazing strategies should be keyed to the condition of the riparian vegetation.

Second, when riparian systems are grazed, moderate use during late fall and winter, or short-term use in spring, will be less damaging than continuous or growing-season grazing. Nevertheless, fall–winter grazing should be carefully controlled to ensure the maintenance of residual plant cover.

Third, degraded riparian habitats may require complete rest from livestock grazing to initiate the recovery process. Four years after cattle removal from riparian habitat in Arizona, understory vegetation and Neotropical migrants showed dramatic increases in abundance (Krueper 1993). In systems requiring long-term rest, the necessary period will be highly variable depending upon the extent of damage and growth rate of regenerating plant species (Clary and Webster 1989). Damaged riparian areas should be rehabilitated by revegetating with native species.

Coniferous Forest

Land managers and field biologists have an unparalleled opportunity to provide information where none currently exists concerning the impacts of grazing on Neotropical migrants in western coniferous forests. Monitoring of migratory landbirds both during the breeding season and in migration, with attention to matched forested habitats differing in grazing regimes or grazing histories could supply much-needed data. Explicit quantitative assessment of grazing pressure and grazing histories, in conjunction with the collection of appropriate vegetation data will be critically important for assessing the relationships between grazing and Neotropical migrants. For breeding season studies. emphasis should be placed on species that nest and/or forage on or near the ground. Migration-period studies should be focused more broadly on the entire suite of species that utilize coniferous forest habitats as staging areas and for foraging activities.

ACKNOWLEDGMENTS

The authors appreciate the helpful suggestions of A. Cruz, W. P. Clary, C. R. Groves, D. Lee, and C. H. Trost. We thank F. L. Knopf, D. M. Finch, H. A. Kantrud, and an anonymous reviewer for constructive comments on earlier versions of this manuscript. B. Fuller provided invaluable assistance in word processing.

LITERATURE CITED

Agnew, W., D. W. Uresk, and R. M. Hansen. 1986. Flora and fauna associated with prairie dog colonies and adjacent ungrazed mixed-grass prairie in western South Dakota. J. Range Manag. 39:135–139.

Allred, B. W. 1941. Grasshoppers and their effect on sagebrush on the Little Powder River in Wyoming and Montana. Ecology 22:387–392.

American Ornithologists' Union. 1983. Check-list of North American birds, 6th ed. Allen Press, Lawrence, KS.

Anderson, R. C. 1982. An evolutionary model summarizing the roles of fire, climate, and grazing animals in the origin and maintenance of grasslands: an end paper. Pp. 297–308 *in* Grasses and grasslands; systematics and ecology (J. R. Estes, R. J. Tyrl, and J. N. Brunken, eds). University of Oklahoma Press, Norman, OK.

Archer, S. 1989. Have southern Texas savannas been converted to woodlands in recent history? Amer. Natur. 134:545–561.

Axelrod, D. I. 1985. Rise of the grassland biome, central North America. Botan. Rev. 51:163–201.

Bahre, C. J. 1991. A legacy of change. University of Arizona Press, Tucson, AZ.

Bailey, R. G. 1976. Ecoregions of the United States (map). USDA Forest Service, Intermountain Region, Ogden, UT.

Bailey, R. G. 1978. Description of the ecoregions of the United States. USDA Forest Service, Intermountain Region, Ogden, UT.

Baker, D. L., and F. S. Guthery. 1990. Effects of continuous grazing on habitat and density of ground-foraging birds in south Texas. J. Range Manag. 43:2–5.

Barbour, M. G. 1988. Californian upland forests and woodlands. Pp. 131–164 *in* North American terrestrial vegetation (M. G. Barbour, and W. D. Billings, eds). Cambridge University Press, New York.

Behnke, R. J. 1979. Values and protection of riparian ecosystems. Pp. 164–167 *in* The mitigation symposium: a national workshop on mitigating losses of fish and wildlife habitats (G. A. Swanson, tech. coord.). USDA Forest Service Gen. Tech. Rep. RM-65.

Behnke, R. J., and R. F. Raleigh. 1979. Grazing and the riparian zone: impact and management perspectives. Pp. 263–267 *in* Strategies for protection and management of flood plain wetlands and other riparian ecosystems (R. R. Johnson, and J. F. McCormick, eds). US Forest Serv. Gen. Tech. Rep. WO-12.

Best, L. B. 1972. First-year effects of sagebrush control on two sparrows. J. Wildl. Manag. 36:534–544.

Blaisdell, J. P., R. B. Murray, and E. D. McArthur. 1982. Managing Intermountain rangelands: sagebrush-grass ranges. USDA Forest Serv. Gen. Tech. Rep. INT-134.

Bock, C. E., and J. H. Bock. 1987. Avian habitat occupancy following fire in a Montana shrubsteppe. Prairie Natur. 19:153–158.

Bock, C. E., and J. H. Bock. 1988. Grassland birds in southeastern Arizona: impacts of fire, grazing, and alien vegetation. Pp. 43–58 *in* Ecology and conservation of grassland birds (P. D. Goriup, ed.). International Council for Bird Preservation, Publ. no. 7, Cambridge, England.

Bock, C. E., and J. H. Bock. 1993. Effects of livestock grazing on cover of perennial grasses in southeastern Arizona. Conserv. Biol. 7:371–377.

Bock, C. E., and B. Webb. 1984. Birds as grazing indicator species in southeastern Arizona. J. Wildl. Manag. 48:1045–1049.

Bock, C. E., J. H. Bock, W. R. Kenney, and V. M. Hawthorne. 1984. Responses of birds, rodents, and vegetation to livestock exclosure in a semidesert grassland site. J. Range Manag. 37:239–242.

Bock, C. E., J. H. Bock, and H. M. Smith. 1993. Proposal for a system of federal livestock exclosures on public rangelands in the Western United States. Conserv. Biol. 7:731–733.

Bock, J. H., and C. E. Bock. 1984. Effects of fires on woody vegetation in the pine–grassland ecotone of the southern Black Hills. Amer. Midland Natur. 112:35–42.

Braun, C. E., M. F. Baker, R. L. Eng, J. S. Gashwiler, and M. H. Schroeder. 1976. Conservation committee report on effects of alteration of sagebrush communities on the associated avifauna. Wilson Bull. 88: 165–171.

Brougham, R. W., and W. Harris. 1967. Rapidity and extent of changes in genotypic structure induced by grazing in a rye grass population. N. Zealand J. Agric. Res. 106–112.

Buffington, L. C., and C. H. Herbel. 1965. Vegetation changes on semidesert grassland range from 1858 to 1963. Ecol. Monogr. 35:139–164.

Butler, D. C. 1979. Effects of a high-density population of ungulates on breeding bird communities in deciduous forest. MS thesis, Colorado State University, Fort Collins, CO.

Carothers, S. W. 1977. Importance, preservation and management of riparian habitat: an overview. Pp. 2–4 *in* Symposium on management of forest and range habitats for nongame birds. USDA Forest Serv. Gen. Tech. Rep. RM-43.

Carothers, S. W., and R. R. Johnson. 1975. Water management practices and their effects on nongame birds in range habitats. Pp. 210–222 *in* Symposium on management of forest and range habitats for nongame birds (D. R. Smith, ed.). USDA Forest Serv. Gen. Tech. Rep. WO-1, Washington, DC.

Castrale, J. S. 1982. Effects of two sagebrush control methods on nongame birds. J. Wildl. Manag. 46:945–952.

Chaney, E., W. Elmore, and W. S. Platts. 1990. Livestock grazing on western riparian areas. Produced for the Environmental Protection Agency by Northwest Resource Information Center, Inc., Eagle, ID.

Clary, W. P., and B. F. Webster. 1989. Managing grazing of riparian areas in the Intermountain Region. USDA Forest Serv. Gen. Tech. Rep. INT-263, Intmt. Res. Sta.

Cody, M. L. 1985. Habitat selection in grassland and open-country birds. Pp. 191–226 *in* Habitat selection in birds (M. L. Cody, ed.). Academic Press, Inc., New York.

Cooper, C. F. 1960. Changes in vegetation, structure, and growth of southwestern pine forests since white settlement. Ecol. Monogr. 30:129–164.

Cooper, H. W. 1953. Amounts of big sagebrush in plant communities near Tensleep, Wyoming, as influenced by grazing treatment. Ecology 34:186–189.

Coppock, D. L., J. E. Ellis, J. K. Detling, and M. I. Dyer. 1983. Plant–herbivore interactions in a North American mixed-grass prairie. Oecologia 56:10–15.

Costello, D. F., and G. T. Turner. 1941. Vegetation changes following exclusion of livestock from grazed ranges. J. For. 39:310–315.

Cottam, W. P., and G. Stewart. 1940. Plant succession as a result of grazing and of meadow desiccation by erosion since settlement. Ecology 34:613–626.

Craddock, G. W., and C. L. Forsling. 1938. The influence of climate and grazing on spring–fall sheep range in southern Idaho. USDA Tech. Bull. 600.

Crouch, G. L. 1982. Wildlife on ungrazed and grazed bottomlands on the South Platte River, Northeastern Colorado. Pp. 188–197 *in* Wildlife–livestock relationships symposium. Forest Wildlife and Range Experiment Station, Moscow, ID.

Crumpacker, D. W. 1984. Regional riparian research and a multi-university approach to the special problem of livestock grazing in the Rocky Mountains and Great Plains. Pp. 413–423 *in* California riparian systems: ecology, conservation, and productive management (R. E. Warner, and K. M. Hendrix, eds). University of California Press, Berkeley, CA.

Daubenmire, R. 1956. Climate as a determinant of vegetation distribution in eastern Washington and northern Idaho. Ecol. Monogr. 26:131–154.

Daubenmire, R. 1970. Steppe vegetation of Washington. Wash. Agric. Exp. Sta. Bull. 62.

Daubenmire, R. 1978. Plant geography. Academic Press, New York.

Dealy, J. E., D. A. Leckenby, and D. M. Concannon. 1981. Wildlife habitats in managed rangelands—The Great Basin of Southeastern Oregon: Plant communities and their importance to wildlife. USDA Forest Serv. Gen. Tech. Rep. PNW-120.

DeGraaf, R. M., V. E. Scott, R. H. Hamre, L. Ernst, and S. H. Anderson. 1991. Forest and rangeland birds of the United States. Agriculture Handbook 688.

Dobkin, D. S. 1993. Neotropical migrants landbirds in the Northern Rockies and Great Plains: a handbook for conservation management. USDA Forest Serv. North. Reg. Publ. no. R1-93-94, Missoula, MT.

Dobkin, D. S., and B. A. Wilcox. 1986. Analysis of natural forest fragments: riparian birds in the Toiyabe Mountains, Nevada. Pp. 293–299 *in* Wildlife 2000: modeling habitat relationships of terrestrial vertebrates (J. Verner, M. L. Morrison, and C. J. Ralph, eds). University Wisconsin Press, Madison, WI.

Duebbert, H. F., and J. T. Lokemoen. 1977. Upland nesting of American bitterns, marsh hawks, and short-eared owls. Prairie Natur. 9:33–40.

Dunning, J. B., Jr, and J. H. Brown. 1982. Summer rainfall and winter sparrow densities: a test of the food limitation hypothesis. Auk 99:123–129.

Dunwiddie, P. W. 1977. Recent tree invasion of subalpine meadows in the Wind River Mountains, Wyoming. Arct. Alp. Res. 9: 393–399.

Dyer, M. I. 1979. Consumers. Pp. 73–86 *in* Grassland ecosystems of the world: analysis of grasslands and their uses (R. T. Coupland, ed.). Cambridge University Press, Cambridge, England.

Ehrlich, P. R., D. S. Dobkin, and D. Wheye. 1988. The birder's handbook—a field guide to the natural history of North American birds. Simon & Schuster Inc., New York.

Ellison, L. 1960. Influence of grazing on plant succession in rangelands. Botan. Rev. 26:1–78.

England, R. C., and A. Devos. 1969. Influence of animals on pristine conditions on the Canadian grasslands. J. Range Manag. 22: 87–94.

Fautin, R. W. 1946. Biotic communities of the northern desert shrub biome in western Utah. Ecol. Monogr. 16:251–310.

Feinsinger, P. 1983. Coevolution and pollination. *In* Coevolution (D. J. Futyma and M. Slatkin, eds). Sinauer Associates, Inc., Sunderland, MA.

Feldhamer, G. A. 1979. Vegetative and edaphic factors affecting abundance and distribution of small mammals in southeast Oregon. Great Basin Natur. 39:207–218.

Franklin, J. F. 1988. Pacific Northwest forests. Pp. 103–130 *in* North American terrestrial vegetation. (M. G. Barbour, and W. D. Billings, eds). Cambridge University Press, New York.

Gano, K. A., and W. H. Rickard. 1982. Small mammals of a bitterbrush–cheatgrass community. Northwest Sci. 56:1–7.

Gauthreaux, S. A. 1992. Preliminary lists of migrants for *Partners in Flight* neotropical migratory bird conservation program. Partners Flight Newslett. 2 (1) (Winter 1992): 30.

Gibson, D. J., and L. C. Hulbert. 1987. Effects of fire, topography and year-to-year climate variation on species composition in tallgrass prairie. Vegetatio 72:175–185.

Gleason, R. S. 1978. Aspects of the breeding biology of burrowing owls in southeastern Idaho, MS thesis, University of Idaho, Moscow, ID.

Glinski, R. L. 1977. Regeneration and distribution of sycamore and cottonwood trees along Sonoita Creek, Santa Cruz County, Arizona. Pp. 116–123 *in* Importance, preservation and management of riparian habitat: a symposium (R. R. Johnson, and D. A. Jones, tech. coords.). USDA Forest Serv. Gen. Tech. Rep. RM-43.

Goebel, C. J., and G. Berry. 1976. Selectivity of range grass seeds by local birds. J. Range Manag. 29:393–395.

Good, E. E., and C. A. Dambach. 1943. Effect of land use practices in breeding bird populations in Ohio. J. Wildl. Manag. 7(3):291–297.

Goodman, T., G. B. Donart, H. E. Kiesling, J. L. Holecheck, J. P. Neel, D. Manzanares, and K. E. Severson. 1989. Cattle behavior with emphasis on time and activity allocations between upland and riparian habitats. Pp. 95–102 *in* Practical approaches to riparian resource management: an educational workshop (R. E. Gresswell, B. A. Barton, and J. L. Kershner, eds). USDI Bureau of Land Management, Billings, MT.

Grayson, D. K. 1977. Paleoclimatic implications of the Dirty Shame Rockshelter mammalian fauna. Tebiwa Misc. Pap. Idaho State Univ. Mus. Nat. Hist. 9.

Gregory, S. V., F. J., Swanson, W. A. McKee, and K. W. Cummins. 1991. An ecosystem perspective of riparian zones. BioScience 41:540–551.

Graul, W. D. 1975. Breeding biology of the mountain plover. Wilson Bull. 87:6–31.

Graul, W. D., and L. E. Webster. 1978. Breeding status of the mountain plover. Condor 78: 265–267.

Gustafson, C. E. 1972. Faunal remains from the Marmes Rockshelter and related archaeological sites in the Columbia Basin. PhD dissertation, Washington State University, Pullman, WA.

Hanley, T. A., and W. W. Brady. 1977. Feral burro impact on a Sonoran Desert range. J. Range Mang. 30:374–377.

Harrison, H. H. 1979. A field guide to western birds' nests. Houghton Mifflin Co., Boston, MA.

Heitschmidt, R. K., S. L. Dowhower, and S. W. Walker. 1986. 14- vs. 42-paddock rotational grazing: aboveground biomass dynamics, forage production, and harvest efficiency. J. Range Manag. 40:216–223.

Henke, M., and C. P. Stone. 1979. Value of riparian vegetation to avian populations along the Sacramento River system. Pp. 228–235 *in* Strategies for protection and management of floodplain wetlands and other riparian ecosystems (R. R. Johnson and J. F. McCormick, eds). USDA Forest Serv. Gen. Tech. Rep. WO-12.

Herrera, C. M. 1984. A study of avian frugivores, bird-dispersed plants, and their interaction in mediterranean scrublands. Ecol. Monogr. 54:1–23.

Hofslund, P. B. 1957. Cowbird parasitism of the Northern Yellow-throat. Auk 74:42–48.

Holechek, J. L., R. D. Piper, and C. H. Herbel. 1989. Range mangement: principles and practices. Prentice-Hall, Englewood Cliffs, NJ.

Humphrey, R. R. 1987. 90 years and 535 miles.

Vegetation changes along the Mexican border. University of New Mexico Press, Albuquerque, NM.

Hutto, R. L. On ecological uniqueness of post-fird bird communities in northern Rocky Mountain forests (unpublished MS).

Johnsgard, P. A. 1979. Birds of the Great Plains. University of Nebraska Press, Lincoln, NB.

Johnsen, T. N., Jr. 1962. One-seed juniper invasion of North Arizona grasslands. Ecol. Monogr. 32:187–207.

Johnson, R. R., L. T. Haight, and J. M. Simpson. 1977. Endangered species *vs.* endangered habitats: a concept. Pp. 68–79 *in* Importance, preservation and management of riparian habitat: a symposium (R. R. Johnson and D. A. Jones, eds). USDA Forest Serv. Gen. Tech. Rep. RM-43.

Kantrud, H. A. 1981. Grazing intensity effects on the breeding avifauna of North Dakota native grasslands. Can. Field Natur. 95:404–417.

Kantrud, H. A., and K. F. Higgins. 1992. Nest and nest site characteristics of some ground-nesting, non-passerine birds of northern grasslands. Prairie Natur. 24:67–84.

Kantrud, H. A., and R. L. Kologiski. 1982. Effects of soils and grazing on breeding birds of uncultivated upland grasslands of the northern Great Plains. USDI Fish Wildlife Serv. Wildl. Res. Rep. no. 15, Washington, DC.

Kauffman, J. B., and W. C. Krueger. 1984. Livestock impacts on riparian ecosystems and streamside management implications: a review. J. Range Manage. 37(5):430–438.

Kauffman, J. B., W. C. Krueger, and M. Vavra. 1982. Impacts of a late season grazing scheme on nongame wildlife habitat in a Wallowa Mountain riparian ecosystem. Pp. 208–220 *in* Wildlife–livestock relationships symposium: proceedings (J. M. Peek and P. D. Dalke, eds). University of Idaho, Forest, Wildl. Range Exp. Sta., Moscow, ID.

Kauffman, J. B., W. C. Krueger, and M. Vavra. 1983. Effects of late season cattle grazing on riparian plant communities. J. Range Manag. 36(6):685–691.

Kennedy, P. B., and S. B. Doten. 1901. A preliminary report on the summer ranges of western Nevada sheep. Nevada Agr. Exp. Sta. Bull. 51.

Kirsch, L. M., and K. F. Higgins. 1976. Upland sandpiper nesting and management in North Dakota. Wildl. Soc. Bull. 4:16–20.

Knopf, F. L. 1985. Significance of riparian vegetation to breeding birds across an altitudinal cline. Pp. 105–111 *in* Riparian ecosystems and their management: reconciling conflicting uses. First North American riparian conference. (R. R. Johnson, C. D. Ziebell, D. R. Patton, P. F. Ffolliott, and R. H. Hamre, tech. coords.). USDA Forest Serv. Gen. Tech. Rep. RM-120.

Knopf, F. 1988. Conservation of steppe birds in North America. Pp. 27–42 *in* Ecology and conservation of grassland birds (P. Goriup, ed.). Internation Council for Bird Preservation, publication no. 7, Cambridge, England.

Knopf, F. L. 1994. Avian assemblages on altered grasslands. Stud. Avian Biol. No. 15:247–257.

Knopf, F. L. 1995. Perspectives on grazing nongame bird habitats. Pp. xx–xx *in* Rangeland wildlife (P. Krausman, ed.). Society of Range Management (in press).

Knopf, F. L., and R. W. Cannon. 1982. Structural resilience of willow riparian community to changes in grazing practices. Pp. 198–207 *in* Wildlife–livestock relationships symposium: proceedings (J. M. Peek and P. D. Dalke, eds.). University of Idaho, Forest. Wildl. Range Exp. Sta., Moscow, ID.

Knopf, F. L., and F. B. Samson. 1994. Scale perspectives on avian diversity in Western riparian ecosystems. Conserv. Biol. 8:669–676.

Knopf, F. L., R. R. Johnson, T. Rich, F. B. Samson, and R. C. Szaro. 1988a. Conservation of riparian ecosystems in the United States. Wilson Bull. 100:272–284.

Knopf, F. L., J. A. Sedgwick, and R. W. Cannon. 1988b. Guild structure of a riparian avifauna relative to seasonal cattle grazing. J. Wildl. Manag. 52(2):280–290.

Knowles, C. J., C. J. Stoner, and S. P. Gieb. 1982. Selective use of black-tailed prairie dog towns by mountain plovers. Condor 84:71–74.

Kochert, M. N. 1989. Responses of raptors to livestock grazing in the western United States. Pp. 194–203 *in* Proc. Western Raptor Manag. Symp. Workshop. National Wildlife Federation, Washington, D.C.

Kovalchik, B. L., and W. Elmore. 1992. Effects of cattle grazing systems on willow-dominated plant associations in central Oregon. Pp. 111–119 *in* Proceedings symposium on ecology and management of riparian shrub communities (W. P. Clary, E. D. McArthur, D. Bedunah, and C. L. Wambolt, (comps). USDA Forest Serv. Gen. Tech. Rep. INT-289.

Krueper, D. J. 1993. Effects of land use practices on western riparian ecosystems. Pp. 321–330 *in* Status and management of neotropical migratory birds (D. M. Finch and P. W. Stangel eds). USDA Forest Serv. Gen. Tech. Rep. RM-229.

Kuchler, A. W. 1964. Potential natural vegetation of the conterminous United States. Am.

Geogr. Soc. Spec. Publ. no. 36, 154 pp. + map.

Larson, F. 1940. The role of the bison in maintaining the short grass plains. Ecology 21:113–121.

Laycock, W. A. 1967. How heavy grazing and protection affect sagebrush-grass ranges. J. Range Manag. 20:206–213.

Laycock, W. A. 1983. Evaluation of management as a factor in the success of grazing systems. Pp. 166–171 *in* Managing Intermountain rangelands—improvement of range and wildlife habitats (S. B. Monsen and N. Shaw, eds). USDA Forest Serv. Gen. Tech. Rep. INT-157.

Leopold, A. S. 1972. The wildlife of Mexico. University of California Press, Berkeley, CA.

Lewis, M. E. 1989. National Grasslands in the Dust Bowl. Geograph. Rev. 79:161–171.

Little, E. L., Jr. 1977. Research in the pinyon–juniper woodland. Pp. 8–19 *in* Ecology, uses and management of pinyon–juniper woodlands (E. F. Aldon and T. J. Loring, eds). USDA Forest Serv. Gen. Tech. Pap. RM-39.

Littlefield, C., S. Thompson, and B. Ehlers. 1984. History and present status of Swainson's Hawk in southeastern Oregon. Raptor Res. 18:1–5.

Lokemoen, J. T., and H. F. Duebbert. 1976. Ferruginous hawk nesting ecology and raptor populations in northern South Dakota. Condor 78:464–470.

MacCracken, J. G., L. E. Alexander, and D. W. Uresk. 1983. An important lichen of southeastern Montana rangelands. J. Range Manag. 36:35–37.

Mack, R. N., and J. N. Thompson. 1982. Evolution in steppe with few large, hooved mammals. Amer. Natur. 119:757–773.

Madany, M. H., and N. E. West. 1983. Livestock grazing–fire regime interactions within montane forests of Zion National Park, Utah. Ecology 64:661–667.

Maher, W. J. 1979. Nesting diets of prairie passerine birds at Matador, Saskatchewan, Canada. Ibis 121:437–452.

Martin, J. W. 1987. Behavior and habitat use of breeding Northern Harriers in southwestern Idaho. J. Raptor Res. 21:57–66.

Martin, T. E. 1993. Nest predation among vegetation layers and habitat types: revising the dogmas. Amer. Natur. 141:897–913.

Martin, T. E., and P. Li. 1992. Life history traits of open- vs. cavity nesting birds. Ecology 73:579–592.

Mayfield, H. F. 1965. The Brown-headed Cowbird, with old and new hosts. Living Bird 4: 13–28.

McArthur, E. D., and B. L. Welch (eds). 1986. Symposium on the biology of *Artemisia* and *Chrysothamnus*. USDA Forest Serv. Gen. Tech. Rep. INT-200, Provo, UT.

McGee, J. M. 1982. Small mammal populations in an unburned and early fire successional sagebrush community. J. Range Manag. 35:177–180.

McKnight, T. L. 1958. The feral burros in the United States: distribution and problems. J. Wildl. Manag. 22:163–179.

McNaughton, S. J. 1979. Grassland–herbivore dynamics. Pp. 46–81 *in* Serengeti: dynamics of an ecosystem. (R. E. Sinclair and M. Norton-Griffiths (eds). University Chicago Press, Chicago, IL.

McNaughton, S. J. 1986. On plants and herbivores. Amer. Natur. 128:765–770.

Medin, D. E. 1986. Grazing and passerine breeding birds in a Great Basin low-shrub desert. Great Basin Natur. 46:567–572.

Medin, D. E. 1990. Birds of a shadscale (*Atriplex confertifolia*) habitat in east central Nevada. Great Basin Natur. 50:295–298.

Medin, D. E., and W. P. Clary. 1990. Bird and small mammal populations in a grazed and ungrazed riparian habitat in Idaho. USDA Forest Serv. Intermt. Res. Sta. Res. Pap. INT-425.

Medin, D. E., and W. P. Clary. 1991. Breeding bird populations in a grazed and ungrazed riparian habitat in Nevada. USDA Forest Serv. Intermt. Res. Sta. Res. Pap. INT-441.

Mengel, R. M. 1970. The North American Central Plains as an isolating agent in bird speciation. Pp. 280–340 *in* Pleistocene and recent environments of the central Great Plains. (W. Dort and J. K. Jones, eds). University of Kansas Press, Lawrence, KS.

Milchunas, D. G., O. E. Sala, and W. K. Lauenroth. 1988. A generalized model of the effects of grazing by large herbivores on grassland community structure. Amer. Natur. 132:87–106.

Mosconi, S. L., and R. L. Hutto. 1982. The effects of grazing on land birds of a western Montana riparian habitat. Pp. 221–233 *in* Wildlife–livestock relationships symposium (J. M. Peek and P. D. Dalke, eds). University of Idaho, Forest Wildl. Range Exp. Sta., Moscow, ID.

Mueggler, W. F. 1950. Effects of spring and fall grazing by sheep on vegetation of the upper Snake River Plains. J. Range Manag. 3:308–315.

Neilson, R. P. 1986. High-resolution climatic analysis and Southwest biogeography. Science 232:27–34.

Nydegger, N. C., and G. W. Smith. 1986. Prey populations in relation to *Artemisia* vegetation types in southwestern Idaho. Pp. 152–156 *in*

Symposium on the biology of *Artemisia* and *Chrysothamnus* (E. D. McArthur and B. L. Welch, eds). USDA Forest Serv. Gen. Tech. Rep. INT-200. Provo, UT.

Ohmart, R. D., and B. W. Anderson. 1986. Riparian habitat. Pp. 169–199 *in* Inventory and monitoring of wildlife habitat (A. Y. Cooperrider, R. J. Boyd, and H. R. Stuart, eds). USDI Bureau of Land Management, Service Center, Denver, CO.

Olson, R. A. 1974. Bird populations in relation to changes in land use *in* Curlew Valley, Idaho and Utah. MS thesis, Idaho State University, Pocatello, ID.

Owens, R. A., and M. T. Myres. 1973. Effects of agriculture upon populations of native passerine birds of an Alberta fescue grassland, Can. J. Zool. 51:697–713.

Page, J. L., N. Dodd, T. O. Osborne, and J. A. Carson. 1978. The influence of livestock grazing on non-game wildlife. Cal. Nev. Wildl. 1978:159–173.

Parrish, T. L., S. H. Anderson, and W. F. Oelklaus. 1993. Mountain Plover habitat selection in the Powder River Basin, Wyoming. Prairie Natur. 25:219–226.

Peet, R. K. 1988. Forests of the Rocky Mountains. Pp. 63–101 *in* North American terrestrial Vegetation. (M. G. Barbour, and W. D. Billings, eds). Cambridge University Press, New York.

Pickford, G. D. 1932. The influence of continued heavy grazing and of promiscuous burning on spring–fall ranges in Utah. Ecology 13:159–171.

Piemiesel, R. L. 1938. Changes in weedy plant cover on cleared sagebrush land and their probable causes. USDA Tech. Bull. 654, Washington, DC.

Platts, W. S. 1981. Effects of livestock grazing. *In* W. R. Meehan (tech. ed.). Influence of forest and rangeland management on anadramous fish habitat in western North America. USDA Forest Serv. Gen. Tech. Rep. PNW-124.

Platts, W. S. 1991. Livestock grazing. Influences of forest and rangeland management on salmonid fishes and their habitats Amer. Fisheries Soc. Spec. Publ. 19:389–423.

Platts, W. S., and R. L. Nelson. 1985. Streamside and upland vegetation use by cattle. Rangelands 7:5–7.

Pond, F. W. 1960. Vigor of Idaho fescue in relation to different grazing intensities. J. Range Manag. 13:28–30.

Popper, F. J., and D. E. Popper. 1991. The reinvention of the American Frontier. Amicus J. 13:4–8.

Poulton, C. E. 1955. Ecology of the non-forested vegetation in Umatilla and Morrow Counties, Oregon, PhD dissertation, Washington State University, Pullman, WA.

Pulliam, H. R., and T. A. Parker, III. 1979. Population regulation in sparrows. Fortschrift Zool. 25:137–147.

Rauzi, F., and C. L. Hanson. 1966. Water uptake and runoff as affected by intensity of grazing. J. Range Manag. 35: 19:351–356.

Reynolds, T. D. 1980. Effects of some different land management practices on small mammal populations. J. Mammalogy 61:558–561.

Reynolds, T. D., and T. D. Rich. 1978. Reproductive ecology of the sage thrasher (*Oreoscoptes montanus*) on the Snake River Plain in south-central Idaho. Auk 95:580–582.

Reynolds, T. D., and C. H. Trost. 1980. The response of native vertebrate populations to crested wheatgrass planting and grazing by sheep. J. Range Manag. 33:122–125.

Reynolds, T. D., and C. H. Trost. 1981. Grazing, crested wheatgrass, and bird populations in southeastern Idaho. Northwest Sci. 55: 225–234.

Rice, J., R. D. Ohmart, and B. W. Anderson. 1983. Turnovers in species composition of avian communities in contiguous riparian habitats. Ecology 64:1444–1455.

Rich, T. D. 1978. Cowbird parasitism of Sage and Brewer's Sparrows. Condor 80:348.

Rich, T. D. 1986. Habitat and nest-site selection by burrowing owls in the sagebrush steppe of Idaho. J. Wildl. Manag. 50:548–555.

Rich, T. D., and S. I. Rothstein. 1985. Sage thrashers reject cowbird eggs. Condor 87: 561–562.

Richard, W. H., and B. E. Vaughan. 1988. Plant communities: characteristics and responses. Pp. 109–179 *in* Shrub-steppe: Balance and change in a semi-arid terrestrial ecosystem (W. H. Rickard, L. E. Rogers, B. E. Vaughan, and S. F. Liebetrau, eds). Elsevier Science Publishers, Amsterdam.

Risser, P. G., E. C. Birney, H. D. Blocker, S. W. May, W. J. Parton, and J. A. Wiens. 1981. The true prairie ecosystem. Hutchinson Ross Publishing Company, Stroudsburg, PA.

Robbins, C. S., J. R. Sauer, and B. G. Peterjohn. 1993. Population trends and management opportunities for Neotropical migrants. Pp. 17–23 *in* Status and management of Neotropical migratory birds (D. M. Finch and P. W. Stangel, eds). USDA Forest Serv. Gen. Tech. Rep. RM-229.

Robertson, J. H. 1971. Changes on a sagebrush–grass range in Nevada ungrazed for 30 years. J. Range Manag. 31:28–34.

Robinson, S. K., J. A. Grzybowski, S. I. Rothstein, M. C. Brittingham, L. J. Petit, and F. R. Thompson. 1993. Management implications of cowbird parasitism on Neotropical migrant songbirds. Pp. 93–102 *in* Status and management of Neotropical migratory birds (D. M. Finch and P. W. Stangel, eds). USDA Forest Serv. Gen. Tech. Rep. RM-229.

Rogers, L. E., and J. D. Hedlung. 1980. A comparison of small mammal populations occupying three distinct shrub-steppe communities in eastern Oregon. Northwest Sci. 54:183–186.

Rogers, L. E., and W. H. Rickard. 1988. Introduction: shrub-steppe lands. Pp. 1–12 *in* Shrub-steppe: balance and change in a semi-arid terrestrial ecosystem (W. H. Rickard, L. E. Rogers, B. E. Vaughan, and S. F. Liebetrau, eds). Elsevier Science Publishers, Amsterdam.

Rotenberry, J. T., and S. T. Knick. 1992. Passerine surveys on the Snake River Birds of Prey Area. Pp. 220–228 *in* Snake River Birds of Prey Area, 1991 Annual Report (K. Steenhof, ed.). USDI Bureau of Land Management, Boise, ID.

Rotenberry, J. T., and J. A. Wiens. 1980a. Temporal variation in habitat structure and shrubsteppe bird dynamics. Oecologia (Berlin) 47:1–9.

Rotenberry, J. T., and J. A. Wiens. 1980b. Habitat structure, patchiness, and avian communities in North American steppe vegetation: a multivariate approach. Ecology 61:1228–1250.

Rothstein, S. I., J. Verner, and E. Stevens. 1980. Range expansion and diurnal changes in dispersion of the Brown-headed Cowbird in the Sierra Nevada. Auk 97:253–267.

Rothstein, S. I., J. Verner, and E. Stevens. 1984. Radio-tracking confirms a unique diurnal pattern of spatial occurrence in the parasitic Brown-headed Cowbird. Ecology 65:77–88.

Rummell, R. S. 1951. Some effects of livestock grazing on ponderosa pine forests and range in central Washington. Ecology 32:594–607.

Ryder, R. A. 1980. Effects of grazing on bird habitats. Pp. 51–66 *in* Management of western forests and grasslands for nongame birds (R. M. Degraff and N. G. Tilghman; comps). USDA Forest Serv. Gen. Tech. Rep. INT-86. Intermountain Forest and Range Experiment Station, Ogden, UT.

Saab, V., and C. Groves. 1992. Idaho's migratory landbirds: description, habitats, and conservation. Idaho Wildl. 12:11–26.

Saab, V. A., and D. R. Petit. 1992. Impact of pasture development on winter bird communities in Belize, Central America. Condor 94:66–71.

Sabadell, J. E. 1982. Desertification of the United States. USDI Bureau of Land Management, Washington, DC.

Sauer, C. O. 1950. Grassland, climax, fire, and man. J. Range Manag. 3:16–22.

Savory, A. 1988. Holistic resource management. Island Press, Washington, DC.

Savory, A., and S. D. Parsons. 1980. The Savory grazing method. Rangelands 2:234–237.

Schlesinger, W. H., J. F. Reynolds, G. L. Cunningham, L. F. Huenneke, W. M. Jarrell, R. A. Virginia, and W. G. Whitford. 1990. Biological feedbacks in global desertification. Science 247:1043–1048.

Schroeder, M. H., and D. L. Sturges. 1975. The effect on the Brewer's sparrow of spraying big sagebrush. J. Range Manag. 28:294–297.

Schroedl, G. F. 1973. The archaeological occurrence of bison in the southern plateau. PhD dissertation. Washington State University, Pullman, WA.

Schulz, T. T., and W. C. Leininger. 1991. Nongame wildlife communities in grazed and ungrazed montane riparian sites. Great Basin Natur. 51:286:292.

Sedgwick, J. A., and F. L. Knopf. 1987. Breeding bird response to cattle grazing of a cottonwood bottomland. J. Wildl. Manag. 51:230–237.

Sedgwick, J. A., and F. L. Knopf. 1988. A high incidence of Brown-headed Cowbird parasitism of Willow Flycatchers. Condor 90:253–256.

Sedgwick, J. A., and F. L. Knopf. 1991. Prescribed grazing as a secondary impact in a western riparian floodplain. J. Range Manag. 44:369–373.

Short, H. L. 1986. Rangelands. Pp. 93–122 *in* Inventory and monitoring of wildlife habitat (A. Y. Cooperrider, R. J. Boyd and H. R. Stuart, eds). USDI Bureau Land Manag. Serv. Center, Denver, CO.

Short, H. L., and K. P. Burnham. 1982. Technique for structuring wildlife guilds to evaluate impacts on wildlife communities. USDI Fish Wildl. Serv. Spec. Sci. Rep. Wildl. 244.

Skinner, R. M. 1975. Grassland use patterns and prairie bird populations of Missouri. Pp. 171–180 *in* Prairie: a multiple view (M. K. Wati, ed.). University of North Dakota Press, Grand Forks, ND.

Skovlin, J. M. 1984. Impacts of grazing on wetlands and riparian habitat: a review of our knowledge. Pp. 1001–1102 *in* Developing strategies for rangeland management. National Research Council/National Academy of Sciences. Westview Press, Boulder, CO.

Smith, D. A., and E. M. Schmutz. 1975. Vegetative changes on protected versus grazed desert grassland ranges in Arizona. J. Range Manag. 28:453–458.

Smith, D. R. 1967. Effects of cattle grazing on a ponderosa pine – bunchgrass range in

Colorado. USDA Forest Serv. Tech. Bull. 1371.

Snyder, N. F., and H. A. Snyder. 1976. Raptors in range habitat. Pp. 190–209 *in* Symposium on management of forest and range habitats for nongame birds. USDA Forest Service, Gen. Tech. Rep. WO-1. Washington, DC.

Springfield, H. W. 1976. Characteristics and management of southwestern pinyon–juniper ranges: the status of our knowledge. USDA Forest Serv. Res. Pap. RM-160.

Stebbins, G. L. 1981. Coevolution of grasses and herbivores. Ann. Missouri Botan. Garden 68:75–86.

Steinauer, E. M., and T. B. Bragg. 1987. Ponderosa pine (*Pinus ponderosa*) invasion of Nebraska Sandhills Prairie. Amer. Midland Natur. 118:358–365.

Stevens, L. E., B. T. Brown, J. M. Simpson, and R. R. Johnson. 1977. The importance of riparian habitat in migrating birds. Pp. 156–164 *in* (R. R. Johnson and D. A. Jones, tech. coords), USDA Forest Serv. Gen. Tech. Rep. RM-43.

Stoddart, L. A., A. D. Smith, and T. W. Box. 1975. Range management. McGraw-Hill Book Co., New York.

Szaro, R. C. 1980. Factors influencing bird populations in southwestern riparian forests. Pp. 403–418 *in* Workshop proceedings management of western forests and grasslands for nongame birds (R. M. DeGraff, tech. coord.). Intermt. Res. Sta. Gen. Tech. Rep. INT-86. USDA Forest Service, Salt Lake City, UT, 11–14 February.

Szaro, R. C., and J. N. Rinne. 1988. Ecosystem approach to management of southwestern riparian communities. Tras. 53rd North Amer. Wildl. Nat. Resources Conf.:502–511.

Taylor, C. A., Jr., N. E. Garza, Jr, and T. D. Brooks. 1993. Grazing systems on the Edwards Plateau of Texas: are they worth the trouble? Rangelands 15:53–60.

Taylor, D. M. 1986. Effects of cattle grazing on passerine bird nesting in riparian habitat. J. Range Manag. 39:254–258.

Taylor, D. M., and C. D. Littlefield. 1986. Willow Flycatcher and Yellow Warbler response to cattle grazing. Amer. Birds 40:1169–1173.

Terborgh, J. 1989. Where have all the birds gone? Princeton University Press, Princeton, NJ.

Tester, J. R., and W. H. Marshall. 1961. A study of certain plant and animal interrelations on a native prairie in northwestern Minnesota. University of Minnesota Museum of Natural History, Occ. Pap. No. 8.

Thilenius, J. F. 1975. Alpine range management in the western United States—principles, practices, and problems: the status of our knowledge. USDA Forest Serv. Res. Pap. RM-157.

Thomas, J. W., C. Maser, and J. E. Rodiek. 1979. Wildlife habitats in managed rangelands—the great basin of southeastern Oregon. USDA Forest Serv. Gen. Tech. Rep. PNW-80.

Tisdale, E. W. 1947. The grasslands of the southern interior of British Columbia. Ecology 28:346–382.

Tisdale, E. W. 1961. Ecological changes in the Palouse. Northwest Sci. 35:134–138.

Tisdale, E. W., and M. Hironaka. 1981. The sagebrush–grass region: a review of the ecological literature. For. Wildl. Range Exp. Sta. Bull. 33, University of Idaho, Moscow, ID.

Tisdale, E. W., M. Hironaka, and M. A. Fosberg. 1969. The sagebrush region in Idaho: a problem in range resource management. Agr. Exp. Sta. Bull. 512, University of Idaho, Moscow, ID.

Udvardy, M. D. F. 1958. Ecological and distributional analysis of North American birds. Condor 60:50–66.

Uresk, D. W. 1984. Black-tailed prairie dog food habits and forage relationships in western South Dakota. J. Range Manag. 37:325–329.

Uresk, D. W., and D. D. Paulson. 1988. Estimating carrying capacity for cattle competing with prairie dogs, and forage utilization in western South Dakota. Pp. 387–390 *in* Management of amphibians, reptiles, and small mammals in North America (R. C. Szaro, K. E. Severson, and D. R. Patton, tech. coord.). USDA Forest Serv. Gen. Tech. Rep. RM-166. Rocky Mt. Forest Range Exp. Sta. Fort Collins, CO.

Urness, P. J. 1979. Wildlife habitat manipulation in sagebrush ecosystems. Pp. 169–178 *in* The sagebrush ecosystem: a symposium. Utah State University Logan, UT.

Vale, T. R. 1975. Presettlement vegetation in the sagebrush-grass area of the Intermountain West. J. Range Manag. 28:32–36.

Vale, T. R. 1977. Forest change in the Warner Mountains, California. Ann. Assoc. Amer. Geogr. 67:28–45.

Van Horne, B. 1983. Density as a misleading indicator of habitat quality. J. Wildl. Manag. 47:893–901.

Vankat, J. L., and J. Major. 1978. Vegetation changes in Sequoia National Park, Cal. J. Biogeogr. 5:377–402.

Verner, J. 1980. Bird communities of mixed-conifer forests of the Sierra Nevada. Pp. 198–223 *in* Workshop proceedings: management of Western forests and grasslands for nongame birds (R. M. DeGraff and N. G. Tilghman, eds) USDA Forest Serv. Gen. Tech. Rep. INT-86.

Verner, J. 1984. The guild concept applied to management of bird populations. Environ. Manag. 8:1–14.

Visher, S. S. 1966. Climatic atlas of the United States. Harvard University Press, Cambridge, MA.

Warren, J. T., and I. Mysterud. 1991. Summer habitat use and activity patterns of domestic sheep on coniferous forest range in southern Norway. J. Range Manag. 44:2–6.

Weatherhead, P. J. 1989. Sex ratios, host-specific reproductive success, and impact of Brown-headed Cowbirds. Auk 106:358–366.

Weltz, M., and M. K. Wood. 1986. Short duration grazing in central New Mexico: effects on infiltration rates. J. Range Manag. 39:365–368.

Webb, E. A., and C. E. Bock. 1990. Relationship of the Botteri's sparrow to sacaton grassland in Southeastern Arizona. Pp. 199–209 *in* Managing Wildlife in the Southwest (P. R. Krausman and N. S. Smith, eds). Arizona Chapter of the Wildlife Society, Phoenix. AZ.

West, N. E. 1979. Basic synecological relationships of sagebrush-dominated lands in the Great Basin and the Colorado Plateau. Pp. 33–41 *in* The sagebrush ecosystem: a symposium. Utah State University, Logan, UT.

West, N. E. 1984. Successional patterns and productivity potentials of pinyon–juniper ecosystems. Pp. 1301–1332 *in* Developing strategies for range management. Westview Press, Boulder, CO.

West, N. E. 1988. Intermountain deserts, shrub steppes, and woodlands. Pp. 209–230 *in* North American terrestrial vegetation (M.G. Barbour and W. D. Billings, eds). Cambridge University Press, New York.

West, T. 1990. USDA Forest Service management of the National Grasslands. Agric. Hist. 64:86–98.

Wiens, J. A. 1973. Pattern and process in grassland bird communities. Ecol. Monogr. 43:237–270.

Wiens, J. A. 1974. Climatic instability and the "ecological saturation" of bird communities in North American grasslands. Condor 76: 385–400.

Wiens, J. A. 1985. Habitat selection in variable environments: shrub-steppe birds. Pp. 227–251 *in* Habitat selection in birds (M. Cody, ed.). Academic Press, London, England.

Wiens, J. A., and M. I. Dyer. 1975. Rangeland avifaunas: their composition, energetics and role in the ecosystem. Pp. 146–182 *in* Symposium on management of forest and range habitats for nongame birds. USDA Forest Serv. Gen. Tech. Rep. WO-1.

Wiens, J. A., and J. T. Rotenberry. 1981a. Censusing and the evaluation of avian habitat occupancy. Pp. 522–532 *in* Estimating numbers of terrestrial birds (C. J. Ralph and J. M. Scott, eds). Stud. Avian Biol. 6.

Wiens, J. A., and J. T. Rotenberry. 1981b. Habitat associations and community structure of birds in shrubsteppe environments. Ecol. Monogr. 51:21–41.

Wiens, J. A., and J. T. Rotenberry. 1985. Response of breeding passerine birds to rangeland alteration in a North American shrubsteppe locality. J. Appl. Ecol. 22:655–668.

Wiens, J. A., J. T. Rotenberry, and B. Van Horne. 1986. A lesson in the limitations of field experiments: shrubsteppe birds and habitat alteration. Ecology 67:365–376.

Willard, E. E. 1989. Use and impact of domestic livestock in whitebark pine forests. Pp. 201–207 *in* Symposium on whitebark pine ecosystems: ecology and management of a high-mountain resource (W. C. Schmidt and K. J. McDonald, eds). USDA Forest Serv. Gen. Tech. Rep. INT-270.

Woodbury, A. M. 1947. Distribution of pigmy conifers in Utah and northeastern Arizona. Ecology 28:113–126.

Yensen, D. L. 1981. The 1900 invasion of alien plants into southern Idaho. Great Basin Natur. 41:176–183.

Young, J. A., and B. A. Sparks. 1985. Cattle in the cold desert. Utah State University Press, Logan, UT.

Young, J. A., R. E. Eckert, Jr, and R. A. Evans. 1979. Historical perspectives regarding the sagebrush ecosystem. Pp. 1–13 *in* The sagebrush ecosystem: a symposium. Utah State University, Logan, UT.

APPENDIX: REVIEW OF GRAZING SYSTEMS USED IN THE WESTERN UNITED STATES

Grazing systems are defined by the season, duration, and intensity of use by a particular type of livestock. Timing, stocking rate (number of animals per unit area), and kind of livestock may have differential effects on habitats used by Neotropical migrants. The following are descriptions of the most commonly used grazing systems.

Continuous Season-Long

This system involves grazing a pasture annually during the growing season, which could be year-round depending on the location. Plants and microhabitats preferred by livestock receive excessive use even with light stocking rates (Holechek et al. 1989). At moderate rates, continuous grazing may have the least impact on the shortgrass steppe of the Great Plains but overall, this grazing system is the most damaging to plant communities (e.g., Kauffman and Krueger 1984, Platts 1991).

Rest Rotation

In this scheme, one pasture receives 12 months of nonuse while other selected pastures are grazed. The period of nonuse is rotated among (usually four) pastures over the cycle (Holechek et al. 1989). An advantage of this system is that the nonuse period allows for plant regeneration of preferred forbs and grasses. Benefits accrued to the ungrazed pastures may be nullified by the higher forage use that occurs on the grazed pastures (Holechek et al. 1989, Platts 1991).

Deferred Rotation

At least one pasture is ungrazed during part of the grazing season and this deferment is rotated among pastures in succeeding years. This scheme is typically used to graze one pasture during the early part of the growing season and remaining pastures later in the season (Holechek et al. 1989). Increased amounts of fencing and moving livestock are required for this system.

Seasonal Suitability

This method involves partitioning an area into pastures based on the condition and type of vegetation (Holechek et al. 1989). The pasture that provides the best nutrition within a particular season is used that time of the year (Platts 1991).

Short Duration

Under this system, pastures are grazed at very high intensities for short durations (Savory and Parsons 1980, Savory 1988). Applying this method in North America has been extremely controversial because it was developed for rangelands in Africa. This system seems particularly inappropriate in the Intermountain West and desert Southwest, where plants evolved in the absence of large herds of grazing ungulates (e.g., Mack and Thompson 1982).

PART V

SCALE PERSPECTIVES

13

HABITAT FRAGMENTATION IN THE TEMPERATE ZONE

JOHN FAABORG, MARGARET BRITTINGHAM, THERESE DONOVAN, AND JOHN BLAKE

INTRODUCTION

The term "habitat fragmentation" has become a buzzword among conservationists and managers in recent years. Although it may sound like a new concept, some wildlife biologists suggest that it is simply a new term for an old concept; they argue that wildlife management has always focused on such factors as habitat distribution, juxtaposition, and edge, which are some of the major factors associated with habitat fragmentation (e.g., Leopold 1933).

Modern studies of habitat fragmentation, however, are very different from classic wildlife management in several ways. First, they deal with many more species than the game species that wildlife managers historically attempted to manipulate. Secondly, these new species of concern often are those whose habitat requirements include late successional habitat types as opposed to classical management, which focused on early successional habitats where game species were abundant. Finally, today's approaches vary greatly in matters of scale; classical wildlife management generally dealt with nonmigratory species, such as quail, turkey, or deer, whose populations could be manipulated on a scale of a few hundred or thousand acres. By contrast, fragmentation studies include species with a variety of migratory strategies and dispersal abilities, generally low fecundity, and often longer life spans (Greenberg 1980). Managing for these species in a particular site may require knowledge about the ecology of the species in a region that not only includes several states, but that also may include two continents, if wintering grounds and migratory pathways are considered—as they must be (Sherry and Holmes 1993).

Modern approaches to habitat fragmentation began with the MacArthur–Wilson equilibrium theory of island biogeography (MacArthur and Wilson 1963, 1967). Although their model was intended as an explanation for species–area relationships on true islands, they noted that similar relationships might exist on mainlands, as habitats that were once large and continuous were broken into smaller isolated pieces (fragments). Within a few years following presentation of this model, various studies tried to describe effects of habitat fragmentation on animal species, often using the MacArthur–Wilson model to explain observed patterns (see references in Rosenfield et al. 1992). Numerous articles reviewing concepts of applied biogeography and fragmentation have appeared in books (Soulé and Wilcox 1980, Soulé 1986, Verner et al. 1986, Simberloff 1988) and recent journals (Saunders et al. 1991). Several books have dealt almost exclusively with fragmentation (Harris 1984, Forman and Godron 1986, Saunders et al. 1987, Shafer 1990, Saunders and Hobbs 1991a). Additionally, reviews dealing with the occurrence and causes of declines in Neotropical migrant populations often have included reviews of fragmentation (e.g., Wilcove et al. 1986, Askins et al. 1990, Wilcove and Robinson 1990, Finch 1991, Robinson and Wilcove 1994). Here, we offer a bibliography listing over 300 fragmentation references, mostly dealing with birds.

In this chapter, we limit our discussion to effects of habitat fragmentation on Neotro-

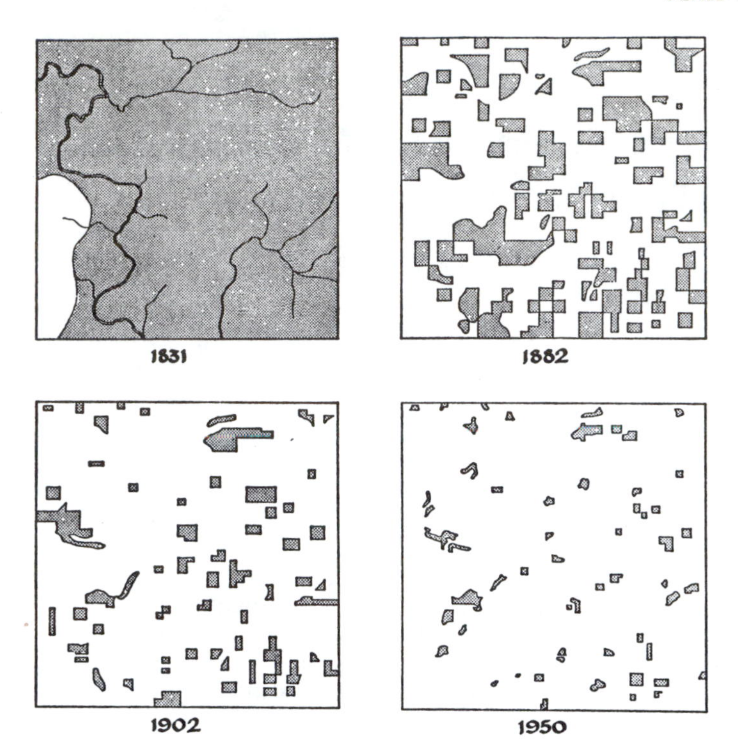

Figure 13-1. Fragmentation of forest habitat in Cadiz Township, Wisconsin, over time. After Curtis (1956).

pical migrants. We focus on permanent habitat fragmentation, where isolated remnants of once broadly occurring vegetation types exist within a matrix of dramatically altered habitat. Although small habitat disturbances within still-homogeneous vegetation types (such as a clearcut in a forested landscape) may show some similarities with fragments, generally we will not deal with such relatively temporary disturbances. We suggest the reader consult the chapter on silviculture in this volume. Other factors associated with fragmentation but covered in more detail elsewhere include cowbird parasitism (Chapter 15), avian demography (Chapter 4), and landscape ecology (Chapter 14).

WHAT IS HABITAT FRAGMENTATION?

Despite its widespread use, "habitat fragmentation" is almost never defined formally; this may be because it is basically a simple concept. Lord and Norton (1990, p. 197) defined fragmentation as "simply the disruption of continuity," which allowed it to apply to any spatial scale. As generally used, habitat fragmentation occurs when a large, fairly continuous tract of a vegetation type is converted to other vegetation types or land uses such that only scattered fragments of the original vegetation type remain. These remnants (also called fragments, isolates, habitat islands, patches, etc.) occupy less area of habitat than the initial condition, are of variable size, shape and location, and are separated by habitats that differ from the original condition. The classic example of fragmentation (Fig. 13-1) shows how large, uniform areas of forest and prairie were broken into small isolates through development of farms and cities (Curtis 1956).

Although island biogeography provided the framework to describe effects of fragmentation, there are some major differences between how birds respond to real islands and habitat islands. For example, the number of land birds on an oceanic island appears to be a function of the size, isolation and habitat characteristics of the island itself, with the factors affecting these species ending abruptly at the coastline (Abbott 1980, Williamson 1981). Birds that occur along the coast generally have completely different ecological requirements from, and do not biologically interact with, interior landbirds. In contrast, habitat islands are surrounded by other terrestrial habitat types, which can support birds that may interact with the birds on habitat islands; further, there are often species of birds adapted to living on the interface (edge) between the two habitat types (Martin 1981a, Blake 1983). Thus, populations adapted to living in large, nonfragmented habitats not only have to survive loss of area as habitat is fragmented, they also must deal with a different set of biological and physical conditions as they come into closer contact with habitat edge. These conditions may include competing bird species, higher levels of nest predation or parasitism, changed microclimatic factors, among others (Ambuel and Temple 1983, Brittingham and Temple 1983, Wilcove 1985, Saunders et al. 1991). Thus, managers dealing with habitat fragmentation not only have to be concerned with reduction in amount of habitat available for the remaining population (quantitative change), but also may have to deal with a whole new set of interactions facing this remnant population (qualitative change). Additionally, persistence of bird populations on a particular fragment may depend upon its proximity to other fragments that might serve as sources of colonists (Chapter 14, this volume).

Thus, although a tract of Missouri oak–hickory forest surrounded by row crops is easily identified as a habitat fragment, many factors interact to determine what species occupy that fragment. Our goal in this chapter is to describe some of the patterns and processes characteristic of bird populations in habitat fragments and to discuss why these occur, how they vary, and what managers can do to minimize detrimental effects. Because these interactions are so complex, we do not have detailed recommendations that apply to all areas. In fact, recent insights into both landscape ecology (Freemark et al., Chapter 14, this volume) and the effects of location within a species' range on its demography (Maurer and Villard 1994, Lawton 1993) suggest that management options will be strongly dependent upon factors that vary within each species' range.

PROBLEMS ASSOCIATED WITH HABITAT FRAGMENTATION

Habitat fragmentation results in both a quantititative and qualitative loss of habitat for the species originally dependent on that habitat type (Temple and Wilcox 1986). As a consequence, abundance and diversity of species originally present in that habitat often declines; losses are greatest in smaller fragments (see Askins et al. 1990, for a review of references). Below we discuss mechanisms that may be responsible for these changes, focusing on forest fragmentation because most work has been done in that habitat. However, fragmentation effects have been noted in other terrestrial habitats as well; these will be noted when appropriate.

Habitat Loss

The most obvious effect of habitat fragmentation is outright quantitative loss of the original habitat type. Organisms directly affected by habitat loss through fragmentation include species with large home ranges or territories, species that depend on specific microhabitats, and species with poor dispersal abilities (Willis 1979, Wilcove et al. 1986, Wilcove 1988, Bierregaard and Lovejoy 1989). When the home range or territory requirements of a species are greater than fragment size, the species may disappear. This may be a factor for raptors, such as the Northern Goshawk or galliforms like the Greater Prairie-Chicken (Herkert et al. 1993); however, most Neotropical migrants have relatively small home ranges and often do not use apparently available habitat on smaller fragments (Wenny et al. 1993). Perhaps the loss of a general habitat type through fragmentation may result in the loss of specific microhabitats within the remaining fragments (see Karr 1982a). As a consequence, species dependent on those microhabitats also may decline or disappear, despite the occurrence of what appears to be available habitat. For example, the Louisiana Waterthrush requires a range of microhabitats, which are generally lacking in small fragments (Wilcove et al. 1986).

Although forest size may be the strongest predictor of species abundance on fragments (Blake and Karr 1987, Askins et al. 1990), it is often not clear how reduced habitat area per se might cause decline in numbers or disappearance from small fragments. Most Neotropical migrants have small territories (< 2 ha), but disappear from fragments tens or even hundreds of times larger than territory size (see section on "Minimum area requirements" below). This suggests that fragmentation produces important qualitative changes in the remaining habitat (Temple and Wilcox 1986).

Increase in Edge Habitat and Edge Effects

As an area is fragmented, the amount of edge relative to interior area and any "edge effects" increase (Temple 1986). Edge is defined as the junction between two dissimilar habitat types or successional stages (Anderson 1991). "Edge effects" are ecological characteristics associated with this junction that positively or negatively affect species living there (Harris 1988, Yahner 1988, Saunders et al. 1991). Edge effects are generally negative for Neotropical migrants that prefer or require forest interior habitats; edge effects reduce the quality of habitat within fragments for these species (Paton 1994).

Determining effects of edge on Neotropical migrants is confounded by the fact that different species respond to edge in very different ways. Some migrants are purely edge species, occurring only where two habitat types or successional stages come together; it is obviously difficult to talk about negative edge effects for species that require edge as breeding habitat. Negative effects are shown more frequently for habitat-interior (i.e., forest interior or prairie interior) species, which have evolved in contiguous habitats that have now become fragmented (Whitcomb et al. 1976). Some of these show edge avoidance or occur at low densities near the habitat edge (Kroodsma 1984, Van Horn et al. 1995), at least in parts of their ranges. Some species apparently use both edge and interior habitats with equal frequency. Thus, creation of edge by fragmentation produces a variety of both positive and negative effects on the abundance of Neotropical migrants even within the same species.

Adverse Effects of Edge for Neotropical Migrants

Traditionally, edge effect has been defined as an increase in abundance and diversity of wildlife along the boundary between two habitat types (Leopold 1933). Many game species are more abundant near edges and, as a consequence, wildlife managers believed that "edge" was good for wildlife; wildlife management often was considered synonymous with creating edge habitat (Harris 1988).

Our concepts of edge and edge effect have changed for several reasons. Most important, our definition of wildlife has expanded to include "nongame" species, many of which evolved within extensive areas of unfragmented habitat away from habitat edges (Temple and Cary 1988). In addition, our method of determining edge effect has changed; instead of simply looking at number and abundance of species, we are using demographic parameters such as reproductive success. This is important because misleading conclusions can be reached if abundance is used as the only measure of habitat quality (Van Horne 1983). For example, a consequence of increased abundance and diversity of wildlife along edges is an increase in adverse biotic interactions, such as predation, brood parasitism, and competition. Below, we discuss specifically how these changes can negatively affect Neotropical migrants.

1. Microclimate or microhabitat change. Alteration of habitats bordering habitat fragments could potentially affect microclimatic conditions along the fragment edge, which could then affect such habitat-related characteristics as succession, habitat structure, etc. Detailed evidence of such microclimatic change has been documented within tropical rainforest study sites

(Lovejoy et al. 1986), where temperature and evaporation rates were higher next to openings than within the continuous forest. Such changes have been suggested to extend less than 30 m into a temperate forest (Wilcove et al. 1986, Saunders et al. 1991), but we know of no research looking at how fragmentation might affect such long-term phenomena as successional characteristics.

2. *High Rates of Nest Predation.* Several species of mammalian and avian nest predators are more abundant along the forest edge than in the forest interior (Bider 1968, Forsyth and Smith 1973, Robbins 1979, Whitcomb et al. 1981). A variety of studies using both artificial and natural nests have been conducted in both forested areas (Gates and Gysel 1978, Wilcove 1985, Small and Hunter 1988, Yahner and Scott 1988) and on prairie fragments (Johnson and Temple 1986, Burger et al. 1994). Most of these tested the idea of Gates and Gysel (1978) that edges may serve as "ecological traps" to some species by offering an enticing distribution of habitat characteristics but exposing the nesting bird to higher predation rates; some purport to have refuted it (Yahner and Wright 1985, Angelstam 1986, Ratti and Reese 1988, and others), but generalizing from these studies is difficult.

A recent review examined existing empirical evidence critically regarding edge effects and predation (Paton 1994). The majority of studies (ten of 14 artificial nest studies and four of seven natural nest studies) found depredation rates higher near edges. The evidence that edge effects usually occur within 50 m of forest edge was strong, whereas studies suggesting that increased depredation rates extend much further into the forest were less convincing.

3. *High Rates of Brood Parasitism.* The Brown-headed Cowbird often is more abundant along forest edges than in forest interior and rates of parasitism usually are higher near forest edges (Brittingham and Temple 1983, also see Robinson et al., Chapter 15, this volume). This is also true for prairie fragments (Johnson and Temple 1986). Edge-related changes in parasitism rates have been reported most frequently from the Midwest in areas where forests have been fragmented by agriculture practices and cowbird abundance is greatest (Brittingham and Temple 1983, Hoover and Brittingham 1993), but they also occur in the west as forests are cleared and livestock operations appear (Rothstein et al. 1980).

Cowbird densities and parasitism rates are not, however, always highest near the edge; these factors may vary depending on the landscape context in which a fragment is situated. For example, Robinson et al. (Chapter 15, this volume) detected high cowbird densities and parasitism rates throughout his forest study sites. Parasitism rates are, in contrast, so low in the Ozarks of southeast Missouri, where cowbird densities are quite low, that any edge effect would be difficult to detect (J. Faaborg, personal observation). Cowbird densities in the fragmented agricultural lands of central Missouri are intermediate to the previous two examples, and are highest around edge and in clearcuts (O'Conner and Faaborg 1992). In a review of published work, Paton (1994) found three of five studies showing clearly that parasitism rates were highest near edge.

Cowbird densities vary with the amount of forest area or perimeter/area ratio on a variety of scales, from local (Donovan et al. 1995) to regional (Hoover and Brittingham, 1993); such variation could affect the strength of cowbird response to edges. At the continental scale, cowbird abundance is positively related to distance from the center of the cowbird's distribution (Thompson et al. 1995). At the regional level, cowbird abundance is related to amount of forest cover and amount of edge available. At the local scale, cowbird abundance is positively related to density of potential hosts. These relationships suggest that a focus on local edge effects oversimplifies patterns of cowbird density and parasitism.

4. *High Rates of Interspecific Competition.* Ambuel and Temple (1983) hypothesized that edge species may outcompete forest interior songbirds. However, no quantita-

tive data currently exist to support this hypothesis. This will, in fact, be a difficult hypothesis to test properly, particularly because some forest interior Neotropical migrant species seem to either avoid edge or have lower densities near edge (Kroodsma 1984, Van Horn et al. 1995).

5. *Reductions in Pairing Success.* Ovenbirds in forest fragments in Missouri, Ontario, and Minnesota had lower chances of attracting a mate when in smaller fragments (Gentry 1989, Gibbs and Faaborg 1990, Villard et al. 1993) and when near the edge rather than in the interior of larger forests (Van Horn et al. 1995). Similarly, lower pairing success was found in the Wood Thrush and Red-eyed Vireo, where selective cutting had created clearings in a formerly continuous forest (Ziehmer 1993).

The apparent avoidance of edge for breeding by females of several species in Missouri may be adaptive, given that this region contained a heterogeneous mix of forest and prairie at the time of European settlement. The greatly reduced pairing success of males on fragments, and the observation that those males on fragments that did attract mates did so later in the summer (Gentry 1989), is more difficult to explain. Females may fill the "best" territories first, using both size of forest and distance from edge as cues. Males that never attract mates then may reflect actual skews in adult sex ratios within these species, perhaps due to a high rate of loss of females on the nest during predation events or simply greater female mortality due to breeding effort. It may reflect female dispersal following nest failure (Jackson et al. 1989). The sex-ratio skew may be a function of sexual differences in dispersal characteristics, compounded by the fragmented nature of the central Missouri habitat. It is assumed that the female is the dispersing sex in most passerine birds (Greenwood and Harvey 1982); if true, it is possible that some females may lose a considerable time in the spring attempting to find one of the few fragments in central Missouri large enough (>300 ha) to support a population of males among the thousands of fragments that exist. If males were more faithful to a particular site, they would simply return to the same site year after year.

6. *Reductions in Nesting Success.* Reduced pairing success and increased predation and parasitism rates can be devastating demographically to Neotropical migrants living in fragments, whether or not these effects are edge related. Temple and Cary (1988) found that only 18% of nests within 100 m of forest edge in Wisconsin were successful, whereas 70% of nests >200 m from an edge were successful. Work by Hoover et al. (1995) and Porneluzi et al. (1993) on SE Pennsylvania forest fragments that ranged in size from 9.2 to >500 ha found that probability of nest success was correlated with forest area. Studies currently underway in the Shawnee Forest of southern Illinois and in other forests in central Missouri are also finding that reproductive rates in many fragments are far below those needed to balance mortality. Data from the Hoosier National Forest of Indiana suggest that edge effects are important, with higher parasitism and nest predation rates around both external edges (where the forest meets agricultural land) and internal edges (such as clearcuts; D. Whitehead, unpublished data). Birds nesting away from these edges achieve high nesting success, as do most birds within the extensive Missouri and Arkansas Ozarks (Martin 1993, Faaborg unpublished data).

How Far Do Edge Effects Extend?

As the above material shows, edge effects are variable in both their occurrence and severity. We will not suggest a universal threshold distance beyond which edge effects diminish because doing so ignores effects of factors that operate at landscape or regional levels (Freemark et al., Chapter 14, this volume). Rather, we suggest that local studies are needed to determine which edge-related factors may occur within a site and at which distance this edge effect is most likely to be felt.

Isolation Effects with Habitat Fragmentation

Relative isolation of a fragment may be detrimental to population survival because dispersal capabilities and characteristics of the habitat separating fragments limit movement. All other things being equal, theory suggests that isolated fragments will support fewer species or lower densities than less isolated fragments (MacArthur and Wilson 1967, Shafer 1990). However, Neotropical migrants seem to have excellent dispersal capabilities and few quantitative data exist that address the role of habitat isolation in population declines (Askins et al., 1990, Opdam 1991, but see Freemark et al., Chapter 14, this volume). More isolated woodlots in the northeast US had fewer Neotropical migrant species than less isolated woodlots of comparable size and vegetation (Lynch and Whigham 1984). Ongoing studies at the Connecticut Arboretum (Askins and Philbrick 1987, Askins et al. 1987) found recent increases in Neotropical migrant numbers, presumably because old fields that once isolated this site had grown into forest. This effect might also occur as a function of increased area of forest throughout the Connecticut region (Askins et al. 1987). On the other hand, several measures of isolation did not account for much variation in number of species in woodlots in east-central Illinois (Blake and Karr 1987), and extremely isolated woodlots in Illinois contained more forest-interior migrants than expected (Robinson 1992). In the fragmented situation found in much of Illinois, migrants might have little choice but to pack into the few remaining forests that occur. Woodlots in South Dakota also showed no isolation effects (Martin 1981a).

The ability of corridors of habitat to connect fragments and to thereby reduce the negative effects of isolation has been hotly debated in the literature (Noss 1987, Simberloff and Cox 1987, Hobbs 1992). Corridors have the potential to increase movement rates between fragments, although few data exist that document such movements for birds (Wegner and Merriam 1979). On the other hand, corridors by their nature are long strips of edge habitat with all their associated problems (Harris and Scheck 1991). Most likely, detailed studies will show that corridors may be critical to species with limited dispersal abilities, such as salamanders, small rodents, etc. (Bennett 1990a,b, Lynch and Saunders 1991) but of less value to good dispersers, e.g., birds and bats.

A better understanding of patterns of Neotropical migrant dispersal is needed to appreciate the role of habitat isolation in determining distributional patterns in fragmented regions. To date, bird banding has been the chief means of determining avian movements, but it has provided few data. A better system for measuring long-distance bird movements is needed.

AVIAN RESPONSES TO HABITAT FRAGMENTATION

The above problems associated with habitat fragmentation do not paint a favorable picture for the success of Neotropical migrants on fragments, but different bird species respond to fragmentation in different ways. Here, we briefly examine patterns by which different species respond to fragmentation in an attempt to gain insight into how habitat management might minimize detrimental effects of fragmentation.

Species–Area Relationships

Recognizing the relationship between species–area patterns on oceanic islands and the equilibrium model, researchers in habitat fragmentation carried out censuses of birds on habitat fragments of varying size within a region (e.g., Bond 1957, Galli et al. 1976, Whitcomb et al. 1976, Robbins 1979, Martin 1980, 1981a, Hayden et al. 1985, Robbins et al. 1989). They found that the number of species on habitat islands increased with increasing habitat size and verified that area per se had the greatest influence on the number of species in a given area; factors such as habitat heterogeneity, degree of isolation, and vegetation structure were relatively unimportant (see Askins et al. 1990). Amount of core area (area of forest more than 100 m from an edge) provided an even better predictor than area because large fragments that lack interior habitat have a species composition similar to small fragments (Temple 1986).

The slope of the species–area curve for a particular habitat may vary regionally, perhaps depending upon where the study was done in relation to the centers of ranges of the species involved (Freemark and Collins 1992). This variation is considered in Chapter 14 of this volume on landscape ecology by Freemark et al. In all situations, however, the number of species increases arithmetically while fragment size increases logarithmically. Thus, it often takes a tenfold increase in habitat size to achieve a doubling in species number.

Area-Sensitive Species

Studies in many areas, including Illinois (Blake and Karr 1984), Missouri (Hayden et al. 1985), Wisconsin (Ambuel and Temple 1983), Maryland (Robbins et al. 1989), New Jersey (Galli et al. 1976), and Ontario (Freemark and Merriam 1986), have shown that forest-dwelling species are not randomly distributed with regard to fragment size. As a group, long-distance (Neotropical) migrants are more abundant on large fragments than on small fragments, and species richness of Neotropical migrants increases more rapidly with fragment size than species richness of short-distance migrants or permanent residents (Whitcomb et al. 1981, Blake and Karr 1984). Those species categorized as requiring forest interior habitat were also more area sensitive than edge or interior/edge species (Whitcomb et al. 1981).

On a species by species basis, the use of either an incidence function approach (Diamond 1975b, Martin 1981a) or an analysis of "nested subsets" of species (Blake 1991) has shown that distributions of species with respect to fragment size are strongly nonrandom. Species that are more abundant on large fragments tend to be Neotropical migrants rather than short-distance migrants or permanent residents, open nesters rather than cavity nesters, and single brooded rather than multiple brooded (Whitcomb et al. 1981, Martin 1988).

Minimum Area Requirements and Source–Sink Dynamics

With the recognition that some species seemed to "require" large areas, attempts were made to determine the minimum area requirement (MAR) of each species within a region. The MAR was defined in a variety of ways, ranging from the "size class of habitat at which the frequency of occurrence undergoes a sharp decline" (Robbins 1979) to "the area in which young can be produced in sufficient numbers to replace adult attrition under the poorest conditions of weather, food availability, competition from other wildlife, and other disturbances" (Robbins et al. 1989).

Published estimates of MARs for a given species can vary widely, depending upon the study technique used and the region under study. For example, the MAR for the Ovenbird was suggested to be as little as 4 ha in New Jersey (Galli et al. 1976), around 300 ha in central Missouri (Hayden et al. 1985), and 2650 ha in eastern Maryland (Robbins 1979).

Most published estimates of MARs were based on presence–absence data from bird censuses (e.g., Hayden et al. 1985, Robbins et al. 1989), without taking nesting success into account. Recent data from a number of regions suggest that many, if not most, populations of area-sensitive species that occur on minimally sized fragments are not producing young at a rate that matches natural mortality rates (Temple and Cary 1988, Robinson 1992). Biologists also did not appreciate the ability of species to colonize fragments continually where production was low or nonexistent. Thus, a true MAR must include information on both demography and dispersal, such that a particular patch of habitat within a given region can maintain a population with some degree of probability over a particular period of time (Shaffer 1981).

To understand regional dynamics of populations in fragments, a variety of source–sink models have been developed (Brown and Kodric-Brown 1977, Pulliam 1988, Howe et al. 1991). A sink population is one that does not produce enough young to balance adult mortality, and which exists because of continued colonization from elsewhere—the "rescue effect" of Brown and Kodric-Brown (1977). A source produces enough young to exceed the number needed to replace the annual mortality of breeding adults. Excess young produced in source populations could populate other fragments through dispersal. It is critical that future studies determine whether or not particular areas within a region are functioning as sources or sinks over time, so that areas of a region that are truly supporting populations can be determined and protected. Managing sink populations, however, may help retain a valuable store house of genetic variation if birds on sinks can successfully reproduce at some point in their lifetime (Howe et al. 1991).

Unfortunately, the theory of source–sink dynamics is well ahead of our empirical knowledge; further insight requires both an understanding of population demography in a wide variety of habitats and knowledge of dispersal characteristics for each species. Suffice it to say that we are a long way from estimating true MARs for species.

HABITAT FRAGMENTATION: MANAGEMENT GUIDELINES

Selecting Fragments to Protect

Despite our lack of knowledge about the details of avian demography and ecology necessary to provide quantitative management guidelines, we can provide general guidelines for providing quality breeding habitat for Neotropical migrant songbirds.

Single Large or Several Small (SLOSS)

One of the first questions that arises is whether, given the choice, it is better to protect one large reserve or several small reserves, whose total area equals the large reserve (Simberloff and Abele 1976). The SLOSS debate initially focused on total number of species preserved and found, in theory, that more species might be preserved on several small reserves than on one large one, given certain assumptions about species distributions (Simberloff 1986). However, the occurrence of area-sensitive species only in habitats of large size makes most of these arguments invalid for Neotropical migrants (e.g., Blake 1991). Rather, the general concensus when managing for breeding Neotropical migrants is that a single large reserve is better than several small reserves (Forman et al. 1976, Wilcove et al. 1986, Askins et al. 1987, 1990).

Size of Reserve

Once a manager has decided to direct conservation efforts towards preserving large fragments, the next question is how large do these fragments need to be. Area requirements vary among species, and even within the same species requirements vary regionally and with surrounding landscape (Temple 1986, Askins et al. 1987, Robbins et al. 1989; see also Chapter 14, this volume, on landscape effects). In a region that still contains large amounts of habitat, knowledge of the minimum fragment size that could support source populations would allow a manager to direct protection efforts at the most important fragments. In a situation where few large fragments exist, that knowledge may be essential in choosing a fragment of the proper size or, in the worst case, managing to make a fragment large enough to support source, or at least stable, populations.

Published MARs might not serve as adequate guidelines to determine the size of fragment required for a region. As noted above, often these are based on presence–absence data. In analyzing a set of 14 area-sensitive species, Robbins et al. (1989) determined that 3000 ha fragments were the minimum size that would retain all species of area-sensitive forest birds in the Mid-Atlantic states. Fragments of 1000 ha in mid-Missouri contain all the expected species of Neotropical migrants found in that region (J. Faaborg, personal observation).

Preliminary data suggest, however, that even 1000 ha fragments in central Missouri consist of sink populations (Donovan, unpublished data). The vast Ozark forests of Missouri (over 2 million acres of contiguous forest) support large Neotropical migrant breeding populations with low parasitism and nest predation rates; this area may be the source for many migrant populations occupying the fragmented parts of Missouri. To truly understand the minimum area required to support a stable or source population necessitates examining fitness components such as reproduction or survival in relation to habitat area (Martin 1992, Hoover et al. 1995).

Although large areas may be needed for the maintenance of all species where annual reproduction exceeds annual mortality, smaller reserves are not without value (Forman et al. 1976, Wilcove et al. 1986, Blake and Karr 1987, Howe et al. 1991). Some migrant species can successfully breed at least occasionally on small fragments, and edge and edge–interior species make extensive use of these fragments (Forman et al. 1976, Blake and Karr 1987). Small fragments are important as stopover and foraging sites during migration (Forman et al. 1976, Blake 1986). Small fragments may serve as important reservoirs for that part of the population that may be unable to find space to breed within source habitats in a given year and thereby contribute to the overall size and stability of the larger metapopulation (Howe et al. 1991). For the goals of many nature reserves, the presence of particular Neotropical migrant species may be of value even if successful reproduction is not occurring, because their presence may increase public awareness of these species.

Shape of Reserve

Fragment shape determines the ratio of edge to interior, with the ratio largest for long narrow fragments and smallest for fragments that are circular or square. Because the reproductive success of many Neotropical migrants is highest within forest interior, the quality of habitat can be strongly influenced by fragment shape. Temple (1986) compared the distribution of area-sensitive migrants in forest fragments of different sizes and shapes. For each area, he recorded total area and core area (area >100 m from a forest edge) and compared species distribution as a function of these two variables. He found that core area was a better predictor of species occurrence than area alone (Fig. 13-2). Consequently, compact shapes that maximize core area are favored over narrow shapes where edge habitat predominates (Temple 1986, Wilcove et al. 1986).

Managing Fragments

In many cases, managers will not be able to choose size, shape, or location of fragments being managed. Instead, the concern may be how to best manage existing fragments. Many of the guidelines for managing fragments and the rationale behind those guidelines are discussed in other chapters. Below, we outline management guidelines and suggest chapters the reader should refer to for more details.

Maximize Area and Amount of Interior

Throughout this chapter and others we have emphasized the importance of forest area and the interior. Guidelines for managing large fragments are similar to guidelines used to manage contiguous forests (see Chapters 7–9, this volume, on silviculture practices). In general, a manager should minimize disturbance within the forest interior. Openings, including roads and power lines, should be concentrated along existing habitat edges. The size of small fragments can be increased by allowing reforestation to occur either through natural regeneration or through planting trees and shrubs. Where possible, managers should select areas for afforestation that will maximize the amount of forest interior. Obviously, the option of afforestation will depend on land use and the ownership adjacent to fragments.

Maximize Vertical Diversity

Forest birds forage and nest at different heights within forest and, in general, the diversity of species present increases with an increase in vertical foliage diversity (MacArthur and MacArthur 1961, MacArthur et al. 1962). Managers can enhance vertical diversity in a number of ways, including planting trees and shrubs, "releasing" smaller trees and shrubs by selectively removing

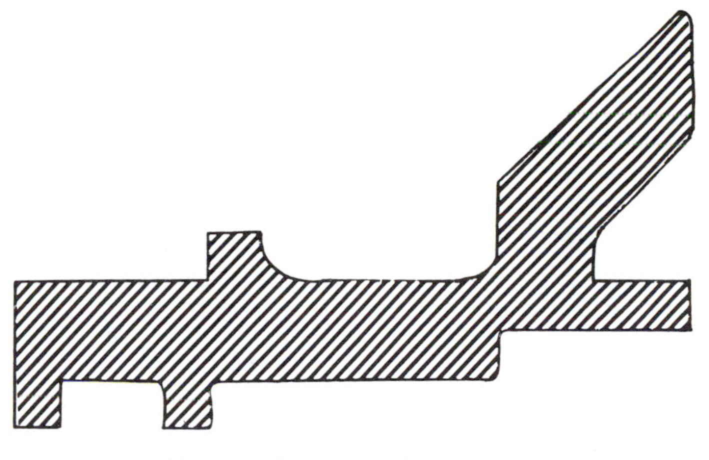

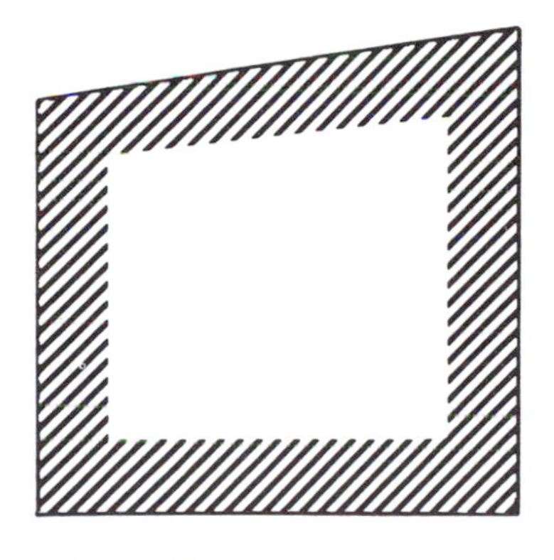

Figure 13-2. Amount of edge and interior habitat on two forest patches of similar size but different shapes. From Temple (1986).

larger trees in direct competition with them, protecting the fragment from grazing by livestock, and by maintaining deer populations at a level which does not result in extensive browsing (see Chapter 7–9, this volume, on silvicultural practices, and Chapter 12, this volume, on livestock grazing).

Use Insecticides Cautiously

Neotropical migrants are primarily insectivorous during the breeding season and the number of young produced has been correlated with the available food supply (see Chapter 4, this volume). Consequently, insecticides that substantially reduce insect

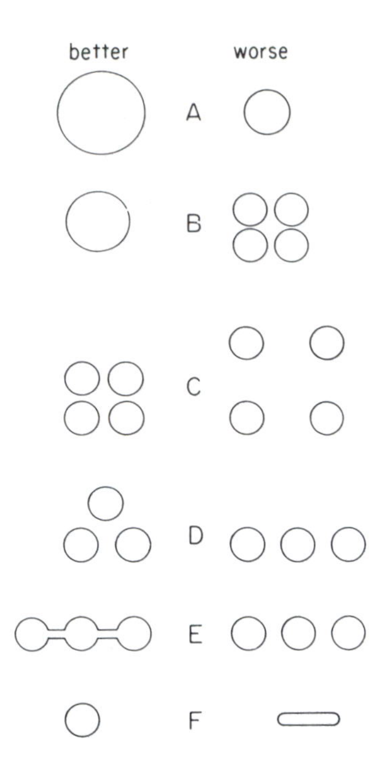

Figure 13-3. Suggested qualitative principles for the selection and management of nature reserves, showing better and worse options with regard to extinction rates. From Diamond (1975a).

populations can have a major impact on subsequent reproductive performance and should not be used on fragments managed for Neotropical migrants (see Chapter 11, this volume, pesticides and environmental contaminants).

Concentrate Recreational Use into Designated Areas

Heavy recreational use and unrestrained pets can have a negative impact on nesting birds, particularly those that nest on or near the ground. When possible, managers should concentrate recreational use along edges and maintain interior areas with little disturbance. Unrestrained dogs should be forbidden during the breeding season.

General Summary

The general guidelines for nature reserve design provided by Diamond (1975a; Fig. 13-3) still best summarize the qualitative approach managers should take in selecting and managing habitat remnants in fragmented environments. All other things being equal, bigger is better than smaller, compact shapes are better than narrow shapes, and reserves closer together or close to a source area are better than widely spaced reserves. As researchers discover the details of Neotropical migrant demography and ecology, they will be able to provide more quantitative suggestions about the details of minimum area requirements and the extent of edge effects on a regional basis.

ACKNOWLEDGMENTS

The authors would like to thank the many students and colleagues who have reviewed this manuscript, including Tom Martin, Deborah Finch, Deanna Dawson, Dirk Burhans, Don Dearborn, Wendy Gram, Paul Porneluzi, and George Wallace. We also thank Scott Robinson for access to his accumulation of fragmentation references.

BIBLIOGRAPHY

Abbott, I. 1980. Theories dealing with the ecology of landbirds on islands. Adv. Ecol. Res. 11:329–371.

Alverson, W. S., D. M. Waller, and S. L. Solheim. 1988. Forest too deer: edge effects in northern Wisconsin. Conserv. Bio. 2:348–358.

Ambrose, J. P., and S. P. Bratton. 1990. Trends in landscape heterogeneity along the borders of Great Smoky Mountains National Park. Conserv. Biol. 4:135–143.

Anderson, S. H. 1991. Managing our wildlife resources, 2nd ed. Prentice-Hall, Englewood Cliffs, NJ.

Ambuel, B., and S. A. Temple. 1983. Area-dependent changes in the bird communities and vegetation of southern Wisconsin forests. Ecology 64:1057–1068.

Anderson, S. H., K. Mann, and H. H. Shugart, Jr. 1977. The effect of transmission-line corridors on bird populations. Amer. Midland Natur. 97:216–221.

Andren, H., and P. Angelstam. 1988. Elevated predation rates as an edge effect in habitat islands: experimental evidence. Ecology 69: 544–547.

Angelstam, P. 1986. Predation on ground-nesting birds' nests in relation to densities and habitat edge. Oikos 47:365–373.

Arguedes, N. 1992. Genetic variation and differentiation in the Ovenbird (*Seiurus aurocapillus*) in central Missouri and Puerto Rico. Unpublished MA thesis, University of Missouri—Columbia, Columbia, MO.

Arnold, G. W., R. A. Maller, and R. Lichfield. 1987. Comparison of bird populations in remnants of wandoo woodland and in adjacent farmland, Austral. Wildl. Res. 14:331–341.

Arnold, G. W., D. E. Steven, and J. R. Weeldenburg. 1993. Influences of remnant size, spacing pattern and connectivity on population boundaries and demography in euros *Macropus robustus* living in a fragmented landscape. Biol. Conserv. 64:219–230.

Ashley, M. V., D. J. Melnick, and D. Western. 1990. Conservation genetics of the Black Rhinoceros (*Diceros bicornis*), I: Evidence from the mitrochondrial DNA of three populations. Conserv. Biol. 4:71–77.

Askins, R. A., and M. J. Philbrick. 1987. Effects of changes in regional forest abundance on the decline and recovery of a forest bird community. Wilson Bull. 99:7–21.

Askins, R. A., M. J. Philbrick, and D. S. Sugeno. 1987. Relationships between the regional abundance of forest and the composition of forest bird communities. Biol. Conserv. 39: 129–152.

Askins, R. A., J. F. Lynch, and R. Greenberg. 1990. Population declines in migratory birds in eastern North America. Curr. Ornithol. 7: 1–57.

Bennett, A. F. 1987. Conservation of mammals within a fragmented forest environment: the contributions of inslular biogeography and autecology. Pp. 41–52 *in* Nature conservation: the role of remnants of native vegetion (D. A. Saunders, G. W. Arnold, A. A. Burbidge, and A. J. M. Hopkins, eds). Surrey Beatty and Sons Pty, Limited, Chipping Norton, Australia.

Bennett, A. F. 1990a. Habitat corridors: their role in wildlife management and conservation. Department of Conservation and Envionment, Melbourne, Australia.

Bennett, A. F. 1990b. Habitat corridors and the conservation of small mammals in a fragmented forest environment. Landscape Ecol. 4:109–122.

Bennett, A. F. 1990c. Land use, forest fragmentation and the mammalian fauna at Naringal, south-western Victoria. Austral. Wildl. Res. 17:325–347.

Bider, J. R. 1968. Animal activity in uncontrolled terrestrial communities as determined by a sand transect technique. Ecol. Monogr. 38: 269–308.

Bierregaard, R. O., Jr, and T. E. Lovejoy. 1989. Effects of forest fragmentation on Amazonian understory bird communities. Acta Amazon. 19:215–241.

Billington, H. L. 1991. Effect of population size on genetic variation in a dioecious conifer. Conserv. Biol. 5:115–119.

Blake, J. G. 1983. Trophic structure of bird communities in forest patches in east-central Illinois. Wilson Bull. 95:416–430.

Blake, J. G. 1986. Species–area relationship of migrants in isolated woodlots. Wilson Bull. 98:291–296.

Blake, J. G. 1987. Species–area relationships of winter resident in isolated woodlots. Wilson Bull. 99:243–252.

Blake, J. G. 1991. Nested subsets and the distribution of birds on isolated woodlots. Conserv. Biol. 5:58–66.

Blake, J. G., and J. R. Karr. 1984. Species composition of bird communities and the conservation benefit of large versus small forests. Biol. Conserv. 30:173–187.

Blake, J. G., and J. R. Karr. 1987. Breeding birds of isolated woodlots: area and habitat relationships. Ecology 68:1724–1734.

Bleich, V. C., J. D. Wehausen, and S. A. Holl. 1990. Desert-dwelling mountain sheep: conservation implications of a naturally fragmented distribution. Conserv. Biol. 4:383–390.

Blouin, M. S., and E. F. Conner. 1985. Is there a best shape for nature reserves? Biol. Conserv. 32:277–288.

Boecklen, W. J. 1986. Effects of habitat heterogeneity on the species–area relationships of forest birds. J. Biogeog. 13:59–68.

Boecklen, W. J., and C. W. Bell. 1987. Consequences of faunal collapse and genetic drift for the design of nature reserves. Pp. 141–149 *in* Nature conservation: the role of remnants of native vegetation (D. A. Saunders, G. W. Arnord, A. A. Burbidge, and A. J. M. Hopkins, eds). Surrey Beatty and Sons, Chipping Norton, Australia.

Boecklen, W. J., and D. Simberloff. 1986. Area-based extinction models in conservation. Pp. 247–276 *in* Dynamics of extinction (D. K. Elliott, ed.). John Wiley & Sons, New York.

Bolger, D. T., A. C. Alberts, and M. E. Soulé. 1991. Occurrence patterns of bird species in habitat fragments: sampling, extinction, and nested species subsets. Amer. Natur. 137: 155–166.

Bond, R. R. 1957. Ecological distribution of breeding birds in the upland forests of southern Wisconsin. Ecol. Monogr. 27:351–384.

Briggs, S. A., and J. H. Criswell. 1978. Gradual silencing of spring in Washington. Atlant. Natur. 32:20–26.

Brittingham, M. C., and S. A. Temple. 1983. Have cowbirds caused forest songbirds to decline? BioScience 33:31–35.

Brown, J. H., and A. Kodric-Brown. 1977. Turnover rates in insular biogeography: effect of immigration on extinction. Ecology 58: 445–449.

Burger, L. D. 1988. Relations between forest and prairie fragmentation and depredation of artificial nests in Missouri. Unpublished MA thesis, University Missouri–Columbia, Columbia, MO.

Burger, L. D., L. W. Burger, and J. Faaborg. 1994. Effects of prairie fragmentation on predation rates of artificial nests. J. Wildl. Manag. 58:249–254.

Burgess, R. L., and D. M. Sharpe, eds. 1981. Forest island dynamics in man-dominated landscapes. Springer-Verlag, New York.

Burkey, T. V. 1989. Extinction in nature reserves: the effect of fragmentation and the importance of migration between reserve fragments. Oikos 55:75–81.

Butcher, G. S., W. A. Niering, W. J. Barry, and R. H. Goodwin. 1981. Equilibrium biogeography and the size of nature preserves: an avian case study. Oecologia 49:29–37.

Carpenter, J. R. 1935. Forest edge birds and exposures of their habitats. Wilson Bull. 47:106–108.

Case, T. J., and M. L. Cody. 1987. Testing theories of island biogeography. Amer. Sci. 75:402–411.

Chasko, G. G., and J. E. Gates. 1982. Avian habitat sutability aong a transmission-line corridor in an oak–hickory forest region. Wildl. Monogr. 82:1–41.

Cieslak, M. 1985. Influence of forest size and other factors on breeding bird species number. Ekol. Polska 33:103–121.

Conner, R. N., and D. C. Rudolph. 1991. Forest habitat loss, fragmentation, and Red-cockaded Woodpecker populations. Wilson Bull. 103:446–457.

Connor, E. F., and E. D. McCoy. 1979. The statistics and biology of the species–area relationship. Amer. Natur. 113:791–833.

Corlett, R. T. 1988. Bukit Timah: the history and signifcance of a small rain-forest reserve. Environ. Conserv. 15:37–44.

Cowling, R. M., and W. J. Bond. 1991. How small can reserves be? An empirical approach in Cape Fynbos, South Africa. Biol. Conserv. 58:243–256.

Cox, J. 1988. The influence of forest size on transient and resident bird species occupying maritime hammocks of northeastern Florida. Florida Field Natur. 16:25–34.

Crawford, H. S., R. G. Hooper, and R. W. Titterington. 1981. Songbird population response to silvicultural practice in central Appalachian hardwoods. J. Wildl. Manag. 45:680–692.

Curtis, J. T. 1956. The modification of mid-latitude greasslands and forests by man. Pp. 721–736 *in* Man's role in changing the face of the earth (W. L. Thomas, ed.). University Chicago Press, Chicago, IL.

Date, E. M., H. A. Ford, and H. F. Recher. 1991. Frugivorous pigeons, steeping stones, and weeds in northern New South Wales. Pp. 241–245 *in* Nature conservation 2: the role of corridors (D. A. Saunders and R. J. Hobbs, eds). Surrey Beatty & Sons, Chipping Norton, Australia.

Diamond, J. M. 1975a. The island dilemma: lessons of modern biogeographic studies for the design of nature reserves. Biol. Conserv. 7:129–146.

Diamond, J. M. 1975b. Assembly of species communities. Pp. 342–444 *in* Ecology and evolution of communities (M. L. Cody and J. M. Diamond, eds). Harvard University Press, Cambridge, MA.

Diamond, J. M. 1976a. Relaxation and differential extinction on landbridge islands: applications to natural preserves. Proc. Internat. Ornithol. Cong. 16:616–628.

Diamond, J. M. 1976b. Island biogeography and conservation: strategy and limitations. Science 193:1027–1029.

Diamond, J. M. 1984. "Normal" extinctions of isolated populations. Pp. 191–246 *in* Extinctions (M. H. Nitecki, ed.). University Chicago Press, Chicago, IL.

Diamond, J. M., K. D. Bishop, and S. Van Balen. 1987. Bird survival in an isolated Javan woodland: island or mirror? Conserv. Biol. 1:132–142.

Dirzo, R., and A. Miranda. 1990. Contemporary Neotropical defaunation and forest structure, function, and diversity—a sequel to John Terborgh. Conserv. Biol. 4:444–447.

Donovan, T. M., F. Thompson, III, and J. Faaborg. 1995. Ecological trade-offs and the influence of scale on Brown-headed Cowbird distribution in fragmented and unfragmented Missouri landscapes. *In* Ecology and management of cowbirds (T. L. Cook, S. Robinson, S. Rothstein, J. Smith, and S. Sealy, eds). University of Texas Press (in press).

Dowsett, R. J. 1980. Extinctions and colonisations in the African montane island avifaunas. Proc. Pan-African Ornithol. Cong. 4:185–197.

East, R. 1981. Species–area curves and populations of large mammals in African savanna reserves. Biol. Conserv. 21:111–126.

East, R., and G. R. Williams. 1984. Island biogeography and the conservation of New Zealand's indigenous forest-dwelling avifauna. N. Zealand J. Ecol. 7:27–35.

Faaborg, J. 1979. Qualitative patterns of avian extinction on Neotropical land-bridge islands: lessons for conservation. J. Appl. Ecol. 16: 99–107.

Faeth, S. H., and E. F. Connor. 1979. Supersaturated and relaxing island faunas: a critique of the species–age relationship. J. Biogeog. 6:311–316.

Fahrig, L., and G. Merriam. 1985. Habitat patch connectivity and population survival. Ecology 66:1762–1768.

Finch, D. M. 1991. Population ecology, habitat requirements, and conservation of Neotropical migratory birds. Gen. Tech. Rep. RM-205, USD Ag Forest Serv. Rocky Mt. Forest Range Exp. Sta., Fort Collins, CO.

Fleming, C. A. 1975. Scientific planning of reserves. Forest Bird 196:15–18.

Forman, R. T. T. 1981. Interaction among landscape elements: a core of landscape ecology. Perspectives in landscape ecology. Proc. 1981 Symp. Netherlands Soc. Landscape Ecol. Veldhoven Pudoc, Wegeningen, The Netherlands.

Forman, R. T. T., and M. Godron. 1986. Landscape ecology. John Wiley and Sons, New York.

Forman, R. T. T., A. E. Galli, and C. F. Leck. 1976. Forest size and avian diversity in New Jersey woodlots with some land use implications. Oecologia 26:1–8.

Forsyth, D. J., and D. A. Smith. 1973. Temporal variability in home ranges of eastern chipmunks (*Tamias striatus*) in a southeastern Ontario woodlot. Amer. Midland Natur. 90:107–117.

Fowler, N. E., and R. W. Howe. 1987. Birds of remnant riparian forests in northeastern Wisconsin. West. Birds 18:77–83.

Freemark, K. E., and B. Collins. 1992. Landscape ecology of birds breeding in temperatue forest fragments. Pp. 443–454 *in* Ecology and conservation of Neotropical migrant landbirds (J. M. Hagan, III and D. W. Johnston, eds). Smithsonian Institution Press, Washington, DC.

Freemark, K. E., and H. G. Merriam. 1986. Importance of area and habitat heterogeneity to bird assemblages in temperate forest fragments. Biol. Conserv. 36:115–141.

Friedmann, H. 1929. The cowbirds: a study in the biology of social parasitism. C. C. Thomas, Springfield, IL.

Galli, A. E., C. F. Leck, and R. T. Forman. 1976. Avian distribution in forest islands of different sizes in central New Jersey. Auk 93:356–364.

Gaston, A. J. 1984. Is habitat destruction in India and Pakistan beginning to affect the status of endemic passerine birds? J. Bombay Natural Hist. Soc. 81:636–641.

Gates, J. E., and L. W. Gysel. 1978. Avian nest dispersion and fledging success in field–forest ecotones. Ecology 59:871–883.

Gentry, R. M. 1989. Variable mating success of the Ovenbird (*Seiurus aurocapillus*) within remnant forest tracts of central Missouri. Unpublished MA thesis, University Missouri–Columbia, Columbia, MO.

Gibbs, J. P. 1991. Avian nest predation in torpical wet forest: an experimental study. Oikos 60:155–161.

Gibbs, J. P., and J. Faaborg. 1990. Estimating the viability of Ovenbird and Kentucky Warbler populations in forest fragments. Conserv. Biol. 4:193–196.

Gilbert, F. S. 1980. The equilibrium theory of island biogeography: fact or fiction? J. Biogeog. 7:209–235.

Gilpin, M. E. 1988. A comment on Quinn and Hastings: extinction in subdivided habitats. Conserv. Biol. 2:290–292.

Gilpin, M. E., and J. M. Diamond. 1980. Subdivision of nature reserves and the maintenance of species diversity. Nature 285:567–569.

Gotelli, N. J., and G. R. Graves. 1990. Body size and the occurrence of avian species on land-bridge islands. J. Biogeog. 17: 315–325.

Gradwohl, J., and R. Greenberg. 1989. Conserving nongame migratory birds: a strategy for monitoring and research. Pp. 297–328 *in* Audubon wildlife report 1989/1990 (W. J. Chandler, ed.). Academic Press, Inc., San Diego, CA.

Greenberg, R. 1980. Demographic aspects of long-distance migration. Pp. 493–504 *in* Migrant birds in the Neotropics: ecology, behavior, distribution, and conservation (A. Keast and E. S. Morton, eds). Smithsonian Institution Press, Washington, DC.

Greenwood, P. J., and P. H. Harvey. 1982. The natal and breeding dispersal of birds. Annu. Rev. Ecol. System. 13:1–21.

Grumbine, R. E. 1990. Viable populations, reserve size, and federal lands management: a critique. Conserv. Biol. 4:127–134.

Haila, Y. 1981. Winter bird communities in the Aland archipelago: an island biogeographic point of view. Holarctic Ecol. 4:174–183.

Haila, Y. 1985. Birds as a tool in reserve planning. Ornis Fennica 62:96–100.

Haila, Y. 1986. North Euopean land birds in forest fragments: evidence for area effects? Pp. 315–319 *in* Wildlife 2000: Habitat relationships of terrestrial vertebrates (J. Verner, M. L. Morrison, and C. J. Ralph, eds). University Wisconsin Press, Madison, WI.

Haila, Y., and I. K. Hanski. 1984. Methodology for studying the effect of habitat fragmentation on land birds. Ann. Zool. Fennici 21:393–397.

Haila, Y., I. K. Hanski, and S. Raivo. 1987. Breeding bird distribution in fragmented coniferous taiga in southern Finland. Ornis Fennica 64:90–106.

Hall, J. B. 1981. Ecological islands in south-eastern Nigeria. Afr. J. Ecol. 19:55–72.

Halme, E., and J. Niemela. 1993. Carabid beetles in fragments of coniferous forest. Ann. Zool. Fennici 30:17–30.

Hamel, P. B., W. P. Smith, and J. W. Wahl. 1993. Wintering bird populations of fragmented forest habitat in the Central Basin, Tennessee. Biol. Conserv. 66:107–115.

Hansen, L., and G. O. Batzli. 1978. The influence of food availability on the white-footed mouse: populations in isolated woodlots. Can. J. Zool. 56:2530–2541.

Hansson, L. 1983. Bird numbers across edges between mature conifer forest and clearcuts in central Sweden. Ornis Scand. 14:97–103.

Harris, L. D. 1984. The fragmented forest: island biogeographic theory and the preservation of biotic diversity. University Chicago Press, Chicago, IL.

Harris, L. D. 1988. Edge effects and conservation of biotic diversity. Conserv. Biol. 2:330–332.

Harris, L. D., and J. Scheck. 1991. From implications to applications: the dispersal corridor principle applied to conservation of biolgoical diversity. Pp. 189–220 *in* Nature conservation the role of corridors (D. A. Saunders and R. J. Hobbs, eds). Surrey Beatty and Sons, Chipping Norton, Australia.

Hayden, T. J., J. Faaborg, and R. L. Clawson. 1985. Estimates of minimum area requirements for Missouri forest birds. Trans. Missouri Acad. Sci. 19:11–22.

Helle, P. 1984. Effects of habitat area on breeding communities in northeastern Finland. Ann. Zool. Fennici 21:421–425.

Helle, P. 1985. Effects of forest fragmentation on bird densities in northern boreal forests. Ornis Fennica 62:35–41.

Herkert, J. R., R. E. Szafoni, V. M. Kleen, and J. E. Schwegman. 1993. Habitat establishment, enhancement and management for forest and grassland birds in Illinois Department of Conservation, Springfield, IL.

Higgs, A. J. 1981. Island biogeography theory and nature reserve design. J. Biogeog. 8:117–124.

Higgs, A. J., and M. B. Usher. 1980. Should nature reserves be large or small? Nature 285:568.

Hobbs, R. J. 1987. Disturbance regimes in remnants of natural vegetation. Pp. 233–240 *in* Nature conservation: the role of remnants of native vegetation (D. A. Saunders, G. W. Arnold, A. A. Burbidge, and A. J. M. Hopkins, eds). Surrey Beatty & Sons, Chipping Norton, Australia.

Hobbs, R. J. 1992. The role of corridors in conservation: solution or bandwagon. Trends Recent Ecol. Evol. 77:389–392.

Hoobs, R. J. 1993. Effects of landscape fragmentation on ecosystem processes in the Western Australian wheatbelt. Biol. Conserv. 64: 193–201.

Hobbs, R. J., and A. J. M. Hopkins. 1990. From frontier to fragments: European impact on Australia's vegetation. Proc. Ecol. Soc. Austral. 16:93–114.

Hobbs, R. J., and A. J. M. Hopkins. 1991. The role of conservation corridors in a changing climate. Pp. 281–290 *in* Nature conservation 2: the role of corridors, (D. A. Saunders and R. J. Hobbs, eds). Surrey Beatty & Sons, Chipping Norton, Australia.

Hoover, J. 1992. Nesting success of Wood Thrush in a fragmented forests. Unpublished MS thesis, Pennsylvania State University, University Park, PA.

Hoover, J. P., and M. C. Brittingham. 1993. Regional variation in cowbird parasitism of Wood Thrush. Wilson Bull. 105:228–238.

Hoover, J. P., M. C. Brittingham, and L. J. Goodrich. 1995. Effect of forest patch size on nesting success of Wood Thrushes. Auk (in press).

Howe, H. F. 1984. Implications of seed dispersal by animals for tropical reserve management. Biol. Conserv. 30:261–281.

Howe, R. W. 1979. Distribution and behaviour of birds on small islands in northern Minnesota. J. Biogeog. 6:379–390.

Howe, R. W. 1984. Local dynamics of bird assemblages in small forest habitat islands in Australia and North America. Ecology 65: 1585–1601.

Howe, R. W., and G. Jones. 1977. Avian utilization of small woodlots in Dane County, Wisconsin. Passenger Pigeon 39:313–319.

Howe, R. W., G. J. Davis, and V. Mosca. 1991. The demographic significance of "sink" populations. Biol. Conserv. 57:239–255.

Humphreys, W. F., and D. J. Kitchener. 1982. The effect of habitat utilization on species—area curves: implications for optimal reserve area. J. Biogeog. 9:391–396.

Hutto, R. L. 1988. Is tropical deforestation responsible for the reported declines in neotropical migrant populations? Amer. Birds 42:375–379.

Hutto, R. L. 1989. The effect of habitat alteration on migratory land birds in a west Mexican tropical deciduous forest: a conservation perspective. Conserv. Biol. 3: 138–148.

Jackson, W. M., S. Rohwer, and V. Nolan, Jr. 1989. Within-season breeding dispersal in Prairie Warblers and other passerines. Condor 91:233– 241.

Janzen, D. 1987. Habitat sharpening. Oikos 48:3–4.

Janzen, D. 1988. There are differences between tropical and extra-tropical national parks. Oikos 51:121–123.

Janzen, D. H. 1983. No part is an island: increase in interference from outside as park size decreases. Oikos 41:402–410.

Janzen, D. H. 1986. The eternal external threat. Pp. 286–303 *in* Conservation biology: the science of scarcity and diversity (M. E. Soulé, ed.). Sinauer Associates, Inc., Sunderland, MA.

Järvinen, O. 1982. Conservation of endangered plant populations: single large or several small reserves? Oikos 38:301–307.

Jennersten, O. 1988. Pollination in *Dianthus deltoides* (Caryophyllaceae): effects of habitat fragmentation on visitation and seed set. Conserv. Biol. 2:359–366.

Johnson, N. K. 1975. Controls of number of bird species on montane islands in the Great Basin. Evolution 29:545–567.

Johnson, R. G., and S. A. Temple. 1986. Assessing habitat quality for birds nesting in fragmented tallgrass prairie. Pp. 245–250 *in* Wildlife 2000: modeling habitat relationships of terrestrial vertebrates (J. Verner, M. L. Morrison, and C. J. Ralph, eds). University Wisconsin Press, Madison, WI.

Johnson, W. C., D. M. Sharpe, D. L. DeAngelis, D. E. Fields, and R. J. Olson. 1981. Modeling seed dispersal and forest island dynamics. Pp. 215–239 *in* Forest island dynamics in man-dominated landscapes (R. L. Burgess and D. M. Sharpe, eds). Springer-Verlag, New York.

Jones, K. B., L. P. Kepner, and T. E. Martin. 1985. Species of reptiles occupying habitat islands in Western Arizona: a deterministic assemblage. Oecologia 66:595–601.

Karieva, P. 1987. Habitat fragmentation and the stability of predator–prey interactions. Nature 326:388–390.

Karr, J. R. 1982a. Avian extinction on Barro Colorado Island, Panama: a reassessment. Amer. Natur. 119:220–239.

Karr, J. R. 1982b. Population variability and extinction in the avifauna of a tropical land bridge island. Ecology 63:1975–1978.

Karr, J. R. 1990. Avian survival rates and the extinction process on Barro Colorado Island, Panama. Conserv. Biol. 4:391–397.

Keller, J. K. 1990. Using aerial photography to model species–habitat relationships: the importance of habitat size and shape. N. York State Mus. Bull. 471:34–46.

Kemp, A. C. 1980. The importance of the Kruger National Park for bird conservation in the Republic of South America. Koedoe 23: 99–122.

Kinnaird, M. F., and T. G. O'Brien. 1991. Viable populations for an endangered forest primate, the Tana River Crested Mangabey (*Cercocebus galeritus galeritus*). Conserv. Biol. 5:203–213.

Kirkpatrick, J. B. 1983. An iterative method for establishing priorities for the selection of nature reserves: an example from Tasmania. Biol. Conserv. 25:127–134.

Kitchener, D. J. 1982. Predictors of vertebrate species richness in nature reserves in the Western Australian wheatbelt. Austral. Wildl. Research 9:1–7.

Kitchener, D. J., J. Dell, B. G. Muir, and M. Palmer. 1982. Birds in Western Australian wheatbelt reserves—implications for conservation. Biol. Conserv. 22:127–163.

Klein, B. C. 1989. Effects of forest fragmentation on dung and carrion beetle communities in central Amazon. Ecology 70:1715–1725.

Kroodsma, R. L. 1982a. Edge effect on breeding forest birds along a power-line corridor. J. Appl. Ecol. 19:361–370.

Kroodsma, R. L. 1982b. Bird community ecology on power-line corridors in east Tennessee. Biol. Conserv. 23:79–94.

Kroodsma, R. L. 1984. Ecological factors associated with degree of edge effect in breeding birds. J. Wildl. Manag. 48:418–425.

Kushlan, J. A. 1979. Design and management of continental wildlife reserves: lessons from the Everglades. Biol. Conserv. 15:281–290.

Laan, R., and B. Verboom. 1990. Effects of pool size and isolation on amphibian communities. Biol. Conserv. 54:251–262.

Lauga, J., and J. Joachim. 1992. Modelling the effects of forest fragmentation on certain species of forest-breeding birds. Landscape Ecol. 6: 183–193.

Laurance, W. F. 1990. Comparative responses of

five arboreal marsupials to tropical forest fragmentation. J. Mammal. 71:641–653.

Laurance, W. F. 1991a. Ecological correlates of extinction proneness in Australia tropical rain forest mammals. Conserv. Biol. 5:79–89.

Laurance, W. F. 1991b. Predicting the impacts of edge effects in fragmented habitats. Biol. Conserv. 55:77–92.

Laurance, W. F. 1991c. Edge effects in tropical forest fragments: application of a model for the design of nature reserves. Biol. Conserv. 57:205–219.

Lawton, J. H. 1993. Range, population abundance and conservation. Trends Ecol. Evol. 8:409–413.

Leck, C. F., B. G. Murray, Jr, and J. Swinebroad. 1988. Long-term changes in the breeding bird populations of a New Jersey forest. Biol. Conserv. 46:145–157.

Leopold, A. 1933. Game management. Scribners, New York.

Lewin, R. 1984. Parks: how big is big enough? Science 225:611–612.

Lewis, D., G. B. Kaweche, and A. Mwenya. 1990. Wildlife conservation outside protected areas—lessons from an experiment in Zambia. Conserv. Biol. 4:171–180.

Loiselle, B. A., and W. G. Hoppes. 1983. Nest predation in insular and mainland lowland rainforest in Panama. Condor 85:93–95.

Loman, J., and T. Von Schantz. 1991. Birds in a farmland—more species in small than in large habitat island. Conserv. Biol. 5:176–188.

Lomolino, M. V. 1986. Mammalian community structure on islands: the importance of immigation, extinction and interactive effects. Biol. J. Linnean Soc. 28:1–21.

Lord, J. M., and D. A. Norton. 1990. Scale and the spatial concept of fragmentation. Conserv. Biol. 4:197–202.

Lovejoy, T. E., and D. C. Oren. 1981. The minimum critical size of ecosystems. Pp. 7–12 *in* Forest island dynamics in man-dominated landscapes (R. L. Burgess and D. M. Sharpe, eds). Springer-Verlag, New York.

Lovejoy, T. E., R. O. Bierregaard, J. M. Rankin, and H. O. R. Schubart. 1993. Ecological dynamics of tropical forest fragments. Pp. 377–384 *in* Tropical rain forest: ecology and management (S. L. Sutton, T. C. Whitmore, and A. C. Chadwick, eds). Blackwell Scientific Publishers, Oxford, England.

Lovejoy, T. E., R. O. Bierregaard, Jr, A. B. Rylands, J. R. Malcolm, C. E. Quintela, L. H. Harper, K. S. Brown, Jr, A. H. Powell, G. V. N. Powell, H. O. R. Schubart, and M. B. Hays. 1986. Edge and other effects of isolation on Amazon forest fragments. Pp. 257–285 *in* Conservation biology: the science of scarcity and diversity (M. E. Soulé, ed.). Sinauer Associates, Sunderland, MA.

Lovett, J. C., and G. W. Norton. 1989. Afromontane rainforest on Malundwe Hill in Mikumi National Park, Tanzania. Biol. Conserv. 48:13–19.

Loyn, R. H. 1987. Effects of patch area and habitat on bird abundances, species numbers and tree health in fragmented Victorian forests. Pp. 65–77 *in* Nature conservation: the role of remnants of native vegetation (D. A. Saunders, G. W. Arnold, A. A. Burbidge, and A. J. M. Hopkins, eds). Surrey Beatty and Sons, Chipping Norton, Australia.

Lynch, J. F. 1987. Responses of breeding bird communities to forest fragmentation. Pp. 123–140 *in* Nature conservation: the role of remnants of native vegetation (D. A. Saunders, G. W. Arnold, A. A. Burbidge, and A. J. M. Hopkins, eds). Surrey Beatty and Sons, Chipping Norton, Australia.

Lynch, J. F., and D. A. Saunders. 1991. Responses of bird species to habitat fragmentation in the wheatbelt of Western Australia: interiors, edges and corridors. Pp. 143–158 *in* Nature conservation 2: the role of corridors (D. A. Saunders and R. J. Hobbs, eds). Surrey Beatty and Sons, Chipping Norton, Australia.

Lynch, J. F., and D. F. Whigham. 1984. Effects of forest fragmentation on breeding bird communities in Maryland, USA. Biol. Conserv. 28:287–324.

MacArthur, R. H., and J. W. MacArthur. 1961. On bird species diversity. Ecology 42:594–598.

MacArthur, R. H., and E. O. Wilson. 1963. An equilibrium theory of insular biogeography. Evolution 17:373–387.

MacArthur, R. H., and E. O. Wilson. 1967. The theory of island biogeography. Princeton University Press, Princeton, NJ.

MacArthur, R. H., J. W. MacArthur, and J. Preer. 1962. On bird species diversity. Part 2, Prediction of bird census from habitat measurements. Amer. Natur. 96:167–174.

Mader, H.-J. 1984. Animal habitat isolation by roads and agricultural fields. Biol. Conserv. 29:81–96.

Mankin, P. C., and R. E. Warner. 1992. Vulnerability of ground nests to predation on an agricultural habitat island in east-central Illinois. Amer. Midland Natur. 128:281–291.

Mannan, R. W., and E. C. Meslow. 1984. Bird population and vegetation characteristics in managed and old-growth forests, northeastern Oregon. J. Wildl. Manag. 48:1219–1238.

Margules, C. R. 1992. The Wog Wog habitat

fragmentation experiment. Environ. Conserv. 19:316–325.

Margules, C., A. J. Higgs, and R. W. Rafe. 1982. Modern biogeographic theory: are there any lessons for nature reserve design? Biol. Conserv. 24:115–128.

Margules, C. R., A. O. Nicholls, and R. L. Pressey. 1988. Selecting networks of reserves to maximise biological diversity. Biol. Conserv. 43:63–76.

Martin, T. E. 1980. Diversity and abundance of spring migratory birds using habitat islands on the Great Plains. Condor 82:430–439.

Martin, T. E. 1981a. Limitation in small habitat islands: chance or competition? Auk 98: 715–734.

Martin, T. E. 1981b. Species-area slopes and coefficients: a caution on their interpretation. Amer. Natur. 118:823–837.

Martin, T. E. 1988. Habitat and area effects on forest bird assemblages: is nest predation an influence? Ecology 69:74–84.

Martin, T. E. 1992. Breeding productivity considerations: what are the appropriate habitat features for management. Pp. 455–473 *in* Ecology and conservation of Neotropical migrant landbirds (J. M. Hagan III and D. W. Johnston, eds). Smithsonian Inst. Press, Washington, DC.

Martin, T. E. 1993. Nest predation among vegetation layers and habitat types: revising the dogmas. Amer. Natur. 141:897–913.

Mauer, B. A., and M.-A. Villard. 1994. Population density. Nat. Geog. Res. Explor. 10:306–317.

May, R. M. 1975. Island biogeography and the design of wildlife preserves. Nature 254: 177–178.

McCoy, E. D. 1982. The application of island-biogeographic theory to forest tracts: problems in the determination of turnover rates. Biol. Conserv. 22:217–227.

McCoy, E. D. 1983. The application of island-biogeographic theory to patches of habitat: how much land is enough? Biol. Conserv. 25:53–61.

McLellan, C. H., A. P. Dobson, D. S. Wilcove, and J. F. Lynch. 1986. Effects of forest fragmentation on New- and Old-World bird communities: empirical observations and theoretical implications. Pp. 305–313 *in* Wildlife 2000: modeling habitat relationships of terrestrial vertebrates (J. Verner, M. L. Morrison, and C. J. Ralph, eds). University Wisconsin Press, Madison, WI.

Menges, E. S. 1991. Seed germination percentage increases with population size in a fragmented prairie species. Conserv. Biol. 5:158–164.

Merriam, G. 1991. Corridors and connectivity: animal populations in heterogeneous environments. Pp. 133–142 *in* Nature conservation 2: the role of corridors (D. A. Saunders and R. J. Hobbs, eds). Surrey Beatty and Sons, Chipping Norton, Australia.

Merriam, G., and A. Lanoue. 1990. Corridor use by small mammals: field measurement for three experimental types of *Peromyscus leucopus*. Landscape Ecol. 4:123–131.

Miller, R. I., and L. D. Harris. 1977. Isolation and extirpations in wildlife reserves. Biol. Conserv. 12:311–315.

Miller, R. I., S. P. Bratton, and P. S. White. 1987. A regional strategy for reserve design and placement based on an analysis of rare and endangered species' distribution patterns. Biol. Conserv. 39:255–268.

Möller, A. P. 1987. Breeding birds in habitat patches: random distribution of species and individuals? J. Biogeog. 14:225–236.

Möller, A. P. 1988. Nest predation and nest site choice in passerine birds in habitat patches of different size: a study of magpies and blackbirds. Oikos 53:215–221.

Moore, N. W., and M. D. Hooper. 1975. On the number of bird species in British woods. Biol. Conserv. 8:239–250.

Murphy, D. D., and B. A. Wilcox. 1986. On island biogeography and conservation. Oikos 47: 385–389.

Murphy, D. D., and B. A. Wilcox. 1986. Butterfly diversity in natural habitat fragments: a test of the validity of vertebrate-based management. Pp. 287–292 *in* Wildlife 2000: Modeling habitat relationships of terrestrial vertebrates (J. Verner, M. L. Morrison, and C. J. Ralph, eds). University of Wisconsin Press, Madison, WI.

Myers, J. P., R. I. G. Morrison, P. Z. Antas, B. A. Harrinton, T. E. Lovejoy, M. Sallaberry, S. E. Senner, and A. Tarak. 1987. Conservation strategy for migratory species. Amer. Sci. 75:19–26.

Naiman, R. J., H. Decamps, and M. Pollock. 1993. The role of riparian corridors in maintaining regional biodiversity. Ecol. Applic. 3:209–212.

Newmark, W. D. 1987. A land-bridge island perspective on mammalian extinctions in western North American parks. Nature 325: 430–432.

Newmark, W. D. 1991. Tropical forest fragmentation and the local extinction of understory birds in the eastern Usambara Mountains, Tanzania. Conserv. Biol. 5:67–78.

Nilsson, S. G. 1986. Are bird communities in small biotope patches random from communities in large patches? Biol. Conserv. 38:179–204.

Noon, B. R., and K. Young. 1991. Evidence of continuing worldwide declines in bird populations: insights from an international conference in New Zealand. Conserv. Biol. 5:141–143.

Norment, C. J. 1991. Bird use of forest patches in the subalpine forest–alpine tundra ecotone of the Beartooth Mountains, Wyoming. Northwest Sci. 65:1–9.

Noss, R. F. 1983. A regional landscape approach to maintain diversity. BioScience 33:700–706.

Noss, R. F. 1987. Corridors in real landscapes: a reply to Simberloff and Cox. Conserv. Biol. 1:159–164.

Noss, R. F. 1991. Effects of edge and internal patchiness on avian habitat use in an old-growth Florida hammock. Natural Areas J. 11:34–47.

O'Conner, R. J., and J. Faaborg. 1992. The relative abundance of the Brown-headed Cowbird (*Molothrus ater*) in relation to exterior and interior edges in forests of Missouri. Trans. Missouri Acad. Sci. 26:1–9.

Opdam, P. 1991. Metapopulation theory and habitat fragmentation: a review of holarctic breeding bird studies. Landscape Ecol. 5:93–106.

Opdam, P., D. van Dorp, and C. J. F. ter Braak. 1984. The effect of isolation on the number of woodland birds in small woods in the Netherlands. J. Biogeog. 11:473–478.

Opdam, P., G. Rijsdijk, and F. Hustings. 1985. Bird communities in small woods in an agricultural landscape: effects of area and isolation. Biol. Conserv. 34:333–352.

Osborne, P. 1984. Bird numbers and habitat characteristics in farmland hedgerows. J. Appl. Ecol. 21:63–82.

Overcash, J. L., J. L. Roseberry, and W. D. Klimstra. 1989. Wildlife openings in the Shawnee National Forest and their contribution to habitat change. Trans. Illinois Acad. Sci. 82:137–142.

Paton, P. W. C. 1994. The effect of edge on avian nest success: how strong is the evidence? Conserv. Biol. 8:17–26.

Patterson, B. D. 1987. The principle of nested subsets and its implications for biological conservation. Conserv. Biol. 1:323–334.

Patterson, B. D. 1990. On the temporal development of nested subset patterns of species composition. Oikos 59:330–342.

Patterson, B. D. 1991. The integral role of biogeographic theory in the conservation of tropical forest diversity. Pp. 124–149 *in* Latin American mammalogy: history, biodiversity, and conservation (M. A. Mares and D. J. Schmidly, eds), University Oklahoma Press, Norman, OK.

Patterson, B. D., and W. Atmar. 1986. Nested subsets and the structure of insular mammalian faunas and archipelagos. Biol. J. Linnnean Soc. 28:65–82.

Patterson, B. D., and J. H. Brown. 1991. Regionally nested patterns of species composition in granivorous rodent assemblages. J. Biogeog. 18:395–402.

Pickett, S. T. A., and J. N. Thompson. 1978. Patch dynamics and the design of nature reserves. Biol. Conserv. 13:27–37.

Picton, H. D. 1979. The application of insular biogeographic theory to the conservation of large mammals in the northern Rocky Mountains. Biol. Conserv. 15:73–79.

Picton, H., and R. J. Mackie. 1980–1981. Single species island biogeography and Montana mule deer. Biol. Conserv. 19:41–49.

Porneluzi, P., J. C. Bednarz, L. J. Goodrich, N. Zawada, and J. Hoover. 1993. Reproductive performance of territorial Ovenbirds occupying forest fragments and a contiguous forest in Pennsylvania. Conserv. Biol. 7:618–622.

Pulliam, H. R. 1988. Sources, sinks, and population regulation. Amer. Natur. 132:652–661.

Quinn, J. F., and A. Hastings. 1987. Extinction in subdivided habitats. Conserv. Biol. 1:198–209.

Quinn, J. F., and A. Hastings. 1988. Extinction in subdivided habitats: reply to Gilpin. Conserv. Biol. 2:293–296.

Quinn, J. F., C. L. Wolin, and M. L. Judge. 1989. An experimental analysis of patch size, habitat subdivision, and extinction in a marine intertidal snail. Conserv. Biol. 3:242–251.

Rafe, R. W., M. B. Usher, and R. G. Jefferson. 1985. Birds on reserves: the influence of area and habitat on species richness. J. Appl. Ecol. 22:327–335.

Ranney, J. W., M. C. Bruner, and J. B. Levenson. 1981. The importance of edge in the structure and dynamics of forest islands. Pp. 67–95 *in* Forest island dynamics in man-dominated landscapes (R. L. Burgess and D. M. Sharpe, eds). Springer-Verlag, New York.

Rapoport, E. H., G. Borioli, J. A. Monjeau, J. E. Puntieri, and R. D. Oviedo. 1986. The design of nature reserves: a simulation trial for assessing specific conservation value. Biol. Conserv. 37:269–290.

Ratti, J. T., and K. P. Reese. 1988. Preliminary test of the ecological trap hypothesis. J. Wildl. Manag. 52:484–491.

Recher, H. F., J. Shields, R. Kavanagh, and G. Webb. 1987. Retaining remnant mature forest for nature conservation at Eden, New South Wales: a review of theory and practice. Pp. 177–194 *in* Nature conservation: the role of remnants of native vegetation (D. A.

Saunders, G. W. Arnold, A. A. Burbidge, and A. J. M. Hopkins, eds). Surrey Beatty and Sons, Chipping Norton, Australia.

Robbins. C. S. 1979. Effect of forest fragmentation on bird populations. Pp. 198–212 *in* Management of northcentral and northeastern forests for nongame birds (R. M. DeGraaf and K. E. Evans, eds). Gen. Tech. Rep. NC-51, USDA Forest Serv. North Central Forest Exp. Sta. St Paul, MI.

Robbins, C. S., D. K. Dawson, and B. A. Dowell. 1989. Habitat area requirements of breeding birds of the Middle Atlantic States. Wildl. Monog. 103:1–34.

Robbins. C. S., B. A. Dowell, D. K. Dawson, J. Colon, F. Espinoza, J. Rodriguez, R. Sutton, and T. Vargas. 1987. Comparison of Neotropical winter bird populations in isolated patches versus extensive forest. Acta Ecol./Oecol. General. 8:285–292.

Robinson, S. K. 1992. Population dynamics of breeding birds in a fragmented Illinois landscape. Pp. 408–418 *in* Ecology and conservation of Neotropical migrant landbirds (J. M. Hagan, III and D. W. Johnston, eds). Smithsonian Institution Press, Washington, DC.

Robinson, S. K., and D. S. Wilcove. 1994. Forest fragmentation in the temperate zone and its effects on migratory songbirds. Bird Conserv. Int. 4:233–249.

Rolstad, J., and P. Wegge. 1987. Distribution and size of Capercaille leks in relation to old forest fragmentation. Oecologia 72:389–394.

Rosenberg, K. V., and M. G. Raphael. 1986. Effects of forest fragmentation on vertebrates in Douglas-fir forests. Pp. 263–272 *in* Wildlife 2000: modeling habitat relationships of terrestrial vertebrates (J. Verner, M. L. Morrison, and C. J. Ralph, eds). University of Wisconsin Press, Madison, WI.

Rosenfield, R. N., C. M. Morasky, J. Bielefeldt, and W. L. Loope. 1992. Forest fragmentation and island biogeography: a summary and bibliography. Tech. Rep. NPS/NRUW/NRTR-92/08, US Dept Interior, National Park Service, Denver, CO.

Rothstein, S. I., J. Verner, and E. Stevens. 1980. Range expansion and diurnal changes in dispersion of the Brown-headed Cowbird in the Sierra Nevada. Auk 97:253–267.

Saunders, D. A. 1989. Changes in the avifauna of a region, district and remnant as a result of fragmentation of native vegetation: the wheatbelt of Western Australia. A case study. Biol. Conserv. 50:99–135.

Saunders, D. A. 1990. Problems of survival in an extensively cultivated landscape: the case of Carnaby's Cockatoo *Calyptorhynchus funereus latirostris*. Biol. Conserv. 54:277–290.

Saunders, D. A. 1993. A community-based observer scheme to assess avian responses to habitat reduction and fragmentation in south-western Australia. Biol. Conserv. 64:203–218.

Saunders, D. A., and C. P. de Rebeira. 1991. Values of corridors to avian populations in a fragmented landscape. Pp. 221–240 *in* Nature conservation 2: the role of corridors (D. A. Saunders and R. J. Hobbs, eds). Surrey Beatty and Sons, Chipping Norton, Australia.

Saunders, D. A., and R. J. Hobbs (eds). 1991a. Nature conservation 2: the role of corridors. Surrey Beatty and Sons, Chipping Norton, Australia.

Saunders, D. A., and R. J. Hobbs. 1991b. The role of corridors in conservation: what do we know and where do we go? Pp. 421–427 *in* Nature conservation 2: the role of corridors (D. A. Saunders and R. J. Hobbs, eds). Surrey Beatty and Sons, Chipping Norton, Australia.

Saunders, D. A., G. W. Arnold, A. A. Burbidge, and A. J. M. Hopkins (eds). 1987. Nature conservation: the role of remnants of native vegetation. Surrey Beatty and Sons, Chipping Norton, Australia.

Saunders, D. A., R. J. Hobbs, and C. R. Margules. 1991. Biological consequences of ecosystem fragmentation: a review. Conserv. Biol. 5:18–32.

Schofield, E. K. 1989. Effects of introduced plants and animals on island vegetation: examples from the Galapagos Archipelago. Conserv. Biol. 3:227–238.

Schwarzkopf, L., and A. B. Rylands. 1989. Primate species richness in relation to habitat structure in Amazonian rainforest fragments. Biol. Conserv. 48:1–12.

Shafer, C. L. 1990. Nature reserves: Island theory and conservation practice. Smithsonian Institution Press, Washington, DC.

Shaffer, M. 1981. Minimum population sizes for species conservation. BioScience 31:131–134.

Sherry, T. W., and R. T. Holmes. 1993. Are populations of Neotropical migrant birds limited in summer or winter? Implications for management. Pp. 47–57 *in* Status and management of Neotropical migratory birds (D. M. Finch and P. W. Stangel, eds). USDA Forest Serv. Gen. Tech. Rep. RM-229.

Simberloff, D. 1976. Species turnover and equilibrium island biogeography. Science 194:572–578.

Simberloff, D. 1986. Are we on the verge of a mass extinction in tropical rain forests? Pp. 165–180 *in* Dynamics of extinction (D. K. Elliot, ed.). John Wiley & Sons, New York.

Simberloff, D. S. 1988. The contribution of

population and community biology to conservation science. Ann. Rev. Ecol. System. 19:473–511.

Simberloff, D., and L. G. Abele. 1976. Island biogeography theory and conservation practice. Science 191:285–286.

Simberloff, D., and L. G. Abele. 1982. Refuge design and island biogeographic theory: effects of fragmentation. Amer. Natur. 120: 41–50.

Simberloff, D., and L. G. Abele. 1984. Conservation and obfuscations: subdivision of reserves. Oikos 42:399–401.

Simberloff, D., and J. Cox. 1987. Consequences and costs of conservation corridors. Conserv. Biol. 1:63–71.

Small, M. F., and M. L. Hunter. 1988. Forest fragmentation and avian nest predation in forested landscapes. Oecologia 76:62–64.

Solheim, S. L., W. S. Alverson, and D. M. Waller. 1987. Maintaining biotic diversity in national forests: the necessity for large blocks of mature forest. Endangered Species 4:1–3.

Soulé, M. E. (ed.). 1986. Conservation biology: the science of scarcity and diversity. Sinauer Associates, Inc., Sunderland, MA.

Soulé, M. E., and D. Simberloff. 1986. What do genetics and ecology tell us about the design of nature reserves? Biol. Conserv. 35:19–40.

Soulé, M. E., and B. A. Wilcox, (eds). 1980. Conservation biology: an evolutionary–ecological perspective. Sinauer Associates, Inc., Sunderland, MA.

Soulé, M. E., B. A. Wilcox, and C. Holtby. 1979. Benign neglect: a model of faunal collapse in the game reserves of East Africa. Biol. Conserv. 15:259–272.

Soulé, M. E., D. T. Bolger, A. C. Alberts, J. Wright, M. Sorice, and S. Hill. 1988. Reconstructed dynamics of rapid extinctions of chaparral-requiring birds in urban habitat islands. Conserv. Biol. 2:75–92.

Stamps, J. A., M. Buechner, and V. V. Krishnan. 1987. The effects of edge permeability and habitat geometry on emigration from patches of habitat. Amer. Natur. 129:533–552.

Telleria, J. L., and T. Santos. 1992. Spatiotemporal patterns of egg predation in forest islands: an experimental approach. Biol. Conserv. 62: 29–33.

Temple, S. A. 1981. Applied island biogeography and the conservation of endangered island birds in the Indian Ocean. Biol. Conserv. 20:147–161.

Temple. S. A. 1986. Predicting impacts of habitat fragmentation on forest birds: a comparison of two models. Pp. 301–304 *in* Wildlife 2000: modeling habitat relationships of terrestrial vertebrates (J. Verner, M. L. Morrison, and C. J. Ralph, eds). University of Wisconsin Press, Madison, WI.

Temple, S. A., and J. R. Cary. 1988. Modeling dynamics of habitat–interior bird populations in fragmented landscapes. Conserv. Biol. 2:340–347.

Temple, S. A., and B. Wilcox. 1986. Predicting effects of habitat patchiness and fragmentation. Pp. 261–262 *in* Wildlife 2000: modeling habitat relationships of terrestrial vertebrates (J. Verner, M. L. Morrison, and C. J. Ralph, eds). University of Wisconsin Press, Madison, WI.

Terborgh. J. 1989. Where have all the birds gone? Princeton University Press, Princeton, NJ.

Thiollay. J.-M. 1988. Forest fragmentation and the conservation of raptors: a survey on the island of Java. Biol. Conserv. 44:229–250.

Thiollay, J.-M. 1989. Area requirements for the conservation of rain forest raptors and game birds in French Guiana. Conserv. Biol. 3:128–137.

Thompson, F. R., S. K. Robinson, and J. Faaborg. 1995. Biogeographic, landscape, and local constraints on cowbirds: the importance of scale to managing brood parasitism. *In* The ecology and management of cowbirds (T. Cook, S. Robinson, S. Rothstein, J. Smith, and S. Sealy, eds). University of Texas Press, Austin, TX (in press).

Van Horn, M. A. 1990. The relationship between edge and the pairing success of the Ovenbird. Unpublished MA thesis, University of Missouri–Columbia, Columbia, MO.

Van Horn, M. A., R. M. Gentry, and J. Faaborg. 1995. Patterns of pairing success of the ovenbird (*Seiurus aurocapillus*) within Missouri forest fragments. Auk (in press).

Van Horne, B. 1983. Density as a misleading indicator of habitat quality. J. Wildl. Manag. 47:893–901.

Verboom, J., A. Schotman, P. Opdam, and J. A. J. Metz. 1991. European nuthatch metapopulations in a fragmented agricultural landscape. Oikos 61:149–156.

Verner, J., M. L. Morrison, and C. J. Ralph. 1986. Wildlife 2000: modeling habitat relationships of terrestrial vertebrates. University of Wisconsin Press, Madison, WI.

Villard, M. A., K. Freemark, and G. Merriam. 1992. Metapopulation dynamics as a conceptual model for Neotropical migrant birds: an empirical investigation. Pp. 474–482 *in* Ecology and conservation of Neotropical migrant landbirds (J. M. Hagan, III and D. W. Johnston, eds). Smithsonian Institution Press, Washington, DC.

Villard, M. A., P. R. Martin, and C. G. Drummond.

1993. Habitat fragmentation and pairing success in the Ovenbird (*Seiurus aurocapillus*). Auk 110:759–768.

Weaver, M., and M. Kellman. 1981. The effects of forest fragmentation on woodlot tree biota in Southern Ontario. J. Biogeog. 8:199–210.

Wegner, J. F., and G. Merriam. 1979. Movements by birds and small mammals between a wood and adjoining farmland habitats. J. Appl. Ecol. 16:349–357.

Wenny, D. G., R. L. Clawson, J. Faaborg, and S. L. Sheriff. 1993. Population density, habitat selection and minimum area requirements of three forest-interior warblers in central Missouri. Condor 95:968–979.

Western, D., and J. Semakula. 1981. The future of the savannah ecosystems: ecological islands or faunal enclaves? Afr. J. Ecol. 19:7–19.

Whitcomb, R. F. 1977. Island biogeography and "habitat islands" of eastern forest. Amer. Birds 31:3–23, 91–93.

Whitcomb, R. F. 1987. North American forests and grasslands: biotic conservation. Pp. 163–176 *in* Nature conservation: the role of remnants of native vegetation (D. A. Saunders, G. W. Arnold, A. A. Burbidge, and A. J. M. Hopkins, eds). Surrey Beatty and Sons, Chipping Norton, Austraiia.

Whitcomb, R. F., J. F. Lynch, P. A. Opler, and C. S. Robbins. 1976. Island biogeography and conservation: strategy and limitations. Science 193:1030–1032.

Whitcomb, R. F., C. S. Robbins, J. F. Lynch, B. L. Whitcomb, K. Klimkiewicz, and D. Bystrak. 1981. Effects of forest fragmentation on avifauna of the eastern deciduous forest. Pp. 125–205 *in* Forest island dynamics in man-dominated landscapes (R. L. Burgess and D. M. Sharpe, eds). Springer-Verlag, New York.

Wiens, J. A., C. S. Crawford, and J. R. Gosz. 1985. Boundary dynamics: a conceptual framework for studying landscape ecosystems. Oikos 45:421–427.

Wilcove, D. 1985. Nest predation in forest tracts and the decline of migratory songbirds. Ecology 66:1211–1214.

Wilcove, D. S. 1988. Changes in the avifauna of the Great Smoky Mountains: 1947–1983. Wilson Bull. 100:256–271.

Wilcove, D. S., and S. K. Robinson. 1990. The impact of forest fragmentation on bird communities in eastern North America. Pp. 319–331 *in* Biogeography and ecology of forest bird communities (A. Keast, ed.). SPB Academic Publishing, The Hague, The Netherlands.

Wilcove, D. S., and J. W. Terborgh. 1984. Patterns of population declines in birds. Amer. Birds 38:10–13.

Wilcove, D. S., C. H. McLellan, and A. P. Dobson. 1986. Habitat fragmentation in the temperate zone. Pp. 237–256 *in* Conservation biology: the science of scarcity and diversity (M. E. Soulé, ed.). Sinauer Associates, Inc., Sunderland, MA.

Wilcox, B. A. 1980. Insular ecology and conservation. Pp. 95–117 *in* Conservation biology: an evolutionary–ecological perspective (M. E. Soulé and B. A. Wilcox, eds). Sinauer Associates, Inc., Sunderland, MA.

Wilcox, B. A., and D. D. Murphy. 1985. Conservation strategy: the effects of fragmentation on extinction. Amer. Natur. 125:879–887.

Wilcox, B. A., D. D. Murphy, P. R. Ehrlich, and G. T. Austin. 1986. Insular biogeography of the montane butterfly faunas in the Great Basin: comparison with birds and mammals. Oecologia 69:188–194.

Williamson, M. 1981. Island populations. Oxford University Press, Oxford, England.

Willis, E. O. 1979. The composition of avian communities in remanescent woodlots in southern Brazil. Papeis Avulsos Zool. 33: 1–25.

Willis, E. O. 1984. Conservation, subdivision of reserves and the anti-dismemberment hypotheis. Oikos 42:396–398.

Willson, M. F., and S. W. Carothers. 1979. Avifauna of habitat islands in the Grand Canyon. Southwest. Natur. 24:563–576.

Wilson, E. O., and E. O. Willis. 1975. Applied biogeography. Pp. 522–534 *in* Ecology and evolution of communities (M. L. Cody and J. M. Diamond, eds). Belknap Press, Cambridge, MA.

Witkowski, Z., and P. Plonka. 1984. The species/area relationship in plants and birds on protected and unprotected isolates in Poland. Bull. Polish Acad. Sci. Biol. Sci. 32:7–8.

Wu, J., J. L. Vankat, and Y. Barlas. 1993. Effects of patch connectivity and arrangement on animal metapopulation dynamics: a simulation study. Ecol. Model. 65:221–254.

Wylie, J. L., and D. J. Currie. 1993. Species–energy theory and patterns of species richness: I. Patterns of bird, angiosperm, and mammal species richness on islands. Biol. Conserv. 63:137–144.

Yahner, R. H. 1987. Short-term avifaunal turnover in small even-aged forest habitats. Biol. Conserv. 39:39–47.

Yahner, R. H. 1988. Changes in wildlife communities near edges. Conserv. Biol. 2:333–339.

Yahner, R., and D. P. Scott. 1988. Effects of forest fragmentation on depredation of artificial nests. J. Wildl. Manag. 52:158–161.

Yahner, R. H., and A. L. Wright. 1985. Depredation on artificial ground nests: effects of edge and plot age. J. Wildl. Manag. 49:508–513.
Ziehmer, R. L. 1993. Effects of uneven-aged timber management on forest bird communities. Unpublished MS thesis, University of Missouri—Columbia, Columbia, MO.
Zimmerman, B. L., and R. O. Bierregaard. 1986. Relevance of the equilibrium theory of island biogeography and species-area relations to conservation with a case from Amazonia. J. Biogeog. 13:133–143.

14

A LANDSCAPE ECOLOGY PERSPECTIVE FOR RESEARCH, CONSERVATION, AND MANAGEMENT

KATHRYN E. FREEMARK, JOHN B. DUNNING, SALLIE J. HEJL, AND JOHN R. PROBST

INTRODUCTION

It is becoming increasingly clear that the distribution and population dynamics of Neotropical migratory birds cannot be understood solely from processes occurring within individual habitat patches. Effects from the surrounding landscape also have to be considered. The need for a landscape perspective in conservation and management has been explicitly recognized in the ecosystem management approach being developed by the USDA Forest Service (Kessler et al. 1992 and related papers in the same issue) and is currently being evaluated in the Pacific Northwest (Franklin 1989, Hansen et al. 1991, Bormann et al. 1994).

Landscape ecology represents a renewed interest in the development and dynamics of landscape mosaics, the effects of landscape patterns on species, biotic interactions and ecological processes, and how landscape heterogeneity can be managed to benefit society (Risser et al. 1984, Naveh and Lieberman 1984, Allen and Hoekstra 1992). An emphasis on spatial heterogeneity, human influences, and spatio-temporal dynamics distinguishes landscape ecology from other types of ecological investigation.

Landscape ecology arose initially in Europe from traditions of regional geography and vegetation science (Naveh and Lieberman 1984, Turner and Gardner 1991). In North America, it evolved primarily from the theory of island biogeography (MacArthur and Wilson 1963, 1967), which used spatial pattern to explain species distributions (Burgess and Sharpe 1981, Risser et al. 1984, Burel 1992), and engendered much of the recent work on effects of habitat fragmentation (see Faaborg et al. 1993, and Chapter 13, this volume).

Landscape ecology has emerged as a recognized discipline only relatively recently. The International Association of Landscape Ecology (IALE) was founded in 1982 in Europe (Bunce and Jongman 1993). Turner (1987) and Moss (1988) provide overviews of landscape ecological research presented at the first meetings of the United States chapter of IALE and the Canadian Society of Landscape Ecology and Management, respectively. The journal *Landscape Ecology* was first published in 1987. The following references provide a good introduction to both the European and North American perspectives of landscape ecology for those readers not familiar with the subject: Naveh and Leiberman (1984), Risser et al. (1984), Forman and Godron (1986), Urban et al. (1987), Turner (1989), Zonneveld and Forman (1990), Turner and Gardner (1991), Hansen and diCastri (1992) and Bunce et al. (1993).

In this paper, we highlight key concepts of landscape ecology important to the research, conservation, and management of Neotropical migratory birds. We then review empirical studies related to the landscape ecology of Neotropical migratory birds in forests, farmland, wetlands, riparian habitats and urban habitats of temperate breeding areas, and to a more limited extent, on migration stopover areas and Neotropical overwintering areas. Research, conservation, and management implications for Neotropical migratory birds arising from a landscape perspective are then discussed.

A Landscape Perspective

To wildlife such as Neotropical migratory birds, a landscape is a heterogenous mosaic of habitat patches in which individuals live and disperse (Dunning et al. 1992). Landscape studies of wildlife examine patterns in the mosaic of habitat patches in the landscape and how they influence the distribution and dynamics of individuals, populations and communities (Kotliar and Wiens 1990, Barrett 1992). The size of a landscape and the way its spatial heterogeneity is perceived (i.e., how a patch is defined) varies among organisms (Turner 1989, Wiens 1989, Karr 1994, Pearson et al. 1995). For wildlife, patches can be viewed as relatively discrete areas of land of relatively homogeneous environmental conditions between which fitness prospects differ. "Grain" is the smallest scale at which individuals perceive and respond to patch structure, and varies with different physiological and perceptual abilities among species (Kotliar and Wiens 1990, Pearson et al. 1995). "Extent" is the coarsest scale of spatial heterogeneity to which individuals respond. For a given individual, extent is determined by lifetime home range. Extent varies among individuals and species. For a given species, extent at the population level exceeds that of individuals. Thus patches can be defined hierarchically in scales ranging between the grain and extent for the individual, population, or range of each species of interest (Kotliar and Wiens 1990, McGarigal and Marks 1995). For small or relatively immobile species, such as some insects, a heterogeneous mosaic of habitat patches might be included in a landscape that covers only a few hectares (Fig. 14-1). For Neotropical migratory birds, landscapes occupy the spatial scales intermediate between an individual's territory or home range (typically one to a few hundred hectares), and a species' distribution over larger areas (e.g., physiographic region). Within this spatial range (typically 1–1000 km^2), landscapes should be delineated based on the scale of the patch mosaics, their larger regional contexts, and

Figure 14-1. A multiscale view of landscape pattern from an organism-centered perspective. From McGarigal and Marks (1995).

the openness of the landscape mosaic in relation to the population(s) of interest (McGarigal and Marks 1995).

By focusing on intermediate spatial scales, the study of landscape patterns and processes forms a bridge in the spatial hierarchy between habitat studies at local (within-patch) scales that have dominated avian ecology, and larger-scale regional and biogeographical studies (Allen and Hoekstra 1992, Dunning et al., 1995). To better understand the population status of Neotropical migratory birds however, landscape studies need to be integrated with those at smaller and larger spatial scales over breeding, wintering and migratory seasons (Probst and Crow 1991, Freemark et al. 1993). During the breeding season for example, the distribution of migratory bird populations is determined by interactions among local habitat factors and the landscape, regional and continental contexts of habitat distributions, species' distributions and population levels (Behle 1978, Wiens et al. 1987, Holmes and Sherry 1988, Mauer and Heywood 1993, Probst and Weinrich 1993).

Habitat selection by Neotropical migratory birds involves responses to patch structure at a series of hierarchical levels including territory, patch and landscape scales (Wiens et al. 1987, Gutzwiller and Anderson 1992, Steele 1992, Probst and Weinrich 1993). The array of habitats ultimately selected varies in response to local, landscape and regional population dynamics because higher quality habitat is usually preferentially selected to lower quality habitat (Zimmerman 1982, Lanyon and Thompson 1986, Hill 1988, Probst 1988). When populations are large relative to the availability of preferred habitat, both higher and lower quality habitats are selected. When populations are small relative to the availability of preferred habitat, only higher quality habitat is selected. As a consequence, higher quality habitats are more consistently occupied than lower quality habitats (Howe et al. 1991). Furthermore, birds usually leave lower quality habitats to colonize more preferred habitats that become available. However, if availability of preferred habitat is substantially larger than existing populations, lags in reproductive output can constrain selection of preferred habitats (Probst and Weinrich 1993). Population variability over space and time is the major reason why habitat suitability models derived solely from local habitat variables fail to predict species abundances adequately (Van Horne and Wiens 1991).

Metapopulations

Many, if not most populations are patchily distributed. This is especially the case in landscape mosaics that are fragmented by natural processes or human activities. Within individual habitat patches, populations can decline, become extinct, and become re-established by the dispersal of individuals from other patches in the landscape. Sets of local populations that interact through the dispersal of individuals have been termed metapopulations (Merriam 1988, Opdam 1990, 1991). Because of spatial and temporal dynamics in local populations, the distribution pattern of a metapopulation shifts over time (Opdam 1990, 1991, Villard et al. 1992). For the metapopulation to persist, it is important that local populations do not fluctuate synchronously, since this increases the probability of simultaneous extinction of all local populations (Gilpin 1990, Hanski 1991, Fahrig and Merriam 1994). Landscape structure coupled with the habitat-specific demography (e.g., survival, productivity) and dispersal characteristics of organisms influence the number of patch populations that can interact, the size of those patch populations, their temporal variability, and, ultimately, the survival of the metapopulation (Merriam 1988, Opdam 1991, McKelvey et al. 1993). For example, if landscape structure restricts dispersal, patch populations will remain extinct for longer and patches that have suffered population extinctions will cover more area. To persist, the metapopulation must then cover a larger area including more patches (Hansson 1991, Fahrig and Merriam 1994).

Within a metapopulation, the probability of local extinction is inversely related to the size of a patch population, which in turn is proportional to patch size and quality. In smaller and lower quality patches, populations are more likely to suffer local extinctions from stochastic processes, adverse biotic interactions and movement of individuals to more preferred habitats that become available elsewhere. Recolonization rates are proportional to the regional availability of dispersing individuals, the proximity of suitable habitat patches, the quality of intervening patch types, and the density and quality of dispersal routes through the landscape (Opdam 1990, Fahrig and Freemark 1995).

Although dispersal is key to metapopulation dynamics, only dispersal distances for adult birds have been well documented. The distribution of dispersal distances generally resembles an exponential or skewed-normal function; most individuals move short distances, a few move long distances (Opdam 1991). For Neotropical migratory passerines, median distances are reportedly shorter than 350 m (Villard et al. 1995). Little is known about the dispersal of birds returning to breed for the first time. Low recapture rates of first-year birds suggest high overwinter mortality and/or substantially longer dispersal movements than those of adults (Villard et al. 1995). The influence of landscape structure on the movement of dispersing individuals has not been well studied for birds. Simulation models and field studies of other taxa indicate that the presence, quality and spatial configuration of dispersal corridors (e.g., wooded fencerows) and "stepping-stone" habitats are potentially important (reviewed by Fahrig and Freemark 1995). Woody vegetation in fields and fencerows appears to concentrate movements of birds between forests and the surrounding agricultural mosaic (MacClintock et al. 1977, Wegner and Merriam 1979, McDonnell and Stiles 1983, Johnson and Adkisson 1985).

Some authors have suggested that metapopulations exist in a "source–sink" fashion (Pulliam 1988, Howe et al. 1991). Offspring disperse from populations in source patches where productivity exceeds mortality to populations in sink patches, which, in the absence of immigration, would go locally extinct. Computer simulation models show that sink areas can be occupied by a large fraction of the metapopulation and can make a significant contribution to the size and longevity of the metapopulation (Pulliam 1988, Howe et al. 1991) as long as sink areas are not preferentially selected over source areas (Best 1986, McKelvey et al. 1993). A given patch may oscillate between acting as a source or a sink with environmental variation (Stacey and Taper 1992). When every patch can experience an extinction, the spatial configuration of patches is particularly important to assure recolonization (Fahrig and Merriam 1994).

The potential genetic consequences of metapopulation dynamics may have dramatic impact on the both the ecological and evolutionary viability of species (Lande and Barrowclough 1987, Gilpin 1991, Manicacci et al. 1992, Holsinger 1993, McCauley 1993). In most cases, metapopulation dynamics is expected to reduce a species' ability to respond adaptively to environmental change. Theoretical models suggest that genetic variation is lost from within local populations and from the metapopulation as a whole even in species that are fairly abundant (McCauley 1993), and especially in species that were formerly widespread and common prior to fragmentation of their habitat (Holsinger 1993). The effect is greatest when the number of local populations remaining is small, extinction rates are high, and the number of individuals in a founding group is small. Using the limited information available on gene flow in plants, Holsinger (1993) hypothesized a typology of evolutionary consequences based on the proximity of suitable habitat patches relative to a species' dispersal distance, its migration rate (number of migrants per generation), and the size of the remaining population fragments. A scenario of large interpatch distances among small fragments with low immigration rates would likely maintain genetic diversity among populations but compromise metapopulation persistence because of lower mean fitness in all populations from loss of within-population genetic diversity. At present, there is no empirical evidence for birds (or other vertebrates) to test current theories of gene flow in metapopulations.

Landscape Structure

The composition and spatial configuration of a landscape can independently or in combination affect ecological processes including species' distributions and biotic interactions (Dunning et al. 1992). Landscape composition includes the variety and abundance of patch types within a landscape, but not the location or relative placement of patches within the mosaic. It is well known that the species richness, composition and abundance of Neotropical migratory birds vary among habitat types (Keast and Morton 1980, DeGraaf and Rudis 1986, Verner et al. 1986, DeGraaf et al. 1992, 1993, Hagan and Johnston 1992, Rodenhouse et al. 1993, Chapter 10, this volume). Although most bird species use more than one habitat type, the extent to which individual species, particularly Neotropical migrants, require a mosaic of habitat types in a given season or are affected by variation in habitat conditions between seasons and years has not been as well studied (but see Karr and Freemark 1983, 1985, Rodenhouse and Best 1983, Warner 1984, Warner et al. 1984, Rosenberg and Raphael 1986, Vander Haegen et al. 1989).

What has also been less apparent until relatively recently is that landscape configuration (i.e., the spatial character and placement of habitat patches within the mosaic) is also important. For many Neotropical migratory birds, smaller and relatively isolated habitat patches support fewer species, and a depauperate subset of species compared to patches that are larger and closer to other patches of similar habitat (Brown and Dinsmore 1986, Robbins et al. 1989, Gibbs et al. 1991, Herkert 1991, Freemark and Collins 1992). The geographical orientation of habitat patches in relation to migratory pathways can also affect the number of migrants that settle in a patch (Gutzwiller and Anderson 1992). The composition, variety, abundance, and productivity of Neotropical migratory birds within a landscape can also be significantly affected by the nature and intensity of biotic interactions with species from patches in the "ecological neighborhood" (*sensu* Addicott et al. 1987) of a given patch (Szaro and Jakle 1985, Temple and Cary 1988, Dunning et al. 1992). The nature of structural or functional boundaries created by the juxtapostion of different patch types is also important (Hansen and diCastri 1992). The presence of corridors may facilitate the movement of organisms across boundaries or through intervening, inhospitable patches (MacClintock et al. 1977, Wegner and Merriam 1979, Harris 1984, Harris and Gallagher 1989, Saunders and Hobbs 1991, Beier 1993). However, the conservation value of corridors has been questioned (Simberloff et al. 1992) and has not been well studied for Neotropical migratory birds.

Both landscape composition and configuration are dynamic over time as a result of natural processes (e.g., fire, flooding, succession) and human activities (e.g., forestry, agriculture, urbanization). As a consequence, responses of individual organisms and populations to landscape structure are also dynamic in both space and time (Pickett and White 1985). In order to understand the current distribution of species, it may be necessary to consider both the structure of past as well as present landscapes (Burel 1992).

EMPIRICAL STUDIES OF LANDSCAPE ECOLOGY OF BIRDS

Recent work has greatly improved the conceptual basis and quantitative methods for landscape studies (e.g., Turner and Gardner 1991, Hansen and DiCastri 1992 and references therein). Empirical studies will now be reviewed to develop a landscape perspective for the research, conservation, and management of Neotropical migratory birds.

Northeastern and Central Hardwood Forests

The realization that a landscape perspective is necessary for conservation and management of Neotropical migratory birds evolved largely from studies of northeastern and central hardwood forests fragmented by agriculture and suburban development. Forest fragmentation has had an array of effects on Neotropical migratory birds as a result of habitat loss, smaller patch size, reduced proximity of forest patches, more edge, and negative interactions with species from surrounding nonforest patches (Table 14-1). Much of the existing literature has been comprehensively reviewed by others (Robbins

Table 14-1. Effects of landscape structure on Neotropical migratory species breeding in hardwood forests within agricultural and suburbanized landscapes of northeastern and central North America. For extensive reviews of the literature see Robbins et al. (1989), Askins et al. (1990), Wilcove and Robinson (1990), Finch (1991) and Faaborg et al. (1993, Chapter 13, this volume).

Landscape Structure	Effects	Selected References[a]
Landscape composition		
Forest type	Successional age, plant-species composition, vegetation structure and habitat heterogeneity influence presence and abundance of species, particularly those with specific habitat requirements	1–9
Forest cover	More area-sensitive species per area with more forest in surrounding landscape	4, 6, 10, 11
Habitat proportions	Higher rates of nest parasitism in landscapes with lower and more fragmented forest cover, and greater extent of habitats suitable for cowbird feeding	12
Landscape configuration		
Patch size	More species with greater area	1, 2, 5, 11, 13
	Species in smaller areas limited subset of species in larger areas	1, 6, 10, 11, 13–15
	Higher probability of local extinction of breeding populations in smaller areas	16
	Lower mating success of males in small fragments compared to extensive forests	17–19
	Small areas important to forest interior–edge species breeding in landscapes with limited forest cover	5, 11
Patch shape	Area-sensitive species more abundant in areas with lower perimeter to area ratios (i.e., more "core" area)	20
	Abundance of species increase with core area	6
Interpatch distance	More species in areas less distant from nearby forest	1, 21, 22
	Higher probability of population recolonization in patches closer to other occupied patches	16
	Higher density and mating success of males in areas less isolated from nearby forest	18
Nonforest edge	Lower reproductive success near nonforest edges from higher rates of nest predation and nest parasitism	12, 23–25
Habitat juxtaposition/ interspersion	Higher rates of nest parasitism in landscapes where forest cover more interspersed with habitats suitable for cowbird feeding	12

[a] References: 1, Howe (1984); 2, Freemark and Merriam (1986); 3, DeGraaf and Rudis (1986); 4, Askins et al. (1987); 5, Blake and Karr (1987); 6, Robbins et al. (1989); 7, DeGraaf et al. (1992); 8, DeGraaf et al. (1993); 9, Dickson et al. (1993); 10, Askins and Philbrick (1987); 11, Freemark and Collins (1992); 12, Robinson et al. (1993, Chapter 15, this volume); 13, Blake and Karr (1984); 14, Blake (1991); 15, Whitcomb et al. (1981); 16, Villard et al. (1995); 17, Gibbs and Faaborg (1990); 18, Villlard et al. (1993); 19, Van Horn et al. (1995); 20, Temple (1986); 21, MacClintock et al. (1977); 22, Lynch and Whigham (1984); 23, Temple and Cary (1988); 24, Robinson (1992); 25, Faaborg et al. (1993, Chapter 13, this volume).

1989, Askins et al. 1990, Wilcove and Robinson 1990, Finch 1991, Faaborg et al. 1993, Chapter 13 this volume). Only key points and selected references are presented here.

It is well known that successional stage, plant-species composition and vegetation structure influence the suitability of hardwood forests to Neotropical migratory species, particularly those with specific habitat requirements (Table 14-1). For mature hardwoods remaining in agricultural and suburbanized landscapes, the number of bird species increases with forest size. Smaller forests support fewer species and only a limited subset of species characteristic of larger forests (Table 14-1). Species uncommon in small forests (see Appendix for details) typically have large territories (>5 ha), overwinter in the Neotropics, preferentially nest in forest interior habitats, nest on or near the ground, build open cup nests, have low reproductive potential (i.e., small and few clutches), glean insects from bark or foliage, and are regionally uncommon. Species–area relationships within and among regions are similar for total numbers of bird species but vary significantly for subsets of species

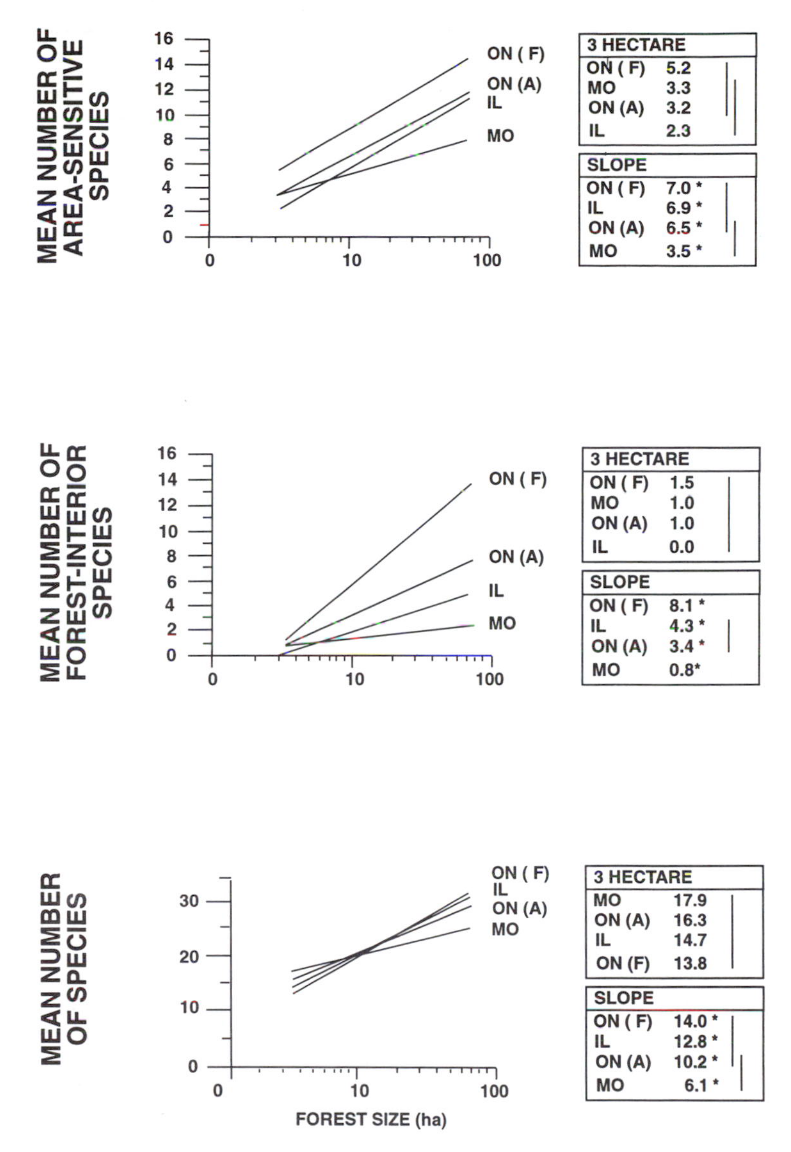

Figure 14-2. Regressions of mean number of species on forest size for: (bottom) all bird species; (middle) species that preferentially use forest-interior habitat; and (top) species that were found to be more frequent in larger forests. Study areas were in Ontario [ON(F) and ON(A)], Missouri (MO) and Illinois (IL). Total forest cover estimated at 30%, 19%, 22%, and 2% for ON(F), ON(A), MO and IL, respectively. Regression slopes are given in the "slope" box. Regression estimates for a 3 ha forest are given in the "3 hectare" box. In each box, a solid line joins regression values that are not significantly different ($P > 0.05$); * denotes nonzero regression slope ($P < 0.05$). Abstracted from Freemark and Collins (1992).

known to be intolerant of forest area reduction (Fig. 14-2). About 72% of the 47 species showing positive area sensitivity (in at least some studies) overwinter in the Neotropics (Fig. 14-3). Many have breeding territories of less than 2 ha (Whitcomb et al. 1981, Freemark and Merriam 1986). For these species, area per se has the greatest influence on species number. Factors such as plant-species composition, vegetation structure,

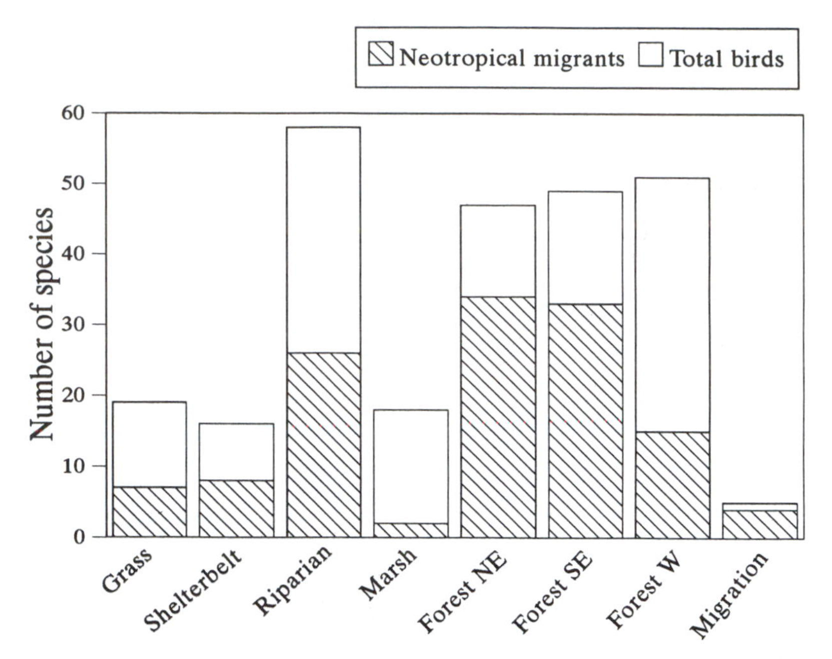

Figure 14-3. Numbers of bird species exhibiting positive area-sensitivity in at least one habitat type or region based on lower density, frequency of occurrence, or nest success with forest fragmentation or reduction in habitat area. Species are listed in the Appendix.

habitat heterogeneity, and patch isolation can be important but relatively less so than forest size. Differences in the slope of the species–area relationships appear to reflect differences in the amount and landscape configuration of forests within and among regions (Freemark and Collins 1992). For example, the number of forest-interior species increases with greater area more rapidly in agricultural landscapes where forest cover is more extensive and less isolated (Fig. 14-2). Small forests (< 10 ha) in agricultural regions support few, if any, area-sensitive or forest-interior species but can provide important habitat for bird species during migration and for some breeding species (e.g., forest interior–edge generalists), particularly when forest cover is severely limited in the landscape (Table 14-1).

Analyses of presence–absence data indicate that the minimum area requirement of many forest species is much greater than their territory size and varies among geographic regions (Martin 1992). While instructive, the use of presence–absence data alone to assess and compare minimum area requirements (or other bird species patterns) within and among regions can be problematical. Population demography within patches can vary with the quantity and quality of different habitats within a patch, patch size and shape, the spatial configuration of patches, and the location of studies in relation to the composition, regional population dynamics and geographic ranges of the species involved. Single-brooded, low-nesting, Neotropical migratory species are particularly vulnerable to the higher rates of predation and nest parasistism associated with nonforest edges. As a result, mated individuals often breed less successfully in small forests or larger forests with high ratios of nonforest edge to interior (Table 14-1). The magnitude of nest parasitism by the Brown-headed Cowbird varies within and among regions, apparently in response to landscape-level variation in forest fragmentation and availability of habitat suitable for cowbird feeding (Table 14-1). Landscape-level management to maintain large areas of contiguous forests and eliminate cowbird feeding areas has been recommended as the most effective approach to reduce the impact of cowbirds on Neotropical migrants (Robinson et al. 1993).

For at least some Neotropical migratory species, the density and mating success of males is significantly higher in extensive forests than in forest patches isolated from nearby woodland (Table 14-1). The magnitude of the effect varies among geographical areas and, for the Ovenbird, is particularly severe at the periphery of its breeding range where it is less abundant (Villard et al. 1993). Empirical evidence is needed to test the various hypotheses proposed to explain reduced pairing success in forest fragments including alteration of dispersal dynamic and habitat selection by females (Gibbs and Faaborg 1990, Villard et al. 1993).

Temporal dynamics in species distributions and population demography also need to be considered in assessing minimum area requirements, and other aspects of habitat suitability. Turnover in species occurrence within forests can average as high as 18% between years irrespective of forest size or habitat heterogeneity (Whitcomb et al. 1977, Blake and Karr 1987, Freemark 1989). Villard et al. (1992, 1995) observed year-to-year changes in the distribution of Ovenbird, Wood Thrush, Black-and-white Warbler and Scarlet Tanager among forest fragments in agricultural landscapes of eastern Ontario. Patch populations of each species exhibited local extinctions and recolonizations between years. In contrast, the proportion of occupied patches in each landscape did not change significantly, a pattern consistent with metapopulation theory. Patches that exhibited population turnovers between years were smaller and tended to be further from other occupied patches. Observed numbers of population turnovers were best approximated by a hypothetical distribution of spring migrants assuming a combination of passive sampling proportional to area, dispersal according to proximity to previously occupied patches, resettlement according to vegetation structure within a patch, and site fidelity of successful breeders (Villard et al. 1995). Other studies have shown that habitat occupancy can also vary in response to regional population dynamics. Smaller and other lower quality patches within the landscape may only be occupied by species in years when regional populations are high relative to the availability of more suitable habitat elsewhere (Probst and Weinrich 1993).

A few studies have considered the size and number of hardwood forests needed to conserve area-sensitive bird species within agricultural and suburbanized landscapes in northeastern North America. Forest fragments under 10 ha are unsuitable for many forest-dwelling Neotropical migratory species. Forests as large as 3000 ha may be required to retain all species of the forest-interior avifauna (Robbins et al. 1989). In suburbanized landscapes of the mid-Atlantic States, the probability of detecting even one-half of the regional pool of area-sensitive species in a 50 ha forest appears to be very low, although several 50 ha forests can contain the species complement of a single, 3000 ha forest (Robbins et al. 1989). In contrast, at least one-half of the regional pool of area-sensitive species was detected annually in a single 54–65 ha forest in agricultural landscapes compared among Ontario, Missouri, and Illinois (Freemark and Collins 1992). Area sensitivity of many forest birds species appears to be less pronounced when the surrounding landscape is at least 30% forested (Robbins et al. 1989, Freemark and Collins 1992). To date, analyses have been based on point-count data only and without consideration of metapopulation dynamics. As argued above, information on habitat- and patch-specific demographic information is needed before the conservation value of different landscapes can be more accurately assessed.

Landscape-level considerations for Neotropical migratory birds in regions of extensive hardwood forests in central and northeastern hardwoods have recently been reviewed by DeGraaf et al. (1993) and Dickson et al. (1993). The focus of attention in the relatively few studies available has been on the maintenance of diverse landscape mosaics of forest types and successional age classes, wetland and aquatic habitats, and native, nonforest habitats, and on how the mosaic of forested landscapes is impacted by human activities, particularly silvicultural

Table 14-2. Effects of landscape structure on selected species breeding in southeastern forests.

Landscape Structure	Effects	References[a]
Landscape composition		
Forest type	Successional age, plant-species composition and vegetation structure influence presence and abundance of species, particularly those with specific habitat requirements	1–5
Forest cover	Number and abundance of forest-interior species decline with loss of suitable forest cover	6, 9, 10
Landscape configuration		
Patch size	More species with greater area	3, 7, 8
	Fewer species and birds with greater area of cypress ponds	5
	Species in smaller areas limited subset of species in larger areas	3, 7, 8
	Only a few species show a positive relationship with area of cypress ponds	5
Interpatch distance	More isolated patches less likely to be occupied by species of resident birds	9, 10
Nonforest edge	Species' abundances increase, decrease or do not change near nonforest edge	5, 11

[a] References: 1, Johnson and Landers (1982); 2, Dickson et al. (1993); 3, Hamel et al. (1982); 4, Hirth et al. (1991); 5, O'Meara (1984); 6, Burdick et al. (1989); 7, Harris (1989); 8, Harris and Wallace (1984); 9, Dunning and Watts (1990); 10, Conner and Rudolph (1991); 11, Noss (1991).

practices. Isolation effects have not been regarded as important but remain largely unstudied. Turnover in species occurrence within plots has been observed among years in extensively forested areas (Wilcove 1988). In the Adirondack Mountains, smaller habitat patches and those more distant from other occupied patches were unoccupied by Spruce Grouse (Fritz 1979). Higher rates of nest predation have been observed in forested areas surrounded by nonforest habitats but not in relation to distance from the edge (Small and Hunter 1988). Assessing the cowbird effect is currently an active area of research (DeGraaf et al. 1993; Robinson et al. 1993, Chapter 15, this volume).

Southeastern Forests

Neotropical migrants are prominent components of the breeding avifaunas of southeastern forests. Successional stage, plant-species composition, and vegetation structure influence the presence and abundance of Neotropical migratory species, particularly those with specific habitat requirements (Table 14-2). Neotropical migratory species make up five of the ten bird species that are dependent on mature forests in the southeastern United States (Hamel et al. 1982) and are common breeders in field and shrub-dominated habitats (J. B. Dunning, personal observation).

Other than habitat type, little work has been done on the influence of landscape structure on Neotropical migratory birds breeding in southeastern forests (Table 14-2). Both the number and densities of Neotropical migratory species characteristic of interiors of bottomland hardwood forests have decreased with successive clearing across a region. Populations of resident species such as Bachman's Sparrow and the endangered Red-cockaded Woodpecker have also declined as the extent of suitable forest habitat decreased in the landscape (Dunning and Watts 1990, Conner and Rudolph 1991). Except for cypress ponds (O'Meara 1984), smaller forests support fewer bird species and a limited subset of species characteristic of larger forests. Neotropical migratory species make up 67% of the 49 species considered to be area sensitive in at least some studies of southeastern forests (Fig. 14-3).

Abundance patterns in old-growth hammocks and cypress ponds in Florida show that Neotropical migratory species respond in a variety of ways to nonforest edge. Some species are more abundant near edge (e.g., Yellow-billed Cuckoo, White-eyed Vireo, Northern Parula, Summer Tanager, Blue-gray Gnatcatcher), some are less abundant (e.g., Red-eyed Vireo, Hooded Warbler, Acadian Flycatcher, Wood Thrush), and others are unaffected (e.g., Yellow-throated Vireo, Great Crested Flycatcher, Black-and-white Warbler, Ovenbird). At least for resident species such as Bachman's Sparrow and the Red-cockaded Woodpecker, more isolated patches of pine woodlands are less likely to be occupied than patches closer to nearby forest.

Southern pinelands and bottomland hardwoods are of special management interest. Pinelands are intensively managed for their economic value, especially commodity (timber) production. Bottomland forests are valued both for their economic (commodity) value and because they are centers of regional biodiversity (Burdick et al. 1989). Management options such as stand size, harvest rotation, and the spatial dispersion of harvest operations can have a tremendous impact on wildlife populations by affecting the composition and spatial configuration of habitat patches throughout managed landscapes (Dickson et al. 1993, Chapter 9, this volume).

Neotropical migratory birds are especially common breeders in newly planted pine stands (1–5 years postclearing) and mature pine stands over 100 years old (Johnson and Landers 1982, Hirth et al. 1991, Dickson et al. 1993, Dunning unpublished data). Intermediate-aged, monoculture pine forest support very few birds of any species, migratory or resident, although a few migrant species such as Pine Warbler do occupy these stands. Unfortunately, common management strategies maintain the majority of southeastern pine woodlands in these sterile, intermediate-aged classes. To date, other landscape effects on Neotropical migrants in southeastern pine woodlands have not been studied.

Bottomland hardwood forests support many species of migratory warblers, vireos and flycatchers. Bottomland forests often exist as relatively narrow strips of habitat following rivers and streams, surrounded by nonforest habitats. The bottomlands are therefore particularly prone to increasing isolation and decreasing connectivity from local land-use decisions that cumulatively fragment extensive bottomland forests into small patches (Burdick et al. 1989, Harris 1989). Thus, a regional perspective needs to be adopted when reviewing permit applications for harvesting individual stands (Burdick 1989). Since many landscape impacts are cumulative, the value of any particular stand in maintaining wildlife populations cannot be assessed without considering how the surrounding landscape would be affected in space and time.

Harris (1989) recommends that mature forest patches should be at least 30 ha to retain some forest interior habitat. Long and narrow patches, such as bottomland riparian forests, need to be larger than more rectangular patches to include interior habitat. Harris (1989) also recommends that mature forest tracts be surrounded by buffer management areas consisting of low-intensity managed forest to ensure that more of the mature core patch is distant from sharp edges caused by the juxtaposition of dissimilar habitats. The benefits of buffer areas, however, need to be carefully evaluated in relation to potential metapopulation dynamics in the species of concern. Buffer areas that create sink habitat can jeopardize metapopulation persistence, if sink habitats are preferentially used over source habitats (McKelvey et al. 1993).

Timber management strategies also create habitat for species that prefer shrub- or grass-dominated habitats such as the Prairie Warbler and the Yellow-breasted Chat. These habitats are found primarily in the early stages of succession in regeneration stands (1–10 years postplanting). When harvest operations are scattered throughout a forest region, the patches of open habitat may resemble an archipelago of relatively isolated patches. Prairie Warblers and other Neotropical migrants that breed in open country have shown significant population declines in the southeast (Odum and Turner 1987). The role of landscape effects in these declines is unknown. Managers should monitor open-habitat species to determine whether populations in their region are stable, or, if declines are detected, whether these declines are associated with changes in habitat composition and configuration across landscapes.

Western Forests

Many western landscapes have a fragmented pattern of natural origin. Western North America is dominated by forests, intermountain deserts, shrubsteppe, grasslands, and warm deserts (Barbour and Billings 1988). The Rocky Mountains, Sierra Nevada, and Cascade Mountains are large areas primarily covered by forests of different tree species and ages. Grasslands, urban areas, agricultural lands, shrubsteppe, desert, and isolated mountains are located in valleys between and within these large mountain ranges. Narrow riparian areas, often naturally fragmented by flood regimes, follow streams and rivers throughout all habitats. The mosaic of natural landscape patterns are especially dynamic in forested areas of high relief and frequent disturbances by fire, wind, and other physical processes (Swanson et al. 1990, Hejl 1992). Forestry practices (e.g., staggered setting clearcuts in the Pacific Northwest) introduce further spatial patterning and subsequent disturbance from factors such as increased windthrow (Franklin and Forman 1987, Swanson et al. 1990, Hansen et al. 1991, Ripple et al. 1991, Hejl et al., Chapter 8, this volume).

Investigations of landscape structure have been different in western forests compared to forests in the east (Verner and Larson 1989, Hansen et al. 1992). Many studies in the west have been conducted primarily in extensively forested landscapes fragmented by silvicultural activities (Table 14-3) rather than in agricultural and urbanized landscapes as is largely the case in the east (Table 14-1). In the west, moist forests are often punctuated by clearcuts that at first break continuous forest into interconnected forest fragments (Swanson et al. 1990). With time, forests usually grow back and the edges become less distinct. Landscape heterogeneity probably initially increases in these cases. Drier forests are sometimes naturally interspersed with grasslands and are logged in various ways, often resulting in managed forests with distinct boundaries on the natural grasslands but not along the cutting treatment (Hejl 1992). Landscape heterogeneity may actually decrease when these drier forests are logged (for an example from the Sierra Nevada, see McKelvey and Johnston 1992). Today, a great deal of privately owned forest lands in the West has been almost entirely cutover, creating various landscape patterns.

Relatively few studies have been conducted on the effects of landscape structure on birds in western forests (Table 14-3). Successional age, plant-species composition, and vegetation structure of western forests influence the presence and abundance of species, particularly those with specific habitat requirements. Other effects of landscape structure are most well known for a resident species, the Spotted Owl (Thomas et al. 1990). Dramatic population declines of the owl are probably due to the combined effects of habitat loss, increasing edge effects, and increasing habitat isolation on factors such as home-range size, the ability to find appropriate nest locations, and dispersal of juvenile owls (Forsman et al. 1984, Carey et al. 1990, Murphy and Noon 1992).

While the Spotted Owl, a predator with a large home range, is known to be affected by changes in landscape structure associated with forest fragmentation, effects on songbirds are much less clear. The importance of patch size appears to vary among regions in the west (Table 14-3). Patch size was a highly significant factor for birds in aspen groves in Saskatchewan, forested drainages in the southern Rockies, and subalpine spruce–fir groves in Wyoming. Verner and Larson (1989) were uneasy with their finding that patch size was the single best predictor of bird-species richness in mixed conifer forests of the Sierra Nevada in California, primarily because of the interconnectedness of their forest sites. Other studies in mountainous areas consisting of interconnected forests have found only a few species significantly related to patch size (Table 14-3).

Half of the landscape studies of western birds have been conducted in fragmented or unfragmented areas, not in isolated forest patches. Fragmented areas usually consist of interconnected forests of varying ages interdigitated with clearcuts or some other logged forests. Some bird species, both migratory and nonmigratory, appear to be associated with areas of either fragmented or unfragmented forest (Table 14-3). In northwestern

Table 14-3. Effects of landscape structure on Neotropical migratory species breeding in western forests.

Landscape Structure	Effects	References[a]
Landscape composition		
Forest type	Successional age, plant-species composition and vegetation structure influence presence and abundance of species, particularly those with specific habitat needs	1–4, 12, 14, 15
Forest cover	Abundance of some species higher in landscapes with more forest cover	1, 5
Habitat proportions	Brown-headed Cowbird (a nest parasite) more abundant in landscapes with more grassland and agricultural fields	5
Landscape configuration		
Patch size	Greater bird species richness with greater area of aspen groves, forested drainages, subalpine spruce–fir groves, and mixed conifer forests	6, 11–14
	Species in smaller aspen groves, forested drainages and subalpine spruce–fir groves limited subset of species in larger areas	6, 12, 13
	Fewer species in guilds prone to nest predation in smaller forested drainages	12
	Only a few species related to patch size in areas of interconnected forests	1, 2, 11
	Some species associated with areas of fragmented forest	1, 2, 7, 8, 10, 15
	Some species associated with areas of unfragmented forest	1, 2, 7, 8, 10, 15
Interpatch distance	More forest interior species in less isolated aspen groves	6
	Abundance higher for four and lower for one Neotropical migrant species in less isolated aspen groves	6
Nonforest edge (e.g., clearcut)	Forest species both positively and negatively associated with nonforest edge	1, 13
	No forest species show preference or avoidance of nonforest edge	7
Habitat juxtaposition/ interspersion	Many species more abundant in Douglas-fir stands adjacent to hardwoods	1
	A few forest species occur more often at nonforest edge	1
	Cowbird abundance higher in forests near open land	9
	Avian nest predators more abundant in old-growth stands interrdigitated with recent clearcuts than in continuous old-growth forest	8

[a] References: 1, Rosenberg and Raphael (1986); 2, Lehmkuhl et al. (1991); 3, Hansen et al. (1991); 4, Bunnell and Kremsater (1990); 5, Hejl (unpublished data); 6, Johns (1993); 7, Keller and Anderson (1992); 8, Hejl and Paige (1994); 9, Hejl (1992); 10, Tobalske et al. (1991); 11, Aney (1984); 12, Martin (1988); 13, Norment (1991); 14, Verner and Larson (1989); 15, Wetmore et al. (1985).

California, for example, abundances of three Neotropical migrant species (Wilson's Warbler, MacGillivray's Warbler, House Wren) were higher in areas of fragmented Douglas-fir (*Pseudotsuga menziesii*) forest (Rosenberg and Raphael 1986). In contrast, 51 species were found to be intolerant of fragmentation or area reduction in a least one study of western forests; 29% are Neotropical migrants (Fig. 14-3).

The influence of patch isolation on Neotropical migratory birds breeding in western forests is less known (Table 14-3). In Saskatchewan parkland, less isolated aspen groves had more forest-interior species, higher abundances of four Neotropical migratory species (Least Flycatcher, Veery, Ovenbird, Connecticut Warbler) and lower abundances of one other (Clay-colored Sparrow). Isolation of montane forests may be important at the regional scale (Johnson 1975, Behle 1978; but see Brown 1978, for an opposing viewpoint).

The effect of nonforest edge (e.g., clearcut) on birds in western forests is mostly unknown (Table 14-3). In northwestern California, abundances of six Neotropical migrant species increased and two decreased significantly with greater length of forest/clearcut edge in 1000 ha blocks of Douglas-fir forests (Rosenberg and Raphael 1986). Species responding positively to greater edge included

Olive-sided Flycatcher, Western Wood Pewee, Hammond's Flycatcher, House Wren, Warbling Vireo, and Wilson's Warbler. Western Tanager and Black-headed Grosbeak responded negatively to greater edge. No such effects were observed in Wyoming (Keller and Anderson 1992). Many different types of edges exist in western forest landscapes and need to be considered in future studies.

The nature and extent of adjacent habitat might be more important for most species than the existence of forest edge *per se* (Table 14-3). Twelve species (including three Neotropical migrants: Warbling Vireo, Western Tanager, Black-headed Grosbeak) were more abundant in Douglas-fir stands adjacent to hardwoods (Rosenberg and Raphael 1986). Stand abundances of many bird species, including Neotropical migrants, also vary in relation to the extent of forest cover in the landscape. Brown-headed Cowbird tends to be more abundant in forest stands in landscapes with more agricultural fields and grasslands (Hejl, unpublished data) which presumably provide good foraging areas (Hejl 1992). Cowbirds were mainly centered around human-based sources of supplemental food in the Sierra Nevada (Verner and Ritter 1983). Probably no host species is currently threatened by cowbird parasitism in the Sierra Nevada, but the problem could worsen as human developments become more widespread in the mountains.

Predation may also affect the number of species occupying habitat patches of different sizes (Table 14-3). In the southern Rockies, species (including many Neotropical migrants) that were absent from small forested drainages tended to belong to guilds that are more vulnerable to nest predation (e.g., foliage nesting vs. cavity nesting) (Martin 1988). Preliminary results in the Northern Rockies indicate that some avian nest predators are more abundant in old-growth forest fragments interdigitated with recent clearcuts than in continuous old-growth forest (Hejl and Paige 1994).

No strong conclusions can be made about the effects of landscape structure on Neotropical migratory birds in western forests. Most information is from short-term studies of bird distributions. Habitat use and habitat-specific demographics in different landscapes have not been studied. As discussed in previous sections, assessments of habitat suitability based solely on distributional data can be misleading. In the Pacific Northwest, for example, species may be packing into remaining fragments of Douglas-fir forest, thereby living and reproducing less successfully than they would in less crowded conditions (Rosenberg and Raphael 1986, Lehmkuhl et al. 1991). If so, distributional data (e.g., higher abundances in fragmented forests) grossly underestimate long-term impacts of habitat fragmentation in these landscapes. Current assessments of forest fragmentation effects on Pacific Northwest birds (Lehmkuhl and Ruggiero 1991, Hansen and Urban 1992, Hansen et al. 1992, 1993) need to be validated with more extensive and intensive information.

Studies of bird distributions, habitat use, reproductive success, dispersal patterns, and other life-history traits in landscapes throughout the West are needed before any solid conclusions can be made about the effects of different landscape structures on birds in western forests (Martin 1992). Compared to species in remnants of northeastern and central hardwood forests, birds in western forests appear not to have responded strongly (at least based on their distributions) to current levels of fragmentation. This may be because: (1) habitat fragmentation is a relatively recent phenomenon in western forests (Rosenberg and Raphael 1986); (2) habitat fragmentation has rarely resulted in isolated forest patches; (3) there is a time-lag in bird responses (Rosenberg and Raphael 1986, Lehmkuhl et al. 1991); and (4) many western forests are naturally fragmented (especially by fire) and human-induced fragmentation has not yet created sufficiently different landscape patterns (except for species with large home ranges such as the Spotted Owl) to affect many changes in birds (Hejl 1992). This last hypothesis is less likely to be true for birds in naturally extensive forests as in the Pacific Northwest, and may be more relevant for species in areas such as the Rocky Mountains, where fire has historically maintained naturally diverse landscape mosaics.

Northern Conifer/Boreal Forest

Few studies are currently available to evaluate effects of landscape structure on Neotropical migratory birds in northern conifer or boreal forests of North America. Probst and Weinrich (1993) argue that population trends of the endangered Kirtland's Warbler are primarily related to effects of landscape composition (availability of stands of jack pine (*Pinus banksiana*) of suitable age and density) and configuration (habitat areas larger than 200 ha) on nonbreeding birds, and dispersal among fragmented areas of marginal quality rather than reduced average nesting success of breeding pairs.

Based on analyses of long-term trends in forest-bird populations of northern Finland, Helle (1985, 1986) concluded that habitat fragmentation was responsible for declines of several species, especially cavity-nesting, nonmigratory species. In a recent review, Hunter (1992) concluded that sedentary, stereotypic species, especially those that prefer large tracts of old forest, are most likely to be vulnerable to changes in boreal landscapes. He specifically noted concerns about Spruce Grouse (Fritz 1979) and Red Crossbill (Benkman 1989). His conclusion that Neotropical migratory species are likely to be relatively insensitive because of their behavioral plasticity, mobility and small home-range requirements was qualified by the need for better data on impacts of habitat fragmentation from human activities in boreal landscapes.

Farmland

A large portion of temperate North America is farmed: 52% of the land area of the contiguous USA and 11% of Canada (Rodenhouse et al. 1993, and Chapter 10, this volume). Neotropical migrants represent the majority of bird species using farmland (Freemark et al. 1991; Rodenhouse et al. 1993 and Chapter 10, this volume). Despite this, the landscape ecology of birds, especially Neotropical migratory species, in farmland has not been well studied in North America (Table 4). Much of our current understanding is based on work conducted in Britain and Europe (O'Connor and Shrubb 1986, Eijsackers and Quispel 1988, Bunce 1990, Firbank et al. 1991, Spellerberg et al. 1991, Bunce et al. 1993). Existing studies suggest that loss of crop diversity, increasing field size, and reduction, degradation, and increasing isolation of noncrop habitats (such as native grasslands, shelterbelts, wooded fencerows) pose potentially significant adverse consequences for the dispersal, abundance, species richness, and regional survival of farmland biota, including Neotropical migratory (and resident) birds. Management practices, such as mowing, grazing, and use of pesticides, further aggravate effects on birds through direct impacts, and modification of vegetation and food resources (Rands 1985, Best 1986, Grue et al. 1986, O'Connor and Shrubb 1986, Bollinger and Gavin 1992, Fry 1991, Warner 1992, 1994, Freemark and Boutin 1995).

While migratory birds use a wide variety of croplands, species richness and abundance is greatest in grasslands, pasture, early-successional habitats, and uncultivated edge vegetation, such as grassed waterways, roadside verges and, especially, shelterbelts and wooded fencerows (Table 14-4). Because they support a unique complement of species, remnant woodlands, riparian areas and marshes are particularly important components of farmland structure (Best et al. 1995). Since they are considered elsewhere in this chapter, they are not discussed further here.

The size of farmland habitats affects their use by birds (Table 14-4). The density and species richness of nesting birds increase with greater area of roadsides, grasslands and, at least in some studies, shelterbelts. In Kansas, shelterbelts that support forest-interior species are larger, taller, and wider, and have higher snag densities and more diverse vegetation structure than those without forest-interior species (Schroeder et al. 1992). Sixteen species (50% Neotropical migrants) appear to be intolerant of a reduced area of shelterbelts (Fig. 14-3). In grasslands, 19 species (37% Neotropical migrants) show positive area sensitivity (Fig. 14-3). Six species observed by Herkert (1991) were never found in grassland fragments less than 10 ha, despite the fact that four of the

Table 14-4. Effects of landscape structure on Neotropical migratory species breeding in temperate farmland. See related reviews by Rodenhouse et al. (1993, Chapter 10, this volume) and Freemark (1995).

Landscape Structure	Effects	References[a]
Landscape composition		
Habitat type	Plant-species composition and vegetation structure influence the pressence and abundance of species in crop, noncrop strip cover, and native habitats in farmland, particularly species with specific habitat needs	1–8, 16
	Lower productivity in fields and noncrop habitats subject to management activities	2, 4, 8–10
Habitat proportions	Species abundances increase or decrease with extent of cultivation in the landscape	15
Landscape configuration		
Patch size	Fewer species and limited subset of species with smaller area of shelterbelts	6, 11, 17, 18
	Species richness not related to shelterbelt area	5
	Number of species and nests of grassland passerines increase with roadside width	4
	Fewer species and limited subset of species with smaller area of grasslands or prairies	12, 19, 20
	Higher rates of nest predation on smaller prairie remnants	13, 22
Patch shape	Species' abundances related to length and perimeter of shelterbelts	5, 18
	Rates of nest predation high in narrow habitats	2, 8
Habitat edge	Higher rates of nest parasitism and nest predation for grassland species nesting near wooded edge	13, 22
	Higher bird densities in rowcrop fields near edges	14
Interpatch distance	Distance to nearest other wooded habitat (at <500 m) unrelated to species richness in shelterbelts	6, 18
	Distances between prairie remnants occupied by a resident endangered species significantly smaller (mean = 14 km, $n = 12$) than those between unoccupied remnants (mean = 81 km, $n = 3$)	20
	Fewer species on an isolated prairie remnant compared to a similar-sized area in a cluster of remnants	21
Habitat juxtaposition/ interspersion	Species richness and species' abundances in shelterbelts influenced by adjacent land-uses in Minnesota (5) but not Kansas (6)	5, 6
	Density of grassland passerine nests higher in areas with more heterogeneous mosaics of crop and noncrop habitats	8

[a] References: 1, Howe et al. (1985); 2, Rodenhouse et al. (1993, Chapter 10, this volume); 3, Freemark et al. (1991); 4, Warner (1992); 5, Yahner (1983); 6, Schroeder et al. (1992); 7, Best (1983); 8, Warner (1994); 9, Bollinger and Gavin (1992); 10, Best (1986); 11, Martin (1981a); 12, Herkert (1991); 13, Johnson and Temple (1990); 14, Best et al. (1990); 15, Smutz (1987); 16, Best et al. (1995); 17, Martin and Vohs (1978); 18, Martin (1981b); 19, Vickery et al. (1994); 20, Samson (1980a); 21, Samson (1980b); 22, Burger et al. (1994).

six typically had territories less than 2.5 ha. Rates of nest predation appear to be higher in smaller areas (4–32 ha) compared to larger areas (up to 571 ha) of grassland (Johnson and Temple 1990, Burger et al. 1994). Grasslands of 100 to 300 ha may be necessary to support area-sensitive species (Samson 1980a,b, Herkert 1991, Vickery et al. 1994). Compared to patch size, vegetation heterogeneity and floristics of grasslands are of relatively minor importance (Samson 1980a, Warner 1994).

In contrast to noncrop habitats, bird species richness and abundance in croplands is likely to decrease as field size increases because more birds and species use the perimeter of fields compared to the center (Best et al. 1990). Narrower and/or more irregularly shaped fields may support more birds because they have proportionately more perimeter. In Britain, O'Connor and Shrubb (1986) found that densities for 23 of the 57 most common farmland species decreased as field size increased.

Reproductive success of grassland species is also affected by patch shape because rates of nest predation and nest parasitism are higher near (<45–60 m) the edge, especially wooded edges (Table 14-4). Rates of nest predation are also high in narrow, strip-cover habitats, such as fencerows and roadsides. Shelterbelt shape affects the abundance of many species during the breeding season.

Effects of isolation on birds using farmland habitats have not been well studied (Table 4). Samson (1980a,b) observed more species and higher occupancy by the Greater Prairie Chicken in prairie remnants closer to other prairie remnants. In Europe, networks of field-edge habitats, such as hedgerows, support more species and greater abundances of birds and other biota than their extent alone would suggest (Lack 1988, Burel and Baudry 1990, Fry 1991).

Interspersion of crop and noncrop habitats is important to both resident and migratory bird species (Table 14-4). The density of passerine nests (seven species, three Neotropical migrants) is higher in areas of Illinois with more grassland (primarily small grains), fewer rowcrop fields and more grassy strip cover (Warner 1994). The decline of the Ring-necked Pheasant in Illinois is related, at least in part, to replacement of hay and oats by corn and soybeans (Warner 1984, 1994, Warner et al. 1984). Nest predation is higher with low grass cover and brood survival is lower on less diverse farms because reduced insect abundance forces birds to move more while foraging, resulting in greater energy expenditures and increased vulnerability to predation. Interspersion of corn fields, pasture and woodlands was needed to provide escape and roosting cover in close association with food required by Wild Turkey to survive overwinter in the eastern USA (Vander Haegen et al. 1989). In central Canada, the abundance of Red-winged Blackbird increased as hayfield hectarages declined and agricultural landscape diversity increased (Clark et al. 1986). In prairie Canada, the nesting density of Ferruginous Hawks declined with increasing cultivation in the landscape (Smutz 1987). In contrast, Swainson's Hawks were more abundant in areas of moderate cultivation than in grassland or in areas of extensive cultivation, partly because they were better able to utilize the prey available in crop fields and were more tolerant of agricultural activity than Ferruginous Hawks.

Wetlands and Riparian Habitats

Plant-species composition and vegetation structure of wetlands and riparian habitats influence the presence and abundance of Neotropical migratory species, particularly those with specific habitat requirements (Table 14-5). Habitat size is also important. Species richness increases with marsh area and with the area or width of riparian forests. Neotropical migratory species account for 45% of 58 area-sensitive bird species in riparian habitats and 11% of 18 species in marshes (Fig. 14-3).

Patch shape and isolation are also important (Table 14-5). Isolated wetlands support fewer species and lower abundances of some migratory species than wetlands in complexes. Isolation appears to be more important than area when wetlands are small (Gibbs et al. 1991). Abundances of some migratory species are higher in the interior compared to the edge of riparian habitats. Development of riparian forests on the Great Plains during the last 100 years has facilitated the movement of eastern bird species into Colorado (Knopf 1986). In contrast, riparian habitats in western coniferous forests may not serve as corridors for species associated with upland areas (McGarigal and McComb 1992).

The interaction of patch size, shape, and geographical orientation in the landscape may also influence the relative importance of riparian habitats to Neotropical migratory birds (Table 14-5). Species abundances were significantly different between west-slope and east-slope canyons in Nevada (Dobkin and Wilcox 1986). In Wyoming, larger patches of riparian woodland oriented perpendicularly to the northerly migration path have more species and higher nest abundances of cavity-nesting species (Gutzwiller and Anderson 1992). By manipulating landscape structure, species richness and abundance during the breeding season could be changed by maximizing (or minimizing, if effects are detrimental) the interception of dispersing or migrating birds (and other taxa).

Habitat juxtaposition and interspersion can affect birds in riparian habitats as well as the surrounding landscape (Table 14-5). Carothers et al. (1974) attributed higher bird abundances in riparian forests adjacent to

Table 14-5. Effects of landscape structure on Neotropical migratory species breeding in wetlands and riparian habitats.

Landscape Structure	Effects	References[a]
Landscape composition		
Patch type	Plant-species composition and vegetation structure influence the presence and abundance of species, particularly those with specific habitat needs	1–6, 19
Habitat proportions	Species richness and abundance of some species higher in wetlands with more additional wetland nearby (≤5 km)	3, 6, 7
	Development of riparian forest on the Great Plains facilitates movement of eastern bird species into Colorado	8
Landscape configuration		
Patch size	Species richness increases with wetland area	3, 6, 7, 9
	Species richness increases with area or width of riparian forests	1, 2, 18
	Species in smaller wetlands and riparian forests limited subset of species in larger areas	2, 3, 7, 9, 10, 18
Patch shape	Abundance of migrants higher in riparian interior	11
Geographic orientation	Riparian woodlands oriented perpendicularly to migration pathways have more nests of migrant species	12
	Species abundances in riparian forests different between west-slope and east-slope canyons	2
Habitat juxtaposition/ interspersion	Bird abundance higher in riparian woodland surrounded by agricultural habitats	13
	Riparian birds use adjacent habitats	11, 14
	Bird abundances higher in upslope than streamside habitats in coniferous forest in Oregon	15
	Abundances higher in riparian–wetland habitats in undisturbed than disturbed landscapes	16, 17

[a] References: 1, Stauffer and Best (1980); 2, Dobkin and Wilcox (1986); 3, Gibbs et al. (1991); 4, Strong and Bock (1990); 5, Bull and Skovlin (1982); 6, Craig and Beal (1992); 7, Brown and Dinsmore (1986); 8, Knopf (1986); 9, Tyser (1983); 10, Gutzwiller and Anderson (1987); 11, Szaro and Jakle (1985); 12, Gutzwiller and Anderson (1992); 13, Carothers et al. (1974); 14, Hehnke and Stone (1978); 15, McGarigal and McComb (1992); 16, Croonquist and Brooks (1991); 17, Croonquist and Brooks (1993); 18, Keller et al. (1993); 19, Farley et al. (1994).

farmland in Arizona to the productivity of nearby agricultural and second-growth fields and pastures. For example, Western Kingbird and Cassin's Kingbird, both Neotropical migratory species, had high densities in riparian woodlands but foraged in nearby agricultural habitats. Both of these species were absent in another reach of the same river with a different landscape setting (Stamp 1978). Riparian species can also supplement the avifauna in adjacent habitats (Table 14-5). In California, 32% fewer species and 95% fewer birds used agricultural habitats from which adjacent riparian vegetation had been removed than on agricultural habitats in association with riparian vegetation (Hehnke and Stone 1978). In contrast to riparian habitats in semi-arid areas, most bird species were more abundant in adjacent, upslope habitats than along streams in coniferous forests in the Pacific Northwest (McGarigal and McComb 1992).

Croonquist and Brooks (1991) assessed the cumulative effects of human activities on birds in riparian-wetland habitats in central Pennsylvania by comparing among landscapes in a protected watershed and in a watershed disturbed by agricultural and residential development. Forty per cent of all birds in the undisturbed watershed and undisturbed portion of the disturbed watershed were Neotropical migrants. Their numbers dropped by half in the disturbed portion of the disturbed watershed. They attributed differences in Neotropical migrants to differences in landscape patterns between the two watersheds. In the disturbed watershed, Neotropical migrants concentrated along the riparian fringe where some natural vegetation remained (Croonquist and Brooks 1993). Many Neotropical migrants were only noted during migration at disturbed sites. Despite limited and only seasonal use by Neotropical migrants of disturbed areas, Croonquist and

Brooks (1993) suggest that riparian corridors, whether undisturbed or partially disturbed, are vital feeding, resting and migrating corridors for uncommon sensitive species and warrant protection. Temporal variability in habitat use can complicate the identification of critical habitat. For example, the Snail Kite, a wandering species, only uses small wetlands during drought in Florida (Takekawa and Beissinger 1989). Although these small wetlands are little used in nondrought years, they are nonetheless, critical habitats for the conservation of this species.

Urban

Substantial work in cities, suburbs, and towns has characterized urban bird populations of North America (e.g., Emlen 1974, Lancaster and Rees 1979, Williamson and DeGraaf 1981, Beissinger and Osborne 1982, Gotfryd and Hansell 1986, Mills et al. 1989). Many native birds are common in urban habitats, especially in areas where native plants are used for plantings around homes, businesses and parks (Mills et al. 1989). Neotropical migratory birds can be relatively rare in urban settings. For example, in a survey of 19 woodlots in neighborhoods of four cities in southern Ontario, only one of the ten most common species was a neotropical migrant, while nine of the ten rarest species recorded were migratory birds (Gotfryd and Hansell 1986). Still, some migratory birds can be reasonably common in urban neighborhoods (Geibert 1980, Gotfryd and Hansell 1986, Mills et al. 1989).

Rosenberg et al. (1987) have suggested that urban environments can be used to mitigate the loss of endangered habitats if native plants are used in neighborhood vegetation plantings. The importance of such plantings to Neotropical migratory birds will depend on the spatial distribution of suitable habitat patches across the urban region; such effects of landscape structure in urban environments are unstudied.

Urban environments can be particularly hazardous for breeding birds. Nest predators such as jays and chipmunks are often common, as are nest parasites (cowbirds). In addition, human impacts on the environment (Matlack 1993) and the birds themselves can be intense. In general, species that are common in urban environments are those that have adaptations that allow them to overcome these hazards. For instance, cavity nesters, such as chickadees and titmice, are common in many North American towns and cities; these birds have lowered nest predation compared to most open-cup nesters. Landscape changes that enhance predator or nest-parasite populations will exacerbate these interspecific interactions.

A few management recommendations have been proposed to retain native bird populations in urban settings. The most important factor is undoubtably the retention of native vegetation (both plant species and vegetation structure) in landscaping (Mills et al. 1989). It is especially important to retain a natural distribution of vegetation in the ground and shrub layers to retain native birds that depend on resources in these layers (Geibert 1980, Mills et al. 1989). To minimize human impacts on suburban forests, Matlack (1993) recommends that land-use planning should restrict road access, avoid creating fragments with a diameter of less than ca. 150 m, and use well-maintained footpaths to channel pedestrian traffic away from floristically sensitive areas and to affect positioning of campsites by reducing the perceived isolation of the fragment interior.

Fewer management recommendations have been offered for landscape-level patterns. Gotfryd and Hansell's (1986) study of Toronto woodlots suggested the following guidelines: (1) wooded areas should be about 7 ha in size to maximize the number of native bird species (both resident and migratory); (2) undulating edges between patches, perceptible at the 5–10 m scale, are more attractive to birds than straight edges; and (3) a dense ground layer of vegetation should be retained. Gotfryd and Hansell also recommended urban woodlots should have long perimeters to maximize edge habitat. However, the species that were most commonly detected in this study were resident edge specialists, not Neotropical migrants. Maximization of edge habitat is not a recommended goal for urban wildlife habitat because increases in edge would exacerbate edge-related problems such as predation and nest parasitism.

Migration

Habitat use during migration has not been studied extensively. However, availability of woody habitats appears to be an important component of landscape structure for birds on migration. During both spring and fall migration, species richness and abundance was higher in forest edges and grassed waterways bounded by wooded hedgerows than in either habitat alone (Morgan and Gates 1982). At the peak of spring trans-Gulf migration, significantly more birds than expected use scrub/shrub, pine forest and relic dune habitats on the Mississippi coast, with scrub/shrub having the highest species richness and abundance of the five habitat types available (Moore et al. 1990).

Patch size continues to be an important aspect of landscape structure during migration. The number of bird species occupying maritime hammocks in Florida during spring migration was higher in large (>20 ha) than in small (<5 ha) areas (Cox 1988). Five species (four are Neotropical migrants) show a significant preference for larger habitat size during migration (Fig. 14-3). No Neotropical migrants have shown a significant preference for small hammocks. Numbers of species and birds during migration increase with greater area of shelterbelts (Martin 1980), woodlots (Blake 1986), and riparian woodlands (Willson and Carothers 1979). When measured, isolation and habitat diversity were relatively less important. In Illinois, species richness in woodlots was greater during migration than the breeding season due to the presence of transient species and to the occurrence of species (e.g., warblers) in woodlots smaller than those typically required for breeding (Blake 1986). However, Blake (1986) cautions that, while small woodlots may increase the ability of migrants to traverse highly disturbed landscapes, they may not be sufficient as refuges for migrants if larger blocks of forest are not also available.

The interaction of patch size, shape, and geographical orientation may also influence the relative importance of different habitat patches in the landscape to migrating birds. Large, oblong patches of riparian woodland oriented perpendicular to migration (or dispersal) paths, intercept more migrating (or dispersing) birds (among other taxa), and thereby influence species richness and abundance during the breeding season (Gutzwiller and Anderson 1992).

Neotropics

Neotropical migrants use a wide variety of natural and disturbed habitats in overwintering areas (Keast and Morton 1980, Hagan and Johnston 1992). Some migratory species are associated mainly with mature forest while others reach their highest abundance in early-successional habitats, agricultural fields and pastures (Greenberg 1992, Lynch 1992, Robbins et al. 1992). For a given migratory (or resident) species, patterns in habitat use can vary from one geographic area to another (Hutto 1992), within and among years (Karr and Freemark 1983) and diurnally (Staicer 1992). As a consequence, a number of authors have concluded that a diverse complement of habitats within the regional mosaic is required for the temporal persistence of diverse assemblages of migratory (and resident) species (Karr and Freemark 1983, Kricher and Davis 1992, Petit et al. 1992, Staicer 1992). Within these mosaics, mature forest is expected to function as a source habitat for many migratory (and resident) species. Habitat loss and degradation pose a serious threat for overwintering migrants. Diamond (1991) estimated that 31 Neotropical migratory species (representing over half of all migratory species that breed in Canadian forests) are likely to lose more than 25% of the winter habitat they had in 1985 by the year 2000. Eleven of these are expected to lose 50% or more. Potential impacts of degradation of remaining habitat from practices such as overgrazing, intensive management or use of pesticides have been acknowledged (Lynch 1992, Robbins et al. 1992) but remain unstudied.

Askins et al. (1992) found that the number of migratory species was significantly higher at points in extensive tracts of moist forest on St John than in small remnant patches of moist forest on St Thomas in the US Virgin

Islands. The abundance of migrants was significantly higher in all habitats on St John than comparable habitats on St Thomas. They suggested that the relatively low density and species richness of migratory birds on St Thomas was related to the widespread destruction and degradation of forest on this island. In contrast, Robbins et al. (1992) found that many (but not all) of 12 species analyzed (data aggregated for seven countries) used isolated forests fragments (5–50 ha) at densities comparable to those in extensive forest. Greenberg (1992) found that even small patches (i.e., 8–10 or more small trees) of native forest increase use of agricultural habitats by some migratory species. Staicer (1992) observed that mature trees in successional habitat or pasture were important roost sites for migrants foraging in adjacent second-growth scrub.

CONSERVATION AND MANAGEMENT IMPLICATIONS

It is becoming increasingly recognized that the conservation and management of Neotropical migratory birds needs to include a landscape perspective. However, our current understanding is largely based on landscape studies during the breeding season of Neotropical migratory birds in central and northeastern hardwood forests, and the resident Spotted Owl in western coniferous forests. More detailed and geographically extensive landscape studies of Neotropical migratory birds are needed to provide regionally specific information and to compare among regions. For example, it is difficult to generalize from landscape studies in western and eastern forests because landscape studies in the west have been conducted primarily in extensively forested landscapes fragmented by silvicultural activities, while those in the east have concentrated on forest remnants in agricultural and suburbanized landscapes. Interfacing studies are needed to investigate potential differences in responses of Neotropical migratory birds to changes in landscape structure along the fragmentation gradient from farmland to extensive forest both within and between these regions. Neotropical migratory species may be particularly sensitive to changes in landscape composition and configuration where habitat conditions are marginal such as at the periphery of their range (Martin 1992, Villard et al. 1993), and in intensively managed landscapes, such as farmland.

Statistically rigorous approaches for conducting comparative landscape studies need to be developed. Poorly designed studies can generate meaningless and misleading relationships, miss important interactions, and confound time or space scales affecting ecological patterns (Wiens 1989, Burel 1992, Pace 1993). An important part of comparative landscape research will be to test hypotheses using independent data and by altering landscape structure experimentally (Pace 1993). In this regard, landscape experiments (cf. Lovejoy et al. 1986, Franklin 1989, Robinson et al. 1992, Rodenhouse et al. 1992), demonstration areas and adaptive management strategies (Bormann et al. 1994) should be useful. Longer-term observation and smaller scale experiments embedded in larger-scale studies should help to distinguish among multiple pathways of effects. To be successful, however, experiments will likely require substantial manipulations that create strong contrasts in the manipulated state variable(s) or input(s), and long-term durations relative to the return time of the ecological processes under study (Carpenter et al. 1993). Both experimental and empirical approaches would benefit from better development and application of scaling theory (Wiens 1989, Gardner and Turner 1991, Allen and Hoekstra 1992).

To date, landscape studies of Neotropical migratory birds have been largely based on short-term studies of species distributions during the breeding season. The influence of landscape structure on migrating birds needs much more study. The current focus on landscape composition in overwintering areas in the Neotropics should be continued, but needs to be extended to consideration of landscape configuration and potential impacts of management activities (e.g., pesticide use).

A cursory analysis of distributional data suggest that a total of 158 bird species are intolerant of forest fragmentation or a reduction in habitat area; 43% of these apparently "area-sensitive" species overwinter in the Neotropics (see Appendix). To be useful in a management context, more detailed analyses are needed to investigate apparent differences in area sensitivity of species, particularly within a region. More importantly, however, recent work in central and northeastern hardwood forests during the breeding season clearly shows that more detailed information on habitat- and patch-specific demography is needed to assess the influence of patch size and other aspects of landscape structure on neotropical migratory birds more accurately.

Comparative studies of habitat use are still needed in most regions, particularly western forests, boreal forests, and farmland. The extent to which individual species require a mosaic of habitat types in a given season or are affected by variation in habitat conditions between seasons and years has not been well studied in any region. Impacts of human activities on habitat quality also need to be more explicitly considered, particularly in intensively managed landscapes, such as farmland. Effects of patch size and isolation on birds breeding in and adjacent to croplands warrants more study in North America. Much more information is needed on dispersal characteristics of Neotropical migratory species (especially first-time breeders), particularly in relation to interpatch distance and dispersal routes (e.g., corridors) in the landscape. Corridor function will be best quantified by comparing the intensities of dispersal flows through the corridor versus adjacent habitats (Opdam 1990). The need for metapopulation studies on temperate breeding areas is discussed at length below.

Metapopulations

Understanding the relationship between landscape structure, management practices, species' distributions, population demographics, and metapopulation dynamics is an important prerequisite for developing and implementing effective conservation and management plans for Neotropical migratory birds. There is some evidence that forest bird populations (including Neotropical migratory species) function as metapopulations within landscapes (Freemark 1989, Stacey and Taper 1992, Villard et al. 1992, 1995, McKelvey et al. 1993, Probst and Weinrich 1993) and among landscapes in a region (Temple and Cary 1988, Robinson 1992). Field data in support of "source–sink" metapopulation structure for Neotropical migratory birds are currently limited to temperate forests in the east (Villard et al. 1992, 1993, 1995) and in the midwest (Temple and Cary 1988, Gibbs and Faaborg 1990, Robinson 1992, Probst and Weinrich 1993).

Metapopulation theory provides an important context for developing conservation and management strategies based on nodes and networks of breeding habitat within and among landscapes (Dyer and Holland 1991, Hudson 1991, Murphy and Noon 1992). For at least some Neotropical migratory bird species, understanding metapopulation dynamics may be essential if viable regional populations are to be maintained (Temple and Cary 1988, Robinson 1992, Probst and Weinrich 1993). Without considering metapopulation dynamics, land managers may misinterpret immigration to and local extinction in sink areas as a population response to management actions. Buffer areas that create sink habitat can jeopardize metapopulation persistence if sink habitats are preferentially used over source habitats (McKelvey et al. 1993). Cumulative impacts of habitat alterations may be underestimated, particularly for productive source areas. For example, actions that reduce the abundance and size of suitable habitat below extinction thresholds for the metapopulation may lead to the regional extirpation of a species even if some habitat of suitable quality remains (Davis and Howe 1992, Lamberson et al. 1992). The failure of existing bird-habitat models to predict population density adequately among different locations and times is related, at least in part, to such landscape-level phenomenon (Van Horne and Wiens 1991).

Effective conservation of Neotropical migratory birds may require the preservation of suitable but intermittently unoccupied habitat (Takekawa and Beissinger 1989). Efforts to identify critical habitat areas and landscapes need to consider differences in population demography and variability, and species-specific dispersal characteristics as well as population density (Van Horne 1983, Pulliam 1988, Murphy and Noon 1992, Probst and Weinrich 1993). In the absence of such information, management plans should protect the diversity of habitats and landscapes used by a species, not just where the species is most common. In some situations, a diversity of habitats and landscapes may be maintained by attempting to mimic the composition and geometry of presettlement landscapes (Thomas et al. 1990, Hunter 1992, Hejl 1992).

A great deal of study is needed on the genetic consequences of metapopulation dynamics (Holsinger 1993, McCauley 1993). The mode of evolution in metapopulations remains largely unstudied. Better theoretical analyses of metapopulation processes that explicitly deal with ecological dynamics of gene flow, local population extinction, and recolonization are needed. Ecological circumstances governing dynamics of local population extinction and recolonization need to be better understood. Empirical evidence is needed to test theories of gene flow for birds (and other vertebrates) and to demonstrate effects of population turnover on genetic structure. The relationship between genetic structure, genetic variability and short- and long-term viability of populations and species needs to be more clearly established. For management purposes, particular attention needs to be paid to transient dynamics as changes in landscape structure fragment previously continuous populations.

Landscape Analysis

Landscape analysis considers the influence of landscape structure on ecological characteristics, such as Neotropical migratory birds. To do this, landscape structure needs to be measured in ways that have ecological meaning to the organism(s) or process(es) of interest (O'Neill et al. 1988, Turner 1989, Turner and Gardner 1991, Pearson et al. 1995). One difficulty in quantifying effects of landscape structure stems largely from the inability to distinguish, a priori, important relationships from merely interesting ones (Gardner and Turner 1991).

For Neotropical migratory birds, ecological relevance is best achieved if the grain and extent of the population(s) of interest closely match the grain and extent of the defined landscape. In the latter context, extent is defined as the area included within the landscape boundary. Grain is the size of the individual units of observation (e.g., pixel size for raster data). In analyses of landscape structure, extent and grain define the upper and lower limits of spatial resolution, and contrain the statistical inferences that can be made about scale dependency in the population of interest (Wiens 1989). In practice, extent and grain are often dictated by the scale of the landscape data available (e.g., aerial photography scale), logistical constraints imposed by field surveys, technical capabilities of the computing environment, and sample sizes of landscapes required for robust statistical analyses. Attempts to compare landscapes of different grain and extent need to be done cautiously until more is known about the sensitivity of measures of landscape structure to changes in scale (McGarigal and Marks 1995).

For Neotropical migratory birds, relevant measures of landscape composition include the proportion of the landscape in each patch type, patch richness, patch evenness, and patch diversity (Table 14-6). Many of these indices have direct counterparts in the indices used to quantify habitat variation at local scales. For instance, patch richness (the number of different patch types within a landscape) is analogous to plant species richness (the number of plant species in a patch).

Many measures of landscape configuration have been developed to quantify the spatial character and arrangement of patches. Metrics can be formulated for individual patches, different habitat types, or the entire landscape mosaic depending on the question of concern (Table 14-6). For example, patterns of habitat fragmentation can be summarized for different habitat types, or the whole landscape by the mean and variance of patch sizes. Quantifying the total area and mean patch size of "core" habitat in the landscape could provide better measures of landscape suitability for species that are sensitive to adverse edge effects from surrounding patches. Patch dominance allows one to

Table 14-6. Measures useful in quantifying landscape structure important to neotropical migratory birds. See Turner (1989) and McGarigal and Marks (1995) for a more complete listing and additional details.

Measure	Method of calculation	Level
Patch type	Area within the landscape	Habitat type[c]
Patch size	Area of a patch	Patch Habitat type[a] Landscape[b]
Core area	Area within patch(es) > specified distance from the patch perimeter	Patch Habitat type[a,c] Landscape[b,d]
Patch shape index	Perimeter/Area^2 compared to a circular standard	Patch Habitat type[a] Landscape[b]
Fractal dimension (d)	$d = 2 \log$ (patch perimeter)/log (patch area)	Patch Habitat type[a] Landscape[b]
Nearest-neighbor distance	Distance to nearest neighboring patch of same type measured edge-to-edge or center-to-center	Patch Habitat type[a] Landscape[b]
Proximity index	Sum of patch area divided by (nearest edge-to-edge distance)^2 between a patch and a focal patch of the same patch type whose edges are within a specified distance	Patch Habitat type[a] Landscape[b]
Patch dominance (largest patch index)	Area of the largest patch divided by total landscape area	Habitat type Landscape
Patch density	Number of patches divided by total landscape area	Habitat type[c] Landscape[d]
Edge density	Sum of lengths of all edge segments divided by total landscape area	Habitat type[c] Landscape[d]
Contrast-weighted edge density	Sum of lengths of edge segments multiplied by corresponding contrast weight divided by total landscape area, where contrast weight based on difference between habitat types	Habitat type[c] Landscape[d]
Habitat richness	Number of different habitat types	Landscape
Habitat diviersity	An index of proportional abundance of each habitat type (e.g., Shannon's, Simpson's)	Landscape
Habitat evenness	An index of distribution of proportional abundances among habitat types	Landscape
Interspersion, juxtaposition, or contagion	Distribution of adjacencies among habitat types	Landscape

[a] Averaged over all patches of the same habitat type in the landscape.
[b] Averaged over all patches in the landscape.
[c] Summed over all patches of the same habitat type in the landscape.
[d] Summed over all patches in the landscape.

separate mainland-island metapopulations (which have a large dominance index) from classic-style or source-sink metapopulations (which have a smaller patch dominance index). The use of fractal dimension to represent relative edge influence can more closely reflect the actual details of patch perimeter (at a given scale), and thus may be a more accurate predictor of edge-to-interior ratios than other shape indices (Milne 1991). Some metrics, such as the nearest-neighbor distance and proximity index, measure the placement of individual patches relative to other patches of the same patch type. Measures of juxtaposition, interspersion, and contagion may be important to species that require close proximity of resources provided by different patch types or are adversely affected by the edges created by adjacency of certain patch types. Some landscape metrics, such as patch density, edge density, and contrast-weighted edge density, represent a combination of landscape composition and configuration that can be important to Neotropical migratory birds.

Until recently, most efforts to quantify landscape structure have been tailored to meet the needs of specific research objectives and have performed analyses with user-generated computer programs limited to particular hardware environments. McGarigal and Marks (1995) have published a versatile computer program, which offers a comprehensive choice of landscape metrics for raster or vector images. Their supporting documentation presents a cogent discussion of important concepts and definitions critical to landscape analysis, fully describes each metric, and discusses its ecological application and limitations. At present, there is a strong movement to emphasize simple indices with clear interpretations as measures of landscape structure (S. Pearson, personal communication). Further study is needed to evaluate the performance of metrics both within and across a variety of spatial and temporal scales (McGarigal and Marks 1995) and to incorporate metrics for narrow habitats (e.g., fencerows, riparian areas). Developing biologically relevant measures of patch isolation and connectivity remains a challenge. Approaches for characterizing landscape structure using a suite of metrics need to be developed and evaluated in relation to their significance to population, community, and ecosystem dynamics across landscapes (Turner and Gardner 1991, Dunning et al. 1992). Use of spatially explicit, computer simulation models may help to identify aspects of landscape structure that warrant field study.

Computer Simulation Studies and Geographic Information Systems

Part of the difficulty in establishing the importance of landscape-level effects is the impracticality of performing manipulative experiments at the spatial scales involved. Few managers have the resources, time, and land required to do a trial run of a management plan and determine the effects on bird communities across a large area. Spatially explicit computer models can be used to help focus research, to design monitoring and management efforts, and to simulate short- and long-term impacts on Neotropical migratory birds of current and alternative management strategies for a landscape (Shaffer 1985, Lande 1988, Murphy and Noon 1992, Dunning et al. 1995, Turner et al. 1995, Pearson et al. 1995). For the greatest accuracy in developing management strategies, models should be tailored to meet specific local conditions of landscape structure and should use locally accurate information of habitat-specific demography and dispersal behavior (Hansen et al. 1992, Pulliam et al. 1992). Such information is rare, however, and research on habitat-specific demography and on dispersal in complex landscapes should be a priority.

To create locally accurate models of populations in landscapes, a major innovation is the linkage of simulation models with geographic information systems (GIS). Management agencies are increasingly turning to GIS technology to map their holdings, and these databases are ideal for generating current and future landscapes maps. By linking these maps to a population simulation model, a manager can observe the impact of a change in management strategy (e.g., habitat type/interspersion objectives, cropping/rotation patterns, pesticide use) on the actual landscapes where the changes have been proposed (Pulliam et al. 1992, Dunning et al. 1995).

Spatially explicit population models have been developed for several nonmigratory bird species, most notably the Bachman's Sparrow in the southeastern United States (Pulliam et al. 1992), and the Spotted Owl in the Pacific Northwest (McKelvey et al. 1993). These models demonstrate that bird populations can be affected by management practices that change the dispersion of suitable habitat patches across the landscape. In particular, population size decreases and the probability of local extinction increases, as suitable patches become more isolated, and as the total amount of suitable habitat decreases in the landscape, especially habitat types that commonly produce dispersing individuals.

Simulation models that use hypothetical landscapes, instead of GIS-derived maps, have shown how bird populations in general may be affected by changes in the landscape (Urban and Shugart 1986, Lande 1988). One Neotropical migratory bird model using hypothetical landscapes has been published. With this model, Temple and Cary (1988) demonstrated that decreases in reproductive success associated with edge effects in highly fragmented landscapes can produce population declines similar to those reported for migratory birds in eastern temperate forests. Thus, this model supports the position that landscape changes on the breeding grounds can account for much of the population declines reported for these species.

To date, no spatially explicit models using GIS-derived landscapes have been developed specifically for Neotropical migrants. The existing models developed for resident species could be modified easily to meet the life-history characteristics of a migratory bird. The major change required would be in the dispersal subroutines. Resident populations are generally assumed to recruit new breeders from locally produced offspring. Thus, the previous year's reproductive success can be used to estimate numbers of dispersers in a model of a resident bird species. Migratory populations, on the other hand, receive dispersers from a much wider region. The disperser pool for a migratory bird model will have to be estimated with different techniques than are currently used in models of resident birds.

Spatially explicit models are extremely data intensive because they require tremendous amounts of locally accurate, habitat-specific information on reproductive success, survivorship and dispersal. Such data are unavailable for most birds, although detailed information can be found in the literature for some species. The new Birds of North America series (e.g., Confer 1992) provides excellent summaries of known information on natural history, demography, and dispersal. This information will need to be supplemented by local studies in most cases. In particular, better field data on dispersal, especially for juveniles, are needed (Pulliam et al. 1992, Villard et al. 1992, 1993, 1995).

Hierarchical Conservation and Management Planning

By focusing on intermediate spatial scales, landscape studies form a bridge in the spatial hierarchy between habitat studies at local (within-patch) scales and larger scale regional and biogeographical studies (Allen and Hoekstra 1992, Dunning et al. 1995). To incorporate landscape patterns and processes, however, conservation and management plans must be developed and implemented over larger spatial and longer temporal scales than in the past (Pickett and White 1985, Turner et al. 1995). It is also critical to interpret and integrate landscape studies in relation to studies done at other spatio-temporal scales during breeding, wintering, and migratory seasons in order to plan future research, and to develop and implement more comprehensive conservation and management strategies (Probst and Crow 1991, Freemark et al. 1993). A logical consequence of the multiscale view of "landscape" from an organism-centered perspective is a mandate to manage wildlife habitats across the full range of spatial scales because each scale will likely be important for a subset of species whether Neotropical migratory birds, other animal taxa, or plants (McGarigal and Marks 1995).

In a previous paper, we presented a hierarchical approach to understanding population processes of Neotropical migratory birds and other biota (Freemark et al. 1993). In that approach (Fig. 14-4), studies at different spatio-temporal scales and levels of resolution are linked together for broad considerations about distribution patterns and their causative mechansms. In this way, it should be possible to predict and test multiscale interrelationships by working progressively from: (1) geographic distribution patterns; (2) habitat relationships; (3) variability in landscape-local distribution; (4) indirect productivity measurements, (5) direct productivity measurements; (6) survivorship estimates; and finally to (7) assessments of viability and population modeling.

The first step is to document continental distribution from existing information, such as field guides, breeding bird atlases, and other available sources. This step helps determine centers of distribution and importance of various regions to each species [see regional status reports in Finch and Stangel (1993)]. Range mapping can easily be extended to physiographic regions, landforms and major patterns of land ownership. Such maps provide a first approximation of the area available for conservation assessments, and potential problems and opportunities associated with each ownership.

A further refinement for predicting species distribution ordinates species along gradients, such as latitude, moisture, elevation, succes-

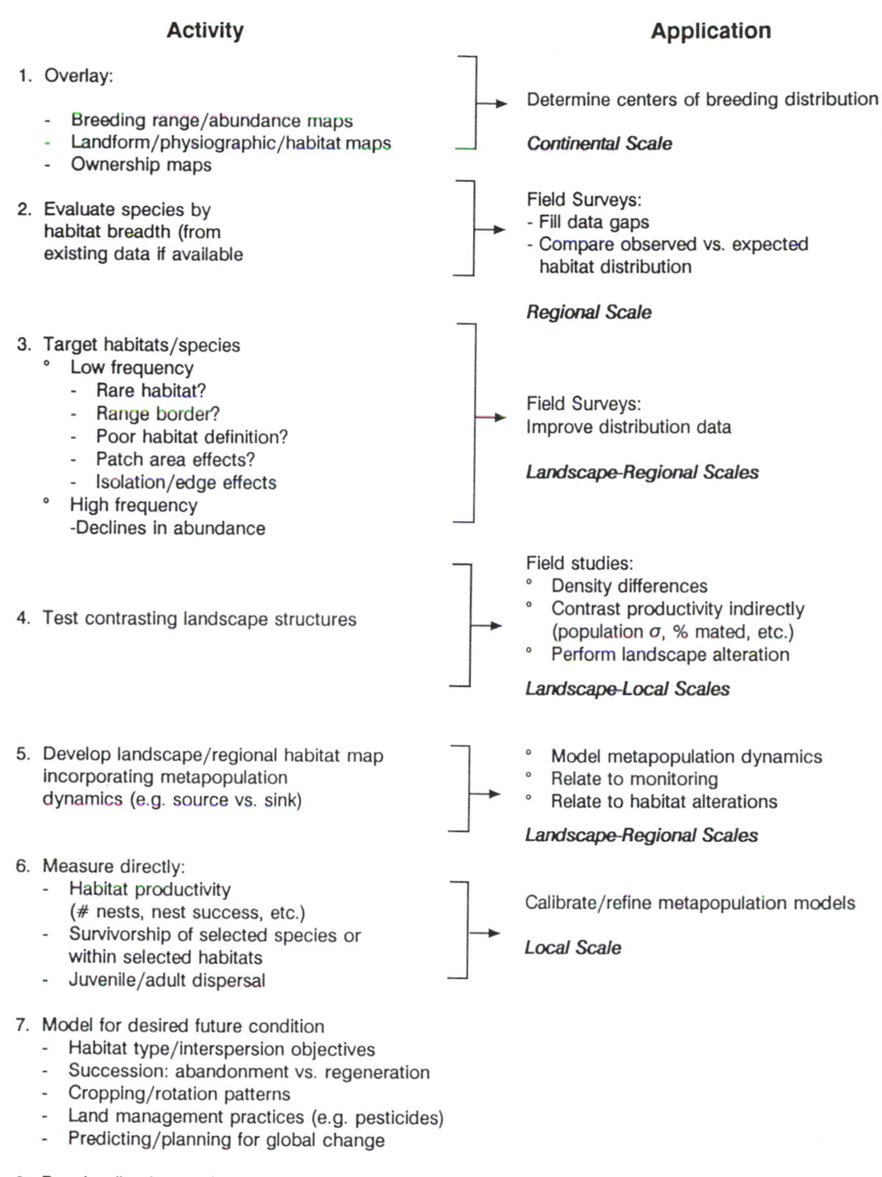

Figure 14-4. A spatially hierarchical framework for developing conservation and management plans for Neotropical migratory bird species (and other biota). Relevant spatial scales are indicated in bold. From Freemark et al. (1993).

sion or life form of vegetation. Using multiple gradients allows for both individual and joint consideration of species without assuming that communities are surrogates for component species. At this stage, field surveys can contrast observed versus expected distributions to modify existing species–habitat relationships. Initial surveys can lead to questions that can be addressed through targeted surveys of specific species or habitats. At this level, frequency data can be used to approximate relative quality of habitats, landscape contexts or geographic regions. Variability in frequency of use within years (e.g., Zimmerman 1982, Probst 1988) and between years (Howe et al. 1991) can add further refinement to assessments of habitat quality by assuming that preferred habitat is selected preferentially.

Habitat quality, biogeographic effects and metapopulation structure can be documented further by higher resolution studies on actual population densities or indirect estimates of productivity, such as pairing success (e.g., Probst and Hayes 1987), or proportion of young relative to adults shortly after fledging. Finally, direct measurements of productivity (nest studies) represent the highest level of resolution. The most difficult estimate to obtain for widespread species is survivorship. Frequently, it is best to solve for survivorship as an unknown variable (e.g., Probst and Hayes 1987) in an otherwise complete demographic equation.

The approach presented above should be time and cost effective, because successive approximations allow for comprehensive consideration of species and geographic scope. More extensive and general survey work on geographic and habitat use can be combined with intensive and detailed field work on productivity and survivorship. In so doing, the approach facilitates implementation of coarse-filter conservation and management strategies, while progressively finer filters are being developed. To be more comprehensive, the approach needs to be extended to include the full complement of species, and ecological and evolutionary processes characteristic of the region of interest.

Developing conservation and management plans for intensively managed areas (e.g., farmland) will require a strong integration with societal values. Freemark (1995) develops the conceptual and scientific basis for a hierarchical approach to ecological risk-assessment in agricultural landscapes. Such efforts would be improved by development and application of scaling theory (Wiens 1989, Gardner and Turner 1991). In practice, the most common approach to dealing with scale is to compare patterns among several arbitrarily selected points on a scale spectrum that reflects our own perceptions of nature. Nonarbitrary, operational ways of defining and detecting scales are needed (see Turner et al. 1991, Pearson 1993). Wiens (1989) contends that research needs to focus on linkages between domains of scale (i.e., how to translate pattern–process relationships across nonlinear spaces between domains of scale). He argues that discovering these linkage requires recognition that: (1) ecological patterns and processes are scale dependent; (2) scale dependency differs for different ecological systems and for different questions that we ask; (3) ecological dynamics and relationships may be well behaved and orderly within domains of scale, but differ from one domain to another and become seemingly chaotic at the boundaries of these domains, and (4) an arbitrary choice of scales of investigation will do relatively little to define these scaling relationships.

Monitoring

There is a particular need to integrate survey, monitoring, and research in the hierarchical planning approach. Current thinking tends to separate research from trends analysis based on survey and monitoring. In addition, there should be a clear distinction between survey (the observation phase) and monitoring (repeated surveys for rigorous hypothesis testing). Currently, almost all coordination on monitoring concerns *standardization* of techniques without first considering the conceptual basis for the sampling strategy. Such a separation of monitoring and research can lead to misinterpretation of monitoring data and delays in problem identification. Deterministic variability, such as metapopulation dynamics, can be confused with stochasticity unless monitoring is properly structured in time and space to answer questions that are specific but broad in scope. By inferring population processes from patterns of distribution, it is possible to test hypotheses about differences in abundance, sex ratio, or age structure, as well as their variability in time.

Recently there has been increased emphasis on population monitoring to verify the status of Neotropical migrants. Some bird monitoring may be misdirected if it is structured on political or ownership boundaries rather than on specific problems (predation, land-use change), species groups, or landscape pattern (habitat distribution or location). For example, there are currently no monitoring plans under development in the Upper Midwest that address major breeding habitats for Cerulean Warbler and several other lowland broadleaf breeding birds. Monitoring based on problems and species' distribution is likely to be very uneven in space in order to focus on problems and comparative situations. It may be unwise to go from inventory to the costly commitment of long-term monitoring before first postulating metapopulation structure (Freemark et al. 1993). When this is done, it is possible to focus efforts on habitats and/or species of concern, which may be quite limited in their distribution.

Metapopulation dynamics may confuse interpretations of monitoring data based solely on abundance. Monitoring of productivity and survivorship is also highly desirable. Indirect measures of productivity such as population variability within and between years (Villard et al. 1992, Probst and Weinrich 1993), mating status of males (Probst and Hayes 1987, Gibbs and Faaborg 1990, Villard et al. 1993) and adult: young ratios (Robinson 1992) are most easily used for more extensive surveys, but need to be supplemented with direct measures of nesting success. Juvenile survival rates, while important, are notoriously difficult to measure because juveniles disperse from their natal areas, and birds that disappear from a study area may survive elsewhere (Pulliam et al. 1992). Estimating adult survival is somewhat easier because most (but not all) adults that nest successfully usually return year after year to the same breeding site.

By understanding the mechanisms underlying population distribution and dynamics, problems associated with nonbreeding habitats (resulting in poor survival) can begin to be separated from problems associated with alteration of breeding habitats (resulting in poor productivity). Such information can by used to evaluate, modify, or design sampling schemes for population or demographic monitoring of specific Neotropical migratory bird species, habitats, geographical areas, or latitudes.

Protected Areas and Cross-boundary Management

Historically, preserving single species has dominated management efforts. Recently, preserving ecosystems has been receiving more attention (Kessler et al. 1992). Approaches cognizant of processes across different spatial and temporal scales are needed for creating or protecting sustainable landscapes (Turner 1989) and preserving biotic diversity, including Neotropical migratory birds (Wilcove 1989, Probst and Crow 1991, Freemark et al. 1993). Hansen and Urban (1992) recommend combining knowledge of species life histories and local landscape dynamics to create management strategies.

The long-term conservation of Neotropical migratory birds (and other biota) is dependent not only on establishment of protected areas, but also on maintaining hospitable environments and viable populations within managed landscapes (Western 1989). Parks and reserves may be important core areas in these landscapes. However, not even the largest national forest or national park is ecologically isolated from activities and conditions in surrounding areas. Furthermore, the viability of species in reserves may often depend on interreserve migration through intervening habitats managed for agricultural or forestry production. Thus conservation efforts for protected areas must also extend to other lands having a variety of uses and ownerships. Public and industrial managers are important to these conservation efforts because they often manage large contiguous tracts of land (Probst and Crow 1991). Such tracts can mitigate the effects of habitat fragmentation in surrounding landscapes.

To be effective, approaches for regional decision-making and cross-boundary management (administratively and on the ground) need to be developed (Headley 1980, Buechner et al. 1992, Schonewald-Cox et al. 1992). Otherwise, the "tyranny of small decisions" (Anglestam 1992) will prevail with many local, relatively unimportant land-use decisions cumulatively resulting in profound landscape changes in the longer term, such as the fragmentation and degradation of southeastern bottomland forests (Burdick et al. 1989, Harris 1989). Approaches will have to include ecological, socio-economic, legal, cultural, ethical, and esthetic considerations (Pimentel and Perkins 1980, Nassauer and Westmacott 1987, Dearden 1988, Hansen et al. 1991, Kessler et al. 1992, Schonewald-Cox et al. 1992). To minimize and resolve conflicts, effective education, communication and carefully designed mechanisms for planning, cooperation and coordination are required (Grumbine 1991, Schonewald-Cox et al. 1992, Slocombe 1993). In this regard, the *Partners in Flight* program and related efforts of its working groups have been important first steps.

Global Change

To predict the effect of global change on Neotropical migratory birds, scientists will need to integrate information on climate, landforms, landscape structure, and dynamics of species' distributions across a hierarchy of spatial and temporal scales (Kareiva et al. 1993). Comparative studies across gradients, regions, or larger geographic areas (e.g., countries, continent, the globe) will be particularly important in predicting the impacts of changes in landscape structure in response to global change and its associated human-driven land-use change (Pace 1993). For example, the International Geosphere–Biosphere Programme is interested in the possible effects of changing the diversity within agricultural and forestry production systems on ecological complexity and function at the regional scale. Agricultural and forestry production systems that are more diverse and complex may not only be more sustainable, but also more conducive to the migration of species among nature reserves and hence, lead to reduced rates of extinction.

Quantitative measures of landscape structure derived from remote-sensing technology can provide appropriate metrics for monitoring regional ecological changes in response to factors such as global change (Gardner and Turner 1991). Potential effects of global change on Neotropical migratory birds (and other biota) can then be inferred from contemporary landscape studies. Use of spatially explicit models should help to focus related research, monitoring and conservation activities in relation to global change. If landscape structure can be linked to population demographics then spatially explicit models could be used to simulate impacts of global change on species, such as Neotropical migratory birds (Dunham 1993, Lodge 1993). Spatially explicit multispecies models also need to be developed to understand expected changes in biotic interactions at broad spatial and temporal scales (Clark 1993, Murdoch 1993).

ACKNOWLEDGMENTS

This paper benefited substantially from comments by Jim Karr, Bob Howe, Eric Preston, Tom Martin and two anonymous reviewers. Funding for K. Freemark was provided by an interagency agreement (DWCN935524) and cooperative agreement (CR821795-01-0) between Environment Canada and the US Environmental Protection Agency (EPA). Research by J. Dunning was supported by Contract DE-AC09-76SR00-819 between the US Department of Energy and the Savannah River Ecology Laboratory at the University of Georgia and also by grants from the EPA (CR820668-01-2) and the USDA Forest Service (12-11-008-876). This paper has undergone EPA peer and administrative review, and has been approved for publication.

LITERATURE CITED

Addicott, J. F., J. M. Aho, M. F. Antolin, J. S. Richardson, and D. A. Soluk. 1987. Ecological neighborhoods: scaling environmental patterns. Oikos 49:340–346.

Allen, T. F. H., and Hoekstra, T. W., 1992. Toward a unified ecology. Columbia University Press, New York.

Aney, W. C. 1984. The effects of patch size on bird communities of remnant old-growth pine stands in western Montana. MS thesis, University of Montana, Missoula, MT.

Anglestam, P. 1992. Conservation of communities —the importance of edges, surroundings and landscape mosaic structure. Pp. 9–70 *in* Ecological principles of nature conservation: applications in temperate and boreal environments (L. Hansson, ed.). Elsevier, New York.

Askins, R. A., and M. J. Philbrick. 1987. Effect of changes in regional forest abundance on the decline and recovery of a forest bird community. Wilson Bull. 99:7–21.

Askins, R. A., M. J. Philbrick, and D. S. Sugeno. 1987. Relationship between the regional abundance of forest and the composition of forest bird communities. Biol. Conserv. 39: 129–152.

Askins, R. A., J. F. Lynch, and R. Greenberg. 1990. Population declines in migratory birds in eastern North America. Curr. Ornithol. 7: 1–57.

Askins, R. A., D. N. Ewert, and R.L. Norton 1992. Abundance of wintering migrants in fragmented and continuous forests in the US Virgin Islands. Pp. 197–206 *in* Ecology and conservation of Neotropical migrant landbirds (J. M. Hagan, III and D. W. Johnston, eds). Smithsonian Institution Press, Washington, DC.

Barbour, M. G., and W. D. Billings. 1988. North American terrestrial vegetation. Cambridge University Press, New York.

Barrett, G. 1992. Landscape ecology: designing sustainable agricultural landscapes. J. Sustainable Agr. 2:83–103.

Behle, W. H. 1978. Avian biogeography of the Great Basin and Intermountain Region. Great Basin Natur. Mem. 2:55–80.

Beier, P. 1993. Determining minimum habitat areas and habitat corridors for cougars. Conserv. Biol. 7:94–108

Beissinger, S. R., and D. R. Osborne. 1982. Effects of urbanization on avian community organization. Condor 84:75–83.

Benkman, C. W. 1989. On the evolution and ecology of island populations of crossbills. Evolution 43:1324–1330.

Best, E. B. 1983. Bird use of fencerows: implications for contemporary fencerow management practices. Wildl. Soc. Bull. 11: 343–347.

Best, L. B. 1986. Conservation tillage: ecological traps for nesting birds? Wildl. Soc. Bull. 14:308–317.

Best, L. B., R. C. Whitmore, and G. M. Booth. 1990. Use of cornfields by birds during the breeding season: The importance of edge habitat. Amer. Midl. Natur. 123:84–99.

Best, L. B., K. E. Freemark, J. J. Dinsmore, and M. Camp. 1995. A review and synthesis of habitat use by breeding birds in agricultural landscapes of Iowa. Amer. Midl. Natur. (in press).

Blake, J. G. 1986. Species–area relationship of migrants in isolated woodlots in east-central Illinois. Wilson Bull. 98:291–296.

Blake, J. G., 1991. Nested subsets and the distribution of birds of isolated woodlots. Conserv. Biol. 5:58–66.

Blake, J. and J. R. Karr. 1984. Species composition of bird communities and the conservation benefit of large versus small forests. Biol. Conserv. 30:173–187.

Blake, J. G., and J. R. Karr. 1987. Breeding birds of isolated woodlots: area and habitat relationships. Ecology 68:1724–1734.

Bollinger, E. K., and T. A. Gavin. 1992. Eastern Bobolink populations: Ecology and conservation in an agricultural landscape. Pp. 483–496 *in* Ecology and conservation of Neotropical migrant landbirds (J. M. Hagan, III and D. W. Johnston, eds). Smithsonian Institution Press, Washington, DC.

Bormann, B. T., M. H. Brookes, E. D. Ford, A. R. Kiester, C. D. Oliver, and J. F. Weigand. 1994. Eastside forest ecosystem health assessment. Vol. v: A framework for sustainable-ecosystem management. USDA Forest Serv. Gen. Tech. Rep. PNW-GTR-331, Pacific Northwest Res. Sta., Portland, OR.

Brown, J. H. 1978. The theory of insular biogeography and the distribution of boreal birds and mammals. Great Basin Natur. Mem. 2:209–227.

Brown, M., and J. J. Dinsmore. 1986. Implications of marsh size and isolation for marsh bird management. J. Wildl. Manag. 50:392–397.

Buechner, M., C. Schonewald-Cox, R. Sauvajot, and B. A. Wilcox. 1992. Cross-boundary issues for national parks: What works "on the ground." Environ. Manag. 16:799–809.

Bull, E. L., and J. M. Skovlin. 1982. Relationships between avifauna and streamside vegetation. Trans. North Amer. Wildl. Nat. Resources Conf. 47:497–506.

Bunce, R. G. H. (ed.). 1990. Species dispersal in agricultural habitats. Belhaven Press, c/o CRC Press Boca Raton, FL.

Bunce, R. G. H., and R. H. G. Jongman. 1993. An introduction to landscape ecology. Pp. 3–10 *in* Landscape ecology and agroecosystems

(R. G. H. Bunce, L. Ryszkowski, and M. G. Paoletti, eds). Lewis Publications, Boca Raton, FL.

Bunce, R. G. H., L. Ryszkowski, and M. G. Paoletti (eds). 1993. Landscape ecology and agroecosystems. Lewis Publications, Boca Raton, FL.

Bunnell, F. L., and L. L. Kremsater. 1990. Sustaining wildlife in managed forests. Northwest Environ. J. 6:243–269.

Burdick, D. M., D. Cushman, R. Hamilton and J. G. Gosselink. 1989. Faunal changes and bottomland hardwood forest loss in the Tensas Watershed, Louisiana. Conserv. Biol. 3:282–292.

Burel, F. 1992. Effect of landscape structure and dynamics on species diversity in hedgerow networks, Landscape Ecol. 6:161–174.

Burel, F. and J. Baudry. 1990. Hedgerow networks as habitats for forest species: Implications for colonising abandoned agricultural land. Pp. 238–255 *in* Species dispersal in agricultural habitats (R. G. H. Bunce, ed.). Belhaven Press, c/o CRC Press, Boca Raton, FL.

Burger, L. D., L. W. Burger, Jr, and J. Faaborg. 1994. Effects of prairie fragmentation on predation of artificial nests. J. Wildl. Manag. 58:249–254.

Burgess, R. L., and D. M. Sharpe (eds). 1981. Forest island dynamics in man-dominated landscapes. Springer-Verlag, New York.

Carey, A. B., J. A. Reid, and S. P. Horton. 1990. Spotted Owl home range and habitat use in southern Oregon coast ranges. J. Wildl. Manag. 54:11–17.

Carothers, S. W., R. R. Johnson, and S. W. Aitchison. 1974. Population structure and social organization of southwestern riparian birds. Amer. Zool. 14:97–108.

Carpenter, S. R., T. M. Frost, J. F. Kitchell, and T. R. Kratz. 1993. Species dynamics and global environmental change: a perspective from ecosystem experiments. Pp. 267–279 *in* Biotic interactions and global change (P. M. Kareiva, J. G. Kingsolver, and R. B. Huey, eds). Sinauer Associates, Sunderland, MA.

Clark, J. S. 1993. Paleoecological perspectives on modeling broad-scale responses to global change. Pp. 315–332 *in* Biotic interactions and global change (P. H. Kareiva, J. G. Kingsolver, and R. B. Huey, eds). Sinauer Associates, Sunderland, MA.

Clark, R. G., P. Weatherhead, H. Greenwood, and R. D. Titman. 1986. Numerical responses of Red-winged Blackbird populations to changes in regional land-use patterns. Can. J. Zool. 64:1944–1950.

Confer, J. L. 1992. Golden-winged Warbler. Pp. 1–15 *in* The birds of North America, no. 20 (A. Poole, P. Stettenheim, and F. Gill, eds). The Academy of Natural Sciences, Philadelphia, PA/The American Ornithologists' Union, Washington, DC.

Conner, R. N., and D. C. Rudolph. 1991. Forest habitat loss, fragmentation, and Red-cockaded Woodpecker populations. Wilson Bull. 103:446–457.

Cox, J. 1988. The influence of forest size on transient and resident bird species occupying maritime hammocks of northeastern Florida. Florida Field Natur. 16:25–34.

Craig, R. J., and K. G. Beal. 1992. The influence of habitat variables on marsh bird communities of the Connecticut River estuary. Wilson Bull. 104:295–311.

Croonquist, M. J., and R. P. Brooks. 1991. Use of avian and mammalian guilds as indicators of cumulative impacts in riparian–wetland areas. Environ. Manag. 15:701–714.

Croonquist, M. J., and R. P. Brooks. 1993. Effects of habitat disturbance on bird communities in riparian corridors. J. Soil Water Conserv. 48:65–70.

Davis, G. J., and R. W. Howe. 1992. Juvenile dispersal, limited breeding sites and the dynamics of metapopulations. Theor. Pop. Biol. 45:184–207.

Dearden, P. 1988. Landscape aesthetics, tourism and landscape management in British Columbia. Pp. 183–190 *in* Landscape ecology and management (M. R. Moss ed.). Polyscience Publications, Montreal, Canada.

DeGraaf, R. M., and D. D. Rudis. 1986. New England wildlife: habitat, natural history, and distribution. USDA Forest Serv. Gen. Tech. Rep. NE-108, Northeastern Forest Exp. Sta., Broomall, PA.

DeGraaf, R. M., M. Yamasaki, W. B. Leak, and J. W. Lanier. 1992. New England wildlife: management of forested habitats. USDA Forest Serv. Gen. Tech. Rep. NE-144, Northeastern Forest Exp. Sta., Radnor, PA.

DeGraaf, R. M., M. Yamasaki, and W. B. Leak. 1993. Management of New England northern hardwoods, spruce-fir and eastern white pine for Neotropical migratory birds. Pp. 363–373 *in* Status and management of Neotropical migratory birds (D. M. Finch and P. W. Stangel, eds). USDA Forest Serv. Gen. Tech. Rep. RM-229, Rocky Mt. Forest Range Exp. Sta., Fort Collins, CO.

Diamond, A. W. 1991. Assessment of the risks from tropical deforestration to Canadian songbirds. Trans. North Amer. Wildl. Nat. Resources Conf. 56:177–194.

Dickson, J. G., F. R. Thompson, III, R. N. Conner, and K. E. Franzreb. 1993. Effects of silviculture on Neotropical migratory birds in central and southeastern oak pine forests. Pp. 374–385 *in* Status and management of Neotropical migratory birds (D. M. Finch and P. W. Stangel, eds). USDA Forest Serv. Gen. Tech. Rep. RM-229, Rocky Mt. Forest Range Exp. Sta., Fort Collins, CO.

Dobkin, D. S., and B. A. Wilcox. 1986. Analysis of natural forest fragments: riparian birds in the Toiyabe Mountains, Nevada. Pp. 293–299 *in* Wildlife 2000: modeling habitat relationships of terrestrial vertebrates (J. Verner, M. L. Morrison, and C. J. Ralph, eds). University of Wisconsin Press, Madison, WI.

Dunham, A. E. 1993. Population responses to environmental change: operative environments, physiologically structured models, and population dynamics. Pp. 95–119 *in* Biotic interactions and global change (P. M. Kareiva, J. G. Kingsolver, and R. B. Huey, eds). Sinauer Associates, Sunderland, MA.

Dunning, J. B., and B. D. Watts. 1990. Regional differences in habitat occupancy by Bachman's Sparrow. Auk 107:463–472.

Dunning, J. B., B. J. Danielson, and H. R. Pulliam. 1992. Ecological processes that affect populations in complex landscapes. Oikos 65:169–175.

Dunning, J. B., D. J. Stewart, B. J. Danielson, B. R. Noon, T. L. Root, R. H. Lamberson, and E. E. Stevens. 1995. Spatially explicit population models: current forms and future uses. Ecol. Applic. 5: 3–11.

Dyer, M. I., and M. M. Holland. 1991. The biosphere-reserve concept: needs for a network design. BioScience 41:319–325.

Eijsackers, H., and A. Quispel (eds). 1988. Ecological implications of contemporary agriculture. Ecol. Bull. (Copenhagen) 39.

Emlen, J. T. 1974. An urban bird community in Tucson, Arizona: derivation, structure, regulation. Condor 76:184–197.

Faaborg, J., M. Brittingham, T. Donovan, and J. Blake. 1993. Habitat fragmentation in the temperate zone: a perspective for managers. Pp. 331–338 *in* Status and management of Neotropical migratory birds (D. M. Finch and P.W. Stangel, eds). USDA Forest Serv. Gen. Tech. Rep. RM-229, Rocky Mt. Forest Range Exp. Sta., Fort Collins, CO.

Fahrig, L., and K. E. Freemark. 1995. Landscape-scale effects for toxic events for ecological risk assessment. Pp. 193–208 *in* Ecological toxicity testing: scale, complexity and relevance (J. Cairns Jr. and B. R. Niederlehner, eds). Lewis Publishers, Boca Raton, FL.

Fahrig, L., and G. Merriam. 1994. Conservation of fragmented populations. Conserv. Biol. 8:50–59.

Farley, G. H., L. M. Ellis, J. N. Stuart, and N. J. Scott Jr. 1994. Avian species richness in different-aged stands of riparian forest along the middle Rio Grande, New Mexico. Conserv. Biol. 8:1098–1108.

Finch, D.M. 1991. Population ecology, habitat requirements, and conservation of Neotropical migratory birds. USDA Forest Serv. Gen. Tech. Rep. RM-205. Rocky Mt. Forest Range Exp. Sta., Fort Collins, CO.

Finch, D. M. and P. W. Stangel (eds). 1993. Status and management of Neotropical migratory birds. USDA Forest Serv. Gen. Tech. Rep. RM-229, Rocky Mt. Forest Range Exp. Sta., Fort Collins, CO.

Firbank, L. G., N. Carter, J. F. Darbyshire, and G. R. Potts. 1991. The ecology of temperate cereal fields. Blackwell Scientific Publications, Boston, MA.

Forman, R. T. T., and M. Godron. 1986. Landscape ecology. John Wiley & Sons, New York.

Forsman, E. D., E. C. Meslow, and H. M. Wight. 1984. Distribution and biology of the Spotted Owl in Oregon. Wildl. Monogr. 84:1–64.

Franklin, J. 1989. Toward a new forestry. Amer. For. Nov/Dec: 37–44.

Franklin, J. F., and R. T. T. Forman. 1987. Creating landscape patterns by forest cutting: ecological consequences and principles. Landscape Ecol. 1:5–18.

Freemark, K.E. 1989. Landscape ecology of forest birds in the northeast. Pp. 7–12 *in* Is forest fragmentation a management issue in the Northeast? (R. M. DeGraaf and W. M. Healy, comps). USDA Forest Serv. Gen. Tech. Rep. NE-140, Northeastern Forest Exp. Sta., Radnor, PA.

Freemark, K. 1995. Assessing effects of agriculture on terrestrial wildlife: Developing a hierarchical approach for the US EPA. Landscape Urban Plan. 31:99–115.

Freemark, K., and C. Boutin. 1995. Impacts of agricultural herbicide use on terrestrial wildlife in temperate landscapes: a review with special reference to North America. Agr. Ecosyst. Environ. 52:67–91.

Freemark, K. E., and B. Collins. 1992. Landscape ecology of birds breeding in temperate forest fragments. Pp. 443–454 *in* Ecology and conservation of Neotropical migrant landbirds (J. M. Hagan, III and D. W. Johnston, eds). Smithsonian Institution Press, Washington, DC.

Freemark, K. E., and H. G. Merriam. 1986. Importance of area and habitat heterogeneity to bird assemblages in temperate forest fragments. Biol. Conserv. 36:115–141.

Freemark, K. E., H. Dewar, and J. S. Saltman. 1991. A literature review of bird use of farmland habitats in the Great Lakes–St

Lawrence region. Tech. Rep. Ser. no. 114, Canadian Wildlife Service, Environment Canada, Ottawa, Canada.

Freemark, K. E., J. R. Probst, J. B. Dunning, and S. J. Hejl. 1993. Adding a landscape ecology perspective to conservation and management planning. Pp. 346–352 *in* Status and management of Neotropical migratory birds (D. M. Finch and P. W. Stangel, eds). USDA Forest Ser. Gen. Tech. Rep. RM-229, Rocky Mt. Forest Range Exp. Sta., Fort Collins, CO.

Fritz, R. S. 1979. Consequences of insular population structure: distribution and extinction of spruce grouse populations. Oecologia 42: 57–65.

Fry, G. L. A. 1991. Conservation in agricultural ecosystems. Pp. 415–443 *in* The scientific management of temperate communities for conservation (I. F. Spellerberg, F. B. Goldsmith, and M. G. Morris, eds). Blackwell Scientific Publications, Oxford, England.

Gardner, R. H., and M. G. Turner. 1991. Future directions in quantitative landscape ecology. Pp. 519–525 *in* Quantitative methods in landscape ecology (M. G. Turner and R. H. Gardner, eds). Springer-Verlag, New York.

Gauthreaux, S. A. 1992. Preliminary lists of migrants for *Partners in Flight* Neotropical migratory bird conservation program. 1991 annual report. *Partners in Flight* 2(1):30.

Geibert, E. H. 1980. Songbird diversity along an urban powerline right-of-way in Rhode Island. Environ. Manag. 4:205–213.

Gibbs, J. P., and J. Faaborg. 1990. Estimating the viability of Ovenbird and Kentucky Warbler populations in forest fragments. Conserv. Biol. 4:193–196.

Gibbs, J. P., J. R. Longcore, D. G. McAuley, and J. K. Ringelman. 1991. Use of wetland habitats by selected nongame water birds in Maine. US Fish Wildl. Serv. Fish Wildl. Res. 9, Washington, DC.

Gilpin, M. 1991. The genetic effective size of a metapopulation. Biol. J. Linnean Soc. 42:165–175.

Gilpin, M. E. 1990. Extinction of finite metapopulations in correlated environments. Pp. 177–186 *in* Living in a patchy environment (B. Shorrocks and I. R. Swingland, eds). Oxford Science Publications, Oxford, England.

Gotfryd, A., and R. I. C. Hansell. 1986. Prediction of bird-community metrics in urban woodlots. Pp. 321–326 *in* Modeling habitat relationships of terrestrial vertebrates (J. Verner, M. L. Morrison and C. J. Ralph, eds). University of Wisconsin Press, Madison, WI.

Greenberg, R. 1992. Forest migrants in non-forest habitats on the Yucatan Peninsula. Pp. 273–286 *in* Ecology and conservation of Neotropical migrant landbirds (J. M. Hagan, III and D. W. Johnston, eds). Smithsonian Institution Press, Washington, DC.

Grue, C. E., L. R. DeWeese, P. Mineau, G. A. Swanson, J. R. Foster, P. M. Arnold, J. N. Huckins, P. J. Sheehan, W. K. Marshall, and A. P. Ludden. 1986. Potential impacts of agricultural chemicals on waterfowl and other wildlife inhabiting prairie wetlands: an evaluation of research needs and approaches. Trans. North Amer. Wildl. Nat. Resources Conf. 51:357–383.

Grumbine, R. E. 1991. Co-operation or conflict? Interagency relationships and the future of biodiversity of US parks and forests. Environ. Manag. 15:27–37.

Gutzwiller, K. J., and S. H. Anderson. 1987. Multiscale associations between cavity-nesting birds and features of Wyoming streamside woodlands. Condor 89:534–548.

Gutzwiller, K. J., and S. H. Anderson. 1992. Interception of moving organisms: influences of patch shape, size, and orientation on community structure. Landscape Ecology 6:293–303.

Hagan, J. W., III and D. W. Johnston (eds). 1992. Ecology and conservation of Neotropical migrant landbirds. Smithsonian Institution Press, Washington, DC.

Hamel, P. B., H. E. LeGrand, M. R. Lennartz, and S. A. Gauthreaux. 1982. Bird–habitat relationships on southeastern forest lands. USDA Forest Serv. Gen. Tech. Rep. SE-22, Southeast Forest Exp. Sta., Asheville, NC.

Hansen, A. J., and F. diCastri (eds). 1992. Landscape boundaries: consequences for biotic diversity and ecological flows. Ecol. Stud. 92, Springer-Verlag, New York.

Hansen, A. J., and D. L. Urban. 1992. Avian response to landscape pattern: the role of species life histories. Landscape Ecol. 7:163–180.

Hansen, A. J., T. A. Spies, F. J. Swanson, and J. L. Ohmann. 1991. Conserving biodiversity in managed forests. BioScience 41:382–392.

Hansen, A. J., D. L. Urban, and B. Marks. 1992. Avian community dynamics: The interplay of landscape trajectories and species life histories. Pp. 170–195 *in* Landscape boundaries: consequences for biotic diversity and ecological flows (A. J. Hansen and F. diCastri, eds). Ecol. Stud. 92, Springer-Verlag, New York.

Hansen, A. J., S. L. Garman, and B. Marks. 1993. An approach for managing vertebrate diversity across multiple-use landscapes. Ecol. Applic. 3:481–496.

Hanski, I. 1991. Single-species metapopulation

dynamics: concepts, models and observations. Biol. J. Linnean Soc. 42:17–38.

Hansson, L. 1991. Dispersal and connectivity in metapopulations. Biol. J. Linnean Soc. 42:89–103.

Harris, L. D. 1984. The fragmented forest. Island biogeography theory and the preservation of biotic diversity. University of Chicago Press, Chicago, IL.

Harris, L. D. 1989. The faunal significance of fragmentation of southeastern bottomland forests. Pp. 126–134 *in* The forested wetlands of the southern United States (D. D. Hook and R. Lea, eds). USDA Forest Serv. Gen. Tech. Rep. SE-5, Southeastern Forest Exp. Sta., Asheville, NC.

Harris, L. D., and P. B. Gallagher. 1989. New initiatives for wildlife conservation. The need for movement corridors. Pp. 11–34 *in* In defense of wildlife: preserving communities and corridors (G. Mackintosh, ed.). Defenders of Wildlife, Washington, DC.

Harris, L. D., and R. D. Wallace. 1984. Breeding bird species in Florida forest fragments. Proc. Annu. Conf. Southeast. Assoc. Fish Wildl. Agencies 38:87–96.

Headley, J. C. 1980. The economic milieu of pest control: have past priorities changed? Pp. 81–97 *in* Pest control: cultural and environmental aspects (D. Pimentel, and J. H. Perkins, eds). Westview Press, Boulder, CO.

Hehnke, M., and C. P. Stone. 1978. Value of riparian vegetation to avian populations along the Sacramento River system. Pp. 228–235 *in* Strategies for protection and management of floodplain wetlands and other riparian ecosystems (R. R. Johnson and J. F. McCormick, tech. coords). USDA Forest Serv. Gen. Tech. Rep. WO-12.

Hejl, S. J. 1992. The importance of landscape patterns to bird diversity: a perspective from the Northern Rocky Mountains. Northwest Environmental J. 8:119–137.

Hejl, S. J., and L. C. Paige. 1994. A preliminary assessment of birds in continuous and fragmented forests of western red cedar/western hemlock in northern Idaho. Pp. 189–197 *in* Interior cedar–hemlock–white pine forests: ecology and management. Proceedings of a symposium held March 2–4, 1993, in Spokane, WA. Washington State University Cooperative Extension, Pullman, WA.

Helle, P. 1985. Effects of forest fragmentation on bird densities in northern boreal forests. Ornis, Fenn. 62:35–41.

Helle, P. 1986. Bird community dynamics in a boreal forest reserve: the importance of large-scale regional trends. Ann. Zool. Fennici 23:157–166.

Herkert, J. R. 1991. Prairie birds of Illinois: population response to two centuries of habitat change. Illinois Natural Hist. Surv. Bull. 34:393–399.

Hill, G. E. 1988. Age, plumage brightness, territory quality, and reproductive success in the Black-headed Grosbeak. Condor 90:379–388.

Hirth, D. H., L. D. Harris, and R. F. Noss. 1991. Avian community dynamics in a penisular Florida longleaf pine forest. Florida Field Natur. 19:33–48.

Holmes, R. T., and T. W. Sherry. 1988. Assessing population trends of New Hampshire forest birds: local vs regional patterns. Auk 105: 756–768.

Holsinger, K. E. 1993. The evolutionary dynamics of fragmented plant populations. Pp. 198–216 *in* Biotic interactions and global change (P. M. Kareiva, J. G. Kingsolver and R. B. Huey, eds). Sinauer Associates, Sunderland, MA.

Howe, R. W. 1984. Local dynamics of bird assemblages in small forest habitat islands in Australia and North America. Ecology 65: 1585–1601.

Howe, R. W., D. M. Roosa, J. P. Schaufenbuel, and W. R. Silcock. 1985. Distribution and abundance of birds in the loess hills of western Iowa. Proc. Iowa Acad. Sci. 92:164–175.

Howe, R. W., G. J. Davis, and V. Mosca. 1991. Demographic significance of sink populations. Biol. Conserv. 57:239–255.

Hudson, W. (ed.). 1991. Landscape linkages and biodiversity. Island Press, Washington, DC.

Hunter, M. L., Jr. 1992. Paleoecology, landscape ecology, and conservation of Neotropical migrant passerines in boreal forests. Pp. 511–523 *in* Ecology and conservation of Neotropical migrant landbirds (J. M. Hagan, III and D. W. Johnston, eds). Smithsonian Institution Press, Washington, DC.

Hutto, R. L. 1992. Habitat distributions of migratory landbird species in western Mexico. Pp. 211–239 *in* Ecology and conservation of Neotropical migrant landbirds (J. M. Hagan, III and D. W. Johnston, eds). Smithsonian Institution Press, Washington, DC.

Johns, B. W. 1993. The influence of grove size on bird species richness in aspen parklands. Wilson Bull. 105:256–264.

Johnson, A. S., and J. L. Landers. 1982. Habitat relationships of summer resident birds in slash pine flatwoods. J. Wildl. Manag. 46:416–428.

Johnson, N. K. 1975. Controls of number of bird species on montane islands in the Great Basin. Evolution 29:545–567.

Johnson, R. G., and S. A. Temple. 1990. Nest predation and brood parasitism of tallgrass prairie birds. J. Wildl. Manag. 54:106–111.

Johnson, W. C., and C. S. Adkisson. 1985. Dispersal of beech nuts by Blue Jays in fragmented landscapes. Amer. Midland Natur. 113:319–324.

Kareiva, P. M., J. G. Kingsolver, and R. B. Huey (eds). 1993. Biotic interactions and global change. Sinauer Associates, Sunderland, MA.

Karr, J. R. 1994. Landscapes and management for ecological integrity. Pp. 229–251 *in* Biodiversity and landscape: a paradox for humanity (R. C. Kim and R. D. Weaver, eds). Cambridge University Press, New York.

Karr, J. R., and K. E. Freemark. 1983. Habitat selection and environmental gradients: dynamics in the "stable" tropics. Ecology 64:1481–1494.

Karr, J. R., and K. E. Freemark. 1985. Disturbance and vertebrates: An integrative perspective. Pp. 153–168 *in* The ecology of natural disturbance and patch dynamics (S. T. A. Pickett and P. S. White, eds). Academic Press, New York.

Keast, A., and E. S. Morton (eds). 1980. Migrant birds in the Neotropics: ecology, behavior, distribution and conservation. Smithsonian Institution Press, Washington, DC.

Keller, M. E., and S. H. Anderson. 1992. Avian use of habitat configurations created by forest cutting in southeastern Wyoming. Condor 94:55–65.

Keller, C. M. E., C. S. Robbins, and J. S. Hatfield. 1993. Avian communities in riparian forests of different widths in Maryland and Delaware. Wetlands 13:137–144.

Kessler, W. B., H. Salwasser, C. W. Cartwright, Jr, and J. A. Caplan. 1992. New perspectives for sustainable natural resources management. Ecol. Applic. 2:221–225.

Knopf, F. L. 1986. Changing landscapes and the cosmopolitism of the eastern Colorado avifauna. Wildl. Soc. Bull 14:132–142.

Kotliar, N. B., and J. A. Wiens. 1990. Multiple scales of patchiness and patch structure: a hierarchical framework for the study of heterogeneity. Oikos 59:253–260.

Kricher, J. C., and W. E. Davis, Jr. 1992. Patterns of avian species richness in disturbed and undisturbed habitats in Belize. Pp. 240–246 *in* Ecology and conservation of Neotropical migrant landbirds (J. M. Hagan, III and D. W. Johnston, eds). Smithsonian Institution Press, Washington, DC.

Lack, P. C. 1988. Hedge intersections and breeding bird distribution in farmland. Bird Study 35:133–136.

Lamberson, R. H., R. McKelvey, B. R. Noon, and C. Voss. 1992. The effects of varying dispersal capabilities on the population dynamics of the Northern Spotted Owl. Conserv. Biol. 6:1–8.

Lancaster, R. K., and W. E. Rees. 1979. Bird communities and the structure of urban habitats. Can. J. Zool. 57:2358–2368.

Lande, R. 1988. Demographic models of the Northern Spotted Owl (*Strix occidentalis caurina*). Oecologia 75:601–607.

Lande, R., and G. B. Barrowclough. 1987. Effective population size, genetic variation, and their use in population management. Pp. 87–123 *in* Viable populations for conservation (M. E. Soulé, ed.). Cambridge University Press, New York.

Lanyon, S. M., and C. F. Thompson. 1986. Site fidelity and habitat quality as determinants of settlement patterns in male Painted Bunting. Condor 88:206–210.

Lehmkuhl, J. F., and L.F. Ruggiero. 1991. Forest fragmentation in the Pacific Northwest and its potential effects on wildlife. Pp. 35–46 *in* Wildlife and vegetation of unmanaged Douglas-fir forests (L. F. Ruggiero, K. B. Aubry, A. B. Carey, and M. H. Huff, tech. coords). USDA Forest Serv. Gen. Tech. Rep. PNW-285, Portland, OR.

Lehmkuhl, J. F., L. F. Ruggiero, and P. A. Hall. 1991. Landscape-scale patterns of forest fragmentation and wildlife richness and abundance in the southern Washington Cascade Range. Pp. 425–442 *in* Wildlife and vegetation of unmanaged Douglas-fir forests (L. F. Ruggiero, K. B. Aubry, A. B. Carey, and M. H. Huff, tech. coords). USDA Forest Serv. Gen. Tech. Rep. PNW-285, Portland, OR.

Lodge, D. M. 1993. Species invasions and deletions: community effects and responses to climate and habitat change. Pp. 367–387 *in* Biotic interactions and global change (P. M. Kareiva, J. G. Kingsolver, and R. B. Huey, eds). Sinauer Associates, Sunderland, MA.

Lovejoy, T. E., R. O. Bierregaard, Jr, A. B. Rylands, J. R. Malcolm, C. E. Quintela, L. H. Harper, K. S. Brown, Jr, A. H. Powell, G. V. N. Powell, H. O. R. Schubart, and M. B. Hays. 1986. Edge and other effects of isolation on Amazon forest fragments. Pp. 257–285 *in* Conservation biology: the science of scarcity and diversity (M. E. Soulé, ed.). Sinauer Associates, Sunderland, MA.

Lynch, J. F. 1992. Distribution of overwintering Nearctic migrants in the Yucatan Peninsula, II: use of native and human-modified vegetation. Pp. 178–196 *in* Ecology and conservation of Neotropical migrant landbirds (J. M. Hagan, III and D. W. Johnston, eds). Smithsonian Institution Press, Washington, DC.

Lynch, J. F., and D. F. Whigham. 1984. Effects of

forest fragmentation on breeding bird communities in Maryland, USA. Biol. Conserv. 28:287–324.

MacArthur, R. H., and E. O. Wilson. 1963. An equilibrium theory of insular zoogeography. Evolution 17:373–387.

MacArthur, R. H., and E. O. Wilson. 1967. The theory of island biogeography. Princeton University Press, Princeton, NJ.

MacClintock, L., R. F. Whitcomb, and B. L. Whitcomb. 1977. Island biogeography and "habitat islands" of eastern forest. II. Evidence for the value of corridors and minimization of isolation in preservation of biotic diversity. Amer. Birds 31:6–16.

Manicacci, D., I. Olivieri, V. Perrot, A. Atlan, P.-H. Gouyon, J.-M. Prosperi, and D. Couvet. 1992. Landscape ecology: population genetics at the metapopulation level. Landscape Ecol. 6: 147–159.

Martin, T. E. 1980. Diversity and abundance of spring migratory birds using habitat islands on the Great Plains. Condor 82:430–439.

Martin, T. E. 1981a. Species–area slopes and coefficients: a caution on their interpretation. Amer. Natur. 118:823–837.

Martin, T. E. 1981b. Limitation in small habitat islands: Chance or competition? Auk 98:715–734.

Martin, T. E. 1988. Habitat and area effects on forest bird assemblages: is nest predation an influence? Ecology 69:74–84.

Martin, T. E. 1992. Landscape considerations for viable populations and biological diversity. Trans. North Amer. Wildl. Nat. Resource. Conf. 57: 283–291.

Martin, T. E., and P. A. Vohs. 1978. Configuration of shelterbelts for optimum utilization by birds. Great Plains Agr. Council Publ. 87:79–88.

Matlack, G. R. 1993. Sociological edge effects: Spatial distribution of human impact in suburban forest fragments. Environ. Manag. 17:829–835.

Mauer, B. A., and S. G. Heywood. 1993. Geographic range fragmentation and abundance in Neotropical migratory birds. Conserv. Biol. 7: 501–509.

McCauley, D. E. 1993. Genetic consequences of extinction and recolonization in fragmented habitats. Pp. 217–233 *in* Biotic interactions and global change (P. M. Kareiva, J. G. Kingsolver, and R. B. Huey, eds). Sinauer Associates, Sunderland, MA.

McDonnell, J. J., and E. W. Stiles. 1983. The structural complexity of old field vegetation and the recruitment of bird-dispersed plant species. Oecologia 56:109–116.

McGarigal, K., and B. J. Marks. 1995. FRAGSTATS: spatial pattern analysis program for quantifying landscape structure. USDA Forest Serv. Gen. Tech. Rep., Pacific Northwest Res. Sta., Corvallis, OR.

McGarigal, K., and W. C. McComb. 1992. Streamside versus upslope breeding bird communities in the central Oregon Coast Range. J. Wildl. Manag. 56:10–23.

McKelvey, K. S., and J. D. Johnston. 1992. Historical perspectives on forests of the Sierra Nevada and the Transverse Ranges of southern California: forest conditions at the turn of the century. Pp. 225–246 *in* The California Spotted Owl: a technical assessment of its current status (J. Verner, K. S. McKelvey, B. R. Noon, R. J. Gutierrez, G. I. Gould, Jr, and T. W. Beck, tech. coords), USDA Forest Serv. Gen. Tech. Rep. PSW-GTR-133.

McKelvey, K., B. R. Noon, and R. H. Lamberson. 1993. Conservation planning for species occupying fragmented landscapes: The case of the Northern Spotted Owl. Pp. 424–450 *in* Biotic interactions and global change (P. M. Kareiva, J. G. Kingslover, and R. B. Huey, eds). Sinauer Associates, Sunderland, MA.

Merriam, G. 1988. Landscape dynamics in farmland. Trends in Ecol. Evol. 3:16–20.

Mills, G. S., J. B. Dunning, and J. M. Bates. 1989. Effects of urbanization on breeding bird community structure in southwestern desert habitats. Condor 91:416–428.

Milne, B. 1991. Lessons from applying fractal models to landscape patterns. Pp. 199–235 *in* Quantitative methods in landscape ecology (M. G. Turner, and R. H. Gardner, eds). Spring-Verlag, New York.

Moore, F. R., P. Kerlinger and T. R. Simons. 1990. Stopover on a Gulf coast barrier island by spring trans-Gulf migrants. Wilson Bull. 102:487–500.

Morgan, K. A., and J. E. Gates. 1982. Bird population patterns in forest edge and strip vegetation at Remington farms, Maryland. J. Wildl. Manag. 46:933–944.

Moss, M.R. (ed.). 1988. Landscape ecology and management. Polyscience, Montreal, Canada.

Murdoch, W. W. 1993. Individual-based models for predicting effects of global change. Pp. 147–164 *in* Biotic interactions and global change (P. M. Kareiva, J. G. Kingsolver, and R. B. Huey, eds). Sinauer Associates, Sunderland, MA.

Murphy, D. D., and B. R. Noon. 1992. Integrating scientific methods with habitat conservation planning: reserve design for Northern Spotted Owls. Ecol. Applic. 2:3–17.

Nassauer, J. I., and R. Westmacott. 1987. Progressiveness among farmers as a factor in heterogeneity of farmed landscapes. Pp. 199–210 *in* Landscape heterogeneity and disturbance (M. G. Turner, ed.). Springer-Verlag, New York.

Naveh, Z., and A. S. Lieberman. 1984. Landscape ecology: theory and application. Springer-Verlag, New York.

Norment, C. J. 1991. Bird use of forest patches in the subalpine forest-alpine tundra ecotone of the Beartooth Mountains, Wyoming. Northwest Science 65:1–9.

Noss, R. F. 1991. Effects of edge and internal patchiness on avian habitat use in an old-growth Florida hammock. Natural Areas J. 11:34–47.

O'Connor, R. J., and M. Shrubb. 1986. Farming and birds. Cambridge University Press, London, England.

Odum, E. G., and M. G. Turner. 1987. The Georgia landscape: a changing resource. Kellogg Phys. Res. Task Force Final Rep. Institute of Ecology, University of Georgia, Athens, GA.

O'Meara, T. E. 1984. Habitat–island effects on the avian community in cypress ponds. Proc. Annu. Conf. Southeast. Assoc. Fish Wildl. Agencies 38:97–110.

O'Neill, R. V., J. R. Krummel, R. H. Gardner, G. Sugihara, B. Jackson, D. L. DeAngelis, B. T. Milne, M. G. Turner, B. Zygmunt, S. W. Christensen, V. H. Dale, and R. L. Graham. 1988. Indices of landscape pattern. Landscape Ecol. 1:153–162.

Opdam, P. 1990. Dispersal in fragmented populations: the key to survival. Pp. 3–17 *in* Species dispersal in agricultural habitats (R. G. H. Bunce and D. C. Howard, eds). Belhaven Press, New York.

Opdam, P. 1991. Metapopulation theory and habitat fragmentation: a review of holarctic breeding bird studies. Landscape Ecol. 5:93–106.

Pace, M. L. 1993. Forecasting ecological responses to global change: the need for large-scale comparative studies. Pp. 356–363 *in* Biotic interactions and global change (P. M. Kareiva, J. G. Kingsolver, and R. B. Huey, eds). Sinauer Associates, Sunderland, MA.

Pearson, S. M. 1993. The spatial extent and relative influence of landscape-level factors on wintering bird populations. Landscape Ecol. 8:3–18.

Pearson, S. M., M. G. Turner, R. H. Gardner, and R. V. O'Neill, 1995. An organism-based perspective of habitat fragmentation. *in* Biodiversity in managed landscapes: theory and practice (R. C. Szaro, ed.). Oxford University Press (in press).

Petit, D. R., L. J. Petit, and K. G. Smith. 1992. Habitat associations of migratory birds overwintering in Belize, Central America. Pp. 247–256 *in* Ecology and conservation of Neotropical migrant landbirds (J. M. Hagan, III and D. W. Johnston, eds). Smithsonian Institution Pres, Washington, DC.

Pickett, S. T. A., and P. S. White (eds). 1985. The ecology of natural disturbance and patch dynamics. Academic Press, New York.

Pimentel, D., and J. H. Perkins (eds). 1980. Pest control: cultural and environmental aspects. Amer. Assoc. Adv. Sci. Selected Symp. 43, Westview Press, Boulder, CO.

Probst, J. R. 1988. Kirtland's Warbler breeding biology and habitat management. Pp. 28–35 *in* Integrating forest management for wildlife and fish (T. W. Hoekstra and J. Capp, comps). USDA Forest Service, Gen. Tech. Rep. NC-122.

Probst, J. R., and T. R. Crow. 1991. Integrating biological diversity and resource management: an essential approach to productive, sustainable ecosystems. J. For. 89:12–17.

Probst, J. R., and J. P. Hayes. 1987. Pairing success of Kirtland's warblers in marginal versus suitable habitat. Auk 104:234–241.

Probst, J. R., and J. Weinrich. 1993. Relating Kirtland's warbler population to changing landscape composition and structure. Landscape Ecol. 8:257–271.

Pulliam, H. R. 1988. Sources, sinks, and population regulation. Amer. Natur. 132:652–661.

Pulliam, H. R., J. B. Dunning, and J. Liu. 1992. Population dynamics in complex landscapes: a case study. Ecol. Applic. 2:165–177.

Rands, M. R. W. 1985. Pesticide use on cereals and the survival of Grey Partridge chicks: a field experiment. J. Appl. Ecol. 22:49–54.

Ripple, W. J., G. A. Bradshaw, and T. A. Spies. 1991. Measuring forest landscape patterns in the Cascade range of Oregon, USA. Biol. Conserv. 57:73–88.

Risser, P. G., J. R. Karr and R. T. T. Forman. 1984. Landscape ecology: directions and approaches. Illinois Natural Hist. Surv. Spec. Publ. 2, Champaign, IL.

Robbins, C. S., D. K. Dawson, and B. A. Dowell. 1989. Habitat area reguirements of breeding forest birds of the Middle Atlantic States. Wildl. Monogr. 103:1–34.

Robbins, C. S., B. A. Dowell, D. K. Dawson, J. A. Colon, R. Estrada, A. Sutton, R. Sutton, and D. Weyer. 1992. Comparison of Neotropical migrant landbird populations wintering in tropical forest, isolated fragments, and agricultural habitats. Pp. 207–210 *in* Ecology and conservation of Neotropical migrant landbird (J. M. Hagan, III and D. W. Johnston, eds). Smithsonian Institution Press, Washington, DC.

Robinson, G. R., R. D. Holt, M. S. Gaines, S. P. Hamburg, M. L. Johnson, H. S. Fitch, and E.

A. Martinko. 1992. Diverse and contrasting effects of habitat fragmentation. Science 257:524–526.

Robinson, S. K. 1992. Population dynamics of breeding Neotropical migrants in a fragmented Illinois landscape. Pp. 408–418 *in* Ecology and conservation of Neotropical migrant landbirds (J. M. Hagan, III and D. W. Johnston, eds). Smithsonian Institution Press, Washington, DC.

Robinson, S. K., J. A. Grzybowski, S. I. Rothstein, M. C. Brittingham, L. J. Petit, and F. R. Thompson. 1993. Management implications of cowbird parasitism on Neotropical migrant songbirds. Pp. 93–102 *in* Status and Management of Neotropical migratory birds (D. M. Finch and P. W. Stangel, eds). USDA Forest Serv. Gen. Tech. Rep. RM-229, Rocky Mt. Forest Range Exp. Sta., Fort Collins, CO.

Rodenhouse, N. L., and L. B. Best. 1983. Breeding ecology of Vesper Sparrows in corn and soybean fields. Amer. Midl. Natur. 110:265–275.

Rodenhouse, N. L., G. W. Barrett, D. M. Zimmerman, and J. C. Kemp. 1992. Effects of uncultivated corridors on arthropod abundances and crop yields in soybean agroecosystems. Agr. Ecosyst. Environ. 38:179–191.

Rodenhouse, N. L., L. B. Best, R. J. O'Connor, and K. Bollinger. 1993. Effects of temperate agriculture on Neotropical migrant landbirds. Pp. 280–295 *in*: Status and management of Neotropical migratory birds (D. M. Finch and P. W. Stangel, eds). USDA Forest Serv. Gen. Tech. Rep. RM-229, Rocky Mt. Forest Range Exp. Sta., Fort Collins, CO.

Rosenberg, K. V., and M. G. Raphael. 1986. Effects of forest fragmentation on vertebrates in Douglas-fir forests. Pp. 263–272 *in* Wildlife 2000: modeling habitat relationships of terrestrial vertebrates (J. Verner, M. L. Morrison, and C. J. Ralph, eds). University of Wisconsin Press, Madison, WI.

Rosenberg, K. V., S. B. Terrill, and G. H. Rosenberg. 1987. Value of suburban habitats to desert riparian birds. Wilson Bull. 99:642–654.

Samson, F. B. 1980a. Island biogeography and the conservation of prairie birds. Pp. 293–299 *in* Proc. 7th North Amer. Prairie Conf. (C. L. Kucera, ed). Southwest Missouri State University, Springfield, MO.

Samson, F. B. 1980b. Island biogeography and the conservation of nongame birds. Trans. North Amer. Wildl. Nat. Res. Conf. 45:245–251.

Saunders, D. A., and R. J. Hobbs. 1991. Nature conservation: the role of corridors. Surrey Beatty and Sons, Chipping Norton, Australia.

Schonewald-Cox, C., M. Buechner, R. Sauvajot, and B. A. Wilcox. 1992. Environmental auditing cross-boundary management between national parks and surrounding lands: a review and discussion. Environ. Manag. 16:273–282.

Schroeder, R. L., T. T. Cable, and S.L. Haire. 1992. Wildlife species richness in shelterbelts: test of a habitat model. Wildl. Soc. Bull. 20:264–273.

Shaffer, M. L. 1985. The metapopulation and species conservation: the special case of the Northern Spotted Owl. Pp. 86–99 *in* Ecology and management of the Spotted Owl in the Pacific Northwest (R. J. Gutierrez and A. B. Carey, eds). USDA Forest Serv. Gen. Tech. Rep. PNW-185, Pacific Northwest Forest Range Exp. Sta., Portland, OR.

Simberloff, D., J. A. Farr, J. Cox, and D. W. Mehlman. 1992. Movement corridors: conservation bargains or poor investments? Biol. Conserv. 6:493–504.

Slocombe, D. S., 1993. Implementing ecosystem-based management. BioScience 43:612–622.

Small, M. F., and M. L. Hunter. 1988. Forest fragmentation and avian nest predation in forested landscapes. Oecolgia 76:62–64.

Smutz, J. K. 1987. The effect of agriculture on Ferruginous and Swainson's hawks. J. Range Manag. 40:438–440.

Spellerberg, I. F., F. B. Goldsmith, and M. G. Morris (eds). 1991. The scientific management of temperate communities for conservation. Blackwell Scientific Publication, Cambridge, MA.

Stacey, P. B., and M. Taper. 1992. Environmental variation and the persistence of small populations. Ecol. Applic. 2:18–29.

Staicer, C. A. 1992. Social behavior of the Northern Parula, Cape May Warbler, and Prairie Warbler wintering in second-growth forest in southwestern Puerto Rico. Pp. 308–320 *in* Ecology and conservation of Neotropical migrant landbirds (J. M. Hagan III and D. W. Johnston, eds). Smithsonian Institution Press, Washington, DC.

Stamp, N. E. 1978. Breeding birds of riparian woodland in south-central Arizona. Condor 80:64–71.

Stauffer, D. F., and L. B. Best. 1980. Habitat selection by birds of riparian communities: evaluating effects of habitat alterations. J. Wildl. Manag. 44:1–15.

Steele, B. B. 1992. Habitat selection by breeding Black-throated Blue Warblers at two spatial scales. Ornis Scand. 23:33–42.

Strong, T. R., and C. E. Bock. 1990. Bird species distribution patterns in riparian habitats in southeastern Arizona. Condor 92:866–885.

Swanson, F. J., J. F. Franklin, and J. R. Sedell. 1990. Landscape patterns, disturbance, and management in the Pacific Northwest, USA. Pp. 191–213 *in* Changing landscapes: an ecological perspective (I. S. Zonneveld and R. T. T. Forman, eds). Springer-Verlag, New York.

Szaro, R. C. and M. D. Jakle. 1985. Avian use of a desert riparian island and its adjacent scrub habitat. Condor 87:511–519.

Takekawa, J. E., and S. R. Beissinger. 1989. Cyclic drought, dispersal, and the conservation of the Snail Kite in Florida: lessons in critical habitat. Conserv. Biol. 3:302–311.

Temple, S. A. 1986. Predicting impacts of habitat fragmentation on forest birds: a comparison of two models. Pp. 301–304 *in* Wildlife 2000: modeling habitat relationships of terrestrial vertebrates (J. Verner, M. L. Morrison, and C. J. Ralph, eds). University of Wisconsin Press, Madison, WI.

Temple, S. A., and J. R. Cary. 1988. Modeling dynamics of habitat–interior bird populations in fragmented landscapes. Conserv. Biol. 2:340–347.

Thomas, J. W., E. D. Forsman, J. B. Lint, E. C. Meslow, B. R. Noon, and J. Verner. 1990. A conservation strategy for the Northern Spotted Owl. Report of the Interagency Scientific Committee to address the conservation of the Northern Spotted Owl. US Government Printing Office 1990-791-171/20026, Washington, DC.

Tobalske, B. W., R. C. Shearer, and R. L. Hutto. 1991. Bird populations in logged and unlogged western larch/Douglas-fir forest in northwestern Montana. USDA Forest Serv. Res. Pap. INT-442, Intermt. Res. Sta., Ogden, UT.

Turner, M. G. (ed.) 1987. Landscape heterogeneity and disturbance. Springer-Verlag, New York.

Turner, M. G. 1989. Landscape ecology: the effect of pattern on process. Annu. Rev. Ecol. Syst. 20:171–197.

Turner, M. G., and R. H. Gardner (eds). 1991. Quantitative methods in landscape ecology. Springer-Verlag, New York.

Turner, M. G., G. J. Arthaud, R. T. Engstrom, S. J. Hejl, J. Liu, S. Loeb, and K. McKelvey. 1995. Usefulness of spatially explicit population models in land management. Ecol. Applic. 5:12–16.

Turner, S. J., R. V. O'Neill, W. Conley, M. R. Conley and H. C. Humphries. 1991. Pattern and scale: statistics for landscape ecology. Pp. 17–50 *in* Quantitative methods in landscape ecology (M. G. Turner and R. H. Gardner, eds). Springer-Verlag, New York.

Tyser, R. W. 1983. Species–area relations of cattail marsh avifauna. Passenger Pigeon 45: 125–128.

Urban, D. L., and H. H. Shugart, Jr. 1986. Avian demography in mosaic landscapes: modeling paradigm and preliminary results. Pp. 273–279 *in* Wildlife 2000: modeling habitat relationships of terrestrial vertebrates (J. Verner, M. L. Morrison, and C. J. Ralph, eds). University of Wisconsin Press, Madison, WI.

Urban, D. L., R. V. O'Neill, and H. H. Shugart. 1987. Landscape ecology. BioScience 37:119–127.

Vander Haegen, W. M., M. W. Sayre, and W. E. Dodge. 1989. Winter use of agricultural habitats by wild turkeys in Massachusetts. J. Wildl. Manag. 53:30–33.

Van Horn, M. A., R. M. Gentry, and J. Faaborg, 1995. Patterns of pairing success of the Ovenbird (*Seiurus aurocapillus*) within Missouri forest fragments. Auk (in press).

Van Horne, B. 1983. Density as a misleading indicator of habitat quality. J. Wildl. Manag. 47:89–101.

Van Horne, B. and J. A. Wiens. 1991. Forest bird habitat suitability models and the development of general habitat models. US Fish Wildl. Serv. Fish Wildl. Res. 8, Washington, DC.

Verner, J., and T. A. Larson. 1989. Richness of breeding bird species in mixed-conifer forests of the Sierra Nevada, California. Auk 106:447–463.

Verner, J. and L. V. Ritter. 1983. Current status of the Brown-headed Cowbird in the Sierra National Forest. Auk 100:355–368.

Verner, J., M. L. Morrison, and C. J. Ralph (eds). 1986. Wildlife 2000: modeling habitat relationships of terrestrial vertebrates. University of Wisconsin Press, Madison, WI.

Vickery, P. D., M. L. Hunter, Jr., and S. M. Melvin. 1994. Effects of habitat area on the distribution of grassland birds in Maine. Conserv. Biol. 8:1087–1097.

Villard, M.-A., K. E. Freemark, and H. G. Merriam. 1992. Metapopulation theory and Neotropical migrant birds in temperate forests: an empirical investigation. Pp. 474–482 *in* Ecology and conservation of Neotropical migrant landbirds (J. M. Hagan, III and D. W. Johnston, eds). Smithsonian Institution Press, Washington, DC.

Villard, M.-A., P. R. Martin and C. G. Drummond. 1993. Habitat fragmentation and pairing success in the Ovenbird (*Seiurus aurocapillus*). Auk 110:759–768.

Villard, M.-A., G. Merriam, and B. A. Maurer. 1995. Dynamics in subdivided populations of Neotropical migratory birds in a fragmented temperate forest. Ecology 76:27–40.

Warner, R. E. 1984. Effects of changing agriculture

on Ring-necked Pheasant brood movements in Illinois. J. Wildl. Manag. 48:1014–1018.

Warner, R. E. 1992. Nest ecology of grassland passerines on road rights-of-way in central Illinois. Biol. Conserv. 59:1–7.

Warner, R. E. 1994. Agricultural land use and grassland habitat in Illinois: future shock for midwestern birds? Converv. Biol. 8: 147–156.

Warner, R. E., S. L. Etter, G. B. Joselyn, and J. A. Ellis. 1984. Declining survival of Ring-necked Pheasant chicks in Illinois agricultural ecosystems. J. Wildl. Manag. 48:82–88.

Wegner, J. F., and G. Merriam. 1979. Movements by birds and small mammals between a wood and adjoining farmland habitats. J. Appl. Ecol. 16:349–357.

Western, D. 1989. Conservation without parks: wildlife in the rural landscape. Pp. 158–165 *in* Conservation for the twenty-first century (D. Western and M. C. Pearl, eds). Oxford University Press, New York.

Wetmore, S. P., R. A. Keller, and G. E. J. Smith. 1985. Effects of logging on bird populations in British Columbia as determined by a modified point-count method. Can. Field Natur. 99:224–233.

Whitcomb, B. L., R. F. Whitcomb, and D. Bystrak. 1977. Island biogeography and "habitat islands" of eastern forest. III. Long-term turnover and effects of selective logging on the avifauna for forest fragments. Amer. Birds 31:17–23.

Whitcomb, R. F., C. S. Robbins, J. F. Lynch, B. L. Whitcomb, K. Klimkiewicz, and D. Bystrak. 1981. Effects of forest fragmentation on avifauna of the eastern deciduous forest. Pp. 125–205 *in* Forest island dynamics in man-dominated landscapes (R. L. Burgess and D. M. Sharpe, eds). Springer-Verlag, New York.

Wiens, J. A. 1989. Spatial scaling in ecology. Funct. Ecol. 3:385–397.

Wiens, J. A., J. T. Rotenberry, and B. Van Horne. 1987. Habitat occupancy patterns of North American shrubsteppe birds: the effects of spatial scale. Oikos 48:132–147.

Wilcove, D. S. 1988. Changes in the avifauna of the Great Smoky Mountains: 1947–1983. Wilson Bull. 100:256–271.

Wilcove, D. S. 1989. Protecting biodiversity in multiple-use lands: lessons from the US Forest Service. Trends Ecol. Evol. 4:385–388.

Wilcove, D. S., and S. K. Robinson, 1990. The impact of forest fragmentation on bird communities in eastern North America. Pp. 319–331 *in* Biogeography and ecology of forest bird communities (A. Keast, ed). SPB Academic Publications, The Hague, The Netherlands.

Williamson, R. D., and R. M. DeGraaf. 1981. Habitat associations of ten bird species in Washington, DC. Urban Ecol. 5:125–136.

Willson, M. F., and S. W. Carothers. 1979. Avifauna of habitat islands in the Grand Canyon. Southwest. Natur. 24:563–576.

Yahner, R. H. 1983. Seasonal dynamics, habitat relationships, and management of avifauna in farmstead shelterbelts. J. Wildl. Manag. 47: 74–84.

Zimmerman, J. L. 1982. Nesting success of Dickcissels in preferred and less preferred habitats. Auk 99:292–298.

Zonneveld, I. S., and R. T. T. Forman (eds). 1990. Changing landscapes: an ecological perspective. Springer-Verlag, New York.

APPENDIX (*follows on pages 422–427*)

APPENDIX

Bird species exhibiting possitive area sensitivity based on lower density, frequency of occurrence or nest success with forest fragmentation or reduction in habitat area ($P < 0.10$ used when given). **Bold type** indicates species that overwinter in the Neotropics (according to Gauthreaux (1992: List A) or analyses of range maps).

	Breeding Season							Migration
	Forest			Grassland	Marsh	Riparian	Shelter-belt	
	NE	SE	W					
Least Bittern					**13, 14**			
American Bittern					13, 14(26)[a]			
Black-crowned Night Heron					13			
Great Blue Heron					26			
Green-backed heron					14(26)			
American Swallow-tailed Kite		**3**						
Mississippi Kite		**3**						
Osprey					26			
Bald Eagle					26			
Northern Goshawk						15		
Sharp-shinned Hawk			7			(15)		
Broad-winged Hawk		**3**						
Red-shouldered Hawk	1(31)	3, 4						
Northern Harrier				11, 24	26	15		
American Kestrel			(23)			25(15)		
Ruffed Grouse		3	7(23)					
Wild Turkey		3						
Greater Prairie Chicken				24				
Northern Bobwhite	(31)	4	(11)			(29)		
Virginia Rail					14, 26(13)			
Sora					13(14, 26)			
Upland Sandpiper				**11, 24, 30**				
Spotted Sandpiper					26	15		
Common Snipe					26			
Belted Kingfisher					26	15		
Band-tailed Pigeon						**15**		
Mourning Dove	(1)	4	6			(15, 29)	27	(19)

Black-billed Cuckoo	**2(1)**	**3**	**6**			**27**	
Yellow-billed Cuckoo	**1**[b]	**4, 5**			**29(16)**		**(19)**
Great Horned Owl	(31)	4					
Barred Owl	(31)	3, 4	23				
Spotted Owl			7, 10				
Rufous Hummingbird			**28(21, 23)**				
Ruby-throated Hummingbird	**31(1)**	**4**	**6**		**(29)**		
Northern Flicker	(1)		(6, 7, 22, 23, 28)		15(25, 29)	27	
Pileated Woodpecker	1[b]	3, 4(5)	7, 20(23)				(19)
Red-bellied Woodpecker	1	4(5)			16(29)		(19)
Red-cockaded Woodpecker		3					
Red-headed Woodpecker	(1)				25		
Acorn Woodpecker			7				
Red-breasted Sapsucker			7				
Yellow-bellied Sapsucker			(6)		15		
Downy Woodpecker	1[b]	4(5)	6(7)		15, 25, 29(16, 29)		
Hairy Woodpecker	1	3	6(7, 21–23)		15, 16(29)		
Eastern Kingbird	**(1)**					**27**	
Scissor-tailed Flycatcher				**24**			
Eastern Wood-Pewee	**31(1)**				**29(16)**	**27**	
Western Wood-Pewee			**(7, 22)**		**15**		
Say's Phoebe			28				
Eastern Phoebe					29(29)		
Acadian Flycatcher	**1**	**3**			**29**		
Least Flycatcher	**1**		**6**				
Great Crested Flycatcher	**1**[b]	**(4, 5)**	**6**		**16(29)**		**(19)**
Horned Lark				24			
Violet-green Swallow					**15**		
Blue Jay	31(1)	4			(16, 29)	17, 27	(19)
Pinyon Jay					15		
American Crow	1	(4, 5)			(29)		
Common Raven	(31)		7(22, 23)				
Boreal Chickadee			28				
Black-capped Chickadee	1[b]		6(22, 23)		(16)	17(18)	(18)
Chestnut-backed Chickadee			7(21, 23, 28)				

(*continued*)

	Breeding Season							Migration
	Forest			Grassland	Marsh	Riparian	Shelter-belt	
	NE	SE	W					
Mountain Chickadee			8, 9(22, 28)			(15)		
Tufted Titmouse	1[b]	4(5)				16(29)		(19)
White-breasted Nuthatch	1(32)		7, 9			15(16)		
Red-breasted Nuthatch			8, 9, 28 (7, 21–23, 28)					
Brown Creeper	1[b]		8, 23, 28 (7, 21, 22, 28)			15		
Marsh Wren					13, 14			
Sedge Wren				24(11)				
House Wren	**1**[b]		**6(7)**			**15, 25**	**27**	
Winter Wren			7, 21, 23 (21, 22, 28)					
American Dipper						15		
Ruby-crowned Kinglet			9(8, 22, 28)					(19)
Golden-crowned Kinglet			23(7, 8, 21, 22, 28)					
Blue-gray Gnatcatcher	**1**	**4(5)**				**16**		**(19)**
Gray Catbird	**1**[b]		**(6)**			**(29)**	**(18)**	**(19)**
Brown Thrasher	(1)		(6)	(30)		16	17, 27(18)	(18)
Sage Thrasher						15		
Cedar Waxwing	(1)		6					
Eastern Bluebird				24				
Mountain Bluebird			(6, 9, 22, 23)			15		
American Robin	(1)		7, 8(6, 21–23, 28)			15(29)	(18, 27)	
Veery	**1**	**3**	**(6)**					
Hermit Thrush	1	3	6, 8, 9, 28 (7, 21–23, 28)			15		
Swainson's Thrush		**3**	**23(21)**			**15**		
Wood Thrush	**1**	**3**				**16, 29**		

Varied Thrush			22, 28(21, 23)			
European Starling	(1)		(6)	25(29)	(18)	(18)
Solitary Vireo	(31)	3	(7, 22, 23)			
Yellow-throated Vireo	1[b]	3				
Red-eyed Vireo	1[b]	3(4)	6	16, 29		(19)
Warbling Vireo			6(7, 22, 23)	16(15)	27	
White-eyed Vireo	(1)	4(5)		29(29)		(19)
Orange-crowned Warbler			7(23)	15		
Virginia's Warbler				15		
Bachman's Warbler		3				
Northern Parula	1	3(4, 5)				19
Yellow-rumped Warbler			8, 9, 28(7, 22, 23)	15		(18, 19)
Yellow warbler			(6, 23)	15		
Chestnut-sided Warbler	1[b]					
Magnolia Warbler	2(1)					
Black-throated Blue Warbler	1					
Black-throated Gray Warbler			7(21, 28)	(15)		
Townsend's Warbler			22, 28(21, 23, 28)			
Black-throated Green Warbler	1[b]	3				
Blackburnian Warbler	2	3				
Yellow-throated Warbler	(31)	3				
Cerulean Warbler	1	3				
Black-and-white Warbler	1	3				19
American Redstart	1(31)		6	16		
Prothonotary Warbler	(31)	3		29		
Worm-eating Warbler	1	3				
Swainson's Warbler		3				
Pine Warbler	31(1)	3(5)				
Ovenbird	1	3	(6)	16		19
Northern Waterthrush	1					
Louisiana Waterthrush	1	3		29		
Kentucky Warbler	1[b]	3		29		
Mourning Warbler	1					
Hooded Warbler	1[b]	3				

(continued)

	Breeding Season							Migration
	Forest			Grassland	Marsh	Riparian	Shelter-belt	
	NE	SE	W					
Canada Warbler	**1**							
Common Yellowthroat	(1)	**5**		**(11, 30)**	**14**	**29**	**17, 27**	**(19)**
Summer Tanager	**1**	**3, 4**						**19**
Scarlet Tanager	**1**	**3**				**16, 29**		
Rose-breasted Grosbeak	**1**[b]	**3**						
Black-headed Grosbeak			**7(28)**			**15**		
Rufous-sided Towhee	(1)		6(7)			16(15)		
Clay-colored Sparrow			**(6)**	**12**				
Brewer's Sparrow						**15**		
Vesper Sparrow			6(9)	24, 30(11)				
Savannah Sparrow			(28)	11, 12, 24, 30				
Grasshopper Sparrow				**11, 12, 24, 30**				
Henslow's Sparrow				11, 24				
Fox Sparrow			(22, 23)			15		
Song Sparrow	(1)			(11, 30)	14	15	27(18)	(18)
Lark Sparrow				**24**				
Swamp Sparrow					13, 14			
Field Sparrow	(1)			30(11, 30)		(29)		
White-crowned Sparrow			(9)			15		
Dark-eyed Junco			9, 21 (7, 8, 22, 23, 28)			15		(18)
Dickcissel				**24(11)**				
Bobolink				**11, 12, 30**				
Eastern Meadowlark				30(11, 24)				
Western Meadowlark				12				
Brewer's Blackbird						15		
Brown-headed Cowbird	(1)		6(7)			15, 29(29)	27	(18, 19)
Common Grackle	(1)	(5)			14(13)	(29)	(17, 18, 27)	18(17)

Northern Oriole	**(1)**		**6**				**18, 27**	
Orchard Oriole						**(29)**	**27**	
Pine Siskin			9, 28 (7, 8, 21–23, 28)			15		
Evening Grosbeak			28(21)					
Pine Grosbeak			9(8)					
White-winged Crossbill			9					
Purple Finch			28(7)					
American Goldfinch	(1)		(6, 28)	24(11)		(29)	27	(18, 19)
Area-sensitive species								
Number (Total 158 species)	47	49	51	19	18	58	16	5
Number Neotropical migrants (68 of total)	**34**	**33**	**15**	**7**	**2**	**26**	**8**	**4**
% Neotropical migrants (43% of total)	**72**	**67**	**29**	**37**	**11**	**45**	**50**	**80**

[a] No positive area-sensitivity found in studies listed in parentheses.
[b] Disagreement among studies reviewed.
References: 1, Studies reviewed by Freemark and Collins (1992); 2, Freemark (unpublished data); 3, Hamel et al. (1982); 4, Harris and Wallace (1984); 5, O'Meara (1984); 6, Johns (1993); 7, Rosenberg and Raphael (1986); 8, Keller and Anderson (1992); 9, Norment (1991); 10, Thomas et al. (1990); 11, Herkert (1991); 12, Johnson and Temple (1990); 13, Brown and Dinsmore (1986); 14, Tyser (1983); 15, Dobkin and Wilcox (1986); 16, Stauffer and Best (1980); 17, Martin and Vohs (1978); 18, Yahner (1983); 19, Cox (1988); 20, Aney (1984); 21, Lehmkuhl et al. (1991); 22, Tobalske et al. (1991); 23, Hejl and Paige (1994); 24, Samson (1980a,b); 25, Gutzwiller and Anderson (1987); 26, Gibbs et al. (1991); 27, Martin (1981b); 28, Wetmore et al. (1985); 29, Keller et al. (1993); 30, Vickery et al. (1994); 31, Robbins et al. (1989); 32, Temple (1986).

15

ECOLOGY AND BEHAVIOR OF COWBIRDS AND THEIR IMPACT ON HOST POPULATIONS

SCOTT K. ROBINSON, STEPHEN I. ROTHSTEIN, MARGARET C. BRITTINGHAM, LISA J. PETIT, AND JOSEPH A. GRZYBOWSKI

INTRODUCTION

Cowbirds of the genus *Molothrus* are brood parasites that lay their eggs in the nests of other (host) species. Because many hosts raise cowbird young instead of their own, brood parasitism can substantially reduce host breeding productivity, and high levels of brood parasitism can threaten some host populations. Population declines and the endangered status of several species have been linked to high levels of cowbird parasitism (see below). Neotropical migrant landbirds may be particularly vulnerable to cowbird parasitism, especially in landscapes where cowbird feeding habitat has been created or enhanced by human activities and cowbird numbers are high (Brittingham and Temple 1983). The purpose of this chapter is to review the natural history of North American cowbirds with special emphasis on the Brown-headed Cowbird, their effects on host population dynamics, and to identify the areas where research is most needed. Methods of cowbird management and control are discussed in detail in Robinson et al. (1993).

SYSTEMATICS AND PLUMAGE

There are five species of parasitic cowbirds (Friedmann 1929, Lanyon 1992), two of which, the Screaming and Giant Cowbirds, are found only in the Neotropics. All three of the remaining species now breed in North America. Among the two exclusively North American species, the Bronzed Cowbird breeds mainly south of the US from Mexico to Panama, but also occurs in Florida, the Gulf States and the Southwest, and the Brown-headed Cowbird is widespread throughout North America, with breeding-season records from every state and province except Hawaii. The Shiny Cowbird, which was previously confined to South America, has colonized the Caribbean, where it may be responsible for major population declines in some host species (Post and Wiley 1977) and has invaded the southeastern United States within the last 7 years (Cruz et al. 1989, Grzybowski and Fazio 1991).

Adult males of all five parasitic species are entirely glossy black except in the case of the Brown-headed Cowbird. Females are a dull grayish brown in all species except for the Screaming Cowbird and one race of the Shiny Cowbird, which show little or no sexual dimorphism in plumage. The dull, nondescript plumage of most female cowbirds may be an adaptation that makes them relatively inconspicuous to hosts (Payne 1977). Except for the distinctive juvenile (fledgling) plumage of the Screaming Cowbird, juveniles in all species have female-like plumage. Juvenile Brown-headed Cowbirds, however, can nearly always be distinguished from adult females by their streaked breasts and, especially, the white-to-buffy colors on the edges of nearly all back and wing feathers.

GEOGRAPHIC DISTRIBUTION AND ABUNDANCE OF BROWN-HEADED COWBIRDS

Historical Range

Prior to European settlement, the Brown-headed Cowbird is generally thought to have been largely confined to the short-grass prairie region west of the Mississippi (Friedmann 1929, Mayfield 1965). Cowbirds feed by walking on the ground in areas of short grass and bare ground. They originally followed the herds of bison present in the prairies, and are likely to have used scattered trees and riparian woodlands within prairie areas for nest searching and display perches. Cowbirds may have been absent from the large expanse of forest that covered eastern North America (although they may have been present in eastern grasslands) and from arid western states.

Range Expansion Eastward

European settlement was beneficial to Brown-headed Cowbirds and enabled them to expand their range eastward. As the eastern forests were cut by settlers and land was cleared for farming, open habitat was expanded, which provided opportunities for an eastward expansion of the cowbird's range. Livestock, introduced by the settlers, replaced the bison, and cowbirds were reported associated with herds of cattle (Friedmann 1929, Mayfield 1965). Mayfield (1965) provided historic records for the range expansion of the cowbird eastward. As early as 1790, cowbirds were reported at scattered sites throughout eastern North America. By the late 1800s cowbirds were widespread in eastern North America, but were not abundant and were found primarily in cultivated areas. At that time, they were still uncommon in forest habitat (Bendire 1895, Friedmann 1929), and their impact on forest host species was probably minimal.

Since the early 1900s, the range of the cowbird within eastern North America has not changed substantially, but the percentage of Christmas counts on which cowbirds were detected has increased dramatically (Fig. 15-1) (Brittingham and Temple 1983). An increase in winter food supply and wintering habitat

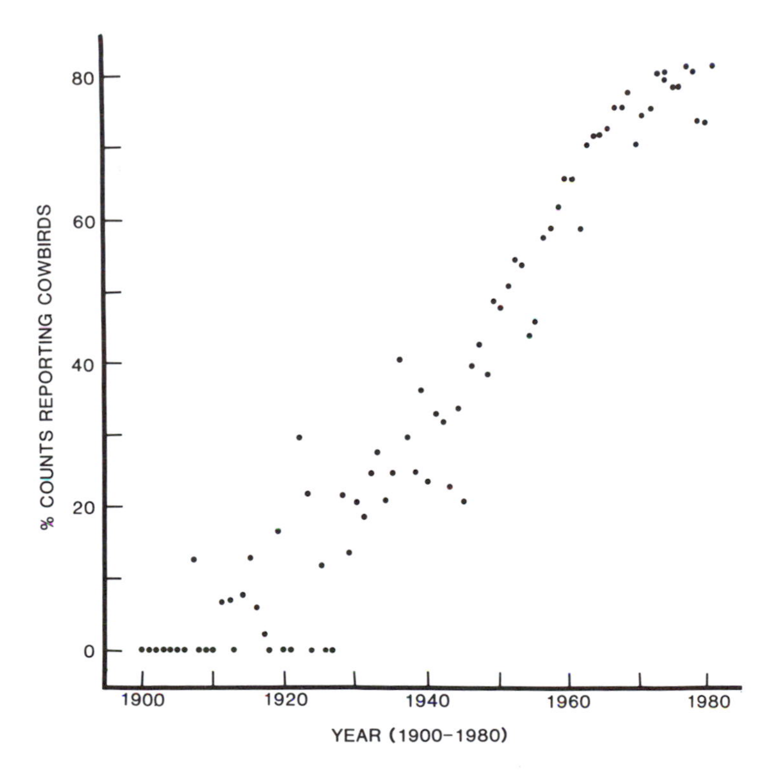

Figure 15-1. An index to cowbird distribution from 1900 to 1980 taken from Audubon Christmas bird count records ($r = 0.96$, $P < 0.01$) (Brittingham and Temple 1983).

may have also contributed to increasing cowbird populations. Cowbird reproductive success may also have increased as contact was made with new hosts (Mayfield 1965). As cowbirds increased in abundance, their impact on host species undoubtedly increased.

Range Expansion Westward

As European settlers moved westward, they also improved habitat conditions for cowbirds (Mayfield 1965, 1977a). The prairies and arid intermountain regions were converted to farm land, and domestic livestock were introduced, which provided cowbirds with an abundant and readily accessible foraging habitat. Trees planted within the prairies provided perches for displaying and nest searching.

The most detailed account of a range expansion for cowbirds occurred in the far western United States in the mountains of the Sierra Nevada (Rothstein et al. 1980). As recently as the 1940s, cowbirds were absent or extremely rare throughout the Sierra Nevada. By the 1970s, however, they were one of the most abundant species in the area. Rothstein and colleagues (Rothstein et al. 1980, 1984, Verner and Rothstein 1988) showed how human settlements consisting of pack stations, corrals, and range cattle provided cowbirds with foraging opportunities that enabled cowbirds to invade areas where they were formerly excluded. Rothstein (1994) also provided a detailed account of the cowbird's spread throughout the far west where it became well established in coastal southern California by the 1920s, in Oregon west of the Cascades in the 1940s, and in Washington and southern British Columbia by the 1950s. In addition, cowbirds have become more abundant in the Great Basin and Intermountain Region since the late 1800s.

Current Geographic Distribution and Future Changes

The center of greatest abundance for cowbirds in the United States today extends from North Dakota to Oklahoma and coincides with the historic distribution of the cowbird, which includes southcentral Canada. The Midwest has the next highest abundance of cowbirds. Numbers of cowbirds drop off gradually in the east and more abruptly in the West (Fig. 15-2) but are continuing to expand their ranges to the north in the Yukon and northern Alberta

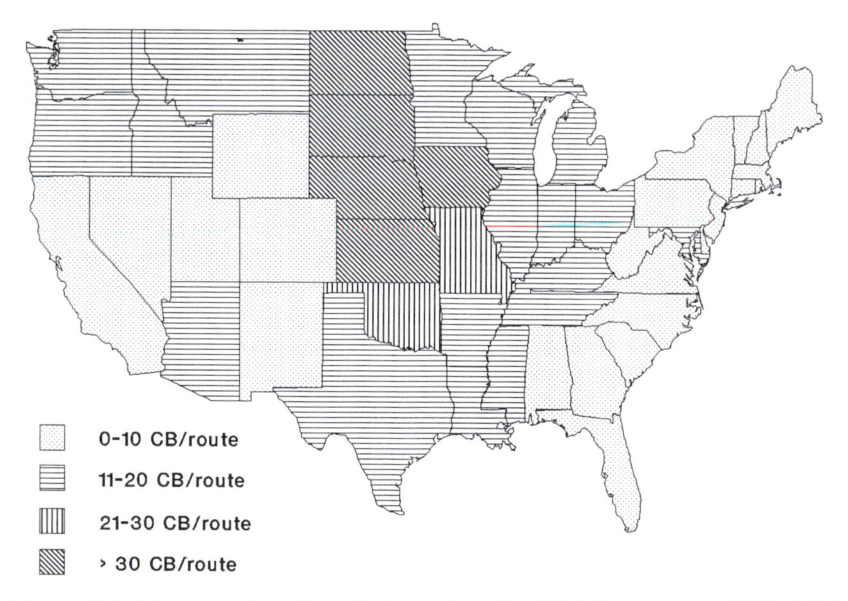

Figure 15-2. Mean number of cowbirds per Breeding Bird Survey route (1980–1989).

Table 15-1. Areas where cowbird parasitism levels are consistently high in most host species.

Geographical Area	Evidence	References
Fragmented Midwestern forests (Wisconsin, Illinois, Missouri)	Wood Thrush (80–100% parasitism levels); high community-wide levels in migrant songbirds	Robinson (1992), Brittingham and Temple (1983), Robinson et al. (1995)
Northern Great Plains Grasslands (North Dakota, Manitoba)	20–70% parasitism levels in center of Brown-headed Cowbird abundance	R. Koford and D. Johnson (personal communication), S. Sealy (personal communication)
Kansas tallgrass prairie (especially pastures)	Multiple parasitism is the rule in some prairie birds (e.g., Dickcissel)	Elliott (1978)
Southern California Riparian woodlands	Endangered status of Least Bell's Vireo and Southwestern Willow Flycatcher	Whitfield (1990), Franzreb (1989)
Puerto Rican woodlands and scrub (Shiny Cowbird)	Endangered status of Yellow-shouldered Blackbird following recent colonization of cowbirds	Cruz et al. (1989)

(J. N. M. Smith, personal communication), and are locally abundant in the Far West (Trail and Baptista 1993).

According to Breeding Bird Survey (BBS) data (Robbins et al. 1986), cowbirds are decreasing slightly in the Northeast and increasing in abundance in all other regions, with the greatest rates of increase in the Southeast. The cowbird has only recently become established in Florida and other southeastern states. Host species in these states may be particularly vulnerable because the Shiny Cowbird is simultaneously expanding its range northward into Florida and other southern states (Cruz et al. 1989, Grzybowski and Fazio 1991).

Geographic Variation in Parasitism Levels

Although cowbirds occur across most of the United States and southern Canada, at least five areas show particularly high levels of cowbird parasitism (Table 15-1). In fragmented midwestern woodlands and Kansas tallgrass prairies, multiple parasitized nests are the rule rather than the exception. Data from the northern Great Plains also show high levels of parasitism. Two areas where cowbirds have only recently invaded, southern California and Puerto Rico, now have species that are endangered at least partly as a result of cowbird parasitism. No doubt further studies will reveal new areas where parasitism levels are high enough to threaten entire communities.

Levels of cowbird parasitism within species also vary geographically, sometimes as a result of changes in cowbird or host abundance, but in other cases for unknown reasons. Geographic variation is most evident for host species that nest over wide geographic areas. For example, the Wood Thrush nests throughout the eastern deciduous forest and is a common cowbird host. Researchers in the midwest have consistently reported levels of parasitism exceeding 80% in all but the most heavily forested sections of the Midwest (Brittingham and Temple 1983, Robinson 1992 Trine et al. in press, Robinson et al 1995). In the Eastern United States, however, local levels of parasitism are generally under 20% (Hoover 1992, Roth and Johnson 1993). Hoover and Brittingham (1993) found that levels of parasitism differed significantly among regions, and were positively correlated with cowbird abundance and negatively correlated with Wood Thrush abundance.

Red-winged Blackbirds are frequently parasitized and their breeding range coincides with the cowbird's breeding range. Freeman et al. (1990) provided data on parasitism levels of red-wings from a variety of studies conducted throughout North America. Levels of parasitism were variable, ranging from 0 to >40% of observed nests, as shown previously by Friedmann (1963). Levels of parasitism were highest in the central parts of North America where cowbird numbers were highest and were lowest in states where cowbird numbers were low, although there was a substantial amount of local variation. Engstrom (unpublished data) obtained over 6000 red-wing blackbird nest records from the Cornell nest record program. The highest levels of parasitism were clustered in the Midwest where cowbird numbers were highest but there also was within-region variation in parasitism levels. In Illinois and southern California, for example, red-wings are rarely parasitized even though cowbirds are abundant (S.I. Rothstein and S. K. Robinson, unpublished data). Parasitism levels of grassland-nesting birds are also much lower in Illinois than in the Great Plains (J. Herkert and R. Koford, personal communication).

HABITAT USE

The cowbird's parasitic habit enables it to breed in a wider range of habitats than perhaps any other North American passerine. Cowbirds were recorded on a higher percentage of 1988–1992 breeding bird censuses listed in the *Journal of Field Ornithology* (Vol. 60–64) than any other bird species (Rothstein 1994). Because cowbirds do not need to tend to their own offspring, their two main activities during the breeding season, reproduction and feeding, can be uncoupled and carried out in different locations. Thus, cowbirds commonly occupy habitats that fulfill only one of these needs (Rothstein et al. 1984), although in some areas they breed and feed in the same sites (J. N. M. Smith, personal communication). By contrast, most nonparasitic birds must remain close to their nests and are therefore limited to habitats that meet the food needs of adults as well as the requirements of nesting. Areas that provide foraging opportunities for cowbirds often contain few hosts relative to surrounding areas, and it is habitats with high passerine density and, possibly, diversity, that are preferred as breeding sites (Rothstein et al. 1984). In most regions of North America, cowbirds reach their greatest population densities in mixed habitats containing grassy areas and scattered bushes and trees, such as old-field and forest-meadow ecotones where breeding and feeding areas are close together. Riparian areas, freshwater marshes, and orchards are also centers of abundance in California. The only terrestrial habitats in which cowbirds rarely breed are treeless areas in high mountains and deserts, although they were abundant in desert riparian corridors (Grinnel 1914, Rosenberg et al. 1991).

Local factors also affect the distribution and abundance of cowbirds, and levels of parasitism within landscapes (Table 15-2). In the following section, we elaborate on some of the factors listed in Table 15-2 that have been associated with high levels of parasitism and cowbird abundance.

Internal and External Openings

Numbers of cowbirds and levels of parasitism within the eastern deciduous forest vary with distance from openings (Gates and Gysel 1978, Brittingham and Temple 1983). Within an extensive area of contiguous forest in Wisconsin, Brittingham and Temple (1983) found that 65% of the nests they found within 99 m of an edge or opening were parasitized, and levels of parasitism declined as distance from an edge or opening increased. Parasitism levels fell to 18% of observed nests at >300 m from an edge or opening, although some species were heavily parasitized even 400 m into the forest interior. Because forest fragmentation increases the ratio of forest edge to interior, higher levels of parasitism appear to be an indirect consequence of fragmentation.

Recent research suggests that the magnitude of the "parasitism edge effect" may vary as a function of the local landscape and cowbird abundance (Thompson et al., in press) Robinson and Wilcove (1994), for example, working in Illinois where forests are highly fragmented and cowbird populations are high, found few edge-related changes in parasitism. Instead, levels of parasitism were high (>50% of observed nests) throughout

Table 15-2. Local factors that affect cowbird habitat use and levels of parasitism.

Factor	Locale	Parasitism Frequency	Source
Internal openings (≥0.2 ha)	Southern Wisconsin	30–80%; higher near openings	Brittingham and Temple (1983)
Powerline corridors	Maryland	Higher near corridor	Chasko and Gates (1982)
Field-forest edges	Maryland	Higher near edges	Gates and Gysel (1978)
	Illinois	Parasitism levels high everywhere	Robinson and Wilcove (1994) Thompson et al. (in press)
Clearcuts	Indiana	Higher near clearcuts	D. Whitehead (unpublished data)
	Missouri	Cowbird abundance highest near clearcuts	Thompson et al. (1992)
Tract size	Illinois	Parasitism levels highest (70–80%) in tracts less than 200 ha	Robinson et al. (in press)
Streams	Maryland	Cowbirds use streams as travel lanes	Gates and Giffen (1991)
Woody edges of graslands	Wisconsin	Higher near woody corridors and edges where cowbirds perch	Johnson and Temple (1990)
Proximity to pasture and feedlots	California	Cowbird abundance higher near pasture and feedlots	Verner and Ritter (1983)
Host density	Illinois	Cowbird abundance proportional to host density within regions	Robinson and Wilcove (1994)

the forest even at distances >300 m from an edge or opening (Thompson et al., in press).

In the West, cowbirds occur regularly within coniferous forests but at lower densities than in other habitats in the same region such as meadows and riparian zones (Rothstein et al. 1980, 1984, Verner and Ritter 1983, Airola 1986). Because some western coniferous forests are more open than eastern forests, it is unclear whether or not western and eastern cowbirds differ in their preferences for forests, or if host distribution or some other factors influence habitat occupancy by cowbirds.

There is little information on differences between the effects of external agricultural edges and internal edges created by forestry practices. Clearcuts do not provide prime foraging habitat for cowbirds, but they may be used for displaying and nest searching. Whitehead (personal communication) has found higher rates of parasitism near clearcuts and agricultural edges than in the forest interior (see also Thompson et al. 1992). How different types of openings influence the impact of cowbirds on host populations is a research question that needs to be addressed.

Corridors

Corridors such as powerlines within forest habitats create internal edges. Gates and his colleagues have looked at whether numbers of cowbirds and rates of parasitism are higher near these openings than away from them, and compared these results with the effects of natural corridors created by streams running through forest habitat (Chasko and Gates 1982, Gates and Giffen 1991). They found numbers of cowbirds and rates of parasitism were higher near both types of corridors than away from them. Because host density is also higher along these corridors, they suggested that cowbirds are attracted to the high density of nests. Gates is continuing his research on cowbird use of corridors.

Open corridors within forest habitat are not the only type of corridor that attracts cowbirds. Johnson and Temple (1990) found that rates of cowbird parasitism were higher near woody corridors and edges within tall grass prairie habitat than within the prairie interior. Friedmann (1963), and Peck and James (1987) showed that obligate ground-nesting grassland species were rarely parasitized. Cowbirds may only have been able to find nests near the edges of grasslands, where trees and shrubs were available as elevated perches, where cowbirds could watch hosts and search for host nests.

Forest Fragmentation

Studies of parasitism rates in midwestern oak–hickory forests show clear effects of forest fragmentation on parasitism levels. In small (<100 ha) woodlots located in a sea of corn and soybeans in central Illinois, 75% of the nests of Neotropical migrants contained an average of 3.3 cowbird eggs per nest (Robinson 1992). In the Shawnee National Forest of southern Illinois, a region that is 40–60% forested, parasitism levels of nests of Neotropical migrants averaged 50–60% (Robinson and Wilcove 1994, Robinson et al., in press). In the Hoosier National Forest of southern Indiana, an area of >80% forest, less than 30% of host nests were parasitized (Robinson et al. 1995). In the central Missouri Ozarks, where over 90% of the landscape is forested, most neotropical migrants suffered parasitism of less than 5% of their nests (Robinson et al. 1995). Similarly, Holmes et al. (1986) have never found a parasitized nest in the Hubbard Brook Experimental Forest of New Hampshire, an area that is virtually entirely forested. Within Illinois forests, parasitism levels generally decrease as tract size increases, but remain high (20–50%) even in the interior of the largest (up to 5000 ha) tracts (Robinson et al. in press).

Livestock and Other Human-based Feeding Sites

Cowbirds show a remarkable ability to find and exploit food resources created by human activities. They forage at livestock corrals where they feed on insects attracted by the mammals, and on hay and grain provided for livestock (Rothstein et al. 1980, 1987, Verner and Ritter 1983). Where livestock are unavailable, cowbirds exploit bird feeders and food scraps at campgrounds, where their tameness often attracts attention and additional food. Numbers of cowbirds and levels of parasitism are highest near these food sources and decline with distance from these sources of disturbance (Verner and Ritter 1983, S. K. Robinson, unpublished data).

Radio-tracking studies showed that cowbirds in the Sierra Nevada, Missouri Ozarks, and southern Illinois forests commuted on a daily basis between morning breeding ranges and afternoon feeding sites up to 7 km away (Rothstein et al. 1984, Thompson 1994), although most flights were much shorter. This commuting behavior means that a single human-based feeding site, such as horse corrals at a pack station, can make large forest tracts accessible to cowbirds.

"Commuting" cowbirds shifted from largely antisocial behavior in the morning during which individuals showed considerable aggression towards conspecifics of the same sex to highly social feeding in the afternoon when little overt aggression occurred. Feeding sites such as horse corrals in the Sierras that typically had 10–30 cowbirds in the early afternoon normally had five or fewer in the first 2–4 hours after sunrise. Conversely, cowbirds were rarely detected in the afternoon in breeding habitat where they occurred in the morning in all three sites (Rothstein et al. 1984). While most passerines are more active and therefore more easily detected in the morning, none of 11 other common species showed such a large afternoon decline in the frequency with which they were seen or heard in breeding habitat (Rothstein et al. 1984).

Although most cowbirds radiotracked in the Sierra Nevada made one trip per day to feeding sites (where they generally stayed from early afternoon to shortly before sundown), some individuals whose breeding ranges were within 1–2 km of feeding sites made occasional brief trips to feeding sites during the morning. Similarly, cowbirds in Illinois commuted several times a day to feeding sites (F. Thompson, unpublished data). This shows the extreme flexibility of cowbird activity patterns and the need for further studies in other landscapes. If feeding sites are near or within suitable breeding areas, cowbirds will make use of them during the morning (Rothstein et al. 1986). Such proximity or intermixing of feeding and breeding sites occurs over much of North America.

Host Density

In Illinois, where cowbirds are abundant in all woodlots studied to date, cowbird abundance is strongly correlated with host density within study areas (Robinson and Wilcove, 1994, Robinson et al., in press). In areas where hosts are abundant, cowbirds are also generally abundant and the ratio of cowbird to host abundance is a good predictor of community-wide parasitism levels (Thompson et al., in press, Robinson et al., in press). Tract size, proximity to cowbird feeding areas and, possibly, forest type also appear to influence cowbird abundance.

ANNUAL CYCLE AND BREEDING BEHAVIOR OF BROWN-HEADED COWBIRDS

Annual Cycle

Cowbirds begin to appear on their breeding grounds in late March to early May depending upon latitude and elevation. As with many birds, males appear one to several weeks earlier than females (Friedmann 1929). When they first return to a region, cowbirds are likely to occur in small flocks centered around food sources. Whereas these flocks contain local breeders, they may also contain many birds that are still migrating northwards (Yokel 1986a,b). Over several weeks, the early season flocks gradually break down, and cowbirds spend more of each day in the actual sites where breeding occurs. In most regions male song becomes conspicuous by early April to early May.

The peak period of cowbird egg-laying is May in most regions, although this is delayed until June in the Sierra Nevada and presumably in other high-elevation areas of the West, and probably in northern areas. Even in low-elevation areas, egg-laying begins later in the West than in the East, as is evident from egg dates (Bent 1958, Lowther 1993). In the Northeast, cowbirds often lay during the second week of April in nests of Eastern Phoebes (S. I. Rothstein, personal observation). In Kansas, most parasitism of this species occurs in mid-to-late April (Klaas 1975). Even in parts of the West with a more moderate climate such as southern California, coastal British Columbia, and Arizona, little laying occurs before May (Finch 1982, 1983) even though suitable hosts such as Song Sparrows, Abert's Towhees, and Brewer's Blackbirds begin to nest in March or April. In eastern Washington, parasitism occurred at 7.0% of 415 nests in which Brewer's Blackbirds began laying between 10 April and 9 May, but at 55.7% of 422 nests begun between 10 May and 20 June (Friedmann et al. 1977). Freeman et al. (1990) reported a similar trend in Washington for Red-winged Blackbirds. By contrast, cowbirds in Connecticut begin to lay 1–3 weeks earlier than redwings (S.I. Rothstein, personal observation).

Late breeding by cowbirds allows some western passerine populations to escape parasitism almost completely during their period of peak productivity. The extent to which this occurs may vary from year to year. In 1990 and 1991, White-crowned Sparrows in the San Francisco area experienced parasitism at 45% of 121 nests (Trail 1992) but this declined to 30% of 66 nests in 1992 when the sparrows began to breed about 2 weeks earlier (D. Bell, personal communication). It is possible that cowbird parasitism is selecting for early breeding in some hosts. Because cowbirds may time parasitism to coincide with the maximum period of host nest availability, early and late-nesting hosts may escape parasitism. In contrast to the West, cowbird parasitism on some hosts in the East and Great Plains declines in late May and June (Klaas 1975, Petit 1991).

Cowbird egg-laying declines rapidly in early July and ends in most regions by mid-July (Friedmann 1929, Scott 1963), although eggs are sometimes laid as late as the first few days in August in the Sierra Nevada (Friedmann et al. 1977). In some regions, cowbirds begin to form flocks in late June (Scott 1963, Payne 1973a). These are composed mainly of males and may consist of males that have failed to secure mates, as males outnumber females in all regions by from 1.5–3 to one (Darley 1971, Beezely and Reiger 1987, Yokel 1989). Most passerines become inconspicuous and hard to find after the breeding season in late July and August, and this is especially pronounced in cowbirds. The scarcity of cowbirds in some places during this period results from migration, as in the Sierra Nevada Mountains, where nearly all adults leave for lower elevations by mid-to-late July (Rothstein et al. 1980, Verner and Ritter 1983). In some areas such as southern California where cowbirds are year-round residents, adult cowbirds are very difficult to find in August, although juveniles may be readily found around livestock. It is likely that adults form large flocks at this time while undergoing the postnuptial molt.

Most cowbirds that migrate from low-elevation areas do not do so until October or November, well after the peak of passerine migration, although some depart Canada in August (J. N. M. Smith, personal observation). Cowbirds, however, seem to be exploiting man-made food sources in the winter in some northern areas, such as eastern Washington, which is enabling them to become year-round residents. Wintering aggregations or roosts of cowbirds in the South often reach huge numbers in the tens of thousands or even millions (Meanley 1975, Johnson et al. 1980). Frequently, other blackbird species or starlings occur in these roosts, which typically exploit man-made feeding sites, such as agricultural fields and livestock concentrations. Banding records confirm that these roosts contain cowbirds that breed over disparate locations (Crase et al. 1972, Coon and Arnold 1977, Dolbeer 1982). Even in regions where the local breeding populations are not migratory, such as southern California, most cowbirds present in the winter are probably migrants from elsewhere (S. Rothstein, personal observation).

Social System

Social systems and other aspects of the cowbird's behavioral ecology are reviewed in Rothstein et al. (1986). Cowbirds are highly social year-round, although during the breeding season social aggregations occur mostly in the afternoon and at night. These latter groups contain breeding birds because about 70% of the females in them have eggs in their oviducts (Scott and Ankney 1983, Fleischer et al. 1987), i.e., eggs scheduled to be laid the next morning. Although they are not strictly territorial in the morning because they do not occupy exclusive breeding ranges, cowbirds of both sexes seek out and threaten members of the same sex whenever they are detected in the morning. This is confirmed by experimental playback of both female chatter calls and male flight whistles (Dufty 1982, 1985, Rothstein et al. 1988, Yokel 1989). This intrasexual aggression may result in some individuals being excluded from areas of prime breeding habitat. In addition, individuals of both sexes are attracted to vocalizations of the other sex (Rothstein et al. 1988), undoubtedly for purposes of mate assessment.

There is strong male–male competition for mates because males outnumber females, with the male bias more extreme in the West than in the East (Rothstein et al. 1986, Yokel 1986a, 1989 and references therein). J. N. M. Smith (personal communication) saw one male kill another by forcing it into the water and pecking it repeatedly on the head. Monogamy is the usual mating system in New York, Ontario, and California, but polygyny occurs regularly at low rates (Darley 1978, 1982, Dufty 1982, Teather and Robertson 1986). The most conclusive study was based on over 100 observed copulations in California and showed that matings outside the pair bond are rarer in cowbirds than in most nonparasitic passerines (Yokel 1986b, Yokel and Rothstein 1991). The lack of extra-pair copulations may result partly from mate guarding by males, but female choice may also be involved because females typically refuse to mate with these other mates even when their mate is not present. The only exception to this pattern of monogamy is a Kansas study in which a small sample of copulations involving marked birds demonstrated promiscuity (Elliott 1980), i.e., no fixed mating relationships. Cowbird density at the Kansas site was far higher and the sex ratio more even than at the other sites, and behavioral ecology theory would predict promiscuity under such circumstances (Yokel 1989). It is unknown whether promiscuity is typical throughout the Great Plains where cowbirds are especially abundant (Van Velzen 1972) as the

Kansas study site has an unusually high cowbird density, even for this region (R. C. Fleischer, personal communication).

In California, virtually all pair bonds involve adult males at least 2 years old (Yokel 1986a, Rothstein et al. 1986), whereas in Ontario and New York, adult and yearling males have equal mating success (Darley 1978, Dufty 1982). The adult–yearling contrast in mating success in California may be related to song development. Yearling males in California have smaller song repertoires and share few song types with adults (O'Loghlen and Rothstein 1993), whereas yearling and adult males in New York have equivalent song repertoires (Dufty 1985). Yearling females arrive on the breeding grounds and begin to lay about a week later than adults (Fleischer et al. 1987, see also Holford and Roby 1993).

Parasitic Habits

Female cowbirds generally sneak into host nests and lay their eggs about 10–25 min before sunrise (Scott 1991). Related blackbirds lay no earlier than shortly after sunrise and some other passerines lay as late as 5 h after sunrise. Cowbirds do not parasitize nests at random but instead appear to adjust their egg-laying so as to parasitize nests while the host is laying its eggs (Hann 1941). Cowbird eggs laid earlier have increased chances of being rejected by some hosts (Clark and Robertson 1981, Sealey 1992) or of simply being passively buried by nesting material (Rothstein 1986). Conversely, eggs laid too late after the host's laying period are unlikely to hatch. Cowbirds find host nests and probably estimate a nest's readiness for parasitism by secretively watching hosts build nests (Hann 1941, Wiley 1988). On occasion, cowbirds also find nests by actively searching through vegetation or by attempting to flush birds from well-concealed nests (Norman and Robertson 1975, Wiley 1988).

There is an extensive literature on the fecundity of female cowbirds (reviewed in Rothstein et al. 1986). The most reliable and simplest method for quantifying laying rates is to determine the proportion of females with oviductal eggs, which is equal to the proportion of females that will lay an egg the next morning. This is equivalent to the probability that any female will lay and also to the number of eggs each female lays per day. At the mid-point of the cowbird's breeding period, a laying rate of 0.7–0.8 eggs per day has been found in all regions where data have been gathered—Ontario, lowland California, Sierra Nevada Mountains, and Michigan (Fleischer et al. 1987). This constancy is remarkable given the variation among these regions in host species and densities, the availability of cowbird feeding sites and other energetic considerations.

The only major factors that seem to affect the total number of eggs that a female lays per year is the length of the breeding season and female age. Where egg-laying begins late, as in the Sierra Nevada, females average 30.5 eggs per season whereas in southern Ontario where the moderate climate allows cowbirds to breed earlier, the average is at least 40 (Scott and Ankney 1980). No other wild bird is known to lay so many eggs, including other brood parasites such as cuckoos (Payne 1973b), which has prompted Scott and Ankney (1980) to dub the cowbird as the "passerine chicken." Captive females have laid as many as 77 eggs in a season (D. Roby, personal communication). The Shiny Cowbird may lay more than 100 eggs per season (Kattan 1993).

Thirty to 40 eggs a year amounts to a combined mass of at least 90–120 g, which is roughly 2.5–4.0 times the mass of a female cowbird. Obviously then, the material for eggs comes primarily from a female's daily foraging intake rather than from long-term energy stores. A detailed study of nutrient reserves failed to show any nutritional cost of egg-laying in female cowbirds (Ankey and Scott 1980) but another study found low red blood cell levels in female cowbirds, which may represent a cost of egg laying (Keys et al. 1986). A cost is also suggested by mortality rates that are equivalent in males and females for the first year of life when neither have yet bred, but are much higher in females in subsequent years (Darley 1971). Calcium also limits egg production in cowbirds (Holford and Roby 1993).

It should not be assumed that each of a cowbird's 30–40 plus eggs results in the loss of host young. Some host species are "ejecters" and routinely remove cowbird eggs from their nests (Rothstein 1976, Scott 1977). Other hosts commonly desert some or most parasitized nests, and often escape parasitism when they renest (Clark and Robertson 1981, Burgham and Picman 1989, Sealy 1992). In addition, some cowbird eggs are not timed properly and result in little harm to hosts, and eggs in multiply parasitized nests have lower success rates than those in singly parasitized nests (C. L. Trine, unpublished data). The conditions under which cowbirds "waste" eggs (e.g., scarcity of hosts relative to cowbird abundance) and whether or not they avoid laying in nests of ejectors are questions worth further research. Nor do we know what proportion of eggs are wasted.

EFFECTS OF COWBIRD PARASITISM ON HOST SPECIES

Cowbirds have been known to parasitize 240 species of birds (Friedmann and Kiff 1985). There is some evidence that individual females also appear to parasitize several species (Fleischer 1985). A lack of host specificity by individual cowbirds is potentially very damaging to rare species because cowbird reproduction is not dependent on these rare hosts. Even if rare species decline, there may still be a nearly constant number of cowbirds attempting to parasitize them, if most of the recruitment into cowbird populations is provided by common hosts.

Factors that Lower the Success of Parasitized Hosts

Cowbird parasitism reduces host reproductive success in several ways. Female cowbirds usually remove one host egg from about 33% to as many as 90% of all nests they parasitize (Friedmann 1963, Weatherhead 1989, Sealy 1992). This is nearly always done on a visit other than the one during which the female lays her eggs (Mayfield 1961, Nolan 1978, Sealy 1992). Cowbirds have unusually thick and therefore strong eggshells (Blankespoor et al. 1982, Spaw and Rohwer 1987), which can result in the breakage and loss of occasional host eggs while the cowbird is laying, or during incubation when the cowbird eggs knock against host eggs (Roskaft et al. 1990, but see Weatherhead 1991).

The potential for more extreme damage comes from cowbird nestlings. Cowbird eggs have a short incubation period of 10–12 days, whereas the eggs of many host species require 12–14 days of incubation (Nice 1953, Friedmann 1963, Briskie and Sealy 1990). Hosts begin to feed cowbirds as soon as they hatch. Cowbird nestlings double their mass within 24 h after hatching (Norris 1947, Hatch 1983), and adult cowbirds usually parasitize hosts smaller than themselves. Thus, at hatching, host young of many species are small compared with cowbird nestlings, which are usually at least a day or two older. Birds generally feed the nestling that provides the strongest stimulus, and this is normally the much larger cowbird. Often a cowbird nestling is so much larger that the host young are covered by it and not even visible. All of this is exacerbated by the fact that cowbird nestlings beg more loudly than most host young (Friedmann 1929), have larger relative mouth sizes (Ortega and Cruz 1991), and probably develop more quickly (Weatherhead 1989, Hatch 1983, Briskie et al. 1990).

Cowbird parasitism is especially damaging when two or more parasitic eggs are laid in a nest, which occurs in at least a third of all parasitized nests (Friedmann 1963). The majority of parasitized nests in some communities contain multiple cowbird eggs (e.g., Elliott 1978, Robinson 1992). When such multiple parasitism occurs, even species that normally raise a single cowbird without any losses among their own nestlings often fail to raise any of their own young. Host species that normally accept a single cowbird egg are just as likely to accept a clutch with two parasitic eggs or one made up entirely of cowbird eggs (Rothstein 1982, 1986, Sealy 1992).

Because they beg so loudly, it is possible that cowbird nestlings make a nest more conspicuous to predators. Nest predation can benefit some hosts that raise none of their own young when parasitized. Nest losses in these species often result in renesting, which gives the host another chance to escape parasitism. Nevertheless, there is no consistent evidence that parasitized nests are more likely to be depredated. Parasitized and unparasitized nests were destroyed at equal rates in three well-studied host species (Mayfield 1960, Klaas 1975, Southern 1958). In another common host, the Song Sparrow, one study found that parasitized nests were destroyed at a higher rate than unparasitized ones (Nice 1937) whereas Smith (1981, personal communication) found no difference. Finch (1983) found increased mortality of parasitized Abert's Towhee nests. This issue is very difficult to resolve because both parasitism and predation often vary seasonally and spatially and only predation rates on nests with nestlings can be used to test the hypothesis that cowbird begging increases predation rates.

Effects of Parasitism on Host Population Dynamics

Population dynamics of hosts are determined by the number of young produced (births), and the mortality of those young and of adults (deaths). The impact of parasitism on host populations depends directly on how it influences these parameters, which are used to calculate intrinsic growth rates of populations. For birds, the three critical parameters to measure are: (1) seasonal fecundity, i.e., the number of young a female host can produce over the course of an entire breeding season given renesting; and (2) juvenile, and (3) female survivorship.

Annual Fecundity

Many studies have documented the impact of cowbird nest parasitism on brood reduction in passerine hosts (e.g., Walkinshaw 1961, Zimmerman 1982, 1983, Finch 1983, Marvil and Cruz 1989). In some cases, this brood reduction is substantial (e.g., Mayfield 1960, Nolan 1978, Weatherhead 1989). Multiple nestings, however, can occur, and some hosts abandon parasitized nests and renest. Some species may nest before or after egg-laying periods of cowbirds. Thus, brood reduction is only an indirect measure of the consequences of brood parasitism on seasonal fecundity. Nolan (1978), for example, found that Prairie Warblers, a Neotropical migrant vulnerable to cowbird parasitism, produced 3.4 and 0.9 young per unparasitized and parasitized nest, respectively, a reduction of 74%. Assuming this level of reduction, a parasitism level of 27% of all nests (i.e., 0.74×0.27) (Nolan 1978), should result in a 20% reduction in seasonal fecundity. The actual impact of parasitism, however, was only a 13% reduction because Prairie Warblers tend to abandon parasitized nests and renest. Nolan's (1978) calculations illustrate the complexity and difficulty of determining the impact of brood parasitism on the host population dynamics. Similarly, Sedgwick and Knopf (1988) followed Willow Flycatchers throughout their breeding season and showed that the impact of cowbird parasitism was much less on seasonal fecundity per pair than in the productivity of individual nests.

Few studies have measured most of the variables needed to quantify the impact of cowbird parasitism on host population dynamics (May and Robinson 1985, but see Trail and Baptista 1993). In addition, estimates of levels of parasitism based on percentages of parasitized nests found may underestimate actual levels of parasitism if parasitized nests are abandoned or depredated at higher rates than unparasitized nests (e.g., Nolan 1978, Finch 1983).

Models of host population dynamics (e.g., May and Robinson 1985, Temple and Cary 1988, Pease and Grzybowski, 1995) incorporate several key variables to assess seasonal fecundity (Table 15-3). Pease and Grzybowski (1995) have developed a model using the parameters in Table 15-3 that makes possible calculations of seasonal productivity per female even in the absence of data on productivity of color-marked females throughout the entire breeding season. By using estimates of the timing of events in the nesting cycle (e.g., egg-laying, incubation, and nestling times), the length of the breeding season, a set of observations of the changes in status for a sample of individual nesting attempts, parasitism and predation rates, fecundity of parasitized and unparasitized nests, and probability of abandonment, average seasonal fecundity per female can be calculated (Table 15-3).

Table 15-3. Parameters needed to estimate effects of cowbird parasitism on host populations.[a]

Instantaneous parasitism rate (per day)
Instantaneous nest predation rate (per day)
Probability a parasitized nest is abandoned
Time to initiation of egg-laying[b]
Time to initiation of incubation[b]
Time to fledging[b]
Time to when successful females can renest[b]
Time to when the non-renesting parent terminates parental care[b]
Time from when first and last nesting cycles are initiated
Fecundity of unparasitized and successful nests
Fecundity of unabandoned and successful parasitized nests
Seasonal fecundity

[a] Adapted from Pease and Grzybowski (1995).
[b] Times from initiation of nesting attempt.

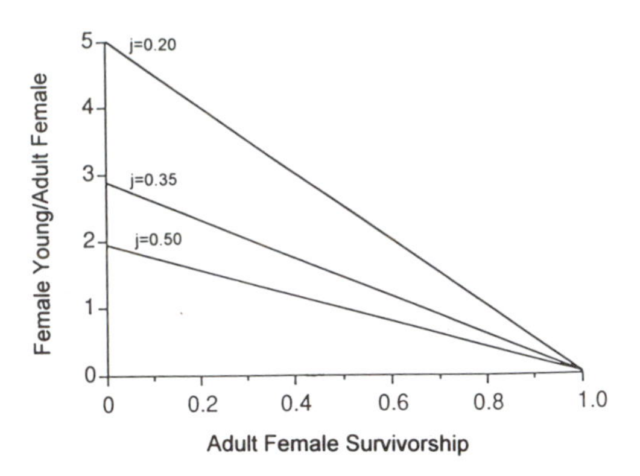

Figure 15-3. Relationship of juvenile survival (j), adult survivorship and fecundity necessary to sustain local populations. Fecundity is defined as the number of female young produced per adult female. Lines for differing j represent points where intrinsic growth rates are stable.

Many of these model parameter values can be obtained from the extant literature.

The Pease and Grzybowski model confirms that seasonal fecundity is strongly affected by daily predation and parasitism levels, and the length of the breeding season relative to the length of the nest cycle. Potentially, brood parasitism may have a greater impact on host productivity than nest predation even when it is much less frequent than predation because hosts usually renest after nest predation, but may not have time to renest after raising a cowbird to independence (Rothstein 1990). Nevertheless, predictions for the effects of parasites on host population dynamics will be limited until we have better data on adult and juvenile survival rates. Figure 15-3 shows how important both adult and juvenile survivorship are for estimates of fecundity needed to sustain local populations.

Adult and Juvenile Survival Rates and Lifetime Fecundity

The product of seasonal fecundity and juvenile survival rates must be equal to or greater than adult mortality for a closed population to endure. Because females are the important component of population models, adult survival rates should be expressed as female survival rates (May and Robinson 1985). Estimates of the threshold between growing and declining populations are strongly affected by variation in adult and juvenile survival rates.

Annual adult survival estimates of small landbirds generally range from 30% to 65% (Ricklefs 1973), but are uncertain because of the difficulty of distinguishing death from breeding dispersal. Roberts (1971) reported slightly higher estimates for several species of warblers, including an estimate of 85% for Ovenbirds. In many cases, however, survival rates of females are less than those of males (e.g., Graber 1961, Nolan 1978, Payne 1989).

Most estimates of survival rates are generated from band-return data (often skewed by more abundant data for males). Even detailed studies usually take place on plots small relative to the dispersal potential of the birds. Whereas males of many passerines maintain high site fidelity, females may switch territories between and within breeding seasons (Jackson et al. 1989). Because juveniles of most Neotropical migrants do not return to their natal areas, estimates of juvenile survival rates are particularly crude and perhaps more subject to inter-year variation than rates for adults. Usually, juvenile survival rates are assumed to be half that of adults (May and Robinson 1985).

Parasitism may also increase adult mortality if parasitized females expend more energy during breeding than unparasitized females. Some indirect evidence for this possibility is provided for Black-capped Vireos in Texas where ratios of males to females have climbed from 1:0.73, at pre-cowbird removal parasitism levels in excess of 80%, to 1:0.89, at parasitism levels reduced to less than 30% of all nests (Grzybowski unpublished data). Nest predation rates were similar during this period. Alternatively, females may disperse from areas where cowbirds are abundant. While annual male return rates were relatively constant over a period of 4 years (64–74%), female return rates increased from 27% following a year in which 45% of all nests were parasitized to 89% following a year when only 4% of nests were parasitized (Grzybowski 1991a). Smith (1981), however, found no differences in survival of parasitized and unparasitized female Song Sparrows in a resident island population.

EFFECTS OF HOST LIFE HISTORY ON HOST VULNERABILITY

Accepters and Ejecters

Most North American passerines can be placed into one of two groups on the basis of their responses to cowbird eggs. Among "accepter species," all or nearly all (85–100%) individuals accept cowbird eggs, regardless of whether a clutch contains a single cowbird egg or consists only of cowbird eggs. Conversely, nearly all individuals of "ejecter species" remove cowbird eggs (Rothstein 1975a, 1977, 1982). Ejecters are less adversely affected by cowbird parasitism than accepters. Ejecters most commonly rid themselves of cowbird eggs by ejecting the eggs from their nests and often do so only minutes after a cowbird egg has been placed in a nest (Rothstein 1975b, 1977, Scott 1977, Finch 1982). This means that considerable amounts of cowbird parasitism may go undetected because hosts remove some cowbird eggs before observers ever see them. In addition, this "unseen parasitism" means that for ejecter species the naturally occurring parasitism that is most likely to be observed are cases of acceptance, even if acceptance is a rare response. There is then a potential bias when inferring host responses solely from natural parasitism; experimental parasitism (placement of real or artificial cowbird eggs into nests) is often needed to assess a species' responses (Rothstein 1975a, 1976, 1977). Nevertheless, a host that accepts natural parasitism in at least a considerable minority (>25%) of its nests, and rarely or never ejects such parasitism, is almost certain to be an accepter given the all-or-none nature of host responses in most species.

Some hosts, including many accepter species, often desert naturally parasitized nests and renest (Rothstein 1975a, Graham 1988), although it is not clear that nest abandonment is always an antiparasite adaptation (S. Sealy, personal communication). Desertion in response to experimental addition of cowbird eggs has been found in only one species, so the proximate stimulus that triggers desertion in most species is unclear. Desertion may occur when hosts detect an adult parasite at or near its nest, a factor known to be important to some cuckoo hosts (Davies and Brooke 1988, Davies et al. 1989, Moksnes and Roskaft 1989). Clay-colored Sparrows and Yellow Warblers, however, are not more likely to abandon nests following detection of cowbirds (Hobson and Sealy 1989, S. Sealy, unpublished data).

At least two hosts do not fit easily into the accepter–rejecter dichotomy as both frequently show acceptance and desertion of naturally parasitized nests. The Yellow Warbler typically deserts its nest or buries over the cowbird egg with a new nest floor if a cowbird egg is laid before it begins to lay. Rejection occurs about 50% of the time if a cowbird egg is laid during the first half of the warbler's own laying period after which rejection declines to near zero (Clark and Robertson 1981, Burgham and Picman 1989, Sealy 1992). Secondly, the Prothonotory Warbler is likely to desert parasitized nests only if there are additional nest sites on its territory (Petit 1991). However, this species is an obligate cavity nester and there are few other cowbird hosts that are limited in their nest site to the same extent as this species (but see Briskie et al. 1992, Sealy 1992).

Both of these warblers and all ejecter species show no tendency towards increased ejection when parasitized naturally (Graham

1988, Petit 1991) or experimentally (Rothstein 1982) with a single versus two or more cowbird eggs. This is somewhat anomalous because multiple parasitism is extremely deleterious to host success. Why hosts such as Yellow Warblers would evolve a response to early but not to multiple cowbird eggs is unclear.

Timing of the Breeding Cycle

The impact of parasitism is determined in part by the timing and length of various components of the reproductive cycle, particularly the length of the breeding season. Breeding seasons generally decrease in length in more northerly climates and at high altitudes. In essence, this means that potential hosts in such areas have less opportunity for renesting after a parasitism or predation event, allowing these potential hosts less flexibility. Shorter breeding seasons greatly reduce seasonal fecundity for a fixed set of model parameters. As cowbirds expand their range both northward and to higher elevations in the mountains of western North America (Snyder 1957, Mayfield 1965, Hanka 1985), their impact on host populations may therefore increase.

In contrast, species with long breeding seasons might buffer their populations from the effects of brood parasitism because of the higher production generated from the greater number of possible renesting attempts possible, and because such hosts can sometimes evade brood parasitism by nesting before or after the egg-laying periods of cowbirds. Host productivity may be elevated to such a high level that brood parasites have only a minor role in regulating host numbers. The Northern Cardinal in southern Ontario, for example, incurs high levels of parasitism (0.8 of all nests), yet maintains higher seasonal fecundity (2.8 young per female; Scott 1963, Scott and Lemon unpublished data) than Neotropical migrants such as the Black-capped Vireo and Prairie Warbler at moderate parasitism levels (Nolan 1978, Grzybowski unpublished data). Some resident passerine species can evade parasitism by nesting before or after the egg-laying periods of cowbirds (e.g., Abert's Towhees; Finch 1983; and Indigo Buntings; Carey 1982). Many Neotropical migrants, however, have breeding seasons constrained by the requirements of migration.

Incubation Time

Another life-history trait affecting host fecundity from parasitized nests is incubation time. Hosts with incubation times greater than the 10–12 day period of the cowbird may suffer substantial brood reduction. This is the case among several currently endangered passerines affected by cowbird nest parasitism (Walkinshaw 1983, Goldwasser et al. 1980, Grzybowski et al. 1986). Kirtland's Warbler and Least Bell's Vireo each have incubation times of about 14 days whereas that of the Black-capped Vireo is 14–17 days (Graber 1961). In general, incubation times of many vireos and *Empidonax* flycatchers are 13–14 days (Ehrlich et al. 1988, Briskie and Sealy 1987), putting them at great risk in heavily parasitized populations. The incubation period of many warblers is 12–13 days, only slightly longer than that of cowbirds.

Host Size

Small hosts with long incubation periods usually fail to raise any of their own young if a single cowbird egg hatches, even though cowbird nestlings do not overtly attack host young. Species in this category are small flycatchers such as *Empidonax* spp. (Briskie and Sealy 1987) and phoebes, small vireos such as Black-capped and Bell's, small warblers such as the Prairie Warbler and the Common Yellowthroat, and small sparrows such as the Chipping and Clay-colored. At the other extreme, raising a cowbird nestling usually results in the loss of no host nestlings in medium-sized hosts such as the Song Sparrow (Smith 1981), and large ones such as Red-winged and Brewer's blackbirds (Friedmann et al. 1977, Weatherhead 1989, Roskaft et al. 1990) with relatively short incubation periods. However, parasitized individuals of these species still average about one lost egg as a result of egg removal and breakage.

Table 15-4. Number of Neotropical migratory species ($N = 98$) categorized as common hosts (C), uncommon hosts (U), or rare hosts or never parasitized (R) grouped by vertical placement of nests and general habitat type. Categorization of relative parasitism frequency is based upon that of Ehrlich et al. (1988). Only species that could be considered potentially suitable hosts (i.e., breeding range overlaps the range of the Brown-headed Cowbird, body weight < 50 g, insectivorous nestling diet, nonaerial foragers) are included. Numbers in parentheses are percentages of the total number of species within each cell.

Nest Placement[a]	Habitat[b]				
	Deciduous Forest	Forest Edge/ Second growth	Grassland/Steppe/ Desert	Open Forest/ Riparian	Coniferous Forest
Ground	C = 6 (55) U = 2 (18) R = 3 (27)	C = 1 (50) U = 1 (50) R = 0	C = 1 (33) U = 2 (67) R = 0	C = 0 U = 0 R = 0	C = 1 (20) U = 2 (40) R = 2 (40)
Shrub	C = 3 (50) U = 2 (33) R = 1 (17)	C = 12 (71) U = 4 (24) R = 1 (6)	C = 1 (17) U = 3 (50) R = 2 (33)	C = 2 (29) U = 1 (14) R = 4 (57)	C = 1 (25) U = 1 (25) R = 2 (50)
Subcanopy	C = 4 (57) U = 3 (43) R = 0	C = 1 (100) U = 0 R = 0	C = 0 U = 0 R = 0	C = 2 (29) U = 2 (29) R = 3 (43)	C = 1 (14) U = 1 (14) R = 5 (71)
Canopy	C = 0 U = 1 (33) R = 2 (67)	C = 0 U = 0 R = 0	C = 0 U = 0 R = 0	C = 1 (50) U = 0 R = 1 (50)	C = 0 U = 2 (50) R = 2 (50)
Cavity	C = 1 (50) U = 0 R = 1 (50)	C = 0 U = 0 R = 1 (100)	C = 0 U = 0 R = 3 (100)	C = 0 U = 0 R = 0	C = 0 U = 0 R = 0

[a] Vertical habitat stratum in which the nest is typically located: ground = on ground or within herb layer (e.g., usually < 0.25 m in height); shrub = in woody vegetation > 0.25–3.0 m; subcanopy = > 3.0–10.0 m; canopy = > 10.0 m.

[b] General habitat where most individuals of a species are typically found: deciduous forest = mature (> 20 years old); forest edge/second growth = forest–field ecotones and forest < 20 years old; grassland/steppe/desert = shrub or grassland areas lacking a subcanopy/canopy layer; open forest/riparian = mature forests with relatively low canopy closure (e.g., willow forests, parks, orchards, western riparian corridors); coniferous forest = mature forest in which conifers comprise $\geq 30\%$ of tree species.

Nest-site Placement

Type and vertical location of the nesting substrate can influence host vulnerability to cowbird parasitism by affecting both detectability and accessibility of the nest to female cowbirds. Obvious differences in vulnerability to parasitism exist between open- and hole-nesting species, and approximately 86% of all hole-nesting passerines in North America are rarely or never parasitized (Friedmann 1929, 1963, 1966). Of the six species of hole-nesting Neotropical migratory passerines that are potentially suitable hosts, two species are never parasitized, and four are rare hosts (Table 15-4).

Hole-nesting species may escape parasitism for a variety of reasons unrelated to nesting substrate per se, including aggressive behavior (e.g., wrens and flycatchers), early nesting time (parids), or because they are otherwise unsuitable hosts (e.g., aerial insectivores such as swallows; but see Mason 1986). Hole nests are probably no less (and may be more) detectable to a cowbird than are open nests. Hole nests, however, presumably are usually less accessible to cowbirds either because the small entrance hole and deep nest restrict the cowbird, or there is greater difficulty in determining presence of the host female on the nest and thus greater risk of discovery to the parasite. The Prothonotary Warbler is the only hole-nesting species in North America that is a common cowbird host, perhaps because of this warbler's habit of using shallow nest cavities or filling deeper cavities with moss to elevate the nest almost to the entrance hole (Friedmann 1929, Petit 1989, 1991). Evidence from studies of Eastern Bluebirds suggests that the size of the entrance hole can affect the probability of parasitism, as relatively high rates of parasitism have occurred in several studies where entrance holes of natural and artificial sites were uncharacteristically large (Musselman 1946, Pinkowski 1974, Woodward and Woodward 1979). House Wren nest boxes with large entrance holes were heavily parasitized by Shiny Cowbirds in Colombia (Kattan 1993).

For open-nesting species, the relationship between nest concealment and the probability of parasitism remains unclear, as few studies have compared microhabitat features of unparasitized and parasitized nests. To date, most evidence suggests that the probability of parasitism is not significantly related to increased (subjectively ranked) nest concealment (Anderson and Storer 1976, Best 1978, Best and Stauffer 1980, Smith 1981, Buech 1982; but see DellaSala 1985). Lack of relationship between concealment and parasitism frequencies is perhaps not surprising, however, given that female cowbirds locate many or most nests by monitoring movements of female hosts (Friedmann 1929, Hann 1937, Norman and Robertson 1975). Brittingham and Temple (unpublished data), however, compared the structure of the vegetation around parasitized and nonparasitized nests. Structural features of the vegetation associated with high rates of parasitism included both an open subcanopy and a dense shrub layer. The open subcanopy may aid in nest searching by allowing cowbirds perched in the canopy to obtain an unobstructed view of host nest-building activity while dense shrub cover provides nest sites for host species.

Cowbirds are frequently observed perched or displaying at the top of dead snags. Anderson and Storer (1976), working within relatively open jack pine (*Pinus banksiana*) habitat, tested whether parasitism rates of Kirtland's warblers were higher when a dead snag was in close proximity to the nest. They found that parasitism rates were significantly higher near dead snags and suggested that cowbirds were using the exposed perches for both displaying and nest searching. Robbins (1979) hypothesized that a similar situation might occur within the deciduous forest. Brittingham and Temple (unpublished data) tested his hypothesis but found no relationship between proximity of dead snags and probability of parasitism. The difference in results between the two studies may result from differences in the general habitat structure. In deciduous woods, snags below the canopy do not provide a better observation perch than exposed branches of living trees, and snags above the canopy do not provide females with a view of host nest-building activity below. On the other hand, in the more open jack-pine habitat, snags may provide an extensive view of nest-building activities.

The structure of the vegetation may also influence how successful cowbirds are in selecting appropriate host nests to parasitize (Martin 1993). Freeman et al. (1990) studied cowbird parasitism on Red-winged Blackbirds in marshes of eastern Washington. They divided the marshes into those with: (1) no trees around the perimeter; (2) 1–4 trees around the perimeter; and (3) more than four trees on the edge of the marsh. Tree density did not affect overall rates of parasitism but cowbirds laid fewer eggs in inappropriate (inactive, abandoned, etc.) nests when trees were available than when they were absent. Female cowbirds used the trees as perches to locate nests and observe host behavior.

The probability of parasitism might also be related to the habitat type and vertical position of the nest within the habitat. A greater proportion (71%) of shrub-nesting species inhabiting forest-edge/second-growth habitats tend to be common or frequent cowbird hosts compared with species nesting in other vegetation strata and habitats (Table 15-4). This result is corroborated by evidence summarized from the literature (Martin 1992) that shrub-nesting species occupying shrub/grassland habitats suffer a greater percentage of nest losses as a result of cowbird parasitism than do species occupying other strata and habitats. However, log-linear analysis of relative parasitism frequencies in host species occupying different vegetation strata across five general habitat types indicated that relative frequency of parasitism was significantly related to habitat type ($\chi^2 = 19.56$, df $= 4$, $P = 0.0006$; Table 15-5), but not to nest placement within the habitat (i.e., ground, shrub, subcanopy/canopy; $\chi^2 = 0.77$, df $= 2$, $P = 0.68$; Table 15-6). In addition, no significant interaction existed between effects of nest height and habitat type on parasitism frequency ($\chi^2 = 2.29$, df $= 5$, $P = 0.81$). Thus, habitat type appears to be most important in influencing relative parasitism frequencies among species, with a significantly greater

Table 15-5. Relative parasitism frequencies (based on Ehrlich et al. 1988) for 92 species of open-nesting Neotropical migratory passerines occupying different general habitat types. Numbers represent the number of species (%) for each frequency cateogy.

Habitat[a]	Parasitism Frequency		
	Common	Uncommon	Rare/Never
Deciduous forest	13 (48)	8 (30)	6 (22)
Forest edge/second growth	14 (70)	5 (25)	1 (5)
Grassland/steppe/desert	2 (22)	5 (56)	2 (22)
Open forest/riparian	5 (31)	3 (19)	8 (50)
Conifrerous forest	3 (15)	6 (30)	11 (55)

[a] See Table 15-4 for descriptions of habitats.

Table 15-6. Relative parasitism frequencies (based on Ehrlich et al. (1988) for 92 species of open-nesting Neotropical migratory passerines occupying different vertical habitat strata. Numbers represent the number of species (%) for each frequency category.

Nest Placement[a]	Parasitism Frequency		
	Common	Uncommon	Rare/Never
Ground	9 (43)	7 (33)	5 (24)
Shrub	19 (48)	11 (28)	10 (25)
Subcanopy/canopy[b]	9 (29)	9 (29)	13 (42)

[a] See Table 15-4 for description of nest placement categories.
[b] Categories were lumped together for statistical analyses.

proportion of species that inhabit forest edge/second growth being common hosts (Table 15-5). In Illinois, however, parasitism levels did not differ consistently among nesting strata and were lowest in species of grassland and shrubland habitats (Robinson et al., in press).

Within species, there is some evidence that nest height can affect the probability of parasitism, although this evidence is conflicting among studies and may be confounded by shifts in both nest parasitism and nest-height levels within seasons. Fleischer (1986) found that parasitized nests of Red-winged Blackbirds and Dickcissels in marsh and grassland habitats were significantly higher in the vegetation than unparasitized nests($\bar{X}$ = 0.94 vs 0.73 m for Red-wings, 0.65 vs 0.34 m for Dickcissels). Newman (1970), however, found that parasitism tended to be more frequent in ground nests (63.6%) than in above-ground nests (36.4%) of Lark Sparrows. Other studies of sparrows (Song Sparrow: Smith 1981; Field Sparrow: Best 1978, Buech 1982; Chipping Sparrow: Buech 1982: and Clay-colored Sparrow: Buech 1982) have also found no differences in heights of parasitized and unparasitized nests. Finally, DellaSalla (1985) found that parasitized nests of Yellow Warblers were significantly lower ($\bar{X} = 1.51 \pm 0.10$ m) than unparasitized nests ($\bar{X} = 2.78 \pm 0.61$ m) (see also Briskie et al. 1990).

CASE HISTORIES

Species that are endangered at least partly as a result of cowbird parasitism have several features in common (Table 15-7). All have restricted geographical breeding ranges in which most or all populations are exposed to parasitism. In all cases, remaining habitat within their breeding ranges is under severe threat as a result of habitat destruction, fragmentation, and/or fire suppression, which compounds problems with cowbird parasitism. In three of four cases, cowbirds are fairly recent colonists in the breeding areas. Encouragingly, cowbird control efforts and habitat management have been success-

Table 15-7. Summary of case histories of hosts endangered by Brown-headed Cowbird parasitism.

Species	Breeding Range	Habitat	Parasitism	Cowbird Control Efforts	Population Trend
Kirtland's Warbler	Very narrow (Michigan)	Specialized (Jack Pine) fire-dependent	50–75% (pre-control)	Yes	Stable to increasing following cowbird control and prescriptive burn
Least Bell's Vireo	Small (Southern California)	Riparian corridors (habitat loss and fragmentation have been severe)	50–60% (pre-control)	Yes	Increasing in cowbird control areas
Southwestern Willow Flycatcher	Southwestern US, Mexico	Riparian corridors (habitat loss and fragmentation)	50–60% (Kern River)	Yes (Kern River)	Local increases after control
Black-capped Vireo	Southcentral US, Northern Mexico	Oak scrub habitats (fire-dependent successional stage)	Up to 95%	Yes	Populations increasing in cowbird control areas

ful in stabilizing or increasing host populations. Below we elaborate on these case histories.

Kirtland's Warbler

Probably the best known endangered host of the Brown-headed Cowbird, the Kirtland's Warbler, was driven to the brink of extinction at least partly because of the detrimental effects of brood parasitism. Populations of the warbler have never been large; the peak population size in the late 1800s was probably only a few thousand individuals (Mayfield 1977b). The breeding area, discovered in 1903, is restricted to a small portion of northern lower Michigan where multiple critical habitat (vegetation and soil) characteristics are present (Mayfield 1960). Suitable nesting habitat is restricted to young jack pine stands (generally 2–6 m in height) during a narrow window of succession after fire. Breeding habitat was relatively abundant in the late 1800s and early 1900s due to logging practices and lack of fire control.

Cowbird populations, and thus rates of nest parasitism, steadily increased in the breeding range of Kirtland's Warbler during the 1900s. Parasitism occurred in a mean of 55% of warbler nests between 1944 and 1957 (Mayfield 1960), rising to a mean of 69% of nests between 1966 and 1971 (Walkinshaw 1983). Mean number of young fledged per nest (including successful and depredated nests) during these latter 6 years was 0.8. Breeding success of the warbler has been improved dramatically through the highly successful cowbird removal program (Shake and Mattson 1975, Kelly and DeCapita 1982, Walkinshaw 1983). Immediately following implementation of cowbird control in 1972, frequency of parasitism dropped to 6%, and remained at a mean of 3.4% over the first 10 years of the program (Kelly and DeCapita 1982). Warbler reproductive success also showed an immediate increase, and the mean number of young fledged per nest between 1972 and 1977 was 2.7 (Walkinshaw 1983).

Yet, in spite of the successful control of cowbird parasitism, and a concomitant rise in potential recruitment, total number of singing Kirtland's Warbler males remained relatively constant at a mean of approximately 200 (Mayfield 1983, Walkinshaw 1983, Weinrich 1989) until 1989 when it began to increase, reaching 488 in 1993 (M. E. DeCapita, personal communication). Fluctuations in numbers of singing males appears to be tied directly to abundance of suitable breeding habitat (Mayfield 1983, Weinrich 1989) and may also be due to an inability of juveniles to migrate successfully and particularly to locate the relatively tiny areas of breeding and wintering habitat (Mayfield 1960, 1983). A large proportion of males occupying suboptimal habitat apparently remain unmated (Probst and Hayes 1987);

successful reproduction therefore may be almost entirely limited to high-quality breeding habitat. Further research on relative impacts of habitat availability and quality on breeding and wintering grounds, and on juvenile survival is needed to understand fully the conservation needs of the Kirtland's Warbler.

Least Bell's Vireo

The Least Bell's Vireo was initially common in riparian woodland throughout the Central Valley and coastal parts of southern California. It also occurred at scattered desert locations with riparian habitats and in the western part of northern Baja California. A decline was noticeable starting about 1930 (Grinnell and Miller 1944). By the early 1970s, it was extirpated from the Central Valley (Goldwasser et al. 1980), the region where the largest populations initially occurred. It was designated as an endangered species by the US Fish and Wildlife Service in 1982 (Franzreb 1987). Today, the remaining populations breed in riparian habitats mostly in coastal southern California counties from Santa Barbara to San Diego, and in the inland county of Riverside as well as northern Baja California of Mexico. Most (84.8%) of the 440 territorial males estimated to be in the United States in 1987 were in San Diego County (Franzreb 1989). Among these, most were in four river drainages with small numbers of individuals along about 10 other creeks and rivers.

Cowbirds began to invade southern California at about 1900 (Grinnel 1909) and reached the San Francisco area by 1922 (La Jeunesse 1923, Rothstein et al. 1980, Rothstein 1994, Laymon 1987). The first reported cases of parasitism were from the southern end of the Central Valley in 1907 (Linton 1908). Within a decade or two, most observed vireo nests in southern California were parasitized (Franzreb 1989). (Unfortunately, there are no comparable nesting data for the northern part of the vireo's range.) Thus, there is a clear correlation between the decline of vireos and the arrival of cowbirds. When studies of the remnant vireo populations began in the late 1970s, most were found to be experiencing parasitism in about 50% of their nests (Goldwasser et al. 1980, Franzreb 1989). As parasitized Bell's Vireos generally raise only a cowbird (Pitelka and Koestner 1942, Mumford 1952), it is obvious that cowbirds currently have an enormous effect on population growth of this species. It is possible that southern California vireos have persisted because they began to breed earlier than Central Valley populations and therefore partially escaped parasitism of early nesting attempts due to later breeding by cowbirds.

It is clear that cowbird parasitism is not the only factor in the vireo's decline. The Central Valley has lost 95% of its riparian vegetation in this century to agriculture and other anthropogenic factors (Smith 1977) and a drive along southern California's freeways will quickly demonstrate that habitat loss there has also been massive (Collins et al. 1989). Even in places where natural, dense, riparian habitat remains, flood-control programs have sometimes made it unsuitable. The vireos prefer willow thickets 4–7 years old that used to be generated periodically when flooding rivers scoured out vegetation. Vireos originally bred in the Owens Valley, a place where cowbirds were probably found under primeval conditions (Rothstein et al. 1980). Their disappearance there seems directly attributable to the City of Los Angeles, whose aqueducts have drained the Owens River and its tributaries to the point where most of the valley's once lush vegetation has died. A second western subspecies, *Vireo bellii arizonae*, breeds along the lower Colorado River, where it has been virtually extirpated. Even though heavy parasitism was noted along the Colorado River as early as 1900 (Brown 1903), vireos did not begin to decline until the 1950s when much of their riparian habitat was converted to agriculture and flooding was reduced by dam construction (Rosenberg et al. 1991). Thus, there is reason to believe that large vireo populations in extensive suitable habitat can maintain themselves in the face of cowbird parasitism.

Although it is unlikely that vireos would have declined so drastically if cowbirds had been the only new challenge confronting them, it seems certain that parasitism will cause the current remnant populations to go extinct without human intervention (Laymon 1987). Removal of cowbirds (Beezely and Rieger 1987) from all of the major drainages where the vireos breed has greatly increased vireo productivity (Franzreb 1989). Parasitism occurred in 47% of 93 nests along the Santa Margarita River in 1982. A cowbird trapping program began in 1983 and resulted in the removal of 244 cowbirds and an observed 10% parasitism of 86 nests. Productivity went from 2.08 fledglings per pair in 1982 to 2.86 in 1983. Demographic analyses and mortality data indicate that the latter productivity rate, but not the former, should allow the populations to persist and experience moderate growth rates. In another southern California area, the Prado Basin, a cowbird removal program, which began in 1986, seems responsible for a dramatic rise in the number of breeding vireos from 19 in 1986 to 122 in 1993 (L. R. Hays, personal communication).

Even cowbird control, however, may not sustain all extant populations as indicated by the fate of the most northern and remote one, which occurs in Santa Barbara County (Greaves 1987). It was the third largest population in 1986 (Franzreb 1989) but has decreased from 57 territorial males to about 15 at present (J. Greaves personal communication). This is the most sizable vireo population in a remote area away from extensive human development, and cowbird parasitism was never above 15–20%. Greaves (personal communication) has documented parent–offspring matings, so inbreeding depression or a failure of juveniles to settle in this relatively isolated vireo population may be responsible for the decline.

The goal of a long-term management plan is to establish a self-sustaining population of at least 4000 pairs and to reintroduce vireos to the Central Valley (Franzreb 1989). Organizations such as The Nature Conservancy and governmental agencies are re-establishing riparian vegetation in the Central Valley and southern California. Franzreb (1990) assessed various introduction methods to establish new populations, something that has never been done before for an endangered migratory passerine. The favored approach at present is to capture juveniles at the end of summer or first-time breeders at the beginning of the breeding season and to release birds in the Central Valley after a minimal time in captivity. Franzreb (1990) proposed testing these procedures on other riparian passerines in California or on other races of the vireo, although the stage at which a juvenile becomes attached to habitat and geographical area still needs to be established.

Willow Flycatcher

This species has experienced an enormous decline in the southwestern part of its distribution, especially in California and Arizona, where it nests along riparian corridors dominated by willows. Grinnell and Miller (1944) described it as common in lowland parts of California and sporadic in montane localities up to 8000 feet. Today, it is apparently absent from the Central Valley where it was probably widespread early in this century. The entire Californian population was estimated at less than 150 pairs in the mid 1980s (Harris et al. 1987). Unitt (1987) reviewed its current occurrence in the Southwest from southern California to western Texas and found it absent at numerous sites where it had once occurred. Although a 1993 survey found that many historical sites in Arizona had at least one singing male, only 45–55 territories were found in the entire state (S. Sferra, personal communication). As with Bell's Vireo, habitat destruction and cowbird parasitism appear to be major problems. In both of these species, acceptance of a single cowbird egg nearly always results in the loss of the host's entire brood. Most other riparian hosts in California do not suffer complete reproductive loss when parasitized, and have not declined as drastically as the vireo and flycatcher. Despite similarities, there are three critical differences between these two declining species in the West.

First, there was considerable taxonomic confusion regarding the western subspecies of the Willow Flycatcher. Aldrich (1951) proposed a new subspecies, *Empidonax traillii extimus*, and Unitt (1987) confirmed that *extimus* is a valid subspecies on the basis of plumage color and wing formula (relative lengths of various primaries), and that it occurred from the Los Angeles basin and the extreme southern Sierra Nevada eastward to southern Nevada and Utah, Arizona, New Mexico, and southwestern Texas. This is precisely the region in which Willow Flycatchers have declined the most. Had *extimus* been recognized earlier on as a valid taxon, it might have received endangered species status at least several years ago. Unitt (1987) estimates a total *extimus* population of no more than 500 pairs.

In 1990, the Willow Flycatcher was listed as an endangered species by the California Fish and Game Commission. This listing includes *extimus* as well as the two other subspecies that occur in California, *brewsteri* and *adastus*. The latter two subspecies have also been virtually extirpated from California but may be maintaining reasonably healthy populations further north from Oregon and British Columbia east to the Rocky Mountains. In July 1993, a formal proposal was made by the US Fish and Wildlife Service to list the only globally threatened subspecies, *extimus* (the "Southwestern Willow Flycatcher"), as an endangered species. Workers in Oregon to British Columbia west of the Cascades should closely monitor population trends of the Willow Flycatcher. This is a region that lacked cowbirds until the 1940s and 1950s (Rothstein 1994). If Willow Flycatchers decline there, it may be due largely to cowbird parasitism, given the widespread nature of suitable habitat in this mesic region.

A second difference between the Least Bell's Vireo and Willow Flycatcher is that only the latter occurs at high and low elevations. In California, the causative agents for the flycatcher's decline appear to be different at low and high elevations. At elevations below about 1000 m, destruction of riparian habitat and cowbird parasitism seem to be responsible for the decline of *extimus* and of *brewsteri*. As with the case of the Bell's Vireo, most nests of the flycatcher seemed to be parasitized within 10–20 years of the cowbird's arrival in southern California. However, a less clear picture emerges further east, where the cowbird has been present throughout recorded history and some samples show little parasitism (data in Unitt 1987), while one showed a parasitism level of 50% (Brown 1988). Because Willow Flycatchers have declined in all of these regions, it is likely that both habitat destruction and cowbird parasitism are factors. Nevertheless, it is probable that cowbird parasitism will cause the complete extirpation of many remnant flycatcher populations if it is left unchecked.

The first cowbird control program under the auspices of recovery efforts for the Willow Flycatcher began in 1993 in a Nature Conservancy preserve along the South Fork of the Kern River where the largest California population of *extimus* exists. This population experiences about a 50–55% rate of parasitism (Whitfield 1990). A cowbird control program undertaken to boost Least Bell's Vireo productivity along the Santa Margarita River in San Diego County, however, may be responsible for a steady rise in flycatcher numbers there from five territorial birds in 1981 to 17 in 1986 (Unitt 1987).

Cowbird parasitism at elevations above 1000–1500 m is absent to slight, even in places where cowbirds occur in California (Stafford and Valentine 1985, Flett and Sanders 1987). In the mountains of California, suitable habitat was probably always limited in the mountains of California, occurring in patchily distributed moist meadows and streams with stands of willows surrounded by forest and sagebrush. Because of their lush grass and water, these meadows are heavily used by range cattle which knock over nests and degrade the habitat by consuming the lower foliage of willows. This problem is exacerbated by the fact that many meadows within national forests are privately owned because they were favored by early homesteaders. Thus, federal agencies do not have a free hand in limiting cattle grazing, which pilot efforts have shown to be effective in boosting flycatcher productivity (Valentine et al. 1988). Although cowbird parasitism may not now be a major factor in high elevation parts of California, the situation is probably different in the Rocky Mountains (Sedgewick and Knopf 1988). Furthermore, cowbird parasitism may be the chief cause of the flycatcher's complete extirpation from Yosemite Valley (Gaines 1988), which is at an intermediate elevation of 1200 m.

The last difference between the vireo and flycatcher is the late breeding of the latter, which usually begins about 1 June and reaches its peak in mid-June even in the warm climate of lowland Southern California; extreme dates for 187 clutches of *extimus* in museums range from 24 May to 30 July (Unitt 1987). Because of this late breeding, the entire laying season of the flycatcher overlaps that of the cowbird. Although cowbird activity in California declines noticeably by late June (Payne 1973a,b), cowbirds in Santa Barbara County typically show signs of breeding such as courtship and male–male aggression until the first few days in August (S. I. Rothstein, personal observation). This complete overlap in the breeding seasons of the cowbird and flycatcher may explain why the latter has declined even more drastically than the Least Bell's Vireo, some of whose early nests escape parasitism.

Black-capped Vireo

The Black-capped Vireo formerly nested in scrub habitats from south-central Kansas, through central Oklahoma and central Texas, and central Coahuila, and possibly Nuevo Leon and Tamaulipas in Mexico (Graber 1961, AOU 1983). However, it has not been reported in Kansas since 1953 (Tordoff 1956, Graber 1961) and occurs at only two isolated localities in Oklahoma (Grzybowski et al. 1986, Grzybowski 1991b, US Fish and Wildlife Service 1991). Furthermore, the species appears to be disappearing in a random pattern through much of its range in the northern and central portions of Texas (Marshall et al. 1985).

A number of factors may be contributing to the decline of the vireo, including fire suppression, but one that is undoubtedly making a significant impact is nest parasitism by Brown-headed Cowbirds. Nest parasitism of vireo nests by cowbirds has been documented at greater than 70% for most years in many vireo-breeding areas in Oklahoma and Texas unprotected by cowbird removal (Grzybowski unpublished data, Tazik and Cornelius unpublished data). In one area in Oklahoma, where parasitism was documented at 100%, the population of vireos has been extirpated (Grzybowski 1991 b). Production of unprotected vireos in Oklahoma from 1983–1987 was 0.4 young/ pair/year, 0.7 for a series of study sites in central Texas, far below that expected to maintain their populations without immigration. With cowbird removal, this production has been increased to generally greater than 2.0 young/female/year, and as high as 3.78 (Grzybowski 1990).

The Black-capped Vireo situation differs from the other case histories described here because the vireo's breeding distribution is within the historical range of the Brown-headed Cowbird. However, the reduction in buffalo and introduction of cattle in the southern Great Plains has likely made the distribution of cowbirds in this region more homogeneous. Rather than travel with large buffalo herds that were highly clumped and mobile, cowbirds now use livestock and human-associated habitats that are more widespread and thus continuously provide access to cowbirds.

As with the Least Bell's Vireo and Willow Flycatcher, the Black-capped Vireo has suffered from a reduction of suitable habitat to smaller and more widely distributed patches. Regional fire suppression has allowed the vireo's preferred early successional habitat to mature or to occur in small patches (Grzybowski et al. 1994). These smaller groupings of vireos may be more vulnerable to cowbird parasitism, perhaps because the effective size of vireo populations is small relative to regional cowbird populations.

COWBIRD MANAGEMENT

The cowbird's abundance and parasitic life history makes it one of the most severe potential threats to Neotropical migrants and other host species. Because cowbirds have been shown to commute up to 7 km between breeding and feeding areas, attempts to manage cowbird populations must consider landscape-level mosaics of habitats when deciding on management practices (Robinson et al. 1993). At the regional scale, cowbirds are most likely to be a problem in the Great Plains, the Midwest and in the far southwest, including California (Fig. 15-1). Within regions, the landscape composition also needs to be considered. In extensively forested areas, cowbirds may be limited by foraging opportunities. In these landscapes, cowbird management may be unnecessary or focused on just a few cowbird foraging sites. In areas with fragmented habitats where cowbird feeding sites are abundant, management of cowbirds may depend upon a potential combination of local and broad-scale options. We still need a better understanding of dynamic interactions between host and cowbird populations, but when cowbird populations become limited by hosts, a serious problem may occur. Management options may include elimination of cowbird feeding sites within forest fragments and the reduction of feeding sites by the consolidation of existing forest patches. In areas with locally endangered hosts, intensive trapping and removal of cowbirds may be the best immediate strategy. Although controversial, a case could also be made for eradication of large numbers of cowbirds in regions where cowbird parasitism is a severe problem in all remaining habitats. These management recommendations are described in greater detail in Robinson et al. (1993).

The recent range expansion of the Shiny, Bronzed, and Brown-headed Cowbirds into the Southeast is also of particular concern. The Shiny Cowbird has already had a detrimental impact on hosts in recently colonized islands of the Caribbean (Post and Wiley 1976). As cowbirds continue to increase in numbers and expand their range, detrimental effects on additional hosts are likely to be reported. Anecdotal evidence suggests, for example, that the recent declines in highland species in the Monteverde Cloud Forest Preserve of Costa Rica might be a result of parasitism by the newly colonized Bronzed Cowbird (G. Powell, personal communication). Baseline data on host-nesting success are needed to evaluate the impact of cowbird range expansion and population increases.

Nevertheless, because cowbirds benefit from intensive human activity, the ubiquity of the cowbird problem is likely to persist. A network of large (> 20,000 ha), unfragmented habitat patches in each major geographical region might meet the requirements of Neotropical migrants as well as those of other animals that require large habitat blocks. Identifying potential refuges remains a high research priority as does the development of predictive models of the dynamics of host and cowbird populations.

RESEARCH NEEDS

Virtually all aspects of the cowbird–host interactions described in this paper require further research. Here we summarize some of the more pressing research needs that we have identified. Our focus is on conservation-related problems.

1. Host population dynamics and predictive models. Our understanding of the demographic parameters used in modeling host demographics (e.g., May and Robinson 1985, Temple and Cary 1988, Pease and Grzybowski 1995; Table 15-1) is primitive at best. Estimating the variables in Table 15-1 or directly measuring productivity of color-marked birds (e.g., Nolan 1978, Smith 1981, Holmes et al. 1992, Sherry and Holmes 1992), is extremely labor intensive and has only been done for a few species. Our estimates of adult and juvenile survivorship are also crude in part because juveniles rarely return to their natal areas to breed (Whitcomb et al. 1981), but also because the between-year return rates of adults to the same plots is a minimal estimate of adult survival rates (Greenberg 1980). If adults disperse following losses of nests to predators (e.g., Robinson 1985, Jackson et al. 1989), the low return rates reported for Neotropical migrants (Greenberg 1980) may seriously underestimate adult survival. The high return rates reported by Roberts (1971) and Roth and Johnson (1993) may therefore not represent local artifacts as hypothesized by Greenberg (1980). Much uncertainty still remains in these estimates, but these parameters form the basis for assessing dynamics of parasitized host populations of individual species and the level of parasitism these hosts can tolerate. Once a more solid empirical framework is established for demographics of migrants,

models can become a powerful tool for managers to predict the impacts of their actions on host populations.

2. Habitat fragmentation and cowbird parasitism. As more community-wide studies of brood parasitism are initiated, we should be able to predict more accurately the effects of landscape-level features on cowbird populations and parasitism. Cowbird parasitism appears not to be a significant problem in mostly forested landscapes (e.g., Missouri Ozarks: Robinson et al. 1995; or in the White Mountains of New Hampshire: Holmes et al. 1992, Sherry and Holmes 1992). In contrast, cowbirds appear to have reached saturation levels in the forests of Illinois, which are all situated in a mostly agricultural matrix (Robinson 1992, Robinson and Wilcove 1994, Thompson et al., in press). In both of these landscapes, features such as edges around clearcuts and wildlife openings may have little impact on cowbird distribution and parasitism. The edge effects observed by Brittingham and Temple (1983) may therefore only be pronounced in moderately fragmented landscapes where cowbird populations are not limited either by a lack of foraging areas (as in mostly forested areas) or the scarcity of hosts (as in small forests in mostly agricultural landscapes). It is in these moderately fragmented landscapes that management to minimize edge and reduce cowbird feeding habitats has the greatest chance to succeed.

3. Cowbird parasitism and community composition. Cowbird parasitism may strongly affect local avian community organization. In Illinois, most of the species that suffer high levels of parasitism are either rare (e.g., Hooded Warbler) or locally distributed and declining (e.g., Wood Thrush) (Robinson 1992, unpublished data). Although cause and effect are difficult to disentangle, it is possible that parasitism is selectively reducing the population densities of some but not all species. Community-level data on parasitism rates and long-term censuses might detect population trends that are related to parasitism levels. Predictive models of these interactions, however, could prove more useful than correlational studies, especially in areas where cowbird populations are increasing or have only recently invaded.

4. Cowbird removals. An alternative way to measure the effect of cowbirds on host populations is to manipulate parasitism rates by trapping cowbirds. To date, cowbird removal has only been used to protect specific populations of endangered species. Therefore, we have no data on the effectiveness of cowbird removals on the community level or the potential effectiveness of cowbird trapping in landscapes that vary in the extent of fragmentation. Furthermore, none of the removal programs have involved experimental controls, i.e., matched host populations from which local cowbirds are not removed. Thus, the effects of cowbird parasitism cannot be proven with the same degree of rigor that ecologists usually seek. Whereas it may not be practical to set aside as controls some populations of endangered species, controls should be incorporated into removal programs involving more widespread hosts.

5. Energetic costs to hosts of raising cowbirds. If hosts expend more energy raising cowbirds than they would if they were only raising their own young, the cost of parasitism may be even greater than we currently estimate. Studies of host energetics using doubly labeled water for adults with parasitized and unparasitized nests could determine if cowbirds are able to "trick" the hosts into working harder.

6. Calcium limitation in cowbirds. Because cowbirds lay such large clutches, it is possible that their populations in some areas are limited by the supply of dietary calcium needed for egg shells. Examination of stomach contents of cowbird corpses retrieved from trapping programs might provide useful indications of how cowbirds find calcium.

7. Habitat restoration and cowbird abundance. As management programs are implemented to restore contiguous forest in the East and Midwest, and to expand the width of riparian forest corridors in the Southwest, research involving long-term monitoring will determine if cowbird populations decline and parasitism levels decrease in the interior of the tracts.

Experimental elimination of local feeding areas should be tried to determine its effect on cowbird parasitism levels.

8. Timber management and cowbird abundance. One of the most crucial questions to be addressed is the effects of various timber management practices (clearcuts, shelterwood cuts, single-tree selection, group selection) on cowbird abundance and levels of parasitism both within and among regions. It is difficult to develop guidelines for forestry practices because we do not know how cowbirds and host communities are influenced by different management practices (Thompson et al. 1993). Once again, predictive models of how cowbird and host populations, and parasitism levels respond to different logging practices would be particularly beneficial to managers.

9. Host responses to cowbird eggs. Although we know a great deal about parasitism levels in most species, we still have a lot to learn about host defenses, especially in species that may abandon parasitized nests. Geographically varying rates of parasitism may reflect regional differences in egg rejection or other behavioral tactics. S. Sealy (personal communication), for example, found that Warbling Vireos are puncture ejecters in some parts of their range.

10. Host population densities and cowbird abundance. Because host populations may be more abundant along some edges, we still do not know if the edge effect reported by Brittingham and Temple (1983) results from a preference by cowbirds for edges, for areas of high host density, or for areas with particular host species. Robinson (unpublished data) found a strong correlation between host and cowbird abundance in a forest where cowbirds apparently saturate the available landscape and there is no detectable edge effect. Similar studies in landscapes with varying cowbird abundances will shed light on the cues used by cowbirds to locate breeding areas as will playback experiments that test cowbird responses to augmented apparent host population densities.

11. Other cowbirds. As the Shiny Cowbird expands its range into North America, its host selection should be monitored. Shiny Cowbirds may select different hosts from the Brown-headed Cowbird, or they may compete directly for the same hosts, as they do with the Bronzed Cowbird (Carter 1986). A. Cruz's (personal communication) studies of pre-invasion host population dynamics of such species as Prairie Warbler, Black-whiskered Vireo, and Yellow Warbler in peninsular Florida should be particularly useful.

12. Cowbird parasitism and microhabitat features. Few studies have addressed the effects of microhabitat features such as shrub density, ground cover, and vegetation density around the nest on cowbird parasitism rates. Correlative and manipulative studies might provide managers with specific vegetation management guidelines for some species.

ACKNOWLEDGMENTS

We would especially like to thank Jamie Smith, Spencer Sealy, Jeff Brawn, and Dan Petit for carefully reading the manuscript and providing many useful comments, and bringing additional references and unpublished data to our attention. S. I. Rothstein was supported by a Faculty Research Grant from the University of California and by a grant from the US Forest Service. Mark Holmgren and James Greaves provided useful comments on Bell's Vireos and Willow Flycatchers, and Loren Hays, Susan Sferra, and Douglas Bell allowed us access to unpublished data, for which we are grateful.

LITERATURE CITED

Airola, D. A. 1986. Brown-headed Cowbird parasitism and habitat disturbance in the Sierra Nevada. J. Wildl. Manag. 50:571–575.

Aldrich, J. W. 1951. A review of the races of the Traill's Flycatcher. Wilson Bull. 63:192–197.

American Ornithologists' Union. 1983. Check-list of North American birds, 6th ed. Allen Press, Lawrence, Kansas. 877 pp.

Anderson, W. L., and R. W. Storer. 1976. Factors influencing Kirtland's Warbler nesting success. Jack-Pine Warbler 54:105–115.

Ankney, C. D., and D. M. Scott. 1980. Changes in

nutrient reserves and diet of breeding Brown-headed Cowbirds. Auk 97:684–696.

Beezley, J. A., and J. P. Rieger. 1987. Least Bell's Vireo management by cowbird trapping. West. Birds 18:55–62.

Bendire, C. E. 1895. Life histories of North American birds, Vol. 2. US Natl Mus. Spec. Bull. 3.

Bent, A. C. 1958. Life histories of North American blackbirds, orioles, tanagers, and allies. US Natl Mus. Bull. 211.

Best, L. B. 1978. Field Sparrow reproductive success and nesting ecology. Auk 95:9–22.

Best, L. B., and D. F. Stauffer. 1980. Factors affecting nesting success in riparian bird communities. Condor 82:149–158.

Blankespoor, G. W., J. Oolman, and C. Uthe. 1982. Eggshell strength and cowbird parasitism of Red-winged Blackbirds. Auk 99:363–365.

Briskie, J. V., and S. G. Sealy. 1987. Responses of Least Flycatchers to experimental inter- and intraspecific brood parasitism. Condor 89: 899–901.

Briskie, J. V., and S. G. Sealy. 1990. Evolution of short incubation periods in the parasitic cowbirds, *Molothrus* spp. Auk 107:789–794.

Briskie, J. V., S. G. Sealy, and K. A. Hobson. 1990. Differential parasitism of Least Flycatchers and Yellow Warblers by the Brown-headed Cowbird. Behav. Ecol. Sociobiol. 27:403–410.

Briskie, J. V., S. G. Sealy, and K. A. Hobson. 1992. Behavioral defenses against avian brood parasitism in sympatric and allopatric host populations. Evolution 46:334–340.

Brittingham, M. C., and S. A. Temple. 1983. Have cowbirds caused forest songbirds to decline? BioScience 33:31–35.

Brown, B. T. 1988. Breeding ecology of a Willow Flycatcher population in Grand Canyon, Arizona. West. Birds 19:25–33.

Brown, H. 1903. Arizona bird notes. Auk 20:43–50.

Buech, R. R. 1982. Nesting ecology and cowbird parasitism of Clay-colored, Chipping and Field Sparrows in a Christman tree plantation. J. Field Ornithol. 53:363–369.

Burgham, M. C. J., and J. Picman. 1989. Effect of Brown-headed Cowbirds on the evolution of Yellow Warbler anti-parasite strategies. Anim. Behav. 38:298–308.

Carey, M. 1982. An analysis of factors governing pair-bonding period and the onset of laying in Indigo Buntings. J. Field Ornithol. 53:240–248.

Carter, M. D. 1986. The parasitic behavior of the Bronzed Cowbird in south Texas. Condor 88: 11–25.

Chasko, G. G., and J. E. Gates. 1982. Avian habitat suitability along a transmission-line corridor in an oak–hickory forest region. Wildl. Monogr. 82: 1–41.

Clark, K. L., and R. J. Robertson. 1981. Cowbird parasitism and evolution of antiparasite strategies in the Yellow Warbler. Wilson Bull. 93:249–358.

Collins, C. T., L. R. Hays, M. Wheeler, and D. Willick. 1989. The status and management of the Least Bell's Vireo within the Prado Basin, California, during 1989. Orange County Water District, Fountain Valley, CA., 53 pp.

Coon, D. W., and K. A. Arnold. 1977. Origins of Brown-headed Cowbird populations wintering in central Texas. North Amer. Bird Bander 2:7–11.

Crase, F. T., R. W. Dehaven, and P. P. Woronecki. 1972. Movements of Brown-headed Cowbirds banded in the Sacramento Valley, California. Bird Banding 43:197–204.

Cruz, A. C., J. W. Wiley, T. K. Nakamura, and W. Post. 1989. The Shiny Cowbird *Molothrus bonariensis* in the West Indian Region—biographical and ecological implications. Pp. 519–540 *in* Biogeography of the West Indies—past, present, and future (C. A. Woods, ed.). Sandhill Crane Press, Gainesville, FL.

Darley, J. A. 1971. Sex ratio and mortality in the Brown-headed Cowbird. Auk 88: 560–566.

Darley, J. A. 1978. Pairing in captive Brown-headed Cowbirds (*Molothrus ater*). Can. J. Zool. 56:249–252.

Darley, J. A. 1982. Territoriality and mating behavior of the male Brown-headed Cowbird. Condor 84:15–21.

Davies, N. B., and N. De L. Brooke. 1988. Cuckoos versus Reed Warblers: adaptations and counteradaptations. Anim. Behav. 36:262–284.

Davies, N. B., A. F. G. Bourke, M., and N. De L. Brooke. 1989. Cuckoos and parasitic ants: interspecific brood parasitism as an evolutionary arms race. Trends Ecol. Evol. 4:274–278.

DellaSalla, D. A. 1985. The Yellow Warbler in southeastern Michigan: factors affecting its productivity. Jack-Pine Warbler 63:52–60.

Dolbeer, R. A. 1982. Migration patterns for age and sex classes of blackbirds and starlings. J. Field Ornithol. 53:28–46.

Dufty, A. M., Jr. 1982. Movements and activities of radio-tracked Brown-headed Cowbirds. Auk 99:316–327.

Dufty, A. M., Jr. 1983. Variation in the egg markings of the Brown-headed Cowbird. Condor 85:109–111.

Dufty, A. M., Jr. 1985. Song sharing in the

Brown-headed Cowbird (*Molothrus ater*). Z. Tierpsychol. 69:177–190.

Ehrlich, P. R., D. D. Dobkin, and D. Wheye. 1988. The Birder's Handbook; A field guide to the natural history of North American birds. Simon and Schuster, New York.

Elliott P. F. 1978. Cowbird parasitism in the Kansas tallgrass prairie. Auk 95:161–167.

Elliott, P. F. 1980. Evolution of promiscuity in the Brown-headed Cowbird. Condor 82:138–141.

Finch, D. M. 1982. Rejection of cowbird eggs by Crissal Thrashers. Auk 99:719–724.

Finch, D. M. 1983. Brood parasitism of the Abert's Towhee: timing, frequency and effects. Condor 85:355–359.

Fleischer, R. C. 1985. A new technique to identify and assess the dispersion of eggs of individual brood parasites. Behav. Ecol. Sociobiol. 17: 91–99.

Fleischer, R. C. 1986. Brood parasitism by Brown-headed Cowbirds in a simple host community in eastern Kansas. Bull. Kansas Ornithol. Soc. 37:21–29.

Fleischer, R. C., A. P. Smyth, and S. I. Rothstein. 1987. Temporal and age-related variation in the laying rate of the Brown-headed Cowbird in the eastern Sierra Nevada, CA. Can. J. Zool. 65:2724–2730.

Flett, M. A., and S. D. Sanders. 1987. Ecology of a Sierra Nevada population of Willow Flycatchers. West. Birds 18:37–42.

Franzreb, K. E. 1987. Endangered status and strategies for conservation of the Least Bell's Vireo (*Vireo bellii pusillus*) in California. West. Birds 18:43–49.

Franzreb, K. E. 1989. Ecology and conservation of the endangered Least Bell's Vireo. US Fish Wildl. Serv. Biol. Rep. 89(1), 17 pp.

Franzreb, K. E. 1990. An analysis of options for reintroducing a migratory, native passerine, the endangered Least Bell's Vireo, *Vireo bellii pusillus*, in the Central Valley, California. Biol. Conserv. 53:105–123.

Freeman, S., D. F. Gori, and S. Rohwer. 1990. Red-winged Blackbirds and Brown-headed Cowbirds: some aspects of a host–parasite relationship. Condor 92:336–340.

Friedmann, H. 1929. The cowbirds: a study in the biology of social parasitism. C. Thomas, Springfield, IL.

Friedmann, H. 1963. Host relations of the parasitic cowbirds. US Natl Mus. Bull. 233.

Friedmann, H. 1966. Additional data on the host relations of the parasitic cowbirds. Smithson. Misc. Coll. 149:1–12.

Friedmann, H., and L. F. Kiff. 1985. The parasitic cowbirds and their hosts. Proc. West. Found. Zool. 2:226–304.

Friedmann, H., L. F. Kiff, and S. I. Rothstein. 1977. A further contribution to knowledge of the host relations of the parasitic cowbirds. Smithsonian Contrib. Zool. 235.

Gaines, D. 1988. Birds of Yosemite and the east slope. Artemisia Press, Lee Vining, CA.

Gates, J. E., and N. R. Giffen. 1991. Neotropical migrant birds and edge effects at a forest stream ecotone. Wilson Bull. 103:204–217.

Gates, J. E., and L. W. Gysel. 1978. Avian nest dispersion and fledging success in field–forest ecotones. Ecology 59:871–883.

Goldwasser, S., D. Gaines, and R. S. Wilbur. 1980. The Least Bell's Vireo in California: a de facto endangered race. Amer. Birds 34:742–745.

Graber, J. W. 1961. Distribution, habitat requirements, and life history of the Black-capped Vireo (*Vireo atricapilla*). Ecol. Monogr. 31: 313–336.

Graham, D. S. 1988. Responses of five host species to cowbird parasitism. Condor 90:586–591.

Greaves, J. M. 1987. Nest-site tenacity of Least Bell's Vireos. Western Birds 18:50–54.

Greenberg, R. 1980. Demographic aspects of long-distance migration. Pp. 493–504 *in* Migrant birds in the Neotropics: ecology, behavior, distribution, and conservation (A. Keast and E. S. Morton, eds). Smithsonian Institution Press, Washington, DC.

Grinnell, J. 1909. A new cowbird of the genus *Molothrus*, with a note on the probable genetic relationship of the North American forms. Univ. Calif. Publ. Zool. 5:275–281.

Grinnell, J. 1914. An account of the mammals and birds of the lower Colorado valley. Univ. Calif. Publ. Zool. 12:51–194.

Grinnell, J., and A. H. Miller. 1944. The distribution of the birds of California. Pacific Coast Avif. 27:1–608.

Grzybowski, J. A. 1990. Population and nesting ecology of the Black-capped Vireo—1990. Resource Protection Division, Texas Parks and Wildlife Department, Austin, TX, 39 pp.

Grzybowski, J. A. 1991a. Survivorship, dispersal, and population structure of the Black capped Vireo at the Kerr Wildlife Management Area, Texas. Resource Protection Division, Texas Parks and Wildlife Department, Austin, TX, 42 p.

Grzybowski, J. A. 1991b. Ecology and management of isolated populations of the Black-capped Vireo (*Vireo atricapillus*) in Oklahoma. Oklahoma Dept. Conserv. Perform. Rep. E-1-6. 27 p.

Grzybowski, J. A., and V. W. Fazio. 1991. Shiny cowbird reaches Oklahoma. Amer. Birds 45:50–52.

Grzybowski, J. A., R. B. Clapp, and J. T. Marshall, Jr. 1986. History and population status of the

Black-capped Vireo in Oklahoma. Amer. Birds 40:1151–1161.

Grzybowski, J. A., D. J. Tazik, and G. D. Schnell. 1994. Regional analysis of Black-capped vireo breeding habitats. Condor 96:512–544.

Hanka, L. R. 1985. Recent altitudinal range expansion of the Brown-headed Cowbird in Colorado. Western Birds 16:183–184.

Hann, H. W. 1937. Life history of the Oven-bird in southern Michigan. Wilson Bull. 49:145–237.

Hann, H. W. 1941. The cowbird at the nest. Wilson Bull. 53:211–221.

Harris, J. H., S. D. Sanders, and M. A. Flett. 1987. Willow Flycatcher surveys in the Sierra Nevada. West. Birds 18:27–36.

Hatch, S. A. 1983. Nestling growth relationships of Brown-headed Cowbirds and Dickcissels. Wilson Bull. 95:669–671.

Hobson, K. A., and S. G. Sealy. 1989. Responses of yellow warblers to the threat of cowbird parasitism. Anim. Behav. 38:510–519.

Holford, K. C., and D. D. Roby. 1993. Factors limiting fecundity of captive Brown-headed Cowbirds. Condor 95:536–545.

Holmes, R. T., T. W. Sherry, and F. W. Sturges. 1986. Bird community dynamics in a temperate deciduous forest: Long-term trends at Hubbard Brook, Ecol. Monogr. 50:201–220.

Holmes, R. T., T. W. Sherry, P. P. Marra, and K. E. Petit. 1992. Multiple brooding and productivity of a Neotropical migrant, the Black-throated Blue Warbler (*Dendroica caerulescens*) in an unfragmented temperate forest. Auk 109:321–333.

Hoover, J. P. 1992. Factors influencing Wood Thrush (*Hylocichla mustelina*) nesting success in a fragmented forest. MS thesis, The Pennsylvania State University, University Park, PA.

Hoover, J. P., and M. C. Brittingham. 1993. Regional variation in cowbird parasitism of Wood Thrushes. Wilson Bull. 105:228–238.

Jackson, W. M., S. Rohwer, and V. Nolan, Jr. 1989. Within-season breeding dispersal in Prairie Warblers and other passerines. Condor 91: 233–241.

Johnson, D. M., G. L. Stewart, M. Corley, R. Ghrist, J. Hagner, A. Ketterer, B. McDonnell, W. Newsom, E. Owen, and P. Samuels. 1980. Brown-headed Cowbird (*Molothrus ater*) mortality in an urban winter roost. Auk 97:299–320.

Johnson, R. G., and S. A. Temple. 1990. Nest predation and brood parasitism of tallgrass prairie birds. J. Wildl. Manag. 54:106–111.

Kattan, G. H. 1993. Reproductive strategy of a generalist brood parasite, The Shiny Cowbird, in the Cauca Valley, Columbia. PhD dissertation, University of Florida, Gainesville, FL.

Kelly, S. T., and M. E. Decapita. 1982. Cowbird control and its effect on Kirtland's Warbler reproductive success. Wilson Bull. 94:363–365.

Keys, G. C., R. C. Fleischer, and S. I. Rothstein. 1986. Relationships between elevation, reproduction and the hematocrit level of Brown-headed Cowbirds. Comp. Biochem. Physiol. 83A:765–769.

Klaas, E. E. 1975. Cowbird parasitism and nesting success in the Eastern Phoebe. Occ. Pap. Mus. Natural Hist. Univ. Kansas 41:1–18.

Lack, D. 1968. Ecological Adaptations for Breeding in Birds. Methuen, London.

La Jeunesse, H. V. 1923. Dwarf Cowbird nesting in Alameda County, California. Condor 25:31–32.

Lanyon, S. M. 1992. Interspecific brood parasitism in blackbirds (*Icterinae*): a phylogenetic perspective. Science 255:77–79.

Laymon, S. A. 1987. Brown-headed Cowbirds in California: historical perspectives and management opportunities in riparian habitats. West. Birds 19:63–70.

Linton, C. B. 1908. Notes from Buena Vista Lake, May 20 to June 16, 1907. Condor 10:196–198.

Lowther, P. E. 1993. Brown-headed Cowbird. Pp. 1–28 *in* The Birds of North America, no. 47 (A. Poole and F. Gill, eds). The Academy of Natural Science, Philadelphia, PA.

Marshall, J. T., Jr, R. B. Clapp, and J. A. Grzybowski. 1985. Status report: *Vireo atricapillus* Woodhouse, Black-capped Vireo. Rep. Offi. Endangered Species US Fish Wildl. Serv. Albuquerque, NM, 55 pp.

Martin, T. E. 1992. Breeding productivity considerations: what are the appropriate habitat features for management? Pp. 455–473 *in* Ecology and conservation of Neotropical migrant landbirds (J. M. Hagan and D. W. Johnston, eds). Smithsonian Institution Press, Washington, DC.

Martin, T. E. 1993. Nest predation among vegetation layers and habitat types: revising the dogmas. Amer. Natur. 141:897–913.

Marvil, R. E., and A. Cruz. 1989. Impact of Brown-headed Cowbird parasitism on the reproductive success of the Solitary Vireo. Auk 106:476–480.

Mason, P. 1986. Brood parasitism in a host generalist, the Shiny Cowbird: I. The quality of different species of hosts. Auk 103:52–60.

May, R. M., and S. K. Robinson. 1985. Population dynamics of avian brood parasitism. Amer. Natur. 126:475–494.

Mayfield, H. F. 1960. The Kirtland's Warbler. Cranbrook Institute, Bloomfield Hills, MI.

Mayfield, H. F. 1961. Vestiges of a proprietary interest in nests by the Brown-headed Cowbird parasitizing the Kirtland's Warbler. Auk 78:162–167.

Mayfield, H. F. 1965. The Brown-Headed Cowbird with Old and New Hosts. Living Bird 4:13–28.

Mayfield, H. F. 1977a. Brown-headed cowbird: agent of extermination? Amer. Birds 31:107–113.

Mayfield, H. F. 1977b. Brood parasitism: reducing interactions between Kirtland's Warblers and Brown-headed Cowbirds. Pp. 85–91 *in* Endangered birds: management techniques for preserving threatened species. (S. A. Temple, ed.). University of Wisconsin Press, Madison, WI.

Mayfield, H. F. 1983. Kirtland's Warbler, victim of its own rarity? Auk 100:974–976.

Meanley, B. 1975. The blackbird–starling roost problem. Atlantic Natur. 30:107–109.

Moksnes, A., and E. Roskaft. 1989. Adaptations of Meadow Pipits to parasitism by the Common Cuckoo. Behav. Ecol. Sociobiol. 24:25–30.

Mumford, R. 1952. Bell's Vireo in Indiana. Wilson Bull. 64:224–233.

Musselman, T. E. 1946. Some interesting nest habits of the Eastern Bluebird (*Sialia sialis sialis*). Bird Banding 17:60–63.

Newman, G. A. 1970. Cowbird parasitism and nesting success of Lark Sparrows in southern Oklahoma. Wilson Bull. 82:304–309.

Nice, M. M. 1937. Studies in the life history of the Song Sparrow, Part 1. Trans. Linn. Soc. N. York 4:1–247.

Nice, M. M. 1953. The question of ten-day incubation periods. Wilson Bull. 65:81–93.

Nolan, V., Jr. 1978. The ecology and behavior of the Prairie Warbler *Dendroica discolor*. Ornithological monographs no. 26. American Ornithologists' Union, Washington, DC.

Norman, R. F., and R. J. Robertson. 1975. Nest-searching behavior in the Brown-headed Cowbird. Auk 92:610–611.

Norris, R. T. 1947. The cowbirds of Preston Frith. Wilson Bull. 59:83–103.

O'Loghlen, A. L., and S. I. Rothstein. 1993. An extreme example of delayed vocal development: song learning in a population of wild Brown-headed Cowbirds. Anim. Behav. 46:293–304.

Ortega, C. P., and A. Cruz. 1991. A comparative study of cowbird parasitism in Yellow-headed Blackbirds and Red-winged Blackbirds. Auk 108:16–24.

Payne, R. B. 1973a. The breeding season of a parasitic bird, the Brown-headed Cowbird, in central California. Condor 75:80–99.

Payne, R. B. 1973b. Individual laying histories and the clutch size and number of eggs of parasitic cuckoos. Condor 75:414–438.

Payne, R. B. 1977. The ecology of brood parasitism in birds. Ann. Rev. Ecol. Syst. 8:1–28.

Payne, R. B. 1989. Indigo bunting. Pp. 153–172 *in* Lifetime reproduction in birds (I. Newton, Ed.). Academic Press, London.

Pease, C. M., and J. A. Grzybowski. 1995. A model for assessing the consequences of brood parasitism on seasonal fecundity in passerine birds. Auk (in press).

Peck, G. K., and R. D. James. 1987. Breeding birds of Ontario. Nidology and distribution, Vol. 2: passerines. Royal Ontario Museum, Toronto, Canada.

Petit, L. J. 1989. Breeding biology of Prothonotary Warblers nesting in riverine habitat in Tennessee. Wilson Bull. 101:51–61.

Petit, L. J. 1991. Adaptive tolerance of cowbird parasitism by prothonotary warblers: a consequence of nest-site limitation? Anim. Behav. 41:425–432.

Pinkowski, B. C. 1974. Cowbird parasitism of a bluebird nest. Jack-Pine Warbler 52:45.

Pitelka, F. A., and E. J. Koestner. 1942. Breeding behavior of Bell's Vireo in Illinois. Wilson Bull. 54:97–106.

Post, W., and J. W. Wiley. 1976. The yellow-shouldered blackbird—present and future. Amer. Birds 30:13–20.

Post, W., and J. W. Wiley. 1977. The Shiny Cowbird in the West Indies. Condor 79:119–121.

Probst, J. R., and J. P. Hayes. 1987. Pairing success of Kirtland's Warblers in marginal vs. suitable habitat. Auk 104:234–241.

Ricklefs, R. E. 1973. Fecundity, mortality and avian demography. *In* Breeding biology of birds (D. S. Farner ed.). National Academy of Science, Washington, DC.

Robbins, C. S. 1979. Effects of forest fragmentation on bird populations. Pp. 198–213 *in* Management of Northcentral and Northeastern Forests for nongame birds (R. M. DeGraaf and K.E. Evans, eds). US Forest Serv. Gen. Tech. Rep. NC-51.

Robbins, C. S., D. Bystrak, and P. H. Geissler. 1986. The breeding bird survey: Its first fifteen years, 1965–1979. USFWS Res. Publ. 157, Washington, DC.

Roberts, J. O. L. 1971. Survival among some North American wood warblers. Bird banding 42:165–184.

Robinson, S. K. 1985. Coloniality as a defense against nest predators of the Yellow-rumped Cacique. Auk 102:509–519.

Robinson, S. K. 1992. Population dynamics of breeding Neotropical migrants in a fragmented Illinois landscape. Pp. 408–418 *in* Ecology and conservation of Neotropical migrant landbirds (J. M. Hagan, III and D. W. Johnston, eds). Smithsonian Institution Press, Washington, DC.

Robinson, S. K., and D. S. Wilcove. 1994. Forest fragmentation in the temperate zone and its effects on migratory songbirds. Bird Conserv. Int. 4:233–249.

Robinson, S. K., J. Grzybowski, S. I. Rothstein, L. J. Petit, M. C. Brittingham, and F. R. Thompson. 1993. Management implications of cowbird parasitism on neotropical migrant songbirds. Pp. 93–102 *in* Status and management of Neotropical migratory birds (D. M. Finch and P. W. Stangel, eds). USDA Gen. Tech. Rep. RM-229.

Robinson, S. K., F. R. Thompson III, T. M. Donovan, D. R. Whitehead, and J. Faaborg. 1995. Regional forest fragmentation and the nesting success of migratory birds. Science 267:1987–1990.

Robinson, S. K., J. P. Hoover, R. Jack, and J. L. Herkert. In press. Effects of tract size, habitat, cowbird density, and nesting stratum on cowbird parasitism levels. *In* Ecology and management of cowbirds (T. Cook, S. K. Robinson, S. I. Rothstein, S. G. Sealy, and J. N. M. Smith, eds). University of Texas Press, Austin, TX.

Rosenberg, K. V., R. D. Ohmart, W. C. Hunter, and B. W. Anderson. 1991. Birds of the Lower Colorado River Valley. University of Arizona Press, Tucson, AZ.

Roskaft, E., G. H. Orians, and L. D. Beletsky. 1990. Why do Red-winged Blackbirds accept eggs of Brown-headed Cowbirds? Evol. Ecol. 4:35–42.

Roth, R. R., and R. K. Johnson. 1993. Long-term dynamics of a Wood Thrush population breeding in a forest fragment. Auk 110:37–48.

Rothstein, S. I. 1975a. An experimental and teleonomic investigation of avian brood parasitism. Condor 77:250–271.

Rothstein, S. I. 1975b. Mechanisms of avian egg-recognition: do birds know their own eggs? Anim. Behav. 23:268–278.

Rothstein, S. I. 1976. Cowbird parasitism of the Cedar Waxwing and its evolutionary implications. Auk 93:498–509.

Rothstein, S. I. 1977. Cowbird parasitism and egg recognition of the Northern Oriole. Wilson Bull. 89:21–32.

Rothstein, S. I. 1982. Successes and failures in avian egg and nestling recognition with comments on the utility of optimality reasoning. Amer. Zool. 22:547–560.

Rothstein, S. I. 1986. A test of optimality: egg recognition in the Eastern Phoebe. Anim. Behav. 34:1109–1119.

Rothstein, S. I. 1990. A model system for coevolution: avian brood parasitism. Annu. Rev. Ecol. Syst. 21:481–508.

Rothstein, S. I. 1994. The cowbird's invasion of the Far West: history, causes and consequences experienced by host species. *In* A century of avifaunal change in western North America (J. R. Jehl, Jr. and N. R. Johnson, eds). Stud. Avian Biol. no. 15.

Rothstein, S. I., J. Verner, and E. Stevens. 1980. Range expansion and diurnal changes in dispersion of the Brown-headed Cowbird in the Sierra Nevada. Auk 97:253–267.

Rothstein, S. I., J. Verner, and E. Stevens. 1984. Radio-tracking confirms a unique diurnal pattern of spatial occurrence in the parasitic Brown-headed Cowbird. Ecology 65:77–88.

Rothstein, S. I., D. A. Yokel, and R. C. Fleischer. 1986. Social dominance, mating and spacing systems, female fecundity, and vocal dialects in captive and free-ranging Brown-headed Cowbirds. Curr. Ornithol. 5:127–185.

Rothstein, S. I., J. Verner, E. Stevens, and L. V. Ritter. 1987. Behavioral differences among sex and age classes of the Brown-headed Cowbird and their relation to the efficacy of a control program. Wilson Bull. 99:322–327.

Rothstein, S. I., D. A. Yokel, and R. C. Fleischer. 1988. The agonistic and sexual functions of vocalizations of male Brown-headed Cowbirds, *Molothrus ater*. Anim. Behav. 36:73–86.

Scott, D. M. 1963. Changes in the reproductive activity of the Brown-headed Cowbird within the breeding season. Wilson Bull. 75:123–239.

Scott, D. M. 1977. Cowbird parasitism on the Gray Catbird at London, Ontario. Auk 94:18.

Scott, D. M. 1991. The time of day of egg laying by the Brown-headed Cowbird and other icterines. Can. J. Zool. 69:2093–2099.

Scott, D. M., and C. D. Ankney. 1980. Fecundity of the Brown-headed Cowbird in southern Ontario. Auk 97:677–683.

Scott, D. M., and C. D. Ankney. 1983. The laying cycle of Brown-headed Cowbirds: passerine chickens? Auk 100:583–593.

Sealy, S. G. 1992. Removal of Yellow Warbler eggs in association with cowbird parasitism. Condor 94:40–54.

Sedgwick, J. A., and F. L. Knopf. 1988. A high incidence of Brown-headed Cowbird parasitism of Willow Flycatchers. Condor 90:253–256.

Shake, W. F., and J. P. Mattsson. 1975. Three years of cowbird control: an effort to save the Kirtland's Warbler. Jack-Pine Warbler 53: 48–53.

Sherry, T. W., and R. T. Holmes. 1992. Population fluctuations in a long-distance Neotropical migrant: demographic evidence for the importance of breeding season events in the American Redstart. Pp. 431–442 *in* Ecology and conservation of Neotropical migrant landbirds (J. M. Hagan, III and D. W. Johnston, eds). Smithsonian Institute Press, Washington, DC.

Smith, F. 1977. A short review of the status of riparian forests in California. Pp. 1–2 *in* Riparian forests in California: their ecology and conservation (A. Sands, ed.). Institute for Ecology Publication 15, 122 p.

Smith, J. N. M. 1981. Cowbird parasitism, host fitness, and age of the host female in an island Song Sparrow population. Condor 83:152–161.

Snyder, L. L. 1957. Changes in the avifauna of Ontario. Pp. 26–42 Changes in the fauna of Ontario (F. A. Urquhart, ed.). University Toronto Press, Toronto, Canada.

Southern, W. E. 1958. Nesting of the Red-eyed Vireo in the Douglas Lake region, Michigan. Jack-Pine Warbler 36:105–130, 185.

Spaw, C. D., and S. Rohwer. 1987. A comparative study of eggshell thickness in cowbirds and other passerines. Condor 89:307–318.

Stafford, M. D., and B. E. Valentine. 1985. A preliminary report on the biology of the Willow Flycatcher in the central Sierra Nevada. Calif. Nev. Wildl. Trans. 1985:66–77.

Teather, K. L., and R. J. Robertson. 1986. Pair bonds and factors influencing the diversity of mating systems in Brown-headed Cowbirds. Condor 88:63–69.

Temple, S. A., and J. R. Cary. 1988. Modeling dynamics of habitat–interior bird populations in fragmented landscapes. Conserv. Biol. 2:340–347.

Thompson F. R., III. 1994. Temporal and spatial pattern of breeding in Brown-headed Cowbirds in the midwestern United States. Auk (in press).

Thompson, F. R., III, W. D. Dijak, T. G. Kulowiec, and D. A. Hamilton. 1992. Breeding bird populations in Missouri Ozark forests with and without clearcutting. J. Wildl. Manag. 56:23–30.

Thompson, F. R. III, J. R. Probst, and M. G. Raphael. 1993. Silvicultural options for Neotropical migrants. Pp. 353–362 *in* Status and management of Neotropical migratory birds (D. M. Finch and P. W. Stangel, eds). USDA Forest Serv. Gen. Tech. Rep. RM-229, Rocky Mt. Forest Range Exp. Sta., Fort Collins, CO.

Thompson, F. R., III, S. K. Robinson, T. M. Donovan, J. Faaborg, and D. R. Whitehead. In press. Biogeographic, landscape, and local factors affecting cowbird abundance and host parasitism levels. *In* Ecology and management of cowbirds (T. Cook, S. K. Robinson, S. I. Rothstein, S. G. Sealy, and J. N. M. Smith, eds). University of Texas Press, Austin, TX.

Tordoff, H. B. 1956. Checklist of the birds of Kansas. Univ. Kansas Mus. Natural Hist. Publ. 8:307–359.

Trail, P. 1992. Nest invaders. Pacific Discovery. Spring:32–37.

Trail, P. W., and L. F. Baptista. 1993. The impact of Brown-headed Cowbird parasitism on populations of the Nuttall's White-crowned Sparrow. Conserv. Biol. 7:309–315.

Trine, C. L., W. D. Robinson, and S. K. Robinson. In press. Consequences of cowbird parasitism for host population dynamics. *In* Avian brood parasitism (S. I. Rothstein and S. K. Robinson, eds). Oxford University Press, New York.

Unitt, P. 1987. *Empidonax traillii extimus*: an endangered subspecies. West. Birds 18:137–162.

US Fish and Wildlife Service. 1991. Black-capped Vireo recovery plan. US Fish and Wildl. Serv. Region 2. Albuquerque, NM, 74 pp.

Valentine, B. E., T. A. Roberts, S. B. Boland, and A. P. Woodman. 1988. Livestock management and productivity of Willow Flycatchers in the Central Sierra Nevada. Trans. West. Sect. Wildl. Soc. 24:105–114.

Van Velzen, W. T. 1972. Distribution and abundance of the Brown-headed Cowbird. Jack-Pine Warbler 50:110–113.

Verner, J., and L. V. Ritter. 1983. Current status of the Brown-headed Cowbird in the Sierra National Forest. Auk 100:355–368.

Verner, J., and S. I. Rothstein. 1988. Implications of range expansion into the Sierra Nevada by the parasitic Brown-headed Cowbird. Pp. 92–98 *in* Proceedings, State of the Sierra Symposium (D. Bradley, ed.). Pacific Publishing Company, San Francisco, CA.

Walkinshaw, L. H. 1961. The effect of parasitism by the Brown-headed Cowbird on *Empidonax* flycatchers in Michigan. Auk 78:266–268.

Walkinshaw, L. H. 1983. Kirtland's Warbler. Cranbrook Institute of Science, Bloomfield Hills, MI.

Weatherhead, P. J. 1989. Sex ratios, host-specific reproductive success, and impact of Brown-headed Cowbirds. Auk 106:358–366.

Weatherhead, P. J. 1991. The adaptive value of thick-shelled eggs for Brown-headed Cowbirds. Auk 108:196–198.

Weinrich, J. A. 1989. Status of Kirtland's Warbler, 1988. Jack-Pine Warbler 67:69–72.

Whitcomb, R. F., C. S. Robbins, J. F. Lynch, B. L. Whitcomb, K. Klimkiewicz, and D.

Bystra. 1981. Effects of forest fragmentation on avifauna of the eastern deciduous forest. Pp. 125–205 *in* Forest island dynamics in man-dominated landscapes (R. L. Burgess and D. M. Sharpe, eds), Springer-Verlag, New York.

Whitfield, M. J. 1990. Willow Flycatcher reproductive response to Brown-headed Cowbird parasitism. Master's thesis, California State University, Chico, CA.

Wiley, J. W. 1988. Host selection by the Shiny Cowbird. Condor 90:289–303.

Woodward, P. W., and J. C. Woodward. 1979. Brown-headed Cowbird parasitism on Eastern Bluebirds. Wilson Bull. 91:321–322.

Yokel, D. A. 1986a. Monogamy and brood parasitism: an unlikely pair. Anim. Behav. 34:1348–1358.

Yokel, D. A. 1986b. The social organization of the Brown-headed Cowbird in the Owens Valley, California in Natural History of the White-Inyo Range. White Mt. Res. Sta. Univ. Calif., Los Angeles, CA.

Yokel, D. A. 1989. Intrasexual aggression and the mating behavior of Brown-headed: their relation to population densities and sex ratios. Condor 91:43–51.

Yokel, D. A., and S. I. Rothstein. 1991. The basis for female choice in an avian brood parasite. Behav. Ecol. Sociobiol. 29:39–45.

Zimmerman, J. L. 1982. Nesting success of dickcissels (*Spiza americana*) in preferred and less preferred habitats. Auk 99:292–298.

Zimmerman, J. L. 1983. Cowbird parasitism of dickcissels in different habitats and at different nest densities. Wilson Bull. 95:7–22.

16

SINGLE-SPECIES VERSUS MULTIPLE-SPECIES APPROACHES FOR MANAGEMENT

WILLIAM M. BLOCK, DEBORAH M. FINCH, AND LEONARD A. BRENNAN

INTRODUCTION

Neotropical migratory birds are major components of the avifauna in most North American terrestrial ecosystems. Over 150 species of Neotropical migratory birds are known to breed in North America (Finch 1991a). Given the large number of species, developing effective management strategies for Neotropical migratory birds is a monumental task because each species exploits a unique niche, and thus requires different considerations for the management of its habitats and populations. Management is complicated further by temporal and spatial variations in resource-use patterns by many species. Thus, detailed knowledge of a species' habitat and population ecology from one place and time might have little relevance to other locations or periods.

Managers have a continuum of options for managing wildlife resources. This continuum ranges from management for one or a few featured species to the management of entire communities, landscapes, ecosystems, or regions. Each approach has advantages and disadvantages, and the choice of one approach over another represents a series of tradeoffs. Resource managers must decide which approach meets their objectives and the level of risk that is acceptable.

We present a conceptual framework of alternative approaches to wildlife management, with special emphasis on the management of Neotropical migratory birds. In particular, we discuss the use of single-species approaches, management- and ecological-indicator species, the guild concept and some of its permutations, and ecosystem approaches. We outline these approaches and weigh their merits and limitations. We also provide examples of how each has been and might be used in resource management.

CONCEPTUAL FRAMEWORK OF THE MANAGEMENT CONTINUUM

Single-species Approaches

Traditionally, the management of wildlife resources has emphasized single species. Initially, such management emphasized game, with the assumption that managing for game would provide suitable habitat for numerous other species as well (American Game Policy, 1930). Whereas this assumption is undoubtedly valid for some situations, the effects of game management also reduce habitat quality and quantity for populations of numerous other species. For example, managing for a game species that relies on early successional habitats will reduce habitat availability for species of Neotropical migratory birds that require late successional stages. The following case studies provide examples of how single-species management approaches can potentially influence populations of Neotropical migratory birds.

Case Studies

Northern Bobwhite and Red-cockaded Woodpecker

Management of game and Neotropical migratory birds is not mutually exclusive. As an example, we present preliminary results of an ongoing study by Brennan et al. (1995) of the effects of management actions for Red-cockaded Woodpeckers and Northern

Bobwhites on the habitats of Neotropical migratory birds in the southeast. In pine forests of the southeastern United States, situations exist where Northern Bobwhite and Red-cockaded Woodpecker management is complementary (Brennan and Fuller 1993). Management for either species has implications for Neotropical migratory birds. Therefore, this case study consists of two parts: (1) general effects of Northern Bobwhite habitat management on Neotropical migratory birds, and (2) how management actions beneficial to both Northern Bobwhites and Red-cockaded Woodpeckers may or may not benefit Neotropical migratory birds.

Most habitat management for the Northern Bobwhite is done in two types of environment: (1) old fields and field margins near cropland, and (2) pine and mixed pine–hardwood forests. The key to Northern Bobwhite habitat management is the frequent, periodic disturbance of vegetation in small, patchy mosaics (Stoddard 1931, Rosene 1969). In old-field habitats, the disturbance is accomplished by disking, prescribed fire, or combinations thereof. Widespread agricultural plantings of small (generally <0.5 ha) patches of corn, millet, milo, etc., are used in many bobwhite management efforts. In pine forests, prescribed fire and the reduction of tree basal area are primary bobwhite management tools. Regardless of the habitat type, the goal of the manager is to maintain approximately 70% of an area in understory plant communities that are 1–3 years of age. The remaining 30% is left as permanent cover areas, usually in the form of "cover blocks" of habitat that are allowed to develop more advanced seral stages of pine forests, or hedges and fencerows in old fields.

Understory and ground-cover plants are of the utmost importance to bobwhites. These plants produce seeds and provide substrates for insects needed by bobwhites for survival. This vegetation also provides escape cover for protection from predators. In pine forests, the tree canopy must remain open (<50% cover) to allow sufficient light to reach the ground and stimulate the growth of vegetation that produces food and cover for quail. One key factor in the recent bobwhite population decline in the southeastern United States is the proliferation of high-density pine plantations (Brennan 1991). This silvicultural practice results in a sterile understory that is dominated by decaying pine needles, and provides virtually no food resources for quail or other ground-foraging birds. Therefore, effective bobwhite management in pine forests requires the maintenance of an open-canopied forest with an understory that is disturbed on a frequent (1–3 years) basis.

Pine-forest habitats managed for the Northern Bobwhite will favor Neotropical migratory birds that are considered edge and open-country species, and will be detrimental to those that require dense, continuous, closed-canopy forests. In the southeastern United States, bobwhite management in pine forests will benefit species such as the Blue-gray Gnatcatcher, Common Yellowthroat, Eastern Wood-pewee, Indigo Bunting, and Great Crested Flycatcher. Species that would be likely to be impacted negatively by bobwhite management in pine forests include Wood Thrush, Black-and-White Warbler, Hooded Warbler, Red-eyed Vireo, and Yellow-throated Vireo.

Old-field and agricultural habitats managed for bobwhites will benefit Neotropical migratory species such as the Eastern Meadowlark, Mourning Dove, Indigo Bunting, and Yellow-breasted Chat. It is unknown whether bobwhite management in old, fallow-field environments has a negative impact on particular species of Neotropical migratory birds. With the recent widespread decline in Northern Bobwhite populations, efforts at habitat management will most likely increase because interest in quail hunting remains high, and membership of private organizations such as Quail Unlimited is increasing. Many people are interested in bobwhite management, and vast areas in the quail "plantation country" of southern Georgia and northern Florida continue to be managed for bobwhite.

The link between habitat management for the Northern Bobwhite and the endangered Red-cockaded Woodpecker presents a unique example of management for both game and endangered species (Brennan 1991). Habitat management for the woodpecker in pine-dominated systems entails

Table 16-1. Point counts of Neotropical migratory birds in forest stands managed for Red-cockaded Woodpeckers and unmanaged stands of similar age (>50 years) at the Bienville National Forest and Noxubee National Wildlife Refuge, Mississippi, May–June 1992 (Brennan et al. 1995).

Species	Bienville		Noxubee	
	Managed	Unmanaged	Managed	Unmanaged
Mourning Dove	1[a]	0		
Yellow-billed Cuckoo	0	1	0	2
Ruby-throated Hummingbird	2	0		
Great Crested Flycatcher	1	5	5	2
Eastern Wood-pewee	1	6	20	5
Blue-gray Gnatcatcher			11	0
Wood Thrush	8	7	0	14
Gray Catbird			0	1
White-eyed Vireo	2	7	1	3
Yellow-throated Vireo	0	2	1	13
Red-eyed Vireo	0	1	1	14
Black-and-white Warbler	5	6	2	11
Yellow-rumped Warbler			0	2
Pine Warbler	57	63	46	59
Palm Warbler	2	0		
Kentucky Warbler	6	15	2	5
Hooded Warbler	1	20	0	13
Common Yellowthroat	0	1	45	0
Yellow-breasted Chat	16	2	59	0
Indigo Bunting	14	1	52	2
Scarlet Tanager			0	2
Summer Tanager	7	0	9	14

[a] Total number of birds detected from seven replicate surveys of seven points in each stand type (managed and unmanaged).

maintainance of low (generally $<14\ m^2/ha$) basal area, and control of hardwood midstory by frequent fire, mechanical, or chemical means (Richardson and Smith 1992). Such techniques have also been widely successful in sustaining abundant populations of Northern Bobwhites at a variety of locations (Rosene 1969).

The effects of management in loblolly pine (*Pinus taeda*) forests for Red-cockaded Woodpeckers on other nontarget forest vertebrates, including Neotropical migratory birds, has been assessed at Bienville National Forest and Noxubee National Wildlife Refuge in Mississippi (Brennan et al., 1995). Unlike the Pacific Northwest, where interest in the Spotted Owl spurred comprehensive research efforts on terrestrial vertebrates in Douglas-fir (*Pseudotsuga menziesii*) forests (Ruggiero et al. 1991), wildlife researchers in the South have not conducted similar comprehensive assessments of the impacts of Red-cockaded Woodpecker habitat management on nontarget vertebrates.

At Bienville National Forest, Brennan et al. (1995) found 14 species of Neotropical migratory birds in stands actively managed for Red-cockaded Woodpeckers (Table 16-1). Three of these species (Great Crested Flycatcher, Indigo Bunting, and Yellow-breasted Chat) were apparently favored by Red-cockaded habitat management. Hooded Warbler, Kentucky Warbler, and Summer Tanager were detected most frequently in mature pine stands that were not managed for the woodpecker (Table 16-1). At Noxubee, 14 species of Neotropical birds were detected in stands managed for the woodpecker (Table 16-1). Five of these species (Blue-gray Gnatcatcher, Common Yellowthroat, Eastern Wood-pewee, Indigo Bunting, and Yellow-breasted Chat) were apparently favored by Red-cockaded Woodpecker management. The Black-and-white Warbler, Hooded Warbler, Red-eyed Vireo, Summer Tanager, Wood Thrush, and Yellow-throated Vireo were detected most frequently in mature pine stands that were not managed for woodpeckers at Noxubee (Table 16-1).

The differences in the Neotropical migratory birds at Bienville and Noxubee may be partly a function of how the habitat is managed at each site (Brennan et al., 1995). At Noxubee, the hardwood midstory is sheared with a v-blade mounted on a bulldozer, then controlled by burning on 2–4 year intervals (Richardson and Smith 1992). At Bienville, the hardwood midstory is treated with herbicides and then controlled by burning at 5–7 year intervals (Brennan et al., 1995). These two different management approaches result in different vegetation structures, influencing Neotropical migratory birds differently at each site (Brennan et al., 1995).

Two major lessons can be learned from the case history described above. First, it is possible to integrate the management of endangered species, Neotropical migratory birds, and game in the pine-dominated forests of the southeastern coastal plain. Second, the particular way that endangered or game species are managed can have a major influence on populations and habitats of Neotropical migratory birds.

Several other issues have important implications for Neotropical migratory birds in the context of management for Red-cockaded Woodpeckers and Northern Bobwhites. Whether the patterns observed by Brennan et al. (1995) in loblolly-pine forests apply to wildlife–habitat relationships in longleaf pine (*Pinus longirostris*) is unknown and urgently needs to be investigated. Also unknown are the effects of the timing of prescribed fire on Neotropical migratory bird populations. Most prescribed fire in pine systems of the southeastern coastal plain has been conducted during February and March over the past 60 years rather than from May though August when most natural lightning-caused fires occur. The effects of this departure from the natural fire regime must be assessed as managers strive to return to a more "natural" use of fire in an ecosystem context. Clearly, research on how variations in Red-cockaded Woodpecker habitat and differences in management practices influence Neotropical migratory birds should be a high priority in the future.

Mountain Quail

Another study that compared habitats of Neotropical migratory birds with that of a game species requiring early successional vegetation was described by Block et al. (1991). They found extensive overlap in the habitats of ground-foraging birds (including Lazuli Bunting, Rufous-sided Towhee, Green-tailed Towhee, Chipping Sparrow) with that of the Mountain Quail in northern California (Fig. 16-1). Typically, Mountain Quail are found in early-successional montane brushfields that result from even-aged forestry practices (such as clearcutting) or stand-replacing fires.

Northern Spotted Owl

As populations of many species of plants and animals have declined to the point of concern, and even extinction, the public has become more concerned with conserving the extant biota. Enactment of the Endangered Species Act provided a legal framework under which such conservation efforts could be undertaken. Contained within the Endangered Species Act is a provision for the maintenance and enhancement of "critical habitat;" that is, "...physical or biological features (I) essential to the conservation of the species and (II) which may require special management considerations or protection" (US Government Printing Office 1983, p. 2). Because enactment of management plans for threatened, endangered, and sensitive species is directed towards those species, effects on Neotropical migratory birds will be a byproduct of these management actions.

Frequently, management of habitat for threatened, endangered, and sensitive species requires setting land in reserved status and prohibiting activities other than those directed towards the target species. For example, the recovery plan for the Northern Spotted Owl sets aside large blocks of land as Designated Conservation Areas (Bart et al. 1992). These areas include existing owl habitat as well as large blocks of forest that have the capability of maturing into suitable owl habitat. The developers of the recovery plan recognized that its implementation would affect numerous other species, and

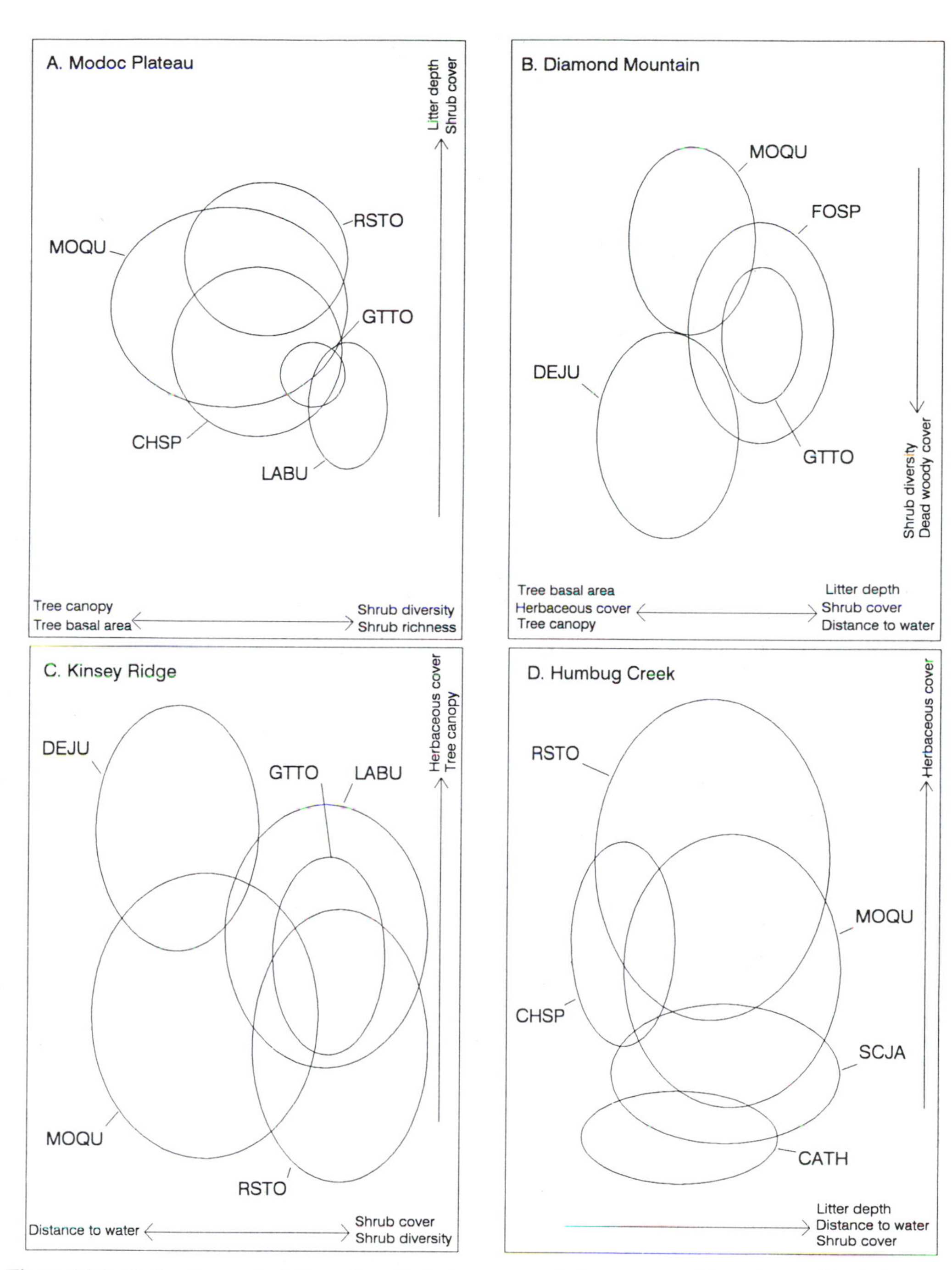

Figure 16-1. Ordinations depicting the first two canonical axes resulting from discriminant analyses of the habitats of ground-foraging birds found at four locations in northern California. Ellipses include 95% of all observations. Species codes are: MOQU, Mountain Quail; CATH, California Thraher; SCJA, Scrub Jay; RSTO, Rufous-sided Towhee; GTTO, Green-tailed Towhee; CHSP, Chipping Sparrow; FOSP, Fox Sparrow; DEJU, Dark-eyed Junco; and LABU, Lazuli Bunting (from Block et al. 1991).

Table 16-2. Threatened, endangered, sensitive, candidate, and old-growth-associated birds within the range of the Northern Spotted Owl (from Bart et al. 1992).

Species	Status[a]			Old Forest Association[b]		
	WA	OR	CA	WA	OR	CA
Northern Goshawk	C	SC	SC	+	+	+
Flammulated Owl	C	SC		+	+	+
Vaux's Swift	C			*	*	?
Pacific-slope Flycatcher				+	+	*
Hammond's Flycatcher						*
Willow Flycatcher			E			
Hermit Thrush						*
Warbling Vireo						+
Hermit Warbler						*
Wilson's Warbler				+		+

[a] WA (Washington): C = candidate; OR (Oregon): SC = sensitive (critical); CA (California): SC = species of concern.
[b] + = old-growth associate; * = close old-growth associate; ? = insufficient data.

they attempted to evaluate those effects. The recovery team identified 23 bird species of special concern [e.g., threatened, endangered, and sensitive (TES) species, those whose distributions are endemic within that of the Northern Spotted Owl, those with greatest abundance in older forests, or those with greatest need of conservation], including ten Neotropical migratory birds, that would benefit from implementation of the recovery plan (Bart et al. 1992; Table 16-2). In addition to these species, numerous others that use mature and old-growth forest would also stand to benefit from this recovery plan—see Ruggiero et al. (1991) for papers that document species associated with mature forests in the Pacific Northwest.

In conclusion, single-species approaches to wildlife management contain many direct and indirect, positive and negative effects on other species. Usually, effects on other species are considered simply as byproducts of the intent of the management actions for the target species or are not considered at all. Only recently, with considerations of biological diversity and resultant conservation efforts for all native species, have single-species approaches (e.g., the Northern Spotted Owl) evaluated the needs of other species as well.

INDICATOR SPECIES

Indicator species mark a transition from single-species to multiple-species approaches in wildlife management. Although a single species is emphasized, this species is used to index or represent specific environmental conditions or the population status of other, ecologically similar species.

Indicator species can be divided into two major categories: ecological indicators and management indicators. The concept of ecological indicators was formally proposed by Clements (1920) to explain plant distributions based on specific environmental conditions, most notably edaphic and moisture regimes. Birds are also tied to specific environmental conditions, as this provides the basis for describing species' habitats (Block and Brennan 1993). Birds, particularly migratory birds, are highly vagile and consequently can adjust to greater variations in environmental conditions than most plant species. Thus, the relationship between environmental conditions and the presence of a particular bird species is less predictable than for many plants. For this reason, the predictive value of birds as indicators of environmental conditions may be extremely limited (Morrison 1986).

Management Indicator Species

Regardless of these limitations, resource managers have endorsed the use of avian indicator species for land-use planning and decisions. For example, the USDA Forest Service has promoted the use of Management Indicator Species (MIS) in the forest-planning process. Four broad categories of MIS are commonly used: (1) recovery species; (2) featured species; (3) specific habitat indicators; and (4) ecological indicators (Salwasser et al. 1982). Recovery species are those that are managed to increase their populations because they are recognized as threatened or endangered. Migratory species such as the Whooping Crane, Golden-cheeked Warbler, and Kirtland's Warbler are examples of recovery species. Featured species include those managed for consumptive purposes or those that are valued for nonconsumptive recreational use. Examples of featured migratory species are ducks, geese, Mourning Dove, and Elegant Trogon. Specific habitat indicator species are those with potentially limiting habitat needs that might be affected adversely by land-management practices. Forest-interior species might be regarded as a group of species with specific habitat needs (Robbins et al. 1989). Of course, it could be argued that all species have specific needs, and consequently might be regarded as specific habitat indicators.

Ecological indicator species are those whose populations can be used to index habitat quality and population status of other species. Guild-indicator species provide a typical example of how ecological indicators have been used. Essentially, a guild indicator is one species of a guild (discussed beyond) whose population and habitat can be monitored as an index for other members of the guild. This concept was initially proposed by Severinghaus (1981), but the validity of guild indicators has been questioned by many, including Verner (1983), Szaro (1986), and Block et al. (1987).

The value of birds, particularly migratory raptors such as the Peregrine Falcon and waterbirds, as indicators of contaminants and related environmental problems is well documented (Hickey 1969). The banning of dichlorodiphenyltrichloroethane (DDT) can be attributed largely to birds as indicators through eggshell thinning and nestling deformity. Presently, unregulated use of pesticides on the wintering grounds may be contributing to population declines in migratory birds. If so, population declines noted on the breeding grounds may be indicators of factors occurring on the wintering grounds, such as the use of contaminants or deforestation.

Typically, indicator species are used to monitor population and habitats of ecologically similar species. Their use in these instances has been the subject of much criticism (Morrison 1986, Block et al. 1987, Landres et al. 1988). Morrison (1986) questioned the use of birds as environmental indicators because many factors affect birds. Thus, assessing cause–effect relationships may be muddled by the vast number of complicating factors affecting the species. Landres et al. (1988) felt that the use of indicators to assess population statuses and habitat suitabilities for other species was indefensible unless basic research had been done to document that the chosen species indicated the population and habitat status of other species. This is not surprising given that habitat requirements and population ecologies are specific to each species (Mannan et al. 1984, Block et al. 1987).

Standards for Indicator Species

Landres et al. (1988) concluded that, if indicators were used, they should meet the following rigorous standards. First, criteria to assess whether or not goals have been met should be stated clearly. Second, indicators should be used only when absolutely necessary. Third, selection of indicator species should be by clear and unambiguous rules. Subjectivity in the selection of indicator species should be minimized. In this regard, we know of no clear guidelines for the selection of indicator species. Odum (1953), Hill et al. (1975), and Graul and Miller (1984) suggested that indicators should be species that tolerated a narrow range of conditions, whereas Szaro and Balda (1982) felt that indicators should represent a wide range of conditions. Further, few investigators present quantitative methods for selecting indicators (cf. Hill et al. 1975, Szaro and Balda 1982), and most methods commonly used are qualitative (e.g., Salwasser et al. 1982). Fourth, all appropriate species must be included in the assessment process, and selection criteria for indicators must not be modified for convenience. Fifth, investigators must have a thorough knowledge of the biology of the organism selected as the indicator. Sixth, all assessment programs that use indicator species must be reviewed and evaluated, particularly with reference to assessment design, data collection, and data analysis. Seventh, investigators must consider natural fluctuations in population parameters and incorporate concepts from landscape ecology when developing monitoring strategies. From the recommendations of Landres et al. (1988), it is clear that indicator species should not be used in an ad hoc, haphazard manner. Detailed evaluations must be undertaken prior to using indicators, and they should be used only when alternative strategies are not possible.

GUILDS

Guilds represent the most basic approach to true multiple-species management. The basis of the guild concept can be traced to Salt (1953, 1957), who grouped species of birds into functional units based on their general foraging ecologies. Salt used the functional-unit concept to compare general ecological attributes of avifaunas from different locations. Bock and Lynch (1970) applied this grouping procedure to explain change in a bird-community structure following forest fire in the Sierra Nevada. Root (1967), in his monograph on the niche of the Blue-gray Gnatcatcher, defined a guild as a group of species that exploit the same class of environmental resources in a similar fashion. He stressed that guilds were not restricted taxonomically to closely related species, but that guild membership should transcend taxonomic boundaries to include species representative of grossly different taxa such as avian and mammalian, or vertebrate and invertebrate. In practice, however, guilds are generally restricted within a taxonomic class (e.g., Aves, Mammalia) and frequently to birds (Verner 1984). Failure to include species from different taxa prompted Jaksić (1981) to coin a different term, the taxonomic assemblage, to reflect the common deviation from Root's intent of what species should comprise a guild. To our knowledge few investigators have defined guilds across taxonomic boundaries; Jaksić and his coworkers are among the few researchers to have included species from different major taxa in their analyses of a predatory guild. Even Root did not apply the guild concept as he defined it. His foliage-gleaning guild included only birds that foraged by gleaning insects from the foliage of oaks (*Quercus* spp.).

Root's analysis of his foliage-gleaning guild clearly demonstrated that each species within the guild differed from the others, to varying degrees, in exactly how they obtained food. For example, species differed in foraging mode, foraging substrate, and the general location within a tree where they obtained food. He concluded that, although these species overlapped in their general foraging ecology, they differed when one examined aspects of their foraging in greater detail (i.e., a finer scale). Similar intraguild analyses have led to similar conclusions (e.g., Noon 1981, Block et al. 1991). This should come as no surprise because each species, even within a guild, has unique morphology and exhibits a unique pattern of extracting resources from the environment (Verner 1984, Block et al. 1991). Pianka (1978) concluded that species in a guild should exhibit relatively intense interspecific competition, forcing each species to diverge in its niche-utilization patterns. Thus, even though species in a guild share some general mode of exploiting resources, their intrinsic species-specific properties entail varying degrees of divergence in their ecologies.

Guild Delineation

Placing species in a guild is not a simple matter (Jaksić 1981, MacMahon et al. 1981). Frequently, investigators group species a priori according to some investigator-defined notion of resource use (MacMahon et al. 1981). Thus, guilds are essentially human constructs that hopefully have some relevance to ecological similarity of the species contained therein. Generally, these groupings are based on foraging (Verner 1984), although other aspects of resource use can provide equally valid bases for grouping. Unfortunately, a priori groupings may have little relevance to how and what resources

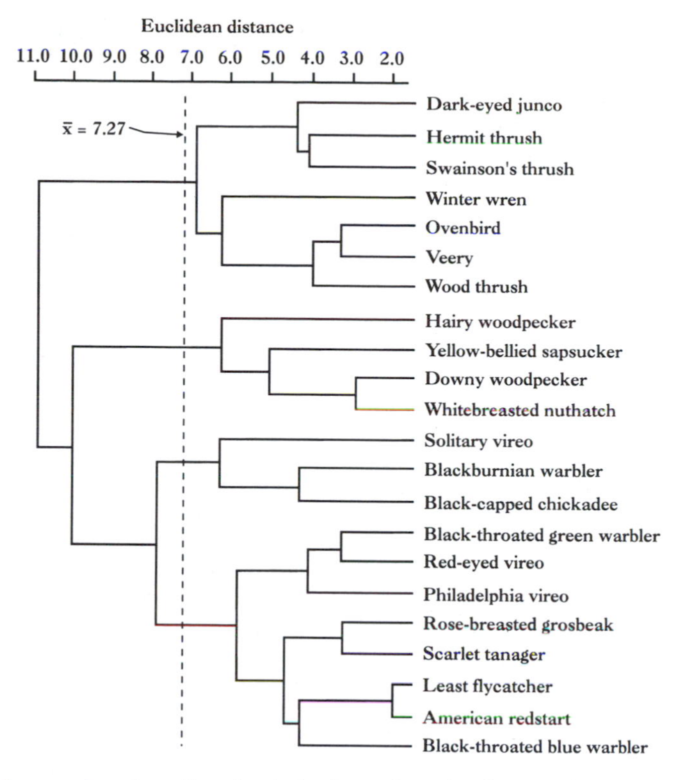

Figure 16-2. Dendogram based on foraging behaviors of species from Hubbard Brook Experimental Forest, New Hampshire, used to group species into foraging guilds (from Holmes et al. 1979).

are actually used at a specific location or during a given time. Further, these groupings are generally based on expert opinion or on data obtained at different places or times. Temporal variations in ecologies of species occur within and among both seasons and years; spatial variations occur as well (cf. Block 1990, Hejl and Verner 1990, Szaro et al. 1990). Thus, guilds defined a priori may not be valid for the specific time or location where they are intended for use.

A more desirable approach is to group species into guilds based on time- and site-specific data (i.e., a posteriori groupings). Although this approach is still somewhat subjective (Morrison et al. 1992), it offers a more objective approach to developing guilds than those developed a priori. Holmes et al. (1979) provided an example of how to use empirical data to group species into guilds based on cluster analysis (Fig. 16-2). As Morrison et al. (1992) noted, however, subjectivity still is involved in defining the level of similarity or dissimilarity for grouping or separating species into guilds. The primary difficulty with developing guilds a posteriori, however, is that it requires a great deal of field work to obtain adequate data upon which guilds can be defined. Resource managers rarely have the time, personnel, or monetary resources to obtain such data and are thus faced with the dilemma of relying on little, and perhaps inappropriate, information for defining guilds, or simply not using guilds at all. We believe that there is no correct solution to this problem, except to employ more rigorous a posteriori procedures for grouping species whenever possible.

Guild Dimensionality

Another important consideration is that guilds are generally based on few (usually one or two) resource-use dimensions. Thus, even if species overlap greatly in one aspect of their niche utilization patterns, it is possible that they will differ substantially in other aspects of niche utilization (Schoener 1974, Pianka 1978). For example, species might forage in the same general fashion but have completely different nest requirements (Martin 1991). Consequently, management actions whose goals are to provide suitable foraging habitat may fail to provide or even decrease the value of the habitat for other life history needs.

Management Applications

The concept of the guild was viewed by many as a potential tool for the management of wildlife populations (Johnston 1981, Severinghaus 1981, Short and Burnham 1982, Verner 1984). Short and Burnham (1982) introduced the concept of "guild-blocking." This method involved developing a matrix based on the foraging and nesting location (according to height strata) and grouping species that appeared in the same matrix block. Short and Burnham recognized that some species nested and/or foraged within multiple-height strata allowing some species to be assigned to more than one guild. Again, a major problem with this approach was that species were assigned to guilds a priori, based on expert opinion and not on site-specific, empirical information.

Severinghaus (1981) suggested that guilds could be used to assess environmental impacts. He defined guilds based on general diet composition, foraging mode, activity patterns, and gross habitat structure. Amazingly, many of the guilds that he derived (Figs 1 and 2 in Severinghaus 1981) consisted of species that were allopatric rather than sympatric. Thus, his guilds deviated from the basic premise of guild analysis, namely that of sympatry (Verner 1984). Regardless, Severinghaus (1981) conjectured that environmental impacts that affected one member of a guild should affect other members of the guild similarly. This assumption, the basis for guild indicators, was discredited by Mannan et al. (1984), who observed that species in the guilds that they studied of the Coast Ranges and Blue Mountains of Oregon exhibited different population responses to environmental change. Further, Block et al. (1987) noted that species within a ground-foraging guild differed in their use of microhabitat. They concluded that subtle changes in the environment would affect the habitat of each species differently.

Szaro (1986) and Verner (1984) proposed grouping species according to similar population responses to environmental perturbations. This approach has some merit when considering the effects of pronounced habitat change (Knopf et al. 1988, Finch 1991b, Morrison et al. 1992). For example, one would expect populations of most canopy-dwelling birds to decline following the removal of a substantial percentage of trees as might result from crown fires, clearcuts, shelterwood cuts, or seedtree cuts. However, population responses of birds to less pronounced habitat changes may not be so easy to predict with precision (Mannan et al. 1984). Thus, response guilds probably provide managers with only the most basic information of population responses to the most severe forms of disturbance. Even then, managers rarely have information detailing population responses to even these severe disturbances. Thus, research that defines species' responses and groups species into guilds could be useful.

Regardless of the plethora of criticisms, resource managers have embraced the concepts of guilds and of guild indicators in their management plans. We do not contend that these approaches are without merit. Their utility, however, may be limited to situations of predicting effects of drastic and pronounced environmental perturbations. Their use in predicting the effects of more subtle environmental change may be limited. Reliance on guilds as the sole management tool may be extremely misleading and potentially deleterious to populations of many of the species contained within the guild. Guilds can be useful tools when reliable knowledge of joint relationships among habitats or species exists. However, when guilds and related concepts are used in ad hoc fashions and functional relationships have not been established, the risk of making ill-informed management decisions is high.

Managers should be aware of the limitations of guilds in land-use planning and assessment, and they should recognize that guilds are but one of many potential tools at their disposal.

MANAGEMENT ASSEMBLAGES

We define management assemblage as a grouping of species to meet specific management objectives (e.g., Verner 1984). Management assemblages differ from guilds in that the groupings need not be based on patterns of resource use. Frequently, it may be useful to group species according to other, more broad-based criteria for specific management purposes. For Neotropical migratory birds, a number of potential groupings may be useful such as long- and short-distance migrants, lowland and upland birds, or forest-interior and edge species. These are but a few examples of perhaps limitless possibilities.

The initial step in using this approach is to define the specific objective or rationale for grouping species in such a manner. For example, long-distance migrants might experience a different set of limiting factors than short-distance migrants, and thus should be managed differently. O'Connor (1991) suggested that short-distance migrants were more eurytypic in habitat use than were long-distance migrants, and thus were less vulnerable to habitat change. This was partly supported by the findings of Hussell et al. (1991) who noted that a greater proportion of long-distance migrants exhibited trends of declining populations than short-distance migrants. Hagan et al. (1991) noted that population declines during winter were greater for short-distance migrants that wintered in temperate regions than for long-distance migrants wintering in the tropics, although Hagan et al (1991) were unable to document the underlying causal factors. Determining the common factors that affect members of such assemblages in a similar way is the critical step in developing appropriate management strategies. Obviously, this step is one of the most difficult to implement but, without it, management of these types of assemblages would be tenuous, at best.

A potential problem with management assemblages is that they include species that respond to different ecological conditions. As noted in our discussion of guilds, species with different ecologies are likely to respond differently to environmental perturbations (Mannan et al. 1984, Block et al. 1987). Thus, changes that may cause a positive population response in some species might decrease populations of others. This does not necessarily negate the efficacies of this approach, but it does stress the need for detailed and accurate biological information about the species comprising the assemblage.

Certainly, use of the management assemblages holds some potential for Neotropical migratory birds. Exactly what that potential might entail is unknown. We feel that the development and testing of the use of these assemblages might provide some novel, yet useful, tools for the future management of Neotropical migratory birds. Management assemblages may be particularly useful as conceptual tools to focus on common links among species that can provide insights into limiting factors and management approaches to sustain their populations.

ECOSYSTEM APPROACHES

Neotropical migratory birds are just one component of the avian portion of the biotic community at any particular location or during a given time. This holds true on the breeding grounds, along spring and fall migration routes, and on the wintering grounds. Although Neotropical migratory birds frequently comprise a major portion of an avian community (Block et al. 1992; Table 16-3), management must account for residents and short-distance migrants as well.

Because an avian community can include a large number of species, management approaches cannot be concerned with species-specific habitat requirements. Rather, the approach that we advocate is one of providing diverse environmental conditions to meet the needs of a multitude of species. Such management cannot be limited to stand

Table 16-3. Numbers of resident and Neotropical migratory species found in the mountains of southeast Arizona during the 1991 breeding season and the oak woodlands of the Tehachapi Mountains, California, during the 1986, 1987, and 1988 breeding seasons.

Species	Arizona Mountains			Tehachapi Mountains
	Mixed conifer	Pine–oak	Oak	
Resident	33	43	31	24
Neotropical migrant	44	54	41	63

management but must consider broad-scaled landscape approaches (cf. Reynolds et al. 1992). Perhaps the most appealing aspect of the management of communities of Neotropical migratory birds is that it is consistent with the current trends towards ecosystem management for the maintenance and enhancement of biological diversity (Wilson 1986, USDA Forest Service 1992).

Ecosystem management is an ecological approach that incorporates both environmental values and the needs of people to sustain natural ecological systems. To implement ecosystem management, reference or desired future conditions must be defined at the onset of the planning process. These conditions must be within the natural ranges of variation, assuming that management within these bounds will sustain the ecosystem structure and function. Management activities are not implemented to meet commodity targets, but to move the ecosystem within these natural ranges.

Ecosystem management and inherent properties of ecosystems (i.e., processes and functions) must be considered along spatial and temporal hierarchies. The birds found within ecosystems at any particular spatial and temporal scale represent only a small part of the system in which they are found. Obviously, ecosystem management is not a direct approach for managing populations and habitats of Neotropical migratory birds. Managers must operate under the assumption that providing appropriate conditions for ecosystem sustainability will also provide adequate conditions for all component systems, including bird populations and communities. Thus, we view the primary role of birds in ecosystem management as monitors of the effectiveness of ecosystem management. By monitoring populations and habitats of birds, managers can assess whether or not the ecosystem is within or approaching the desired target conditions. Admittedly, ecosystem approaches are in the development stage and their true efficacy is unknown. These approaches, however, hold great potential and must include considerations of the avian component in ecosystem evaluations and in applications of ecosystem management practices.

SYNTHESIS: WHERE DO WE GO FROM HERE?

As may be evident from our discussion above, no one management approach is a panacea for managing Neotropical migratory birds. If management is to be directed toward one species or a small subset of species, then we advocate that they be managed singly. If management is to be directed toward multiple species, then single-species approaches are not feasible and alternative methods must be employed. Undoubtedly, no single method will work for the management of large groups of Neotropical birds; rather, multiple approaches should be used simultaneously.

Any one approach contains advantages and disadvantages. For example, management of single species is based on very specific information about that species. We assume that the information used is appropriate for the place and time that it is used, and is in sufficient detail to allow informed and effective management to be initiated. The disadvantage of single-species management is simply that it targets only the species of interest. Further, monitoring population changes of single species requires an extraordinary amount of effort (Verner 1983), generally far beyond what is possible with the logistic constraints faced by most resource management agencies. However, a certain amount of single-species management will continue to be conducted and we must come to terms with the implications of these management schemes for other species.

An advantage of multiple-species approaches is that managers can account for numerous species through management activities at a cost that is equal to or perhaps even less than that incurred for single species (Verner 1983). The primary disadvantage to these approaches is that little species-specific information is available upon which managers can base their decisions. A corollary to this disadvantage is that monitoring habitats and/or populations of species assemblages may mask trends affecting individual species. In the situation where the population of one species might be declining, this represents a potentially large risk. Thus, managers run the risk of invoking strategies that may not benefit, or may even work negatively on, the populations of species that they hope to improve.

Ultimately, the direction taken by managers will depend on their specific objectives. Inherent in this process is that management considers adequate spatial and temporal scales in the planning process. Plans restricted to a forest stand or even a Forest Service District, for example, often are too small for the proper management of certain species. Management considerations may need to be expanded to an entire National Forest, or to even regional or continental scales to manage a particular suite of Neotropical migratory birds adequately.

Presently, management is dictated by single-species management of TES species, and species of great consumptive or nonconsumptive value. Management of these species is either mandated by law, as in the case of TES species, influenced by economic issues (e.g., agriculture, development), or is deeply rooted in the history of wildlife management, particularly with respect to game species. Practices focused on the management of these species will define a set of constraints that will dictate future management. Determining the appropriate strategy for the remaining species is more difficult. Certainly, the cost of managing each of these species singly would be prohibitive. That leaves the manager with alternatives of indicator species, guilds, management assemblages, and ecosystems. We agree with Landres et al. (1988) that the use of indicator species should be avoided if at all possible. As we discussed previously, guilds and management assemblages each contain numerous disadvantages that limit their effectiveness. That leaves ecosystem management as the most viable alternative.

We advocate that management of wildlife resources, including migratory birds, considers the concepts of ecosystem management. Theoretically, ecosystem management is a holistic approach aimed at sustaining natural resources by ensuring that ecosystem processes and functions operate within natural ranges of variation. By providing conditions that sustain ecosystems, avifaunas are sustained as a result. Traditional approaches that veer off from single-species management are compatible with ecosystem management. Single-species approaches are essentially fine-filter strategies to management (Hunter 1991). Ecosystem management involves a coarse-filter approach to evaluate biodiversity and environmental conditions across a landscape (Hunter 1991). Fine-filter approaches are then applied to species or groups of species that pass through the coarse filter; that is, species requiring additional management considerations. The Nature Conservancy estimated that up to 90% of all species would be managed sufficiently by the coarse-filter approach (Hunter 1991). The difference between single-species and ecosystem management is that fine filters are used first in single-species management, and last in ecosystem management.

Ecosystem management, however, cannot be applied to discrete communities in a piecemeal fashion. Rather, it must consider the effects of management along a spatial continuum ranging from the stand to the continental or even global levels. Many aspects of ecosystem management are still being developed but this system does offer possibilities for managing many species over broad, geographic areas. Only through such an approach can management hope to provide for the wide diversity of Neotropical migratory birds found in North America.

ACKNOWLEDGMENTS

We thank I. J. Ball, J. L. Ganey, M. L. Morrison, J. Verner, and an anonymous reviewer for commenting on earlier drafts of the paper.

LITERATURE CITED

American Game Policy. 1930. 17th American Game Conference, American Game Association, New York.

Bart, J., R. G. Anthony, M. Berg, J. H. Beuter, W. Elmore, J. Fay, R. J. Gutiérrez, H. T. Heintz, Jr., R. S. Holthausen, K. Lathrop, K. Mays, R. Nafziger, M. Pagel, C. Sproul, E. E. Starkey, and J. C. Tappenier. 1992. Recovery plan for the Northern spotted Owl—final Draft. US Department of the Interior, Washington, DC.

Block, W. M. 1990. Geographic variation in foraging ecologies of breeding and nonbreeding birds in oak woodlands. Stud. Avian Biol. 13:264–269.

Block, W. M., and L. A. Brennan. 1993. Habitat concept in ornithology: theory and applications. Curr. Ornithol. 11:35–91.

Block, W. M., L. A. Brennan, and R. J. Gutiérrez. 1987. Evaluation of guild-indicator species for use in resource management. Environ. Manage. 11: 265–269.

Block, W. M., L. A. Brennan, and R. J. Gutiérrez. 1991. Ecomorphological relationships of a guild of ground-foraging birds in northern California, USA. Oecologia 87:449–458.

Block, W. M., J. L. Ganey, K. E. Severson, and M. L. Morrison. 1992. Use of oaks by Neotropical migratory birds in the southwest. Pp. 65–70 *in* Ecology and management of oaks and associated woodlands (P. F. Pfolliott, G. J. Gottfried, D. A. Bennett, V. M. Hernandez C., A. Ortega-Rubio, and R. H. Hamre, tech. coords). USDA Forest Serv. Gen. Tech. Rep. RM-215, Fort Collins, CO.

Bock, C. E., and J. F. Lynch. 1970. Breeding bird populations of burned and unburned coniferous forests in the Sierra Nevada. Condor 72:182–189.

Brennan, L. A. 1991. How can we reverse the Northern Bobwhite decline? Wildl. Soc. Bull. 19: 544–555.

Brennan, L. A., and R. S. Fuller. 1993. Bobwhites and red-cockaded woodpeckers: endangered species management helps quail too! Quail Unlimited Mag. 12(3): 16–20.

Brennan, L. A., J. L. Cooper, K. E. Lucas, B. D. Leopold, and G. A. Hurst. 1995. Asessing the influence of Red-cockaded Woodpecker colony site management on non-target forest vertebrates in loblolly pine forests of Mississippi: study design and preliminary results. Pp. 36–47 *in* Red-cockaded Woodpecker Symp. III: Species Recovery, Ecology and Management (D. L. Kulhavey, R. G. Cooper, and R. Costa, eds). Stephen F. Austin State University, Nacogdoches, TX.

Clements, F. E. 1920. Plant indicators. Carnegie Instit. Wash. Publ. 290.

Finch, D. M. 1991a. Population ecology, habitat requirements, and conservation of Neotropical migratory birds. USDA Forest Serv. Gen. Tech. Rep. RM-205, Fort Collins, CO.

Finch, D. M. 1991b. Positive associations among riparian bird species correspond to elevational changes in plant communities. Can. J. Zool. 69:951–963.

Graul, W. D., and A. H. Miller. 1984. Strengthening ecosystem management. Wildl. Soc. Bull. 12:282–289.

Hagan, J. M., III, T. L. Lloyd-Evans, J. L. Atwood, and D. S. Woods. 1991. Long-term changes in migratory landbirds in the northeastern United States: evidence from migration capture data. Pp. 115–130 *in* Ecology and conservation of Neotropical migrant landbirds (J. M. Hagan, III and D. W. Johnston, eds). Smithsonian Institution Press, Washington, DC.

Hejl, S. J., and J. Verner. 1990. Within-season and yearly variations in avian foraging locations. Stud. Avian Biol. 13:202–209.

Hickey, J. J. 1969. Peregrine Falcon populations: their biology and decline. University of Wisconsin Press, Madison, WI.

Hill, M. O., R. G. H. Brunce, and M. W. Shaw. 1975. Indicator species analysis, a divisive polythetic method of classification, and its application to a survey of native pinewoods in Scotland. J. Ecol. 63:597–613.

Holmes, R. T., R. E. Bonney, Jr, and S. W. Pacala. 1979. Guild structure of the Hubbard Brook bird community: a multivariate approach. Ecology 60: 512–520.

Hunter, M. L., Jr. 1991. Coping with ignorance: the coarse filter strategy for maintaining biodiversity. Pp. 266–281 *in* Balancing on the brink of extinction (K. A. Kohm, ed.). Island Press, Washington, DC.

Hussell, D. J. T., M. H. Mather, and P. H. Sinclair. 1991. Trends in numbers of tropical- and temperate-wintering migrant landbirds in migration at Long Point, Ontario, 1961–1988. Pp. 101–114 *in* Ecology and conservation of Neotropical migrant landbirds (J. M. Hagan, III and D. W. Johnston, eds). Smithsonian Institution Press, Washington, DC.

Jaksić, F. M. 1981. Abuse and misuse of the term "guild" in ecological studies. Oikos 37: 397–400.

Johnston, R. A. 1981. Application of the guild concept to environmental impact analysis of terrestrial vegetation. J. Environ. Manage. 13: 205–22.

Knopf, F. L., J. A. Sedgewick, and R. W. Cannon. 1988. Guild structure of a riparian avifauna relative to cattle grazing. J. Wildl. Manag. 52: 280–290.

Landres, P. B., J. Verner, and J. W. Thomas. 1988. Ecological uses of vertebrate indicator species: a critique. Conserv. Biol. 2:316– 328.

MacMahon, J. A., D. J. Schimpf, D. C. Andersen, K. G. Smith, and R. L. Bayn, Jr. 1981. An organism-centered approach to some community and ecosystem concepts. J. Theor. Biol. 88:287– 307.

Mannan, R. W., M. L. Morrison, and E. C. Meslow. 1984. Comment: the use of guilds in forest bird management. Wildl. Soc. Bull. 12:426–430.

Martin, T. E. 1991. Breeding productivity considerations: what are the appropriate habitat features for management? Pp. 455–473 *in* Ecology and conservation of Neotropical migrant landbirds (J. M. Hagan, III and D. W. Johnston, eds). Smithsonian Institution Press, Washington, DC.

Morrison, M. L. 1986. Birds as indicators of environmental change. Curr. Ornithol. 3:429–451.

Morrison, M. L., B. G. Marcot, and R. W. Mannan. 1992. Wildlife–habitat relationships: concepts and applications. University of Wisconsin Press, Madison, WI.

Noon, B. R. 1981. The distribution of an avian guild along a temperate gradient: the importance and expression of competition. Ecol. Monogr. 51:105–124.

O'Connor, R. J. 1991. Population variation in relation to migratory status in some North American birds. Pp. 43–56 *in* Ecology and conservation of Neotropical migrant landbirds (J. M. Hagan, III and D. W. Johnston, eds). Smithsonian Institution Press, Washington, DC.

Odum, E. P. 1953. Fundamentals of ecology. W. P. Saunders, Philadelphia, PA.

Pianka, E. R. 1978. Evolutionary ecology. Harper's, New York.

Reynolds, R. T., R. T. Graham, M. H. Reiser, R. L. Bassett, P. L. Kennedy, D. A. Boyce, G. Goodwin, R. Smith, and E. L. Fisher. 1992. Management recommendations for the Northern Goshawk in the Southwestern United States. USDA Forest Serv. Gen. Tech. Rep. RM-217, Fort Collins, CO.

Richardson, D. M., and D. L. Smith. 1992. Hardwood removal in red-cockaded woodpecker colonies using a shear V-blade. Wildl. Soc. Bull. 20:428–433.

Robbins, C. S., D. K. Dawson, and B. A. Dowell. 1989. Habitat area requirements of breeding forest birds of the middle Atlantic states. Wildl. Monogr. 103.

Root, R. B. 1967. The niche exploitation pattern of the Blue-gray Gnatcatcher. Ecol. Monogr. 37:317–350.

Rosene, W. 1969. The bobwhite quail: its life history and management. Rutgers University Press, New Brunswick, NJ.

Ruggerio, L. F., K. B. Aubry, A. B. Carey, and M. H. Huff (tech. coords). 1991. Wildlife and vegetation of unmanaged Douglas-fir forests. USDA Forest Serv. Gen. Tech. Rep. PNW-285.

Salt, G. W. 1953. An ecological analysis of three California avifaunas. Condor 55: 258–273.

Salt, G. W. 1957. An analysis of avifaunas in the Teton Mountains and Jackson Hole, Wyoming. Condor 59:373–393.

Salwasser, H., I. D. Luman, and D. Duff. 1982. Integrating wildlife and fish into public land forest management. West. Assoc. Fish Wildl. Agen. 62:293–299.

Schoener, T. W. 1974. Resource partitioning in ecological communities. Science 185:27–39.

Severinghaus, W. D. 1981. Guild theory development as a mechanism for assessing environmental impact. Environ. Manag. 5:187–190.

Short, H. L., and K. P. Burnham. 1982. Technique for structuring wildlife guilds to evaluate impacts on wildlife communities. US Fish Wildl. Serv. Spec. Sci. Rep. 244.

Stoddard, H. L. 1931. The bobwhite quail: its habits, preservation, and increase. Charles Scribner's Sons, New York.

Szaro, R. C. 1986. Guild management: an evaluation of avian guilds as a predictive tool. Environ. Manag. 10:681–688.

Szaro, R. C., and R. P. Balda. 1982. Selection and monitoring avian indicator species: an example from a ponderosa forest in the southwest. USDA Forest Serv. Gen. Tech. Rep. RM-89, Fort Collins, CO.

Szaro, R. C., J. D. Brawn, and R. P. Balda. 1990. Yearly variation in resource-use behavior by ponderosa pine forest birds. Stud. Avian Biol. 13:226–236.

USDA Forest Service. 1992. Ecology based multiple-use management. Southwest. Reg. Rocky Mt. Forest Range Exp. Sta.

US Government Printing-Office. 1983. Endangered Species Act. Washington, DC.

Verner, J. 1983. An integrated system for monitoring wildlife on the Sierra National Forest. Trans. North Amer. Wildl. Nat. Resources Conf. 48:355–366.

Verner, J. 1984. The guild concept applied to management of bird populations. Environ. Manag. 8:1–14.

Wilson, E. O. (ed.). 1986. Biodiversity. National Academy Press, Washington, DC.

17

SUMMARY: MODEL ORGANISMS FOR ADVANCING UNDERSTANDING OF ECOLOGY AND LAND MANAGEMENT

THOMAS E. MARTIN

Land managers are often faced with the difficult task of managing for a diversity of resources, including a wide diversity of organisms. Species differ in their space requirements, physiological tolerances, demographies, behaviors, and biological interactions. Added to this complexity is the diversity of land uses, habitat types, and environmental conditions that vary over space and time. This diversity complicates both management strategies and further development of general principles in ecology and evolution. Neotropical migratory birds potentially provide a model set of organisms for aiding our understanding of general principles in ecology and evolution, and for developing general programs of land management because: (1) they are ubiquitous; (2) they include a wide diversity of species with varying ecologies that coexist; (3) they are sensitive to environmental perturbations because they are mobile, short-lived, and differ in their environmental requirements; and (4) behaviors, demographic characters (fecundity, survival), physiology, species interactions, and habitat use can be readily studied for many species. As the chapters in this volume show, Neotropical migratory birds can allow increased understanding of general concepts of population limitation and regulation, ecological processes, habitat selection, spatial scales, and land management.

Sensitivity to environmental modification is indicated by population declines and some Neotropical migrant birds are thought to be declining. However, the number of species, habitat context, and spatial and temporal extent of declines have been debated (e.g., Robbins et al. 1989, Askins et al. 1990, various papers in Hagan and Johnston 1992). Clear delineation of population trends is important from both theoretical and conservation perspectives because it allows identification of species attributes or environmental conditions associated with susceptibility of populations to decline. The issue is also important because the extent and degree of declines indicate the urgency for conservation responses. However, interpretation of trends is complex. This complexity was presented in Part I, where Peterjohn, Sauer, and Robbins (Chapter 1) presented Breeding Bird Survey data on population trends of selected species and guilds, and illustrated that population problems can differ among species in the same geographic region as a function of the kind of habitat they use (e.g., open vs forested habitats), can differ among populations of the same species in different geographic locations throughout its range, or can differ over time (also see James et al. 1992). Interpretation of trends is further complicated by the existence of possible biases created by variation in skills among different observers and changes in observers over time. Peterjohn and his coauthors include such variation in their statistical analyses to reduce the biases. However, statistical analysis has been a further area of debate and this issue is briefly reviewed (also see James et al. 1990, 1992, Sauer and Droege 1990).

Peterjohn and his coauthors acknowledge habitat change as one other source of potential bias that has not yet been possible to evaluate fully or include in statistical models, although it is currently the subject of research investigation. This potential source of bias is important because it begins to address sources and causes of population change; populations may show declines on repeated surveys of a route because preferred habitat has changed from succession or land-use changes. Moreover, if the changes along roadsides are not representative of the region at large, then the surveys do not accurately portray regional changes in populations. Determination of the sources and causes of population declines is critical not only for estimation of population trends, but also for effective management. For example, if species using early successional habitats are declining due to a reduction in these habitats, but forest species are stable, then land-management programs can be developed to increase the necessary habitat types, where appropriate.

Much remains to be learned concerning long-term population trends of Neotropical migrants. Extension of existing monitoring programs to cover underrepresented geographic areas, habitat conditions, and bird species is needed along with further development of statistical and survey methods for reducing potential observer, habitat, and analytical biases. In addition, monitoring programs need to be extended to address breeding productivity and survival to allow determination of population health (i.e., the ability of a population to sustain itself). These programs also need to be closely tied to habitat measurements to allow identification of habitat influences on population trends and habitat conditions required for conservation of healthy populations. The extent of declines among Neotropical migrants is still an open question and needs additional study. Yet, long-term population declines among some Neotropical migrant species or populations are clearly evident and forewarn of the beginning of larger problems. These declines indicate the need to initiate and extend management, research, and monitoring programs to identify causes of declines and allow proactive rather than reactive management responses.

Given the importance of identifying underlying causes of population trends, studies that provide reliable and general conclusions are needed. James and McCulloch (Chapter 2) review problems associated with attempting to determine causes of population trends and the implications for strength of inference. James and McCulloch outline pitfalls and shortcomings of some previous approaches, and suggest approaches for future studies to increase their strength of inference. Their approach is pragmatic, recognizing that the strongest source of inference (experimental manipulations) generally is not possible, but that strength of inference can be markedly increased by careful attention to experimental design and awareness of possible sources of bias.

Determination of causes of population declines are facilitated by understanding when and where populations are most limited. This issue is important theoretically because it allows identification of potentially important periods, and environmental conditions of selection on the evolved niche and phenotype of species (Martin and Karr 1990). Such knowledge has conservation implications: by understanding the ultimate bases of phenotypic attributes of species, we gain insight into determinants of habitat requirements and species coexistence among other things. However, the issue also has even more direct and immediate implications for conservation. Once we know when and where populations are limited, then identification of habitat conditions needed in these locations and periods allows immediate management action to increase and enhance such habitat conditions for Neotropical migratory birds.

Population limitation can operate on two time scales: across years and within years. Rotenberry, Cooper, Wunderle, and Smith (Chapter 3) began Part II by examining across-year influences of natural disturbances or catastrophes (i.e., drought, hurricanes, insect outbreaks, and fire). Some disturbances can affect breeding productivity (e.g., drought, insect outbreaks), while others (e.g., hurricanes, fires) largely change habitat structure and affect birds through habitat-selection processes—see chapters in Cody (1985) for a general overview. Disturbances vary in time and space, causing variation in population sizes in a given location across years, complicating analysis of population trends (see Part I) and management of local populations. Yet, periodic events can shape the evolution of species (e.g., Wiens 1977, Grant and Grant 1989). Moreover, disturbances are a natural component of the dynamics of populations and must be integrated into considerations of population trends and development of land management programs. Indeed, the potential for natural catastrophes to occur means that we need to err on the conservative side (i.e., overestimate) when estimating habitat and population size requirements for management goals. The authors conclude that larger spatial scales (see Part V for more discussion of this issue) need to be considered for management to accomodate such natural disturbances effectively.

The other time scale of concern is across seasons within years; Neotropical migrants are faced with the task (that many humans would like to imitate) of spending summers in North America, winters in Latin America, and migrating long distances in spring and fall, and, in each season, they must interact with entirely different sets of competitors, predators, and environmental conditions. Population limitation may differ among these seasons and much debate has centered on the importance of winter vs breeding season as the main source of population declines (e.g., Wilcove 1985, Hutto 1988, Robbins et al. 1989, Martin 1992). Sherry and Holmes (Chapter 4) examine seasonal patterns of limitation, present a variety of evidence, and conclude that populations are limited in different ways and degrees in all seasons of the year (also see Morse 1980, Martin and Karr 1990). Their conclusions indicate the critical need to understand habitat and ecological requirements, and causes of population limitation in every season for effective conservation of Neotropical migrants.

Probably the least-studied portion of the annual cycle and life history of Neotropical migrants is their ecology and habitat requirements during migration, although migration can clearly be an important period of mortality and selection (Morse 1980, Martin and Karr 1990, Moore and Yong 1991). Moore, Gauthreaux, Kerlinger, and Simons (Chapter 5) outline and review the conceptual basis for understanding habitat selection and ecological requirements during migration. Migration is a difficult period to determine habitat requirements because birds exhibit considerable behavioral plasticity during this period and demographic parameters such as survival cannot be readily measured. By using a conceptual approach that integrates knowledge of habitat use at stopover sites during migration, the authors provide a framework that allows delineation of important stopover habitat on migration pathways for targeted management action.

Much more work has been completed on habitat use during winter in the Neotropics, but habitat preferences are still debated, e.g., the importance of early successional or disturbed habitat vs older undisturbed forest is often debated. One means of resolving the issue is to compare results across a wide array of studies, but the variety of descriptive studies of habitat use mostly have not been summarized and examined for their overall patterns (but see Terborgh 1980). Petit, Lynch, Hutto, Blake, and Waide (Chapter 6) pull together data on habitat use by Neotropical migrants from more than 200 sites in the Neotropics to demonstrate a variety of patterns that support, modify, and counter previous views. Their results indicate that disturbed habitats are often used by more migrants than undisturbed, but that some species of migrants need large blocks of undisturbed habitat. Long-standing notions such as a preference for mid-elevations are not supported. Their synthesis provides a new benchmark from which to work on future patterns. However, a shortcoming of previous studies and our current knowledge is a heavy reliance only on abundance information from survey studies that differ in methodology; the biggest impact of wintering grounds on populations is through their effects on survival. The most effective determination of habitat quality will come from studies of adult survival and abundance rather than abundance alone. Nonetheless, much information is gained from knowledge of distribution and abundance, and much remains to be studied with respect to sex and age differences as well as more thorough coverage across geographical space, habitat types, and elevations.

Most of the remainder of this volume

focuses on the breeding season to provide information for North American land managers, but many of the concepts and approaches are general, and can apply to other organisms and other seasons. Initial concerns about population declines of Neotropical migratory birds arose from concerns about birds of eastern deciduous forest (e.g., Whitcomb et al. 1981, Hall 1984, Robbins et al. 1989) and much attention subsequently focused on forest birds. Forests provide habitats for a substantial proportion of the bird species of North America and they represent a major portion of the land that is under relatively intensive land management. However, forest-management programs differ dramatically in time and space, and can have very different effects on populations. In Part III, Thompson, Probst, and Raphael (Chapter 7) provide an overview of the common kinds of silvicultural treatments that are practiced across the United States. They review habitat influences on forest birds and how these may be related to differing silvicultural systems. Much concern has focused on the potential increase in the amount of edges from silviculture because of the potential for increasing problems from Brown-headed Cowbird parasitism and nest predation. However, concerns about edge were generated from studies in portions of North America with extensive agricultural development rather than in large contiguous blocks of forest, the latter of which is the situation for many public forest lands. The authors argue that edge effects may not be as important as modifications of forest age classes and the consequences for habitat structure and effects on habitat requirements of bird species. They point out the importance of considering forest management from a variety of spatial scales (see Part V for further development of this topic).

Following the overview of silvicultural effects, two case studies are provided as examples of the kinds of detailed information that are needed, but are often lacking. Hejl, Hutto, Preston, and Finch (Chapter 8) review the effects of silvicultural treatments in the Rocky Mountains of the west. A western example is particularly appropriate because: (1) western habitats have received much less study than eastern, even though most public forest land exists in the west; and (2) it brings the focus on to coniferous forests rather than the conventional focus on deciduous forests of the east. In the second case study, Dickson, Thompson, Conner, and Franzreb (Chapter 9) examined silvicultural practices in oak–pine forests of central and southeastern United States. Results of both case studies highlight the difficulty of managing for a diverse array of species as they show that preferences for different age and disturbance conditions differ among bird species. The issue is made even more complex by the action of natural disturbances, such as fires, that historically occurred more frequently and to which some species are particularly adapted. Nonetheless, such approaches identify species that may be most vulnerable to loss of undisturbed forests or changing forest structure. Forest management needs to include consideration of broad-groups of ecological classes of birds, ecological processes, and landscape patterns.

Probably the most detrimental effect of humans has been removal of natural habitat to replace it with human structures or land uses. Grassland birds seem to be showing the most general and extensive population declines of any group of birds. In native grassland and shrubsteppe habitats, agriculture is the biggest source of loss of natural habitats, although agriculture leads to loss of other habitats as well. Rodenhouse, Best, O'Connor, and Bollinger (Chapter 10) begin Part IV by pointing out that replacement of natural habitats with agricultural fields and farmland structures may have had the greatest impact on Neotropical migratory birds of any land-use activity. Agriculture has been detrimental because it has caused direct loss of important natural grassland and woodland habitat, but also by farming practices, such as mowing and tillage, that can cause direct loss of nests. However, in addition, it can influence bird populations in remaining natural habitats in the landscape by affecting distribution and abundance of nest predators and brood parasites. Addition of agriculture and farms can lead to increases in Brown-headed Cowbirds and a variety of nest predators such as raccoons, squirrels, corvids, and skunks, potentially creating dramatic population problems. Yet, population dynamics of birds under differing agricultural practices are poorly studied. Existing information suggests that landscape composition and configuration can affect the presence and abundance of bird species, although effects vary among species. Programs that promote an increase in the proportion of a landscape that is set aside from production (e.g., the Conservation Reserve Program) can generate strong positive benefits to birds and demonstrate the importance of considering large spatial scales.

An indirect effect of agriculture on birds arises from the use of pesticides and their proliferation in the environment at large. Gard and Hooper (Chapter 11) examine not only pesticides but other industrial contaminants in the environment and their potential consequences for birds. Toxicants can be detrimental to birds because they can lead to death. Such lethal effects can be relatively obvious to detect. Less obvious, but equally insidious, contaminants can cause aberrant breeding behavior, decreased fertility and hatching success and increased nestling mortality, all of which can lead to reduced breeding productivity. An indirect mechanism by which contaminants can lead to reduced breeding productivity is through negative effects on the food of birds. Indeed, contaminants can have detrimental effects on a wide array of the components of natural ecosystems, and birds potentially represent an ideal indicator because they are common, widespread, intensively studied, and generally are more sensitive to contaminants than other vertebrates. Yet, effects on birds require detailed study and much more work in this arena is needed.

Another major impact on native grassland and shrubsteppe habitats has been the addition of cattle, although grazing effects are not limited to nonforested habitats. Grazing on public lands is most common in the Western United States and Saab, Bock, Rich, and Dobkin (Chapter 12) review the potential effects of grazing in the West on Neotropical migrants. As with any modification of habitat, grazing effects differ among bird species. Grazing seems to have the greatest negative effects on species that use dense ground cover for nesting or foraging. However, effects of grazing on birds depend on the habitat type and the timing and duration of grazing. Moreover, most studies have been short-term and based on the examination of presence/abundance information; no studies of Neotropical migrants have examined the consequences of grazing for reproductive success or survival. Long-term studies of the effects of differing grazing regimes on breeding productivity and survival among the full diversity of habitat and environmental conditions that are currently grazed are badly needed.

General effects of humans on habitats through use of pesticides and contaminants, conversion of natural habitats to agricultural and farm lands, and modification of habitat structure through grazing are widespread. The evidence indicates that these general effects can affect the abundance and distribution of bird species, but the true consequences for population health and demography are actually poorly studied. Sufficient information on abundance and distribution is available to aid some management decisions, but much greater study of the consequences for population health and, ultimately, for population trends is needed to allow effective long-term management decisions.

Study of demography is important because populations are limited in a proximate sense by reductions in survival, which may be particularly important during winter or migration seasons, or by reductions in breeding productivity. When breeding productivity is insufficient to offset mortality ("sink" populations), then a local population either declines or is maintained by immigration from other populations where breeding productivity exceeds mortality ("source" populations) (Pulliam 1988). As a result, surveys of the presence or abundance of species do not detect general population health. Breeding productivity and survival need to be directly monitored to determine population health. Moreover, study of breeding productivity and survival along with their associated environmental conditions can allow identification of ultimate limits on populations; examination of environmental conditions associated with unhealthy (sink) or healthy (source) populations can aid identification of environmental conditions that need enhancing (unhealthy populations) and environmental conditions to emulate (healthy populations) (Martin 1992). In short, by determining ultimate causes of population problems, we gain important insight into evolutionary influences on ecology and habitat requirements of species over the long term, but in the short term we identify environmental limits on populations and provide immediate targets for management action (Martin 1992).

In most cases, population problems are created when modification of environmental conditions interrupt or modify normal action of ecological processes (e.g., competition, predation, parasitism). Previous chapters repeatedly pointed out the need to consider these effects at multiple spatial scales. Part V considers the importance of differing spatial scales and differing scales of biological organization (species vs assemblages vs ecosystems). One of the most widely recognized examples of spatial influences on ecological processes is habitat fragmentation. Faaborg, Brittingham, Donovan, and Blake (Chapter 13) point out that fragmentation creates increased amounts of edge, which cause changes in microclimate, microhabitats, nest predation, brood parasitism, pairing success, and interspecific competition. All of the latter influence bird abundance and breeding productivity. Nest predation and brood parasitism in particular have received much study, and empirical evidence suggests that fragmentation can lead to increases in both. However, much evidence for nest predation effects have come from experimental studies using artificial nests, and this technique is potentially overused and applied with little regard or study of the many potential biases it can create: (1) artificial nests may not provide cues used by predators for real nests and may only sample subsets of predators that are not representative of predators on real nests; (2) predators may differ among habitat conditions or nest positions and different predators may respond differently to artificial nests; and (3) artificial nests do not provide an opportunity to examine the ability of birds to compensate for nesting mortality through renesting attempts and changes in nest placement. Increased study of real nests is urgently needed. Nonetheless, the few studies on real nests suggest that fragmentation can increase nest predation and brood parasitism, at least in the geographic areas studied so far. In many cases, fragments appear to support sink populations, but the extent of population problems is influenced by the kinds of edges and adjoining vegetation or land uses.

Freemark, Dunning, Hejl, and Probst (Chapter 14) explore the importance of larger spatial considerations by examining the implications of landscape ecology for Neotropical migrants. They point out the importance of landscape composition (i.e., kinds of habitat types and land-use elements and proportion of area they encompass in the landscape) and landscape configuration (i.e., patch size, patch shape, interpatch distance, edge, and habitat juxtaposition/interspersion) to presence, abundance, and demography of birds. These landscape components can affect action of ecological processes such as nest predation and Brown-headed Cowbird parasitism, which can create population sinks. Population sinks are maintained by immigration from other source populations (Pulliam 1988). Links between source and sink populations (i.e., metapopulation dynamics) highlight the need to study demography of populations to determine true habitat suitability and local population health, and that we need to understand and consider spatial scales larger than a local habitat patch. Indeed, landscape-level analyses need to be conducted in every season of the annual cycle and travels of Neotropical migrants, a point made by most of the chapters in this volume. Freemark and her coauthors effectively summarize the existing information on landscape effects on Neotropical migrants and provide management recommendations. Nonetheless, the authors point out that the importance of fragmentation and landscape context is based on geographically and methodologically restricted studies. More studies are needed in a wider array of geographic locations, bird species, and environmental conditions, and they need to be linked with intensive studies of demography and ecological processes to understand the spatio-temporal mosaic affecting populations.

Robinson, Rothstein, Brittingham, Petit,

and Grzybowski (Chapter 15) provide a more detailed look at an ecological process that has been increasing in importance as an impact on Neotropical migrants; they examine Brown-headed cowbird parasitism and its environmental influences. They review evidence showing that both fragmentation and landscape conditions influence the degree and extent of Brown-headed Cowbird parasitism, although vulnerability to parasitism varies among host species depending on their size, geographic range, timing of breeding, and behaviors. Geographic range of vulnerable host species can become important, particularly if the host has a restricted range, but also if its range falls within the main ranges of the Brown-headed Cowbird. Such effects again emphasize the importance of considering many spatial scales in developing management plans. Environmental influences on the presence and abundance of cowbirds at a regional and landscape scale need to be incorporated into management plans. However, more work is needed on the environmental causes and influences of brood parasitism.

Consideration of multiple spatial scales is clearly important to effective management and understanding of bird populations, but management is further complicated by the need to manage for a diversity of species as well. Thus, the scale of biological organization (species, assemblage, community, ecosystem) is another important decision level. Historically, management focused on single species because of a focus on game or endangered species. However, a variety of species are potentially showing population problems, which generally indicate a more general problem with environmental or habitat degradation and, hence, a need to examine large arrays of species. In addition, an increasing recognition of the need to implement conservation programs prior to species becoming threatened or endangered to provide proactive rather than reactive programs will require examination and management of a larger proportion of species in a given management area. Block, Finch, and Brennan (Chapter 16) point out that indicator species and guilds have been used as one means to approach multispecies management, but that this approach has many shortcomings. On the other hand, indicators can be developed using a group of carefully selected species to at least provide an index of the general health of an ecosystem. Block and his coauthors conclude that management of ecosystems holds the greatest promise for the future, but ecosystem management will require much greater knowledge of environmental requirements for species assemblages and a greater understanding of environmental influences on species interactions.

Obviously, our current knowledge of Neotropical migratory birds has many holes. Yet, we cannot wait for more knowledge before we begin conservation programs. Conservation initiatives are needed now and land managers often need access to research information to make decisions. The purpose of this volume was to bring together the majority of information that is available on critical issues in the ecology and management of Neotropical migratory birds. By so doing, land managers have access to the current state of knowledge from which to make the most informed land-management decisions. By reviewing and synthesizing current information, gaps in our knowledge are also highlighted to aid future efforts by researchers. Ultimately, new adaptive management strategies need to be tried and simultaneously evaluated by land managers working together with researchers. Effective proactive conservation requires managers and researchers to share knowledge and work closely together. Hopefully this volume is only a beginning.

ACKNOWLEDGMENTS

I thank C. J. Conway and L. N. Garner for helpful comments and thoughts on an earlier version.

LITERATURE CITED

Askins, R. A., J. F. Lynch, and R. Greenberg. 1990. Population declines in migratory birds in eastern North America. Curr. Ornithol. 7:1–57.

Cody, M. (ed.). 1985. Habitat selection in birds. Academic Press, Inc., New York.

Grant, B. R., and P. R. Grant. 1989. Evolutionary dynamics of a natural population: the large

Cactus Finch of the Galapagos. The University of Chicago Press, Chicago, IL.

Hagan, J. M., III, and D. W. Johnston. 1992. Ecology and conservation of Neotropical migrant landbirds. Smithsonian Institution Press, Washington, DC.

Hall, G. A. 1984. Population decline of Neotropical migrants in an Appalachian forest. Amer. Birds 38:14–18.

Hutto, R.L. 1988. Is tropical deforestation responsible for the reported declines in Neotropical migrant populations? Amer. Birds 42:375–379.

James, F. C., C. E. McCulloch, and L. E. Wolfe. 1990. Methodological issues in the estimation of trends in bird populations with an example: the Pine Warbler. Pp. 84–97 *in* Survey designs and statistical methods for the estimation of avian population trends. (J. R. Sauer and S. Droege, eds). US Fish Wildl. Serv. Biol. Rept. 90(1).

James, F. C., D. A. Wiedenfeld, and C. E. McCulloch. 1992. Trends in breeding populations of warblers: declines in the southern highlands and increases in the lowlands. Pp. 43–56 *in* Ecology and conservation of Neotropical migrant landbirds. (J. M. Hagan, III and D. W. Johnston, eds). Smithsonian Institution Press, Washington, D.C.

Martin, T. E. 1992. Breeding productivity considerations: what are the appropriate habitat features for management? Pp. 455–473 *in* Ecology and conservation of Neotropical migrant landbirds (J. M. Hagan, III and D. W. Johnston, eds). Smithsonian Institution Press, Washington, DC.

Martin, T. E., and J. R. Karr. 1990. Behavioral plasticity of foraging maneuvers of migratory warblers: multiple selection periods for niches? Stud. Avian Biol. 13:353–359.

Moore, F. R., and W. Yong. 1991. Evidence of food-based competition among passerine migrants during stopover. Behav. Ecol. Sociobiol. 28:85–90.

Morse, D. H. 1980. Population limitation: breeding or wintering grounds? Pp. 505–516 *in* Migrant birds in the Neotropics: ecology, behavior, distribution, and conservation (A. Keast and E. S. Morton, eds). Smithsonian Institution Press, Washington, DC.

Pulliam, H. R. 1988. Sources, sinks, and population regulation. Amer. Natur. 132:652–661.

Robbins, C. S., J. R. Sauer, R. S. Greenberg, and S. Droege. 1989. Population declines in North American birds that migrate to the Neotropics. Proc. Natl Acad. Sci. USA 86:7658–7662.

Sauer, J. R., and S. Droege (eds) 1990. Survey designs and statistical methods for the estimation of avian population trends. US Fish Wildl. Serv. Biol. Rept. 90(1).

Terborgh, J. 1980. The conservation status of Neotropical migrants: present and future. Pp. 21–30 *in* Migrant birds in the Neotropics: ecology, behavior, distribution, and conservation (A. Keast and E. S. Morton, eds). Smithsonian Institution Press, Washington, DC.

Whitcomb, R. F., C. S. Robbins, J. F. Lynch, B. L. Whitcomb, M. K. Klimkiewicz, and D. Bystrak. 1981. Effects of forest fragmentation on the avifauna of the eastern deciduous forest. Pp. 125–205 *in* Forest island dynamics in man-dominated landscapes (R. L. Burgess and D. M. Sharpe, eds). Springer-Verlag, New York.

Wiens, J. A. 1977. On competition and variable environments. Amer. Sci. 65:590–597.

Wilcove, D. S. 1985. Nest predation in forest tracts and the decline of migratory songbirds. Ecology 66:1211–1214.

INDEX

Abundance 269
  indices of 6–7, 183–184
Agricultural practices 246, 269–270, 395–397
  burning of crop residues 272
  chemicals 270
  field size 269, 278
  flooding of fields 272
  mowing or harvesting 273
  spatial structure 280, 395–397
  tillage, planting, and cultivation 271
Agroforestry 280
Annual cycle 86–88, 96, 435
Area
  requirements 364–365, 389
  sensitivity 215, 363–364, 386–388, 390, 394, 396, 401–402
Arthropod abundance; *see* Insect outbreaks

Behavioral ecology 146, 174–176, 185
Behavioral plasticity 103, 108, 135–136
Biological diversity 177, 180, 466, 472, 473
Bird–habitat relationships 153, 163, 223
Breeding Bird Survey 3–8, 168
  as a sample of bird populations 31
  comparisons with other survey methods 24–28
  estimation of trends 6–7, 40
  evaluation of efficiency 8
  methods of analysis 32–37
  regional trends 6–7, 47–50
  route trends 6
  sample efficiency 10
  species analyses 10, 47–50
Brown-headed Cowbird 311–312
  breeding distribution 430
  breeding season 435
  effects of agricultural practices on 274, 434, 436, 450, 452
  effects of farmland structure on 280
  egg-laying and fecundity 437
  host specificity 438
  parasitism 145
    edge effects; *see* Edge effects
    effects of corridors 433
    effects of cowbird removal 446, 448–452
    effects of forestry practices 249, 433, 446, 450, 453
    effects of grazing and agriculture 280, 311, 319, 434, 436, 450, 452
    effects of host breeding season 442
    effects of host density 435, 453
    effects on host reproduction success 262, 438–439, 444–445
    effects on host survival 440–441
    geographical variation 431–432
    habitat effects 432–433, 444–445, 453
    host responses to 441, 453
    impacts on host populations 439, 445–451
    on endangered species 442, 445–450
  population changes 431
  population increases 429–430
  range expansion 428–431, 447, 451
  social system 436–437

Captive breeding 108
Catastrophic events 55–84, 165
Climate 157, 171
  El Niño 107
  global climate change 58, 107, 410
  spatial and temporal scales 56–57, 74
  variation and effects on populations 47–49, 56–57
Clutch size 89, 98, 109
Community 161–162, 167, 461, 472
  composition 136, 166–167, 181, 250
  equilibrium 55
  structure 55, 101
Competition 106, 109, 126, 130, 135, 171
  for food 87–88, 91, 95, 146
  for habitat 90, 96–97
  for nest sites 87–88, 91
  interspecific 58, 74, 93–94, 98, 104–105, 361–362
  intraspecific 88, 91, 96–98, 468

Competition (*cont.*)
  migrant/resident 101, 105
  on the breeding grounds 97–98
  on the wintering grounds 97–98
Conservation Reserve Program (CRP) 279
Contaminants 294–305
  effects on reproduction 299–300
Corridors 107, 175–176, 180–181, 363, 384–385
Cowbird; *see* Brown-headed Cowbird
Crops 269, 270, 275, 278
  damage by migrants 281

Deforestation 137, 165, 175, 178, 467
  tropical 49, 145, 400–401
Dispersal 86, 88, 91, 95, 107, 109, 383–385, 389, 392, 402–403, 406
  age-related differences 384
  distances 384
  problems 87–95
  routes 384–385
Disturbance 146, 148, 150, 153–155, 160, 162, 176, 181–182, 462, 470
  human 246, 249
  natural 55–84, 222, 314
Diversity 215
Drought 56–58, 87, 93, 100, 102, 105–106, 108
  effects on food availability 56–57
  effects on populations 56–57
  effects on reproductive success 56–57
  spatial and temporal scales 74
  xeric environments 57, 314

Edge 162, 163, 164, 173, 201, 202, 215, 270, 386, 390–399, 403–405, 471
  area 269, 386–400
  effects 86, 99, 145, 269
    nest parasitism 86, 99, 360, 432–433, 452, 453
    nest predation 86, 99, 360
Edge-sensitive species 215
Elevation 148, 151, 154, 161, 164, 171, 182, 185
Endangered species 45, 442, 445–450, 462, 464, 466, 467, 473
Endangered Species Act 464
Evolution of
  migration 103, 105
  Neotropical migrants 103, 105
Exotic birds 295
Experimental design 40–51
  comparative observational study 41, 42, 44, 47–49, 50
  control group 41, 42
  controlled experiment 41, 43
  correlation and causation 45
  descriptive survey 41
  intervention 41, 42, 44
  multiple time-series design 41, 42, 44, 46–47, 50
  observational study 41, 42
  quasi-experiment 41–44
  randomization 41
  replication 41
  time-series design with intervention 41, 42, 44, 45, 46
  true experiment 41–43
Extinction 55, 56, 69–72, 93, 95, 110, 165, 389, 402, 403, 405, 464

Farmland structure 275
Fecundity 86–89, 93, 94, 102, 105, 106, 109, 111
Fire 74–76, 93, 106, 160, 173, 392, 394, 470
  alternatives to 72, 238
  applications of 67–72, 238, 239, 246
  as a management tool 67–72, 238, 239, 249, 462–464
  effects on birds 68–73, 220, 235
  effects on habitat 68–72, 222, 235, 249, 255, 314
  suppression 67–69, 75, 220, 222, 235, 238, 239, 249, 255
Floaters 88, 89, 96, 97
Floods 106
Food
  acquisition 122, 129–131
  resources 57, 91, 93, 95–100, 104–107, 146, 174–176, 462
Foraging ecology 93, 94, 135, 136, 468
Foraging guilds 176, 468–470
  bark 215
  canopy 216
  foliage-gleaning 468
  ground 216, 470
  understory 216
Forest management; *see* Silviculture
Fragmentation 86, 94, 95, 99, 100, 106, 132, 134, 137, 145, 176, 177, 179, 180, 185, 201, 202, 220, 279, 381, 383–410; *see also* Landscape
  and edge effects 360, 366; *see also* Edge effects
  and species–area relationships 363, 364, 398
  application of island biogeography theory 357
  definition of 358
  effects on cowbird parasitism 86, 99, 361, 432, 434, 451, 452
  effects on interspecific competition 361, 362
  effects on microclimate and microhabitat 360, 361
  effects on Neotropical migrants 74–76, 94, 95, 106, 262, 357–359, 381, 384–399, 403
  effects on nest predation 86, 99, 361
  effects on nesting success 86, 99, 362
  effects on pairing success 362, 389, 408, 409
  fragment shape 366, 386–400
  fragment size 364, 365, 366, 386–388, 390–400, 403, 405

Fragmentation (*cont.*)
habitat patches or islands 123, 132, 136–138, 160, 164, 176, 179, 180, 185, 363, 381–427, 462
management guidelines for 365–368, 401–410
minimum area requirements 364, 365, 389
natural 385, 392, 394
patch occupancy 202, 389

Gene flow 384, 403
Geographic distribution 107, 145, 147–149, 153, 155, 171, 174, 179, 182, 183, 185, 269, 382–384
Grazing 311
effects 311, 314–320, 322–327, 329–338, 340, 341
evaluation of systems 321, 328, 339–341
history 313, 314
Guild 7–10, 461, 467–471, 473
ecological 167, 312
foraging 312; *see also* Foraging guilds
indicator species 467, 470
nesting 312; *see also* Nesting guilds
predatory 467

Habitat
fragmentation; *see* Fragmentation
loss 91, 98–100, 106, 108, 111, 145, 201, 245, 246, 359
quality vs. quantity 86, 89–100, 106, 108–111, 383–385, 398, 391, 402–406
requirements 88–90, 93, 94, 170, 181, 386, 467, 471
saturation 89–90, 95–98
selection 87–98, 102–104, 106, 108, 122, 126, 129, 130, 132–134, 137, 145, 146, 169, 187, 245, 304, 402, 406
age-related patterns 96, 97, 108, 159, 172–174, 185
extrinsic constraints 122, 123, 125–128
intrinsic constraints 122, 123, 126, 129, 130, 133
sex-related patterns 58, 96, 97, 108, 134, 159, 172–174, 185, 389
temporal variation 174, 185, 399
theory, despotic model 89–91
structure 132, 148, 150, 154, 158, 159, 169, 176, 470
suitability models 383
use 87–100, 102–104, 106, 108–110, 126, 129, 132, 136, 145–186, 223–227, 383–384
behavioral aspects of 174–177
breadth 103, 104, 106
during migration 95–96, 121–138
in breeding season 87–100, 102–104, 106, 108–110, 383, 384
in winter 87–100, 102–104, 106, 108–110, 136, 145–186, 383, 384
migrants vs. residents 102, 147, 158, 160, 161
Habitat type
agricultural/residential 137, 138, 149–151, 155, 157–159, 162–164, 169, 171, 183, 185, 394–397, 399, 400, 462
artificial 158–160, 162, 185
coastal dunes 130, 148, 157, 161, 171
deserts 137
disturbed 146, 151, 154, 155, 158–164, 171, 172, 180, 184, 185
forest 422–427
bottomland hardwood 245, 257, 258
coniferous 130, 148, 154, 155, 161–163, 169, 170, 178, 179, 183
deciduous 130, 148, 155, 156, 158, 162, 164, 166–168, 170, 178, 183
early successional 148, 149, 156, 157, 161, 163, 164, 461, 464
late successional 461
mixed 130, 162, 169, 171, 462
monocultures 160, 185
old-growth/primary 161, 163, 168, 169, 174, 178, 466
secondary 161, 163, 164, 169, 173, 174
grasslands 70, 72, 125, 137, 146, 149, 157, 163, 164, 169, 171, 422–427
mangroves 161, 163–165, 169, 171, 174, 178, 183
pastures 151, 158, 160–164, 185
plantation/grove 151, 158, 159, 161–163, 165, 179, 185
riparian 7, 123, 137, 183
scrub 129, 130, 148, 149, 156, 157, 161, 162, 167, 171, 174
shrubsteppe 130, 137, 138
undisturbed 150, 159, 162–165, 168, 176, 183, 185
wetlands 157, 183
Herbicides 302–303
Hurricanes 59–61, 160, 173
Hypothesis testing 40–51

Inbreeding depression 87, 107
Indicator species 167, 466–468, 473
ecological 461, 466, 467
management 461, 466, 467
Insect outbreaks 61–67, 74–76, 93, 98, 104, 106
and bird density 62, 63
and bird population dynamics 66, 67
causes of 62
diet change 62, 63
effects of birds on 62
effects on reproductive success 62, 63, 66, 67
pest species 61
silvicultural management to deter 75
temporal and spatial scales 62
Insularization 148, 151–153, 179

Key factor analysis 88, 100

Land-use planning 466, 470
Land-use practices 160, 162, 165, 174, 176, 177, 184–186, 385–401
Landscape 220, 381–427, 461, 472, 473; *see also* Fragmentation
  and fragmentation 74–76, 94–95, 106, 109, 384–399, 403; *see also* Fragmentation
  approach to conservation 174, 180–182, 185, 401–410
  composition 75, 76, 206, 383–388, 390–393, 403–405
  configuration 74–76, 94, 95, 106, 109, 138, 384–399, 403
    interpatch distance 386–399, 403–405
    interspersion 386–388, 390–399, 403–410
    patch shape 386–400
    patch size 386–388, 390–400, 403–405; *see also* Fragmentation
  ecology 180, 181, 185, 381, 468
  effects on density 237, 384–385, 389
  effects on dispersal 384, 385, 389
  mosaics 382–385
  pattern 381–427
Latitude 146, 148, 151, 153, 155, 157, 158, 166, 173, 182, 183
Life-history traits 88–90, 100–106, 108, 109, 121, 123, 145
  migrants vs. residents 100–106
Life zones 153, 154, 163, 183

Management scales
  assemblages 468, 471, 473
  ecosystems 107, 167, 177, 461, 471–473
  single-species 107, 461, 466, 472, 473
  multiple-species 466, 468, 472, 473
Microhabitat 136, 175, 176, 179, 180
Migration 85, 88, 95, 96, 121, 138
  altitudinal 105, 107
  energetic requirements/costs 121, 122, 130, 132, 135–138
  evolution of 100–103, 105, 106, 122, 136, 138
  geographic patterns 123–126
  habitat and ecological requirements 121, 136, 137, 400, 422–427
  multiple-stage 95, 96
  nocturnal 126, 127, 132, 133
  orientation 122, 127, 128
  resource competition 122, 133
  stopover habitat 121, 122, 127, 129, 130, 132–134, 136–138, 182, 400, 422–427
  weather 59, 122, 123, 126–128, 132, 133, 136, 137
Mixed-species flocks 94, 176, 177, 185
Molt 93, 138
Mortality 86–90, 94–98, 102, 105, 110, 121, 122, 132–136, 145, 174, 175, 180, 384
  adult 86–90, 94–98, 102, 105, 110, 121, 384, 440–441
  yearling 121

Nest 470
  parasitism 86, 99, 106, 109, 361, 388, 389, 394, 396, 399, 428–460
  predation 86, 88, 93, 98, 99, 106, 249, 361, 388, 390, 394, 396–399
  site 91, 93, 99, 102–104, 108
  success 98, 236, 362
Nesting guilds
  cavity 91, 93, 99, 312
  ground 312
  shrub 312

Organochlorine insecticides 294–297
Organophosphorus 297–300

Pairing success 87, 93, 108–110, 362
Pesticides 65, 66, 138, 159, 294–305, 467
  effects on reproduction 299–300
Population 3, 4, 269, 279, 472
  carrying capacity 87, 91–93, 95–98, 104
  declines 137, 145, 146, 154, 165, 168, 179, 201, 269, 270, 467, 471
    breeding season effects 145
    due to weather 56–61
    winter effects 145, 467
  demography 108–111, 262, 383–385, 389, 402, 408–409
    migrants vs. residents 101, 102, 105, 106
    problems with small populations 87, 93, 107, 108
  densities 86, 88–90, 93, 108, 110, 122, 147, 151, 158, 159, 163, 171, 183
  density-dependent regulation 89–93
  density-independent regulation 87
  equilibrium 91, 92
  limitation/regulation 58, 85–111
    summer vs. winter 90–94, 111
  metapopulation 95, 383, 384, 389, 391, 402–406
    dynamics 383
    genetics of 384, 403
  size 202
  small 383–385
  source–sink 95, 109, 110, 137, 249, 364, 365, 384, 402, 404
  stochastic processes 87, 383
  temporal dynamics 74–76, 107, 389, 399, 406, 410
  trends 3, 4, 6–8
    geographic pattern 13, 123
    inferences about causation 40–51
  viability 202, 401, 406

Predation 145, 171
  during migration 130, 132
  nest 262; *see also* Nest predation
Predator 88, 109, 121, 126, 132, 146, 175, 280
  avoidance 122, 130, 177

Recolonization 383–385, 389, 403
Recruitment 98, 109, 110
Removal experiments 88, 89, 96, 97, 446, 448–452
Resource use 171
Resources 58, 74, 86, 91, 93, 98, 101–109, 462, 468, 469

Seasonality 100–107
Sex ratio 97, 109
Silviculture 203
  and fire effects on birds 220, 236, 255–257, 259–261
  effects on birds 228–234, 237–239, 245, 247–249, 251–254, 255–257
  effects on landscape pattern 237, 391, 392
  effects on raptors 229, 233–235
  impacts on habitat 206
    even-aged systems 206, 249, 250
      residual structure 210
      rotation age 210
      spatial distribution and edge effects 207
      stand succession 211
      temporal distribution of forest age-classes 209
    uneven-aged systems 211, 249, 250
      change in stand structure 213
      landscape-level impacts 212
      single- and multi-tree gaps 212
  practices 72, 167, 179, 205, 220, 462, 464
    chaining 229, 232, 233
    clearcuts 228, 236, 238, 246, 247, 250, 258, 261, 470
    group selection 228, 247, 258
    overstory removal 228, 245
    regeneration 205
    related 206
    seedtree 247–250, 470
    shelterwood 220, 247, 250, 470
  systems
    even-aged 223, 245, 464
    uneven-aged 223, 245
  treatments 222
Site fidelity 87, 97, 389
Snags 235, 236, 239
Spatial scales 74–76, 85, 95, 104–106, 145, 381–427, 473
Species richness 104, 130, 147, 150, 157, 161, 165, 245, 385, 386
Species turnover 389, 390
Surveys 3–7, 182–184, 406–409
  methods 147, 170
Survival 86–91, 94–100, 102, 105–106, 109–111, 122, 383, 406, 409
  age-specific 87, 97, 98
  sex-specific 87, 97, 98
  residents vs. migrants 102
  winter 86, 88, 91, 100

Territoriality 88–89, 91, 95–97, 104, 130, 159, 174–176, 382, 383
  interspecific 104

Weather 56–61, 73–76, 86, 88, 93, 95, 97, 98, 104–106, 108, 392
  precipitation 57
  storms 59–61, 87, 104–106, 108
  temperature 57